T0390630

Dimensional Analysis Across the Landscape of Physics

Dimensional Analysis Across the Landscape of Physics

Classic Results, Textbook Examples, and Exploration of Research

Richard W. Robinett
Department of Physics
Penn State University

OXFORD
UNIVERSITY PRESS

Great Clarendon Street, Oxford, OX2 6DP,
United Kingdom

Oxford University Press is a department of the University of Oxford.
It furthers the University's objective of excellence in research, scholarship,
and education by publishing worldwide. Oxford is a registered trade mark of
Oxford University Press in the UK and in certain other countries

Published in the United States of America by Oxford University Press
198 Madison Avenue, New York, NY 10016, United States of America

British Library Cataloguing in Publication Data
Data available

Library of Congress Control Number: 2024938152

ISBN 9780192867551
ISBN 9780192867568 (pbk.)

DOI: 10.1093/oso/9780192867551.001.0001

Printed and bound by
CPI Group (UK) Ltd, Croydon, CR0 4YY

Contents

Part IV **Resources for exploration**

Preface

There can be many purposes for the creation of any book, both from the point of view of the author, and of the intended audience. For STEM (Science, Technology, Engineering, and Mathematics)-focused ones such as this, there are at least several familiar formats. One outcome might be as a textbook for a specific (perhaps frequently offered) course in the standard curriculum, at the introductory (secondary or high school, first or second year of college/university) level, advanced undergraduate, or even graduate level. Such works typically introduce the material (often fairly standardized in terms of content), provide examples to aid reader/student understanding, and (hopefully) include homework problems for further practice in mastering the material, for future use at a more expert level.

An even more advanced format might be a book-length monograph on a research level topic, often written by a true expert, intended for a smaller audience of other experts in the field, designed to document a body of recent progress, perhaps with some introductory material provided to give context to the topic(s) covered.

Given the increasing importance of demonstrating the "broader impact" that STEM education and its applications can have on the (much) larger, non-STEM public, being able to explain the interrelationships among technical subjects across more (or less) familiar topics, can also to help readers "connect the dots" and convey some of the intellectual significance that progress in such fields has on our society, even if the readers themselves do not pursue the topics further after turning the final page.

An analogy for these three approaches in a more liberal arts context might be compare education in (i) art instruction at an introductory level (how to actually do various types of art, painting, sculpture, music, etc.), which could include classroom instruction, but also lessons and studio time (i.e., homework), leading to mastery of specific skills, (ii) art history (documenting the detailed historical provenance or techniques used in works, to codify or advance specialist knowledge in the field), but also (iii) art appreciation (looking for connections among fields of artistic endeavor, and their possible cultural significance and societal importance).

College and university faculty in their careers often balance their time and effort between the teaching, research, and outreach (or public science or service to the community) missions that are typical in the academy, and so might well produce materials in support of any of those professional goals, generalized to their discipline. It can then be a great opportunity for an author with that type of professional background to try to integrate all three aspects of their love for a subject in a single presentation.

My hope is that this book on dimensional analysis (hereafter usually referred to as DA for notational brevity) will have components that may be consistent with all three approaches.

- The tools of DA can certainly be used as one problem-solving technique in physics (and in many other STEM disciplines, as I trust we'll see together) from algebra/trig-level or calculus-based treatments in introductory courses for life science and/or physical science/engineering students, through the advanced undergraduate physics curriculum, and perhaps into the early years of core graduate courses. Even if there may be few (if any) dedicated courses on DA at these levels, the book is designed to be relatively self-contained for use as a reference for classes across much of the breadth and depth of topics encountered in physics.

The first two chapters provide an introduction to the techniques and range of applicability of DA that could be used at the intro (advanced high school or first two years of undergraduate) levels. Part II covers what I (perhaps pretentiously) call the "quadrivium" of core subjects covered in every undergraduate degree program, namely classical mechanics, electricity and magnetism, thermal physics, and quantum mechanics. Each of these separate chapters, with their own collection of self-contained references and extensive set of homework problems, could be used as additional materials in the canonical advanced undergraduate courses on the subject. Those sections might even prove useful at the early graduate level, where some of the same topics may well be revisited, at a much higher level of mathematical sophistication, but often with similar physical content.

Part III focuses on more advanced topics, at the level of senior undergraduate or graduate specialty courses in advanced quantum mechanics (including mention of particle physics and field theory), condensed matter (including soft matter), and astrophysics and gravitation. At my institution, there are senior undergraduate electives in all three of these areas, as well as graduate courses covering the same subjects at a higher level, both of which often touch on research applications, so we expect those chapters to be of use in coverage of these topics as well.

We therefore think that the subtitle *Classic Results and Textbook Examples* is very appropriate. To hopefully make this volume even more useful, a *Solutions Manual* for the approximately 280 end-of-chapter homework problems (many with multiple parts) is in preparation and which should be available to instructors soon after publication of this text.

- A research article from *Science*, *Nature*, or *Physical Review Letters* from this month's edition might be truly new to everyone (save the authors themselves), but a paper presenting a (then) novel experimental or theoretical result from even 100 years ago can still be "new to you" if you've not encountered the topic before. Reading the latest research paper, or reviewing older progress, can pose the same level of challenge in understanding the work presented. There can be a continuum of learning "new" results that starts as early as novice students in intro-level courses (in any field!), culminating in expert-level research deemed worthy of a Nobel Prize, from only a few years ago. Reading a more advanced treatment of a topic in a journal, or listening to a colloquium-level presentation of the latest advances in a field of study, can be challenging tasks, and having as many tools as possible can help anyone interested in pursuing knowledge of a subject make some progress towards that goal. Even if your understanding is not complete, simply knowing how the dimensions relate to each other can be a small step towards mastery, or at least deeper understanding.

  While some advances in physics may be almost purely mathematical in content, the majority of results in our discipline can often be explored by at least checking dimensions to help elucidate the connections between the physical quantities and concepts involved. For this reason, we intentionally include in every section worked-out-examples, homework problems, and references to actual applications of the science involved, taken mostly from the research literature (physics or beyond), to encourage readers/students to intersect with the continuum of realizations of physics they counter, from intro-level coursework to advanced research, wherever their paths cross. This is often accompanied by discussions of the experimental verification (or original discovery) of the dimensional dependences of the physical phenomena in question. These goals are alluded to in the subtitle as *Exploration of Research*.

- Finally, the "Landscape of Physics" metaphor, illustrated by the images on the inside front and back covers, is intended to remind us that practicing physicists (or scientists/engineers) may well work in their own research fields, perhaps sequestered in some corner of that intellectual landscape, with their advances often siloed and not widely shared. We might utilize different skills (mathematical, computational, experimental, statistical, etc.), and use specialized terminology (speaking different dialects, as it were), but ultimately the physical quantities we study can all be "spelled" in terms of their dimensions with just five letters, M, L, T, Q, Θ for mass, length, time, charge, and temperature.

  That statement is, to me, at least as fundamental and consequential as one of the central tenets of biology, namely that all life (on Earth at least) can be "spelled" with four letters, namely A, C, G, T for the nucleotide bases of DNA: adenine, cytosine, guanine, and thymine. The wide applicability of DA across fields of physical science is not only a powerful educational/pedagogical and research tool, but the fact that it exists at all is an amazing organizing principle of Nature, one which may be vastly unappreciated, especially by the general public. One of the "screaming messages" that I trust readers will take away from this book is that **Five Suffice!** as the number of base dimensions needed to "spell" all of the physical properties that they're likely to encounter in their studies. The important background leading to that conclusion is the focus of Chapter 1.

To emphasize the broad utility of DA as a predictive tool, we include examples and/or homework problems from a wide variety of physics-adjacent fields. Instead of including separate, stand-alone chapters on, say geoscience, biophysics, engineering applications, etc., we have striven to incorporate them as examples of the "core" topics whenever possible. I hope that everyone reading this book, wherever they might reside in the "Landscape of Physics", will be able to see themselves represented and can appreciate some interconnectedness with their neighbors.

In common parlance, the phrase ". . .all over the map. . ." can sometimes imply being in a disorganized or disconnected state. In contrast, we argue that the "Landscape of Physics" metaphor advocated here is one attempt to encourage readers/students to ". . .see the big picture. . ." and appreciate how physics is organized, and perhaps sense just how close other physics-adjacent fields are, often proverbially ". . .just over the horizon. . ." off the edges of their corner of the map. Problems in, say fluid mechanics, meteorology, and environmental science almost always touch on fundamental physics concepts and so can make use of DA methods.

To continue metaphorically, over the course of several decades of international travel, my custom has been to purchase travel guides for any country I'm planning to visit. In addition to information on some of the famous "landmark" locations, ones that everyone has heard of and wants to visit, such volumes often cite lesser-known spots of interest or "hidden gems" that might otherwise be missed by a tourist. These volumes always provide guidance on the language(s)/dialect(s) spoken, the currency(ies) used, and suggestions for the best means of "getting around" the local landscape. They often present some of the area's history by way of background, even though the most pleasurable aspects of travel, to me at least, are meeting new people and interacting with the current culture. In that context, we hope that each chapter of this book provides an overview of one relatively connected corner of the physics landscape, offering context and connections to the others, and helpful suggestions of how to navigate the local (intellectual) environment, while getting the most out of the perhaps limited time non-experts spend visiting there. We then encourage readers to explore on their own, via the end-of-chapter problems, or the references to the original literature, to further expand their horizons and be introduced to even more new topics.

Seeing how fields of study are related in the two-dimensional visualizations inside the covers, and how they have evolved in time, is one goal of this text, but a landscape, either in a pictorial/artistic or geophysical sense, can also imply depth (or height). Some actual landscape paintings may well be "flat", but most (since the Renaissance at least) have provided some perspective, allowing us to see deeper into the subject. In somewhat the same way, many areas of study are approached at different levels of rigor/challenge with students "scaling the heights" as they advance in their knowledge, thereby gaining depth and/or perspective, in a different sense. Thus, we also stress the use of DA for use in problem solving at many levels of rigor/complexity in a given discipline, ranging from the first chapter of an intro text to a graduate class on field theory.[1]

Physicists are often used to thinking not only in three spatial dimensions, but also in a connected four-dimensional space-time, and here again the concept of a landscape can be extended. The use of DA techniques stretches back in history for centuries, and seeing those tools applied to historically relevant research documents, in addition to a problem in a recent *Physical Review Letters*, or even in a textbook, can provide more context about the continued value of learning, and using, such tools. We acknowledge this continuity in the time coordinate by citing examples, and providing homework problems, using historical publications (some over a century old) which still provide excellent practice in DA techniques.

We wish to stress, however, that this work is in on way a serious attempt at a *History of Dimensional Analysis*, as that would require the rigorous tools and methods of trained experts in the history of science[2]. Nor is this a *Great Books* approach, of the type advocated by Mortimer Adler and Robert Maynard Hutchins, as we hope to stress the continuum of the utility of DA techniques across fields and their historical development, up to and including last week's journal publications.

While we have attempted to provide as many references to original works we found relevant to each topic as we can, we make no claim for having identified intellectual priorities in their first appearance in the literature. In fact, in searching for original source material for this book, we have ourselves found many cases of the so-called... *zeroth theorem of the history of science*[3] where it often happens "... that a discovery (rule, regularity, or insight) named after someone often did not originate with that person." This has also been called *the Matthew effect in science.*[4] We have, however, been as assiduous as possible in obtaining complete copies (paper or electronic/digital) of any original work (textbooks, journal articles, etc.) that we cite, to ensure that we are quoting the desired content as accurately as possible, and in the most appropriate context. For some sources in other languages (French and German especially), we have used standard online translation tools (mostly for titles and very short quotations), knowing that they are not always infallible, but that they do seem to suitably express the intellectual content, if not the nuanced phrasing, of the original. We have also included, in some cases, overviews by professional historians of science about specific topics by way of background.

We have also focused on listing the most accessible (in two senses) citations that we have encountered. The first sense of accessibility implies, quite literally, references that should be available to someone with access to a university/college library system, including research journals. We hope that many students and academics will be lucky enough to have direct connections to such

[1] Where, of course, one might have to understand units where it's assumed that "$\hbar = c = 1$"! See Section 7.3.

[2] See, e.g., the comments by J. L. Rudolph (2015). "Scholarly Norms for Writing History," *American Journal of Physics* 83, 911.

[3] Part of the title by an article by J. D. Jackson (2008). *American Journal of Physics* 76, 704–719.

[4] The title of a paper by R. K. Merton (1968). *Science* 159, 56–63, which posits that well-known authors receive more credit than they should, simply because they are already famous. The Matthew is an allusion to the biblical Parable of Talents (Matthew 25: 14–30) including the injunction "For to all those who have, more will be given..."

information systems, and that for more general readers, we note that modern search engines can often find content in surprising ways. But this text is ultimately designed to be as self-contained as possible, including a separate final chapter on Ancillary Materials to assist the reader in checking worked-out **Examples** and approaching end-of-chapter problems. The second sense of accessibility relates to our hope that the citations provided are readable and understandable, which is one reason we have included so many from pedagogically focused journals such as the *American Journal of Physics*, the *European Journal of Physics*, and *The Physics Teacher*. I've sometimes cited review articles instead of original discovery papers, as such journals often provide context and historical background (and even more references) with explanations of the impact the work being cited having had on the field, given time to reflect and relate. Given that almost all topics discussed in this work do not contain exhaustive derivations of the phenomena in question (DA, after all, often gets us part of the way to a "complete answer"), we feel that it's important to supply as many relevant references to allow readers to continue to explore these topics in more depth, and at a more rigorous level, if desired.

In terms of the actual formatting of this book, we use sans serif fonts only for the base $\mathsf{M}, \mathsf{L}, \mathsf{T}, \mathsf{Q}, \Theta$ dimensions to distinguish them from other physical quantities. In the body of the text, we often use **boldface** when introducing a new word/concept for the first time, as well as in enumerating the worked out examples (as in **Example 8.1**) and in all of the end-of-chapter homework problems (such as **P3.2**), each of which have a (short) descriptive title to help explain the background for the interested reader. For in-line or displayed equations, the use of bold is limited to vectors, such as $\boldsymbol{F}$ for force or $\boldsymbol{E}, \boldsymbol{B}$ for electric and magnetic fields. The values of exponents in the derivation of the dependence on parameters are almost always lower-case Greek letters, such as $\boldsymbol{F} \propto e^{\alpha} m^{\beta} \omega^{\gamma} \boldsymbol{E}^{\delta} \boldsymbol{\nabla}^{\epsilon}$. In some cases, where the same symbol can be used for different quantities in the same problem, we may use calligraphic font for one, such as E and $\mathcal{E}$ for energy and electric field, or T and $\mathcal{T}$ for tension and torque, or P and $\mathcal{P}$ for pressure and power. Readers should hopefully use contextual clues (and especially dimensions!) to distinguish them in such cases. As mentioned, we also frequently use the abbreviation DA for dimensional analysis to save space.

The references cited in the text are collected separately at the end of each chapter, to make them more self-contained and hopefully more useful in parallel to other topic courses in that disciplinary area. An extended set of **ancillary materials** is presented in Chapter 10 that can be used in confirming the results in each chapter, and in doing the homework problems. These include several template Mathematica© programs, which can be easily morphed (or translated into other coding platforms) to help solve the systems of linear equations present throughout the book.

In the creative universe of literature, authors may generate elaborate alternative universes, sometimes set in exotic landscapes, and often populated by an incredibly diverse cast of characters. In my own, rather low-brow, collection of comic books,[5] there are fictional personalities with powers ranging from almost godlike and superhuman, to "mere" humans, albeit with amazing skill sets. I have been pleased throughout my academic career to be around individuals with incredible intellectual abilities, outstanding experimental skills, and fantastic intuition, traits sometimes so advanced that I could only admire, and not even aspire to. The closest I come to seeing myself in such stories (let specialize to the MCU or DCU[6]) are characters such as Hawkeye or Green Arrow. Both maintain a vast array of specialized tools (in their case, arrows), each with different abilities/purposes, and through training have been able to know how to access them at the right time, to good effect.

[5] I do buy them as trade paperbacks, so perhaps graphic novels is fair.
[6] Marvel Cinematic Universe or DC Universe.

The metaphorical equivalent of developing such a toolkit for many physics students might consist of acquiring an array of both hard and soft skills, including pure mathematics, symbolic manipulation, numerical coding, data analysis including graphing and modeling skills, experimental expertise, orders-of-magnitude thinking, physical intuition, and communication skills (in written, oral, and social media/public science form). I argue that dimensional analysis should rightly be included in the "quiver" of any practicing (or hopeful) physicist, as another option in the quest for mastery of physics, whether for theory problems, or in the lab. DA techniques can be another tool to approach problem-solving in the physics curriculum, at almost every level, and could be included in the array of successful instructional techniques advocated by experts in Physics Education Research (PER).[7]

As an example, Wieman and Perkins[8] raise (and then answer) an important question:

> But what specifically do we mean by *effective* physics instruction? It is instruction that changes the way students think about physics and physics problem solving and causes them to think more like experts—practicing physicists. Experts see the content of physics as a coherent structure of general concepts that describe nature and are established by experiment, and they use systematic concept-based problem-solving approaches that are applicable to a wide variety of situations.

If this text provides one more (metaphorical) arrow for student use, then the extensive array of homework problems offers opportunities for lots of target practice.

In terms of recommended physics background, parts of the book (Chapter 1 especially) can be used starting as early as a first course in mechanics and electricity and magnetism (E&M), even in secondary school, while Part II (Chapters 3–6) is perhaps best appreciated by students in junior–senior (3rd and 4th year undergraduate) level core physics courses. For mathematics preparation, in both the body of the text and the homework problems, we occasionally refer to differential equations, or use simple integrals and series expansions, to obtain "exact" results in straightforward situations, in order to compare to purely DA results in more depth. A (very) brief compilation of **used mathematics**, which covers most of the calculus-level math needed to approach such problems, is included in Section 10.7. For homework problems (or parts of such problems) that do require the use of any such methods, we indicate that with an asterisk, such as **P3.3*** (which requires an integral), **P3.7**(b)* (which asks for the solution of a first-order differential equation), or **P4.37**(b)* (which needs a series expansion). The only real mathematical tool with which readers should be very comfortable (fluent is better!) is in solving systems of linear equations. As early as Chapter 1, we suggest that interested readers/students who want to streamline the workflow of checking worked-out examples, or solving homework problems, might want to make the investment of generating (adapting) very simple software, say from a symbolic manipulation program, to save time. We provide template examples using Mathematica© in Sections 1.4 and 10.8, but encourage users to take advantage of whatever coding expertise and tools to which they have access.

Since dimensional analysis methods can often correctly predict the power-law dependence of one variable on others, via relationships of the form $y = Ax^B$, we argue that another valuable skill very closely related to DA approaches is the ability to analyze data (via curve fitting), especially to determine the power-law exponent B in order to compare experiment to theory. We include some brief examples in Section 10.9 of data-fitting techniques, but also ask in several homework problems for students to extract information from published data sets/figures, using whatever

[7] See, e.g., Redish, E. F (2021) "Using Math in Physics: 1. Dimensional Analysis," *Physics Teacher* 59, 397–400.

[8] See Wieman. C, and Perkins. K (2005) "Transforming Physics Education," *Physics Today* 58(11), 36–41 (2005).

tools with which they are comfortable. For any such homework problems that ask for data analysis, including power-law fits, we denote them with a dagger, such as **P2.1**† or **P3.12**(b)†. Of the over eighty figures included in the book, close to half are taken from published research papers (or review articles) showing experimental data that readers/students are asked to explore using just such methods.

One important role that a preface can play is to allow the author to publicly thank not only the many people who've had an impact on the book in which the preface appears, but also on their personal and professional lives. First and foremost, I thank my late mother, Dr. Betty W. Robinett, for being the best role model I could imagine, both as a parent (and then as Grandma Năinai to my own two children), and as an academic. Her emphasis on the important role that teaching and service play in the triumvirate of academic pursuits (the third being research), and the recognition she received for her classroom contributions and academic administrative achievements, have guided and inspired me throughout my entire career.

I'm grateful for the physics training I've been blessed with receiving, starting in high school (from W. L'Herault, a fellow UM physics BS graduate I learned later, and only a few years my senior!). Thanks as well to all of the faculty, staff, and fellow students at the University of Minnesota, in both the physics and math departments, at both the undergraduate and graduate levels. I especially recognize S. Gasiorowicz for introducing quarks and *SU*(3) symmetry in a modern physics class that helped me decide to add physics as a (then) second/concurrent major, and helped cement my interest in pursuing theoretical physics at the graduate level, mentored by J. Rosner as my excellent PhD adviser. Thanks also to my colleagues and friends in the physics departments at the University of Wisconsin Madison and the University of Massachusetts Amherst for all their support during my postdoctoral appointments there.

Here at Penn State University, I've been continually grateful for the support of generations of departmental/college administrators, physics (and other) faculty colleagues, and especially decades of interactions with students, both graduate and especially undergraduate, over the thirty-seven years of my tenure. During the initial stages of preparing this text, I took seriously the excellent suggestion from my OUP editor to "test run" the material to be covered in this book (topics, level, homework problems, etc.) with undergraduate students and was able to develop (on the fly!) a one-credit special topics course on dimensional analysis, delivered in the Spring 2022 semester. I'm thankful for the students who were brave enough to sign up for the class and who provided continuous real-time feedback on all aspects of the project, especially on material that I often developed on a (late) Tuesday night for presentation to the class on Wednesday! So, special thanks to B. Bankovic, K. Cardona-Martinez, B. Collins, N. Cristello, N. Gluscevich, D. Hartmann, E. Jennerjahn, Y. Jiang, J. Kipiller, C. Kopp, S. McNamara, N. Mitran, G. Sommer, and S. Tobin. I hope that your careers in physics (and beyond) are all very successful. Several professional colleagues have noted to me over the years that it's likely that their biggest impacts on society will be through training and mentoring of students at all levels, including both teaching as well as advising (academic, career, or otherwise), and indeed some of my happiest and proudest moments to this day are reconnecting with former students. I have also very much appreciated discussions (some very lengthy!) with PSU physics faculty colleagues, especially on some technical, research level topics, including C. Hanna, J. Jain, B. Sathyaprakash, and S. Shandera. I also thank V. Karpov (University of Toledo) for an early invitation to present a colloquium based on a pedagogical article[9] that was the seed of this textbook project.

---

[9] Robinett. R (2011) "Dimensional Analysis as the 'Other' Language of Physics," *American Journal of Physics* 83, 353–361.

I acknowledge that the conceit to include the Bern Porter *Being a Map of Physics* image[10] on the inside front cover was prompted by having inherited an intro level textbook[11] co-authored by a former Penn State Physics faculty colleague, Prof. Marsh White, who had included that image, almost seventy years ago. Prof. White was the first PhD in **any** field (not just physics) from Penn State University, and is perhaps best known for his leadership roles in the national Sigma Pi Sigma organization and Society of Physics Students chapters, which spanned six decades! And I very much appreciate the permission given by Dominic Walliman to reproduce a newer, infographic version of a *Map of Physics* on the inside back cover, a design which earned him one of the Kantar Information is Beautiful Awards[12] in 2017. Mr. Walliman's work in visually explaining technical topics is outstanding, and is a great example of scientific communication.

While nominally a textbook, I have striven to document original source materials whenever possible, and having one of the biggest and best university libraries in the country at my disposal was an invaluable resource. The work of the Penn State University library staff was essential to the completion of this volume, especially during the period of "remote everything" caused by the pandemic, so special thanks to N. Butkovich and J. H. Ritchey. I also thank the amazing staff at Oxford University Press, including S. Adlung (who helped with my first OUP book) and especially Giulia Lipparini for her always patient responses to my (many) emails. Thanks as well go to Rajeswari Azayecoche and the production team at Integra Software Services.

And in terms of thanks, last, and most certainly not least (most important, in fact) I am eternally grateful for the love and support bestowed on me by my treasured immediate and extended family, my wife Sarah Malone, our children (James and Katherine Robinett), my daughter-in-law (Diane Pu) and our two granddaughters (Lucy and Penny), and my siblings, official and unofficial, Susie Morton and Francie Isbell, and Mark and Kit Malone. Love you all!

Finally, I hope that readers (casual or serious) of this book may agree with a sentiment associated with Marcel Proust. A commonly found aphorism attributed to him reads:

> The real voyage of discovery consists not in seeking new landscapes, but in having new eyes.

Some online commentators note that this is seemingly a paraphrasing of a more extensive quotation from *Remembrance of Things Past*, Volume 5 (The Captive), Chapter 2, which in translation (the one by C. K. Scott Moncrieff, 1922–1930) reads:

> The only true voyage of discovery, the only fountain of Eternal Youth, would be not to visit strange lands but to possess other eyes, to behold the universe through the eyes of another, of a hundred others, to behold the hundred universes that each of them beholds, that each of them is...

I prefer the more expansive (and eloquent) full quotation, as it emphasizes the role that interacting with others, and benefiting from their diverse experiences, plays in learning about, and appreciating, the world around us. I hope that you enjoy your own time with this book, and I wish you a *bon voyage.*

R. W. Robinett
Professor (Emeritus) of Physics
Department of Physics
Penn State University
December 1, 2023

[10] Thanks to Mark Melnicove, as literary executor of the Porter estate, for his approval to reproduce it in this work.
[11] White, M. W., Manning, K. V., and Weber, R. L. (1955). *Practical Physics*, 2nd ed. (New York McGraw-Hill).
[12] See https://www.informationisbeautifulawards.com/showcase/1793-the-map-of-physics

Inside cover credits

- Inside front cover credit: *Being a Map of Physics* by Bernard H. Porter (1939). Reproduced by permission of Mark Melnicove, literary executor for Bern Porter, <mmelnicove@gmail.com>. Thanks to the staff of the Colby College Library for their help in obtaining a copy.
- Inside back cover credit: *Map of Physics* by Dominic Walliman (2016), reproduced with his permission. For more examples, see YouTube, Domain of Science, and The Map of Physics Video.

Part I

Preparing to explore the landscape of physics

Chapter 1

Getting started

1.1 Introduction to the book

Starting from the youngest of ages, it's easy to be fascinated by the diversity of life around us, starting perhaps with the flora and fauna in our back yards. As we travel beyond our own neighborhoods, either in person (through hiking through forests or visiting zoos) or remotely (via books or other media), we expand our direct experience about the array of living organisms on our planet. We might then learn how to codify our knowledge more concretely as we encounter biological classification and nomenclature schemes (taxonomy) through constructs such as kingdom and phylum through genus and species. These can help orient our thinking about the interconnections among living things, based on morphological, physiological, and eventually genetic characteristics. Ultimately, those molecular connections are explained to us through the language of DNA, with the four nucleobases, adenine (A), cytosine (C), guanine (G), and thymine (T), constituting a four-letter alphabet of sorts, from which information on the organization of life and its interrelationships can be "written." This central tenet is at the core of modern biology and a fundamental organizing principle of that field.

One hopes that many of us are also equally intrigued (initially perhaps) by the diversity of physical phenomena we observe around us ("Why is the sky blue?" or "What causes the seasons?") and some continue to explore the organizing principles codified under the rubric of physics. This could be in terms of traditional subject areas (mechanics, electricity and magnetism, etc.) or through physical phenomena with similar mathematical descriptions (wave motion, diffusion, conduction). It can happen, of course, that as one approaches the subject at each progressively higher level of mathematical sophistication, the technical demands of the math used can be perceived as an obstacle to further progress.

In what follows we argue that there is one organizing principle in physics, across all disciplinary boundaries that remains at a completely accessible level to almost any interested student. The fact that there are five fundamental (or **base**) **dimensions** in physics (and by extension, in other science or engineering applications) is every bit as important and foundational a construct as that leading to DNA. More importantly that simple fact can be used, with a minimum of mathematical technology, in the form of **dimensional analysis** (DA[1]), as a pedagogical and even research tool, when looking to explain connections between known physical properties and phenomena, but also in pursuing research to find new ones.

We start our exploration in the rest of this Chapter 1 by imagining a fly-over (since our tour will initially be quick and high-level) of the main physical concepts and quantities seen in the standard physics curriculum. We will quickly convince ourselves that the dimensions of all of the physical properties likely to be used in practice can be "spelled" with an almost equally small alphabet as

[1] Our mnemonic analogy is then Biology/DNA : Physics/DA.

Dimensional Analysis Across the Landscape of Physics. Richard W. Robinett, Oxford University Press. © Richard W. Robinett (2024).
DOI: 10.1093/oso/9780192867551.003.0001

that in biology: mass (M), length (L), time (T), charge (Q), and temperature (Θ). We then explore how to make use of this fact in a proactive manner to predict new physical connections, starting with research (historical and more recent) examples in Chapter 2, then in traditional curricular areas (Part II, Chapters 3–6), and finally for more advanced, research-intensive topics (Part III, Chapters 7–9). Chapter 10 includes ancillary material designed to make the book as self-contained as possible by providing tools to confirm examples cited in the text and to solve end-of-chapter problems.

1.2 A dimensional analysis overview of the landscape of physics

A metaphorical fly-over of the landscape of physics (perhaps as imagined fancifully on the inside front and back covers) might start by opening up any introductory physics textbook, reviewing the table of contents to get a sense of the main ideas, leafing through its pages to explore the physical concepts, and reviewing how those ideas are translated into symbols, and then relations among them in equation form. For our purposes, just that type of exercise will help establish how many foundational dimensional quantities are needed in physics (actually, in all of the physical sciences) in order to **spell** (metaphorically) all of the measurable quantities that are used in not only fundamental science, but also in engineering applications.

1.2.1 Mechanics

Many texts begin with the study of **kinematics** (how to describe motion), starting in one-dimension, where the relevant quantity is the time-dependent one-dimensional position of an object, $x(t)$, immediately suggesting that **length** (hereafter L) and **time** (or T) are fundamental dimensional constructs. To codify and connect the basic dimensions for any physical quantity, variable, relation, or equation, we will consistently use a square bracket notation, by writing

$$[\chi] \equiv \text{the dimensions of}\,\chi$$

where χ (pronounced 'kai', see Section 1.5 for a listing of the Greek alphabet) is any individual physical quantity, or combinations thereof.

For one-dimensional kinematics, the first two such variables we encounter are clearly then

$$\boxed{[x(t)] = \mathsf{L} \qquad \text{and} \qquad [t] = \mathsf{T}} \tag{1.1}$$

for length and time. Important derived (by repeated differentiation) quantities are then velocity (or speed) and acceleration, given by

$$[\dot{x}(t)] = [v_x(t)] = \left[\frac{dx(t)}{dt}\right] = \frac{[dx]}{[dt]} = \frac{\mathsf{L}}{\mathsf{T}} \tag{1.2}$$

$$[\ddot{x}(t)] = [a_x(t)] = \left[\frac{dv_x(t)}{dt}\right] = \frac{[dv_x]}{[dt]} = \frac{(\mathsf{L}/\mathsf{T})}{\mathsf{T}} = \frac{\mathsf{L}}{\mathsf{T}^2}, \tag{1.3}$$

where we can consider derivatives as ratios of physical (if infinitesimal) quantities, each with the appropriate dimensions. A familiar physical example of Eqn. (1.3) is then the **acceleration of gravity**, g, which has $[g] = \mathsf{L}/\mathsf{T}^2$.

We note for future reference that combinations of dimensional quantities, say A and B, will satisfy

$$\left[\frac{A}{B}\right] = \frac{[A]}{[B]} \qquad \text{and} \qquad [AB] = [A][B], \tag{1.4}$$

and that space and time derivatives can be considered to contribute dimensional factors of

$$\left[\frac{d}{dt}\right] = \frac{1}{[dt]} = \frac{1}{\mathsf{T}} \qquad \text{and} \qquad \left[\frac{d}{dx}\right] = \frac{1}{[dx]} = \frac{1}{\mathsf{L}}, \tag{1.5}$$

regardless of what they act upon. An extension to three dimensions makes use of **position vectors**, which in Cartesian coordinates are given by

$$\boldsymbol{x}(t) = (x(t), y(t), z(t)) \tag{1.6}$$

and we clearly have

$$[\boldsymbol{x}(t)] = [x(t)] = [y(t)] = [z(t)] = \mathsf{L}. \tag{1.7}$$

The notion of dimensionality can therefore be extended to vectors more generally by noting that

$$[\boldsymbol{A}] = [(A_x, A_y, A_z)] = [A_x] = [A_y] = [A_z] = [\,|\boldsymbol{A}|\,] \tag{1.8}$$

are all equal for any vector quantity, and its components, whatever the dimensions of $\boldsymbol{A}$, or its magnitude $|\boldsymbol{A}|$, might be.

Note: Two- and three-dimensional vectors can also be described by cylindrical/polar and spherical (even elliptical) coordinate systems where the vector components might **NOT** share common dimensions, such as position vectors given by $\boldsymbol{r} = (r, \theta, z)$ or $\boldsymbol{r} = (r, \theta, \phi)$. We will always assume that physical quantities represented by vectors, including variables like momentum ($\boldsymbol{p}$), angular momentum ($\boldsymbol{L}$), electric and magnetic fields ($\boldsymbol{E}$ and $\boldsymbol{B}$), and the like, are in Cartesian coordinates, so that the analog of Eqn. (1.8) is valid. We can then generalize Eqns. (1.2) and (1.3) to write

$$[\boldsymbol{x}(t)] = \mathsf{L}, \quad [\boldsymbol{v}(t)] = \frac{\mathsf{L}}{\mathsf{T}}, \qquad \text{and} \qquad [\boldsymbol{a}(t)] = \frac{\mathsf{L}}{\mathsf{T}^2}. \tag{1.9}$$

For numerical calculations, we will most frequently use SI or <u>M</u>K<u>S</u> units, here emphasizing (by underline) the use of the <u>M</u> (meter) and <u>S</u> (second) as specific units of length and time, but we note that scientists working in different domains across the "landscape of physics" may well use their own more specialized sets of units. We will not shy away from discussing problems involving more "parochial" units, noting that there is an extensive translation dictionary in Section 10.4. We think that this strategy not only provides practice in applying dimensional ideas to physical problems in a variety of settings, but recognizes and honors the diversity of expression of the shared basic dimensions upon which all of physics is built. For example, while we may most often utilize *meters per second* for speed or velocity, the use of *miles per hour*, or even *furlongs per fortnight*, should not obscure the fact that we're describing a quantity with the same L/T dimensions.[2]

We note that the assumption implicit in Eqn. (1.7), namely that there is only one length dimensionality needed, even in three spatial dimensions, is consistent with experimental evidence that there is no preferred frame of reference, going back to the Michelson and Morley (1887) experiment probing the isotropy of space. More recent precision tests (such as those by Eisele et al. (2009) and Herrmann et al. (2009)) have extended the validity of this limit, and many related aspects of Lorentz invariance, to the level of 10^{-17} or better. So, in a real sense, the assumptions underpinning DA are themselves subject to rigorous experimental verification.

Turning now to the topic of **dynamics** (the study of how motion changes), the important new constructs are Newton's laws of motion and the concept of **inertial mass** (m), with a base dimension given by

$$\boxed{[m] = \mathsf{M}}\,. \tag{1.10}$$

This adds a third fundamental dimension to L and T, and has SI units of K (kilogram) in the MKS system. Newton's third law, $\boldsymbol{F} = m\boldsymbol{a}$, implies that **force** has dimensions given by

$$[\boldsymbol{F}] = [m\boldsymbol{a}] = [m][\boldsymbol{a}] = (\mathsf{M})\left(\frac{\mathsf{L}}{\mathsf{T}^2}\right) = \frac{\mathsf{ML}}{\mathsf{T}^2} \tag{1.11}$$

for any component of $\boldsymbol{F}$.

Note: We have implicitly assumed that all equations must be dimensionally self-consistent, so that an equation of the form $LHS = RHS$ must also have $[LHS] = [RHS]$. Connections such as those in Eqn. (1.11) also make use of a dimensional identity for mixed scalar (C) and vector ($\boldsymbol{D}$) quantities, namely that $[C\boldsymbol{D}] = [C][\boldsymbol{D}]$, which can then be used, for example, to write $[\boldsymbol{p}] = [m\boldsymbol{v}] = [m][\boldsymbol{v}] = \mathsf{ML/T}$ for the dimensions of **momentum.**

A specific, but important and fundamental, example of a force is Newton's (universal) **law of gravitation**, with $\boldsymbol{F}_G$ acting between two massive objects separated by a distance r, given (dimensionally) by

$$|\boldsymbol{F}_G| = F_G \quad \longrightarrow \quad [F_G] = \left[\frac{Gm_1m_2}{r^2}\right]. \tag{1.12}$$

Inverting this, we can obtain the dimensional "spelling" of **Newton's constant,** G, to be

$$[G] = \frac{[F_G][r^2]}{[m_1m_2]} = \left(\frac{\mathsf{ML}}{\mathsf{T}^2}\right)\frac{\mathsf{L}^2}{\mathsf{M}^2} = \frac{\mathsf{L}^3}{\mathsf{MT}^2}. \tag{1.13}$$

Newton's realization that terrestrial (proverbially the apple) and astronomical (say, the moon) objects obeyed the same law of gravitation was, perhaps, one of the first **unifications** in physics. But for the purposes of DA, it's just as important to note that the $m_{1,2}$ values that appear in Eqn. (1.12) are **gravitational masses** (or m_G) which could, in principle, be distinct from the **inertial masses** (m_I) used in $\boldsymbol{F} = m\boldsymbol{a}$.

[2] Quoting a senior colleague of mine, "Units are the enemy, dimensions are our friends", but we should hopefully be able to appreciate and respect all realizations of physics, in whatever dialect they might be expressed.

The important fact that $m_G = m_I$ is the content of the famous **equivalence principle**[3] (one of the hallmark assumptions of relativity) that has been tested with increasing rigor, both in terrestrial, lab-based experiments (from the early seminal results of Eötvös et al. (1922)) to more recent ones by Wagner et al. 2012) and in space-based settings (see, e.g., Williams et al. 2012, Toublou et al. (2017), and most recently Toublou et al. (2022)), now reaching the level of something like 10^{-15}. For our purposes, this means we can safely assume that only one type of mass dimensionality is needed in any problem involving DA.

Considering now the (scalar) **work** (W) done by a (vector) force ($\boldsymbol{F}$) over a (vector) distance ($\boldsymbol{D}$), we can write

$$[W] = [\boldsymbol{F} \cdot \boldsymbol{D}] = [\boldsymbol{F}][\boldsymbol{D}] = \left(\frac{\mathsf{ML}}{\mathsf{T}^2}\right)\mathsf{L} = \frac{\mathsf{ML}^2}{\mathsf{T}^2} \tag{1.14}$$

or, even more generally, for work done by a variable force over an arbitrary path,

$$[W] = \left[\int dW\right] = \left[\int \boldsymbol{F} \cdot d\boldsymbol{l}\right] = [\boldsymbol{F}][d\boldsymbol{l}] = \frac{\mathsf{ML}^2}{\mathsf{T}^2}. \tag{1.15}$$

This assumes another **distributive property** of dimensions, namely that for two vectors $\boldsymbol{A}, \boldsymbol{B}$, we have for their dot product $[\boldsymbol{A} \cdot \boldsymbol{B}] = [\boldsymbol{A}][\boldsymbol{B}]$. The integral sign ($\int$), representing a summation over infinitesimal (but dimensionful) quantities, plays no role in DA, but the $d\boldsymbol{l}$ line-element certainly does, with $[d\boldsymbol{l}] = \mathsf{L}$.

For many physical systems, the (vector) force $\boldsymbol{F}$ can be derived from a (scalar) potential energy function, $V(\boldsymbol{r})$, via $\boldsymbol{F}(\boldsymbol{r}) = -\nabla V(\boldsymbol{r})$, where ∇ is the **gradient operator**, and when expressed in Cartesian coordinates, we associate its dimensions as

$$[\nabla] = \left[\left(\frac{\partial}{\partial x}, \frac{\partial}{\partial y}, \frac{\partial}{\partial z}\right)\right] = \frac{1}{\mathsf{L}}, \tag{1.16}$$

generalizing the second result in Eqn. (1.5) to three dimensions. This implies that all of the vector calculus operations (such as those discussed in the eponymously titled *Div, Grad, Curl, and All That* by Schey (2005), or in any multi-variable calculus text) where the ∇ is used, can be handled using this dimensional "spelling." The gradient operator is sometimes called **del** or even **nabla.**

All other forms of energy (E), including **kinetic energy** (KE) and **gravitational potential energy** (PE_G) encountered in mechanics (and eventually **heat** in thermodynamics) share the same set of dimensions, since

$$[W] = [E] = [KE] = \left[\frac{1}{2}mv^2\right] = (1)(\mathsf{M})\left(\frac{\mathsf{L}}{\mathsf{T}}\right)^2 = \frac{\mathsf{ML}^2}{\mathsf{T}^2}$$
$$[PE_G] = [mgh] = [m][g][h] = \mathsf{M}\left(\frac{\mathsf{ML}}{\mathsf{T}^2}\right)\mathsf{L} = \frac{\mathsf{ML}^2}{\mathsf{T}^2}. \tag{1.17}$$

Note: Any pure number will be dimensionless, as in

$$\left[\frac{1}{2}\right] = \left[\sqrt{2}\right] = [\pi] = [e] = \cdots = \mathsf{M}^0\mathsf{L}^0\mathsf{T}^0 = 1, \tag{1.18}$$

[3] For a centennial history of the equivalence principle, see Bod et al. (1991).

so that we can write [*dimensionless*] = 1 for such quantities. Such factors will not contribute to the checking of dimensions and so can be ignored, and we will very soon stop writing that step explicitly. Turning this around, however, any proactive use of DA to predict (or at least to constrain) the dependence on the dimensionful parameters in a physical system will never be able to provide any information on purely numerical prefactors, such as the 1/2 above. That said, we can certainly train our intuition on the judicious choice of some combinations of dimension**ful** and dimension**less** parameters through repeated problem solving.

For the description of rotational motion (say about a single axis), the relevant angular variable is often denoted by $\theta(t)$ and assumed to be measured in **radians**, which are, in fact, dimensionless. This gives for the corresponding **angular velocity** ($\omega(t)$) and **angular acceleration** ($\alpha(t)$)

$$[\theta(t)] = 1, \quad [\omega(t)] = \left[\frac{d\theta(t)}{dt}\right] = \frac{1}{\mathsf{T}}, \quad \text{and} \quad [\alpha(t)] = \left[\frac{d\omega(t)}{dt}\right] = \left[\frac{d^2\theta(t)}{dt^2}\right] = \frac{1}{\mathsf{T}^2}, \tag{1.19}$$

with the corresponding connections to linear motion (for rotational motion with fixed radius R) given by

$$s(t) = R\theta(t) \quad ([s] = \mathsf{L}) \quad v(t) = R\omega(t) \quad ([v] = \mathsf{L}/\mathsf{T}) \quad \text{and} \quad a(t) = R\alpha(t) \quad ([a] = \mathsf{L}/\mathsf{T}^2) \tag{1.20}$$

all having the appropriate dimensions.

To capture the **dynamics** of rotational motion, we need the angular analogs of force and momentum, namely **torque** ($\boldsymbol{\tau}$) and **angular momentum** ($\boldsymbol{L}$), given by

$$[\boldsymbol{\tau}] = [\boldsymbol{r} \times \boldsymbol{F}] = [\boldsymbol{r}][\boldsymbol{F}] = \mathsf{L}\left(\frac{\mathsf{ML}}{\mathsf{T}^2}\right) = \frac{\mathsf{ML}^2}{\mathsf{T}^2} \tag{1.21}$$

$$[\boldsymbol{L}] = [\boldsymbol{r} \times \boldsymbol{p}] = [\boldsymbol{r}][\boldsymbol{p}] = \mathsf{L}\left(\frac{\mathsf{ML}}{\mathsf{T}}\right) = \frac{\mathsf{ML}^2}{\mathsf{T}}, \tag{1.22}$$

where we have used the result that for vectors we have $[\boldsymbol{A} \times \boldsymbol{B}] = [\boldsymbol{A}][\boldsymbol{B}]$ for cross-products. The analog of mass for rotational motion is **moment of inertia** (I) which, for a point particle of mass m a fixed distance R away from a rotational axis, is given by

$$[I] = [mR^2] = \mathsf{ML}^2, \tag{1.23}$$

with the analog of Newton's third law being

$$\tau = I\alpha, \tag{1.24}$$

with both sides having the dimensions of $\mathsf{ML}^2/\mathsf{T}^2$.

Note: Despite having the same basic sets of dimensions, namely $\mathsf{ML}^2/\mathsf{T}^2$, energy (in any form) and torque are conceptually radically different and so must be treated in context for any problem in which they appear.

Extending the discussion of the mechanics of point particles to continuous media (fluids such as gases and liquids, solids, but also granular materials), we need the concepts of **mass density** (ρ_m, or mass per unit volume) and **pressure** (P, or force per unit area) with dimensions given by

$$[\rho_m] = \left[\frac{m}{V}\right] = \left[\frac{dm}{dV}\right] = \frac{\mathsf{M}}{\mathsf{L}^3} \quad \text{and} \quad [P] = \left[\frac{F}{A}\right] = \left(\frac{\mathsf{ML}}{\mathsf{T}^2}\right)\left(\frac{1}{\mathsf{L}^2}\right) = \frac{\mathsf{M}}{\mathsf{LT}^2}. \tag{1.25}$$

We include the subscript m for mass density (ρ_m) to distinguish it from **electric charge density** (ρ_e) to be used below in electricity and magnetism (E&M) applications

Another material property is the **bulk modulus**, defined by $B = -V(dP/dV)$, which measures the resistance to volume change caused by an external pressure (a measure of the stiffness or compressibiity of a material), and has dimensions given by (since we can ignore the dimensionless minus sign)

$$[B] = \left[V\frac{dP}{dV}\right] = \left[\frac{V}{dV}\right][dP] = [dP] = \frac{\mathsf{M}}{\mathsf{LT}^2}, \tag{1.26}$$

and so has the same dimensions as pressure, as do other closely related quantities such as the **shear modulus** and **Young's modulus.**

The property of **surface tension** (often denoted S, but also γ), which is sometimes called **surface energy**, can be defined as either the *force per unit length* or *energy per unit area* causing a fluid surface to try to minimize its area. Both descriptions are dimensionally consistent since

$$[S] = \left[\frac{F}{l}\right] = \left(\frac{\mathsf{ML}}{\mathsf{T}^2}\right)\frac{1}{\mathsf{L}} = \frac{\mathsf{M}}{\mathsf{T}^2} = \left(\frac{\mathsf{ML}^2}{\mathsf{T}^2}\right)\frac{1}{\mathsf{L}^2} = \left[\frac{E}{A}\right] = [S]. \tag{1.27}$$

In introductory mechanics, the **coefficient of friction** is defined by the relation $F = \mu N$ where N is the normal force, so that $[\mu] = [F/N]$ is actually dimensionless. For fluids, the corresponding quantity is an important material property called **viscosity** (here labeled η) defined (see Fig. 2.8 for a visualization) by

$$\frac{F}{A} = \eta\frac{\partial v_x}{\partial y} \quad \text{or dimensionally} \quad \left(\frac{\mathsf{ML}}{\mathsf{T}^2}\right)\left(\frac{1}{\mathsf{L}^2}\right) = [\eta]\left(\frac{\mathsf{L/T}}{\mathsf{L}}\right) \quad \text{giving} \quad [\eta] = \frac{\mathsf{M}}{\mathsf{LT}}. \tag{1.28}$$

Note: Many engineering texts use the symbol μ for viscosity (called absolute or dynamic), which should not be confused with the dimensionless coefficient of friction. There are, after all, only so many letters in the Greek and Roman alphabets, so paying attention to disciplinary context is important in deciphering the physical property associated with a particular symbol. A related quantity is the **kinematic viscosity**, defined as $\nu = \mu/\rho_m$, which has dimensions $[\nu] = (\mathsf{M/LT})/(\mathsf{M/L}^3) = \mathsf{L}^2/\mathsf{T}$ and no mass dependence.

Periodic phenomena can be characterized by their **frequency** (f) or **angular frequency** (ω), both with dimensions $[f] = [\omega] = 1/\mathsf{T}$, and their **period** ($\tau$) (with dimension $[\tau] = \mathsf{T}$), all of which are related by

$$\omega = 2\pi f = \frac{2\pi}{\tau}, \tag{1.29}$$

while for waves with spatial regularity, we have the **wavelength** (λ, with dimension $[\lambda] = \mathsf{L}$) or the **wave-number** ($k = 2\pi/\lambda$, with dimension $[k] = 1/\mathsf{L}$.)

An important subset of examples of periodic behavior include waves, and the **wave equation** in one dimension is given by

$$\frac{\partial^2 A(x,t)}{\partial t^2} = v^2 \frac{\partial^2 A(x,t)}{\partial x^2}, \tag{1.30}$$

or in three dimensions by

$$\frac{\partial^2 A(\mathbf{r},t)}{\partial t^2} = v^2 \left\{ \frac{\partial^2 A(\mathbf{r},t)}{\partial x^2} + \frac{\partial^2 A(\mathbf{r},t)}{\partial y^2} + \frac{\partial^2 A(\mathbf{r},t)}{\partial z^2} \right\} = v^2 \left\{ \nabla^2 A(\mathbf{r},t) \right\} \tag{1.31}$$

where v is the **wave speed.** The generalized amplitude, A or $\mathbf{A}$, might be the displacement ($A(x,t) = d(x,t)$) of a stretched string in one dimension, the pressure ($A(\mathbf{r},t) = P(\mathbf{r},t)$) in a three-dimensional sound wave, or the electric field in an electromagnetic (EM) wave ($\mathbf{A}(\mathbf{r},t) = \mathbf{E}(\mathbf{r},t)$). We note that the dimensions of Eqns. (1.30) and (1.31) are consistent, since the quantity $A(x,t)$ (or $A(\mathbf{r},t)$ or $\mathbf{A}(\mathbf{r},t)$) appears linearly on both sides, while

$$\left[\frac{\partial^2}{\partial t^2}\right] = [v^2]\left[\frac{\partial^2}{\partial x^2}\right] \quad \longrightarrow \quad \frac{1}{\mathsf{T}^2} \stackrel{!}{=} \left(\frac{\mathsf{L}}{\mathsf{T}}\right)^2 \frac{1}{\mathsf{L}^2}, \tag{1.32}$$

A simple solution of the wave equation in 1D is $A(x,t) = A_0 \cos(kx - \omega t)$ which can be inserted into Eqn. (1.30) to obtain the relation between wave-number (wavelength), frequency (period), and wave speed, as

$$\omega^2 = v^2 k^2 \quad \text{or} \quad 2\pi f = \underbrace{\omega(k) = vk}_{} = v\frac{2\pi}{\lambda} \quad \text{or} \quad v = f\lambda = \frac{\lambda}{\tau}, \tag{1.33}$$

where the relation between $\omega(k)$ and k is an example of a **dispersion relation**, in this case a simple linear one.

Note: This is our first example (of many) where some component of a DA problem (in this case, the solution) depends in a crucial way on a dimensionless combination of physical parameters, say as the argument of a "pure math function," in this case $\phi(x,t) = kx - \omega t$ and $\cos(\phi(x,t))$ respectively, where

$$[kx] = \left(\frac{1}{\mathsf{L}}\right)\mathsf{L} = 1 \quad \text{and} \quad [\omega t] = \left(\frac{1}{\mathsf{T}}\right)\mathsf{T} = 1. \tag{1.34}$$

An equation we will also encounter multiple times, in many different physical contexts, is the **diffusion equation**, which in one dimension reads

$$\frac{\partial A(x,t)}{\partial t} = D\frac{\partial^2 A(x,t)}{\partial x^2} \quad \text{or in 3D} \quad \frac{\partial A(\mathbf{x},t)}{\partial t} = \nabla^2 A(\mathbf{x},t), \tag{1.35}$$

where D is the **diffusion constant**, with dimensions $[D] = \mathsf{L}^2/\mathsf{T}$.

1.2.2 Electricity and magnetism (E&M)

Even a cursory review of the E&M chapters in an intro-level physics text will show an array of new dimensionful physical quantities (many of which are collected in Table 10.2 in Section 10.1), all of which include only one new base dimension, namely **electric charge**, q, with dimension given by

$$\boxed{[q] = \mathsf{Q}} \tag{1.36}$$

which is the fourth fundamental or base dimension, adding to L, T, M. The derived quantities of **current** (I) and its derivative ($\dot{I}$) satisfy

$$[I] = \left[\frac{dq}{dt}\right] = \frac{\mathsf{Q}}{\mathsf{T}} \quad \text{and} \quad [\dot{I}] = \left[\frac{dI}{dt}\right] = \left[\frac{d^2q}{dt^2}\right] = \frac{\mathsf{Q}}{\mathsf{T}^2}. \tag{1.37}$$

While the quantity $\ddot{q}(t) = \dot{I}(t)$ does not play nearly the same important role in E&M that acceleration ($\ddot{x}(t) = a(t)$) does in mechanics, it does appear in the analysis of circuits involving inductors, so we include it here for completeness.

We will use the Coulomb as the SI unit of charge, with the Ampère then derived as *Coulomb per second* as the unit of current. However, we immediately acknowledge that the MKS<u>A</u> system has the <u>A</u>mpere as the fundamental dimensional unit for E&M, partly because of the historical technical advantages of making precision measurements involving currents. We use this conceptual framework here for what we trust is pedagogical familiarity and simplicity, in lieu of metrological utility and consistency. We note Wikipedia does regularly use MKS<u>A</u> dimensions when discussing units, and that one (very useful!) physics handbook (Cohen et al. (2003, Sec. 1.3) uses LMTI dimensional notation (so, I for current) as base dimensions, logically consistent with MKSA units. We occasionally describe our "hybrid" set of assumed units as MKS(C). See Section 10.10 for a review of current metrological standards and definitions of base units, and especially Section 10.6 for a brief discussion of the alternative system of centimeter-gram-second (CGS)-Gaussian units for E&M.

For problems involving continuous bodies, we need the concepts of **charge density** (ρ_e for charge per unit volume) and **current density** (J_e for current per unit area) with dimensions

$$[\rho_e] = \left[\frac{dq}{dV}\right] = \frac{\mathsf{Q}}{\mathsf{L}^3} \quad \text{and} \quad [J_e] = \left[\frac{I}{A}\right] = \frac{\mathsf{Q}}{\mathsf{TL}^2} \tag{1.38}$$

where we use the subscript e to distinguish charge density from mass density ρ_m.

Perhaps the simplest way to derive the dimensions of the **electric field** ($\boldsymbol{E}$) and **magnetic field** ($\boldsymbol{B}$) is to make use of the **Lorentz force law** for a charged particle, namely

$$\boldsymbol{F} = q\,(\boldsymbol{E} + \boldsymbol{v} \times \boldsymbol{B}) \tag{1.39}$$

which gives

$$[\boldsymbol{E}] = \frac{[\boldsymbol{F}]}{[q]} = \frac{\mathsf{ML}}{\mathsf{QT}^2} \tag{1.40}$$

$$[\boldsymbol{B}] = \frac{[\boldsymbol{F}]}{[\boldsymbol{v}][q]} = \frac{\mathsf{M}}{\mathsf{QT}}, \tag{1.41}$$

with canonical units of N/C (Newton/Coulomb) for $\boldsymbol{E}$ and *Tesla* for $\boldsymbol{B}$.

If we ignore complications arising from quantum mechanics and the intrinsic spin of elementary particles contributing to magnetic fields, so, restricting ourselves to classical physics for the moment, one can fairly say that

> Electric fields are caused by static charges and magnetic fields are caused by moving charges (Griffiths (2022))

and we can use textbook results for Coulomb's law (for an infinitesimal charge, dq) and the Biot–Savart law (for a short length of current I along the direction $d\boldsymbol{l}$) to write

$$d\boldsymbol{E} = \frac{1}{4\pi\epsilon_0}\frac{dq\,\hat{\boldsymbol{r}}}{r^2} \tag{1.42}$$

$$d\boldsymbol{B} = \frac{\mu_0}{4\pi}\frac{Id\boldsymbol{l}\times\hat{\boldsymbol{r}}}{r^2}, \tag{1.43}$$

which can be taken to define the fundamental strengths of the **electro-static** and **magneto-static** interactions via ϵ_0 and μ_0 (along with their conventional dimensionless factors of 4π.) These two constants are respectively referred to as the **vacuum permittivity** (or **permittivity of free space**) and **vacuum permeability** (or **permeability of free space**). We note that the **unit position vector** is dimension<u>less</u> since $[\hat{\boldsymbol{r}}] \equiv [\boldsymbol{r}]/[r] = \mathsf{L}/\mathsf{L} = 1$ and just provides the relevant direction in space.

We can invert Eqns. (1.42) and (1.43) to obtain the dimensions of ϵ_0 and μ_0 as

$$[\epsilon_0] = \frac{[dq]}{[\boldsymbol{E}][r^2]} = \frac{\mathsf{Q}^2\mathsf{T}^2}{\mathsf{ML}^3} \tag{1.44}$$

$$[\mu_0] = \frac{[\boldsymbol{B}][r^2]}{[I][d\boldsymbol{l}]} = \frac{\mathsf{ML}}{\mathsf{Q}^2}, \tag{1.45}$$

and we note (for future reference) the dimensional relation

$$[\mu_0\epsilon_0] = \left(\frac{\mathsf{ML}}{\mathsf{Q}^2}\right)\left(\frac{\mathsf{Q}^2\mathsf{T}^2}{\mathsf{ML}^3}\right) = \frac{\mathsf{T}^2}{\mathsf{L}^2} \quad \text{or} \quad \left[\frac{1}{\sqrt{\epsilon_0\mu_0}}\right] = \frac{\mathsf{L}}{\mathsf{T}}. \tag{1.46}$$

It is clear that Newton's law of gravitation and the Coulomb force law for electrostatics have very similar forms,[4] namely

$$|F_G| = G\frac{m_1 m_2}{r^2} \qquad \text{and} \qquad |F_C| = \frac{1}{4\pi\epsilon_0}\frac{q_1 q_2}{r^2}, \tag{1.47}$$

with G and $1/4\pi\epsilon_0$ playing corresponding roles as the fundamental strengths of their respective interactions. There are, however, several important differences:

- While electric charge can be both positive and negative (thanks, Benjamin Franklin!), gravitational masses (which are, after all, experimentally the same as inertial masses) are only positive and there seemingly is no "antigravity" (Nieto and Goldman (1991)), with matter and antimatter interacting the same way gravitationally.

---

[4] For more mathematical parallelisms, see Section 3.3.

- The proton and electron masses might be considered as possible "quanta of mass" but the simplest bound state of the two (the hydrogen atom) has a mass that is slightly smaller than their combined rest masses, due to their binding energy, so it's perhaps hard to argue that a "mass quantum" is a relevant concept.
- On the other hand, electric charge **is** quantized, with e being the basic unit, with experimental proof going back to Millikan (1913). Fundamental particles such as quarks, with fractional charges of $\pm 1/3e$ or $\pm 2/3e$, are not seen as free particles, only as bound states with net integral values of $\pm e$.
- The charges of the proton and electron, while opposite in sign, are seemingly equal in magnitude, with $|q_{p^+} + q_{e^-}|/|e| \leq 10^{-21}$, so that bulk matter is electrically neutral. (For a review, see Unnikrishan and Gillies 2004), or see experiments by Dylla and King 1973) or more recent/stringent limits from Bressi et al. 2011.)
- Even antihydrogen is observed to be experimentally neutral (see Ahmadi et al. 2106) and the electron and positron have the same mass to within experimental errors (Chu, Mills, and Hall 1984), namely $|m(e^+) - m(e^-)|/m(e^-) \leq 4 \times 10^{-8}$.

We have focused so far on electric fields in empty space, where the vacuum permittivity ϵ_0 is the relevant quantity. In matter, polarization effects can change the "strength" of the $\boldsymbol{E}$ field, which can be described by letting $\epsilon_0 \to \epsilon$ where $\epsilon > \epsilon_0$ in Coulomb's law (and in other applications, such as the calculation of capacitance and energy stored in the electric field in devices). This effect is described variously by writing

$$\epsilon = K\epsilon_0 = \kappa\epsilon_0 = \epsilon_r\epsilon_0 \tag{1.48}$$

where $K = \kappa$ is the **dielectric constant** and ϵ_r is the **relative permittivity**, all of which are equal and, of course, dimensionless. For applications in air, the impact is small, since $K_{air} \approx 1.0006$, while in many materials used in electrostatic applications (say capacitors) $K \approx 2 - 10$. But for a substance as ubiquitous and important as liquid water (say in biophysical applications), $K_{H_2 0} \approx 80$, which can make a large numerical difference. We will generally ignore dielectric constant effects, since DA can provide no guidance on their impact, unless we're being careful with specific examples (such as in **Example 5.4**).

One of the most profound conceptual accomplishments in physics, discussed in all E&M texts, at all levels, is the extension of work by Gauss, Faraday, and Ampère by Maxwell (adding the notion of **displacement current**) resulting in what have come to be known as **Maxwell's equations**. Their standing in intellectual history, and their impact on technology, cannot be overestimated. Feynman's famous quote perhaps expresses it best:

> ...there can be little doubt that the most significant event of the 19th century will be judged as Maxwell's discovery of the laws of electrodynamics. The American Civil War will pale into provincial insignificance in comparison with this important scientific event of the same decade (Feynman et al. 1964), Vol. 2, sections 1–6).

While representing a more mathematically advanced topic than can be covered in intro-level texts with full rigor, requiring a full understanding of vector calculus to appreciate and utilize to solve problems, the dimensionality of Maxwell's equations can be understood perhaps most simply in their differential form, namely

$$\nabla \cdot \boldsymbol{E}(\boldsymbol{r}, t) = \frac{1}{\epsilon_0}\rho_e(\boldsymbol{r}, t) \qquad \text{Gauss's law} \tag{1.49}$$

$$\nabla \cdot \boldsymbol{B}(\boldsymbol{r}, t) = 0 \qquad \text{No magnetic monopoles} \tag{1.50}$$

$$\nabla \times \boldsymbol{E}(\boldsymbol{r}, t) = -\frac{\partial}{\partial t}\boldsymbol{B}(\boldsymbol{r}, t) \qquad \text{Faraday's law} \tag{1.51}$$

$$\nabla \times \boldsymbol{B}(\boldsymbol{r}, t) = \mu_0 \boldsymbol{J}_e(\boldsymbol{r}, t) + \underbrace{\mu_0\epsilon_0\frac{\partial}{\partial t}\boldsymbol{E}(\boldsymbol{r}, t)}_{\text{displacement term}} \qquad \text{Ampère's law (+ Maxwell).} \tag{1.52}$$

Given that we've already encountered the dimensionality of time and space derivatives ($\partial/\partial t$ and ∇), we can confirm the dimensional consistency, of say Gauss's law, by comparing

$$\begin{aligned}
[\nabla \cdot \boldsymbol{E}] &\stackrel{?}{=} \left[\frac{1}{\epsilon_0}\rho_e\right] \\
\left(\frac{1}{\mathsf{L}}\right)\left(\frac{\mathsf{ML}}{\mathsf{QT}^2}\right) &\stackrel{?}{=} \left(\frac{\mathsf{Q}^2\mathsf{T}^2}{\mathsf{ML}^3}\right)^{-1}\left(\frac{\mathsf{Q}}{\mathsf{L}^3}\right) \\
\frac{\mathsf{M}}{\mathsf{QT}^2} &\stackrel{!}{=} \frac{\mathsf{M}}{\mathsf{QT}^2},
\end{aligned} \tag{1.53}$$

and similarly for Eqns. (1.50)–(1.52), as in **P1.4.**

The equivalent versions of Maxwell's equations in integral form are

$$\oint \boldsymbol{E} \cdot d\boldsymbol{A} = \frac{q_{enc}}{\epsilon_0} \qquad \text{Gauss's law} \tag{1.54}$$

$$\oint \boldsymbol{B} \cdot d\boldsymbol{A} = 0 \qquad \text{No magnetic monopoles} \tag{1.55}$$

$$\oint \boldsymbol{E} \cdot d\boldsymbol{l} = -\frac{d}{dt}\int \boldsymbol{B} \cdot d\boldsymbol{A} \qquad \text{Faraday's law} \tag{1.56}$$

$$\oint \boldsymbol{B} \cdot d\boldsymbol{l} = \mu_0 I_{enc} + \underbrace{\mu_0\epsilon_0\frac{d}{dt}\int \boldsymbol{E} \cdot d\boldsymbol{A}}_{\text{displacement term}} \qquad \text{Ampère's law (+ Maxwell)} \tag{1.57}$$

and for dimensional checking, we only require the geometrical results $[d\boldsymbol{A}] = \mathsf{L}^2$ and $[d\boldsymbol{l}] = \mathsf{L}$ for infinitesimal surface areas ($d\boldsymbol{A}$) and line segments ($d\boldsymbol{l}$), and to recall that integral signs (of any kind, $\int$ or $\oint$) are dimensionless, since they just represent (continuous) summations. In this language, the dimensions of Gauss's law are seen to be consistent since

$$\begin{aligned}
\left[\oint \boldsymbol{E} \cdot d\boldsymbol{A}\right] &\stackrel{?}{=} \left[\frac{q_{enc}}{\epsilon_0}\right] \\
\left(\frac{\mathsf{ML}}{\mathsf{QT}^2}\right)(\mathsf{L}^2) &\stackrel{?}{=} \mathsf{Q}\left(\frac{\mathsf{Q}^2\mathsf{T}^2}{\mathsf{ML}^3}\right)^{-1} \\
\frac{\mathsf{ML}^3}{\mathsf{QT}^2} &\stackrel{!}{=} \frac{\mathsf{ML}^3}{\mathsf{QT}^2}.
\end{aligned} \tag{1.58}$$

One of the major implications of Maxwell's equations (especially with the inclusion of the displacement current term) is that (highly correlated) configurations of both $\boldsymbol{E}(\boldsymbol{r},t)$ and $\boldsymbol{B}(\boldsymbol{r},t)$ fields can satisfy the classical **wave equation**, which in free space (with no $\rho_e, \boldsymbol{J}_e$ charge or current density source terms) reads

$$\frac{\partial^2}{\partial t^2}\{\boldsymbol{E}(\boldsymbol{r},t), \boldsymbol{B}(\boldsymbol{r},t)\} = \frac{1}{\mu_0 \epsilon_0}\nabla^2 \{\boldsymbol{E}(\boldsymbol{r},t), \boldsymbol{B}(\boldsymbol{r},t)\}\,. \tag{1.59}$$

Comparing this to the general form of the wave equation in Eqn. (1.31) implies that electric and magnetic fields can support traveling waves characterized by $v = 1/\sqrt{\epsilon_0 \mu_0} = c$ which is, of course, the speed of light, with dimensions matching via Eqn. (1.46). This is another of the great **unifications** in physics, bringing together electricity, magnetism, and optics.

Returning now to electro-statics, the (vector) electric field of a point charge can be derived from a corresponding (scalar) **electric potential**, with

$$\boldsymbol{E}_q(\boldsymbol{r}) = \frac{q}{4\pi\epsilon_0}\frac{\hat{\boldsymbol{r}}}{r^2} \qquad \text{related to} \qquad V_q(\boldsymbol{r}) = \frac{q}{4\pi\epsilon_0}\frac{1}{r} \qquad \text{by} \quad \boldsymbol{E}_q(\boldsymbol{r}) = -\boldsymbol{\nabla} V_q(\boldsymbol{r}), \tag{1.60}$$

where V is measured in *Volts*. This implies that another unit for electric field $\boldsymbol{E}$ is then $V/m = N/C$, with dimensions $[V] = [E][dx] = \mathsf{ML^2/QT^2}$.

An important constitutive relation relevant for many materials is that $\boldsymbol{J}_e = \sigma_e \boldsymbol{E}$, so that an applied electric field can give rise to a current (density), with σ_e called the **electrical conductivity** being an important transport property. This is a version of **Ohm's law** and can also be written as

$$\frac{I}{A} = \underbrace{|\boldsymbol{J}_e| = \sigma_e |\boldsymbol{E}|}_{\text{Ohms' law}} = \sigma_e \left|\frac{dV}{dx}\right| \tag{1.61}$$

which gives the dimensional relation

$$[\sigma_e] = \frac{[I][dx]}{[A][dV]} = \frac{\mathsf{Q^2T}}{\mathsf{ML^3}}. \tag{1.62}$$

The related quantity of **electrical resistivity** is defined as $\rho = 1/\sigma_e$, often shown with the same symbol as mass (ρ_m) or charge (ρ_e) density, but here with the lack of a subscript (so just plain ρ) to help minimize confusion. The important electrical transport relation defined by Eqn. (1.61) can be written in the form

$$\frac{dq}{dt} = I = \sigma_e A \frac{dV}{dx} \tag{1.63}$$

which begs comparison to the corresponding expression for thermal transport (in Eqn. (1.73)) and makes their connection more straightforward. This expression can also be rewritten in terms of the voltage drop across a resistor as

$$\Delta V = dV = I\left(\frac{dx}{\sigma_e A}\right) \equiv IR \tag{1.64}$$

where R is the standard **resistance** of a circuit element in terms of its conductivity and geometry (length dx and cross-section area A).

One can extend the notion of electric potential in Eqn. (1.60) more generally to time-dependent electric and magnetic fields, which can be derived from a combination of **scalar** ($\phi(\boldsymbol{r}, t)$, a generalization of $V(\boldsymbol{r})$) and **vector** ($\boldsymbol{A}(\boldsymbol{r}, t)$) **potentials** via

$$\boldsymbol{E}(\boldsymbol{r}, t) = -\nabla\phi(\boldsymbol{r}, t) - \frac{\partial}{\partial t}\boldsymbol{A}(\boldsymbol{r}, t) \quad \text{and} \quad \boldsymbol{B}(\boldsymbol{r}, t) = \nabla \times \boldsymbol{A}(\boldsymbol{r}, t) \tag{1.65}$$

which imply that

$$[\phi(\boldsymbol{r}, t)] = \frac{\mathsf{ML}^2}{\mathsf{QT}^2} \quad \text{and} \quad [\boldsymbol{A}(\boldsymbol{r}, t)] = \frac{\mathsf{ML}}{\mathsf{QT}}. \tag{1.66}$$

Finally, most intro-level texts (rightly) have a focus on the impact E&M has on practical applications to devices and technologies, including the physical principles underlying basic electronic circuit elements such as **capacitors** (C for capacitance), **resistors** (R for resistance), and **inductors** (L for inductance), whose dimensions are given by

$$[C] = \frac{\mathsf{Q}^2\mathsf{T}^2}{\mathsf{ML}^2}, \quad [R] = \frac{\mathsf{ML}^2}{\mathsf{Q}^2\mathsf{T}}, \quad \text{and} \quad [L] = \frac{\mathsf{ML}^2}{\mathsf{Q}^2}. \tag{1.67}$$

The associated SI units are *farad* (after Michael Faraday), *ohm* (for Georg Ohm), and *henry* (to acknowledge Joseph Henry), while the symbol for inductance being L honors Heinrich Lenz.

Some applications in engineering (and even in condensed matter physics) find it useful to use the inverse of resistance, namely the **conductance** G, with

$$[G] = \frac{1}{[R]} = \frac{\mathsf{Q}^2\mathsf{T}}{\mathsf{ML}^2}. \tag{1.68}$$

1.2.3 Thermal physics

Many advanced undergraduate texts in thermal physics (such as Reif (1965), Kittel and Kroemer (1980), or Schroeder (2000)) or corresponding graduate level treatments on the subject (Landau and Lifschitz (1958), Huang (1987), or Pathria (2006)) introduce the concept of **temperature** in the context of statistical mechanics. In contrast, most introductory-level textbooks (whether for science/engineering or life science students, whether calculus-based or at the algebra–trig level) take a more operational/historical approach and use traditional ideas from thermometry.

For example, citing experiments using ideal gases and constant volume pressure gauge thermometers, many intro-level treatments define the **absolute temperature** (in Kelvin or K) via

$$T \equiv 273.16\,K \lim_{m \to 0}\left(\frac{P}{P_{tp}}\right) = 273.16\,K \lim_{P_{tp} \to 0}\left(\frac{P}{P_{tp}}\right) \tag{1.69}$$

where P_{tp} is the pressure at the **triple point** (where the solid, liquid, and gas phases coexist) and where the limit is taken for increasingly small amounts of gas ($m \to 0$) or low pressure ($P_{tp} \to 0$). The scale factor is then designed to reproduce the "slope" of the Celsius or centigrade scale (with 0 and 100 for the temperature of the ice–water and water–steam transition, in T_C) with $T_K = T_C + 273.15$. Almost all such pedagogical treatments include data (real or imagined) of pressure versus temperature for many ideal gases, showing a $P \propto T$ linear behavior (variously called **Gay-Lussac's law** or sometimes **Amontons' law**), all extrapolating to zero pressure (extrapolated, since real gases eventually condense) at the same $T = 0\,K$ temperature.

Whatever the definition, it's conventional to introduce a new dimension to describe this **absolute temperature** as

$$\boxed{[T] = [\text{absolute temperature}] = \Theta} \tag{1.70}$$

which adds a fifth (and, as we'll see, final) base dimension to our M, L, T, Q set.

One of the simplest (almost trivial) examples where this new base dimension comes into play is for **thermal expansion**, where, for small changes in temperature (ΔT), the change in length (ΔL) is proportional to the original length itself (L) and ΔT, as

$$\Delta L = \alpha_{th} L \Delta T \qquad \text{or} \qquad \frac{\Delta L}{L} = \alpha_{th} \Delta T, \tag{1.71}$$

where α_{th} is the **coefficient of linear expansion**, which therefore has dimensions $[\alpha_{th}] = 1/\Theta$. The related **coefficient of volume expansion** defined by $\Delta V/V = \beta_{th} \Delta T$ clearly has the same dimensional "spelling" and simple geometrical considerations require that $\beta_{th} = 3\alpha_{th}$, with the same dimensions but a different numerical value.

Noting that some authors have cautioned about the "... use and misuse..." (Zemansky (1970)) of the word **heat**, even arguing against having it be used as a noun (Romer (2001)), many textbooks provide carefully worded definitions for this quantity (most often labeled as Q) designed to avoid conceptual sloppiness, with one example being

> ... the energy transferred between a system and the environment as a consequence of a temperature difference between them (Knight 2013), p. 476),

but all agree that its dimensions match those of any form of energy, namely

$$[Q] = [\text{energy}] = \frac{\mathsf{ML}^2}{\mathsf{T}^2}. \tag{1.72}$$

Such definitions are consistent with several important physical concepts involving thermal transport phenomena, such as the transfer of energy due to a temperature difference, which includes the material property known as **thermal conductivity**, κ, given via

$$H \equiv \frac{dQ}{dt} = \kappa A \frac{\Delta T}{\Delta x}, \tag{1.73}$$

where H is the rate of heat flow (so, energy per unit time), $\Delta T/\Delta x$ is the temperature gradient (in one dimension), and A is the cross-sectional area in the geometry defining the flow. (Note again the similarity between this relation and Eqn. (1.63) which is relevant for charge transport.) We can use Eqn. (1.73) to derive the dimensions of κ to be

$$[\kappa] = \left[\frac{dQ}{dt}\right] \frac{[dx]}{[dT]} \frac{1}{[A]} = \left(\frac{\mathsf{ML}^2}{\mathsf{T}^3}\right)\left(\frac{\mathsf{L}}{\Theta}\right)\left(\frac{1}{\mathsf{L}^2}\right) = \frac{\mathsf{ML}}{\mathsf{T}^3\Theta}. \tag{1.74}$$

Another form of thermal energy transfer occurs as the temperature of a material is increased or decreased, giving

$$\Delta Q = mC\Delta T \tag{1.75}$$

for the thermal energy change (ΔQ) related to the temperature change (ΔT) for a given mass (m), which defines the **specific heat** for a given substance, with corresponding dimensions

$$[C] = \frac{[\Delta Q]}{[m][\Delta T]} = \frac{\mathsf{L}^2}{\mathsf{T}^2\Theta}. \tag{1.76}$$

A second important concept for tracking the thermal history of a sample relates to the energy change associated with phase changes, where one has for the solid–liquid (F for fusion) and liquid–vapor (V for vaporization) transitions the corresponding **latent heat**, defined by

$$Q = mL_F \quad \text{and} \quad Q = mL_V \quad \text{with} \quad [L_F] = [L_V] = \frac{\mathsf{L}^2}{\mathsf{T}^2}, \tag{1.77}$$

where Q is the amount of heat released/absorbed when a sample of mass m undergoes a phase transition. Given that these phase changes occur at fixed temperature, some authors generalize the conceptualization of **heat** to be more consistent with such phenomena, by suggesting definitions such as

> Heat is the transfer of energy to a system by thermal contact with a reservoir (Kittel and Kroemer 1980, p. 227).

Moving from classical thermodynamics to making connections to statistical mechanics, more advanced treatments of thermal physics focus on the concept of entropy via the famous **Boltzmann equation**[5]

$$S = k_B \ln(\Omega), \tag{1.78}$$

where Ω is the number of micro-states accessible to the system (and, being a pure number, is therefore dimensionless) and k_B is the **Boltzmann constant.** Relations such as

$$dS = \frac{dU}{T}, \tag{1.79}$$

where U is the energy, can then be used to derive relations such as the **ideal gas law**

$$PV = Nk_BT = nRT, \tag{1.80}$$

where P, V are the pressure and volume of a sample of gas, N is the number of molecules, n the number of moles of gas, and the **gas constant** R is related to k_B by $R = N_Ak_B$, where N_A is **Avogadro's number**, which is dimensionless. The value of k_B is then connected to the conventional, thermometrically defined, Kelvin temperature scale given in Eqn. (1.69), with dimensions and magnitude given by

$$\text{(Boltzmann's constant)} \qquad [k_B] = \frac{\mathsf{ML}^2}{\mathsf{T}^2\Theta} \quad \text{where} \quad k_B = 1.381 \times 10^{-23}\, J/K \tag{1.81}$$

which acts as the fundamental constant of thermal physics.

Given the close relation between thermal energy and temperature, there are many situations where the combination k_BT will appear naturally, one example being the **equipartition theorem**, where the (thermal) average (denoted by $\langle\ \rangle_{th}$) of the kinetic energy (of the translational degrees of freedom) of a gas in one or three dimensions is given by

---

[5] This equation is the simple inscription on his tombstone. See https://www.atlasobscura.com/places/boltzmanns-grave – accessed 11/1/2023.

$$\langle KE \rangle_{th} = \frac{m}{2}\langle v^2 \rangle_{th} = \frac{1}{2}k_B T \quad \text{and} \quad \frac{m}{2}\langle \mathbf{v}^2 \rangle_{th} = \frac{m}{2}\langle v_x^2 \rangle_{th} + \frac{m}{2}\langle v_y^2 \rangle_{th} + \frac{m}{2}\langle v_z^2 \rangle_{th} = \frac{3}{2}k_B T. \tag{1.82}$$

We devote Section 5.2.1 to what we describe as "$k_B T$" physics, exploring physical systems where this combination automatically occurs.

1.2.4 **Quantum mechanics and relativity**

While quantum mechanics is conceptually about as different from classical mechanics as possible (being probabilistic instead of deterministic is only one of many examples), it still describes the behavior (let's not say "motion") of matter, but at a microscopic scale, while incorporating the wave nature of the particles involved. It includes one important new fundamental physical parameter, **Planck's constant**, which was first derived (see a more extensive discussion in Section 6.1) in the context of data-fitting of the blackbody spectrum, and in the original context was labeled h. More frequently it appears in applications in the undergraduate curriculum in the form of the **reduced Planck's constant**, written as $\hbar = h/2\pi$, but of course both have the same dimensions, $[\hbar] = [h/2\pi] = \mathsf{ML}^2/\mathsf{T}$.

One early appearance of h was in Einstein's quantization of photon energy (E_γ) in terms of frequency (f) via

$$E_\gamma = hf = \left(\frac{h}{2\pi}\right)(2\pi f) = \hbar\omega \tag{1.83}$$

when written in terms of $\hbar$ and the angular frequency ω. A second familiar application is the **de Broglie relation**, which associates a wavelength, λ_{dB}, with the momentum of material particles via

$$\lambda_{dB} = \frac{h}{p} \quad \longrightarrow \quad p = \frac{h}{\lambda_{dB}} = \left(\frac{h}{2\pi}\right)\left(\frac{2\pi}{\lambda_{dB}}\right) = \hbar k \tag{1.84}$$

in terms of $\hbar$ and the **wave-number** k. Both of these associations require the same dimensionality, namely that

$$\begin{aligned} [\hbar] = [h] = \frac{[E_\gamma]}{[f]} = \left(\frac{\mathsf{ML}^2}{\mathsf{T}^2}\right)\mathsf{T} &= \frac{\mathsf{ML}^2}{\mathsf{T}} \\ [h] = [\lambda_{dB}][p] = \mathsf{L}\left(\frac{\mathsf{ML}}{\mathsf{T}}\right) &= \frac{\mathsf{ML}^2}{\mathsf{T}}. \end{aligned} \tag{1.85}$$

An important application of quantum mechanics is the derivation of the quantized energy levels of bound state systems, perhaps most notably those of the **hydrogen atom** (which we discuss more extensively in Sections 6.3.4 and 7.2) given by

$$E_n = -\frac{1}{2n^2}\frac{m_e e^4}{\hbar^2(4\pi\epsilon_0)^2} \tag{1.86}$$

where n is a (dimensionless) **quantum number**. We can check that this expression is dimensionally correct by writing

$$\begin{aligned}\frac{\mathsf{ML}^2}{\mathsf{T}^2} = [E_n] &\stackrel{?}{=} \left[-\frac{1}{2n^2}\right]\left[\frac{1}{(4\pi)^2}\right][m_e][e^4][\hbar]^{-2}[\epsilon_0]^{-2} \\ &\stackrel{?}{=} (1)\,(1)\,\mathsf{M}\,\mathsf{Q}^4\left(\frac{\mathsf{ML}^2}{\mathsf{T}}\right)^{-2}\left(\frac{\mathsf{Q}^2\mathsf{T}^2}{\mathsf{ML}^3}\right)^{-2} \\ \frac{\mathsf{ML}^2}{\mathsf{T}^2} &\stackrel{!}{=} \mathsf{M}^{1-2+2}\,\mathsf{L}^{-4+6}\,\mathsf{T}^{2-4}\,\mathsf{Q}^{4-4} = \mathsf{M}^1\,\mathsf{L}^2\,\mathsf{T}^{-2}.\end{aligned} \tag{1.87}$$

Note: We see from Eqn. (1.86) that besides dimensionless fixed numbers/constants (like the 1/2 or 2π factors we've seen before), physical results can depend on "variable" dimensionless quantities, as in this case, where there are quantum numbers, given here by $n = 1, 2, 3, \ldots$. This type of result is not limited to quantum mechanics, but can also appear in the context of normal modes, or in the solutions of eigenvalue problems, in both classical mechanics and E&M, and DA, by itself, does not provide constraints on such values. We also note that DA cannot distinguish between two fundamental pure numbers, namely ± 1, so the **sign** of a physical effect (which can be of paramount importance) is also not necessarily constrained by matching dimensions. In such cases, informed intuition or careful calculations are usually required.

Many introductory texts end with an overview of (special) relativity, likely including some discussions of physical phenomena such as length contraction and time dilation, focusing mostly on the impact of relativity on kinematics (but often not on E&M), but always with the emphasis on the closer relation of space and time than that encountered in classical mechanics. All such texts cite the **speed of light** (c) as a new fundamental constant, and note its role as Nature's ultimate speed limit, and of course we have $[c] = \mathsf{L}/\mathsf{T}$ as its dimensional spelling.

Another famous Einstein result, namely the mass–energy connection, can be written in terms of the **rest mass** of a particle (m_0) and its speed (v) as

$$E = \gamma m_0 c^2 \quad \text{where} \quad \gamma \equiv \frac{1}{\sqrt{1-(v/c)^2}} = \left(1-(v/c)^2\right)^{-1/2}. \tag{1.88}$$

A nice pedagogical discussion of the experimental verification of this connection between the *Speed and kinetic energy of relativistic electrons* is given by Bertozzi (1964). The related expression for relativistic (R) momentum is $\boldsymbol{p} = \gamma m_0 \boldsymbol{v}$.

Note: Dimension**less** combinations of dimension**ful** parameters can appear very naturally, as with γ in this case, ones which DA may or may not necessarily recognize, or be able to constrain, and which can make large numerical differences in predictions for physical results. See **P1.12** for an example of just how large γ factors can be in relevant research settings.

To make contact between Eqn. (1.88) and more familiar non-relativistic (NR) mechanics expressions, we can expand γ for $v << c$ (using a series expansion for $v/c << 1$ or even the binomial theorem), to find

$$\begin{aligned} E = m_0c^2\left(1-\left(\frac{v}{c}\right)^2\right)^{-1/2} &= m_0c^2\left(1+\frac{v^2}{2c^2}+\frac{3v^4}{8c^4}+\cdots\right) \\ &\approx \underbrace{m_0c^2}_{\text{rest energy}} + \underbrace{\frac{1}{2}m_0v^2}_{\text{NR kinetic energy}} + \underbrace{\frac{3}{8}\frac{m_0v^4}{c^2}}_{\text{R correction}} + \cdots \end{aligned} \tag{1.89}$$

which reproduces the rest energy and the classical NR kinetic energy, and provides the first relativistic (R) correction. At the other extreme, for ultra-relativistic particles ($v \lesssim c$) where $E >> m_0c^2$, we can invert Eqn. (1.88) to find

$$\frac{v}{c} = \sqrt{1 - \left(\frac{m_0 c^2}{E}\right)^2} \approx 1 - \frac{1}{2}\left(\frac{m_0 c^2}{E}\right)^2 + \cdots \qquad \text{or} \qquad 1 - \frac{v}{c} \approx \frac{1}{2}\left(\frac{m_0 c^2}{E}\right)^2 \approx \frac{1}{2}\gamma^{-2}, \tag{1.90}$$

which can be used to see just how close one comes to the speed limit value of c.

Returning to the hydrogen atom, a heuristic derivation of the semi-classical speed of the electron in a given quantum state labeled by n (for a Bohr orbit description, at least) corresponding to the quantized energies in Eqn. (1.86) gives

$$v_n = \frac{v_0}{n} \qquad \text{where} \qquad v_0 = \frac{e^2}{4\pi\epsilon_0 \hbar}. \tag{1.91}$$

It's natural to compare this value to c to confirm (to first approximation, at least) that the hydrogen atom, while quantum mechanical, is still an NR problem, since

$$\frac{v_0}{c} = \frac{e^2}{4\pi\epsilon_0 \hbar c} \equiv \alpha_{FS} \approx \frac{1}{137.04} << 1. \tag{1.92}$$

Note: The dimension<u>**less**</u> combination denoted by α_{FS} is called the **fine-structure constant** and plays an important role in the quantum theory of the electromagnetic field (QED or quantum electrodynamics). It is perhaps first seen in the undergraduate curriculum in this context, but gets its name from its appearance in the small corrections to the H-atom spectrum (hence "fine structure") due to including the effects of special relativity (via the last term in Eqn. (1.89) and its impact on the observable energy level structure; we discuss this further in Sections 7.1 and 7.2.3.

If we explore the formalism of special relativity at a slightly more advanced level, say that of a modern physics textbook, we learn that space and time can be considered in a much more consistent and unified manner, sometimes being described by a **four-vector** formalism, where for **space-time** and **energy-momentum** we have

$$\mathcal{X} = (ct, \mathbf{x}) = (ct, x, y, z) \tag{1.93}$$

$$\mathcal{P} = (E/c, \mathbf{p}) = (E/c, p_x, p_y, p_z), \tag{1.94}$$

with $[\mathcal{X}] = \mathsf{L}$ and $[\mathcal{P}] = \mathsf{ML/T}$.

As noted in Eqn. (1.8), we expect (demand!) that the components of a three-vector all be expressed in Cartesian coordinates, and so be able to assume that they share the same dimensions. So it makes sense when extending the notion of $\mathbf{x} \to \mathcal{X}$ from space to space-time (and $\mathbf{p} \to \mathcal{P}$ from momentum to energy-momentum) that we include appropriate factors of c, so that the four-vector components share the same dimensionality. This also means that the **Lorentz transformations** for the components of $(ct, \mathbf{x})$ and $(E/c, \mathbf{p})$, which describe the change in coordinate values when going from one reference frame to another one moving at constant speed (say $V_{rel} = (v, 0, 0)$), will have very similar forms, namely

$$\begin{aligned} ct' &= \gamma\left(ct - \frac{v}{c}x\right) \\ x' &= \gamma\left(x - \frac{v}{c}ct\right) \\ y' &= y \\ z' &= z \end{aligned} \tag{1.95}$$

for space-time, and

$$\frac{E'}{c} = \gamma\left(\frac{E}{c} - \frac{v}{c}p_x\right)$$
$$p'_x = \gamma\left(p_x - \frac{v}{c}\frac{E}{c}\right)$$
$$p'_y = p_y$$
$$p'_z = p_z \tag{1.96}$$

for energy-momentum.

We continue the discussion of the impact of special relativity on E&M in Section 4.3, but note here for completeness that there are two other important four-vectors needed for the study of electromagnetic fields, namely the **four-current** (which contains the (scalar) electric charge density, ρ_e, and the (vector) current density, $\boldsymbol{J}_e$) and the **four-potential**, a relativistic combination of the scalar ($\phi(\boldsymbol{r}, t)$) and vector ($\boldsymbol{A}(\boldsymbol{r}, t)$) potentials in Eqn. (1.65). They are given by

$$\mathcal{J} = (c\rho_e, \boldsymbol{J}_e) = (c\rho_e, J_{e,x}, J_{e,y}, J_{e,z}) \tag{1.97}$$
$$\mathcal{A} = (\phi/c, \boldsymbol{A}) = (\phi/c, A_x, A_y, A_z) \tag{1.98}$$

and both transform in the same way as Eqns. (1.95) and (1.96) and their dimensions are given by $[\mathcal{J}] = \mathsf{Q/TL^2}$ and $[\mathcal{A}] = \mathsf{ML/QT}$.

For the time being, it is most important to see (**P1.12**) that these transformations are all dimensionally consistent. We note for future reference that the quantities $q\phi$ and $q\boldsymbol{A}$ have the dimensions of energy and momentum respectively, namely

$$[q\phi] = \mathsf{Q}\left(\frac{\mathsf{ML^2}}{\mathsf{QT^2}}\right) = \frac{\mathsf{ML^2}}{\mathsf{T^2}} = [E] \quad \text{and} \quad [qA] = \mathsf{Q}\left(\frac{\mathsf{ML}}{\mathsf{QT}}\right) = \frac{\mathsf{ML}}{\mathsf{T}} = [\boldsymbol{p}]. \tag{1.99}$$

Finally, some intro-level textbooks will have chapters (often cited as optional, and so not nearly always covered in detail, or even at all) providing surveys of more advanced topics such as atomic and nuclear structure, molecules and condensed matter physics, or particle physics, astrophysics, and cosmology, and we extend our discussion of the use of DA to those topics in Part III (Chapters 7-9).

This first overview (or fly-over) focused on the canonical topics in what we call the **quadrivium**[6] of physics undergraduate coursework, namely mechanics, E&M, thermal physics, and quantum mechanics, and we turn attention in Part II (Chapters 3–6) on exploring what DA can tell us about those core subjects at a more advanced level. This section then serves mostly to introduce us to the basic assumptions and the "spelling" of many important physical quantities and relationships.

[6] The classical quadrivium of a liberal arts education was said to consist of the subjects of arithmetic, geometry, music, and astronomy. This was supposed to follow the **trivium** of grammar, logic, and rhetoric. Perhaps this Chapter 1 can be thought of in that context, as providing the necessary background on how to read, think, and spell/write/speak dimensionally!

1.3 A posteriori checking versus a priori predictions using dimensional analysis

Now that we are (hopefully) fluent in the practice of identifying the spelling behind the mathematical descriptions of physical phenomena in terms of the basic $\mathsf{M}, \mathsf{L}, \mathsf{T}, \mathsf{Q}, \Theta$ base dimensions, we can practice making use of this technique when reviewing familiar, or novel, results, whether at the textbook or research level.

For example, a standard topic in introductory texts in discussions of fluid mechanics or hydrodynamics is the **Bernoulli equation**, which is essentially the **work-energy** theorem applied to continuous materials. Under several assumptions, including **incompressibility** of the fluid, **isentropy** (no transfer of energy as heat or friction), and the application to a fluid "packet" moving along a **streamline**, we have

$$\underbrace{\frac{1}{2}\rho_m v^2}_{\text{kinetic}} + \underbrace{\rho_m g h}_{\text{potential}} + \underbrace{P}_{\text{pressure}} = \text{constant.} \tag{1.100}$$

It's now easy to check that this formula is self-consistent, as all of the terms share the same dimensionality, since

$$\begin{aligned}
\left[\frac{1}{2}\rho_m v^2\right] &= (1)\left(\frac{\mathsf{M}}{\mathsf{L}^3}\right)\left(\frac{\mathsf{L}}{\mathsf{T}}\right)^2 = \frac{\mathsf{M}}{\mathsf{L}\mathsf{T}^2} \\
[\rho_m g h] &= \left(\frac{\mathsf{M}}{\mathsf{L}^3}\right)\left(\frac{\mathsf{L}}{\mathsf{T}^2}\right)\mathsf{L} = \frac{\mathsf{M}}{\mathsf{L}\mathsf{T}^2} \\
[P] &= \frac{\mathsf{M}}{\mathsf{L}\mathsf{T}^2}.
\end{aligned} \tag{1.101}$$

The same type of dimension checking can then be applied to the mathematical descriptions of physical phenomena which may be far less familiar, say ones from current research, or even from historical sources, such as the **Rayleigh–Plesset equation** of hydrodynamics. This relation describes the dynamics of a spherical bubble (of radius $R(t)$) and depends on the **viscosity** (η) and **density** (ρ_m) of the surrounding fluid, the **surface tension** (here called γ) of the bubble–fluid interface, and the pressure difference ($\Delta P_B(t)$) between the bubble interior and exterior, and is given by

$$\underbrace{\rho_m\left\{R(t)\frac{d^2R(t)}{dt^2} + \frac{3}{2}\left(\frac{dR(t)}{dt}\right)^2\right\}}_{\text{kinetic}} + \underbrace{\frac{4\eta}{R(t)}\frac{dR(t)}{dt}}_{\text{viscosity}} + \underbrace{\frac{2\gamma}{R(t)}}_{\text{surface tension}} + \underbrace{\Delta P_B(t)}_{\text{pressure}} = 0. \tag{1.102}$$

We can confirm that Eqn. (1.102) is dimensionally correct by seeing that all four types of terms satisfy

$$
\begin{array}{lr}
\text{kinetic 1} & \left(\frac{\mathsf{M}}{\mathsf{L}^3}\right)\left(\mathsf{L}\frac{\mathsf{L}}{\mathsf{T}^2}\right) \overset{!}{=} \frac{\mathsf{M}}{\mathsf{LT}^2} \\
\text{kinetic 2} & \left(\frac{\mathsf{M}}{\mathsf{L}^3}\right)\left(\frac{\mathsf{L}}{\mathsf{T}}\right)^2 \overset{!}{=} \frac{\mathsf{M}}{\mathsf{LT}^2} \\
\text{viscosity} & \left(\frac{\mathsf{M}}{\mathsf{LT}}\right)\left(\frac{1}{\mathsf{L}}\right)\left(\frac{\mathsf{L}}{\mathsf{T}}\right) \overset{!}{=} \frac{\mathsf{M}}{\mathsf{LT}^2} \\
\text{surface tension} & \left(\frac{\mathsf{M}}{\mathsf{T}^2}\right)\left(\frac{1}{\mathsf{L}}\right) \overset{!}{=} \frac{\mathsf{M}}{\mathsf{LT}^2} \\
\text{pressure} & \frac{\mathsf{M}}{\mathsf{LT}^2} \overset{!}{=} \frac{\mathsf{M}}{\mathsf{LT}^2}.
\end{array}
\tag{1.103}
$$

Example 1.1 Rayleigh–Plesset bubble dynamics

A specific example of a problem for which Eqn. (1.102) is relevant, first discussed by Besant (1859), and later extended by Rayleigh (1917), has been described as follows:

> An infinite mass of homogeneous incompressible fluid acted upon by no forces is at rest, and a spherical portion of the fluid is suddenly annihilated; it is required to find the instantaneous alteration of pressure at any point of the mass, and the time in which the cavity will be filled up, the pressure at an infinite distance being supposed to remain constant (Rayleigh 1917).

We have underlined the portion of the problem that asks for what we call the **collapse time,** or τ_C. This problem corresponds to the Rayleigh–Plesset equation in the special case of vanishing viscosity and surface tension, namely $\eta = \gamma = 0$, so that only the kinetic and pressure terms are relevant. The solution for τ_C, given that the initial bubble radius is R_0, can be written in the form

$$
\tau_C = \sqrt{\frac{\rho_m R_0^2}{\Delta P}} \underbrace{\left[\frac{1}{\sqrt{6}}\int_0^1 z^{-1/6}(1-z)^{-1/2}\,dz\right]}_{\text{Besant (1859)}} = \underbrace{0.914681}_{\text{Rayleigh (1917)}} \sqrt{\frac{\rho_m R_0^2}{\Delta P}}, \tag{1.104}
$$

with a dimensional combination which is appropriate, since

$$
\tau_C \propto \sqrt{\frac{\rho_m R_0^2}{\Delta P}} \qquad \text{or} \qquad [\tau_C] = [\rho_m]^{1/2}[R_0][\Delta P]^{-1/2} \tag{1.105}
$$

which translates into

$$
\begin{aligned}
\mathsf{T} &\overset{?}{=} \left(\frac{\mathsf{M}}{\mathsf{L}^3}\right)^{1/2} (\mathsf{L}) \left(\frac{\mathsf{M}}{\mathsf{LT}^2}\right)^{-1/2} \\
&\overset{?}{=} \mathsf{M}^{1/2-1/2}\,\mathsf{L}^{-3/2+1+1/2}\,\mathsf{T}^{-2(-1/2)} \\
T &\overset{!}{=} \mathsf{M}^0\,\mathsf{L}^0\,\mathsf{T}^1.
\end{aligned}
\tag{1.106}
$$

While we can use DA to check/confirm the dimensions of an existing result (what we might call a reactive approach), we can make use of the dimensional constraints imposed on physical relationships by the fact that dimensions must match to try to proactively predict the dependence of τ_C on the relevant physical parameters, ρ_m, R_0, and ΔP. This example will be the first of many(!) such cases that we will consider and introduces us to the real power of this technique.

We start by writing

$$\tau_C = C_\tau \rho_m^\alpha R_0^\beta (\Delta P)^\gamma \tag{1.107}$$

with C_τ being a pure number (hence dimensionless, and outside the range of DA methods to predict) and with (at the moment) undetermined powers (α, β, γ, integral or not, in principle, we don't know yet) of the relevant physical quantities. We then insist that the dimensionalities of Eqn. (1.107) match, by putting [ ] brackets around both sides, namely

$$\begin{aligned} [\tau_C] &= [C_\tau][\rho_m]^\alpha [R_0]^\beta [\Delta P]^\gamma \\ \Downarrow & \qquad \Downarrow \\ \mathsf{T} &= (1)\left(\frac{\mathsf{M}}{\mathsf{L}^3}\right)^\alpha (\mathsf{L}^\beta)\left(\frac{\mathsf{M}}{\mathsf{LT}^2}\right)^\gamma . \end{aligned} \tag{1.108}$$

This relation constrains the power-law exponents α, β, γ to satisfy

$$\begin{aligned} \mathsf{M}: &\quad 0 = \alpha + \gamma \\ \mathsf{L}: &\quad 0 = -3\alpha + \beta - \gamma \\ \mathsf{T}: &\quad 1 = -2\gamma, \end{aligned} \tag{1.109}$$

which is a (3×3) system of linear equations that is easily solved yielding $\alpha = 1/2, \beta = 1$, and $\gamma = -1/2$ or $\tau_C \propto \sqrt{\rho_m R_0^2/\Delta P}$, which reproduces (as it should) the dimensional dependences in the exact solution in Eqn. (1.104). The numerical constant ultimately provided by the complete solution, $C_\tau \approx 0.915$, turns out to be remarkably close to unity, but we stress that this is not always the case.

The use of DA can also be seen at the proverbial "next level of mathematical rigor" when solving differential equations, by defining dimensionless quantities to turn a physical model into a "pure math" problem, effectively extracting the DA information in advance. This process is often called **non-dimensionalization** or sometimes **scaling**. For example, the ODE (ordinary differential equation) relevant to the Besant–Rayleigh bubble collapse problem is

$$\rho_m \left\{ R(t)\frac{d^2R(t)}{dt^2} + \frac{3}{2}\left(\frac{dR(t)}{dt}\right)^2 \right\} + \Delta P_B = 0, \tag{1.110}$$

and we can write the time variable as $t = \tau z$ where $[\tau] = \mathsf{T}$ carries the dimensions, leaving z as a now dimensionless (scaled) time variable. Similarly, we can write $R(t) = R_0 y(z = t/\tau)$, as we expect the time-dependent radius to scale as R_0 (which then carries the length dimensionality), and $y(z)$ is a dimensionless function of the scaled time variable z. With these substitutions, Eqn. (1.110) becomes

$$\frac{\rho_m R_0^2}{\tau^2}\left\{ \frac{d^2y(z)}{dz^2} + \frac{3}{2}\left(\frac{dy(z)}{dz}\right)^2 \right\} + \Delta P_B = 0, \tag{1.111}$$

and we can make the entire equation dimensionless by choosing τ to satisfy

$$\frac{\rho_m R_0^2}{\tau^2} = \Delta P \qquad \text{naturally suggesting the time scale} \qquad \tau = \sqrt{\frac{\rho_m R_0^2}{\Delta P_B}}, \tag{1.112}$$

consistent with the "predictive" results above.

Even from this one simple example, we can appreciate that at many stages in the analysis of a physical problem, we can apply DA in a variety of ways, including,

(a) *reactive* or *a posteriori* checking,

(b) *proactive* or *a priori* predicting by matching dimensions, and now

(c) *non-dimensionalization.*

Before proceeding to other examples of predictive DA, let us explore two other examples of non-dimensionalization, both set in the context of one of the most familiar problems in mechanics (both classical and quantum), namely the harmonic oscillator.[7] This problem can be defined by using either the classical force law ($F_{HO}(x) = -kx$) or the corresponding potential energy function ($V_{HO}(x) = kx^2/2$) and either definition determines the dimensions of the **spring constant,** k, to be

$$\frac{\mathsf{M}}{\mathsf{T}^2} = \left(\frac{\mathsf{ML}}{\mathsf{T}^2}\right)\left(\frac{1}{\mathsf{L}}\right) = \frac{[F]}{[x]} = [k] = \frac{[V(x)]}{[x^2]} = \left(\frac{\mathsf{ML}^2}{\mathsf{T}^2}\right)\left(\frac{1}{\mathsf{L}^2}\right) = \frac{\mathsf{M}}{\mathsf{T}^2}. \tag{1.113}$$

In classical mechanics, the oscillator would be explored using Newton's law to find the time-dependent one-dimensional position, given by $x(t)$, determined by

$$m\frac{d^2x(t)}{dt^2} = ma(t) = F = -kx(t), \tag{1.114}$$

subject to initial values for $x(0), \dot{x}(0)$. To get a sense of the natural time scale for the resulting oscillatory motion, without a detailed solution, we can repeat the non-dimensionalization method by writing $t = \tau z$ with $[\tau] = \mathsf{T}$ and $[z] = 1$ being dimensionless, and then let $x(t) = Ay(z = t/\tau)$ where $[A] = \mathsf{L}$ is the amplitude of motion and $[y(z)] = 1$ is dimensionless as well. The equation of motion then reduces to

$$\frac{mA}{\tau^2}\left(\frac{d^2y(z)}{dz^2}\right) = -(kA)y(z), \tag{1.115}$$

and we can make the entire differential equation dimensionless by choosing

$$\frac{mA}{\tau^2} = kA \qquad \text{or} \qquad \tau = \sqrt{\frac{m}{k}}, \tag{1.116}$$

independent of A, the amplitude of the motion. It's easy to confirm that this is dimensionally correct since

$$\mathsf{T} = [\tau] = \left[\sqrt{\frac{m}{k}}\right] = \mathsf{M}^{1/2}(\mathsf{M}/\mathsf{T}^2)^{-1/2} = \mathsf{T}, \tag{1.117}$$

and this result is very similar in spirit to the Besant–Rayleigh bubble problem.

[7] We explore the oscillator in the context of classical mechanics in Section 3.1 and then using quantum mechanics in Section 6.3.2.

The problem of the harmonic oscillator is approached very differently in quantum mechanics, where the relevant physics is encoded in the **time-independent Schrödinger equation**, namely

$$-\frac{\hbar^2}{2m}\frac{d^2\psi_n(x)}{dx^2}+\frac{1}{2}kx^2\psi_n(x)=E_n\psi_n(x), \tag{1.118}$$

where $\psi_n(x)$ and E_n are the eigenfunctions and energy eigenvalues (about which, much more in Section 6.3.2). In this case, we're interested in the natural length scale, so we let $x=\rho y$ where $[\rho]=\mathsf{L}$ and $[y]=1$ is dimensionless, to write

$$-\frac{\hbar^2}{2m\rho^2}\frac{d^2\psi_n(y)}{dy^2}+\frac{1}{2}k\rho^2y^2\psi_n(y)=E_n\psi_n(y) \tag{1.119}$$

or

$$-\frac{d^2\psi_n(y)}{dy^2}+\left(\frac{mk\rho^4}{\hbar^2}\right)y^2\psi_n(y)=\left(\frac{2m\rho^2}{\hbar^2}E_n\right)\psi_n(y)\equiv\mathcal{E}_n\psi_n(y). \tag{1.120}$$

We can then make this equation dimensionless by the choice

$$\left(\frac{mk\rho^4}{\hbar^2}\right)-1 \qquad \text{or} \qquad \rho-\left(\frac{\hbar^2}{mk}\right)^{1/4} \tag{1.121}$$

which also makes $\mathcal{E}_n\equiv(2m\rho^2E_n/\hbar^2)$ dimensionless as well (which you should check). This choice of length scale is easily confirmed to give the correct dimensions since

$$\begin{aligned}\mathsf{L}=[\rho]=\left[\left(\frac{\hbar^2}{mk}\right)^{1/4}\right] &\stackrel{?}{=} \left(\frac{\mathsf{ML}^2}{\mathsf{T}}\right)^{1/2}\mathsf{M}^{-1/4}\left(\frac{\mathsf{M}}{\mathsf{T}^2}\right)^{-1/4}\\ &\stackrel{?}{=} \mathsf{M}^{1/2-1/4-1/4}\,\mathsf{L}^1\,\mathsf{T}^{-1/2-2(-1/4)}\\ \mathsf{L} &\stackrel{!}{=} \mathsf{M}^0\,\mathsf{L}^1\,\mathsf{T}^0.\end{aligned} \tag{1.122}$$

You should be able to reproduce this result (in **P1.16**) by assuming that the harmonic oscillator length scale can be written as $\rho\propto\hbar^\alpha m^\beta k^\gamma$ and matching dimensions to determine α,β,γ.

Example 1.2 Radiation from accelerated charges

Returning to the use of DA to check/confirm or predict dimensional dependences, another result from the nineteenth century, but one which is still prominently featured in many undergraduate and graduate texts on E&M (including Griffiths (1999), Good (1999), Jackson (1999), and Zangwill (2013)), is the calculation of the power radiated by an accelerating charge, first derived by Larmor (1897), which we can write in the following form,

$$P_L=\frac{1}{6\pi}q^2a^2\epsilon_0^{1/2}\mu_0^{3/2}, \tag{1.123}$$

where q, a are the charge and acceleration of the object.

We can check that this expression is dimensionally correct by comparing both sides to confirm that

$$\begin{aligned}
[P_L] &\stackrel{?}{=} [q]^2[a]^2[\epsilon_0]^{1/2}[\mu_0]^{3/2} \\
\left(\frac{\mathsf{ML}^2}{\mathsf{T}^3}\right) &\stackrel{?}{=} \mathsf{Q}^2\left(\frac{\mathsf{L}}{\mathsf{T}^2}\right)^2\left(\frac{\mathsf{Q}^2\mathsf{T}^2}{\mathsf{ML}^3}\right)^{1/2}\left(\frac{\mathsf{ML}}{\mathsf{Q}^2}\right)^{3/2} \\
&\stackrel{?}{=} \mathsf{M}^{-1/2+3/2}\,\mathsf{L}^{2-3/2+3/2}\,\mathsf{T}^{-4+1}\,\mathsf{Q}^{2+1-3} \\
\left(\frac{\mathsf{ML}^2}{\mathsf{T}^3}\right) &\stackrel{!}{=} \left(\frac{\mathsf{ML}^2}{\mathsf{T}^3}\right).
\end{aligned} \tag{1.124}$$

We can also try to more proactively "derive" the dependences on the four dimensionful parameters in the problem by assuming $P_L = C_P q^\alpha a^\beta \epsilon_0^\gamma \mu_0^\delta$ where C_P is a dimensionless constant (which we won't be able to constrain) and then matching dimensions via

$$\frac{\mathsf{ML}^2}{\mathsf{T}^3} = \mathsf{Q}^\alpha\left(\frac{\mathsf{L}}{\mathsf{T}^2}\right)^\beta\left(\frac{\mathsf{Q}^2\mathsf{T}^2}{\mathsf{ML}^3}\right)^\gamma\left(\frac{\mathsf{ML}}{\mathsf{Q}^2}\right)^\delta, \tag{1.125}$$

which requires that

$$\begin{aligned}
\mathsf{M}:&\quad 1 = \quad\quad -\gamma + \delta \\
\mathsf{L}:&\quad 2 = \quad \beta - 3\gamma + \delta \\
\mathsf{T}:&\quad -3 = \quad -2\beta + 2\gamma \\
\mathsf{Q}:&\quad 0 = \alpha \quad + 2\gamma - 2\delta
\end{aligned} \tag{1.126}$$

and this does indeed have the expected solutions, namely $\alpha = \beta = 2$, $\gamma = 1/2$, and $\delta = 3/2$, with the otherwise undetermined value of $C_P = 1/6\pi$. In this case, the dimensionless constant is still within an order-of-magnitude (or so) of unity. We revisit this problem in **Example 4.9.**

Example 1.3 Length scale for quantum gravity

A subject of intense ongoing research interest over the last century has been the pursuit of a mathematically self-consistent theory of **relativistic quantum gravity**, one of the holy grails of theoretical physics, which, of course, must also actually describe nature and agree with experimental observations/constraints. While most of the many conceptual approaches pursued (string theory, loop quantum gravity, causal dynamical triangulations, etc.) include some of the most advanced mathematics in all of science, the same basic dimensional quantities required in such a theory are clear. Already given the phrase "relativistic quantum gravity," we expect such a theory will include the foundational constants of three fields, namely c, $\hbar$, and G.

To estimate the length scale on which all three fundamental concepts must be relevant, we write

$$L_P \sim \hbar^\alpha G^\beta c^\gamma \quad \text{requiring} \quad [L_P] = [\hbar]^\alpha[G]^\beta[c]^\gamma \quad \text{or} \quad \mathsf{L} = \left(\frac{\mathsf{ML}^2}{\mathsf{T}}\right)^\alpha\left(\frac{\mathsf{L}^3}{\mathsf{MT}^2}\right)^\beta\left(\frac{\mathsf{L}}{\mathsf{T}}\right)^\gamma \tag{1.127}$$

to constrain the **Planck length.** Matching dimensions gives

$$\begin{aligned} \mathsf{M}: &\quad 0 = \alpha - \beta \\ \mathsf{L}: &\quad 1 = 2\alpha + 3\beta + \gamma \\ \mathsf{T}: &\quad 0 = -\alpha - 2\beta - \gamma \end{aligned} \tag{1.128}$$

which has solutions $\alpha = \beta = 1/2$ and $\gamma = -3/2$, giving

$$L_P = \sqrt{\frac{\hbar G}{c^3}} \approx 1.6 \times 10^{-35}\, m. \tag{1.129}$$

The actual numerical value of the Planck length (regardless of whether or not we've missed out on some dimensionless constants of $\mathcal{O}(0.1 - 10)$ immediately illustrates the immense challenges facing experimental verification of any theory of quantum gravity. (We explore **Planckian units** in more depth in Section 6.5.2.)

Example 1.4 Sparking the vacuum: Critical electric field strength

The production of new forms of matter in the collisions of known elementary particles has a rich history going back almost a century. From the discovery of the **positron** or antiparticle of the electron (Anderson 1933) and **muon,** the heavy cousin of the electron (Neddermeyer and Anderson 1937, and Street and Stevenson 1937) to the production of the **top quark** (Abe et al. 1995, and Abachi et al. 1995) and more recently the **Higgs boson** (CMS and ATLAS collaborations 2012), the use of the $E = mc^2$ equivalence to produce new massive particles in high-energy collisions has been at the forefront of physics.

A very different effect involving the production of (charged) particle–antiparticle pairs has been predicted by Sauter (1932) and Schwinger (1951) involving pair creation by very strong electric fields. In this quantum effect, a sufficiently strong $\mathbf{E}$ field can produce Q^+Q^- particles from the vacuum, and given that the lightest charged particle is the electron (with its antiparticle the positron, which we know has the equal magnitude but opposite sign charge, and the same mass), predictions for the **critical electric field** (or E_C) focus on e^+e^- production.

Given that this is an effect of QED we expect that E_c would depend on $e, \hbar, c$, and for electron–positron pairs the relevant mass scale is m_e. To use DA to evaluate the field necessary, we write

$$\begin{aligned} E_C &\propto m_e^{\alpha} c^{\beta} e^{\gamma} \hbar^{\delta} \\ \frac{\mathsf{ML}}{\mathsf{QT^2}} &= \mathsf{M}^{\alpha} \left(\frac{\mathsf{L}}{\mathsf{T}}\right)^{\beta} \mathsf{Q}^{\gamma} \left(\frac{\mathsf{ML^2}}{\mathsf{T}}\right)^{\delta} \end{aligned} \tag{1.130}$$

and match dimensions as

$$\begin{aligned} \mathsf{M}: &\quad 1 = \alpha \qquad\qquad + \delta \\ \mathsf{L}: &\quad 1 = \qquad \beta \qquad + 2\delta \\ \mathsf{T}: &\quad -2 = \quad -\beta \qquad - \delta \\ \mathsf{Q}: &\quad -1 = \qquad\quad +\gamma \end{aligned} \tag{1.131}$$

which is easily solved to give $\alpha = 2$, $\beta = 3$, and $\gamma = \delta = -1$, or

$$E_C = \frac{m_e^2 c^3}{e\hbar} \approx 1.3 \times 10^{18}\ V/m. \tag{1.132}$$

This value also sets the scale for when electrodynamics becomes non-linear due to quantum effects, as discussed in Section 7.4. A similar approach can be used to find the **critical magnetic field** or $\boldsymbol{B_c}$ which leads to pair-production; see **P1.14**.

1.4 Automating dimensional analysis calculations: Mathematica© examples

We have already noted that the highest level mathematical tools of which we will make (occasional) use are differentiation, integration, and sometimes series expansions. While we have used (and will continue to use) the symbols of multi-variable calculus (gradient operators, line and surface integrals, etc.), we have only done so to note that equations of those types can still be used in the context of DA problems, without having to master the details of the mathematics.

The one tool that we use regularly (in fact, on almost every page of this book) involves solving systems of linear equations. Most students have some experience with this, even if only using straightforward substitution or elimination techniques, or perhaps the use of matrix formalism (determinants and the like). Given that we regularly require the use of some such methods, it can be helpful to automate and streamline, to the extent possible, the well-defined mathematics involved.[8] (We note that this approach has been advocated by Remillard (1983) and Goth (1986), using early generations of computer software.)

Many symbolic manipulation or computer algebra programs (Maple©, MATLAB©, and Mathematica©) can be used to solve systems of linear equations, and we illustrate below (and in Section 10.8) some examples using Mathematica©, but starting with a very general approach focusing on DA. As many scientists have learned how to program by running someone else's code, changing it slightly, breaking it, and trying again, we provide several working examples related to problems in the text, ones which students/readers can use as templates to adapt, or "translate" into other languages, if they choose to minimize the (admittedly) tedious process of solving sets of linear equations.

The "workflow" we make use of is as follows:

- Identify the physical quantity/variable that you're studying, and note its dimensional "spelling." Since that quantity is often on the "left-hand side" (*LHS*) of a DA consistency equation, you might write, for example, $LHS = \mathsf{M}^a\mathsf{L}^b\mathsf{T}^c\mathsf{Q}^d\Theta^e$ where the exponents a, b, c, d, e are assumed known.
- Do the same for all of the physical variables, say $var_1, var_2, \ldots$, that you've identified as likely candidates for contributing to the dimensions of the *LHS* quantity.

---

[8] As Steve Jobs famously said, ". . . a computer is a bicycle for the mind . . ." and in this context can help us in the efficient ". . . locomotion . . ." of most of the examples or problems in this text.

- Write (conceptually, at least, we won't actually do this)

$$LHS = RHS = (var_1)^{\alpha}(var_2)^{\beta}(var_3)^{\gamma}\ldots \tag{1.133}$$

  where *RHS* is the combination of powers of dimensional quantities on the "right-hand side," with the powers $\alpha, \beta, \gamma, \ldots$ to be determined.
- Since the dimensions on both sides of Eqn. (1.133) must agree, the **ratio** of *RHS/LHS* must be dimensionless, so that the powers/exponents of each of M, L, T, Q, Θ factors in that combination must actually vanish.
- Extracting those powers/exponents and setting them equal to zero gives the required set of linear equations for $\alpha, \beta, \gamma, \ldots$, which can be explored, using, for example, the `Solve[ ]` command in Mathematica©.
- From the cases we've considered so far, we have seen 3×3 and 4×4 systems, with the number of variables matching the number of dimensional constraints, but this need not always be the case.

We demonstrate below a realization of this strategy to approach the Besant–Rayleigh bubble collapse time problem, solved proactively in Eqns. (1.107)–(1.109). The variables required for the RHS are all written in terms of M, L, T dimensions (taken directly from the "dictionary" in the Section 10.1 if necessary) as is the single LHS variable (in this case, the collapse time, τ_C).

```
(* Rayleigh_Bubble.nb *)
(* Finds the dependence of the bubble collapse time on the pressure,
density, and initial radius *)
(* All quantities are given in terms of their M, L, and T base dimensions *)
(* This is what we need for the RHS *)
pressure = M/(L*T^2);                          (* pressure dimensions *)
density = M/L^3;                               (* density dimensions *)
r0 = L;                                        (* initial radius dimensions *)
RHS = density^(alpha) * r0^(beta) * pressure^(gamma);
(* This is what we need for the LHS *)
tauc = T;                                      (* collapse time dimensions *)
LHS = tauc;
ratio = RHS/LHS;
(* The ratio of the RHS to LHS should be dimensionless, as dimensions should
match *)
aa = Exponent[ratio, M];
bb = Exponent[ratio, L];
cc = Exponent[ratio, T];
Solve[{0 == aa, 0 == bb, 0 == cc}, {alpha, beta, gamma}]
(* Press SHIFT+ENTER to compile *)

{{alpha -> 1/2, beta -> 1, gamma -> -(1/2)}}
```

This agrees with the result following Eqn. (1.109). We note that in Mathematica© any text between a (* and a *) pairing is a comment and not compiled. Any line of code that ends in a semi-colon

(`;`) does not produce any output nor is printed, so the only real output is the result of the final `Solve[ ]` command, shown inside the `{{ }}` brackets.

This approach has the benefit that so long as the original var_n quantities and the *LHS* are "spelled" correctly, there is a saving in time and effort, even in the matching of dimensions to set up the system of linear equations, and certainly in their actual solution, with less chance for computational error. Depending on how often you might use these techniques, one can even imagine having a library of dimensional "spellings" already coded in Mathematica© (or insert your favorite language here) for use, translating the dimensional dependences collected in Section 10.1 for easy reference.

We can extend this technique slightly to include the Q base dimension to solve the Larmor problem (power emitted by accelerating particle) considered in **Example 1.2** and Eqns. (1.125)–(1.126), clearly reproducing the answers cited there.

```
(* Larmor_1.nb *)
(* Derives the dependence of the power radiated by an accelerated charge
on q,a, epsilon_0, and mu_0 *)
(* This is the so-called Larmor formula *)
(* All quantities are given in terms of their M, L, T, and Q base
dimensions *)
(* This is what we need for the RHS *)
charge = Q;                           (* particle charge dimensions *)
acc = L/T^2;                          (* particle acceleration
dimensions *)
epsilon0 = (Q^2*T^2)/(M*L^3);         (* permittivity dimensions *)
mu0 = (M*L)/Q^2;                      (* permeability dimensions *)
RHS = charge^(alpha) * acc^(beta) * epsilon0^(gamma) * mu0^(delta);
(* This is what we need on the LHS *)
power = (M*L^2)/T^3;                  (* power dimensions *)
LHS = power;
ratio = RHS/LHS;
(* Ratio of RHS to LHS should be dimensionless, as dimensions should match *)
aa = Exponent[ratio, M];
bb = Exponent[ratio, L];
cc = Exponent[ratio, T];
dd = Exponent[ratio, Q];
Solve[{0 == aa, 0 == bb, 0 == cc, 0 == dd}, {alpha, beta, gamma, delta}]
(* Press SHIFT+ENTER to compile *)

{{alpha -> 2, beta -> 2, gamma -> 1/2, delta -> 3/2}}
```

Section 10.8 provides several more examples, including

- A 5 × 5 problem where all five base dimensions (M, L, T, Q, Θ) are involved (automating the example discussed at length in Section 2.4).
- A 4 × 4 variation of the Larmor problem, where it turns out that <u>**no**</u> solution is possible, illustrating what such a case looks like in Mathematica© (following up on **Example 4.9** and **P4.31** and **P4.32**).

- And perhaps most importantly, an example where there is not a single unique solution, but where information on an important dimension<u>**less**</u> ratio is provided by the method, for use in the analysis of the **Reynolds number** in **Example 3.4.**

1.5 What have we learned about dimensional analysis so far?

Reflecting on our fly-over of the physics landscape as mapped out in many introductory texts, and some applications to research problems (old and new), perhaps the most striking feature of the mathematical descriptions of the phenomena we've encountered so far is that everything can be encoded (we've called it "spelled") in terms of five base dimensions, namely

$$\boxed{\text{Length (L), Time (T), Mass (M), Charge (Q), and Temperature }(\Theta)}\,. \tag{1.134}$$

Some treatments of DA (see e.g. Cohen et al. 2003) add two other SI base dimensions (with corresponding units), namely

Quantity	Base	Unit	Symbol
Amount of substance	N	Mole	*mol*
Luminous intensity	J	Candela	*cd*

and these units are indeed included in the internationally agreed upon SI system of units (see Section 10.10 on metrology). But for the vast majority of problems in the undergraduate and graduate physics curricula, the physics research literature (historical or modern), and even in applications in other science/engineering disciplines, we argue that these two are little-used and so will not consider them further.

Thus, the **screaming message** to be taken away from our 30,000-foot (10,000 *m*) overview is that

FIVE SUFFICE !

and we will henceforth explore the pedagogical utility and predictive power of assuming that the dimensionality of all relevant physical quantities can be "spelled" in the form $\mathsf{L}^{\alpha}\mathsf{T}^{\beta}\mathsf{M}^{\gamma}\mathsf{Q}^{\delta}\Theta^{\epsilon}$, or dimensionless combinations thereof.

We have also noted that the property of dimensionality can be **distributed** (repeatedly) using relations such as Eqn. (1.4) and others, namely

$$[AB] = [A][B], \quad \left[\frac{A}{B}\right] = \frac{[A]}{[B]}, \quad [A\mathbf{C}] = [A][\mathbf{C}] = [A][\,|\mathbf{C}|\,], \quad \text{and} \quad [\mathbf{C}\cdot\mathbf{D}] = [\mathbf{C}\times\mathbf{D}] = [\mathbf{C}][\mathbf{D}], \tag{1.135}$$

whether $A, B, \mathbf{C}, \mathbf{D}$ are scalars or vectors, or combinations thereof (via scalar, dot, or cross product multiplication) and that derivatives have appropriate dimensions, as in Eqns. (1.5) and (1.16).

It's perhaps worthy to recall that in the proverbial "real world" where science and engineering is done on a daily basis, there can be other constraints that limit one's ability to achieve results (do a new experiment, build or market a new product or device, implement a new algorithm, etc.), a limitation that can be quantified: namely how much things cost. We won't discuss this sixth important dimension, save to note that the units in which currency can be measured are perhaps even more diverse than those projected down from our $\mathsf{L}, \mathsf{T}, \mathsf{M}, \mathsf{Q}, \Theta$ dimensions, and you might have purchased this book in \$, £, €, or ¥. At least one engineering handbook (Wood 2007), to its credit, in addition to extensive appendices on *Units and Conversion of Units* and *Dimensionless Groups* (a very complete list that I recommend), has a sixty-page section on *Capital Cost Guidelines* for the purchase of a variety of equipment!

1.6 **Problems for Chapter 1**

Q1.1 Reality check: Since it's always good to take time to reflect on what one has learned, at the start of the Problem section for every chapter, I will always ask the same question, namely, what was the single most important, interesting, engaging, and/or useful thing you learned from this part of the book?

P1.1 Personal experience with DA: Think back on your physics and math (and perhaps other STEM) educational experiences, formal and informal. Do you recall being exposed to/taught some of the concepts/techniques of DA? If so, in what context(s), and did you use them to solve any problems?

P1.2 Your individualized fly-over of the landscape of physics: (a) Find any introductory physics textbook you have laying around (or have an electronic copy of), scan through all of the chapters/sections listed in its Table of Contents, and see if we've missed anything in Section 1.2 that you think would be important for you to know. For example, some physical quantity or variable, or some equation/relation that you think should have been mentioned. If so, can you that add the dimensional *spelling* of that concept/quantity to the list in Section 10.1 for your future use?

(b) If you have access to higher-level texts (say at the advanced undergraduate or graduate level), briefly review those and see if there are topics not listed in Chapters 3–6 (covering mechanics, E&M, thermal physics, or quantum mechanics) or more advanced/specialized content shown in Chapters 7–10 that you might want to explore yourself, using the techniques of DA.

(c) And if your interests lie beyond traditional physics areas of study/research, such as geosciences, life sciences, engineering, or other disciplines, see if textbooks in those fields cite dimensional quantities or relations you'd like to have included.

P1.3 Reviewing the basics: Check the Tables 10.1–10.5 of dimensional "spellings" in Section 10.1 and confirm/reproduce as many of the entries as you can.

P1.4 Checking the dimensions of Maxwell's equations: We confirmed that Gauss's law in both differential (Eqn. (1.49)) and integral form (Eqn. (1.54)) are dimensionally correct. Confirm that the other three Maxwell equations in Eqns. (1.50)–(1.52) and Eqns. (1.55)–(1.57) are also dimensionally consistent.

P1.5 Other variations on the Rayleigh–Plesset equation: There are other limiting cases of Eqn. (1.102) for the dynamics of a bubble where the non-dimensionalization procedure leading to Eqn. (1.111) can be used to extract information on the natural time sales for the collapse process.

(a) Imagine that instead of assuming that $\eta = \gamma = 0$, that we have $\eta = \Delta P = 0$, so that surface tension is the relevant term in the differential equation. Use the substitutions $t = \tau z$ and $R(t) = R_0 y(z = t/\tau)$ and decide what is the natural time scale for collapse, and how it depends on ρ_m, R_0, and γ. Confirm that you obtain the same combination by writing $\tau \propto \rho_m^\alpha R_0^\beta \gamma^\sigma$ (note that we use σ as an exponent here to avoid typographical confusion!) and matching dimensions.

(b) Repeat part (a), but assume that $\gamma = \Delta P = 0$ and find how the natural time scale τ depends on ρ_m, R_0, and now η (the viscosity) using both methods above.

P1.6 Millikan's oil drop experiment: In the final version of Millikan's famous oil drop experiment to determine the charge on the electron, he says:

> As is now well known, the oil-drop method rests originally upon the assumption of Stokes' law and gave a charge e on a given drop through the equation
>
> $$e_n = \frac{4}{3}\pi\left(\frac{9\eta}{2}\right)^{3/2}\left(\frac{1}{g(\sigma-\rho)}\right)^{1/2}\frac{(v_1+v_2)v_1^{1/2}}{F} \tag{1}$$
>
> in which η is the coefficient of viscosity of air, σ the density of the oil, ρ that of the air, v_1 the speed of descent under gravity, and v_2 its speed of ascent under the influence of the electric field of strength F (Millikan 1913).

Looking up the dimensions of viscosity, (mass) density (both σ and ρ), speed, the acceleration of gravity (g), and electric field (called F here, and not E as we have, so use the appropriate dimensions) from Section 10.1, show that Millikan's Eqn. (1) above is dimensionally correct.

P1.7 LRC circuit I—Checking dimensions: Applying Kirchhoff's law for a single loop circuit containing a capacitor (C), resistor (R), and inductor (L) gives an equation of the form

$$\underbrace{L\frac{dI(t)}{dt}}_{\text{inductor}} + \underbrace{I(t)R}_{\text{resistor}} + \underbrace{\frac{q(t)}{C}}_{\text{capacitor}} = 0 \tag{1.136}$$

where $I(t)$ is the current in the circuit and $q(t)$ is the charge on the capacitor. Look up the dimensions of all factors and confirm that all of the terms are dimensionally consistent.

P1.8 LRC circuit II—Non-dimensionalization: The LRC circuit equation in Eqn. (1.136) is a second-order ordinary differential (ODE) for $q(t)$ and so is amenable to an analysis using non-dimensionalization methods. The resulting ODE has three terms, namely

$$\underbrace{L\frac{d^2q(t)}{dt^2}}_{\text{inductor}} + \underbrace{\frac{dq(t)}{dt}R}_{\text{resistor}} + \underbrace{\frac{q(t)}{C}}_{\text{capacitor}} = 0. \tag{1.137}$$

Use the substitutions $t = \tau z$ and $q(t) = q_0 y(z = t/\tau)$ to find the natural time scales for (a) an LR circuit (so no capacitor), (b) an RC circuit, and (c) an LC circuit. What happens if you try to find a logical value for τ for an LRC circuit with all three elements?

P1.9 The cable equation—Non-dimensionalization and DA: A partial differential equation that models the space-time behavior of the voltage signal, $\phi(x,t)$, on a (yes, old-fashioned) telegraphy cable, but which also has applications to signal propagation in neurons, is

$$\sigma_e(2\pi a^2)\frac{\partial^2\phi(x,t)}{\partial x^2} = 2\pi a\left(g\phi(x,t) + c\frac{\partial\phi(x,t)}{\partial t}\right). \tag{1.138}$$

The physical parameters included are the radius (a) of the cylindrical cable, the electrical conductivity (σ_e) of the cable material, the conductance per area ($g = G/A$) of the insulating sheath, and the capacitance per unit area ($c = C/A$) of the cable.

(a) Knowing that $[a] = \mathsf{L}$, look up the dimensions of σ_e, and evaluate those of g and c, and confirm that Eqn. (1.138) is dimensionally consistent.

(b) Introduce new variables $x = \lambda y$ and $t = \tau z$, where λ, τ carry the dimensions of space and time, and evaluate them in terms of the physical parameters so that Eqn. (1.138) becomes non-dimensional.

(c) Try to reproduce part (b) by using just DA, by writing $\lambda \propto a^\alpha \sigma_e^\beta g^\gamma C^\delta$, match dimensions, and solve for the exponents to the extent you can. Are your results consistent with part (b)? Apply the same dimensional analysis approach for τ.

P1.10 Artesian aquifer flows: An equation from a popular geophysics book (Turcotte and Schubert 1982, p. 240) states the relation

$$Q = 7.686\left(\frac{gb}{R'}\right)^{4/7}\left(\frac{\rho_m}{\mu}\right)^{1/7} R^{19/7}, \tag{1.139}$$

where Q is a volume flow rate (volume per time), b, R', R are all distances, while ρ_m and μ are (mass) density and viscosity (again, here called μ and not η), respectively. Looking up the dimensions of all of these quantities, confirm that the equation is dimensionally correct.

Note: The unusual powers (multiples of 1/7) shown here arise in a phenomenological model of fluid flow and not necessarily from fundamental mathematical equations, but nonetheless any such relation should obey dimensional consistency. Compare this to **P1.22** where the powers of 1/5 are a prediction of general relativity!

P1.11 Fly-by power dissipation: A charge (q) moves with constant speed v a distance d parallel to a flat horizontal conducting surface, with the electrical conductivity of the material given by σ_e. The moving charge causes currents to flow in the metal, which, in turn, gives rise to energy losses (Joule heating) and power being dissipated. A popular textbook (Zangwill 2013 pp. 474–75) describes this problem and estimates that the instantaneous power dissipated (P_d) is given by

$$P_d \approx \frac{1}{16\pi}\frac{v^2q^2}{\sigma_e d^3}. \tag{1.140}$$

(a) Show that Eqn. (1.140) is dimensionally correct.

(b) Try to reproduce this equation proactively by writing $P_d \propto q^\alpha v^\beta \sigma_e^\gamma d^\delta$, matching $\mathsf{M, L, T, Q}$ dimensions, and solving for $\alpha, \beta, \gamma, \delta$.

P1.12 Relativistic four-vectors: Show that the four-vectors defined by Eqns. (1.93), (1.94), (1.97), and (1.98) are all dimensionally consistent, that is, with all components having the same dimensions.

P1.13 How big can relativistic γ factors be? Protons can be accelerated to ultra-relativistic ($v \lesssim c$) speeds in experiments at terrestrial accelerators and in astrophysical environments. The protons produced at the LHC (Large Hadron Collider) at CERN have $E_p \approx 6.8\,TeV$ (where $T = tera = 10^{12}$), while the highest energy cosmic ray protons ever measured have $E_p \approx 3 \times 10^{20}\,eV$. Using Eqn. (1.88), evaluate the rest energy, $E_0 = m_0 c^2$, for the proton (in J), convert this to eV energy units, and then evaluate the corresponding γ factors. Using Eqn. (1.90), estimate how close to the speed of light the particles are in the two cases.

P1.14 Critical electric and magnetic fields for pair-production: (a) Use handbook values for $m_e, c, \hbar$, and e in Eqn. (1.132) to evaluate the numerical value of E_c in $Volts/m$.

(b) Repeat the derivation of Eqns. (1.130)–(1.132), but for the **critical magnetic field** by assuming that $B_c \propto m_e^\alpha c^\beta e^\gamma \hbar^\delta$ to find $\alpha, \beta, \gamma, \delta$.

(c) Evaluate B_c in *Tesla* and *Gauss* and compare to the value of the Earth's magnetic field ($B_{earth} \approx 0.5\,Gauss$), to a typical MRI machine ($1 - 3\,Tesla$), and to values found in highly magnetized neutron stars or **magnetars** ($10^{13} - 10^{15}\,Gauss$).

P1.15 Electron–phonon scattering time: A textbook on solid state physics (Snoke 2009) derives an expression for the electron–phonon scattering time, $\tau_{e\text{-}p}$, as

$$\frac{1}{\tau_{e\text{-}p}} \approx \frac{\sqrt{2}}{\pi} \frac{D^2 m^{3/2}}{\rho_m \hbar^4 v^2} (k_B T)^{3/2}, \tag{1.141}$$

where D is a **deformation potential** (the word potential implying it has dimensions of energy), m the electron mass, ρ_m the mass density (mass per unit volume, not ρ_e), v the drift speed, $\hbar$ and k_B Planck's and Boltzmann's constants, and T the temperature. Show that this expression is dimensionally consistent.

P1.16 Quantum harmonic oscillator dimensions: The non-dimensionalization approach to the harmonic oscillator in Eqns. (1.120)–(1.121) shows that the combination $\mathcal{E}_n$ is dimensionless, which then provides the dimensional "spelling" of the energy eigenvalues as $E_n \propto \hbar^2/(m\rho^2)$.

(a) Using the result for ρ in Eqn. (1.121), evaluate E_n in terms of $\hbar, m, k$.

(b) Writing $\rho \propto \hbar^\alpha m^\beta k^\gamma$, match dimensions to solve for α, β, γ and compare to the result in Eqn. (1.121).

(c) Repeat part (b) to find the dimensional dependence of E_n and compare to the results of part (a).

P1.17 Dimensions of Planck's constant: Planck's constant (either h or $\hbar = h/2\pi$) appears in a dizzying array of (seemingly) different contexts in quantum mechanics, and here we ask you to check that the relations below are dimensionally correct:

$$\hat{p}_x = \frac{\hbar}{i}\frac{\partial}{\partial x} \quad \text{(momentum operator)} \tag{1.142}$$

$$\hat{E} = i\hbar\frac{\partial}{\partial t} \quad \text{(energy operator)} \tag{1.143}$$

$$\Delta x \cdot \Delta p \geq \frac{\hbar}{2} \quad \text{(uncertainty principle)} \tag{1.144}$$

$$L_z = m\hbar \quad \text{(quantized angular momentum),} \tag{1.145}$$

The (possibly) new symbols that might need some explanation include the **hat** or **caret** over the quantities $\hat{x}$ and $\hat{p}$, which indicates that they are differential operators, but otherwise have no impact on their dimensions. The values of Δx and Δp are the **uncertainties** in those quantities and so have the same dimensions, and in general $[\Delta S] = [S]$ whatever the variable/quantity S is. For Eqn. (1.145), L_z is an **angular momentum** and m is **NOT** a particle mass, but rather a dimensionless quantum number. (Once again, context is important in DA spelling.) We explore these, and other incarnations/avatars of $\hbar$, extensively in Chapter 6.

P1.18 The equations of magneto-hydrodynamics: Two of the equations governing the behavior of electrically conducting fluids are the **mass continuity equation** and the **Cauchy momentum equation**, given by

$$\frac{\partial \rho_m}{\partial t} + \nabla \cdot (\rho_m \boldsymbol{v}) = 0$$
$$\rho_m \left(\frac{\partial}{\partial t} + \boldsymbol{v} \cdot \nabla \right) \boldsymbol{v} = \boldsymbol{J}_e \times \boldsymbol{B} - \nabla P \tag{1.146}$$

where $\rho_m, \boldsymbol{v}$ are the fluid mass density and local velocity, $\boldsymbol{J}_e, \boldsymbol{B}$ are the current density and local magnetic field, P is the fluid pressure, and ∇ is the gradient operator. Show that Eqns. (1.146) are dimensionally correct.

P1.19 Eddy current brake: A problem taken from an early edition of the famous introductory textbook by Halliday and Resnick (1978 p. 791) describes a magnetic braking problem as follows:

> An electromagnetic "eddy current" brake consists of a disk of conductivity σ and thickness t rotating about an axis through its center with a magnetic field $\boldsymbol{B}$ applied perpendicular to the plane of the disk over a small area a^2 .... If the area a^2 is a distance r from the axis, find an approximate expression for the torque tending to slow down the disk at the instant that its angular velocity equals ω.

Note that we have underlined the word approximate for emphasis as this type of problem is very challenging to solve, even numerically. The answer provided is $\tau = B^2 a^2 r^2 \omega\sigma t$. Show that this expression for the torque τ is dimensionally correct.
Note: We revisit this, and other "magnetic braking" problems, in Section 4.7.

P1.20 Induced currents: Another exercise taken from the same source as **P1.19** (Halliday and Resnick 1978 p. 790) describes a **Faraday's law** problem as follows:

> A uniform magnetic field $\boldsymbol{B}$ is changing in magnitude at a constant rate dB/dt. You are given a mass m of copper which is to be drawn into a wire of radius r and formed into a circular loop of radius R.

Show that the induced current in the loop does not depend on the size of the wire or of the loop and, assuming that **B** perpendicular to the loop, is given by

$$i = \frac{m}{4\pi\rho\,\delta}\frac{dB}{dt}$$

where ρ is the electrical resistivity and δ is the mass density of copper.

Note: The authors had to be careful with symbols for mass density and charge density, versus resistivity, to avoid confusion. We will usually use ρ_m for the first and ρ_e for the second, and "plain" $\rho = 1/\sigma_e$ for the third, but here have transcribed their notation directly.

(a) Show that the equation above is dimensionally correct.

(b) Writing $i \propto m^{\alpha}(dB/dt)^{\beta}\rho^{\gamma}\delta^{\epsilon}$, match dimensions, and solve for $\alpha, \beta, \gamma, \epsilon$ and compare to the result cited above.

P1.21 Using Mathematica© to automate linear algebra: Use the template Mathematica© files in Section 1.4, or those in Section 10.8, or those of your own devising (in any language), to solve for the dimensional dependences in:

(a) The Planck length problem in Eqns. (1.127)–(1.129)

(b) The Schwinger critical electric field in Eqns. (1.130)–(1.132)

(c) The results in **P1.11**, **P1.16**, or **P1.20**.

P1.22 Extracting physics from gravitational wave signals—The chirp mass: One of the major achievements in international science of the twenty-first century is most certainly the discovery of sources of gravitational waves (see Abbott et al. 2016), celebrated by the 2017 Nobel Prize being awarded to Weiss, Thorne, and Drever, and soon thereafter, the Breakthrough Prize being shared by all members of the LIGO/VIRGO collaborations.

The inspiraling of two large masses (either neutron stars or black holes) gives rise to distortions in space-time that propagate at the speed of light and can be detected as a (dimensionless) strain in (extremely!) sensitive detectors on Earth, resulting in signals of the type shown in Fig. 1.1.

A particular combination of the two masses, m_1, m_2, called the **chirp mass** (or $\mathcal{M}_c$), can be extracted directly from the *frequency versus time* dependence of the data. The relation defining this mass, and its dependence on the experimental observables (the instantaneous frequency of the wave form, $f(t)$, and its time derivative, $\dot{f}(t)$, is given by

$$\frac{(m_1 m_2)^{3/5}}{(m_1+m_2)^{1/5}} \equiv \mathcal{M}_c = \frac{c^3}{G}\left\{\left(\frac{5}{96}\right)^3 \pi^{-8} f^{-11} \dot{f}^3\right\}^{1/5}, \tag{1.147}$$

where c and G are the speed of light and Newton's gravitational constant.

(a) Confirm that both equalities in Eqn. (1.147) are dimensionally correct.

(b) A (very rough!) estimate of f and $\dot{f}$ from the frequency–time plot in Fig. 1.2 at $t \approx 0.40\,sec$ gives $f \approx 80\,Hz$ and $\dot{f} \approx 2600\,Hz/sec$ (note the mixed units of $\dot{f}$). Use those data, and handbook values of c and G, to estimate the chirp mass $\mathcal{M}_c$, in both kg, and solar masses ($M_\odot$) and compare to the value cited in Abbott et al. (2017), namely $\mathcal{M}_c \sim (30-40)M_\odot$.

(c) Why do you think that the frequency versus time plot in Fig. 1.2 would suggest the word "chirp?" Noting that the frequencies are clearly in the audible range, can you "reproduce" that data by whistling?

Note: We return to the subject of gravitational waves in Section 9.4.2, where we do our best to try to derive the dimensional dependences in Eqn. (1.147). But learning how to read plots of data, and being able to confirm dimensions, are already important first steps towards understanding such processes.

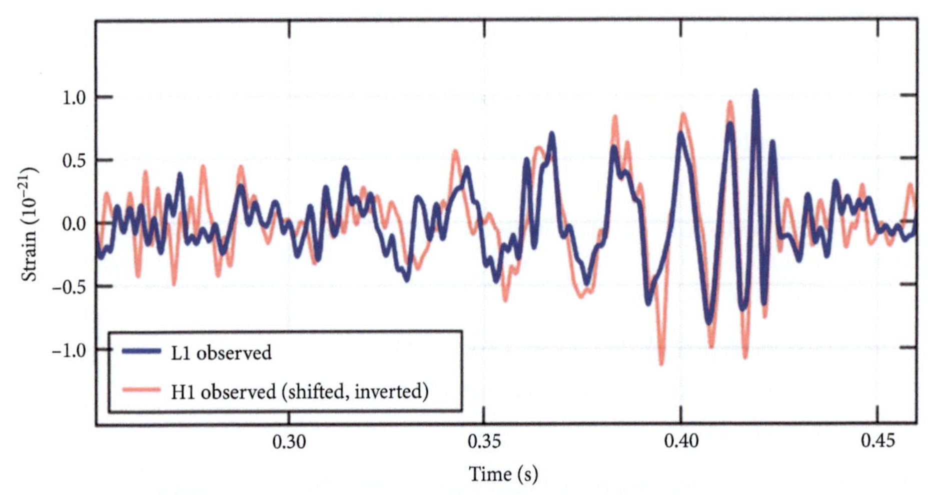

Fig. 1.1 Instrumental strain versus time in the two LIGO detectors, taken from the gravitational wave discovery paper. Reprinted from the review article by Abbot et al. (2017) with permission of John Wiley and Sons.

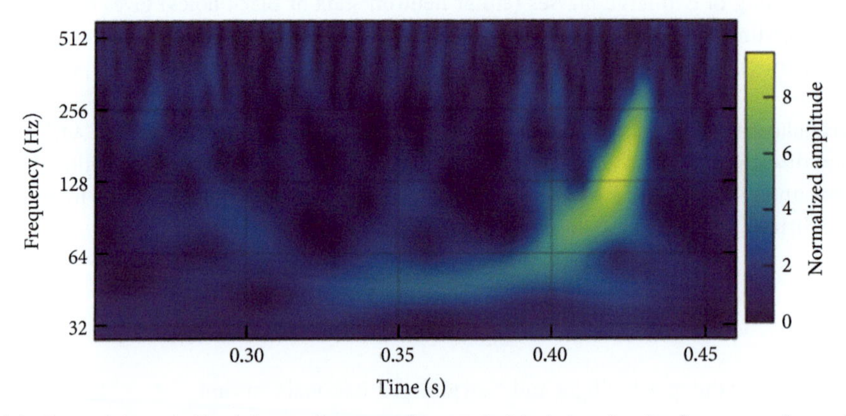

Fig. 1.2 Same data as in Fig. 1.1, over the same time period, but showing the frequency dependence with the characteristic rise in f, or "chirp," near the time of merger. Reprinted from the review article by Abbot et al. (2017) with permission of John Wiley and Sons.

P1.23 Planktonic ecosystem models: As an example of the true diversity of scientific investigations to which DA can be used, we share one related to "... phytoplankton primary production in water ecosystem models" by Golosov et al. (2021).[9] Their parametrization of this process included four-dimensional quantities (whose descriptions and units are reproduced verbatim):

- "Integral photosynthesis μ ($[\mu] = kgC \cdot m^{-3} \cdot s^{-1}$) . . ."
- ". . . energy of the total electromagnetic radiation of the Sun I ($[I] = W \cdot m^{-2}$). . ."
- ". . . the biomass of algae B ($|B| = kg \cdot m^{-3}$) . . ."
- ". . . the thickness of the radiation absorption layer, i.e., water transparency L ($[L] = m$)."

Since context is everything, we note that the C label in the first bullet is **not** for Coulomb, but rather Carbon, and so $kgC = kg$ for our purposes.

(a) Write μ, I, B, and L in terms of their $\mathsf{M}, \mathsf{L}, \mathsf{T}$ dimensions.

(b) Derive a dimensional relation among the variables by writing $\mu \propto I^{\alpha} B^{\beta} L^{\gamma}$ and solve for the three exponents.

P1.24 "Wave turbulence on the surface of a fluid in a high-gravity environment": Those interested in the pursuit of scientific knowledge, from curious amateurs to practicing physicists, may try to keep abreast of recent developments by reviewing the literature, from public science outlets (like *Scientific American* or *Discover*, either in print or online versions) to cutting-edge research journals (such as *Science, Nature*, or *Physical Review Letters*). During the early stages of preparing this book, a paper by Cazaubiel et al. (2019) with the provocative title cited in this problem caught my attention, as it was in a field in which I have very little background, and deals with a topic that is notoriously complex. The abstract began with

> We report on the observation of gravity-capillary turbulence on the surface of a fluid in a high-gravity environment. By using a large-diameter centrifuge, the effective gravity acceleration is turned up to 20 times Earth's gravity.

The authors reported data on the dependence of something called the **power spectrum density**, labeled $S_{\eta}(f)$, as a function of the driving frequency (f) in two different regimes, ones where either surface tension (again labeled γ) or gravitational effects (g, g^*) were dominant (where g^* is the effective acceleration of gravity in the rotating frame), and including the nature of the inertia of the fluid via its density ρ_m. The definition of $S_{\eta}(f)$ was not explicitly stated in the paper, but their figure 4 shows that this variable is plotted with units listed as $m^2 \cdot s$ or $meter^2 \cdot second$, so that the dimensionality must satisfy $[S_{\eta}] = \mathsf{L}^2\mathsf{T}$. The authors cite theoretical predictions for the two regimes, namely

$$S_{\eta}(f) \sim \mathcal{E}^{1/2} \left(\frac{\gamma}{\rho_m} \right)^{1/6} f^{-17/6} \quad \text{when capillary forces, i.e. surface tension, dominate} \tag{1.148}$$

$$S_{\eta}(f) \sim \mathcal{E}^{1/3} g f^{-4} \quad \text{when in the "gravity regime"} \tag{1.149}$$

and where $\mathcal{E} = (\gamma g^* / \rho_m)^{3/4}$ is the **critical energy flux.**

[9] In the spirit of full disclosure, I came across this work since it cites a pedagogical paper I wrote about dimensional analysis, namely Robinett (2015).

Acknowledging that we often work with imperfect and incomplete knowledge of a problem, knowing only the dimensions of all of the quantities involved, let us apply some DA ideas to better understand details of this system.

(a) The critical energy flux, $\mathcal{E}$, is described in a related paper by Falcon (2007) as being "... the energy flux per unit surface and density ...", which they then say implies that $[\mathcal{E}] = (\mathsf{L}/\mathsf{T})^3$. Assuming that this definition means that $\mathcal{E} = (dE/dt)/(A \cdot \rho_m)$, find its dimensions, and then confirm that their dimensional statement is true. Then, compare the dimensions of $(\gamma g^*/\rho_m)^{3/4}$ by inserting the dimensions of γ, g^*, and ρ_m to confirm that they match those of $\mathcal{E}$.

(b) Using the results of part (a), confirm that Eqns. (1.148) and (1.149) are dimensionally consistent.

(c) The frequency at which there is a transition between capillary and gravity behavior is given as

$$f_{gc} = \frac{1}{\sqrt{2\pi}} \left(\frac{\rho_m}{\gamma}\right)^{1/4} (g^*)^{3/4}. \tag{1.150}$$

Confirm that this relation is also dimensionally consistent.

(d) Can you proactively predict the dependence of $\mathcal{E}$ and f_{gc} on γ, g^*, and ρ_m by matching dimensions? For example, write $\mathcal{E} \propto \gamma^\alpha (g^*)^\beta \rho_m^\delta$, match dimensions, and solve for α, β, and γ. Repeat that process for f_{gc} and compare your predictions to the stated results.

(e) A time scale associated with non-linear (hence the subscript nl) behavior in this system is cited as being

$$\tau_{nl} \sim \mathcal{E}^{-2/3} k^{-3/2} (g^*)^{1/2} \tag{1.151}$$

where k is the **wave-number** of the turbulence. Recalling from Section 1.2 that $[k] = 1/\mathsf{L}$, confirm that Eqn. (1.151) is dimensionally correct. Can you predict this dependence by writing $\tau_{nl} \approx \mathcal{E}^\alpha k^\beta (g^*)^\gamma$ and solve for the exponents by matching dimensions? If not, what's gone "wrong?"

(f) Cazaubiel et al. (2019) cite Falcon et al. (2007) who did earlier work in this field, and those authors define $S_\eta(f)$ by saying that it's equal to

> ... the Fourier transform of the autocorrelation function of $\eta(t)$ ...

where $\eta(t)$ is the **wave height** (could be positive or negative, but a height) of the excitations above the equilibrium flat ($\eta(t) = 0$) surface. If you know enough about the words used here, confirm that this definition does suggest that $[S_\eta] = \mathsf{L}^2\mathsf{T}$. If you don't know anything about autocorrelation functions—that's totally fine—as it's a good excuse to do even more research on a new and challenging subject!

Note: I can't think of a better way to end this chapter, as this is a very personal example of using incomplete knowledge about a challenging subject, to make some progress in appreciating and understanding complex physical phenomena, using mostly DA. I hope readers will feel free to use whatever tools they have in their possession in pursuit of their explorations in physics!

References for Chapter 1

Abachi, S. et al. (D0 Collaboration). (1995). "Search for High Mass Top Quark Production in $p\bar{p}$ Collisions at $\sqrt{8} = 1.8$ TeV," *Physical Review Letters* 74, 2422–6.

Abbott B. P., et al. (2016). "Observation of Gravitational Waves from a Binary Black Hole Merger," *Physical Review Letters* 116, 061102.

Abbot B. P., et al. (2017). "The Basic Physics of the Binary Black Hole Merger GW150914," *Annalen der Physik* 529, 1600209.

Abe F., et al. (CDF Collaboration). (1995). "Production in pbar p collisions with the Collider Detector at Fermilab," *Physical Review Letters* 74, 2626.

Ahmadi M., et al. (2016). "An Improved Limit on the Charge of Antihydrogen from Stochastic Acceleration," *Nature* 529, 373-6.

Anderson, C. D. (1933). "The Positive Electron," *Physical Review* 43, 491.

ATLAS Collaboration (2012). "Observation of a New Particle in the Search for the Standard Model Higgs Boson with the ATLAS Detector at the LHC," *Physics Letters B* 32, 1-29.

Bertozzi, W. (1964). "Speed and Kinetic Energy of Relativistic Electrons," *American Journal of Physics* 32, 551-5.

Besant, W. H. (1859). *A Treatise on Hydrostatics and Hydrodynamics* (Cambridge: Deighton, Bell, and Co.).

Bod, L., Fischbach, E., Marx, G., and Maray-Ziegler, N. (1991). "One Hundred Years of the Eötvös Experiment," *Acta Physica Hungarica* 69, 335-55.

Bressi, G., et al. (2011). "Testing the Neutrality of Matter by Acoustic Means in a Spherical Resonator," *Physical Review* 83, 052101.

Cazaubiel, A., et al. (2019). "Wave Turbulence on the Surface of a Fluid in a High-Gravity Environment," *Physical Review Letters* 123, 244501.

Chu, S., Mills, A. P., and Hall, J. L. (1984). "Measurement of the Positronium Interval $1^3S_1 - 2^3S_1$ by Doppler-Free Two-photon spectroscopy," *Physical Review Letters* 52, 1689-92.

CMS Collaboration. (2012). "Observation of a New Boson at a Mass of 125 GeV with the CMS Detector at the LHC," *Physics Letters B* 716, 30-61.

Cohen, E. R., Lide, D. R., and Trigg, G. (2003). *AIP Physics Desk Reference* (Berlin, Springer).

Dylla, H. F., and King, J. G. (1973). "Neutrality of Molecules by a New Method," *Physical Review A* 7, 1224-9.

Eisele, Ch., Nevsky, A., and Schiller, S. (2009). "Laboratory Test of the Isotropy of Light Propagation at the 10^{-17} Level," *Physical Review Letters* 103, 090401.

Eötvös, L., Pekár, D., and Fekete, E. (1922). "Beiträge zum Gesetz der Proportionalität von Trägheit and Gravität (Contributions to the Law of Proportionality of Inertia and Gravity)," *Annalen der Physik* 373, 11-66.

Falcon, A., Laroche, C., and Fauve, S. (2007). "Observation of Gravity-Capillary Turbulence," *Physical Review Letters* 98, 094503.

Feynman, R. P., Leighton, R. B., and Sands, M. (1964). *The Feynman Lectures on Physics* – Vol. II: Mainly Electromagntism and Matter (Reading: Addison-Wesley). See also the (2011) *The New Millennium Edition*, which is freely available on-line at https://www.feynmanlectures.caltech.edu/.

Golosov, S., et al. (2021). "On the Parameterization of Phytoplankton Primary Production in Water Ecosystem Models," *Journal of Physics: Conference Series* 2131, 032079.

Good, R. H. (1999). *Classical Electromagnetism* (Fort Worth, TX: Saunders).

Goth, G. W. (1986). "Dimensional Analysis by Computer," *Physics Teacher* 24, 75-76.

Griffiths, D. (1999). *Introduction to Electrodynamics* 3rd ed. (Englewood Cliffs, NJ: Prentice-Hall).

Griffiths, D. (2022). "Reply to: All Magnetic Phenomena are NOT due to Electric Charges in Motion [Am. J. Phys. 90, 7–8 (2022)]," *American Journal of Physics* 90, 9.

Halliday, D., and Resnick, R. (1978). *Physics, Part Two*, 3rd ed. (New York: Wiley and Sons).

Herrmann, S., et al. (2009). "Rotating Optical Cavity Experiment Testing Lorentz Invariance at the 10^{-17} Level," *Physical Review D* 80, 105011.

Huang, K. (1987). *Statistical Mechanics* 2nd ed. (New York: Wiley and Sons).

Jackson J. D. (1999). *Classical Electromagnetism*, 3rd ed. (New York: Wiley).

Kittel C., and Kroemer, H. (1980). *Thermal Physics* (New York: Freeman).

Knight, R. D. (2013). *Physics for Scientists and Engineers: A Strategic Approach* (Boston: Pearson).

Landau, L. D., and Lifschitz, E. M. (1958). *Statistical Physics* (London: Pergamon).

Larmor, J. (1897). "LXIII: On the Theory of the Magnetic Influence on Spectra; and on the Radiation from Moving Ions," *The London, Edinburgh, and Dublin Philosophical Magazine and Journal of Science* 44, 503–512.

(Lord) Rayleigh. (1917). "VIII: On the Pressure Developed in a Liquid during the Collapse of a Spherical Cavity," *The London, Edinburgh, and Dublin Philosophical Magazine and Journal of Science* 34, 94–8.

Michelson, A. A., and Morley, E. W. (1887). "On the Relative Motion of the Earth and the Luminiferous Ether," *American Journal of Science* XXXIV (No. 203), 333–45.

Millikan, R. A. (1913). "On the Elementary Electrical Charge and the Avogadro Constant," *Physical Review* 2, 109–43.

Neddermeyer, S. H., and Anderson, C. D. (1937). "Note on the Nature of Cosmic Ray Particles," *Physical Review* 51, 884–886.

Nieto, M. M., and Goldman, T. (1991). "The Arguments against 'Antigravity' and the Gravitational Acceleration of Antimatter," *Physics Reports* 205, 221-81.

Pathria, R. K. (2006). *Statistical Mechanics*, 2nd ed. (Amsterdam: Elsevier).

Planck, M. (1899). "Ueber irreversible Strahlungsvorgänge (On Irreversible Radiation Processes)," 306, 69-122.

Reif, F. (1965). *Fundamentals of Statistical and Thermal Physics* (Boston: McGraw-Hill).

Remillard, W. J. (1983). "Applying Dimensional Analysis," *American Journal of Physics* 51, 137–140.

Robinett, R. W. (2015). "Dimensional Language as the *Other* Language of Physics," *American Journal of Physics* 83, 353–61.

Romer, R. (2001). "Heat is Not a Noun," *American Journal of Physics* 69, 107–109.

Sauter, F. (1931). "Über das Verhalten eines Elektrons im homogenen elektrischen Feld nach der relativistischen Theorie Diracs (About the Behavior of an Electron in a Homogeneous Electric Field According to Dirac's Relativistic Theory)," *Zeitschrift für Physik* 69, 742–64.

Schey, H. M. (2005). *Div, Grad, Curl, and All That: An Informal Text on Vector Calculus*, 4th ed. (New York: Norton).

Schroeder, D. (2000). *An Introduction to Thermal Physics* (San Francisco: Addison-Wesley).

Schwinger, J. (1951). "On Gauge Invariance and Vacuum Polarization," *Physical Review* 82, 664–79.

Snoke, D. (2009). *Solid State Physics: Essential Concepts* (San Francisco: Addison-Wesley).

Street, J. C., and Stevenson, E. C. (1937). "New Evidence for the Existence of a Particle of Mass Intermediate between the Proton and the Electron," *Physical Review* 52, 1003–1004.

Touboul, P., et al. (2017). "MICROSCOPE Mission: First Results from a Space Test of the Equivalence Principle," *Physical Review Letters* 119, 231101.

Touboul, P., et al. (2022). "MICROSCOPE Mission: Final Results of the Test of the Equivalence Principle," *Physical Review Letters* 129, 121102.

Turcotte, D. L., and Schubert, G. (1982). *Geodynamics* (New York: John Wiley and Sons).

Unnikrishan, C. S., and Gillies, G. T. (2004). "The Electrical Neutrality of Atoms and Bulk Matter," *Metrologica* 41, S125–S135.

Wagner, T. A., Schlamminger, S., Gundlach, J. H., and Adelberger, E. G. (2012). "Torsion-Balance Tests of the Weak Equivalence Principle," *Classical and Quantum Gravity* 29, 184002.

Williams, J. G., Turyshev, S. G., and Boggs, D. H. (2012). "Lunar Laser Ranging Tests of the Equivalence Principle," *Classical and Quantum Gravity* 29, 184004.

Wood, D. R. (2007). *Rules of Thumb in Engineering Practice* (Weinheim: Wiley-VCH).

Zangwill, A. (2013). *Modern Electrodynamics* (Cambridge: Cambridge University Press).

Zemansky, M. (1970). "The Use and Misuse of the Word 'heat' in Physics Teaching," *Physics Teacher* 8, 295–300.

Chapter 2

Warm-up: Dimensional analysis confronts data

Before embarking on a systematic discussion of the methods of dimensional analysis (sometimes abbreviated as DA for simplicity), and their use in the "quadrivium" of topics in a standard physics curriculum (classical mechanics, electricity and magnetism, thermal physics, and quantum mechanics, in Chapters 3–6), we begin instead with a few extended examples of the use of DA in research areas, old and new, and how their predictions can be confronted very directly with experimental data. We then continue (in Chapters 7–9, but also with examples embedded along the way) with explorations of its use in analyzing research problems (historical and contemporary) in less familiar areas (condensed matter physics, astrophysics, particle physics, geophysics, biophysics, etc.). And, as with all of the chapters in this text, we include a large number of exercises for further practice

2.1 Capillary oscillations of water droplets

Fluid mechanics and hydrodynamics provide a rich array of physical phenomena, many of which are amenable to the application of dimensional analysis, sometimes only needing to know the appropriate dimensionful/physical quantities (and their dimensional "spelling") to proceed. One historical example is the (idealized) case of a spherical droplet of fluid, which is slightly deformed and experiences shape oscillations because of the restoring force due to **surface tension**, that is, **capillary oscillations.**

Surface tension (here labeled S, but sometimes called γ) can be thought of as the "energy cost to create surface area" and so has dimensions determined by $S = E/A$ or $[S] = (\mathsf{ML}^2/\mathsf{T}^2)/(\mathsf{L}^2) = \mathsf{M}/\mathsf{T}^2$. It can conceptually be measured as the *force per unit length* to create new area, as in the scheme shown in Fig. 2.1, and the dimensions still work out to be $[S] = [F/L] = (\mathsf{ML}/\mathsf{T}^2)/(\mathsf{L}) = \mathsf{M}/\mathsf{T}^2$. Another method used to measure surface tension is described as **stalagmometry** (using a geometry more like stalactites than stalagmites), which is to balance the mg weight of a drop of liquid as it falls from a capillary tube of radius r with the surface tension force, via $mg = F = (2\pi r)S$, a method attributed to Tate (1864), and still used in introductory chemistry labs (Worley 1992) or research settings (Vinet et al. 1993) with some corrections.

Given an initially spherical (diameter d) drop of liquid (of density ρ_m and surface tension S), can we predict the dependence of the frequency (f) of capillary oscillations on these three quantities? If we write $f \sim S^\alpha \rho_m^\beta d^\gamma$, then dimensional consistency requires

$$[f] = [S]^\alpha [\rho_m]^\beta [d]^\gamma$$

$$\frac{1}{\mathsf{T}} = \left(\frac{\mathsf{M}}{\mathsf{T}^2}\right)^\alpha \left(\frac{\mathsf{M}}{\mathsf{L}^3}\right)^\beta \mathsf{L}^\gamma \tag{2.1}$$

Dimensional Analysis Across the Landscape of Physics. Richard W. Robinett, Oxford University Press. © Richard W. Robinett (2024).
DOI: 10.1093/oso/9780192867551.003.0002

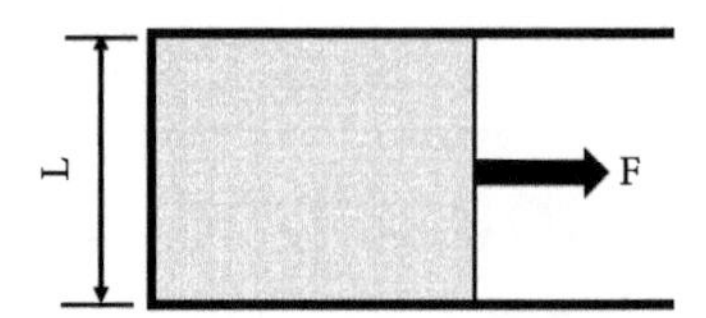

Fig. 2.1 Conceptual measurement of surface tension, as *force per unit length*, giving $S = F/2L$, since there are two surfaces (upper and lower) being created by the force F being applied to the right.

giving

$$\begin{aligned} \mathsf{M}: \quad & 0 = \alpha + \beta \\ \mathsf{L}: \quad & 0 = -3\beta + \gamma \\ \mathsf{T}: \quad & -1 = -2\alpha. \end{aligned} \tag{2.2}$$

This system of equations has solutions $\alpha = 1/2$, $\beta = -1/2$, and $\gamma = -3/2$, so that

$$f \propto S^{1/2} \rho_m^{-1/2} d^{-3/2} = \sqrt{\frac{S}{\rho_m d^3}}. \tag{2.3}$$

This result is one of the "sample problems" posed by Rayleigh (1915) in his *Principle of Similitude* review, where he states the dimensional analysis relation as follows:

> The frequency of vibration of a drop of liquid, vibrating under capillary force, is directly as the square root of the capillary tension and inversely as the square root of the density and as the 1 1/2 power of the diameter

which is exactly what we find, once we translate some of the names ("capillary tension = surface tension"). The "exact" answer to this problem was given by Lamb (1932) and Rayleigh (1896) as

$$f_n = \left[\frac{2n(n-1)(n+2)}{\pi^2}\right]^{1/2} \sqrt{\frac{S}{\rho_m d^3}} \tag{2.4}$$

where n is the (dimensionless) vibrational mode number ($n \geq 2$) (as in normal modes, akin to a quantum number). For the lowest excitation mode with $n = 2$, the dimensionless prefactor is ~ 1.3, thus close to unity as hoped. But we note that we could just have well used the drop radius $r = d/2$ in the analysis, in which case there would be another factor of $2^{3/2} \sim 2.8$, a good reminder that dimensional analysis provides little or no guidance about the ever-present purely numerical constants.

Experiments to explore this behavior were conducted by Nelson and Gokhale (1972) who used wind tunnels to effectively levitate water droplets (so that their default shapes were not precisely spherical). We reproduce in Fig. 2.2 some of their data relating oscillation frequency (for the $n = 2$ lowest mode) versus drop diameter. You are asked in **P2.1** to use this data to confirm the $d^{-3/2}$ dependence, as well as the numerical factors in Eqn. (2.4).

The study of fluid oscillations continues to be of ongoing interest, with a recent publication mentioning historic Rayleigh results, citing data from experiments performed on the International Space Station! See McCraney et al. (2022).

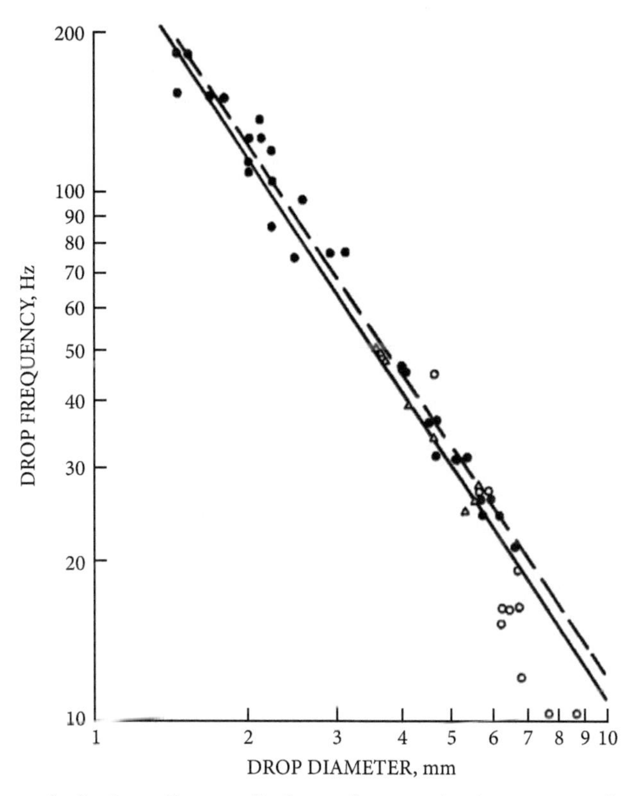

Fig. 2.2 Frequency (*Hz*) of capillary oscillations of water droplets versus size (drop diameter in *mm*). Reproduced from Nelson and Gokhale (1972) with permission by John Wiley and Sons.

2.2 **Blast waves: Atomic bomb explosions**

One of the most famous examples in dimensional analysis is a simplified prediction for some of the main results in a paper studying "*The formation of a blast wave by a very intense explosion I. Theoretical discussion*" by Taylor (1950a). (A similar analysis was done by von Neumann (1941) at about the same time and later by Sedov (1946), so that the resulting self-similarity analysis for this problem is sometimes called the Taylor–von Neumann–Sedov solution.) In his role as a civilian scientist in the Second World War, G. I. Taylor was tasked with doing calculations related to the development of nuclear weapon effects, or in his words:

> The present writer had been told that it might be possible to produce a bomb in which a very large amount of energy would be released by nuclear fission—the name atomic bomb had not then been used—and the work here described represents his first attempt to form an idea of what mechanical effects might be expected if such an explosion could occur.

The first of his two famous papers focused on his original (Taylor 1941) theoretical analysis, which he describes as

> An ideal problem is here discussed. A finite amount of energy is suddenly released in an infinitely concentrated form. The motion and pressure of the surrounding air is calculated.

A follow-up paper (Taylor 1950b) then used (very) recently declassified data of the Trinity bomb tests to test his models of the expansion (namely $R(t)$) of the blast, to compare his results to data of $R(t)$ versus t obtained from published photographs.

To explore this problem using DA, let us first assume that a large amount of energy E_0 is released and "pushes" against the atmospheric pressure (P_0) of the surrounding air as the blast expands in time t. If we assume only these three variables, we might be led to write a possible relation, along with its dimensional constraint, as

$$\begin{aligned} R(t) &\propto E_0^\alpha P_0^\beta t^\gamma \\ \mathsf{L} &= \left(\frac{\mathsf{ML}^2}{\mathsf{T}^2}\right)^\alpha \left(\frac{\mathsf{M}}{\mathsf{LT}^2}\right)^\beta \mathsf{T}^\gamma \end{aligned} \tag{2.5}$$

which implies

$$\begin{aligned} \mathsf{M}: \quad & 0 = \alpha + \beta \\ \mathsf{L}: \quad & 1 = 2\alpha - \beta \\ \mathsf{T}: \quad & 0 = -2\alpha - 2\beta + \gamma\,. \end{aligned} \tag{2.6}$$

These equations have the solution $\alpha = 1/3$, $\beta = -1/3$, and $\gamma = 0$, corresponding to a static result, namely $R(t) \sim (E_0/P_0)^{1/3}$, independent of time, which is clearly observationally wrong.

If instead we assume that the blast "pushes" against the density of air (namely its inertia), then we have $R(t) \sim E_0^\alpha \rho_m^\beta t^\gamma$, giving

$$\mathsf{L} = \left(\frac{\mathsf{ML}^2}{\mathsf{T}^2}\right)^\alpha \left(\frac{\mathsf{M}}{\mathsf{L}^3}\right)^\beta \mathsf{T}^\gamma \tag{2.7}$$

which requires

$$\begin{aligned} \mathsf{M}: \quad & 0 = \alpha + \beta \\ \mathsf{L}: \quad & 1 = 2\alpha - 3\beta \\ \mathsf{T}: \quad & 0 = -2\alpha + \gamma \end{aligned} \tag{2.8}$$

giving $\alpha = 1/5$, $\beta = -1/5$, and $\gamma = 2/5$, so that

$$R(t) \sim E_0^{1/5} \rho_m^{-1/5} t^{2/5} = \left(\frac{E_0 t^2}{\rho_m}\right)^{1/5}. \tag{2.9}$$

This can be compared to Taylor's original (and highly non-trivial) result, which was written in the form $R(t) = S(\gamma)(E_0 t^2/\rho_m)^{1/5}$ where $S(\gamma)$ is a (very complicated and numerically integrated) dimensionless function of the ratio of specific heats of air. Using the value $\gamma = C_P/C_V = 7/5 = 1.4$ (relevant to diatomic gases like N_2 and O_2) gives $S(1.4) \approx 1.03$, so that the dimensional analysis result is actually very close to the rigorous derivation.

Taylor then plotted the available data in a form designed to directly probe the predicted $R(t) \sim t^{2/5}$ dependence and we reproduce his presentation in Fig. 2.3 where there is seen to be

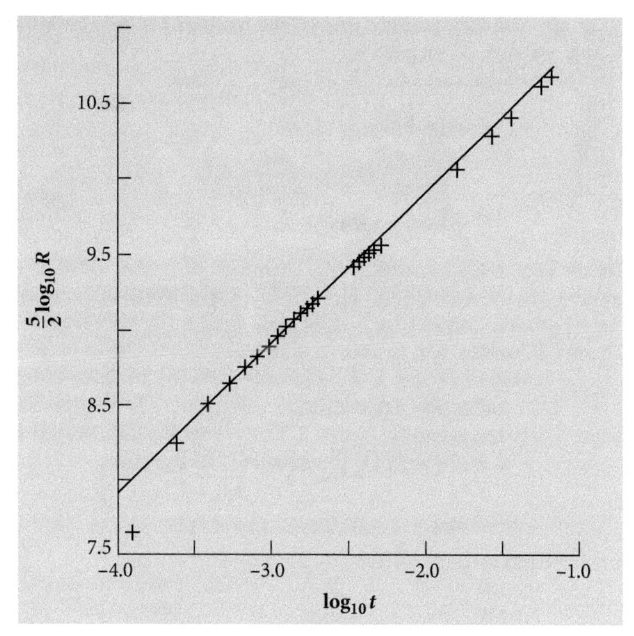

Fig. 2.3 Plot of $5/2\,\log_{10}(R)$ versus $\log_{10}(t)$ for atomic bomb explosion data. Reproduced from Taylor (1950b) with permission from the Royal Society.

good agreement. We provide the original data in **P2.4** and encourage you to plot/analyze it using modern computer methods to visualize/fit the data yourself.

Perhaps most surprisingly, using this simple fit, Taylor was able to extract an approximate value of the bomb "yield," namely E_0. For example, using any of data points in Table 2.3 in **P2.4**, you can invert the result in Eqn. (2.9) to write $E_0 \sim R(t)^5\rho_m/t^2$ to obtain a result of order $E_0 \sim 10^{14}\,J$. Using a somewhat unfamiliar (at least to most physicists) conversion from Joule to "tons of TNT" (a typical way to express weapon yield) namely 1 ton of TNT $\sim 4 \times 10^9\,kJ$, he found an energy release of roughly 20 kton, which was surprisingly close to the actual value.

2.3 **Plasma oscillations**

Plasma or ionized gas (sometimes described as the "fourth state of matter") can show a remarkable array of physical properties, since, in addition to its fluid nature, it can interact with, and create its own, electric and magnetic fields. It can be studied in a wide variety of natural (astrophysical and geophysical) or laboratory (table top experiments to fusion devices) environments. One popular text book (Chen 2018) provides the definition of plasma as

> a quasi-neutral gas of charged and neutral particles which exhibits collective behavior

and one of the simplest examples of such "collective behavior" are plasma oscillations.

If displaced from equilibrium, the electrons in an otherwise neutral plasma will experience a restoring force proportional to their displacement, giving rise to oscillations, which were first studied (both theoretically and experimentally) by Tonks and Langmuir (1929). The physical quantities relevant for this process are the electron properties (charge and mass, e and m_e), their number density (n_e), and, since the restoring force is ultimately electrostatic, we assume it depends on ϵ_0 as well. These quantities have dimensions given by

$$[n_e] = \frac{1}{\mathsf{L}^3}, \qquad [e] = \mathsf{Q}, \qquad [m_e] = \mathsf{M}, \qquad \text{and} \qquad [\epsilon_0] = \frac{\mathsf{Q}^2\mathsf{T}^2}{\mathsf{ML}^3}. \tag{2.10}$$

To explore the dependence of the oscillation behavior on these quantities, we assume that for such oscillatory motion the natural quantity is the angular frequency, ω_P, and if we write

$$\begin{aligned} \omega_P &\propto n_e^{\alpha} e^{\beta} m_e^{\gamma} \epsilon_0^{\delta} \\ \frac{1}{\mathsf{T}} &= \left(\frac{1}{\mathsf{L}^3}\right)^{\alpha} \mathsf{Q}^{\beta} \mathsf{M}^{\gamma} \left(\frac{\mathsf{Q}^2\mathsf{T}^2}{\mathsf{ML}^3}\right)^{\delta}, \end{aligned} \tag{2.11}$$

we can equate dimensions and require

$$\begin{aligned} \mathsf{M}: &\quad 0 = \quad\quad\quad \gamma \quad - \delta \\ \mathsf{L}: &\quad 0 = -3\alpha \quad\quad - 3\delta \\ \mathsf{T}: &\quad -1 = \quad\quad\quad\quad +2\delta \\ \mathsf{Q}: &\quad 0 = \quad\quad \beta \quad + 2\delta \end{aligned} \tag{2.12}$$

and easily find $\alpha = 1/2$, $\beta = 1$, $\gamma = -1/2$, and $\delta = -1/2$ or

$$\omega_P \sim \left(\frac{n_e e^2}{m_e \epsilon_0}\right)^{1/2}. \tag{2.13}$$

This predicted dependence of the plasma frequency on the electron density, namely that $\omega_P^2 \propto n_e$, has been experimentally verified by Looney and Brown (1954) and is illustrated in Fig. 2.4. (The different sets of data points along the "best fit line" correspond to the observation of standing waves in the apparatus, with the corresponding wave patterns show in the inset.)

The result in Eqn. (2.13) turns out to be the exact answer from a first principles calculation, as derived in the original literature, or from standard textbooks (see e.g., Reitz et al. 1993), and you're asked (in **P2.7**) to confirm that the numerical factors in Eqn. (2.13) reproduce the "slope" in Fig. 2.4. We explore plasma physics in more detail in Section 9.1, including reproducing the derivation of ω_P.

As a reality check, please note that we might easily have made different assumptions, for example, using $f = \omega/2\pi$ in this DA approach, or assuming that e^2/ϵ_0 was actually $e^2/4\pi\epsilon_0$, and so could have been off by the typical "order-of-magnitude" factors (values like $2\pi \sim 6$ too big or $1/\sqrt{4\pi} \sim 0.3$ too small) expected from an analysis based purely on dimensions. The use of dimensional analysis benefits, as do most things, from practice helping to inform intuition.

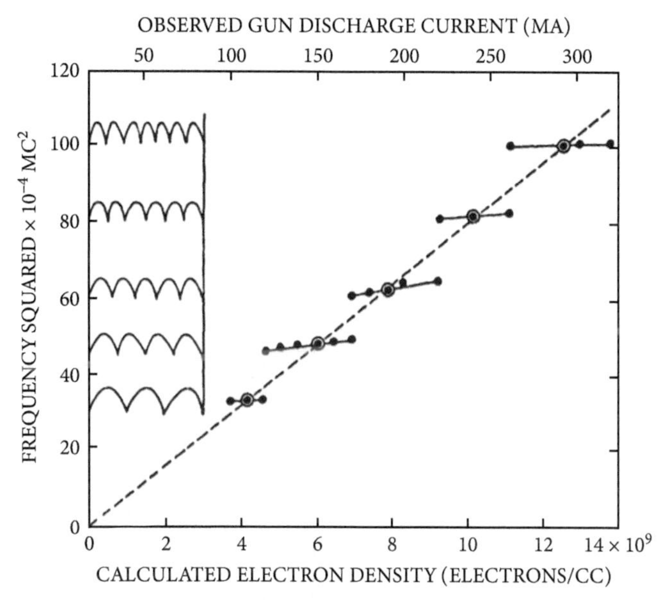

Fig. 2.4 Frequency squared versus plasma density. Reprinted with permission from Looney and Brown (1954). Copyright (1954) by the American Physical Society. Note the units used on both axes, including that of *MC* for *mega-cycle*!

2.4 **Wiedemann–Franz law and the Lorenz number**

The study, both experimental and theoretical, of transport phenomena played, and continues to play, an important role in condensed matter physics and materials science. The two most important quantities which are typically "moved around" in solid systems are

- heat (in a way characterized by the material's thermal conductivity κ) and
- electric charge (described by the electrical conductivity σ_e),

and relations governing their transport can be defined in similar ways. For the conduction of heat, we standardly write

$$\frac{dQ}{dt} = \kappa A \frac{dT}{dx} \tag{2.14}$$

which relates the rate of heat/energy flow (dQ/dt) to the area A and temperature gradient (dT/dx). For electrical conduction, one can start with the relation between current density and electric field (Ohm's law) given by $\boldsymbol{J}_e = \sigma_e \boldsymbol{E}$, but this can be rewritten as

$$\frac{I}{A} = \sigma_e \frac{dV}{dx} \qquad \text{or} \qquad \frac{dq}{dt} = \sigma_e A \frac{dV}{dx}, \tag{2.15}$$

with the analogous quantities to those in Eqn. (2.14) being the rate of charge flow ($I = dq/dt$) and the voltage gradient (dV/dx).

An early result in Franz and Wiedemann's (1853) study of metals was that the ratio of thermal to electrical conductivity (at fixed temperature T) was roughly a constant across a wide range of individual values of κ and σ_e for various materials, as illustrated in Fig. 2.5. An extension of this result was that the combination $\kappa/\sigma_e T$ was actually roughly independent of temperature as well, leading to the statement that

$$\frac{\kappa}{\sigma_e T} \equiv \mathcal{L}_0 \tag{2.16}$$

was approximately a constant for many materials (at least metals, at not too low temperatures). The "constant" $\mathcal{L}_0$ is called the **Lorenz number** and can be modeled using both classical and quantum mechanical theories of electron transport, but here we wish to explore to what extent we can derive its dependence on fundamental constants through dimensional analysis.

If we assume that $\mathcal{L}_0$ depends on the properties of the conduction electrons, we might consider using e and m_e and the number density n_e of carriers, as well as the scattering time τ between collisions (though the validity of Eqn. (2.16) across many materials might suggest no dependence on n_e and τ, both of which can vary appreciably). And given that this clearly involves thermal physics, we would also include Boltzmann's constant k_B. Thus, we are led to try

$$\mathcal{L}_0 = C_{\mathcal{L}} e^{\alpha} m_e^{\beta} k_B^{\gamma} n_e^{\delta} \tau^{\epsilon} \tag{2.17}$$

(with $C_{\mathcal{L}}$ a dimensionless constant) and using the dimensions of σ_e and κ implied by the definitions in Eqns. (2.14) and (2.15), we know that

$$[\kappa] = \frac{\mathsf{ML}}{\mathsf{T}^3\Theta} \quad \text{and} \quad [\sigma_e] = \frac{\mathsf{Q}^2\mathsf{T}}{\mathsf{ML}^3} \quad \text{so that} \quad [\mathcal{L}_0] = \frac{\mathsf{M}^2\mathsf{L}^4}{\mathsf{T}^4\mathsf{Q}^2\Theta^2} \tag{2.18}$$

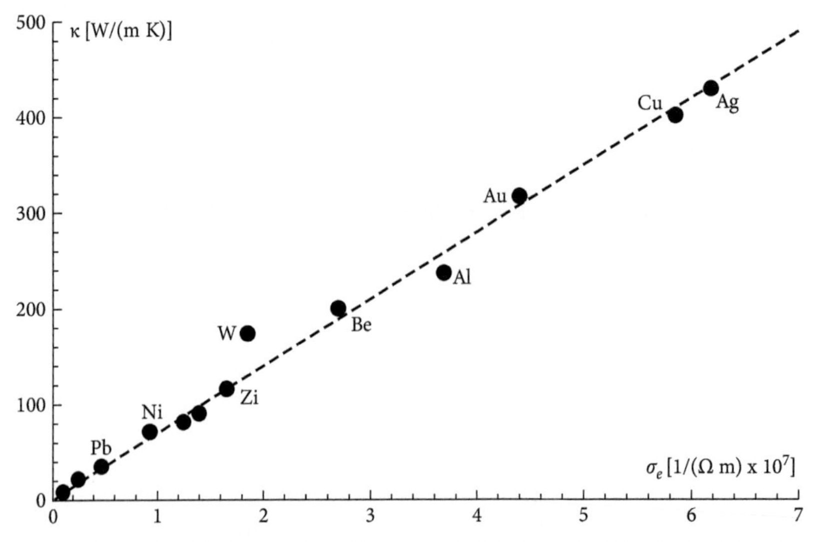

Fig. 2.5 Thermal conductivity (κ in $W/(m \cdot K)$) versus electrical conductivity (σ_e in $1/(\Omega \cdot m) \times 10^7$) for metals at $27\,^{\circ}C = 300\,K$. Data taken from "Thermal and Physical Properties of Pure Elemental Metals," in *CRC Handbook of Chemistry and Physics*, 103rd ed. (Internet version 2002), J. R. Rumble, ed. (Boca Raton, FL: CRC Press/Taylor & Francis). The dashed line is a least-square fit to the data points, which are shown as dots. Only some of the elements are labeled explicitly by their chemical symbols.

giving the dimensional relation

$$\frac{\mathsf{M}^2\mathsf{L}^4}{\mathsf{T}^4\mathsf{Q}^2\Theta^2} = \mathsf{Q}^\alpha \mathsf{M}^\beta \left(\frac{\mathsf{ML}^2}{\mathsf{T}^2\Theta}\right)^\gamma \left(\frac{1}{\mathsf{L}^3}\right)^\delta \mathsf{T}^\epsilon. \tag{2.19}$$

This is one of the few cases where we are able to use constraints from all five of the fundamental dimensions (see also **P2.11**) and find

$$\begin{aligned}
\mathsf{M}: &\quad 2 = \beta + \gamma \\
\mathsf{L}: &\quad 4 = 2\gamma - 3\delta \\
\mathsf{T}: &\quad -4 = -2\gamma + \epsilon \\
\mathsf{Q}: &\quad -2 = \alpha \\
\Theta: &\quad -2 = -\gamma.
\end{aligned} \tag{2.20}$$

These equations have the remarkably simple solution $\alpha = -2$, $\gamma = 2$, and $\beta = \delta = \epsilon = 0$, giving the result $\mathcal{L}_0 \sim (k_B/e)^2$ and indeed the answer is independent of the material-specific carrier properties n_e and τ and even m_e.

Standard textbooks on solid state physics (see e.g., Kittel 1971 or Ashcroft and Mermin 1976) derive the same quantum mechanical result

$$\mathcal{L}_0 = \frac{\pi^2}{3}\left(\frac{k_B}{e}\right)^2 \approx 2.44 \times 10^{-8}\ \text{Watt-Ohm/deg}^2, \tag{2.21}$$

but seemingly have slightly varying statements on the corresponding classical result. Kittel (1971) quotes the classical result as $\mathcal{L}_0 = 3(k_B/e)^2$, while Ashcroft and Mermin (1976) cite half this value and argue that Drude "... erroneously found half the correct result... in extraordinary agreement with experiment." But it is notable that both the classical and quantum mechanical approaches find a result in agreement with dimensional analysis (presumably as they must!). We note that if we replace either n_e or τ by $\hbar$ in the dimensional analysis, we continue to find that $\mathcal{L}_0 \sim (k_B/e)^2$ and you are asked to explore other combinations in **P2.10**.

While the statement of the Wiedemann–Franz (WF) law dates back to the mid-nineteenth century, experiments still use it as the "gold standard" for comparison in electric/thermal transport measurements, with many recent research papers citing an observed *violation* or *breakdown* of the relation in novel systems. See, for example, Principi and Vignale (2015), Crossno et al. (2016), Zarenia et al. (2019), and Robinson et al. (2021).

2.5 **Gravitational bound states of neutrons**

The study of bound states forms an important part of the content of most quantum mechanics texts. The most familiar examples are manifestations of the electromagnetic (EM) force being responsible for the binding, and include atoms (especially the canonical example of the hydrogen atom), molecules, and eventually solids. The strong force (ultimately QCD or quantum chromodynamics) is also responsible for bound states with protons, neutrons, pions, etc., consisting of quarks and gluons, but such systems are perhaps best studied with the machinery of field theory (including computational implementations on the lattice).

There are, of course, bound systems of particles due to gravitational interactions (planets, stars, galaxies, etc.) but there is one (perhaps) under-appreciated example of a bound state due to terrestrial gravity, namely the motion of a particle in the potential due to the Earth's gravitational field, that is,

$$V(z) = \begin{cases} \infty & \text{for } z<0 \\ mgz & \text{for } z>0 \end{cases} . \tag{2.22}$$

Neutrons, despite their short lifetimes (about 15 minutes, or $\tau_n \sim 878\, sec$), have been used to perform some foundational experiments testing many aspects of quantum mechanics (and we will explore other examples in Section 6.4.2). One such result demonstrated the existence of "*Quantum states of neutrons in the Earth's gravitational field*" (see Nesvizhevsky et al. 2002, 2003, 2005), providing evidence for quantized energy states of the neutron "hovering" over a flat surface in the Earth's field (i.e., subject to a classical force mg).

In order to understand some of the results of this series of experiments (and others, such as Jenke et al. 2011) we explore the use of DA to estimate the typical energy and length scales in such systems. We assume that the energies of the neutron quantum states, E_n, (with the subscript n for neutron and NOT a quantum number) depend only on the neutron mass (m_n), the acceleration of gravity (g), and quantum mechanics (via $\hbar$). The required dimensions are given by

$$[E_n] = \frac{\mathsf{ML}^2}{\mathsf{T}^2}, \quad [m_n] = \mathsf{M}, \quad [g] = \frac{\mathsf{L}}{\mathsf{T}^2}, \quad \text{and} \quad [\hbar] = \frac{\mathsf{ML}^2}{\mathsf{T}}. \tag{2.23}$$

We then write

$$\begin{aligned} E_n &\propto m_n^\alpha g^\beta \hbar^\gamma \\ \frac{\mathsf{ML}^2}{\mathsf{T}^2} &= \mathsf{M}^\alpha \left(\frac{\mathsf{L}}{\mathsf{T}^2}\right)^\beta \left(\frac{\mathsf{ML}^2}{\mathsf{T}}\right)^\gamma, \end{aligned} \tag{2.24}$$

and then equating dimensions produces the constraints

$$\begin{aligned} \mathsf{M}: &\quad 1 = \alpha \quad + \gamma \\ \mathsf{L}: &\quad 2 = \quad \beta + 2\gamma \\ \mathsf{T}: &\quad -2 = -\,2\beta - \gamma \end{aligned} \tag{2.25}$$

which gives $\alpha = 1/3$, and $\beta = \gamma = 2/3$, or

$$E_n \sim \left(mg^2\hbar^2\right)^{1/3}. \tag{2.26}$$

Using known values for m_n, g, and $\hbar$ we find that $E_n \sim 1.2 \times 10^{-31}\, J$, or to make contact with the units used in most quantum mechanics experiments, $E_n \sim 0.75 \times 10^{-12}\, eV \sim 0.75\, peV$ where peV is pico-eV or $10^{-12}\, eV$. In contrast, the energy scales for the hydrogen atom problem are of order $10\, eV$, those for van der Waals bonds in solids of order $1-10\, meV$ (milli–eV), while the energy scale for nuclear physics is $1-10\, MeV$, so it may not be immediately clear how any kind of spectroscopy could be used to probe such quantized states (but see **P2.15**).

In a similar way, we can find the dimensional dependence of the typical (vertical) length scale l_n, of these bound states by following the methods above, or by associating $E_n \sim mgl_n$, to find that

$$l_n \sim \frac{E_n}{mg} \sim \left(\frac{\hbar^2}{m^2 g}\right)^{1/3} \sim 7 \times 10^{-6}\, m, \tag{2.27}$$

or length scales of order 10 μm, which is in the mesoscopic range.

In the lead up to the presentation of their experimental results, Nesvizhevsky et al. (2002) show the probability densities for such bound states, obtained by solving the Schrödinger equation for the potential in Eqn. (2.22) (which involves Airy functions), which we reproduce in Fig. 2.6. We note that the energy and length scales cited (namely $E_n \sim \mathcal{O}(peV)$ and $l_n \sim \mathcal{O}(10\,\mu m)$) are exactly the magnitudes suggested by DA, so that a reader of the research paper can appreciate much of the result even before gaining a detailed mastery of the mathematical machinery of quantum mechanics, allowing one to proceed to a deeper understanding of the experimental set-up and data analysis.

The experiment projected cold neutrons horizontally over a flat surface with an upper plate that could be separated, and so was probing the actual spatial extent of the quantum mechanical wave function over lengths of $\mathcal{O}(l_n)$. The transmission as a function of absorber height is shown in Fig. 2.7, and clearly shows structure matching the quantum probability densities shown in Fig. 2.6, on the scales of 10s of microns.

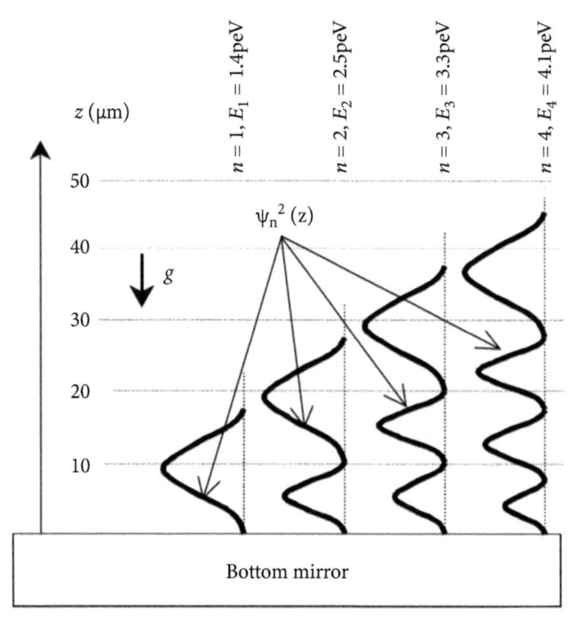

Fig. 2.6 Predicted quantum mechanical probability densities, $|\psi_n(z)|^2$, for the bound states of neutrons in the Earth's gravitational field, along with quantized energy eigenvalues. Reproduced from Nesvizhevsky et al. (2002) with permission of Springer Nature.

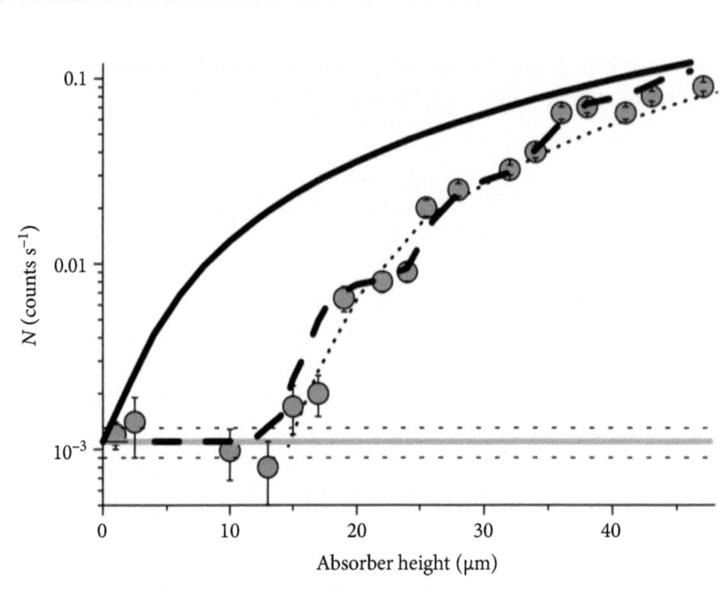

Fig. 2.7 Transmission rate (counts per second) versus absorber height (μm) showing structure near the spatial separations consistent with the quantized wave functions in Fig. 2.6. Reproduced from Nesvizhevsky et al. (2002) with permission of Springer Nature.

2.6 **Nearly perfect fluids: From cold atoms to quark gluon plasma**

Except perhaps for the air we breathe, water (which comprises 70% of our body mass) is the most familiar of all fluids. It is often argued that the special properties of water[1] are largely responsible for life on earth (as we know it) and many of its extraordinary physical behaviors, such as its

- large heat capacity, which mediates temperature extremes,
- excellent solvent properties,
- and temperature dependence of solid versus liquid state (ice floats!)

are all directly related to its molecular structure (see e.g., Brini et al. 2017 or Finney 2004).

Another physical property we use to distinguish fluids is its "slipperiness" or **viscosity** (η), with materials with smaller values of η (as we will see, perhaps scaled to other physical properties) being considered more nearly "*perfect fluids*." Schäfer (2009) described this property as follows:

> A good fluid, such as water, supports complicated flow patterns that decay slowly over time. In contrast, in a "poor fluid", like honey or tar, we cannot observe waves or eddies, and flow processes decay quickly.

and viscosity does play an important role in many classical fluid mechanics and hydrodynamics studies.

---

[1] For a popular science review, see Ball (2001) for a colloquium level presentation, see Finney (2004), or for a more technical discussion, see Brini et al. (2017).

For our first view of its use in a DA approach, we note η can be defined using the geometry shown in Fig. 2.8 by the relation

$$\frac{F}{A} \equiv \eta \frac{\partial}{\partial y}(v_x(y)), \tag{2.28}$$

and where (importantly for us) the dimensions of η are then given by

$$\left[\frac{F}{A}\right] = [\eta]\left[\frac{\partial}{\partial y}v_x\right] \longrightarrow \left(\frac{\mathsf{ML}}{\mathsf{T}^2}\right)\left(\frac{1}{\mathsf{L}^2}\right) = [\eta]\left(\frac{\mathsf{L/T}}{\mathsf{L}}\right) \quad \text{or} \quad [\eta] = \frac{\mathsf{M}}{\mathsf{LT}}. \tag{2.29}$$

The spatial change in velocity described by $\nabla_y(v_x)$ is sometimes described as the **shear rate.**

The MKSA unit of viscosity is then given by $kg/(m \cdot s)$, but is more often written in a form suggested by the definition in Eqn. (2.28), namely $Pa \cdot s$, where $Pa = Newton/m^2$ (for Pascal) is the MKSA unit of pressure. Table 2.1 presents some (approximate) values of viscosity (in those units) for a wide range of materials to which we will refer later.

Returning to our intended focus of applying DA to problems of modern research interest, we note that experimental research in a variety of quantum fluids find a dizzying variation of materials properties, including viscosity, and we show in Table 2.2 values of η for

- liquid water (at atmospheric pressure and just below the boiling point, as well as near the critical) point,
- liquid ^4He (at atmospheric pressure and elsewhere),
- ultra-cold lithium atoms in a Bose–Einstein condensate phase, and
- quark–gluon plasma (QGP) formed by ultra-high nucleus–nucleus collisions at Relativistic Heavy Ion Collider (RHIC),

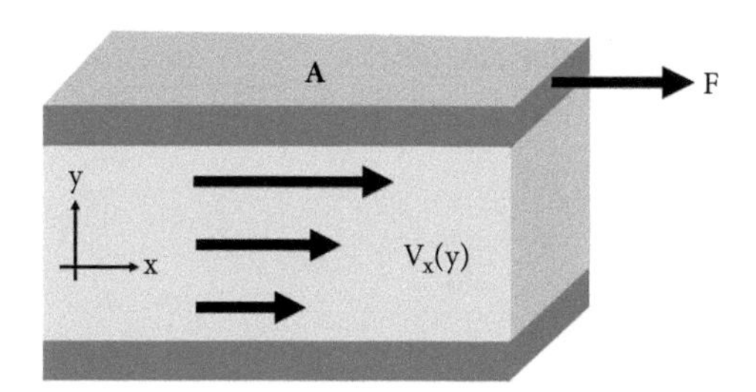

Fig. 2.8 Visualization of the geometry and physical parameters used in the definition of viscosity in Eqn. (2.28).

Table 2.1 Values of viscosity for an assortment of materials.

Material	η ($Pa \cdot s$)	**Material**	η ($Pa \cdot s$)
Air	$2 \cdot 10^{-5}$	Lard (shortening)	10^3
Water	10^{-3}	Lava (wide range)	$10 - 10^{10}$
Vegetable oils	$(3–8) \cdot 10^{-2}$	Pitch (tar)	$10^7 - 10^8$
Honey	2–10	Nuclear matter	10^9
Ketchup/mustard	50	Neutron star	$10^{17} - 10^{18}$
Peanut butter	250	Earth mantle	$10^{19} - 10^{23}$

Table 2.2 Data on pressure ($P\,(Pa)$), temperature ($T\,(K)$), shear viscosity ($\eta\,(Pa \cdot s)$), and ratio of shear viscosity to entropy density (η/s), taken from Schäfer and Teaney (2009). The last column shows η/s scaled by the combination $\hbar/k_B$.

Fluid	$P\,(Pa)$	$T\,(K)$	$\eta\,(Pa \cdot sec)$	$\eta/s\,(sec \cdot K)$	$(\eta/s)/(\hbar/k_B)$
H_20	0.1×10^6	370	2.9×10^{-4}	6.3×10^{-11}	8.2
H_20	22.6×10^6	650	6.0×10^{-5}	1.5×10^{-11}	2.0
^4He	0.1×10^6	2.0	1.2×10^{-6}	1.4×10^{-11}	1.9
^4He	0.22×10^6	5.1	1.7×10^{-6}	5.4×10^{-12}	0.7
^6Li	12.0×10^{-9}	23×10^{-6}	$\leq 1.7 \times 10^{-15}$	$\leq 3.8 \times 10^{-12}$	≤ 0.5
QGP	88.0×10^{-33}	2×10^{12}	$\leq 5.0 \times 10^{11}$	$\leq 3.0 \times 10^{-12}$	≤ 0.4

and we see that the range in typical pressures, temperatures and especially η for these materials can easily span sixteen orders of magnitude, all in controlled laboratory settings.

While there may not seem to be any underlying similarity between these systems, we note that the ratio of viscosity to **entropy density** (entropy per unit volume, S/V), or $\mathcal{R} \equiv \eta/(S/V) = \eta/s$, in each case now covers a range of only a single order of magnitude and might provide a unifying description of these very different systems.

Either of two relations involving entropy provide the important dimensional "spelling" of S, namely the original Boltzmann definition of $S = k_B \ln(\Omega)$ from Eqn. (1.78), or the relation $dS = dU/T$ from Eqn. (1.79), both give $[S] = [U]/[temperature] = \mathsf{ML}^2/(\mathsf{T}^2\Theta)$, so that the entropy density has $[s] = [S/V] = \mathsf{M}/(\mathsf{LT}^2\Theta)$.

Given the similarity in values of $\mathcal{R} = \eta/s$, we might be led to ask whether there is an ultimate limit on this quantity, that is, an **almost perfect fluid**, with a value associated with fundamental constants. Two natural quantities that clearly suggest themselves are $\hbar$ (these are quantum fluids, after all) and k_B (as they all involve thermal processes), but there seems to be no temperature dependence in $\mathcal{R}$, so we ignore the temperature T. This is not a problem obviously related to fundamental gravitational physics (so no G), but perhaps some mass (m_e or m_p or generically m_X) as different particles (atoms, molecules, quarks, etc.) could be involved.

If, for generality, we also assume that $\mathcal{R}$ depends on electromagnetism via e and ϵ_0, we could write for a "basic" unit of $\mathcal{R}_0 = \eta/s$

$$\mathcal{R}_0 \equiv \frac{\eta}{s} \sim \hbar^\alpha k_b^\beta m_X^\gamma \epsilon_0^\delta e^\epsilon \tag{2.30}$$

or matching dimensions

$$[\mathcal{R}_0] = \left[\frac{\eta}{s}\right] = \left(\frac{\mathsf{M}}{\mathsf{TL}}\right)\left(\frac{\mathsf{M}}{\mathsf{LT}^2\Theta}\right)^{-1} = \Theta\mathsf{T} = \left(\frac{\mathsf{ML}^2}{\mathsf{T}}\right)^\alpha \left(\frac{\mathsf{ML}^2}{\mathsf{T}^2\Theta}\right)^\beta \mathsf{M}^\gamma \left(\frac{\mathsf{Q}^2\mathsf{T}^2}{\mathsf{ML}^3}\right)^\delta \mathsf{Q}^\epsilon \tag{2.31}$$

which requires

$$\begin{aligned}
\mathsf{M}: \quad 0 &= \alpha + \beta + \gamma - \delta \\
\mathsf{L}: \quad 0 &= 2\alpha + 2\beta \quad - 3\delta \\
\mathsf{T}: \quad 1 &= -\alpha - 2\beta \quad + 2\delta \\
\mathsf{Q}: \quad 0 &= \qquad\qquad 2\delta + \epsilon \\
\Theta: \quad 1 &= \quad -\beta \qquad\qquad .
\end{aligned} \tag{2.32}$$

This set of equations is easily solved to give $\alpha = 1$, $\beta = -1$, and $\gamma = \delta = \epsilon = 0$, or the very simple result

$$\mathcal{R}_0 \sim \frac{\hbar}{k_B} \sim 7.7 \times 10^{-12}\, sec \cdot K, \tag{2.33}$$

which is within an order of magnitude of the scaled results for $\mathcal{R}$ in Table 2.2 (column 5), motivating the final column, which shows $\mathcal{R}/\mathcal{R}_0$ all being with an order of magnitude of unity.

An amazing insight into the possible origin of such a fundamental lower bound on "perfect fluidity" actually comes from studies of string theory and conformal field theory (see Policastro 2001 and Kovtun 2005), which suggests that the lower-bound for $\mathcal{R}$ might actually be

$$\mathcal{R}_{min} = \frac{1}{4\pi}\mathcal{R}_0 = \left(\frac{1}{4\pi}\right)\frac{\hbar}{k_B} \sim (0.08)\frac{\hbar}{k_B}, \tag{2.34}$$

and the values in Table 2.2 (column 6) can be compared to this. This is a fascinating example of physicists (Kovtun 2005) pursuing some of the most theoretical aspects of one field (in this case, black hole physics) providing insight into any number of seemingly unrelated fields. Looking at the *Map of Physics* (Walliman 2017) on the inside back cover, this is perhaps akin to **The Future** sending us information across the **Chasm of Ignorance** to better inform current understanding, another example of the diversity of subject areas contributing to the unity of physics, all with the common language of dimensions.

2.7 Problems for Chapter 2

Q2.1 Reality check: What was the single most important, interesting, engaging, and/or useful thing you learned from this chapter?

P2.1[†] **Capillary oscillations of water droplets:** (a) Using the data shown in Fig. 2.2, estimate the power-law dependence of f on d given the log–log plot utilized, namely find value of α in a relation like $f = Cd^{\alpha}$. Reminders on how to extract values from such a fit are covered in Section 10.9.
(b) Can you also fit the straight line shown to that same form to also confirm the value of C? Note that the original paper by Nelson and Gokhole used water droplets for which $S = 75$ dynes/cm and $\rho_m = 1$ gr/cc (note the CGS units in both!) and were exciting the $n = 2$ mode. Also note the dimensions used for diameter on the horizontal axis! Is your "fit" accurate enough to distinguish between whether one excited the $n = 2$ or higher modes?

P2.2 Oscillations of charged water droplets: A variation on the oscillating droplet problem of Section 2.1 considers a spherical drop of density ρ_m, radius a_0, and total charge Q.
(a) Ignoring surface tension for the moment, if there are small displacements from the symmetrical equilibrium configuration, what is the natural time scale, call it τ_Q, for the subsequent behavior? Write $\tau_Q \propto Q^{\alpha}\epsilon_0^{\beta}\rho_m^{\gamma}a_0^{\delta}$ and match M, L, T, Q dimensions to find $\alpha, \beta, \gamma, \delta$.
(b) Rayleigh (1896) calculated the frequency of oscillation of a droplet considering both surface tension effects (as in Eqn. (2.4)) and including a net charge and found that

$$f = \sqrt{\frac{n(n-1)}{4\pi^2\rho_m a_0^3}\left\{(n+2)T - \frac{Q^2}{4\pi\epsilon_0 a_0^3}\right\}} \tag{2.35}$$

where T is his notation for surface tension. Show that this result reproduces Eqn. (2.4) for $Q = 0$, and that it is dimensionally consistent with your result from part (a) for $Q \neq 0$.

Table 2.3 Data from Trinity atomic bomb test from Taylor (1950b).

t (msec)	R (m)	t (msec)	R (m)	t (msec)	R (m)	t (msec)	R (m)
0.10	11.1	0.94	36.3	1.79	46.9	4.34	65.6
0.24	19.9	1.08	38.9	1.93	48.7	4.61	67.3
0.38	25.4	1.22	41.0	3.26	59.0	15.0	106.5
0.52	28.8	1.36	42.8	3.53	61.1	25.0	130.0
0.66	31.9	1.50	44.4	3.80	62.9	34.0	145.0
0.80	34.2	1.65	46.0	4.07	64.3	53.0	175.0

(c) How do you interpret the fact that the charge term has the opposite sign from the surface tension term in Eqn. (2.35)? Namely, what would an imaginary frequency look like for the time-dependence?

P2.3 Surface tension vibrations of spherical nuclei: The liquid drop model of nuclei includes a term of the form $E_{ST} = \alpha_{ST}A^{2/3}$, where $A = Z + N$ is the number of nucleons (protons plus neutrons), which is designed to take into account surface tension effects of the nuclear material. The radii of nuclei can be approximated by $R(A) = r_0 A^{1/3}$, which corresponds to the fact that the nuclear density is roughly constant, namely $V(A) = 4\pi[R(A)]^3/3 \propto Ar_0^3$, implying that $\rho_m(A) \propto (Am_p)/(Ar_0^3) \sim m_p/r_0^3$. Use the forms of E_{ST} and $R(A)$ above to write the nuclear energy contribution due to surface tension effects as $E_{ST} = S_{nuc}[4\pi R(A)^2]$ to estimate the surface tension of nuclear matter. Use values of $\alpha_{ST} \approx 20\,MeV$ and $r_0 \approx 1.2\,F$ and express S_{nuc} in terms of MeV/F^2 and then J/m^2 or erg/cm^2 to compare to that of liquid water.

P2.4[†] Trinity atom bomb test data analysis: (a) We reproduce in Table 2.3 the data of blast size (R in meters) versus time after detonation (t in *msec*) from Taylor (1950b). Plot the data either "by hand" (on log–log paper if you know what that is!) or perhaps an Excel spreadsheet, or using computer programs (such as Mathematica©) which can plot and fit data, to confirm that an assumed $R(t) \sim Ct^{\alpha}$ relation gives $\alpha = 2/5 \approx 0.4$. (A rudimentary piece of Mathematica© code that helps plot data to exhibit power-law behavior and find such dependences in included in Section 10.8.)
(b) Use any of the data points, or the results of your fit (i.e., the value of C), to extract the value of E_0 and compare to that cited in the discussion in Section 2.2.

P2.5 Pressure and shock wave velocity after bomb blast: (a) Repeat the analysis of Section 2.2 to explore the time-dependence of the **pressure** of the blast wave as a function of time by assuming that $P(t) \sim E_0^{\alpha}\rho_m^{\beta}t^{\gamma}$ to find α, β, γ and discuss if your solution makes physical sense.
(b) Do the same thing for the **speed** of the shock wave front, first using dimensional analysis directly, and then simply differentiating Eqn. (2.9) with respect to time, checking that the two methods agree.

P2.6 Application to supernovae and other astrophysical blast waves: The methods of Section 2.2 can be applied in a more fundamental research context as applications to astrophysical blasts, such as those accompanying supernovae (for a discussion, see Clarke and Carswell 2007). As an example of the physical parameters relevant to such a problem, assume that a solar mass $M_{\odot}$ of material is expelled from such a blast with a speed of $v \sim 10^4\,km/sec$ and first use this to estimate the initial energy in the explosion, E_0. Then, if the density of the interstellar medium is roughly one hydrogen atom (or proton) per cubic centimeter, show that this is equivalent to

$\rho_m \sim 10^{-21}\ kg/m^3$. Finally, use Eqn. (2.9) to show that the radius and speed of the expanding shell are given by

$$R(t) \sim (0.3\,pc)[t/year]^{2/5} \quad \text{and} \quad v(t) \sim (10^5\ km/s)[t/year]^{-3/5} \tag{2.36}$$

where pc is one parsec. See the table of conversion factors in Section 10.4 if necessary. You can use the results of **P2.5** to find an expression for $v(t)$, or just differentiate your expression for $R(t)$.

P2.7† **Plasma frequency data:** The result in Eqn. (2.13) can be written in the form

$$(f_P)^2 = \left[\frac{e^2}{4\pi^2\epsilon_0 m_e}\right] n_e \equiv C_e n_e,$$

since $\omega_P = 2\pi f_P$, and the data in Fig. 2.4 is plotted to emphasize this relationship. Find the "slope" of the linear $(f_P)^2$ versus n_e plot (keeping in mind the units used on both axes) and compare to the value of C_e obtained by using values of e, ϵ_0, and m_e obtained from the handbook of physical constants in Section 10.3, if needed.

P2.8 AM/FM radio propagation: The ionosphere (as the name implies) is an ionized plasma and is shown in context in the infographic in Fig. 2.9, where it lists typical electron densities in the range $n_e = 10^4 - 10^6\ cm^{-3}$. Evaluate the plasma frequencies (ω_P) for this range and compare to the frequency bands ($f = \omega_P/2\pi$) used in AM (535–1607 kHz) and FM (87–108 MHz) communication. Use your results to comment on the cartoon shown in Fig. 2.10, which implies that AM waves are reflected from the ionosphere while FM waves can propagate through. Have you ever

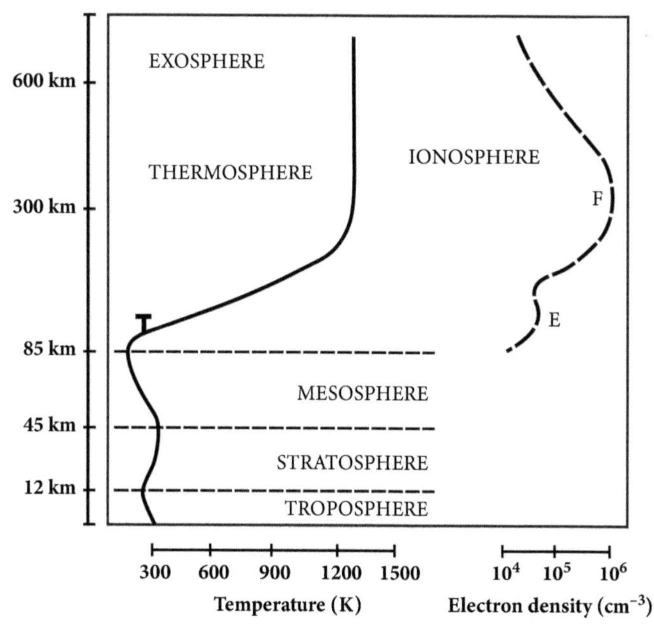

Fig. 2.9 Schematic of the Earth's atmosphere. Note the inset that describes the electron density in the ionosphere as being in the range $n_e = 10^4 - 10^6\ cm^{-3}$. From https://en.wikipedia.org/wiki/Ionosphere. Public domain image authored by Ghamer and updated to SVG by TiZom. Accessed September 19, 2023.

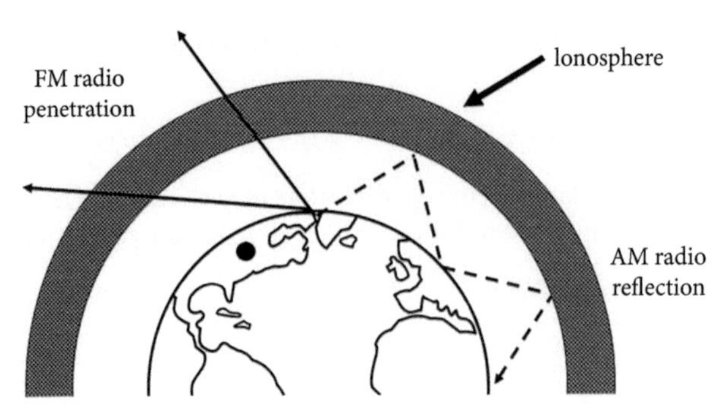

Fig. 2.10 Schematic of AM versus FM signal propagation geometry in the Earth's atmosphere.

much further away than in the daytime? (The answer may be highly dependent on your age!) If so, why do you think that's possible?
Note: We will revisit the subject of plasmas in much more detail in Section 9.1.

P2.9 Wiedemann–Franz Law data analysis: (a) Using handbook data for k_B and e, confirm the numerical value of $\mathcal{L}_0$ shown in Eqn. (2.21), which is the prediction from quantum mechanics. The units cited there are Watt–Ohm/deg^2, which are natural considering the definitions of κ and σ_e. The theory prediction gives dimensions of $\mathcal{L}_0 \propto (k_B/e)^2$ which suggests that the MKS value should be expressible as the square of some basic units. Show that these units also correspond to (Volt/deg)^2, by considering both the left- and right-hand sides of Eqn. (2.21).
(b)† Using the straight-line fit to the κ versus σ_e data shown in Fig. 2.5, extract the experimental value of $\mathcal{L}_0 = \kappa/\sigma_e T$ by (i) finding the slope and then (ii) dividing by the temperature used for this data-set, namely $T = 27\ C = 300\ K$.

P2.10 Wiedemann–Franz law—Checking other possible combinations: Repeat the analysis of the WF law, but this time use as potential variables e, k_B, m_e and instead of n_e and/or τ, use combinations involving $\hbar$ and/or ϵ_0. Do your results change?

P2.11 Seebeck effect: Another thermo-electric phenomena that can be explored using dimensional analysis is the so-called **Seebeck effect.** While evidently first discovered by Alessandro Volta, it is named after Thomas Johann Seebeck and describes the situation where an electric potential difference (and so an electric field, **E**) can be induced in a material subject to a temperature gradient. The formal connection can be written in the form

$$\mathbf{E}_S = -S_B\,\nabla T \tag{2.37}$$

where S_B is the **Seebeck constant.** Assume the same type of dependence we used for the Lorenz number in Section 2.4 for this constant by writing $S_B = e^\alpha m_e^\beta k_B^\gamma n_e^\delta \tau^\epsilon$ and solve for the exponents to the extent you can.
Note: This analysis gives a dimensionally correct combination, but with a value which is two orders of magnitude too large, as it ignores important quantum mechanical effects encoded in the temperature and Fermi energy (E_F); the true answer includes a factor of $k_B T/E_F$ times the answer you'll find, as well as the usual undetermined dimensionless constants. This is one case

where we should have included more variables, with the result that we'd have naturally found the dimensionless ratio $\Pi = k_B T/E_F$. We provide more examples of this in the next chapter.

P2.12 Piezoelectricity: Some materials may generate an electric field in response to being subjected to an external pressure, a phenomenon called **piezoelectricity**, an effect used extensively in many kinds of sensors. The functional relation between the external stress, encoded in the **stress tensor** T (which has the units of pressure), and the induced electric field (E), can be written schematically as

$$\epsilon E \sim d\,T \tag{2.38}$$

where ϵ is the permittivity and d is the material-specific **piezoelectric constant.**

(a) Using the known dimensions of pressure, electric field, and ϵ, find the M, L, T, Q dimensions of d.

(b) Tables of experimental values of d for materials commonly used in applications are cited as being in the $\mathcal{O}(1-10^3)$ pC/N range, where $p = pico = 10^{-12}$. Are those MKSA units consistent with the dimensions you found in part (a)?

Note: The relation in Eqn. (2.38) is described as "schematic" since $\boldsymbol{E}$ is a vector and $\boldsymbol{T}$ is a tensor, so we are ignoring indices, and focusing only on dimensions. A pressure/stress in one direction can give rise to an electric field in another, depending on the details of the crystal structure of the material.

P2.13 Neutron quantum bouncer I: Use the same dimensional analysis methods that were used to analyze the neutron's bound state energy E_n and typical length l_n to find how the neutron (a) speed (v_n), (b) momentum (p_n), and (c) characteristic time (t_n) all depend on $\hbar$, g, and m_n. Confirm that the dimensions of E_n, $mv_n^2/2$, and $p_n^2/2m_n$ all agree, as they must.

P2.14 Neutron quantum bouncer II: Transitions between the quantized energy levels of the neutron in the Earth's field would correspond to the emission of EM radiation of angular frequency $\omega_{ij} = \Delta E_{ij}/\hbar$. Use dimensional analysis to see how these frequencies depend on $\hbar$, g, and m_n. Estimate the photon energy emitted in such a transition and the corresponding wavelength of the emitted photon using $E_\gamma = \hbar\omega_{ij}$ and $E_\gamma = 2\pi\hbar c/\lambda_\gamma$. What type of photon is this and how easy would it be to detect?

P2.15 Neutron quantum bouncer III: Transitions between the quantized neutron bound states have, in fact, been observed, but using more "classical" methods, via resonantly exciting the energy difference ΔE_{ij} **mechanically** by vibrating the entire apparatus. Using the typical value of ω_{ij} found in **P2.13**, estimate the resonant frequency (here it would be $f = \omega/2\pi$) needed and compare to the data from Jenke et al. (2011) reproduced as Fig. 2.11.

P2.16 More oscillations of spherical drops (globes), including viscosity: We began this chapter with a discussion of the oscillation of water droplets due to capillary oscillations, where surface tension was the restoring force. A similar problem (also discussed by Lamb (1932)) asks for the vibrational frequency of a globe of water, where its own gravitational self-attraction is the relevant quantity, driving small perturbations back to equilibrium. (If a "globe of water" sounds like a very artificial problem, imagine instead a spherical mass of fluid, like a star, in the context of astrophysics.)

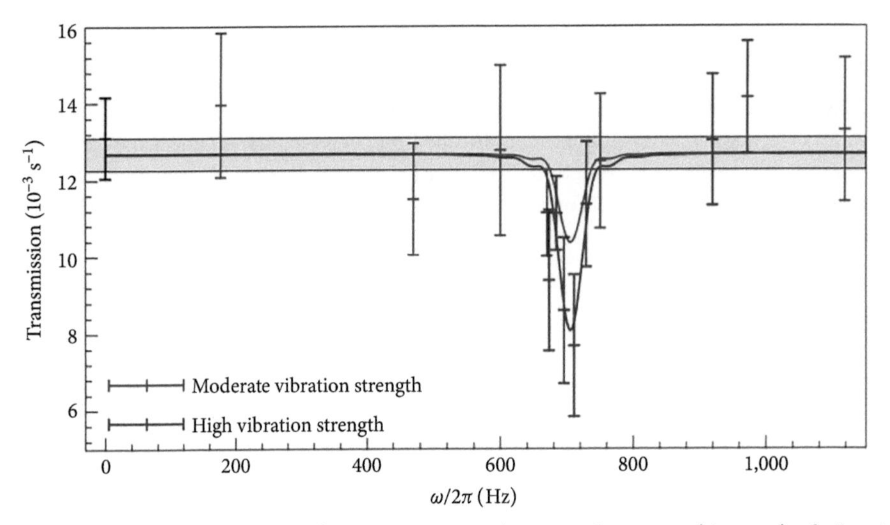

Fig. 2.11 Transmission probability (through apparatus) versus frequency ($f = \omega/2\pi$) of vibration, showing resonant excitation of quantum bound states. Reproduced from Jenke et al. (2011) with permission of Springer Nature.

(a) Assume that f depends on ρ_m and d as before, but now also on G (Newton's constant), write $f \sim G^\alpha \rho_m^\beta d^\gamma$, and solve for α, β, γ. The combination of dimensional factors you've found is present in many astrophysical situations involving gravitational perturbations, determining the time scales for oscillations or collapse.

(b) Whether the water drops (globes) are driven by surface tension or gravity, if there is viscosity present, the oscillations will be exponentially damped, with amplitudes decaying as $\exp(-t/\tau)$. In both cases, the restoring force (S or G) is no longer relevant and the damping time is related to ρ_m, d and now the viscosity η. Write $\tau \sim d^\alpha \rho_m^\beta \eta^\gamma$ and solve for all exponents. Does the dependence on η make physical sense? For the original calculations, see Chandrasekhar (1959) or Reid (1960).

P2.17 Viscosity of nuclear matter: Just as surface tension can play a role in liquid drop models of nuclear structure (as in **P2.3**), so too have some studies considered the relevance of viscosity in large nuclei. One such study (Khokonov 2016) cites a value of the viscosity of nuclear material in "tribal" units used in nuclear physics, namely $\eta \sim (4-8) MeV/(fm^2 c)$ with c being the speed of light and $1\,fm = 10^{-15}\,m = 1\,F$. Confirm that these units do indeed given the appropriate dimensions of viscosity, and convert to the standard $Pa \cdot s$ units to compare to values in Table 2.1.

P2.18 Fundamental limit on $\mathcal{R} = \eta/s$ revisited: Given that some hint of the fundamental $\mathcal{R}_0 = (1/4\pi)(\hbar/k_B)$ limit comes from the study of gravitational physics (Kovtun 2005), reconsider the dimensional analysis argument leading to Eqn. (2.30) and assume instead that $\mathcal{R}_0$ depends on $\hbar, k_b, m_X$, (where m_X is any elementary particle mass) but now G (Newton's constant) and see if your results change.

References for Chapter 2

Ashcroft N. W., and Mermin N. D. (1976). *Solid State Physics* (Australia: Brooks/Cole).

Ball, P. (2001). *Life's Matrix: A Biography of Water* (Berkeley: University of California Press).

Brini, E., et al. (2017). "How Water's Properties are Encoded in its Molecular Structure and Energies," *Chemical Reviews* 117, 12385–414.

Chandrasekhar, S. (1959). "The Oscillations of a Viscous Liquid Globe," *Proceedings of the London Mathematical Society* 9, 141–9.

Chen, F . (2018). *Introduction to Plasma Physics and Controlled Fusion*, 3rd ed. (Heidelberg: Springer).

Clarke, C., and Carswell, B. (2007). *Principles of Astrophysical Fluid Dynamics* (Cambridge: Cambridge University Press).

Crossno, J., et al. (2016). "Observation of the Dirac Fluid and the Breakdown of the Wiedemann–Franz Law," *Science* 351, 1058–61.

Finney, J. L. (2004). "Water? What's So Special about It?" *Philosophical Transactions of the Royal Society of London. Series B, Biological Sciences* 359, 1145–63.

Franz, R., and Wiedemann, G. (1853). "Ueber die Wärme-Leitungsfähigkeit der Metalle (About the Heat Conductivity of Metals)," *Annalen der Physik* 165, 497–531.

Jenke, T., et al. (2011). "Realization of a Gravity-Resonance-Spectroscopy Technique," *Nature Physics* 7, 468–72.

Khokonov, A. Kh. (2016). "Liquid Drop Model of Spherical Nuclei with Account of Viscosity," *Nuclear Physics A* 945, 58–66.

Kittel, C. (1971). *Introduction to Solid State Physics*, 4th ed. (New York: Wiley and Sons).

Kovtun, P. K., Son, D. T., and Starinets, A. (2005). "Viscosity in Strongly Interacting Quantum Field Theories from Black Hole Physics," *Physical Review Letters* 94, 111601.

Lamb, H. (1932). *Hydrodynamics* (Cambridge: Cambridge University Press).

Looney, D. H., and Brown, S. C. (1954). "The Excitation of Plasma Oscillations," Physical Review 93, 965–9.

(Lord) Rayleigh (William Strutt). (1896). *The Theory of Sounds* (New York: McMillan and Co.). See also (Lord) Rayleigh (William Strutt). (1882). "XX: On the Equilibrium of Liquid Conducting Masses Charged with Electricity," *The London, Edinburgh, and Dublin Philosophical Magazine and Journal of Science* 14. 184–6.

(Lord) Rayleigh (William Strutt). (1915). "The Principle of Similitude," *Nature* 95, 66–8.

McCraney, J., et al. (2022). "Oscillations of Drops with Mobile Contact Lines on the International Space Station: Elucidation of Terrestrial Inertial Droplet Spreading," *Physical Review Letters* 129, 084501.

Nelson, A. R., and Gokhale, N. R. (1972). "Oscillation Frequencies of Freely Suspended Water Drops", *Journal of Geophysical Research* 77, 2724–7.

Nesvizhevsky, V. V., et al. (2002) "Quantum States of Neutrons in the Earth's Gravitational Field," *Nature* 415, 297–9.

Nesvizhevsky, V. V., et al. (2003). "Measurement of Quantum States of Neutrons in the Earth's Gravitational Field," *Physical Review D* 67, 10200.

Nesvizhevsky, V. V., et al. (2005). "Study of the Neutron Quantum States in the Gravity Field," Eur. Phys. J. C 40 479-491.

Policastro, G., Son, D. T., and Starinets, A. O. (2001). "Shear Viscosity of Strongly Coupled Supersymmetric Yang-Mills Plasma," *Physical Review Letters* 87, 081601.

Principi, A., and Vignale, G. (2015). "Violation of the Wiedemann–Franz Law in Hydrodynamic Electron Liquids," *Physical Review Letters* 115, 056603.

Reid, W. (1960). "The Oscillations of a Viscous Liquid Drop," *Quarterly of Applied Mathematics* 18, 86–9.

Reitz, J. R., Milford, F. J., and Christy, R. W. (1993). *Foundations of Electromagnetic Theory* (Reading: Addison-Wesley).

Robinson, R., et al. (2021). “Large Violation of the Wiedemann–Franz Law in Heusler, Ferromagnetic, Weyl Semimetal Co_2MnAl,” *Journal of Physics D: Applied Physics* 54, 454001.

Schäfer, T. (2009). “Nearly Perfect Fluidity,” *Physics* 2, 88.

Schäfer, T., and Teaney, D. (2009). “Nearly Perfect Fluidity: From Cold Atomic Gases to Hot Quark Gluon Plasmas,” *Reports on Progress in Physics* 72, 126001.

Sedov, L. I. (1946). “Propagation of Strong Shock Waves,” Prikl. Mat. Mekh. (*Journal of Applied Mathematics and Mechanics*) 10, 241–50 (Pergamon Translations No. 1223).

Tate, T. (1864). “XXX: On the Magnitude of a Drop of Liquid Formed under Different Circumstances,” *The London, Edinburgh, and Dublin Philosophical Magazine and Journal of Science* 27, 176–80.

Taylor, G. I. (1941). *The Formation of Blast Wave by a Very Intense Explosion, Report RC-210, 27 June 1941* (London: Civil Defence Research Committee).

Taylor, G. I. (1950a). “The Formation of a Blast Wave by a Very Intense Explosion. I. Theoretical Discussion,” *Proceedings of the Royal Society of London. Series A, Mathematical and Physical Sciences* 201, 159–74.

Taylor, G. I. (1950b). “The Formation of a Blast Wave by a Very Intense Explosion. II. The Atomic Explosion of 1945,” *Proceedings of the Royal Society of London. Series A, Mathematical and Physical Sciences* 201, 175–86.

Tonks, L., and Langmuir, I. (1929). “Oscillations in Ionized Gases,” *Physical Review* 33, 195–210.

Vinet, B., Garandet, J. P., and Cotella, L. (1993). “Surface Tension Measurements of Refractory Liquid Metals by the Pendant Drop Method under Ultrahigh Vacuum Conditions: Extension and Comments on Tate's law,” *Journal of Applied Physics* 73, 3830–4.

von Neumann, J. (1941). *The Point Source Solution, National Defence Research Committee, Div. B, Report AM-9, June 30, 1941*. (Washington, DC: Civil Defence Research Committee).

Walliman, D. (2017). “The Map of Physics.” Reproduced by permission of author. Wallimann won the 2017 “Kantar Information is Beautiful Award” for this infographic.

Worley, J. D. (1992). “Capillary Radius and Surface Tensions,” *Journal of Chemical Education* 69, 678–80.

Zarenia, M., et al. (2019), “Breakdown of the Wiedemann–Franz Law in AB-Stacked Bilayer Graphene,” *Physical Review Letters* 99, 16104.

Part II

The “quadrivium” of physics

Chapter 3

Classical mechanics: Point particles and continuum systems

Instead of starting with a formal, "theorem-driven" approach to dimensional analysis (DA), we instead first review its use in the context of classical mechanics, using two familiar model systems (the harmonic oscillator and pendulum/rotor) to demonstrate the power (and limitations) of its use in problems that are introduced in the first year of calculus-based physics, but continued through graduate-level coursework. We also include examples from classical Newtonian gravity in advance of exploring more advanced applications of gravitational physics in astrophysical contexts in Chapter 9. We then apply these methods to a variety of continuum mechanics problems across a diverse set of applications (solids, fluids, and even granular materials). So, we begin by focusing on case by case, example upon example, and problem after problem training to encourage students to use any and all methods already into their toolkit as they first approach classic problems and then newer, research-based ones, while maintaining a focus on strategies to use DA whenever possible.

3.1 The harmonic oscillator

The problem of a particle of mass m subject to a linear restoring force, leading to simple harmonic motion, is defined by the force-law, or equivalently by the potential energy function,

$$F_{HO}(x) = -kx \qquad \text{or} \qquad V_{HO}(x) = \frac{1}{2}kx^2 . \tag{3.1}$$

The dimensions of the **spring constant** k are readily determined by either definition to be

$$[k] = \left[\frac{F_{HO}}{x}\right] = \left(\frac{\mathsf{ML}}{\mathsf{T}^2}\right)\left(\frac{1}{\mathsf{L}}\right) \stackrel{!}{=} \frac{\mathsf{M}}{\mathsf{T}^2} \stackrel{!}{=} \left(\frac{\mathsf{ML}^2}{\mathsf{T}^2}\right)\left(\frac{1}{\mathsf{L}^2}\right) = \left[\frac{V_{HO}}{x^2}\right] = [k] . \tag{3.2}$$

Knowing that the motion is periodic, an obvious first question is how the periodicity depends on the physical parameters (m and k) and any initial "stretch" (x_0). We therefore posit that the period has dimensional dependences given by

$$\tau = C_\tau m^\alpha k^\beta x_0^\gamma \tag{3.3}$$

where C_τ is a dimensionless constant which will remain undetermined. Matching dimensions, we require

$$\mathsf{T} = \mathsf{M}^\alpha \left(\frac{\mathsf{M}}{\mathsf{T}^2}\right)^\beta \mathsf{L}^\gamma \tag{3.4}$$

Dimensional Analysis Across the Landscape of Physics. Richard W. Robinett, Oxford University Press. © Richard W. Robinett (2024).
DOI: 10.1093/oso/9780192867551.003.0003

leading to the constraints

$$\begin{aligned} \mathsf{M}: \quad 0 &= \alpha + \beta \\ \mathsf{L}: \quad 0 &= \quad\quad\quad + \gamma \\ \mathsf{T}: \quad 1 &= \quad -2\beta \end{aligned} \tag{3.5}$$

which are easily solved for $\alpha = 1/2$, $\beta = -1/2$, and $\gamma = 0$ or $\tau = C_\tau\sqrt{m/k}$ and we know from introductory-level mechanics that $C_\tau = 2\pi$. The most striking part of this result is that the period does not depend on x_0 and the word **isochronous** is often used to describe the independence of τ on the amplitude of oscillation, for any vibratory system for which this is true. If we had instead asked to solve for the frequency ($f = 1/\tau$) of oscillation, we'd have written $f = C_f m^\alpha k^\beta x_0^\gamma$, matched dimensions, and found that $f = C_f\sqrt{k/m}$, with C_f known from exact solutions to be $1/2\pi$.

This then is one of the easiest examples of the central "conceit" of DA, namely that

- one identifies the relevant physical variables (and their dimensional "spelling") in a given problem involving a new physical quantity;
- one assumes power-law (possibly integral or fractional) dependences, along with a dimensionless constant (C), for the quantity in question;
- one matches dimensions (here only M, L, T are available for mechanics problems) and hope that this constrains them sufficiently that one gets an unambiguous prediction; and
- one finds (it is hoped) that any resulting dimensionless constant (from a closed-form solution, or perhaps from fitting experimental data, or numerical simulations) is within an order-of-magnitude ($C \sim 10 - 0.1$) from unity, compared to some exact (presumably from a more rigorous/mathematical derivation) result. We note that in this case we did indeed find that $C_\tau = 2\pi \sim 6$ and $C_f = 1/2\pi \sim 0.16$.

A similar exercise (in **P3.2**) assuming that τ depends on m, k and v_0 (an initial speed) shows that the period doesn't depend on v_0 and one can explore the dimensional dependences of other quantities such as momentum by assuming $p \sim m^\alpha k^\beta x_0^\gamma$.

For this (very familiar) case, how did we actually know what the period and frequency were, to be able to "quote" values for C_τ, C_f? Depending on the level of sophistication assumed, we might have been able to solve Newton's laws, for which the differential equation is

$$m\ddot{x}(t) = -kx(t) \qquad \text{or} \qquad \ddot{x}(t) = -\omega_0^2 x(t), \tag{3.6}$$

where we have defined $\omega_0 \equiv \sqrt{k/m}$, giving

$$x(t) = A\cos(\omega_0 t) + B\sin(\omega_0 t) \overset{or}{=} C\cos(\omega_0 t + \phi) \tag{3.7}$$

as two forms of the general solution; the complete solution is then determined by enforcing the initial values of $x(0) = x_0$ and $\dot{x}(0) = v_0$. To explicitly demonstrate the periodicity, even in the general case, we can use properties of trig functions to note that

$$x(t+\tau) = C\cos(\omega_0(t+\tau)+\phi) = C\cos(\omega_0 t+\phi+\omega_0\tau) = x(t) \tag{3.8}$$

so long as $\omega_0\tau = n(2\pi)$, which gives a period $\tau = 2\pi/\omega_0 = 2\pi\sqrt{m/k} = 1/f$, which does, in fact, determine C_τ, C_f in the DA approach, and also confirms that τ depends on neither x_0 nor v_0.

It is not an accident that the **angular frequency** (ω) is the natural quantity for which DA gives the "right" answer, as the differential equation in Eqn. (3.7) which determines the motion is one that arises frequently in the analysis of systems displaced (at least by a small amount) from a stable equilibrium. We've seen this once already, in Section 2.3, where the plasma frequency ω_P was correctly given just by DA.

Dynamical systems described by equations for some time-dependent physical quantity $Z(t)$ of the form

$$\mathcal{I}\ddot{Z}(t) = -\mathcal{R}Z(t), \tag{3.9}$$

can have the parameters $\mathcal{I}$ and $\mathcal{R}$ described as generalized **inertia** and **restoring force** respectively, and will have oscillatory (specifically, sinusoidal) solutions with the resulting natural frequency of oscillation given by $\omega_0 = \sqrt{\mathcal{R}/\mathcal{I}}$. This is relevant not only for mechanical systems, but also, for example, in oscillations in an LC circuit where the relevant Kirchhoff's loop equation (balancing voltages) is

$$L\frac{d^2q(t)}{dt^2} + \frac{q(t)}{C} = 0 \qquad \text{or} \qquad \ddot{q}(t) = -\frac{1}{LC}q(t), \tag{3.10}$$

so that $\omega_0 = 1/\sqrt{LC}$.

Returning now to the harmonic oscillator, if we had not been able to solve the differential equation in closed form, could we have also determined the periodicity (both the dimensional dependence and dimensionless constant) in another way? Energy methods (motivated by visualizations where we plot the total energy E_0 versus the potential energy function, $V(x)$, as in Fig. 3.1), already suggest that the particle will bounce "back and forth" between two classical turning points (determined by $V(a) = E_0 = V(b)$) and so clearly exhibit periodic behavior, with one "to" and one "fro" motion each taking half a period, or $\tau/2$.

For an oscillator with an initial displacement x_0, we are thus led to write

$$\frac{m}{2}\left(\frac{dx}{dt}\right)^2 + \frac{k}{2}x^2 = E_0 = \frac{k}{2}x_0^2 \tag{3.11}$$

or

$$\sqrt{\frac{m}{k}}\,dx = \sqrt{x_0^2 - x^2}\,dt, \tag{3.12}$$

and integrating over one "to" or "fro" cycle, we have

$$\sqrt{\frac{m}{k}}\int_{-x_0}^{+x_0}\frac{dx}{\sqrt{x_0^2 - x^2}} = \int_0^{\tau/2} dt = \frac{\tau}{2}. \tag{3.13}$$

With the change of variables, $x = yx_0$, we find that

$$\tau = 2\sqrt{\frac{m}{k}}\left(\frac{x_0}{x_0}\right)\int_{-1}^{+1}\frac{dy}{\sqrt{(1-y^2)}} = 2\pi\sqrt{\frac{m}{k}}, \tag{3.14}$$

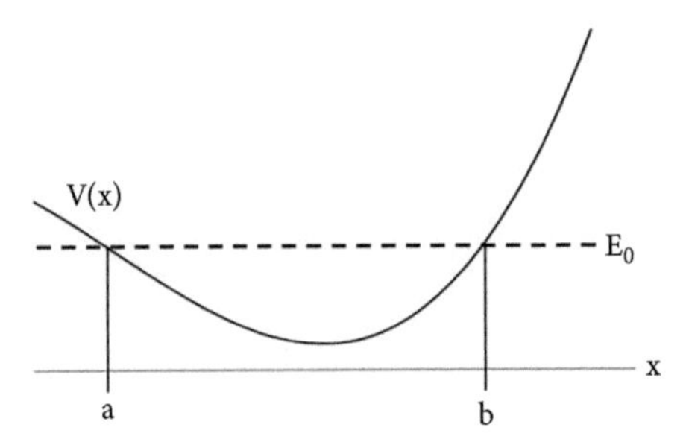

Fig. 3.1 Potential energy function $V(x)$ versus x, with classical turning points defined by $V(a) = E_0 = V(b)$.

and the lack of dependence on x_0 is demonstrated in a different way, the expected dimensions are clearly exhibited, and the dimensionless constant (C_τ) is determined (in this case by a familiar integral). This method can be used in other cases where there is no simple closed-form solution to the Newton's law differential equation (see **P3.4**) to confirm the results of a DA analysis of such problems, and also can be used to reproduce the results in Eqn. (3.7) for the time-dependent sinusoidal solution (see **P3.3**).

Let us now extend the DA analysis of the oscillator to the total energy (E) of the system, to now include a second initial-condition parameter, namely, v_0, by writing

$$E = C_E m^\alpha k^\beta x_0^\gamma v_0^\delta$$
$$\frac{\mathsf{ML}^2}{\mathsf{T}^2} = \mathsf{M}^\alpha \left(\frac{\mathsf{M}}{\mathsf{T}^2}\right)^\beta \mathsf{L}^\gamma \left(\frac{\mathsf{L}}{\mathsf{T}}\right)^\delta . \tag{3.15}$$

In this case we have three equations (for M, L, T), but four variables ($\alpha, \beta, \gamma, \delta$), so we don't expect a unique solution, but instead find

$$\begin{aligned} \mathsf{M}: &\quad 1 = \alpha + \beta \\ \mathsf{L}: &\quad 2 = \qquad\quad +\gamma + \delta \\ \mathsf{T}: &\quad -2 = \quad -2\beta \qquad -\delta, \end{aligned} \tag{3.16}$$

and solving in terms of δ, we have $\alpha = \delta/2$, $\beta = 1 - \delta/2$, and $\gamma = 2 - \delta$, which we can write in the form

$$E = C_E m^{\delta/2} k^{1-\delta/2} x_0^{2-\delta} v_0^\delta = C_E \left(kx_0^2\right)\left(\sqrt{\frac{m}{k}}\frac{v_0}{x_0}\right)^\delta = C_E\,(kx_0^2)\,(\Pi)^\delta \quad \text{with} \quad \Pi \equiv \left(\sqrt{\frac{m}{k}}\frac{v_0}{x_0}\right). \tag{3.17}$$

We note that for the choice of $\delta = 0$ and $C_E = 1/2$, we find the usual result for the energy of a stretched spring ($E = kx_0^2/2$), while for $\delta = 2$ (and $C_E = 1/2$ again) we have the standard form for kinetic energy ($E = mv_0^2/2$), so two familiar limiting cases are easily confirmed.

More importantly, this is our first example where a dimension<u>**less**</u> ratio of important physical parameters, in this case $\Pi = (mv_0^2/kx_0^2)^{1/2}$, is automatically singled out by the DA analysis. Since the power δ is not determined in the DA analysis, we could just have easily written the solution as a linear combination of any number of (let us assume) integral powers, namely

$$E \sim kx_0^2\left(a_0\Pi^0 + a_1\Pi^1 + a_2\Pi^2 + \ldots\right) \quad \sim \quad kx_0^2\sum_{n=0}^{\infty} a_n\Pi^n \sim kx_0^2\, G(\Pi), \tag{3.18}$$

where $G(\Pi)$ is an arbitrary (analytic at least, expressible as a power series) function of the dimensionless parameter Π. This is the first example of many that we will encounter of one aspect of the **Buckingham Pi** (Π) **theorem**, which we discuss in a more formal and systematic way in Section 10.11, but which we continue to use in analyzing problems along the way, as this example illustrates the main result.

In the context of this analysis of the total energy of a harmonic oscillator (with both an initial stretch and speed) we know that

$$E_0 = \frac{1}{2}kx_0^2 + \frac{1}{2}mv_0^2 = kx_0^2\left[\frac{1}{2} + \frac{1}{2}\left(\frac{mv_0^2}{kx_0^2}\right)\right] \qquad \text{or} \qquad G(\Pi) = \frac{1}{2}\left(1 + \Pi^2\right), \tag{3.19}$$

so that $G(\Pi)$ is, in this case, actually a simple polynomial function.

We can see that this Π theorem approach is also consistent with the results for the time-dependent solution of $x(t)$ itself (in a slightly less trivial way), where we can write $x(t) \sim m^\alpha k^\beta x_0^\gamma t^\delta$ requiring

$$\begin{aligned} \mathsf{M}: &\quad 0 = \alpha + \beta \\ \mathsf{L}: &\quad 1 = \qquad\quad + \gamma \\ \mathsf{T}: &\quad 0 = \quad -2\beta \quad + \delta, \end{aligned} \tag{3.20}$$

giving $\alpha = \delta/2$, $\beta = \delta/2$, $\gamma = 1$. This implies that

$$x(t) \sim m^{\delta/2}k^{\delta/2}x_0 t^\delta \sim x_0\left(\frac{\sqrt{k}\,t}{\sqrt{m}}\right)^\delta \sim x_0(\omega_0 t)^\delta \qquad \longrightarrow \qquad x_0\, G(\Pi = \omega_0 t), \tag{3.21}$$

and we know that $G(\Pi) = \cos(\Pi)$ for the exact solution, so that non-trivial (in this case transcendental) functions are allowed by the Π theorem.

The applicability of the Π theorem is not limited to single dimensionless ratios, as can be seen by asking for the most general $x(t)$ solution for the harmonic oscillator, given both arbitrary initial positions and velocities, x_0 and v_0. In this case we write $x(t) \sim m^\alpha k^\beta x_0^\delta v_0^\gamma t^\epsilon$, match M, L, T dimensions, and solving for α, β, δ in terms of γ, ϵ, we find $\alpha = (1-\gamma-\epsilon)/2$, $\beta = (\gamma+\epsilon-1)/2$, and $\delta = 1-\gamma$. For simplicity (wait for it!), we actually choose to use the combination $\sigma = 1-\gamma$ and find

$$\begin{aligned} x(t) &\sim m^{\sigma/2-\epsilon/2}k^{-\sigma/2+\epsilon/2}x_0^{1-\sigma}v_0^\sigma t^\epsilon \\ &\sim x_0\left(\frac{\sqrt{k}t}{\sqrt{m}}\right)^\epsilon\left(\frac{\sqrt{m}v_0}{\sqrt{k}x_0}\right)^\sigma \\ x(t) &\sim x_0\,(\Pi_1)^\epsilon\,(\Pi_2)^\sigma \end{aligned} \tag{3.22}$$

where $\Pi_1 \equiv \omega_0 t$ and $\Pi_2 = (\sqrt{m}v_0/\sqrt{k}x_0)$. Following the same logic leading to Eqn. (3.18), we can generalize Eqn. (3.22) to

$$x(t) = x_0 \sum_{n,k} a_{n,k}(\Pi_1)^n (\Pi_2)^k = x_0 G(\Pi_1, \Pi_2) \tag{3.23}$$

with an arbitrary function of two dimensionless $\Pi_{1,2}$ variables. You should be able to use the most general solution to the harmonic oscillator (as in **P3.5**) to confirm that it can indeed be written in this form, and to evaluate $G(\Pi_1, \Pi_2)$.

Example 3.1 Damped harmonic oscillator

The **damped harmonic oscillator** (a slightly more advanced topic in mechanics) can also be analyzed using DA and compared to well-known exact solutions and provides another example of the Π theorem, with an important new feature. For this problem, in addition to the linear restoring force $-kx(t)$, we consider a velocity-dependent frictional force of the form $-bv(t)$, and explore the dependence of the oscillation frequency ω on m, k, b, x_0 by writing $\omega = m^\alpha k^\beta b^\gamma x_0^\delta$. Using the fact that $[b] = [F/v] = (\mathsf{ML/T^2})/(\mathsf{L/T}) = \mathsf{M/T}$, and matching dimensions, we find that $\delta = 0$ (so once again the period is independent of initial "stretch"), while $\alpha = -1/2 - \gamma/2$, $\beta = 1/2 - \gamma/2$ so that

$$\omega = m^{-1/2-\gamma/2} k^{1/2-\gamma/2} b^\gamma = \sqrt{\frac{k}{m}} \left(\frac{b}{\sqrt{mk}}\right)^\gamma \longrightarrow \omega_0 F(\Pi), \tag{3.24}$$

where DA has identified the dimensionless ratio $\Pi = b/\sqrt{mk}$. We choose to write this result in the entirely equivalent form $\Pi_b \equiv b^2/4mk$ (still dimensionless, after all), solely to facilitate comparison with conventional results for the exact solutions for this problem.

The time-dependence of a damped harmonic oscillator is most often demonstrated by solving the Newton's law equation of motion in the form

$$m\ddot{x}(t) = -kx(t) - b\dot{x}(t), \tag{3.25}$$

and a trial solution of the form $x(t) = Ae^{i\omega t}$ gives a quadratic equation for ω of the form

$$-m\omega^2 + ib\omega + k = 0. \tag{3.26}$$

The oscillation frequencies are then given by the two roots of this quadratic equation as

$$\omega_\pm = \frac{-ib \pm \sqrt{4km - b^2}}{2(-m)} = \sqrt{\frac{k}{m}} \left[\pm\sqrt{1 - \frac{b^2}{4mk}} + i\sqrt{\frac{b^2}{4mk}}\right] = \omega_0 F_\pm(\Pi_b) \tag{3.27}$$

where $F_\pm(\Pi) = \pm\sqrt{1-\Pi} + i\sqrt{\Pi}$.

Depending on the value of Π_b, one has three different "phases," with physically distinct behaviors, namely

- For $0 < \Pi_b < 1$ one has damped oscillations, with a frequency (period) shorter (longer) than for the undamped case;
- For $\Pi_b = 1$ one has critical camping (with no oscillations, since $\tau = 1/\omega \to \infty$);
- For $\Pi_b > 1$ one has over-damped solutions.

This is the first example where a DA approach identifies a dimensionless ratio that can then be used to characterize/classify different types of behaviors in a physical system and also shows that functions of Π

can be real or complex. The next section explores a similar identification of an important dimensionless ratio, leading to a **critical point** in the behavior of another mechanical system.

3.2 The rigid pendulum and rotor

Given the historical interest in "telling time," studies of systems exhibiting periodicity form an important chapter in the early advances of classical mechanics, with the pendulum being one of the most familiar. To explore the full richness of this system, we focus on a point mass (m) at the end of a light (but rigid!) rod (length l), subject to gravity, and with the pivot assumed frictionless, as shown in Fig. 3.2. We want to explore this system where the point mass can not only experience small oscillations about equilibrium (for initial angular displacements $\theta_0 << 1$), but for very high energies (to be quantified below) where it has the limit of a spinning (without friction) rigid rotor, in which case gravity is not relevant.

With our experience that **angular frequency** is the most natural measure of oscillation within a DA analysis, assuming an initial angular displacement of θ_0, we start by writing

$$\begin{aligned} \omega &= C_\omega m^\alpha g^\beta l^\gamma \theta_0^\delta \\ \frac{1}{\mathsf{T}} &= \mathsf{M}^\alpha \left(\frac{\mathsf{L}}{\mathsf{T}^2}\right)^\beta \mathsf{L}^\gamma (1)^\delta \end{aligned} \tag{3.28}$$

and then match dimensions to give

$$\begin{aligned} \mathsf{M}: &\quad 0 = \alpha \\ \mathsf{L}: &\quad 0 = \quad \beta + \gamma \\ \mathsf{T}: &\quad -1 = -2\beta \end{aligned} \tag{3.29}$$

including only those quantities that have dimensions. We immediately note that no information about the dependence on (the dimension<u>**less**</u>) θ_0 will be forthcoming in a DA approach, so we can write $C_\omega = F(\theta_0)$, and we therefore can't assume that C_ω is a constant. In this case it's akin to

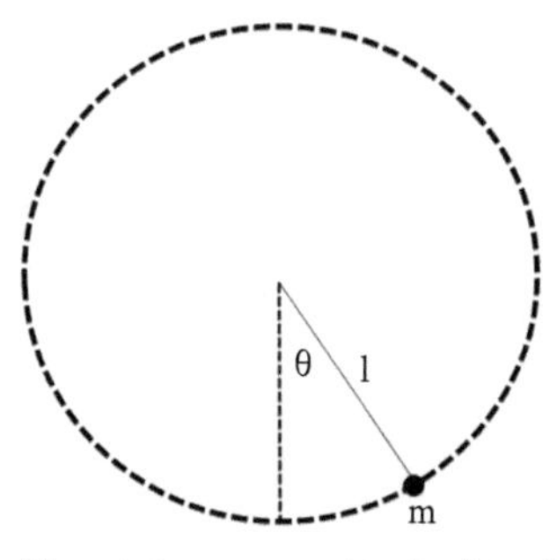

Fig. 3.2 Frictionless pendulum with point mass m attached to rigid and massless rod of length l acting under the influence of gravity.

the Buckingham Π theorem, but where we have already identified an important dimensionless quantity, θ_0, from the beginning.

Matching dimensions, we find that $\alpha = 0$, $\beta = 1/2$, and $\gamma = -1/2$, so that $\omega = \sqrt{g/l}\,F(\theta_0)$. To compare to the exact solution, we use the Newton's law equation of motion for rotational behavior, $I\alpha = \mathcal{T}$, where $I, \mathcal{T}$ are the moment of inertia and applied torque (due to gravity) respectively, and $\alpha \equiv \ddot{\theta}$. With

$$I = ml^2, \qquad V(\theta) = mgl(1 - \cos(\theta)), \qquad \text{and} \qquad \mathcal{T} = -\frac{dV(\theta)}{d\theta} = -mgl\sin(\theta), \tag{3.30}$$

we then have

$$ml^2\ddot{\theta} = -mgl\sin(\theta) \qquad \text{or when } \theta << 1, \qquad ml^2\ddot{\theta} = -mgl\theta \qquad \text{or} \qquad \ddot{\theta} = -\frac{g}{l}\theta. \tag{3.31}$$

Using the general connection in Eqn. (3.9), we find $\omega_0 = \sqrt{g/l}$, so that for $\theta_0 << 1$, we have $F(\theta_0 \approx 0) \rightarrow 1$ and isochronous (period independent of amplitude) motion.

For general values of θ_0 (at least up to $\theta_0 \lesssim \pi$, close to the top of the arc), we can use energy methods to write

$$\frac{1}{2}I\left(\frac{d\theta}{dt}\right)^2 + V(\theta) = E_0 = V(\theta_0), \tag{3.32}$$

and we can equate $d\theta$ and dt via

$$\frac{ml^2}{2}\left(\frac{d\theta}{dt}\right)^2 = mgl[\cos(\theta) - \cos(\theta_0)] \qquad \text{or} \qquad \left(\sqrt{\frac{l}{2g}}\right)\frac{d\theta}{\sqrt{\cos(\theta) - \cos(\theta_0)}} = dt. \tag{3.33}$$

Integrating over a half-cycle, that is, over the angular range $(-\theta_0, +\theta_0)$, we find after rearranging terms

$$\tau = \sqrt{\frac{2l}{g}}\int_{-\theta_0}^{+\theta_0}\frac{d\theta}{\sqrt{\cos(\theta) - \cos(\theta_0)}} = 2\pi\sqrt{\frac{l}{g}}\left[\frac{1}{\sqrt{2}\pi}\int_{-\theta_0}^{+\theta_0}\frac{d\theta}{\sqrt{\cos(\theta) - \cos(\theta_0)}}\right] \equiv 2\pi\sqrt{\frac{l}{g}}\,[G(\theta_0)]. \tag{3.34}$$

For small angles (i.e., when $\cos(\theta) \sim 1 - \theta^2/2 + \ldots$), we have

$$\tau_0 = 2\sqrt{\frac{l}{g}}\int_{-\theta_0}^{+\theta_0}\frac{d\theta}{\sqrt{\theta_0^2 - \theta^2}} = 2\pi\sqrt{\frac{l}{g}} = \frac{2\pi}{\omega_0} \tag{3.35}$$

and $G(\theta_0 \rightarrow 0) \approx 1$.

For values of θ_0 that are not small, one can evaluate $G(\theta_0)$ numerically, but it's also known to be expressible as an **elliptic integral** (Marion and Thornton 2004) and in fact diverges at $\theta_0 \rightarrow \pi$. To visualize that behavior, note that the top of the circle is a point of **unstable equilibrium**, so with $\theta_0 = \pi$ and $\dot{\theta}_0 = 0$ the particle would remain there. The fact that $G(\theta_0)$ depends on θ_0 means that the standard pendulum is only approximately isochronous, and then only for small displacements. (It has been known since the time of Huygens (see Huygens and Blackwell 1986) that one can construct pendula constrained to specific (cycloid) geometries that can have displacement-independent periods, but a DA approach still gives the same $\sqrt{l/g}$ dependence for the period.)

By analogy to the harmonic oscillator problem, where we could give a "kick" to the mass with non-zero values of either x_0 or v_0, we can include possible initial kinetic energy terms to explore the dependence on $\dot{\theta}_0$. But let us instead combine both options by assuming an initial total energy E_0 and write $\tau = C_\tau m^\alpha g^\beta l^\gamma E_0^\delta$ and explore the dependence of τ on E_0. Matching dimensions, we find that

$$\tau \propto m^{-\delta} g^{-\delta-1/2} l^{\delta+1/2} E_0^\delta \quad \propto \quad \sqrt{\frac{l}{g}} \left(\frac{E_0}{mgl}\right)^\delta \quad \rightarrow \quad \sqrt{\frac{l}{g}}\, \mathrm{G}(\Pi), \tag{3.36}$$

where now $\Pi \equiv E_0/mgl$, which is an obvious ratio of total energy to a natural energy scale in the problem.

For $\delta = 0$ we clearly recover the dependence for small oscillations, while in the high-energy limit, where $\Pi >> 1$ (or $E_0 >> mgl$) and the mass rotates rapidly around the pivot (and the effect of gravity is minimal) we have a **rigid rotor** system for which the period is determined by

$$\frac{1}{2} I\omega^2 = E_0 \qquad \text{or} \qquad \frac{2\pi}{\tau} = \omega = \sqrt{\frac{2E_0}{ml^2}} \qquad \text{giving} \qquad \tau = 2\pi\sqrt{\frac{ml^2}{E_0}}, \tag{3.37}$$

and Eqn. (3.36) accommodates this dependence for $\delta = -1/2$, where there is no dependence on g.

Using energy methods once again, we can evaluate the pendulum period for arbitrary values of E_0, appropriately scaled to the small-oscillation limit of $\tau_0 \equiv 2\pi\sqrt{l/g}$, and using the dimensionless variable $z \equiv E_0/mgl$. There are two distinct cases, one where $0 < z < 2$, in which case the pendulum never quite makes it "over the top," and in that range we find $\mathcal{T}(z) \equiv \tau(z)/\tau_0$, where

$$\mathcal{T}^{(-)}(z) = \frac{\sqrt{2}}{\pi} \int_0^{\Theta(z)} \frac{d\theta}{\sqrt{z - 1 + \cos(\theta)}} \tag{3.38}$$

and $\Theta(z) \equiv \cos^{-1}(1 - z) < \pi$ determines the maximum angular range.

When the initial energy satisfies $E_0 > 2mgl$, or $z > 2$, the mass undergoes cyclic motion through complete 2π rotations, and one finds

$$\mathcal{T}^{(+)}(z) = \frac{1}{\sqrt{2}\pi} \int_0^{\pi} \frac{d\theta}{\sqrt{z - 1 + \cos(\theta)}}. \tag{3.39}$$

One can confirm (**P3.9**) that these expressions have the correct limits for $z \to 0, \infty$, namely Eqns. (3.35) and (3.37), respectively.

We plot $\mathcal{T}(z)$ in Fig. 3.3 and we can clearly see the divergence in the period for $z = 2$ (or $E_0 = 2mgl$) corresponding to the unstable equilibrium noted above, as well as the large z behavior corresponding to the rigid-rotor limit (dotted line). For an extensive discussion of the rigid pendulum/rotor system, see Butikov (1999). Once again, the integrals arising in $\mathcal{T}^{(\pm)}(z)$ can be expressed in the form of **incomplete** and **complete elliptic integrals**, but as Feynman (or rather his father) famously (is supposed to have) said, "... knowing the name of something is different than knowing something ...". But that mathematical nomenclature does provide some provenance

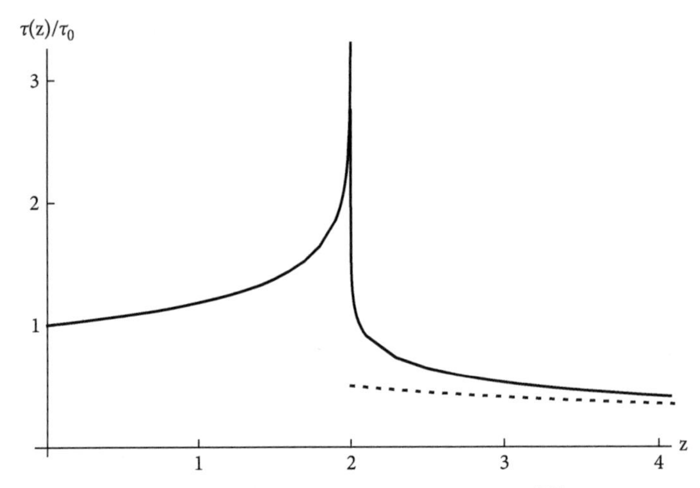

Fig. 3.3 Period of the rigid pendulum/rotor system, scaled to $\tau_0 = 2\pi\sqrt{l/g}$), as a function of $z \equiv E/mgl$. The dotted line corresponds to the high energy ($z >> 1$) limit of $\mathcal{T}(z) = 1/\sqrt{z}$, consistent with Eqn. (3.37).

to find textbook results and/or use computer software to explore its behavior in more numerical detail.[1]

Note: This example shows that the behavior of $G(\Pi)$ in a Buckingham Pi theorem analysis can be generalized well beyond simple rational and transcendental functions to include functions with highly non-trivial behavior, including divergences.

Aside: Despite the fact that the curve in Fig. 3.3 bears a striking resemblance to the λ transition in superfluid helium, and that a famous attempt to calculate the critical behavior of that phase transition (Onsager 1944) used complete elliptical integrals, these two systems are not otherwise connected. The divergence in the period for $z = 2$ is related to the nature of the unstable equilibrium at $\theta_0 = \pi$, where $V(\theta)$ is "flat." One can confirm this by looking at another example, in an extended problem (**P3.10**), where the period obtained using the energy method can be derived using familiar elementary integrals.

3.3 **Newton's law of gravity**

As noted in Section 1.2, Newton's law of gravity is the first of the four fundamental forces of Nature encountered in introductory physics and is, of course, of immense historical and intellectual importance. Students' first exposure to this topic may be at the level of familiar falling body problems, but the array of physical phenomena in terrestrial settings where the acceleration of gravity, g, appears extends far beyond intro-level physics, with applications to fluid mechanics, geophysics, and meteorology, as we'll see in the end-of-chapter problems. Einstein also extended gravitational physics to a general relativistic formulation, which we explore using DA in Section 9.4.1. The topic clearly remains at the forefront of modern research on many fronts (cosmology, dark matter, and most recently the discovery of gravitational waves) and we explore some of these topics in a DA

[1] And Feynman did also say words to the effect that "... it does help to know the names of things if you want to talk to other people ...". Both of these "quotes" by Feynman are actually paraphrases of snippets of videotaped conversations widely available online.

context in Section 9.4.2. Hecht (2021) presents a very readable discussion in *The True Story of Newtonian Gravity*.

In this section (and especially in **P3.11–P3.15**) we present a few examples of the application of DA to problems in the context of classical gravity. But before embarking on a DA-based approach, we want to emphasize the formal similarities between the mathematical formulations of Newtonian gravity and that of electro-statics, at least as described by Coulomb's law, with both being inverse square force laws, and so connected across the landscape of physics. These relations allow us to remind ourselves of less-familiar realizations of Gauss's law for gravitation and we present a number of such analog results, where for electro-statics the parallelism is made even more clear if we use the notation $C_e \equiv 1/4\pi\epsilon_0$.

For example, the force between two point masses, M and m, and that between two point charges, Q and q, can be written as

$$\boldsymbol{F}_G(r) = -\frac{GMm}{r^2}\hat{\boldsymbol{r}} \quad \text{or} \quad \boldsymbol{F}_C(\boldsymbol{r}) = \frac{Qq}{4\pi\epsilon_0 r^2}\hat{\boldsymbol{r}} = \frac{C_e Qq}{r^2}\hat{\boldsymbol{r}}, \tag{3.40}$$

while if one of the objects has finite extension, and mass and charge densities given by $\rho_m(\boldsymbol{r})$ and $\rho_e(\boldsymbol{r})$, one integrates the point mass/charge law to obtain

$$\boldsymbol{F}_G(\boldsymbol{r}') = -Gm\int \frac{\rho_m(\boldsymbol{r}')(\boldsymbol{r}'-\boldsymbol{r})}{|\boldsymbol{r}'-\boldsymbol{r}|^3} \quad \text{and} \quad \boldsymbol{F}_C(\boldsymbol{r}) = C_e q\int \frac{\rho_e(\boldsymbol{r}')(\boldsymbol{r}'-\boldsymbol{r})}{|\boldsymbol{r}'-\boldsymbol{r}|^3}. \tag{3.41}$$

Both Newton's and Coulomb's (vector) force laws can be derived from a (scalar) potential function, by use of the gradient operator, namely

$$V_G(\boldsymbol{r}) = -\frac{GM}{r} \quad \text{and} \quad V_C(\boldsymbol{r}) = \frac{Q}{4\pi\epsilon_0 r} = \frac{C_e Q}{r} \tag{3.42}$$

$$\boldsymbol{g}(\boldsymbol{r}) = -\nabla V_G(r) \quad \text{and} \quad \boldsymbol{E}(\boldsymbol{r}) = -\nabla V_C(\boldsymbol{r}). \tag{3.43}$$

Expressions for the gravitational and electric fields can also be read off directly from Eqn. (3.40) by dividing out the value of any small test mass or charge, namely

$$\boldsymbol{g}(r) = -\frac{GM}{r^2}\hat{\boldsymbol{r}} \quad \text{and} \quad \boldsymbol{E}(r) = \frac{Q}{4\pi\epsilon_0 r^2}\hat{\boldsymbol{r}} = \frac{C_e Q}{r^2}\hat{\boldsymbol{r}}, \tag{3.44}$$

where $\boldsymbol{E}(\boldsymbol{r})$ is the familiar electric field and $\boldsymbol{g}(\boldsymbol{r})$ is its gravitational analog. The value of $|\boldsymbol{g}(R_E)|$ at the Earth's surface is, of course, the familiar acceleration of gravity $g \approx 9.8\ m/s^2$.

Many of the gravitational analogs of the corresponding electricity and magnetism (E&M) expressions are likely familiar to readers of intro-level texts, but one that might be less so is the parallel version of Gauss's law applied to gravity. In their integral form, one has

$$\oint \boldsymbol{g}\cdot d\boldsymbol{A} = -4\pi G M_{enc} \quad \text{and} \quad \oint \boldsymbol{E}\cdot d\boldsymbol{A} = \frac{Q_{enc}}{\epsilon_0} = 4\pi C_e Q_{enc}, \tag{3.45}$$

where M_{enc} and Q_{enc} are the mass and charge (respectively) enclosed by a (generalized) Gaussian surface. The differential forms of Gauss's law of gravity and electricity are then given by

$$\nabla\cdot\boldsymbol{g} = -4\pi G\rho_m(\boldsymbol{r}) \quad \text{and} \quad \nabla\cdot\boldsymbol{E} = \frac{\rho_e(r)}{\epsilon_0} = 4\pi C_e\rho_e(\boldsymbol{r}), \tag{3.46}$$

where the mass and charge densities are again given by $\rho_m(\boldsymbol{r})$ and $\rho_e(\boldsymbol{r})$. In addition to the intrinsic elegance of these analogies, a number of problems of physical interest are actually best approached using Gauss's law of gravitation form in Eqn. (3.45).

Finally, combining the expressions from Eqns. (3.43) and (3.46) gives the **Poisson equations** for the (scalar) gravitational and electric potentials, namely

$$\nabla^2\phi_G(\boldsymbol{r}) = 4\pi G\rho_m(\boldsymbol{r}) \qquad \text{and} \qquad \nabla^2\phi_C(\boldsymbol{r}) = -\frac{\rho_e(\boldsymbol{r})}{\epsilon_0} = -4\pi C_e\rho_e(\boldsymbol{r})\,. \tag{3.47}$$

Many of the standard problems involving the gravitational force involve the motion of a test mass m in fields of various geometries (most famously Kepler's third law, as in **P3.12**), but let us first consider a perhaps less-familiar example. Consider a uniform spherical mass (radius R_* and mass M_*) with a hole drilled through the center from one side to its **antipode**, as shown in Fig. 3.4. We assume that a mass falls through the hole from one side to the other, and then is attracted back and so undergoes some kind of periodic motion.

We can certainly use DA to explore the dependence of the period (τ_G) on the relevant parameters, namely M_*, R_*, and of course G. Based on our experiences so far with periodic systems, we choose to apply DA to find the angular frequency, $\omega_G = 2\pi/\tau_G$, by writing $\omega_G = C_\omega G^\alpha M_*^\beta R_*^\gamma$ (with C_G dimensionless). Recalling that $[G] = \mathsf{L}^3/\mathsf{MT}^2$, we find that $\alpha = \beta = 1/2$ and $\gamma = -3/2$, so that $\omega_G \sim \sqrt{GM_*/R_*^3}$. This result can be compared to that in **P2.16**, where one found $f \sim \sqrt{G\rho_m}$ for a (seemingly) very different fluid mechanics problem, and this combination finds wide relevance in many, much more realistic, astrophysical settings.

In order to solve this problem from first principles, we could certainly find the net force on a point mass m at a given radius r by integrating over the entire volume, using Eqn. (3.41), but given the spherical symmetry inherent in the problem, it is more straightforward to use the gravitational equivalent of Gauss's law in Eqn. (3.45) and integrate over a sphere of radius r, as shown in Fig. 3.4. This gives

$$g(r)(4\pi r^2) = \oint \boldsymbol{g}\cdot d\boldsymbol{A} = -4\pi GM_{enc} = -4\pi GM_*\left(\frac{4\pi r^3/3}{4\pi R_*^3/3}\right) = -4\pi GM_*\frac{r^3}{R_*^3} \tag{3.48}$$

or

$$g(r) = -\frac{GM_*}{R_*^3}r \tag{3.49}$$

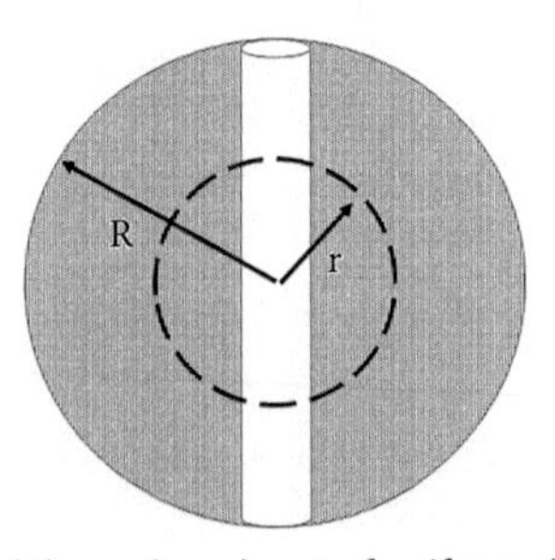

Fig. 3.4 Cartoon model of a tunnel through a planet of uniform density.

which is a linear restoring force in the radial direction. The problem thus reduces to a one-dimensional Newton's law differential equation given by

$$m\frac{d^2r(t)}{dt^2} = ma = F = -\frac{GM_*m}{R_*^3}r(t) \qquad \text{or} \qquad \ddot{r}(t) = -\frac{GM_*}{R_*^3}r(t) = -\omega_0^2 r(t) \tag{3.50}$$

where $\omega_0 \equiv \sqrt{GM_*/R_*^3}$. We see that this three-dimensional problem collapses to the one-dimensional harmonic oscillator problem, where the dimensions of ω_0 (and therefore τ and f) are given correctly by DA and choosing ω_0 as the relevant variable, and we've even confirmed that $C_\omega = 1$ in the DA. Other textbook-level examples are discussed in **P3.11–P3.15**, but we return to the important topic of gravitational physics in much more depth, in the context of modern astrophysical research, in Chapter 9.

3.4 **Continuum mechanics including solids, fluids, and granular materials**

3.4.1 **Waves**

Having explored three examples of the motion of single particles, let us consider as our first example of a continuum system the evaluation of the wave speed (v_s) on a stretched string/wire. We can imagine a length of wire held at a tension F_T, say by having it draped over a pulley with weights suspended at the end (as in an introductory physics lab), or tightened, as a violin string might be, using tuning pegs. The most obvious properties of the wire would be its mass (m) and length (l) and writing the wave speed as $v_s \sim F_T^\alpha m^\beta l^\gamma$, we find $\alpha = \gamma = 1/2$ and $\beta = -1/2$ so that

$$v_s \sim F_T^{1/2} m^{-1/2} l^{1/2} = \sqrt{\frac{F_T l}{m}} = \sqrt{\frac{F_T}{(m/l)}} = \sqrt{\frac{F_T}{\mu}}, \tag{3.51}$$

which depends on the **mass per unit length**, $\mu \equiv m/l$, which is an intensive property of the material (basically a one-dimensional mass density.) In contrast, if we ask about the dimensions of either the frequency (f) or wavelength (λ), the types of quantities one explores in an intro-level lab on standing waves on a wire/string, one finds that DA predicts

$$\lambda \sim l \qquad \text{and} \qquad f \sim \sqrt{\frac{F_T}{ml}} = \frac{1}{l}\sqrt{\frac{F_T}{\mu}}. \tag{3.52}$$

DA clearly has nothing to say about important special cases such as standing waves, where boundary conditions require that where an integral number (n) of half-wavelengths "fit" into the length of wire, imposing the relation $n(\lambda_n/2) = l$, which is a simple (one-dimensional) case of **normal modes.**

The two-dimensional equivalent of a taut string/wire would be to stretch a two-dimensional elastic membrane over a frame of arbitrary shape, which immediately makes this a far richer system in terms of the normal mode patterns one can excite. One can produce the equivalent of tension by hanging weights around the boundary (which would produce a certain *force per unit length*) or tighten it as you would a drum head. Another experimental realization is thin films on a wire frame (as in Section 2.1 and Fig. 2.1) in which case it is also clear that the analogy of **surface tension** S (which has units of *energy per unit area* or *force per unit length*) plays a role here. There are many examples in the pedagogical literature of experiments/demonstrations involving vibrating soap membranes, ranging from simple (circular/square) boundaries (Bergmann 1956) to ones providing analogs of quantum chaos in billiard systems (Arcos et al. 1998). Given our experience with one-dimensional waves, we imagine that the corresponding two-dimensional inertia would be given by the **mass per unit area** (or ν), and assuming only these two parameters we can write $v_s \sim S^{\alpha}\nu^{\beta}$, which requires that $\alpha = -\beta = 1/2$ or $v_s \sim \sqrt{S/\nu}$.

One can compare these DA predictions to textbook solutions for both one-dimensional and two-dimensional problems where one finds (see, e.g., Crawford 1968 for one-dimensional and Morin 2014, or Isenberg 1992 for two-dimensional) the differential equations for the string or membrane/film vibration amplitudes, $\psi(x,t)$ and $\phi(x,y,t)$, namely

$$\mu\left[\frac{\partial^2\psi(x,t)}{dt^2}\right] = F_T\left[\frac{\partial^2\psi(x,y,t)}{dx^2}\right] \tag{3.53}$$

$$\nu\left[\frac{\partial^2\phi(x,y,t)}{dt^2}\right] = S\left[\frac{\partial^2\phi(x,y,t)}{dx^2} + \frac{\partial^2\phi(x,y,t)}{dy^2}\right]. \tag{3.54}$$

These both have the form of the **classical wave equation** (in their respective dimensions) for which functions of the form $\cos(kx \pm \omega t)$, $\sin(kx - \omega t)$, or complex exponential versions, like $\exp(i(kx \pm \omega t)$, provide solutions. For the one-dimensional case, we find the relation between ω, k

$$\mu(\omega^2) = F_T(k^2) \qquad \text{or} \qquad \omega = k\sqrt{\frac{F_T}{\mu}} = v_s k \qquad \text{where} \qquad v_s \equiv \sqrt{\frac{F_T}{\mu}}, \tag{3.55}$$

and if we use the connections $\omega = 2\pi f$ and $k = 2\pi/\lambda$, this reduces to the familiar relation between frequency/wavelength and wave speed, $v_s = \lambda f$. While individual cos, sin, exp solutions are allowed, one can also have a more general linear combination of such solutions, namely

$$\psi(x,t) = \int f(k)\cos(kx - \omega(k)t)\,dk \tag{3.56}$$

by superimposing single ω/k solutions.

A wave for which the relation between $\omega(k)$ and k is linear, as in Eqn. (3.55), is called **non-dispersive** since we can write a solution

$$\begin{aligned}\psi(x,t) &= \int f(k)\cos(kx - \omega(k)t)\,dk \\ &= \int f(k)\cos(k(x - [\omega(k)/k]t)\,dk \\ &= \int f(k)\cos(k[x - v_s t])\,dk \\ \psi(x,t) &= \psi(x - v_s t),\end{aligned} \tag{3.57}$$

and the relative phases between component waves do not depend on k, so any waveform constructed from such solutions will maintain its shape, simply translating to the right (or left), as all of the $\psi_k(x,t)$ components travel at the same speed. The situation where $\omega(k)$ depends on k in a non-linear way leads to a far richer array of phenomena, much of which is readily accessible to DA analysis, as in **P3.18–P3.19.**

The forms of Eqns. (3.53) and (3.54) are somewhat reminiscent of the general harmonic oscillation forms in Eqn. (3.9). Despite having different dimensions, (F_T, S) and (λ, ν), both here play the roles of inertia and restoring forces, $\mathcal{I}$ and $\mathcal{F}$ respectively, giving $v_s = \sqrt{\mathcal{F}/\mathcal{I}} = \sqrt{F_T/\mu} = \sqrt{S/\nu}$ as the wave speed.

Extending these analyses to three-dimensional waves, one finds wave equations of the form

$$\mathcal{I}\left[\frac{\partial^2\psi(\mathbf{r},t)}{dt^2}\right] = \mathcal{F}\left[\nabla^2\psi(\mathbf{r},t)\right] \tag{3.58}$$

and we expect that $\mathcal{I} = \rho_m$ as the three-dimensional mass density to appear. The analog of $\mathcal{F}$ is variously the **elastic** (K), or **bulk** (B), or **shear** (G) modulus, all of which (not surprisingly given the pattern so far) share dimensions of **force per unit area** or **pressure.** For example, the bulk/elastic modulus is defined in terms of the response of a material to an increased external pressure (an applied ΔP) and the corresponding change (reduction) in volume (ΔV) by the connection

$$K = B = -V\frac{\Delta P}{\Delta V} = -\left(\frac{V}{\Delta V}\right)\Delta P \tag{3.59}$$

where the minus sign is conventional, since a positive applied ΔP will result in a negative change in volume ΔV. Note that in terms of dimensions, $[K] = [B] = [P] = \mathsf{M/LT^2}$ as expected, but that the definition involving knowing the form of $P(V)$ as a function of V does have important implications, as we discuss below. For the three-dimensional wave problems we consider, we assume the wave equation is valid and that the wave speeds are given by $v_s = \sqrt{\mathcal{F}/\mathcal{I}} = \sqrt{K/\rho_m}, \sqrt{B/\rho_m}$, but provide specific examples in the problems.

For liquids, we simply use $v_s = \sqrt{B/\rho_m}$, while for gases, we can use more information from the relevant equation of state, namely the form of $P(V)$ or pressure–volume relation. Newton was one of the first to attempt to derive the speed of sound in a gas and he assumed an **isothermal** connection (Boyle's law), namely $PV = nkT$ being constant. With that form, $B = -(dP/dV)V = P$ gives $v_s = \sqrt{P/\rho_m}$, which is roughly 15% smaller than the observed value. Laplace considered instead that the process would be **adiabatic** and so used the relation $pV^\gamma = p_0V_0^\gamma$, where $\gamma = c_P/c_V$ is the (dimensionless) ratio of specific heats and for a diatomic gas (such as those in the atmosphere, namely N_2, O_2) one has $\gamma = 7/5$. With this assumption, one finds $B = -(dP/dV)V = \gamma P$ or $v_s = \sqrt{\gamma P/\rho_m}$. (You're asked to explore these variations in **P3.15.**) We note that DA using just P, ρ_m as the relevant $\mathcal{F}, \mathcal{I}$ attributes does get the dimensions correct, giving the Newtonian/Boyle value, missing the important dimensionless factor of γ in the Laplace/adiabatic exact solution.

We note that neither liquids nor gases support **bending** (meaning no shear forces) while solids do, so the situation is even richer for waves in such media. Including the possibility of a **shear modulus,** G, there turn out to be two different wave speeds in bulk solids, namely

$$v_s = \begin{cases} \sqrt{\frac{K+4G/3}{\rho_m}} & \text{called P or primary waves} \\ \sqrt{\frac{G}{\rho_m}} & \text{called S or secondary waves} \end{cases} . \tag{3.60}$$

In measurements with seismometers following an earthquake, the compressional (P) waves arrive before the shear (S) waves and can help in triangulating the distance to the epicenter, in much the same way that the time-lag between seeing the lightning and hearing the accompanying thunder can be used to estimate the distance to a storm.

3.4.2 Forces, pressures, and power

Students are perhaps most familiar with two approximate conservation laws for fluids, namely the **equation of continuity** and **Bernoulli's equation**, both of which come with important assumptions. If we assume that a fluid is incompressible (no density changes allowed), then conservation of volume follows from conservation of mass since $dm/dt = d(\rho_m V)/dt = \rho_m\, dV/dt$, if ρ_m is constant. This is most often used in fluid flow problems for pipes where we write $dV/dt = Av$ = constant. The version of the **work-energy theorem** used in fluid mechanics associated with Bernoulli is

$$\frac{1}{2}\rho_m v^2 + \rho_m g h + P = \text{constant} \tag{3.61}$$

which assumes that no energy is added/extracted and that dissipation (friction) can be ignored, in which case the system is called **isentropic.** As a simple first exercise in fluid mechanics, we can check that the dimensions of Eqn. (3.61) are consistent, since

$$\begin{aligned}
\left[\rho_m v^2\right] &= \left(\frac{M}{L^3}\right)\left(\frac{L^2}{T^2}\right) = \frac{M}{LT^2} \\
[\rho_m g h] &= \left(\frac{M}{L^3}\right)\left(\frac{L}{T^2}\right)(L) = \frac{M}{LT^2} \\
[P] &= \left[\frac{F}{A}\right] = \left(\frac{ML}{T^2}\right)\left(\frac{1}{L^2}\right) = \frac{M}{LT^2}\,.
\end{aligned} \tag{3.62}$$

Example 3.2 Power from wind energy

Having checked these dimensions, we can apply DA to a first glimpse of a problem of engineering/societal interest, namely the extraction of wind power. Let us consider that a turbine apparatus has characteristic cross-sectional area A (say the area swept out by the turbine blades, $A = \pi R^2$) and that the wind speed and air density are given by v and ρ_m, respectively. Writing power in terms of these as $\mathcal{P} = C_{\mathcal{P}}\rho_m^\alpha A^\beta v^\gamma$ we find that $\alpha = 1$, $\beta = 1$, $\gamma = 3$ or $\mathcal{P} \sim \rho_m A v^3$ and the scaling with v is an important feature.

To confirm this dependence, we can initially assume that the kinetic energy in a thin slab of thickness Δx and area A incident on the turbine is $\Delta E = dm\, v^2/2 = (\rho_m dV)v^2/2 = \rho_m\,(A\Delta x)v^2/2$, and that all of this energy is available for extraction. If this energy intersects the blades over a time Δt, we have $\mathcal{P} = \Delta E/\Delta t = \rho_m A v^2(\Delta x/\Delta t)/2 = \rho_m A v^3/2$ and we find with these assumptions that $C_{\mathcal{P}} = 1/2$ in the DA analysis, as well as confirming the v^3 dependence.

Of course not all of the incident power is available for "capture" as the wind speed after passing through the turbine ($\tilde{v}$) would be reduced to zero, so a full analysis requires a second dimensional parameter related to this final speed. One finds (see Pelka et al. 1978, or Inglis 1979 for details) that the extractable power in terms of $v, \tilde{v}$ is given by

$$P(v,\tilde{v}) = \frac{1}{4}\rho_m A(v+\tilde{v})(v^2-\tilde{v}^2) = \frac{1}{4}\rho_m A v^3(1+\Pi)(1-\Pi^2) \tag{3.63}$$

where the dimensionless ratio $\Pi \equiv \tilde{v}/v$, which is in the form of the Buckingham Π theorem. We discuss the strategy for maximizing this power in **P3.22.**

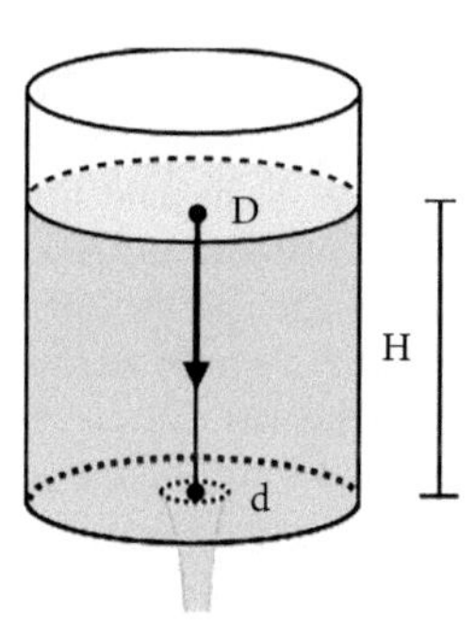

Fig. 3.5 Schematic of liquid (or granular) material flowing from a reservoir.

Example 3.3 Flow of granular materials

The problem of material flowing from a hole in a reservoir (as in Fig. 3.5), whether it's liquid (water, oil, etc.) or granular material (seeds, sand, etc.) is of relevance in a number of industrial applications. The quantity to be studied/predicted is often the **mass flow rate** ($\dot{\mathcal{M}}$ or $d\mathcal{M}(t)/dt$), which clearly has dimensions $[\dot{\mathcal{M}}] = \mathsf{M/T}$. We imagine that this flow rate would depend on the material (via its density ρ_m), certainly gravity (g), and the size of the outflow orifice (d), and perhaps the "overburden" (or depth of fluid/grains) via H.

If we assume the equation of continuity, so that the speed of the surface of the material at the top is related to that which flows out via $\pi(D/2)^2 u = dV/dt = \pi(d/2)^2 v$, we have $u = v(d/D)^2 << v$ for a small outlet size, and so we ignore any dependence on u (or D) at this point. DA proceeds by assuming $\dot{\mathcal{M}} = C_M \rho_m^\alpha g^\beta d^\gamma H^\delta$, and since we have three equations in four unknowns, we can only constrain the result to be

$$\dot{\mathcal{M}} \sim \rho_m g^{1/2} d^\gamma H^{5/2-\gamma}. \tag{3.64}$$

For liquids, we can "solve" the problem by using Bernoulli's equation to equate $v^2 = u^2 + 2gH$, and knowing that $u << v$, we have the approximate "free-fall" result that $v = \sqrt{2gH}$ and

$$\frac{d\mathcal{M}}{dt} = \rho_m \frac{dV}{dt} = \rho_m(Av) = \rho_m \left[\pi\left(\frac{d}{2}\right)^2\right] v = \frac{\pi}{4}\rho_m d^2\sqrt{2gH}, \tag{3.65}$$

which corresponds to $\gamma = 2$ and $C_M = \sqrt{2}\pi/4$ in the general solution above. Thus the flow rate decreases with time since the pressure head is reduced as H gets smaller.

In contrast, it is found that in the *flow of granular solids through orifices* (the title of a paper by Beverloo et al. (1961)) that the flow rate is actually <u>independent</u> of H, so that for such systems $0 = \delta = 5/2 - \gamma$ or $\gamma = 5/2$ giving $\dot{\mathcal{M}} \sim \rho_m g^{1/2} d^{5/2}$ and this dependence has been observed experimentally—you're asked in **P3.22** to explore this further. **Note:** While this result is often associated with the Beverloo et al. (1961) paper, almost identical results (using sand as the granular material) were obtained by Hagen (1852) over a hundred years earlier.

Example 3.4 Reynolds number

Perhaps the most famous example of an important dimension**<u>less</u>** combination appearing in a DA approach to a problem occurs in the study of the force (F_D) on an object either embedded in a moving viscous fluid, or itself moving through such a medium. We assume, for simplicity, a spherical object (diameter $D = 2R$ conventionally) moving at a speed v (or perhaps the stationary object has the fluid move past it with this speed) and the fluid properties are its density and viscosity (ρ_m and η).

When viscosity is negligible, we can consider $F_D \sim \rho_m^{\alpha} v^{\beta} D^{\gamma}$ and matching dimensions we find that $\alpha = 1$ and $\beta = \gamma = 2$ so that $F_D \sim \rho_m v^2 D^2$. This is typically the dominant frictional fluid force for large and fast-moving objects, such as skydivers. Its dependences are consistent with a force–pressure relation, $F = PA$, with $\rho_m v^2$ playing the role of pressure and $A \propto D^2$.

In the limit where viscosity effects dominate inertia (for an example, see **P3.26**) we assume $F_D \sim \eta^{\alpha} v^{\beta} D^{\gamma}$ and find $\alpha = \beta = \gamma = 1$, so that $F_D \sim \eta v D$, which is more relevant for smaller, slower-moving objects in more viscous materials. This last expression is the dimensional part of the well-known **Stokes' law** which gives the frictional force for the special geometry of a spherical shape (diameter D or radius R) as

$$F_S = -3\pi\eta v D \qquad \text{or} \qquad F_S = -6\pi\eta v R, \tag{3.66}$$

where the second form is perhaps more familiar in physics applications. Mentions of Stokes' law (Stokes 1851), which should be familiar from the undergraduate curriculum, include

- the use of F_S in determining the terminal velocity of oil drops in the famous Millikan experiment (**P3.24**), performed and analyzed, one hopes, by almost every physics major, and
- the Einstein–Stokes model for the diffusion constant used in analyses of Brownian motion described by Perrin—with Millikan and Perrin each winning the Nobel prize (in 1923 and 1926, respectively) for research using this fundamental law of hydrodynamics. See Section 5.1.1 and **Example 8.5** for its use.

Viscosity effects (including direct use of Stokes' law) appear in many biophysical and geophysical applications (**P3.26–P3.31**) and in soft matter physics (Section 8.3).

If **both** viscous and inertia effects are relevant, we can include them by writing $F_D \sim \rho_m^{\alpha} \eta^{\beta} v^{\gamma} D^{\delta}$ and solving for β, γ, δ in terms of α we find $\beta = 1 - \alpha$, and $\gamma = \delta = 1 + \alpha$, giving

$$F_D \sim \rho_m^{\alpha} \eta^{1-\alpha} v^{1+\alpha} D^{1+\alpha} = \eta v D \left(\frac{\rho_m v D}{\eta} \right)^{\alpha} = \eta v D \, (\Pi)^{\alpha} \quad \rightarrow \quad \eta v D \, G(\Pi), \tag{3.67}$$

and the two cases considered so far correspond to $\alpha = 1$ and $\alpha = 0$, respectively. We have thus identified the important dimensionless ratio

$$\Pi = \left(\frac{\rho_m v D}{\eta} \right) \equiv \mathcal{R}e \tag{3.68}$$

with $\mathcal{R}e$ being the **Reynolds number.** (See the Mathematica© example in Section 10.8 for how such a ratio naturally appears in the automated solutions we encourage/advocate.)

This ratio provides information on the relative sizes of **inertial** versus **viscous** effects and is extremely useful in scaling arguments, as one expects systems with the same value of $\mathcal{R}e$ to have equivalent flow patterns. This observation is used in the context of **wind** (or **water**) **tunnels** where ρ_m, η are kept constant (same fluid, either air or water) and scale models are used for which the product vD is the same; thus small-scale replica models of airplanes or boats can be placed in high-speed fluid flow environments to test performance. The Reynolds number is one of many such dimensionless ratios in fluid dynamics and the reader is encouraged (see **P3.25**) to acquaint themselves with others.

A common operational approach in engineering applications related to experimental measurements of the force on an object in a moving/viscous fluid is to encode information about the shape of the probe by writing

$$F_D = C_D \frac{\pi}{8} \rho_m v^2 D^2 \tag{3.69}$$

where C_D is a dimensionless constant (undetermined by DA of course), and plotting measurements of F_D scaled via

$$C_D = \frac{F_D}{(\pi/8)\rho_m v^2 D^2}. \tag{3.70}$$

For large values of v, D we expect C_D to be roughly constant (but dependent on the details of the shape) and engineering texts (see e.g., Pritchard 2011) are filled with compilations of **drag coefficient** values (C_D) for various shapes. For small values of v, D, where Stokes' law dominates, we expect (for spherical objects) that

$$C_D = \frac{3\pi\eta v D}{(\pi/8)\rho_m v^2 D^2} = 24\left(\frac{\eta}{\rho_m v D}\right) = \frac{24}{\mathcal{R}e}, \tag{3.71}$$

which suggests that one plots C_D (from F_D measurements) versus $\mathcal{R}e$. One compilation of such measurements for spherical objects (see Brown and Lawler 2003) is shown in Fig. 3.6, where the results of many different experiments (on objects of different sizes and speeds) are seen to "collapse" to one universal curve (albeit with scatter) when plotted as a function of the Reynolds number. Note that the approach to the Stokes' law limit ($C_D = 24/\mathcal{R}e$) for small $\mathcal{R}e$ that is shown in the inset. (You can use the data-fitting methods of Section 10.9.1, or any others you know, and the values from that inset, to extract parameters from a $y = ax^b$ fit to compare to the prediction of $a = 24$ and $b = -1$.)

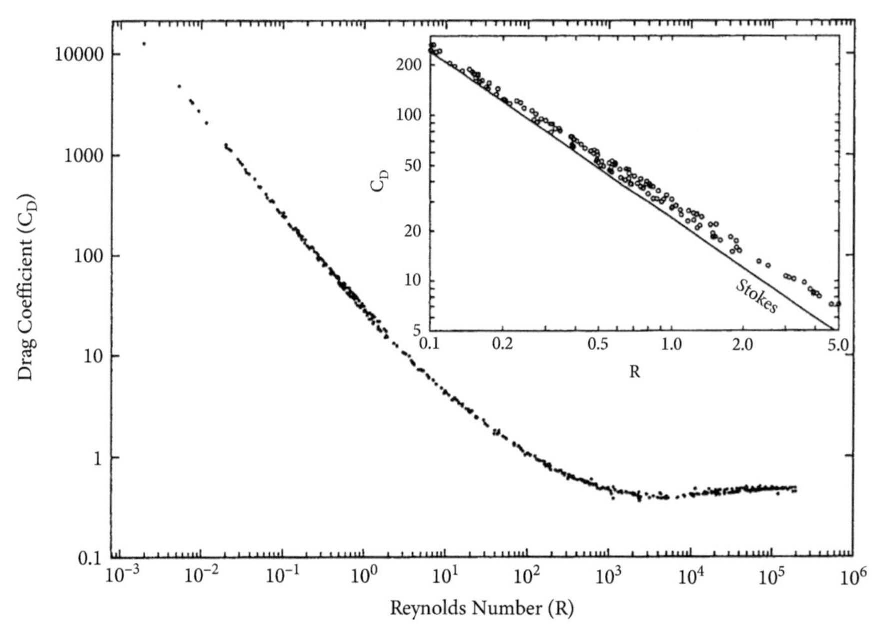

Fig. 3.6 Experimental measurements of the dimensionless drag coefficient C_D in Eqn. (3.71) as a function of Reynolds number $\mathcal{R}$. Used with permission of the ASCE, from Brown and Lawler (2003); permission conveyed through Copyright Clearance Center.

Comparing the data in Fig. 3.6 to the results shown in Fig. 3.3, we see that a large amount of information about the dependence of physical systems on multiple variables can sometimes be encoded in a single plot in terms of dimensionless ratios, ones singled out by a DA approach. One important difference between these two examples, however, is that the idealized pendulum/rotor system is soluble (in terms of known, if perhaps obscure, integrals, but certainly also via simple numerical integration) while the hydrodynamic equations governing the flow of fluids are much more mathematically involved (non-linear differential equations with boundary conditions), and the relatively simple form of the Reynolds number plot has no suggestion of the complex flow patterns, even around simple objects like spheres. (For examples, see Feynman et al. (1964), section 41-1.) In "real-life" systems, the identification of relevant dimensionless ratios (as in **P3.25**) can guide experimentation and save time and money. Performing an experiment ten times, as one varies just the Reynolds number, instead of 10^4 times as one changes all values of ρ_m, v, D, η separately, is clearly an efficient strategy.

One often sees the Reynolds number used in engineering applications, such as dynamic similarity or scaling arguments, or in classifying the types of flow. From the more mathematical physics side, for small values of $\mathcal{R}e$, it can be used to quantify deviations of the simple Stokes' law in Eqn. (3.66) due to the effects of density, when $\mathcal{R}e = \rho_m v D/\eta << 1$. For example, in a *tour de force* calculation, Chester et al. (1969) evaluated higher-order corrections to the drag coefficient in Eqn. (3.71) and found

$$C_D(\mathcal{R}e) = \frac{24}{\mathcal{R}e}\left\{1 + \frac{3}{16}\mathcal{R}e + \frac{9}{160}\mathcal{R}e^2\left[\gamma - \frac{323}{360} + \frac{5}{3}\log(2) + \log\left(\frac{\mathcal{R}e}{2}\right)\right] + \frac{27}{640}\mathcal{R}e^3\log\left(\frac{\mathcal{R}e}{2}\right)\right\} \tag{3.72}$$

where $\gamma \approx 0.577$ is the Euler–Mascheroni constant. An overview of extensions of Stokes' work, with many historical references, has been presented by Dey, Ali, and Padhi (2019). Such dimensionless ratios can sometimes be used to systematically explore perturbative solutions as a natural expansion parameter, as with the fine-structure constant, α_{FS}, of quantum electrodynamics (QED) to be discussed in Section 7.1.

3.5 **Problems for Chapter 3**

Q3.1 Reality check: (a) What was the single most important, interesting, engaging, and/or useful thing you learned from this chapter?

(b) Have you ever used DA in approaching any problem in mechanics?

P3.1 Reflection: (a) After reflecting on the examples presented in this chapter (or working through some of the problems below), what do you consider the most important lessons you've learned so far about the utility and limitations of DA?

(b) Reflect back even further, to all of the problems you've encountered in classical mechanics, and think of one to which you can imagine applying DA; then work it out in as much detail as you can, hopefully comparing it to the "exact" answer, if known.

P3.2 Harmonic oscillator revisited: Re-do the analysis of Section 3.1 for the harmonic oscillator, but assume this time that $x_0 = 0$ and $v_0 \neq 0$ and use DA to find the dependence of the period (τ), energy, and typical distance and momentum scales in terms of k, m, and v_0.

P3.3 Harmonic oscillator solved by energy methods: Use the energy-motivated relation between dx and dt in Eqn. (3.12), namely

$$\sqrt{\frac{m}{k}}\frac{dx}{\sqrt{x_0^2 - x^2}} = dt, \tag{3.73}$$

not over an entire "half-cycle," but rather in the form

$$\sqrt{\frac{m}{k}}\int_{+x_0}^{-x(t)} \frac{dx}{\sqrt{x_0^2 - x^2}} = \int_0^t dt \tag{3.74}$$

to solve for $x(t)$ explicitly and confirm that it reproduces the expected result of $x(t) = x_0\cos(\omega_0 t)$.

P3.4 Quartic oscillator (and beyond) done two ways: A particle of mass m is subject to a quadratic potential, $V(x) = Dx^4$ or a cubic force, $F(x) = -dV(x)/dx = -4Dx^3$.

(a) Using either definition above to find the dimensions of D, evaluate how the period depends on m, D, and an initial "stretch" x_0 by writing $\tau = C_\tau m^\alpha D^\beta x_0^\gamma$ and solve for the exponents.

(b) Use the energy methods in Eqns. (3.11)–(3.14) to find τ, compare to the dimensions found in part (a), and evaluate C_τ in terms of an integral (which you can look up if you wish).

(c) Assume now that the potential is a power law of the form $V^{(P)}(x) = D_P|x|^P$ and redo parts (a) and (b) for a general value of P. You will, of course, have to evaluate the dimensions of D_P.

P3.5 Buckingham Pi ratios for the complete harmonic oscillator solution: Use the most general solution of the harmonic oscillator, namely $x(t) = A\cos(\omega_0 t) + B\sin(\omega_0 t)$, evaluate A, B for arbitrary x_0, v_0, and show that the resulting solution can be written in the form of Eqn. (3.23) and find $G(\Pi_1, \Pi_2)$.

P3.6 Buckingham Pi ratios for motion under a constant force: One of the simplest classical mechanics problems is that of a particle of mass m subject to a constant force F, with initial position and velocity given by x_0, v_0.

(a) Assume that the time-dependent position is given by $x(t) \sim F^\alpha m^\beta x_0^\gamma v_0^\delta t^\epsilon$, solve for β, γ, ϵ in terms of α, δ, and identify two dimensionless combinations Π_1, Π_2.

(b) The exact solution for this motion is well known to be $x(t) = Ft^2/2m + v_0 t + x_0$. Can you show how this solution is consistent with your DA results from part (a) and find the form of $G(\Pi_1, \Pi_2)$?

P3.7 Frictional forces I: A particle of mass m is subject only to a velocity-dependent frictional force given by $F_f = -bv$ and starts at $x_0 = 0$ with initial speed v_0 and eventually comes to rest.

(a) Use DA to find the characteristic time to slow down (τ), and the distance traveled (D), by writing $\tau \sim m^\alpha b^\beta v_0^\gamma$, and similarly for D, matching dimensions and solving for the exponents.

(b)* Solve the Newton's law problem by writing it in the form $mdv(t)/dt = ma(t) = F = -bv(t)$, with $v(0) = v_0$, and solve for $v(t)$. Then integrate $v(t) = dx(t)/dt$ to find $x(t)$, assuming $x(0) = 0$, and compare to your DA results.

(c) Use DA to write $x(t) \sim m^{\alpha} b^{\beta} v_0^{\gamma} t^{\delta}$, find the appropriate dimensionless Π ratio, and compare to your rigorous answer in part (b).

(d)* You can make a connection to the problem of the damped harmonic oscillator in **Example 3.1** by taking the results Eqn. (3.27) and simply letting the spring constant $k \to 0$. Show that the two solutions you obtain in this limit correspond to a general form $x(t) = A + Be^{-bt/m}$ and then apply initial conditions $x(0) = 0$ and $v(0) = v_0$ to compare to your exact result from part (b).

P3.8 Frictional forces II: Repeat **P3.7**(a), (b)*, and (c), but assume a frictional force of the form $F_f = -cv^2$.

P3.9* Low- and high-energy limits of the pendulum/rotor period: Evaluate $\mathcal{T}^{(+)}(z)$ from Eqn. (3.39) in the limit of $z \to \infty$ and confirm that you obtain the high-energy limit of the rigid rotor period in Eqn. (3.37). Do the same for the $z \to 0$ limit of $\mathcal{T}^{(-)}(z)$ to confirm that you recover the small oscillations/isochronous limit.

P3.10 Model potential with a "phase transition" between a low- and high-energy limit: Consider the motion of a particle of mass m in the (admittedly contrived) potential shown in Fig. 3.7,

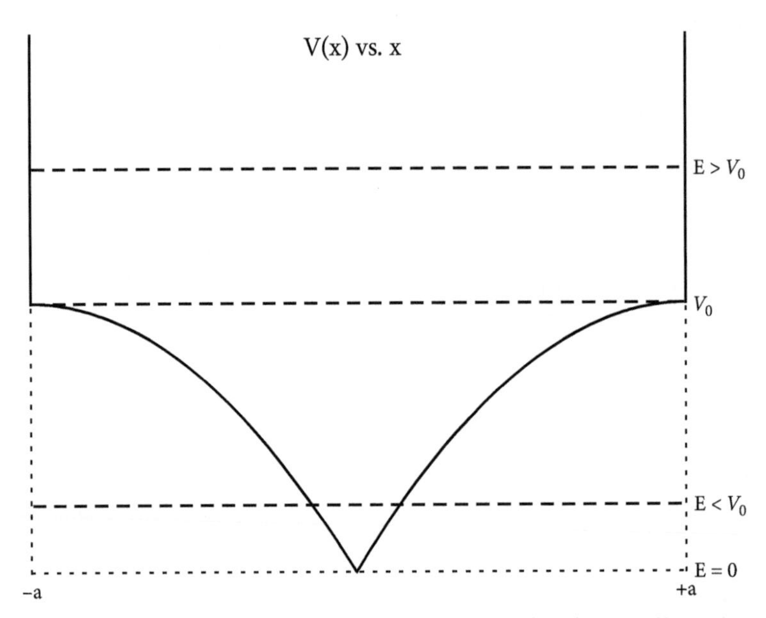

Fig. 3.7 A "toy-model" potential energy, $V(x)$ versus x, from Eqn. (3.75) with different limiting cases for the period $\tau(E)$ for $E << V_0$ and $E >> V_0$, and a divergence in $\tau(E)$ for $E \approx V_0$, as a model of the rigid-pendulum/rotor "phase transition."

defined by

$$V(x) = \begin{cases} \infty & \text{for } x < -a \\ (V_0/a^2)(2a|x| - x^2) & \text{for } -a < x < +a \\ \infty & \text{for } +a < x \end{cases} . \tag{3.75}$$

For $0 < E << V_0$ the potential is effectively $V(x) = F_0|x|$ (where $F_0 \equiv 2V_0/a$), so a symmetric linear potential. For $E >> V_0$ the potential is the familiar infinite square well of quantum mechanics, but treated here as a classical problem.

(a) Before considering the dependence of the actual period on the total energy E, try writing $\tau_0 \propto m^\alpha V_0^\beta a^\gamma$, match dimensions, and solve for the exponents. We refer to this result for τ_0 as a baseline period.

(b) Use DA to evaluate the period for $0 < E << V_0$ where $V(x) = F_0|x|$, assuming that $\tau = C_\tau m^\alpha F_0^\beta E^\gamma$ and solving for α, β, γ. Compare your result to that for τ_0 in part (a).

(c)* Now use the energy methods introduced for the harmonic oscillator to evaluate τ explicitly for $0 < E << V_0$ to confirm the dimensional dependences and to determine C_τ. Recall that $F_0 \equiv 2V_0/a$ and reinsert this combination so that your expression is in terms of m, E, V_0, a.

(d) Repeat part (b), but now assume that $E >> V_0$ (so that V_0 is irrelevant) and assume $\tau = C_\tau m^\alpha a^\beta E^\gamma$ and compare the E dependence to part (b).

(e)* Use the same energy methods in part (c) to evaluate τ exactly in the $E >> V_0$ limit and determine C_τ in this case

(f)* Finally, use the same energy methods, but now for the general case, considering the integrals needed for $0 < E < V_0$ and $E > V_0$ separately, to find $\tau(z = E/V_0)$ for all values of z, writing your answer in terms of τ_0. Confirm that you recover the results of parts (c) and (e) in the appropriate limits.

(f) Plot $\tau(z \equiv E/V_0)/\tau_0$ versus z and compare to Fig. 3.3. Why is there a divergence in $\tau(z)$ at $E = V_0$ in a similar way to that for the pendulum/rotor system at $E = 2mgl$?

P3.11 Gravitational ring: A point mass m is constrained to move along the x-axis and is attracted to a uniform density ring of total mass M (and radius R) that is centered on that axis, as shown in Fig. 3.8. Just as in Section 3.3, the only relevant dimensional quantities that could determine the period or angular frequency are M, G, R, so we expect the same type of dependence as in the "hollow Earth" example. Confirm that one gets the identical expression for the frequency of harmonic motion for small oscillations ($x << R$) by finding the force on the point mass m and solving the resulting Newton's law problem.

P3.12 Keplerian orbits: Kepler's third law relating the period of planetary orbits to their radii (actually their semi-major axes) is another example where DA can be compared to an analytic treatment, and to well-known data.

(a) Assume that the period (τ) of a planetary orbit depends on the (large) mass of the central star (M_*), its orbital radius (R) and of course G. Write $\tau = C_\tau R^\alpha M_*^\beta G^\gamma$ and solve for α, β, γ.

(b)† Use the data below to plot τ versus R and fit the data in such a way to extract α. Presumably a log-log plot and/or say a Mathematica© (or similar) fitting program would help. (See Section 10.9 about analyzing power-law data for reminders if necessary.) Does your answer agree with the results of part (a)? Note that the data below are shown in **years** (so period scaled to the Earth's period) and **AU** or **astronomical units** (so scaled to the Earths' orbital radius). We've also included several solar system objects that aren't on the "canonical" list of planets.

Planet	τ (year)	R (AU)	Planet	τ (year)	R (AU)
Mercury	0.24	0.39	Jupiter	11.86	5.20
Venus	0.60	0.72	Saturn	29.46	9.54
Earth	1.0	1.0	Uranus	84.0	19.19
Mars	1.88	1.52	Neptune	164.82	30.06
(Ceres)	4.6	2.77	(Pluto)	248	38.5
			(Eris)	559	67.9

(c) A simple intro-level derivation of Kepler's law can be obtained by using Newton's laws twice, both $F = ma$ and the law of gravitation, namely

$$m\frac{v^2}{R} = ma_c = ma = F = \frac{GM_* m}{R^2} \tag{3.76}$$

with $v = 2\pi R/\tau$ to solve for τ in terms of R, M_*, G. Does your result agree with the dimensions found in part (a)? Can you use this to also evaluate C_τ?

(d) The data in part (b) were provided in units of **years** for τ and **AU** for R. When expressed in these units (or scaled to these values), show that $C_\tau = 1$!

(e) Show that the period of a satellite in low-Earth orbit ($r = R_E + h$ where $h << R_E$) is the same as the periodicity of the **hollow Earth** problem in Section 3.3.

P3.13[†] **Weighing the massive black hole at the center of our galaxy:** Observations of the orbits of stars around the Super Massive Black Hole (SMBH) Sagittarius A^* at the center of the Milky Way galaxy over the course of more than twenty-five years have produced data on orbital radii and periods that can be used (along with Kepler's third law) to estimate the mass of the central object. The data for six such stars (taken from Gillessen et al. 2017) are shown below, with τ given in years, but R given in *mpc* or *milli–parsecs*.

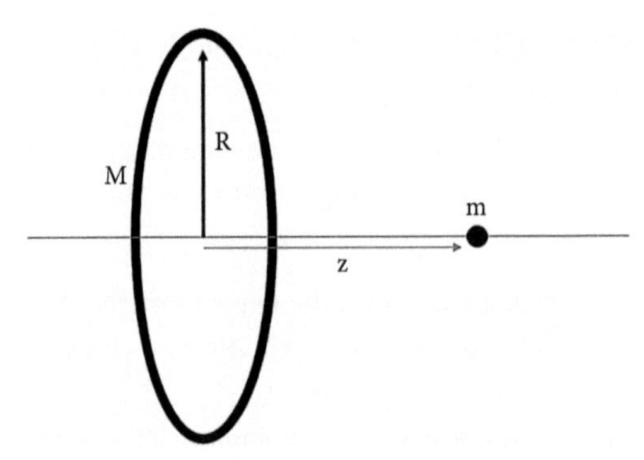

Fig. 3.8 Fixed, uniform ring of mass M and radius R, with test mass m constrained to move along the z-axis.

Star	τ (year)	R (mpc)	Star	τ (year)	R (mpc)
S1	166	23.8	S8	93	16.3
S2	16	5.04	S13	49	10.6
S6	192	26.3	S18	42	9.5

(a) Confirm that the data satisfy Kepler's third law that you found in **P3.12**(c), or looked up. For this part, you don't need to convert τ or R into any particular set of units.

(b) Use the values to "weigh" the central SMBH, in both kg and $M_{\odot}$ (solar masses) units, and for this you will have to look up what a *parsec* is.

P3.14 Gravitational instability and collapse—Jeans's length: Consider a region of gas of size λ and density ρ_m. There are two natural time scales which dictate the stability of such a system under gravitational collapse, namely the **free-fall** time (t_{ff}) due to self-attraction, and the typical time scale for **sound** to propagate across the body (t_s) which is governed by v_s or the local speed of sound. For small λ, perturbations in density can rearrange themselves so long as $t_s < t_{ff}$, while for large λ there is sufficient mass to cause runaway collapse and $t_{ff} < t_s$. A measure of the size at which such a system becomes unstable with respect to gravitational collapse is **Jeans's length** or λ_J (see Jeans 1902 for the original, mathematically sophisticated, paper). Assume that $\lambda_J \sim G^{\alpha}\rho_m^{\beta}v_s^{\gamma}$ and solve for α, β, γ. See also the results of **P2.15**.

P3.15 Pressure in the interiors of planets: For a relatively small planet or moon, where one can consider the mass density (ρ_m) a constant, the pressure varies from a (large) central ($r = 0$) value to zero at the surface $r = R$.

(a) Assuming that the central pressure depends on ρ_m, R, and G, write $\mathcal{P} = C_{\mathcal{P}}\rho_m^{\alpha}G^{\beta}R^{\gamma}$, and evaluate α, β, γ.

(b)* The differential equation governing the variation of pressure, $p(r)$, with radius is

$$\frac{d\mathcal{P}(r)}{dr} = -\rho_m g(r) \qquad \text{and from Eqn. (3.49) we have} \qquad g(r) = \frac{4\pi}{3}\rho_m G r. \qquad (3.77)$$

Integrate this subject to the boundary condition that $\mathcal{P}(r = R) = 0$ to find the exact solution for $\mathcal{P}(r)$ and evaluate $C_{\mathcal{P}}$ from $\mathcal{P}(r = 0)$. (For more details, see Turcotte and Schubert 2002.)

P3.16 Three-dimensional waves in liquids, gases, and solids: Given that the dimensional dependence of the sound speed for liquids, gases, and solids is all quite similar, some practice with "orders of magnitude" in physical systems can be informative. Recall that the MKS unit of pressure is the Pascal Pa (or N/m^2).

(a) For liquid water, the density and bulk moduli are given by $\rho_m \sim 10^3\ kg/m^3$ and $B \sim 2\ GPa$. What is the wave speed given by $v_s = \sqrt{B/\rho_m}$?

(b) The speed of sound in air at standard pressure (STP) is roughly 341 m/s. Air pressure and density at STP are $P \sim 10^5\ Pa$ and $\rho_m \sim 1.2\ kg/m^3$. The Newtonian/Boyle prediction for sound speed is $v_s = \sqrt{P/\rho_m}$, while the Laplacian/adiabatic one is $v_s = \sqrt{\gamma\mathcal{P}/\rho_m}$, where $\gamma = 7/5$ for diatomic molecules. Evaluate both and comment.

(c) Compare v_s for helium and air at the same conditions, given that $\rho_m(He) \sim 0.18\, kg/m^3$ and $\gamma_{He} = 5/3$ (since it's a monatomic gas). Have you ever experienced this difference directly?

(d) The bulk and shear moduli for granite are $K \sim 55\, GPa$ and $G \sim 24\, GPa$, respectively, while its density is $\rho_m \sim 3 \times 10^3\, kg/m^3$. Use Eqn. (3.60) to find the speed of sound for the primary/secondary waves. What is the implication that the secondary wave speed in the outer core (the region between the mantle and inner core) vanishes?

(e) The density of neutron star matter is roughly $\rho_m(NS) \sim (3{-}6) \times 10^{17}\, kg/m^3$. The shear modulus for such material has been estimated (Caplan et al. 2018) to be roughly $G_{NS} \sim 10^{30}\, erg/cc$. Confirm that the units used here do have the appropriate dimensions for a **modulus**, and evaluate the speed of sound predicted by $v_{NS} = \sqrt{G_{NS}/\rho_m(NS)}$.

P3.17 "Ain't no mountain high enough": One model used to explore why mountains on a (let us assume terrestrial) planet have a maximum height (h_{max}) assumes that the limiting factor is that at some point the material (say rock) at the base will crack (slide, shear, break, melt?) because of the gravitationally induced stress due to the material above it. The two material properties cited are the rock density ρ_m and some appropriate *yield stress*, Y, which has the same dimensions as Young's modulus, namely those of pressure, and of course we assume some value of g.

(a) Write $h_{max} \propto Y^{\alpha} \rho_m^{\beta} g^{\gamma}$, match dimensions, and solve for the exponents. Do your results make physical sense?

(b) Instead of using explicit values of Y for some material, consider instead how the h_{max} of a mountain will scale as one moves from planet to planet, assuming that the values of Y, ρ_m are roughly the same, but that the acceleration of gravity changes. Values for g for the Earth and Mars are $9.8\, m/s^2$ and $3.71\, m/s^2$, respectively, while the highest mountains on each body are roughly $8.8\, km$ for Mount Everest (or Sagarmatha) above sea level, or $9.3\, km$ for Mauna Loa (from the ocean floor), with the highest point on Mars being Olympus Mons with a summit of $22\, km$. Does your scaling prediction roughly work?

Notes: One of the few references in the literature to the question of *how high can a mountain be?* (on a rocky planet) that I can find is by Scheuer (1981). The question of how tall a mountain can be on neutron star (it turns out to be a fraction of a mm) is related to the value of its quadrupole moment, which determines its ability to emit gravitational radiation, and so is still a topic of research interest; see, for example, Gittins and Andersson (2021). Scheuer (1981) actually considered this case forty years earlier and finds roughly this more modern value.

P3.18 Gravity versus capillary waves in fluids: Waves can propagate on the surface of fluids and an important quantity is the **dispersion relation** relating the angular frequency ($\omega = 2\pi f$) to the wave number ($k = 2\pi/\lambda$) via $\omega(k)$. As discussed in Section 3.4.1, if the phase velocity $v_p = \omega(k)/k$ is independent of k, then waves are non-dispersive, but otherwise they will spread as they propagate, so we want to explore dispersion effects in two limiting cases, namely surface waves with restoring forces due to gravity (like a pendulum) or surface tension (like the droplet oscillations in Section 1.1). (See the very nice discussion by Gratton and Perazzo (2007) and for this problem we assume that these waves are propagating in deep water, so that depth (h) effects are not relevant.)

(a) Consider first the case where surface tension effects are negligible and assume that the dispersion relation for surface waves depends on the acceleration of g, the fluid density ρ_m, and of course the wave-number k. Write $\omega_g(k) = C_\omega k^\alpha g^\beta \rho_m^\gamma$ and solve for α, β, γ. Are these waves dispersive? The dispersion relation $\omega_g(k)$ for such waves has been measured in laboratory experiments (Taklo et al. 2015) and you can see there to what extent your predicted behavior is reproduced/confirmed.

(b) Now consider the case where gravity is negligible to find the dispersion for **capillary waves** by writing $\omega_c(k) = C_\omega k^\alpha S^\beta \rho_m^\gamma$ where S is the surface tension, and solve again for α, β, γ. Are these waves dispersive? This dispersion relation has also been measured (using laser diffraction techniques) by Zhu et al. (2007) and we reproduce results for ω versus k, for two different fluids, in Fig. 3.9.

(c)† Using the data in Fig. 3.9, for either fluid, can you confirm your prediction for α from part (b)? Using values for water only ($S \sim 73 \times 10^{-3} N/m$ and $\rho_m \sim 10^3\ kg/m^2$), can you estimate the value of the undetermined constant C_ω?

(d) You should have found that the k dependence for the two limiting cases in parts (a) and (b) are different. For what value of $k = k_{gc}$ are the values of $\omega(k_{gc})$ about equal, that is, what is the crossover between gravity and capillary behavior, and what is the corresponding value of ω_{gc}? Do this in two ways, first by equating $\omega_g(k_{gc}) = \omega_c(k_{gc})$ to find the critical value of k_{gc} and then substituting back into either expression, and second by using DA directly by writing $k_{gc}, \omega_{gc} \sim g^\alpha \rho_m^\beta S^\gamma$. How does ω_{gc} depend on ρ_m, S and especially on g?

(e)† I ask specifically about the g dependence, because experiments using high-speed centrifuges are able to explore this behavior of f_{gc} as a function of the **effective acceleration due to**

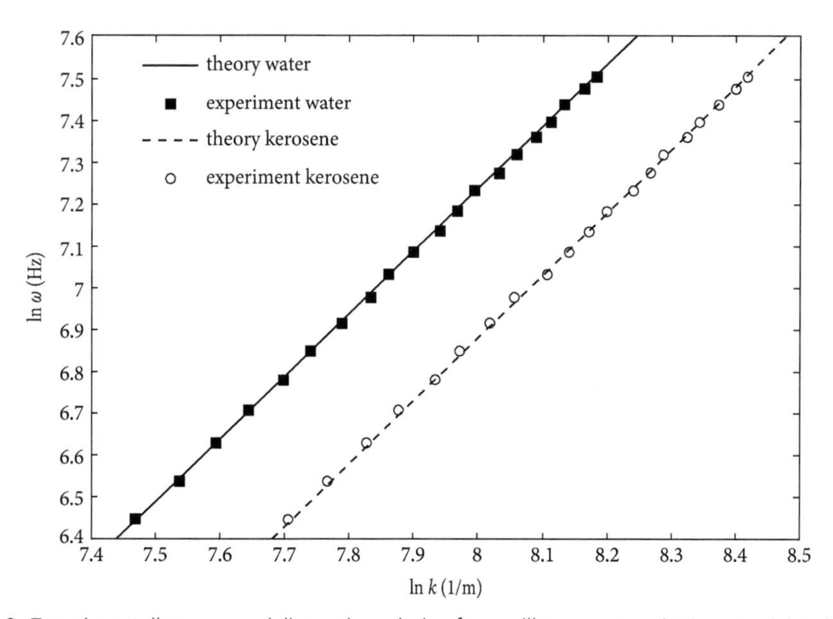

Fig. 3.9 Experimentally measured dispersion relation for capillary waves, ω (Hz) versus k $(1/m)$. Note the log-log scales used. Data (boxes and open circles) and theory predictions (dashed lines) are shown. Reprinted from Zhu et al. (2007), with the permission of AIP publishing.

gravity, $g*$, for values of g^* up to $20g$. The data from one such experiment[2] is plotted in Fig. 3.10 and you should be able to compare your prediction, namely the dependence of $f_{gc} \propto (g^*)^\alpha$ with the solid curve plotted there, which seems to represent the data pretty well.

(f) The insert shows the frequency dependence of the wave spectra in both the capillary and gravity dominated regimes. The predictions for the f^α scaling for each was discussed in **P1.23** as being $f^{-17/6}$ and f^{-4}, respectively. Compare those values (shown as solid lines in the insert) to the experimental results (shown as dashed lines).

(g) Recall that for the damped harmonic oscillator, there was a critical value of the damping, namely $b^2/4mk = 1$, where the oscillations changed from damped but periodic, to over-damped (with no periodicity). For gravity surface waves, when viscosity (η) effects are included, there is a critical value of the wave-number, k^*, which separates two similar types of wave behavior. Assume that $k^* = C_{k^*} g^\alpha \rho_m^\beta \eta^\gamma$ and solve for α, β, γ. Chandrasekhar (1955) considered this problem and found that $C_{k^*} \sim 1.2$, along with the dependences you've predicted.

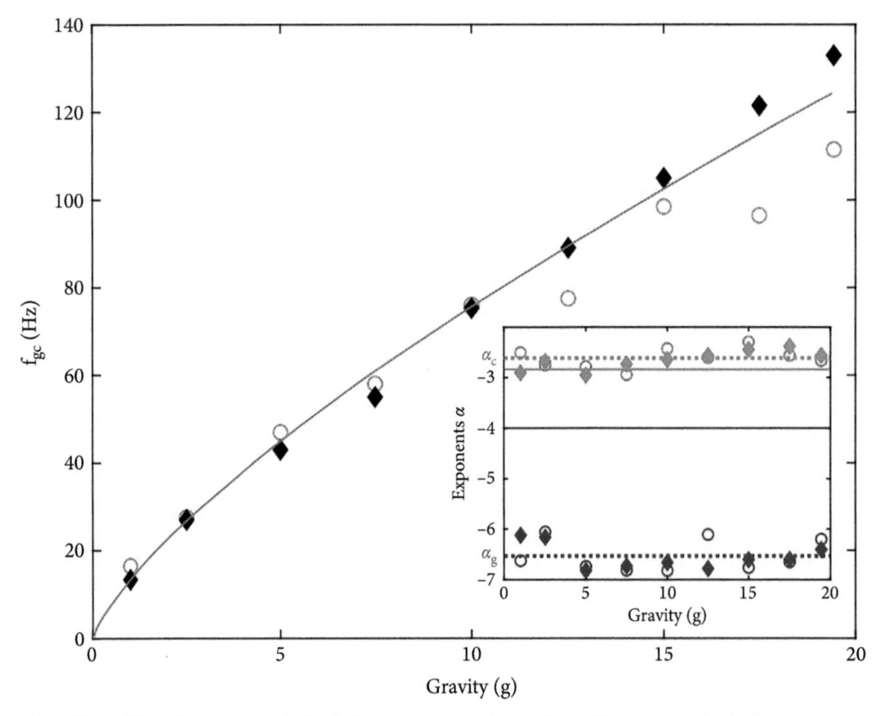

Fig. 3.10 Transition frequency, $f_{gc}(Hz)$, between gravity and capillary wave turbulence, versus g^* (effective acceleration of gravity). Reprinted with permission from Cazaubiel et al. (2019). Copyright (2019) by the American Physical Society. The insert shows the scaling exponent of the frequency dependence f^α for the capillary (top dashed line) and gravity (bottom dashed line) regimes. For use in **P3.18**.

[2] See Cazaubiel et al. (2019) and also the discussion in **P1.23**.

P3.19 Gravity and capillary waves in fluids of any depth: Consider the dispersion relation for surface waves where both gravity (g) and capillary effects (S, ρ_m) are relevant, but now assume a dependence on the depth (h) of the fluid as well.

(a) Write $\omega(k) \sim k^{\alpha} g^{\beta} \rho_m^{\gamma} S^{\delta} h^{\epsilon}$ and solve for α, β, γ in terms of δ, ϵ. Use this to identify two important dimensionless Buckingham Π ratios, Π_1, Π_2, the first involving S and the second involving h. Klemens (1984) has derived the dispersion relation for this general case, which can be written in the form

$$\omega^2 = kg\left(1 + \frac{k^2 S}{\rho_m g}\right)\tanh(kh). \tag{3.78}$$

Is this consistent with your values for Π_1, Π_2

(b) In the limit of deep water, where $kh = 2\pi h/\lambda >> 1$, show that the results (dependence on dimensions) of **P3.18** are recovered in both of gravity and capillary wave limits. Use this exact formula to find C_ω for both cases and comment on how well DA worked there.

(c) Ignoring surface tension, for gravity waves in shallow water, where $kh << 1$, find the dispersion relation, and show that such waves are not dispersive, and find the wave speed given by ω/k.

P3.20 Sloshing modes and seiches—Dimensional <u>and</u> data analysis: If you've ever tried to carry a rectangular pan of water you'll have seen (and perhaps felt) sloshing modes as the water oscillates from one end to the other, something like the picture in Fig. 3.11. A very similar phenomena occurs in a natural setting, often in the context of long and narrow lakes, where they are called **seiches**, where the sloshing is initiated by wind or a sudden change in air pressure.

Location	l (km)	d (m)	τ_O
Lake Earn (Scotland)	10	60	14.5 *min*
Lake George (NSW)	30	5.5	131 *min*
Lake Baikal (Siberia)	665	680	4.64 *hr*

(a) To understand the dependence of the period of oscillation, τ, on the length (l) and depth (d) of the lake (or pan) of these gravity-driven (so we need g) waves, write $\tau = C_\tau g^{\alpha} l^{\beta} d^{\gamma}$ (with C_τ a dimensionless constant). Use DA to constrain the exponents as much as possible, at least as a function of β.

Proudman (1953) collected data on several naturally occurring limnological seiche systems, ones for which a simple hydrodynamic model works well (namely ones which are fairly rectangular, as well as long, narrow, and shallow). We reproduce some of his results in the table below, including values of l and d, as well as the observed period of oscillation, τ_O.

(b)† By taking ratios of your DA result for pairs of data points, find the most likely value of β.

(c)† Using that value of β, use individual data points to find the most likely value of C_τ. The relation you've found using data-fitting and DA is called **Merian's formula.** We can also approach this problem from a more fundamental level as in part (d).

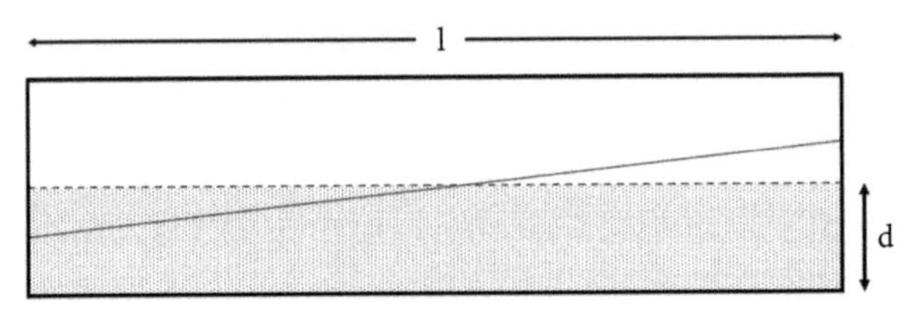

Fig. 3.11 Schematic figure of normal mode oscillation of water leading to seiches

(d) The dispersion relation for gravity waves in shallow water is given by the $kh \to 0$ limit of Eqn. (3.78), where we also let $S = 0$. Using the fact that $\omega/k = v = f\lambda = \lambda/\tau$ and that the fundamental seiche wavelength is given (see again Fig. 3.11) by $l = \lambda/2$, find τ in terms of g, l, h and compare to your "data-fitting" values from parts (b) and (c).

Note: Crawford (1968) suggests exploring this phenomenon as a "Home Experiment," so if you have a rectangular baking dish, a meter/yard stick, and the timer on your phone, you can compare the simple formula you found to data you collect yourself.

P3.21 Dripping faucets, maximum drop size, and pinch-off time: If you very slowly turn on a faucet, you'll see a drop form and get bigger until a point where it "breaks off" and separates.

(a) Presumably the growth/splitting of the drop is governed by a competition between the surface tension (S) holding the drop together and gravity (g), which couples to the fluid density (ρ_m). To get a sense of the maximum size of the drop before it separates, find a typical length (say the radius) by writing $R_{max} \sim S^{\alpha} g^{\beta} \rho_m^{\gamma}$ and solve for α, β, γ. Using values for water ($\rho_m \sim 10^3\ kg/m^3$ and $S \sim 73\ mN/meter$ (where mN is *milli-Newton*) find the typical size of a faucet drop and compare to values of raindrop size you find on the Internet. (The value of R_{max} you found is often called the **capillary length scale** or l_c.)

(b) Does your result make sense when $g \to 0$, a limit relevant for passengers on low-g parabolic plane flights or astronauts on the International Space Station, both of whom have explored water drops (blobs! globes!!) in reduced gravity? See Vollmer and Möllmann (2013) for examples.

(c) Near the time when the drop pinches off, the **neck size** (h_{min}) of the drop shrinks (going to zero when the drop breaks off) and this quantity has been measured to have a power-law dependence on the time (Δt) as measured from the separation. Write $h_{min}(t) \sim S^{\gamma} \rho_m^{\beta} g^{\gamma} \Delta t^{\delta}$ to solve for α, β, γ in terms of δ. One might guess that near the **pinch-off** time (when $\Delta t \to 0$) and the final volume is basically set, that gravity might cease to play a role. If $\gamma = 0$, what then is the predicted Δt dependence? Using the results shown in Fig. 3.12 for $h_{min}(\Delta t)$ versus Δt (for two different liquids), can you confirm the value of δ you've found?

Notes: A nice description of how to perform such experiments in the undergraduate lab, and how to use DA to predict the results, can be found in Keim (2020), and an excellent research level review has been presented by Eggers (1997). A number of important fluid dynamic issues related to modern ink jet printing are discussed by Lohse (2002), so these topics continue to be of relevance in many applications.

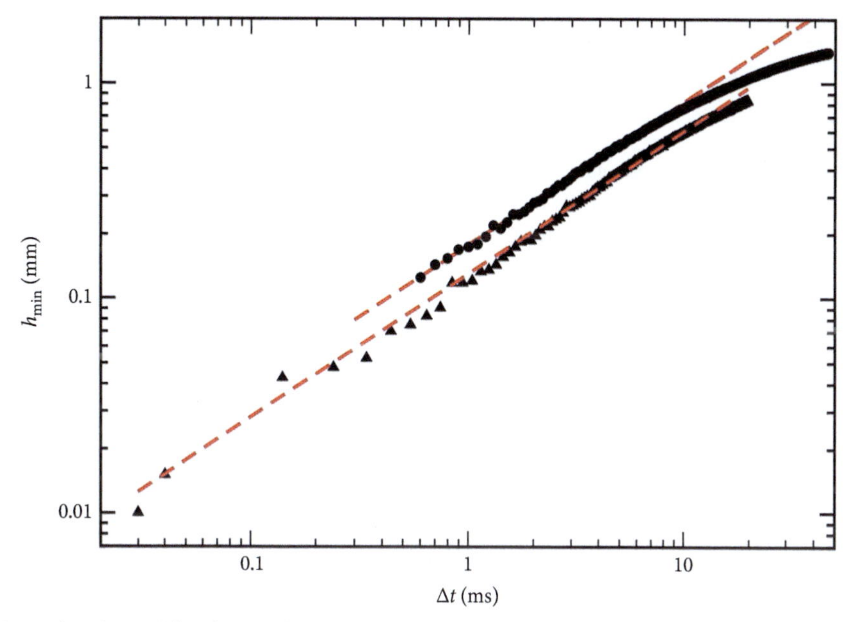

Fig. 3.12 "Neck size" (h_{min}) as a function of Δt for two different liquids, shown on a log-log plot. Reprinted from Sack and Pöschel (2017), with the permission of AIP publishing. Data (circles and triangles) and theory predictions (dashed lines) are shown. Can you extract the power-law exponent α in the relation $h_{min} \propto (\Delta t)^{\alpha}$ from this plot?

P3.22* Maximum power output of a wind turbine: The extractable wind power in Eqn. (3.63) can be written in the form

$$\mathcal{P}(v, \tilde{v}) = \frac{1}{2}\rho_m A v^3 \left[\frac{1}{2}(1 + \Pi)(1 - \Pi^2)\right] \equiv \mathcal{P}_0(v)\, F(\Pi) \tag{3.79}$$

where $\Pi = \tilde{v}/v$ is the ratio of the final (after passing through the turbine) wind speed to the incident value, and we have factored out the naive estimate $\mathcal{P}_0(v)$. Clearly $0 \leq \Pi \leq 1$ (since the wind doesn't speed up through the turbine) and plotting $F(\Pi)$ over that range you should be able to see that the function has a maximum value. Set $dF(\Pi)/d\Pi = 0$ to find the maximum extractable power (compared to $\mathcal{P}_0$) and show that it is approximately 59%. This result was evidently first obtained by Betz (1920).

P3.23† "Like sands through an hourglass": The flow behavior of granular material like sand in a geometry such as in Fig. 3.5 is very different from that of a liquid. One only has to recall that hourglasses are specifically designed so that the flow rate of the material is uniform in time, taking full advantage of the observed independence on that value of the "overburden" H.

(a) We want to analyze data to compare to the result in Eqn. (3.64) with $\gamma = 5/2$, as relevant for granular flow. Data from Beverloo et al. (1961) for the flow rate ($\dot{\mathcal{M}}$) versus orifice diameter (d) for rapeseed is reproduced in the chart below. Can you plot/fit it in such a way to test the predicted dependence on d^{γ}?

Diameter (cm)	0.75	1.0	1.25	1.50	2.00	2.50	3.00
Flow rate (gr/min)	150	405	779	1317	3166	5849	9822

(b) This analysis ignores the fact that the granular materials themselves have a finite size, which can be ignored for liquids where the particles are of molecular dimensions. Beverloo et al. (1961) found that a much better "fit" is obtained by using $d \to d - k D_{seed}$ where k is a dimensionless constant. For their data, $D_{seed} \sim 0.15\, cm$, and they find that $k \sim 1.4$ fits many granular materials well. Try re-doing your plot/fit to see if you agree. Is this phenomenological "fix" an example of the Π theorem?

Notes: Hagen (1852) had earlier found the "finite-size" correction considered here, so such experiments and analyses are not new, but this field of study is still an active research area; see Mankoc et al. (2007), Rao and Nott (2008), and especially Saleh et al. (2018) for many references. Yersel (2000) showed that this type of experiment can be easily done (and analyzed) in an undergraduate lab course, so perhaps you could try this even at home!

P3.24 Millikan oil drop experiment: The famous oil drop experiment that demonstrated quantization of electric charge and found the value of e (Millikan 1913) made use of Stokes' law and the measurement of the terminal velocity of falling drops to measure the size of an individual drop. Assume that for a given drop size that the drop radius R depends on the observed terminal speed v, the density of the oil ρ_m, the acceleration of gravity g, and the viscosity of air η through which the drop is falling.

(a) Write $R = C_R \rho_m^{\gamma} g^{\beta} v^{\gamma} \eta^{\delta}$ and solve for β, γ, δ in terms of α. Given that the first two variables often occur in the combination $\rho_m g$ (so that $\alpha = \beta$), what does that imply about the other exponents?

(b) Find the "exact" answer by balancing the weight of the drop mg (with m expressed in terms of volume and density) with the Stokes' law frictional force to solve for R.

(c) The experiment is done in air, so ρ_{air} is another variable. How do you think this comes into play and does it do so in a manner consistent with the Buckingham Π theorem?

P3.25 Dimensionless numbers in fluid mechanics and hydrodynamics: We've seen that the Reynolds number $\mathcal{R}e$ arises naturally in the study of fluid flow and contains information on the relative importance of inertial versus viscous forces. Do a literature (OK, web) search for other dimensionless numbers relevant for fluid mechanics, perhaps ones named after scientists/engineers, such as Bingham, Bond, Chandrasekhar, Eötvös, Froude, Mach, Morton, Ohnesorge, Prandtl, and Rossby, to see what they encode about the physical systems for which they're used. To see just how many such combinations may be relevant for important modern applications, see Lohse (2022), who lists thirteen different ". . .dimensionless numbers for droplets in inkjet printing. . ." as examples of their relevance.

P3.26 "Life at Low Reynolds Number": In a famous paper (with the same title as that of this problem) Purcell (1977) discusses the motion of microscopic animals (such as bacteria) in water. He notes that "For these animals inertia is totally irrelevant. We know that $F = ma$, but they could scarcely care less", which is because of the importance of viscosity. In this case, the relevant velocity-dependent force is indeed Stokes' law, and in **P3.7** you considered the motion of a mass acted on by a force $F = -bv$ (here with $b = 6\pi\eta R$) and solved for the stopping time (τ) and distance traveled (D) for a mass m and speed v_0 entering a region with a velocity-dependent frictional force of exactly this form. Do (or redo) that problem as a starting point for this one so you can apply those results to this case. You can use either the DA estimates or the actual values from solving the differential equation.

If we assume a spherical shape, the mass of the bacterium would be $m = (4\pi/3)\rho_B R^3$ and other values appropriate for this problem are

$$R \sim 1\,\mu m, \qquad \eta = 10^{-2}\frac{gr}{cm \cdot sec}, \qquad \text{and} \qquad v_0 \sim 30\,\frac{\mu m}{sec} \tag{3.80}$$

and assume that the density of the bacterium (ρ_B) is roughly that of water (1 gr/cc). Use this information to estimate values for b and m, and then the results of **P3.7** to estimate the time (τ) and distance traveled (D) by such a bacterium if it were to coast to a stop.

P3.27 Flow rate of "real fluids" through tubes—Poiseuille's law: The flow of liquids through tubes has obvious applications in engineering problems (from chemical and industrial engineering to oil drilling), but is also important biomedical connections (blood flow). The quantity of interest is usually the volume flow rate ($Q = dV/dt$) and how it depends on the external parameters of the system, namely the radius (R) and length (l) of the pipe and the pressure difference between the two ends (ΔP), along with the relevant fluid property, in this case the viscosity (η).

(a) Write $Q = C_Q \Delta P^{\alpha} R^{\beta} l^{\gamma} \eta^{\delta}$ and solve for α, γ, δ in terms of β. Does your analysis determine any of the exponents unambiguously?

(b)† Simple undergraduate laboratory experiments probing the dependences on all four quantities are easily done (see, e.g., Dolz et al. 2006) and data on the flow rate as a function of ΔP for various pipe lengths are shown in Fig. 3.13. Does the data confirm your prediction for α? Using the data can you find the most likely value of β and therefore for γ? The coefficient C_Q is known from detailed analyses to be $\pi/8$ and the resulting complete formula is known as **Poiseuille's law.**

(c) Repeat this DA problem, but now assume that Q can also depend on the fluid density, ρ_m, so add $Q \sim \rho_m^{\epsilon}$ to the other dependences above. What does the experimental data on the ΔP dependence of the flow rate tell you about ϵ?

Note: The dependence of *Laminar Viscous Flow through Pipes, Related to Cross-section Area and Perimeter Length* has been explored by Lekner (2019), who provides examples for pipes of elliptical and rectangular cross-sections as background for studying if the flow rate is better described as being proportional to A^3/P^2 instead of A^2 where A, P are the area/perimeter of the pipe (both of which have the same final dimensions.)

P3.28 Viscosity in geophysics I—Direct measurement of lava properties: Vulcanologists use a modified version of Stokes's law to make *in situ* measurements of lava viscosity in the field. They use probes (eponymously called **penetrometers**) with shapes for which the viscous force is best approximated by $F_f = -3\pi R\eta v$ and insert them (at the end of a long rod) slowly into flowing/resting lava and measure the applied F_f and resulting v directly. Use the two examples in the chart below of the sample data (from Belousov and Belousova 2018) to estimate the range of viscosities of lava from the Tolbachik (Kamchatka, Russia) volcano, and compare to values in Table 2.1 in Section 2.6, which are obtained from a wide variety of measurements/models. Given the challenges of such measurements, one might not expect complete reproducibility!

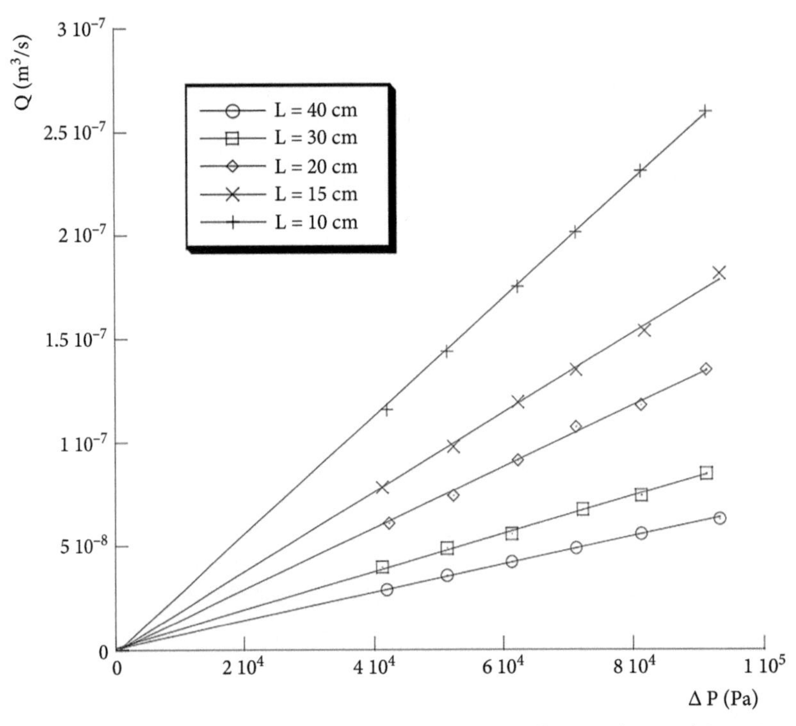

Fig. 3.13 Volume flow rate ($Q\,(m^3/s)$) versus applied pressure difference ($\Delta P\,(Pa)$) for tubes of fixed radius ($R = 0.88\,mm$) and various lengths ($L\,(cm)$). Reproduced from Dolz et al. (2006) by permission of IOP Publishing Ltd. ©European Physical Society.

$F_f\,(N)$	$R\,(cm)$	$v\,(mm/s)$
120	1.4	20.8
30	1.4	13

P3.29 Viscosity in geophysics II—Post glacial rebound: If you're sitting on a cushion for a while and then get up, the surface rebounds on a characteristic time scale, presumably dependent on its material properties. A similar phenomena occurs (over geologic time scales) when a glacier that has been "sitting on" bedrock for a long time melts. Relatively sophisticated geophysical analyses (see, e.g., Heiskanan and Meinesa 1958, Cathles 1975, or Turcotte and Schubert 2002) show that the return to local equilibrium is exponential ($e^{-t/\tau}$) with a characteristic time τ depending on the rock/mantle viscosity (η) and density (ρ_m), the acceleration of gravity (g), and the original lateral size of the glacier (say R).

(a) Use DA to explore the dependencies on these quantities by writing $\tau = C_\tau \eta^\alpha \rho_m^\beta g^\gamma R^\delta$ and solving for β, γ, δ in terms of α. Given that ρ_m and g often appear together as "weight," does this fact uniquely determine the exponents?

(b) The rigorous analyses noted above find that $C_\tau = 4\pi$ (so again, within an order of magnitude of unity). Use experimental data for the Fennoscandia region, namely $R \sim 2400\text{–}3000\,km$,

$\rho_m \sim 3.3\,gr/cm^2$, and $\tau \sim 4400\text{–}5000\,year$ to estimate the value of η for mantle rock and compare to the entries in Table 2.1 in Section 2.6.

P3.30 Viscosity measurements in the lab—Cylindrical viscometers: The viscosity of a liquid can be obtained by measuring the terminal velocity of spherical objects dropped through it. Another method makes use of the original definition from Chapter 2, in Eqn. (2.28), but modifies the conceptually simple parallel-plate visualization used in Fig. 2.8 by wrapping it into a cylindrical geometry, as shown in Fig. 3.14. In this approach, an external torque $\mathcal{T}$ is applied to the inner cylinder, giving a constant angular velocity ω. The region between the (rotating) inner and (fixed) outer cylinders (both of length l and separated by a distance a) is filled with the liquid of viscosity η to be determined, given measurements of $\mathcal{T}$, ω, R, L, and a.

(a) Assume that $\mathcal{T} = C_{\mathcal{T}}\eta^{\alpha}\omega^{\beta}R^{\gamma}l^{\delta}a^{\epsilon}$ and use DA to evaluate $\alpha, \ldots, \epsilon$ to the extent you can. Does your analysis give unambiguous predictions for any of the exponents? Can you use physical arguments to bound γ, δ, ϵ?

(b) Instead of attempting to do this problem all at once, try to do it in stages. For example, the torque $\mathcal{T}$ will be given by a "force times lever arm" and clearly the relevant "lever arm" distance is R. Now try "unwrapping" the viscometer to match the original conceptual visualization in Fig. 2.8 where the definition of viscosity from Eqn. (2.28) suggests that "force = (area) times (viscosity) times (shear rate)", where the relevant area is that of the cylindrical surface, namely $2\pi Rl$, and the shear rate scales as $\nabla_y v_x \sim v_{angular}/a = \omega R/a$. Put these all together to see if you can now confirm your predictions for α, β, determine γ, δ, ϵ, and evaluate $C_{\mathcal{T}}$. Feynman et al. (1964, section 41-2) has a nice discussion of the basics of the rotational viscometer, done with a bit more rigor.

Note: This type of apparatus, with a concentric/coaxial cylinder geometry (often called a **Taylor–Couette viscometer**), has been used for decades in fundamental studies ranging from fluid turbulence (e.g., Taylor 1923, the same G. I. Taylor cited in Section 2.2!) to chaos and irreversibility (e.g., Pine et al. 2005). It is also the basis for one of the most stunning lecture demonstrations in all of physics. Originally proposed by Heller (1960), educational videos of these "unmixing" demos were developed by Taylor (1966) and online versions have regained popularity; see the

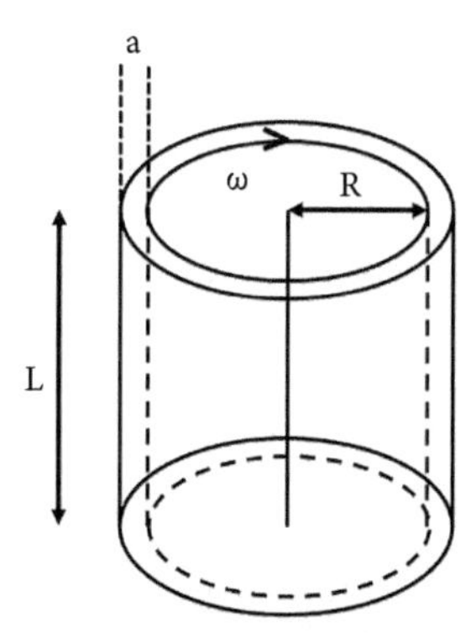

Fig. 3.14 Schematic diagram for cylindrical viscometer with the space between the inner and outer walls of width a filled with a viscous fluid.

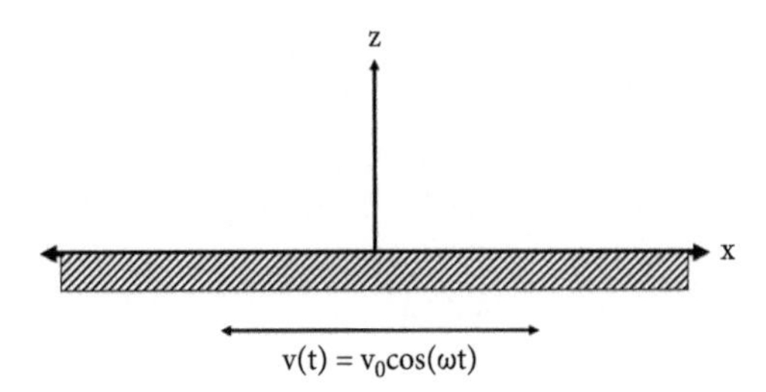

Fig. 3.15 Infinite slab oscillating harmonically in the x direction, with the semi-infinite space ($z > 0$) above filled with material of density ρ_m and viscosity η.

recent review/discussion by Fonda and Sreenivasan (2017). This is one example of the important role played by scientists (across generations) in outreach and "public science".

P3.31 Viscous penetration depth: An idealized (but soluble) problem in fluid dynamics assumes a two-dimensional infinite slab in the $x-y$ plane made to oscillate horizontally with velocity $v_x(t) = v_0 \cos(\omega t)$, with a semi-infinite region (for $z > 0$) above it filled with fluid of density ρ_m and viscosity η, as shown in Fig. 3.15. The fluid in the upper half plane has a velocity profile $v_x(z,t) \propto e^{-z/\delta}$, which is characterized by exponential decay for $z > 0$ with a length scale (δ) determined by the fluid parameters and the oscillation frequency.

(a) Use DA to estimate the **viscous penetration depth** or **skin depth** δ by writing $\delta = C_\delta \eta^\alpha \rho_m^\beta \omega^\gamma$ and solve for α, β, γ. Do the dependences on η, ρ_m, ω seem intuitively correct? For future reference, the combination $\nu \equiv \eta/\rho_m$ is variously called the **kinematic viscosity** or **momentum diffusivity.**

(b)* The differential equation governing the motion of the fluid (Lamb 1945, section 345) is

$$\rho_m \frac{\partial v_x(z,t)}{\partial t} = \eta \frac{\partial^2 v_x(z,t)}{\partial z^2} \tag{3.81}$$

and intuition might suggest a trial solution of the form $v_x(z,t) = v_0 \cos(\omega t - z/\delta)e^{-z/\delta}$, which does satisfy the boundary condition at $z = 0$. Confirm that this functional form does indeed solve the differential equation and evaluate δ from this exact solution to confirm the dimensional dependence found in part (a), as well as evaluating C_δ.

Note: This is one of many examples of the diffusion equation from Eqn. (1.35), where the diffusion constant in this case is $D = \eta/\rho_m = \nu$, hence the name momentum diffusivity.

P3.32 "On the influence of the Earth's rotation on ocean currents"—The Ekman spiral: If the problem in **P3.31** seems somewhat artificial, a physical example with some of the same features, namely an exponential decay in fluid motion as one moves away from a boundary, appears in oceanographic and meteorological applications. In such cases, wind moving over the surface of a fluid provides the "driving force", but the periodicity in time arises from the rotation of the Earth, via the Coriolis force. The equations of motion for the x, y components of velocity, $v_x(z), v_y(z)$ as a function of depth (z) can be written (see, e.g., Stewart 2008, or Holton 2004) as

$$K_z \frac{\partial^2 v_x(z)}{\partial z^2} = -f v_y(z)$$
$$K_z \frac{\partial^2 v_y(z)}{\partial z^2} = +f v_x(z) \tag{3.82}$$

where $f \equiv 2\Omega \sin(\phi)$ is called the **Coriolis parameter** and Ω is the rotation rate of the Earth, $\Omega \equiv 2\pi/\tau$ (with τ being one day) and ϕ is the latitude.

(a) The parameter K_z is called the **eddy viscosity** or **eddy diffusivity** and using the dimensional quantities in the governing differential equations (Eqns. (3.82)) show that it has the same dimensions as the kinematic viscosity or momentum diffusivity $\nu = \eta/\rho_m$ in **P3.31**, namely $[K_z] = \mathrm{L}^2/\mathrm{T}$.

(b)* Show that a trial solution of the form

$$v_x(z) = \cos(z/\delta)\, e^{z/\delta} \qquad \text{and} \qquad v_y(z) = \sin(z/\delta)\, e^{z/\delta} \tag{3.83}$$

solves the coupled differential equations (recall that $z < 0$) and find δ in terms of K_z, f and compare to the results in **P3.31.**

(c) Could you have predicted the dimensional dependence by writing $\delta \propto K_z^{\alpha} f^{\beta}$ (as there are no other dimensionful parameters in the problem) and matching dimensions?

Note: This result was first derived by Ekman (1905) (in a paper with the title of this problem) and is sometimes called the **Ekman spiral** (why spiral?) and has evidently been observed in nature (see, e.g., Chereskin 1995, or Hunkins 1966) and can be demonstrated in the lab (see Beesley et al. 2008).

P3.33 Frictional drag in granular materials: P3.23 showed that the properties of granular materials can be quite different from those of liquids. Another such example involves experiments that

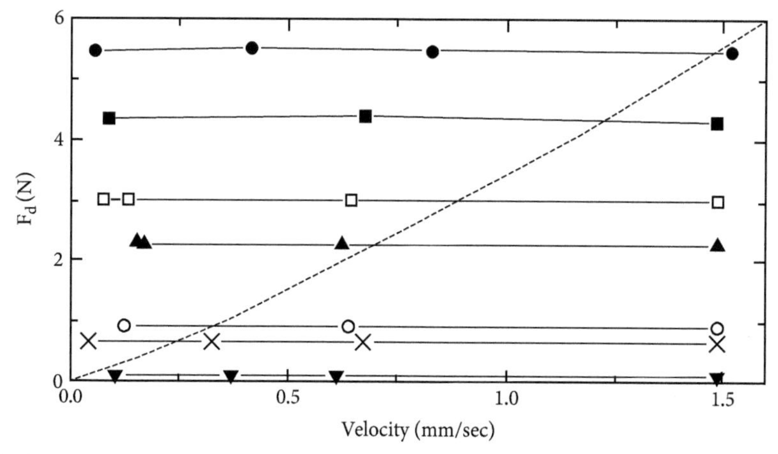

Fig. 3.16 Force required to move a cylindrical probe through granular media as a function of the velocity, for various values of d_{probe}, h_{probe}, showing no dependence on speed. Reprinted with permission from Albert et al. (1999). Copyright (1999) by the American Physical Society. The dashed line corresponds to a Stokes' law ($F \propto v$) prediction.

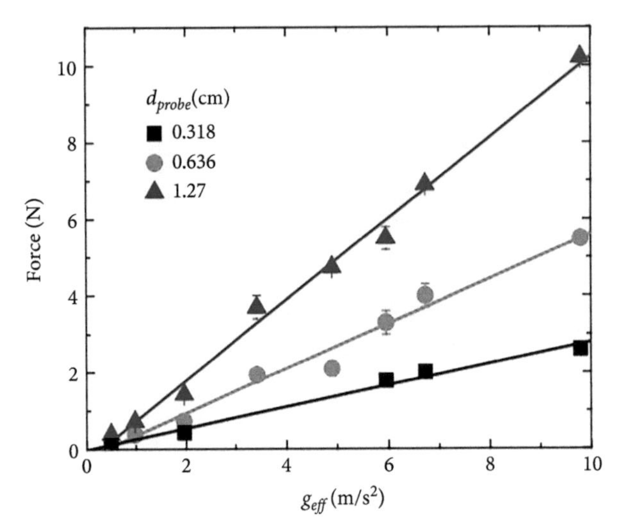

Fig. 3.17 Force required to move a cylindrical probe through granular media as a function of the effective gravitational acceleration (g_{eff}) for various values of the probe diameter, d_{probe}. Reprinted with permission from Costantino et al. (2011). Copyright (2011) by the American Physical Society.

analyze the motion of cylindrical probes moving through granular materials. One can measure the force required to move a probe (F_{probe}) horizontally through glass/polysterene beads of density ρ_m where the cylindrical probe has a diameter d_{probe} and is inserted to a depth h_{probe}. An early version of this experiment (Albert et al. 1999) found that F_{probe} was independent of velocity (so, unlike Stokes' law!) and the data demonstrating this is shown in Fig. 3.16.

A later set of data from the same group arranged that the effective acceleration of gravity could be varied by including various liquids that could produce $g_{eff} = g(1 - \rho_{liquid}/\rho_{grain})$. If we ignore any speed (v) dependence (as already confirmed by experiment) DA suggests that we have

$$F_{probe} = C_g \, \rho_{grain}^{\alpha} \, g_{eff}^{\beta} \, d_{probe}^{\gamma} \, h_{probe}^{\delta} \, . \tag{3.84}$$

(a) Find the constraints on $\alpha, \beta, \gamma, \delta$ from DA, and on γ, δ based on the expected behavior as $d_{probe}, h_{probe} \rightarrow 0$.

(b)† Data from such an experiment (Costantino et al. 2011) shows F_{probe} versus g_{eff} is shown in Fig. 3.17. Is this consistent with your prediction for β? Can you use the data to constrain γ and make a prediction for δ.

Note: Albert et al. (1999) state that this system "... provides the equivalent of a granular Stokes Law for an object moving through a granular medium."

References for Chapter 3

Albert, R., Pfeifer, M. A., Barabási, A-L., and Schiffer, P. (1999). "Slow Drag in a Granular Medium," *Physical Review Letters* 81, 205–208.

Arcos, E., Báez, G., Cuatláyol, P. A., Prian, M. L. H., Méndez-Sánchez, R. A., and Hernández-Saldana, H. (1998). "Vibrating Soap Films: An Analog for Quantum Chaos on Billiards," *American Journal of Physics* 66, 601–607.

Beesley, D., Olejarz, J., Tandom, A., and Marshall, J. (2008). "A Laboratory Demonstration of Coriolis Effects on Wind-Driven Ocean Currents," *Oceanography* 21, 72–76.

Belousov, A., and Belousova, M. (2018). "Dynamics and Viscosity of 'a'ā and Pāhoehoe Lava Flows of the 2012–2013 Eruption of Tolbachik Volcano, Kamchatka (Russia)," *Bulletin of Volcanology* 80, 6.

Bergmann, L. (1956). "Experiments with Vibrating Soap Membranes," *American Journal of Physics* 28, 1043–1047.

Betz, A. (1920). "Das Maximum der theoretisch möglichen Ausnützung des Windes durch Windmotoren," *Zeitschrift für das gesamte Turbinenwesen* 26, 307–309. See the translation by Hamann, H., Thayer, J., and Schaffarczyk, A. P. P. (2013). "The Maximum of the Theoretically Possible Exploitation of Wind by Means of a Wind Motor," *Wind Engineering*, 37 441–6.

Beverloo, W. A., Leninger, H. A., and van de Velde, J. (1961) "The Flow of Granular Solids through Orifices," *Chemical Engineering Science* 15, 260–9.

Brown, P. P., and Lawler, D. F. (2003). "Sphere Drag and Settling Velocity Revisited," *Journal of Environmental Engineering* 129, 222–31.

Butikov, E. (1999). "The Rigid Pendulum: An Antique but Evergreen Physical Model," *European Journal of Physics* 20, 429–41.

Caplan, M. E., Schneider, A. S., and Horowitz, C. J. (2018). "Elasticity of Nuclear Pasta," *Physical Review Letters* 121, 132701.

Cathles, L. M. (1975). *Viscosity of the Earth's Mantle* (Princeton: Princeton University Press).

Cazaubiel, A., et al. (2019). "Wave Turbulence on the Surface of a Fluid in a High-Gravity Environment," *Physical Review Letters* 123, 244501.

Chandrasekhar, S. (1955). "The Character of the Equilibrium of an Incompressible Heavy Viscous Fluid of Variable Density," *Mathematical Proceeding of the Cambridge Philosophical Society* 51, 162–78.

Chereskin, T. K. (1995). "Direct Evidence for an Ekman Balance in the California Current," *Journal of Geophysical Research* 100, C9 18261–9.

Chester, W., Breach, D. R., and Proudman, I. (1969). "On the Flow Past a Sphere at Low Reynolds Number," *Journal of Fluid Mechanics* 37, 751–60.

Costantino, D. J., Bartell, J., Scheidler, K., and Schiffer, P. (2011). "Low-Velocity Granular Drag in Reduced Gravity," *Physical Review E* 83, 011305. (Authors note: I'm very pleased to say that the second and third authors were Penn State undergraduate Physics majors at the time this research was performed.)

Crawford, F. S. (1968). *Waves: Berkeley Physics Course Volume 3* (New York: McGraw-Hill).

Dey, S., Ali, S. Z., and Padhi, E. (2019). "Terminal Fall Velocity: The Legacy of Stokes from the Perspective of Fluvial Hydraulics," *Proceedings of the Royal Society A: Mathematical, Physical and Engineering Sciences* 475, 20190277.

Dolz, M., Hernandez, M. J., Delegido, J., and Casanovas, J. (2006). "A Laboratory Experiment on Inferring Poiseuille's Law for Undergraduate Students," *European Journal of Physics* 27, 1083–1089.

Eggers, J. (1997). "Nonlinear Dynamics and Breakup of Free-Surface Flows," *Reviews of Modern Physics* 69, 865–929.

Ekman, V. W. (1905). "On the Influence of the Earth's Rotation on Ocean Currents," *Archive for Mathematics, Astronomy and Physics* 2, 1–52.

Feynman, R. P., Leighton, R. B., and Sands, M. (1964). *The Feynman Lectures on Physics Vol. II - Mainly Electromagnetism and Matter* (Reading: Addison-Wesley). The famous Feynman lectures are now available free online at https://feynmanlectures.caltech.edu/info/.

Fonda, E., and Sreenivasan, K. R. (2017). "Unmixing Demonstration with a Twist: A Photochromic Taylor–Couette Device," *American Journal of Physics* 85, 796–800.

Gillessen, S., et al. (2017), "An Update on Monitoring Stellar Orbits in the Galactic Center," *Astrophysical Journal* 837, 30.

Gittins, F., and Andersson, N. (2021). "Modelling Neutron Star Mountains in Relativity," *MNRAS* 507, 116–28.

Gratton, J., and Perazzo, C. A. (2007). "Applying Dimensional Analysis to Wave Dispersion," *American Journal of Physics* 75, 158–60.

Hagen, G. H. L. (1852). "Über den Druck und die Bewegung des trocknen Sandes," *Bericht über die zur Bekanntmachung geeigneten Verhandlungen der Königlich Preussischen Akademie der Wissenschaften zu Berlin*, 35–42. Note: This work is more easily accessible as the translation by Tighe, B. P., and Sperl, M. (2007). "Pressure and Motion of Dry Sand: Translation of Hagen's Paper from 1852," *Granular Matter* 9, 141–4.

Hecht, E. (2021). "The True Story of Newtonian Gravity," *American Journal of Physics* 89, 683–92.

Heiskanan, W. A., and Vening Meinesa, F. A. (1958). *The Earth and Its Gravity Field* (New York: McGraw-Hill).

Heller, J. P. (1960). "An Unmixing Demonstration," *American Journal of Physics* 28, 348–53. See also Fonda and Sreenivasan (2017) for a more recent discussion and many online references.

Holton, J. R. (2004). *An Introduction to Dynamic Meteorology*, 4th ed. (Burlington, MA: Elsevier Academic Press).

Hunkins, K. (1966). "Ekman Drift Currents in the Arctic Ocean," *Deep-Sea Research* 13, 607–20.

Huygens, C. (1986). Christiaan Huygens' the Pendulum Clock, or, Geometrical Demonstrations Concerning the Motion of Pendula as Applied to Clocks. Translated by R. J. Blackwell (Ames, IA: Iowa State University Press).

Inglis, D. R. (1979). "A Windmill's Theoretical Maximum Extraction of Power from the Wind," *American Journal of Physics* 47, 416–20.

Isenberg, C. (1992). *The Science of Soap Films and Soap Bubbles* (New York: Dover Publications).

Jeans, J. N. (1902). "The Stability of a Spherical Nebula," *Philosophical Transactions of the Royal Society A: Mathematical, Physical and Engineering Sciences* 199, 1–53.

Keim, N. (2020). "Non-Linear, Granular, and Fluid Physics." In *Experimental Physics: Principles and Practice for the Laboratory*, edited by W. F. Fox, 327–40. (Boca Raton, FL: CRC Press).

Klemens, P. G. (1984). "Dispersion Relations for Waves on Liquid Surfaces," *American Journal of Physics* 52, 451–2.

Lamb, H. L. (1945). *Hydrodynamics*, 6th ed. (New York: Dover Publications).

Lekner, J. (2019). "Laminar Viscous Flow through Pipes, Related to Cross-Section Area and Perimeter Length," *American Journal of Physics* 87, 791–5.

Lohse, D. (2022). "Fundamental Fluid Dynamics Challenges in Inkjet Printing," *Annual Review of Fluid Mechanics* 54, 349–82.

Mankoc, C., et al. (2007). "The Flow Rate of Granular Materials through an Orifice," *Granular Matter* 9. 407–14.

Marion, J. B., and Thornton, S. T. (2004). *Classical Dynamics of Particles and Systems*, 3rd ed. (Boston: Cengage).

Millikan, R. A. (1913). "On the Elementary Electrical Charge and the Avogadro Constant," *Physical Review* 2, 109–43.

Morin, D. (2014). *Problems and Solutions in Introductory Mechanics* (Scotts Valley, CA: CreateSpace).

Onsager, L. (1944). "Crystal Statistics. I. A Two-Dimensional Model with an Order–Disorder Transition," *Physical Review* 65, 117–49.

Pelka D. G., Park R. T., and Singh R. "Energy from the wind," *American Journal of Physics* 46, 495–8.

Pine, D. J., Gollub, J. P., Brady, J. F., and Leshansky, A. M. (2005). "Chaos and Threshold for Irreversibility in Sheared Suspensions," *Nature* 438, 997–1000.

Pritchard, P. J. (2011). *Fox and McDonald's Introduction to Fluid Mechanics*, 8th ed. (Hoboken, NJ: Wiley).

Proudman, J. (1953). *Dynamical Oceanography*, Chapter 11 (London: Methuen).

Purcell, E. M. (1977). "Life at Low Reynolds Number," *American Journal of Physics* 45, 3–11.

Rao, K. K., and Nott, P. R. (2008). *An Introduction to Granular Flow* (Cambridge: Cambridge University Press).

Sack, A., and Pöschel, T. (2017). "Dripping Faucet in Extreme Spatial and Temporal Resolution," *American Journal of Physics* 85, 649–54.

Saleh, K., Golshan, S., and Zarghami, R. (2018). "A Review of Gravity Flow of Free-Flowing Granular Solids in Silos: Basics and Practical Aspects," *Chemical Engineering Science* 192, 1011–1035.

Scheuer, P. A. G. (1981). "How High Can a Mountain Be?", *Journal of Astrophysics & Astronomy* 2, 165–9.

Stewart, R. H. (2008). *Introduction to Physical Oceanography* (Open Textbook Library).

Stokes, G. G. (1851). "On the Effect of Internal Friction of Fluids on the Motion of Pendulums," *Transactions of the Cambridge Philosophical Society* 9, Part II, Number IX, 8–106.

Taklo, T. M. A., Trulsen, K., Gramstad, O., Krogstad, H., and Jensen, A. (2015). "Measurement of the Dispersion Relation for Random Surface Gravity Waves," *Journal of Fluid Mechanics* 766, 326–36.

Taylor, G. I. (1923). "Stability of a Viscous Liquid Contained between Two Rotating Cylinders," *Philosophical Transactions of the Royal Society A: Mathematical, Physical and Engineering Sciences* 223, 289–343.

Taylor, G. I. and Friedman, J. (1966). *Low Reynolds Number Flows* (Cambridge, MA: National Committee on Fluid Mechanics Films, Encyclopedia Britannica Educational Corp.); available as Homsy, G. (ed.). (2000). "G. I. Taylor and Kinematic Reversibility," in *Multi-Media Fluid Mechanics* CD-ROM (Cambridge: Cambridge University Press). See also Taylor, G. I. (1967). *Film Notes for Low Reynolds-Number Flows* (Cambridge, MA: National Committee for Fluid Mechanics Films); http://web.mit.edu/hml/ncfmf/07LRNF.pdf, accessed November 2, 2023.

Turcotte, D. L., and Schubert, G. (2002). *Geodynamics*, 2nd ed. (Cambridge: Cambridge University Press).

Vollmer, M., and Möllmann, K-P. (2013). "Is There a Maximum Size of Water Drops in Nature?", *Physics Teacher* 51, 400–402.

Yersel, M. (2000). "The Flow of Sand," *Physics Teacher* 38, 290–1.

Zhu, F., Miao, R., Xu, C., and Cao, Z. (2007). "Measurement of the Dispersion Relation of Capillary Waves by Laser Diffraction," *American Journal of Physics* 75, 896–8.

Chapter 4

Electricity and magnetism (E&M)

Just as we did for Chapter 3 on classical mechanics, we focus here on the use of dimensional analysis (DA) techniques for the solution (partial though it might be) of standard textbook problems in electricity and magnetism (E&M), but also consider applications in both physics-adjacent fields and research areas beyond the canonical curricular topics. And just as one starts with the topic of kinematics (how to describe motion) in mechanics, we begin our DA-enabled study of electricity and magnetism by exploring some of the many possible motions of charged particles in electric and/or magnetic fields.

4.1 Motion in electric and magnetic fields

4.1.1 Motion in electric fields

Let us first discuss the simplest possible motion of a charged particle in an E or M field, namely the case of a spatially uniform and constant in time electric field, $\boldsymbol{E}_0$, with the initial conditions of the particle given by $\boldsymbol{x}(0) = \boldsymbol{x}_0$ and $\dot{\boldsymbol{x}}(0) = \boldsymbol{v}_0$. There is little real need for a DA treatment here since Newton's law, $q\boldsymbol{E}_0 = \boldsymbol{F} = m\ddot{\boldsymbol{x}}(t)$, is easily solved to give

$$\boldsymbol{x}(t) = \frac{qt^2}{2m}\boldsymbol{E}_0 + \boldsymbol{v}_0 t + \boldsymbol{x}_0, \tag{4.1}$$

exactly analogous to the motion of any other constant force $\boldsymbol{F}$. But we can at least note that Eqn. (4.1) is dimensionally homogeneous, as it must be, with all terms having dimension L.

The case of an oscillating in time, but otherwise spatially uniform, field of the form $\boldsymbol{E}(t) = \boldsymbol{E}_0\cos(\omega t)$ is perhaps the simplest example of a time-dependent field, and is also easily solved (with the same initial conditions leading to Eqn. (4.1)) to give

$$\boldsymbol{x}(t) = \frac{q}{m\omega^2}\{1 - \cos(\omega t)\}\boldsymbol{E}_0 + \boldsymbol{v}_0 t + \boldsymbol{x}_0. \tag{4.2}$$

We see that the coefficients of the $\boldsymbol{E}_0$ terms in both cases have the same dimensions, as they must, but are realized in quite different ways. In the second case of Eqn. (4.2), the charged particle oscillates with (angular) frequency ω around the otherwise uniform motion of the guiding center, $\boldsymbol{x}(t) = \boldsymbol{v}_0 t + \boldsymbol{x}_0$, while in the first there is uniform acceleration with $\boldsymbol{x}(t) \propto t^2$, with the powers of t and $1/\omega$ playing the same role, at least dimensionally.

The dimensional dependence of the first term in Eqn. (4.2) can be explored "from scratch" by writing

$$\begin{aligned} \boldsymbol{x}(t) &\propto q^\alpha m^\beta \omega^\gamma \boldsymbol{E}_0^\delta t^\epsilon \\ \mathsf{L} &= \mathsf{Q}^\alpha \mathsf{M}^\beta \left(\frac{1}{\mathsf{T}}\right)^\gamma \left(\frac{\mathsf{ML}}{\mathsf{QT}^2}\right)^\delta \mathsf{T}^\epsilon, \end{aligned} \tag{4.3}$$

Dimensional Analysis Across the Landscape of Physics. Richard W. Robinett, Oxford University Press. © Richard W. Robinett (2024).
DOI: 10.1093/oso/9780192867551.003.0004

so that matching dimensions gives the constraints

$$\begin{aligned}
\mathsf{M}: &\quad 0 = \quad \beta \quad + \delta \\
\mathsf{L}: &\quad 1 = \qquad\quad +\delta \\
\mathsf{T}: &\quad 0 = \quad -\gamma - 2\delta + \epsilon \\
\mathsf{Q}: &\quad 0 = \alpha \qquad - \delta.
\end{aligned} \tag{4.4}$$

The solutions are then $\alpha = \delta = 1$, $\beta = -1$, and $\gamma = \epsilon - 2$, so that

$$\boldsymbol{x}(t) \propto \frac{q}{m\omega^2}\boldsymbol{E}_0\,(\omega t)^{\epsilon} \quad \longrightarrow \quad \boldsymbol{x}(t) = \frac{q}{m\omega^2}\boldsymbol{E}_0 F\,(\Pi = \omega t)\,, \tag{4.5}$$

so that all of the exponents are fixed by a DA analysis, save for the time-dependence. An appropriate dimensionless ratio, $\Pi = \omega t$, is found, with $F(\Pi) = 1 - \cos(\Pi)$ in the exact solution. It is interesting to note that a purely DA approach also gives the appropriate relation between $\boldsymbol{x}$ and $\boldsymbol{E}_0$, namely that $\boldsymbol{x} \propto \boldsymbol{E}_0$, necessary given that they are the only two vectors in the problem. Example 4.1 shows that such geometric requirements can be used to supplement an analysis based solely on dimensions.

Example 4.1 Ponderomotive force

A more sophisticated problem occurs when there is also a spatial dependence of the time-dependent driving field, namely when $\boldsymbol{E}_0 = \boldsymbol{E}_0(\boldsymbol{r})$. In that case, the particle during an oscillation cycle will find itself in areas of different electric field strengths, and so, averaged over one period (or many cycles), can feel a net attraction or repulsion: this is typically called the **ponderomotive force**. Wikipedia[1] has a succinct description of this effect:

> In physics, a ponderomotive force is a non-linear force that a charged particle experiences in an inhomogeneous oscillating electromagnetic field. It causes the particle to move towards the area of weaker field strength, rather than oscillating around an initial point as happens in a homogeneous field.

While the derivation of the explicit form of the ponderomotive force, $\boldsymbol{F}_p$, is not beyond the level of most E&M books (see, e.g., Chen 2018) we wish to approach this problem using the constraints imposed by DA.

Since this effect is perhaps most prominently used in the acceleration of charged particles, we assume that we are focusing on electrons (with charge $-e$ and mass m_e), with an oscillating (frequency ω) space-dependent external field, $|\boldsymbol{E}_0(\boldsymbol{r})| = E_0(\boldsymbol{r})$. Given that the effect depends on the spatial variation of the electric field ("... towards the area of the weaker field strength ...") we also assume that a gradient operator is present, using only the fact that $\boldsymbol{\nabla}$ has dimensions given by $[\boldsymbol{\nabla}] = 1/\mathsf{L}$. If we then write $\boldsymbol{F}_p \propto e^{\alpha} m_e^{\beta} \omega^{\gamma} \boldsymbol{E}_0^{\delta} (\boldsymbol{\nabla})^{\epsilon}$, we find the dimensional constraint

$$\frac{\mathsf{ML}}{\mathsf{T}^2} = \mathsf{Q}^{\alpha}\mathsf{M}^{\beta}\left(\frac{1}{\mathsf{T}}\right)^{\gamma}\left(\frac{\mathsf{ML}}{\mathsf{QT}^2}\right)^{\delta}\left(\frac{1}{\mathsf{L}}\right)^{\epsilon}, \tag{4.6}$$

[1] See https://en.wikipedia.org/wiki/Ponderomotive_force – accessed July 25, 2023.

which has five unknowns with four constraints (on M, L, T, Q), so we choose to solve for $\alpha, \beta, \gamma, \delta$ in terms of ϵ to find

$$F_p \propto e^{1+\epsilon} m_e^{-\epsilon} \omega^{-2\epsilon} E_0^{1+\epsilon} \nabla^{\epsilon}. \tag{4.7}$$

An obvious special case of this relation is when $\epsilon = 0$, so that the force is simply $\boldsymbol{F} = e\boldsymbol{E}_0$ in the usual way.

If instead we assume that the effect is indeed "non-linear" in the field strength (so that $1 + \epsilon > 1$) and that only one derivative is what's needed ($\epsilon = 1$, again, citing "... towards the area of the weaker field strength ...", we are left with the result

$$F_p \propto \frac{e^2 E_0^2 \nabla}{m_e \omega^2}, \tag{4.8}$$

which uniquely specifies the dependences on the (scalar) quantities e, m_e, and ω. But since this relation is ultimately a vector equation, there are many combinations of ∇ (one factor) and $\boldsymbol{E}_0(\boldsymbol{r})$ (appearing twice) which could be combined to produce a vector force, including $\boldsymbol{E}_0 \times (\nabla \times \boldsymbol{E}_0)$, $(\nabla \cdot \boldsymbol{E}_0)\, \boldsymbol{E}_0$, $(\boldsymbol{E}_0 \cdot \nabla)\, \boldsymbol{E}_0$, or $\nabla(\boldsymbol{E}_0 \cdot \boldsymbol{E}_0)$. It turns out that several of these forms actually do appear at various stages of an explicit first-principles derivation (see Chen 2018), but combine in the final result (including the important sign and dimensional prefactor) to give

$$\boldsymbol{F}_p = -\frac{e^2}{2m_e\omega^2} \nabla \left(\boldsymbol{E}_0 \cdot \boldsymbol{E}_0\right). \tag{4.9}$$

If one imagines (**P4.1**) from the outset that this effect comes from a **ponderomotive potential**, via $\boldsymbol{F}_p = -\nabla \Phi_p(\boldsymbol{r})$, with the (scalar) $\Phi_p(\boldsymbol{r})$ depending only on e, m_e, $\boldsymbol{E}_0(\boldsymbol{r})$, and ω, then one can immediately obtain all the dimensional dependences.

Example 4.2 Charged hoop

A problem where we can profitably use physical intuition in parallel to DA to predict aspects of a particle's motion is illustrated in Fig. 4.1, where a charge Q is spread uniformly on a hoop of radius a. A point charge (q) of mass m is constrained to move along the x-axis and undergoes simple harmonic motion (so that Q and q clearly must have opposite signs, implicitly assumed from the start). To find the period or frequency of this motion (we recall from Chapter 1 that it is often best to focus on the angular frequency in such problems), we can write $\omega = C_\omega q^\alpha Q^\beta \epsilon_0^\gamma m^\delta a^\sigma$ (where we use σ as an exponent to avoid confusion with the permittivity ϵ_0), and matching dimensions we find $\delta = \gamma = -1/2$ and $\sigma = -3/2$, with $\beta = 1 - \alpha$, or

$$\omega \propto \frac{q^\alpha Q^{1-\alpha}}{\sqrt{\epsilon_0 m a^3}}. \tag{4.10}$$

It should be obvious that the product of powers of q and Q must appear symmetrically, in the form qQ, in any expression for the force or electric potential defining this problem, and presumably along with the standard $1/4\pi\epsilon_0$ factors, so we argue that $\alpha = 1 - \alpha$ giving $\alpha = 1/2$, so that

$$\omega = C_\omega \sqrt{\frac{qQ}{4\pi\epsilon_0 m a^3}}, \tag{4.11}$$

and the exact result agrees with this and gives $C_\omega = 1$. (You can compare this result to the similar gravitational problem in **P3.11**.)

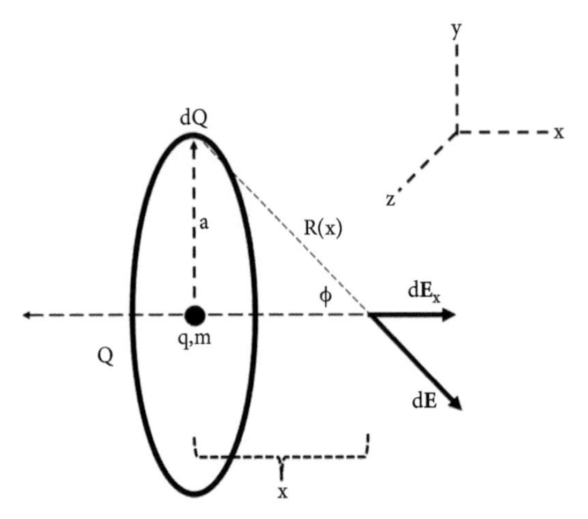

Fig. 4.1 Fixed circular ring, of radius a, with charge Q uniformly applied, with point mass (charge and mass q, m) constrained to move only along the x-axis for **Examples 4.2** and **4.4**. Assume that Q, q have opposite sign charges so that they attract.

4.1.2 **Motion in magneto-static fields**

Turning now to the motion of a charged particle in a spatially uniform (and constant in time) magnetic field, $\boldsymbol{B}_0$, we can fairly easily derive the motions for the parallel and perpendicular (to the field) components of motion. The velocity component parallel to $\boldsymbol{B}_0$ is unaffected (due to the vanishing $\boldsymbol{v} \times \boldsymbol{B}$ cross-product in the Lorentz force law), so that $\boldsymbol{x}_{\parallel}(t) = \boldsymbol{v}_{\parallel,0}t + \boldsymbol{x}_0$.

In the plane perpendicular to the field, the magnetic field provides a radially inward force (as would be felt by a mass attached to a string rotating in a horizontal plane) and there is uniform circular motion at the frequency ω_C (for **cyclotron**) with radius R_L (where L stands for **Larmor**). The dependences of both quantities on q, B_0, m, and $v_\perp$ (since $v_\parallel$ is unaffected by the field) are both amenable to a DA approach, and if we write

$$\omega_C = C_\omega m^{\alpha_1} v_\perp^{\beta_1} q^{\gamma_1} B_0^{\delta_1} \quad \text{and} \quad R_L = C_R m^{\alpha_2} v_\perp^{\beta_2} q^{\gamma_2} B_0^{\delta_2} \tag{4.12}$$

and then match dimensions, we find

$$\omega_C = C_\omega \frac{qB_0}{m} \qquad \text{and} \qquad R_L = C_R \frac{mv_\perp}{qB_0}. \tag{4.13}$$

which are, in fact, the exact results for non-relativistic motion, since the dimensionless constants are $C_\omega = C_R = 1$. You should be able to derive these results (see **P4.4**) from simple Newton's law methods for uniform circular motion. The combination of uniform linear motion in the direction parallel to $\boldsymbol{B}_0$ ($\parallel$) and uniform circular motion in the direction perpendicular ($\perp$) describes a helical trajectory in space.

We note for future reference that the angular frequency (and hence the period of rotation) is independent of the **amplitude** (here meaning the value of R_L, via its $v_\perp$ dependence) similarly to the harmonic oscillator, and this analogy plays a role in the quantum description of the motion of a charged particle in a magnetic field, as discussed in **Example 8.2.**

Example 4.3 Drift velocity

For a charged particle in a uniform magnetic field, the impact of other external forces can often be to add a uniform **drift velocity**, $\mathbf{v}_d$, to the otherwise helical path. For example, if we add a uniform gravitational field given by $\mathbf{g}$ (thereby giving its direction and magnitude, with $[g] = \mathsf{L/T^2}$) for a particle of mass m, we can explore the dimensional dependence of the drift speed by writing $v_d = C_v m^\alpha q^\beta g^\gamma B_0^\delta$. Matching dimensions, the exponents are constrained to be $\alpha = \gamma = +1$ and $\beta = \delta = -1$, so we might naively first write

$$v_d \propto m^1 q^{-1} g^1 B_0^{-1} = \frac{mg}{qB_0}, \tag{4.14}$$

seemingly with a vector quantity in the denominator (at least in terms of dimensions). Given that there are three vector quantities in the problem, $\mathbf{v}_d$, $\mathbf{g}$, and $\mathbf{B}_0$, we also need to ensure that an appropriate combination of vector operations (whether dot-products or cross-products) is utilized (as in **Example 4.1**) and we find that the vectorially (and dimensionally) correct combination is

$$\mathbf{v}_d^{(g)} = \frac{m}{q}\frac{\mathbf{g} \times \mathbf{B}_0}{|\mathbf{B}_0|^2}, \tag{4.15}$$

and once again, fortuitously, $C_v = 1$.

Given that the combination $\mathbf{F}_g = m\mathbf{g}$ appears in Eqn. (4.15), it seems natural that other external constant forces can be substituted, so that we'd have for any given $\mathbf{F}$, and especially for an added electrostatic force $\mathbf{F}_q = q\mathbf{E}$, the results

$$\mathbf{v}_d^{(F)} = \frac{\mathbf{F} \times \mathbf{B}_0}{q|\mathbf{B}_0|^2} \quad \text{and} \quad \mathbf{v}_d^{(E)} = \frac{\mathbf{E} \times \mathbf{B}_0}{|\mathbf{B}_0|^2}. \tag{4.16}$$

The second combination in Eqn. (4.16) is related to the problem of a **velocity selector** or **Wien filter,**[2] which is used to select ions of a specific speed in a mass-spectrometer and in other experimental applications, independent of either the charge (q) or mass (m) of the ions.

An even more important example where the second relation in Eqn. (4.16), $|\mathbf{E}| = |\mathbf{v}||\mathbf{B}|$, is invoked occurs in the **classical Hall** (1879) **effect**, which is illustrated in Fig. 4.2. A long, thin (thickness t and width w, $t << w$) sample of conductor with current I flowing along its length has a magnetic field (B) applied perpendicular to the surface. The moving charge carriers (now known to be electrons, but not understood in Hall's time) are deflected by the $\mathbf{B}$ field, giving rise to a voltage (V_H for Hall) and electric field (E_H) across the sample, such that in the steady state the Lorentz force $\mathbf{F} = q(\mathbf{E} + \mathbf{v} \times \mathbf{B}) = 0$.

The experimentally measured ratio of the observed **Hall voltage** to the current provided is defined as the **Hall resistance.** This quantity could, in principle, depend on the applied field B, the electric charge e, the number density of carriers n_e, and the geometry of the sample (via t, w), so we can write $R_H = C_R B^\alpha e^\beta n_e^\gamma w^\delta t^\epsilon$. DA can only get us so far, as the last three quantities all have dimensions directly related to length (recall that $[n_e] = 1/\mathsf{L}^3$), so matching dimensions gives us the constraint

[2] See Plies et al. (2011) for a brief historical review of this device.

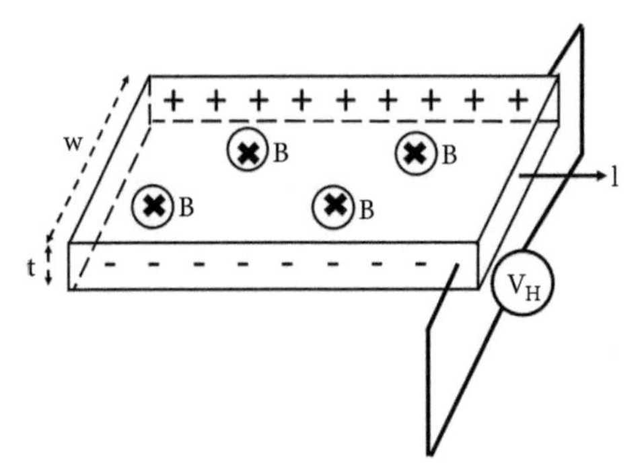

Fig. 4.2 Current-carrying (I) conducting ribbon of width w and thickness t with an external magnetic field B applied perpendicular to the "long" direction. The classical $q\mathbf{v} \times \mathbf{B}$ Lorentz force causes charge to be deflected, building up ± excess on opposite sides. The resulting Hall voltage, V_H, gives an electric field across the strip, which balances the magnetic force.

$$R_H \propto \frac{B}{e} n_e^{\gamma} w^{\delta} t^{2+3\gamma-\delta}. \tag{4.17}$$

Experiments (or a theoretical analysis, as in **P4.7**) find that R_H does not depend on w, so that $\delta = 0$. Since the geometry is quasi-two dimensional ($w >> t$), it can be argued that the result should depend on the two-dimensional number density, that is, on the product $n_e \cdot t$. If so, then we have the additional constraint that $\gamma = \epsilon = 2 + 3\gamma$ or $\gamma = -1$ and

$$\frac{V_H}{I} = R_H = \frac{B}{en_e t}, \tag{4.18}$$

which is the correct answer, with no dimensionless prefactors. Experiments probing the Hall effect can provide important information on the nature of the charge carriers (their number density, n_e, and more notably their sign, confirming that they are electrons) and can be performed even at the undergraduate level; see Armentrout (1990) and **P4.7** for examples. *The Hall Effect and its Applications* by Chien and Westgate (1980) was the result of a symposium celebrating Hall's original 1879 discovery and surveyed why it had become an "... indispensable tool in the studies of many branches of condensed matter physics, especially in metals, semiconductors, and magnetic solids." When similar experiments are done at much higher fields and lower temperatures, novel features are observed, known collectively as the **quantum Hall effect**, to which we return in Chapter 8.

4.2 Sources of electric and magnetic fields

The laws governing the production of electric and magnetic fields by point charges and (short) current loops, namely Coulomb's law and the Biot–Savart equation, are a good starting point as they make use of the fundamental constants ϵ_0 and μ_0, respectively, and have seemingly similar forms. They can then be used to constructively evaluate the $\boldsymbol{E}, \boldsymbol{B}$ fields for almost any configuration of charges and currents, either by direct calculation with closed-form results for simple geometries, or numerical evaluation for others, but at all stages they can be approached using the methods of DA.

4.2.1 Electrostatic fields

A discussion of electric fields arising from (static) configurations of electric charge begins by invoking Coulomb's law for $\boldsymbol{E}$ due to a point charge q, and the related electric potential by writing

$$\boldsymbol{E}_C(\boldsymbol{r}) = \frac{q\boldsymbol{r}}{4\pi\epsilon_0 r^3} = \frac{q\hat{\boldsymbol{r}}}{4\pi\epsilon_0 r^2} \quad \text{and} \quad V_C(\boldsymbol{r}) = \frac{q}{4\pi\epsilon_0 r} \quad \text{with} \quad \boldsymbol{E}_C(\boldsymbol{r}) = -\nabla V_C(\boldsymbol{r}). \quad (4.19)$$

We can then evaluate the field for more complex charge distributions by integrating the contributions $d\boldsymbol{E}$ due to small *cells* of charge given by $dq = \rho_e(\boldsymbol{r})\, dV$ via

$$\boldsymbol{E}(\boldsymbol{r}) = \int \frac{\rho_e(\boldsymbol{r}')(\boldsymbol{r} - \boldsymbol{r}')d\boldsymbol{r}'}{4\pi\epsilon_0 |\boldsymbol{r} - \boldsymbol{r}'|^3}, \quad (4.20)$$

and there are a few cases (usually geometries with a high degree of symmetry) where this integral can be done in closed form, but of course it can also be done numerically.

Instead, let us approach even the Coulomb's law problem via DA, assuming initially that the only three-dimensional quantities in the problem are $q, \epsilon_0, \boldsymbol{r}$, so that there are no other length scales involved, suggestive of the point charge being assumed. We can then expand on this type of analysis to make contact with some of the more familiar exemplary textbook solutions of Eqn. (4.20). If we write $E_C = C_E q^\alpha \epsilon_0^\beta r^\gamma$ (assuming only the magnitudes of $\boldsymbol{E}, \boldsymbol{r}$ are constrained by DA), we would match dimensions to find

$$\frac{ML}{QT^2} = Q^\alpha \left(\frac{Q^2T^2}{ML^3}\right)^\beta L^\gamma, \quad (4.21)$$

which has more constraint equations than powers, and the simple solution $\alpha = 1$, $\beta = -1$, and $\gamma = -2$, and if we assume that the 4π factor comes along with ϵ_0 (so, $C_E = 1/4\pi$) we do reproduce the simple result for a point charge in Eqn. (4.19), and with only two vectors, the fact that $\boldsymbol{E}$ and $\boldsymbol{r}$ are proportional, at least via the $\hat{\boldsymbol{r}}$ term, we've found most of Coulomb's law, even if only as a mnemonic device.

For any example beyond a single point charge, there must be some characteristic length scale (say a) associated with either an extended continuous charge distribution (**Examples 4.4** and **4.5** below), or with an assortment of separated point charges (**Example 4.6**), so that a DA formula would be modified to $E = C_E q^\alpha \epsilon_0^\beta r^\gamma a^\delta$. We can directly extend the result above from $\gamma = -2 \rightarrow \gamma + \delta = -2$, giving

$$E \propto q^1 \epsilon_0 r^{-2-\delta} a^\delta \quad \text{or} \quad E = \left(\frac{q}{4\pi\epsilon_0 r^2}\right)\left(\frac{a}{r}\right)^\delta \rightarrow \left(\frac{q}{4\pi r^2}\right) F\left(\Pi = \frac{a}{r}\right) \quad (4.22)$$

with a Buckingham Pi value given, not surprisingly, by $\Pi = a/r$ or the dimensionless ratio of the two lengths in the problem. Let us next explore two cases (of high symmetry) of an extended charge distribution to see how this ratio can be realized.

Example 4.4 Uniformly charged sphere

Instead of a point charge, let us imagine that the charge q is spread uniformly over the volume of a spherical region of radius a, so that the charge density inside the sphere is given by $\rho_e = q/V = q/(4\pi a^3/3)$.

For values of $r > a$, we can use the spherical symmetry and Gauss's law to evaluate the electric field as

$$
\begin{aligned}
\oint_{r>a} \mathbf{E} \cdot d\mathbf{S} &= \frac{q_{enc}}{\epsilon_0} \\
E(r)(4\pi r^2) &= \frac{q}{\epsilon_0} \\
\Downarrow \quad & \quad \Downarrow \\
E_C(r) &= \frac{q}{4\pi\epsilon_0 r^2},
\end{aligned}
\tag{4.23}
$$

since the electric field $\mathbf{E}$ is radially outward, and so parallel to each surface element $d\mathbf{S}$, giving $\mathbf{E} \cdot d\mathbf{r} = E(4\pi r^2)$, and this reproduces the point-charge result. In contrast, for values of $r < a$, namely inside the sphere, the charge enclosed is a volume fraction of the total charge, giving

$$
\begin{aligned}
\oint_{r<a} \mathbf{E} \cdot d\mathbf{S} &= \frac{q_{enc}}{\epsilon_0} = \frac{1}{\epsilon_0} \int_{r<a} \rho_e (4\pi r^2 \, dr) \\
E(4\pi r^2) &= \frac{1}{\epsilon_0} \left\{ \rho_e \left(\frac{4\pi r^3}{3} \right) \right\} = \frac{q}{\epsilon_0} \frac{r^3}{a^3} \\
\Downarrow \quad & \quad \Downarrow \\
E(r) &= \frac{qr}{4\pi\epsilon_0 a^3} = \left(\frac{q}{4\pi\epsilon_0 r^2} \right) \left(\frac{r^3}{a^3} \right) = E_C \left(\frac{a}{r} \right)^{-3}
\end{aligned}
\tag{4.24}
$$

and $F(\Pi = a/r) = \Pi^{-3}$ is a simple power-law.

Example 4.5 Electric field of charged ring: On-axis geometry

Another tractable example of an extended charge distribution is that of a charge Q spread uniformly around a thin ring of radius a, as shown in Fig. 4.1. The electric field along the axis of symmetry, $\mathbf{E}_x(x)$, can be evaluated by summing the contributions from small dQ elements around the ring. Contributions to $\mathbf{E}_y, \mathbf{E}_z$ will cancel pair-wise, due to symmetry, as any y, z component will have an offsetting contribution from a dQ cell on the opposite side of the ring. The individual contributions to $\mathbf{E}_x$ can be written in terms of $R(x) = \sqrt{a^2 + x^2}$ as

$$
\begin{aligned}
dE_x &= \left(\frac{dQ}{4\pi\epsilon_0 R(x)^2} \right) \cos(\phi) = \frac{dq}{4\pi\epsilon_0 R(x)^2} \left(\frac{x}{R(x)} \right) \\
\Downarrow \quad & \quad \Downarrow \\
E_x(x) &= \frac{Q}{4\pi\epsilon_0} \frac{x}{(a^2 + x^2)^{3/2}} \\
E_x(x) &= \left(\frac{Q}{4\pi\epsilon_0 x^2} \right) \left(1 + (a/x)^2 \right)^{-3/2},
\end{aligned}
\tag{4.25}
$$

where the summation of all of the identical dE_x terms gives $\int dQ = Q$. For $a/x << 1$, we clearly reproduce Coulomb's law, and information on the spatial extent of the charge distribution is lost and the far-field value of $\mathbf{E}_x$ depends only on Q and x, while for $x << a$, the field actually vanishes at the origin for obvious symmetry reasons (where would it even point?) In this case the Buckingham Pi function can be evaluated in closed form and is given by $F(\Pi = a/x) = \left(1 + (a/x)^2\right)^{-3/2}$.

Example 4.6 Electric dipole: Far-field result

A conceptually very important charge configuration consists of a pair of opposite charges, $\pm q$, separated by a distance a giving an **electric dipole moment**, as shown in Fig. 4.3. While the net charge on this system is zero, there is still a non-vanishing electric field, which can be easily approximated[3] for $r >> a$, at least on special axes of symmetry. For example, along the dipole axis we can write

$$\begin{aligned} E_x = E_+ + E_- &= \frac{q}{4\pi\epsilon_0(r - a/2)^2} - \frac{q}{4\pi\epsilon_0(r + a/2)^2} \\ &= \frac{q}{4\pi\epsilon_0 r^2}\left\{\left(1 - \frac{a}{2r}\right)^{-2} - \left(1 + \frac{a}{2r}\right)^{-2}\right\} \\ &= \frac{q}{4\pi\epsilon_0 r^2}\left\{\left(1 + \frac{a}{r} + \cdots\right) - \left(1 - \frac{a}{r} + \cdots\right)\right\} \\ &\approx 2\frac{(qa)}{4\pi\epsilon_0 r^3} \\ E_x &\approx 2\frac{p}{4\pi\epsilon_0 r^3}, \end{aligned} \tag{4.26}$$

where the magnitude of the **electric dipole moment** $\boldsymbol{p}$ is defined by $p = qa$ and therefore has dimensions given by $[p] = [qa] = \mathsf{QL}$. This expression is consistent with Eqn. (4.22) with $F(\Pi = a/r) \propto \Pi^1$ and arises because of the cancellation of the Coulomb field of two oppositely charged objects, when the total (or **monopole**) charge vanishes, but there is a higher-order (in this case **dipole**) field that decreases at large distances faster than the $1/r^2$ Coulomb scaling.

This behavior is not unique to the on-axis field, as we can easily calculate the long-distance electric field along the axis perpendicular to the dipole moment. For example, from Fig. 4.3, if we write $d = \sqrt{r^2 + (a/2)^2} \approx r$, and note that $\sin(\phi) = (a/2)/d$, we can evaluate the field as

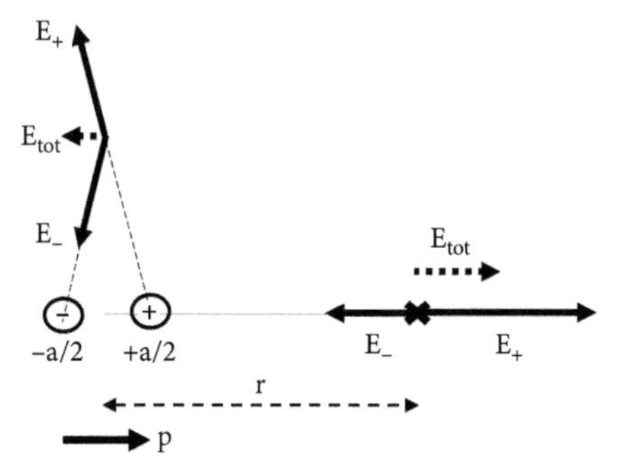

Fig. 4.3 Electric field from a dipole consisting of charges $\pm q$ placed at $(\pm a/2, 0, 0)$. Contributions from the $\pm$ charges to $\boldsymbol{E}_\pm$ are shown separately, as are the total fields ($\boldsymbol{E}_{tot}$) both on-axis, and perpendicular to the dipole moment.

[3] Using say the series expansion in Eqn. (10.26).

$$E_x = -2\left(\frac{q}{4\pi\epsilon_0 d^2}\right)\sin(\phi) = -\frac{qa}{4\pi\epsilon_0 d^3} \approx -\frac{p}{4\pi\epsilon_0 r^3} \tag{4.27}$$

and we can see the (partial, first-order) cancellation between the $\pm q$ field contributions still arise, but in a geometrically different way.

More generally, the electric dipole moment for an array of charges (discrete point charges, or a distribution of charge) can be defined by

$$\boldsymbol{p} = \sum_i q_i \boldsymbol{r}_i \qquad \text{or} \qquad \boldsymbol{p} = \int_V \rho_e(\boldsymbol{r})\, \boldsymbol{r}\, dV, \tag{4.28}$$

both of which clearly have the same dimensions.

A more rigorous analysis of the long-distance electric field from a neutral, but non-trivially distributed charge distribution, is covered in almost all advanced undergraduate textbooks on E&M, where the general expression for the **dipole electric field** is given by

$$\boldsymbol{E}_{dip}(\boldsymbol{r}) = \frac{1}{4\pi\epsilon_0}\left\{\frac{3\hat{\boldsymbol{r}}(\boldsymbol{p}\cdot\hat{\boldsymbol{r}}) - \boldsymbol{p}}{r^3}\right\}, \tag{4.29}$$

where $\hat{\boldsymbol{r}} = \boldsymbol{r}/r$ as always. You should check that this expression reproduces both Eqns. (4.26) and (4.27) for their respective geometries.

A systematic expansion of the electric field of an arbitrary charge distribution is then possible, using higher-order moments, and can be written schematically as

$$E(r >> a) \sim \sum_{n=0}^{\infty} \frac{qa^n}{4\pi\epsilon_0 r^{n+2}}, \tag{4.30}$$

where $n = 0, 1, 2, 3, ...K, ...$ correspond to the **monopole**, **dipole**, **quadrupole**, **octopole**, ..., 2^K**-pole** moments, and this expression is clearly is in the same spirit as Eqn. (4.22). Examples of such higher-order charge configurations are explored in **P4.11** and you can also check that the first-order correction to the Coulomb field (in Eqn. (4.25)) of the charged ring problem in **Example 4.5** is a quadrupolar term, due to the symmetry of the charge distribution.

4.2.2 **Magneto-static fields**

According to the Biot–Savart law, the contribution to the magnetic field from a short length (described by the vector $\boldsymbol{dl}$, which gives both the direction and magnitude) of current (I), with the geometry shown in Fig. 4.4, is given by

$$d\boldsymbol{B} = \frac{\mu_0}{4\pi}\frac{Id\boldsymbol{l}\times\hat{\boldsymbol{r}}}{r^2}, \tag{4.31}$$

where $\hat{\boldsymbol{r}}$ is a unit vector from the current segment to the observation point and r is the distance. This suggests, at face value, that the magnetic field has the same $1/r^2$ dependence as Coulomb's law.

However, a realistic localized current source requires a closed path/loop, which gives rise to cancellations, and we discuss one such example in some detail a little later. But to begin,

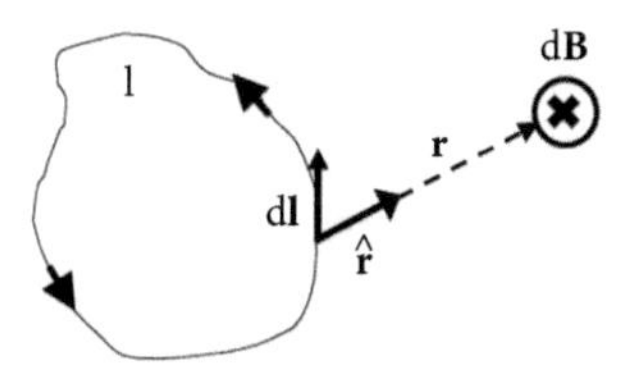

Fig. 4.4 Magnetic field (dB) due to the current (I) segment dl a distance r away. Note the direction of dB determined by the right-hand rule for the cross-product of $dl \times \hat{r}$, where $\hat{r}$ is the unit vector pointing from dl to the observation point.

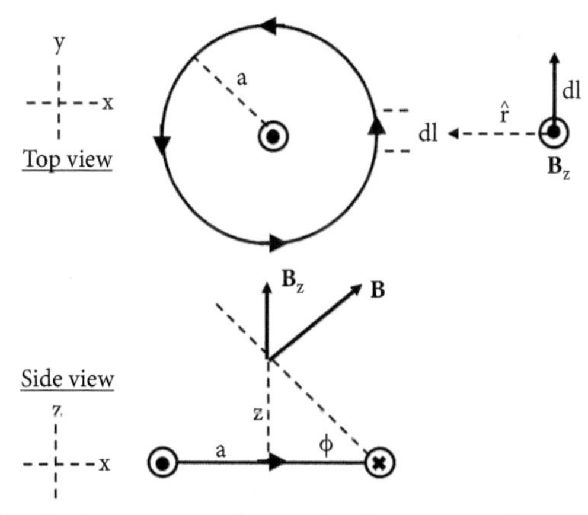

Fig. 4.5 Magnetic field of planar current loop of radius a as a function of z along the axis of symmetry. View from above (top frame) and side view (bottom frame).

we instead explore the dimensional dependence of several well-studied textbook-level problems, with current sources for which there is only **one** relevant length dimension. The most familiar such example by far is the calculation of the magnetic field due to an infinitely long, straight, current-carrying wire, where r is the distance (perpendicular to the wire, of course) to the observation point, and we discuss that geometry in detail in Section 4.2.3. Here we begin with a study of a circular (planar) loop of wire, with fixed radius a, focusing first on the field at the center, then extending the result to along the axis of symmetry, as shown in Fig. 4.5, to compare to the result in the corresponding electric field geometry in **Example 4.5.**

Example 4.7 Current ring

For a planar ring of current, the field at the center must certainly depend on μ_0, and I, and since there is only one length dimension, a, we write $B_{cent} = C_B \mu_0^{\alpha} I^{\beta} a^{\gamma}$. Matching dimensions we find that $\alpha = \beta = +1$ and $\gamma = -1$, so that $B_{cent} \propto \mu_0 I/a$. A constructive calculation using the Biot–Savart formula starts from Fig. 4.1, noting that

$$
\begin{aligned}
d\mathbf{B} &= \frac{\mu_0}{4\pi}\frac{Idl}{a^2}\hat{\mathbf{z}} \\
\mathbf{B} &= \frac{\mu_0 I}{4\pi a^2}\left(\oint dl\right)\hat{\mathbf{z}} \\
\mathbf{B} &= \frac{\mu_0 I}{2a}\hat{\mathbf{z}}
\end{aligned}
\tag{4.32}
$$

since $\mathbf{dl}$ (pointing around the circle) and $\hat{\mathbf{r}}$ (pointing to the center) are orthogonal giving $\mathbf{dl}\times\hat{\mathbf{r}} = dl\,\hat{\mathbf{z}}$, and the (scalar) integral of dl around the circuit just gives the circumference $2\pi a$. This agrees (as it should) with the DA result, giving $C_B = 1/2$, but now with the vector direction specified, about which DA provided no guidance.

To extend this analysis (both in terms of dimensions and explicit calculation) we add a second dimensionful quantity, namely the distance z along the axis of symmetry. A familiar DA approach gives $B(z) \propto (\mu_0 I/a)(z/a)^\delta \to (\mu_0 I/a)F(\Pi = z/a)$, while the Biot–Savart analysis gives

$$
\begin{aligned}
d\mathbf{B}_z &= \frac{\mu_0}{4\pi}\frac{I\,dl}{(a^2+z^2)}\cos(\phi) \\
d\mathbf{B}_z &= \frac{\mu_0}{4\pi}\frac{I\,dl\,a}{(a^2+z^2)^{3/2}} \\
\mathbf{B}_z &= \frac{\mu_0 I a^2}{2(a^2+z^2)^{3/2}} = \frac{\mu_0 I}{2a}F\left(\Pi = \frac{z}{a}\right)
\end{aligned}
\tag{4.33}
$$

since $\cos(\phi) = a/\sqrt{a^2+z^2}$. As usual, this is consistent with a DA analysis with the explicit $F(\Pi = z/a)$ given by $F(\Pi) = (2(1+(z/a)^2)^{3/2})^{-1}$.

Despite the similarity to the "ring of charge" problem in **Example 4.5**, there are important differences, most dramatically in the large z behavior of the magnetic field. For the electrostatic version, the long-distance behavior of the $\mathbf{E}$ field is $E \to q/(4\pi\epsilon_0 z^2)$ and the electric field approaches that of a point charge q, with any information on the structure/distribution of the charge density being lost (at least to leading-order). For the magnetic version, the long-distance behavior ($z >> a$) is given by

$$
B = \frac{\mu_0 I a^2}{2z^3} = \frac{\mu_0 I(\pi a^2)}{2\pi z^3},
\tag{4.34}
$$

which is reminiscent of the **dipole** ($1/z^3$) distance dependence from **Example 4.3**.

While this example is a tractable, exactly calculable result, it doesn't obviously demonstrate the cancellations among circuit elements leading to a dipole dependence. To explore a more general closed-loop geometry of a current loop that does, we next calculate the long-distance behavior of the magnetic field due to a rectangular (planar) geometry, as shown in Fig. 4.6, at two points with $r >> a$, along axes of symmetry, starting with an observation point in the plane of the loop. The contributions to the magnetic field due to the front (f) and back (b) current segments show (we'll add contributions from the sides below) are approximately (for $r >> a$)

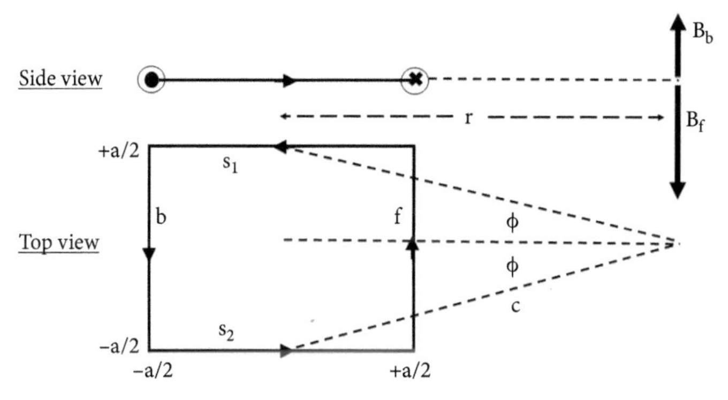

Fig. 4.6 Magnetic field from a square current loop.

$$
\begin{aligned}
B_z = B_{f,z} + B_{b,z} &\approx -\frac{\mu_0 Ia}{4\pi(r - a/2)^2} + \frac{\mu_0 Ia}{4\pi(r + a/2)^2} \\
&\approx -\frac{\mu_0 Ia}{r^2}\left\{\left(1 - \frac{a}{2r}\right)^{-2} - \left(1 + \frac{a}{2r}\right)^{-2}\right\} \\
&\approx -\frac{\mu_0 Ia}{4\pi r^2}\left\{1 + \frac{a}{r} + \cdots - 1 + \frac{a}{r} + \cdots\right\} \\
&\approx 2\frac{\mu_0 \left(Ia^2\right)}{4\pi r^3},
\end{aligned} \tag{4.35}
$$

and indeed the far-field dependence is like the dipole field noted in Eqn. (4.29). More importantly, we can see the cancellation between components in the closed loops (in this case, the front and back ones) clearly, similarly to that in **Example 4.6.**

The quantity here that is analogous to the electric dipole moment is the **magnetic dipole moment,** denoted by m, which (for this specific case) we find to be $m = Ia^2$ and indeed depends on the area of the closed loop, rather than say on the perimeter, necessarily leading to an additional power of r in the denominator. More generally, for a purely planar loop of arbitrary shape, the magnetic dipole moment will be $m = IA$, where A is the enclosed area. The dimensions of such a magnetic moment are clearly $[m] = [IA] = (\mathrm{Q/T})(\mathrm{L}^2) = \mathrm{QL}^2/\mathrm{T}$.

We have considered so far only the two leading terms from the front and back, which illustrate the cancellations inherent in a more general calculation of the far-field for a current loop. For completeness, we can include the contributions from the two sides, which give

$$
B_{z,s} \approx +2\frac{\mu_0 Ia}{4\pi r^2}\sin(\theta) \approx +\frac{\mu_0 Ia^2}{4\pi r^3}, \tag{4.36}
$$

since $\sin(\theta) = (a/2)/r$, which has the same dimensional and spatial dependence as the dominant term arising from the front/back cancellation, but due here to geometrical effects, just as in the second electric dipole example.

In the same way that the electric dipole moment is a vector quantity, the magnetic dipole moment, $\boldsymbol{m}$, has a direction determined by the normal (perpendicular unit vector) to the two-dimensional planar loop, with the familiar **right-hand rule** in play, meaning that wrapping one's fingers around the loop, pointing them along the direction of the current, then has one's (right) thumb pointing in the direction of the vector moment, $\boldsymbol{m}$.

The mathematical expression for the magnetic field of a (classical) point magnetic dipole or the far-field value for a finite loop, is completely analogous to that for the electric dipole field in Eqn. (4.29), namely

$$\boldsymbol{B}_{dip}(\boldsymbol{r}) = \frac{\mu_0}{4\pi}\left\{\frac{3\hat{\boldsymbol{r}}(\boldsymbol{m}\cdot\hat{\boldsymbol{r}}) - \boldsymbol{m}}{r^3}\right\}. \tag{4.37}$$

Adding the results of Eqns. (4.35) and (4.36), we find a dipole (far) field arising from the square loop in Fig. 4.6 to be

$$B_z(f) \approx -2\frac{\mu_0 m}{r^3} + \frac{\mu_0 m}{r^3} = -\frac{\mu_0 m}{r^3}, \tag{4.38}$$

which agrees with the exact expression in Eqn. (4.37) for this geometry, where $\boldsymbol{m} = m\hat{\boldsymbol{z}}$ and $\hat{\boldsymbol{r}} = \hat{\boldsymbol{x}}$, so that $\boldsymbol{m}\cdot\hat{\boldsymbol{r}} = 0$ and only the last term contributes. You should repeat these calculations for the magnetic field at $\boldsymbol{r} = r\hat{\boldsymbol{z}}$ along the z-axis, which is another tractable geometry. You can also confirm that the far-field magnetic field for the circular current loop (along the axis of symmetry) from **Example 4.7** also matches the general form in Eqn. (4.37), with $m = I(\pi a^2)$.

Magnetic fields can arise from the movement of charge in different ways than via current in a circuit, for example, by the rotation of a charged object. Consider a cylindrically symmetric object (such as a sphere or loop or disk) with net charge q that is rotating about its axis at angular velocity ω, and has mechanical properties given by its mass m and radius R. The magnetic moment (here labeled μ to avoid confusion with the mass) arising from such motion can be constrained by writing $\mu \propto q^{\alpha}R^{\beta}\omega^{\gamma}m^{\delta}$ with dimension matching requiring

$$\frac{\mathsf{QL}^2}{\mathsf{T}} = \mathsf{Q}^{\alpha}\mathsf{L}^{\beta}\left(\frac{1}{\mathsf{T}}\right)^{\gamma}\mathsf{M}^{\delta}, \tag{4.39}$$

which defines perhaps the simplest set of linear equations we've encountered so far, and immediately (by inspection) gives $\alpha = \gamma = 1, \beta = 2$, and $\gamma = 0$, namely that $\mu = C_{\mu}qR^2\omega$. The dimensionless C_{μ} depends on the details of the object (shape, mass distribution, etc.)—for a circular hoop, $C_{\mu} = 1$, spherical shell, $C_{\mu} = 1/3$, and for a uniform sphere, $C_{\mu} = 1/5$. The important vector direction of the magnetic moment is given by the standard right-hand rule associated with the angular velocity ω, wrapping ones fingers in the direction of rotation with ones thumb pointing along the vector $\boldsymbol{\omega}$, at least for a positive charge q.

A rotating object of this type will also have a (mechanical) angular momentum $L = C_L m^{\alpha}R^{\beta}\omega^{\gamma}$ and dimensional matching requires

$$\frac{ML^2}{T} = M^{\alpha}L^{\beta}\left(\frac{1}{T}\right)^{\gamma}, \tag{4.40}$$

which also has the trivial solution $\alpha = \gamma = 1$ and $\beta = 2$, so that $L = C_L mR^2\omega$, where again the dimensionless C_L depends on the shape and mass distribution.

It turns out that if the charge and mass densities have the same functional form (at least if the two objects have uniform q and m distributions and the same shapes) then the dimensional prefactors are related, namely $C_{\mu} = C_{\mathcal{L}}/2$. An important quantity in more fundamental physics applications is the **gyromagnetic ratio**, namely $\gamma = \mu/L$, and for the simplest geometry (ring of charge and mass)

$$\gamma = \frac{\mu}{\mathcal{L}} = \frac{qR^2\omega/2}{mR^2\omega} = \frac{q}{2m}. \tag{4.41}$$

4.2.3 **Electric and magnetic fields in "parallel" geometries**

One of the simplest applications of either the Biot–Savart law or Ampère's law is the calculation of the magnetic field of a long (assumed infinite, so one ignores any issues with closing a current loop) straight current-carrying wire, so let us approach that problem using DA, in parallel (no pun intended?) to the calculation of the electric field of a long (again, pretend infinite) line of charge, with linear charge density λ_e, as shown in Fig. 4.7.

We start by assuming very similar forms for the two cases, initially including factors of ϵ_0 and μ_0 for both, writing

$$B(r) \propto I^{\alpha}\epsilon_0^{\beta}\mu_0^{\gamma}r^{\delta} \qquad \text{and} \qquad E(r) \propto \lambda_e^{\overline{\alpha}}\epsilon_0^{\overline{\beta}}\mu_0^{\overline{\gamma}}r^{\overline{\delta}} \tag{4.42}$$

where $[\lambda_e] = \mathsf{Q/(TL)}$ and r is the distance from the wire, the only relevant length scale in the problem. Matching dimensions, we demand that

$$\left(\frac{\mathsf{M}}{\mathsf{QT}}\right) = \left(\frac{\mathsf{Q}}{\mathsf{T}}\right)^{\alpha}\left(\frac{\mathsf{Q^2T^2}}{\mathsf{ML^3}}\right)^{\beta}\left(\frac{\mathsf{ML}}{\mathsf{Q^2}}\right)^{\gamma}(\mathsf{L})^{\delta} \quad \text{and} \quad \left(\frac{\mathsf{ML}}{\mathsf{QT^2}}\right) = \left(\frac{\mathsf{Q}}{\mathsf{L}}\right)^{\overline{\alpha}}\left(\frac{\mathsf{ML}}{\mathsf{Q^2}}\right)^{\overline{\delta}}\left(\frac{\mathsf{Q^2T^2}}{\mathsf{ML^3}}\right)^{\overline{\beta}}(\mathsf{L})^{\overline{\delta}}, \tag{4.43}$$

which are easily solved to find

$$B(r) = c_B\frac{I\mu_0}{r} \qquad \text{and} \qquad E(r) = c_E\frac{\lambda_e}{\epsilon_0 r}. \tag{4.44}$$

We note that this analysis, based purely on dimensions, "knows" that the electric (magnetic) field does not depend on μ_0 (ϵ_0), and gives identical dependences on the charge density/current, their respective fundamental strengths, and their spatial dependence via $1/r$.

The applications of Ampère's law for the magnetic field, and Gauss's law for the electric field, in this geometry are tractable enough that they appear even in many introductory texts, so we reproduce the results here. For the magnetic field we write

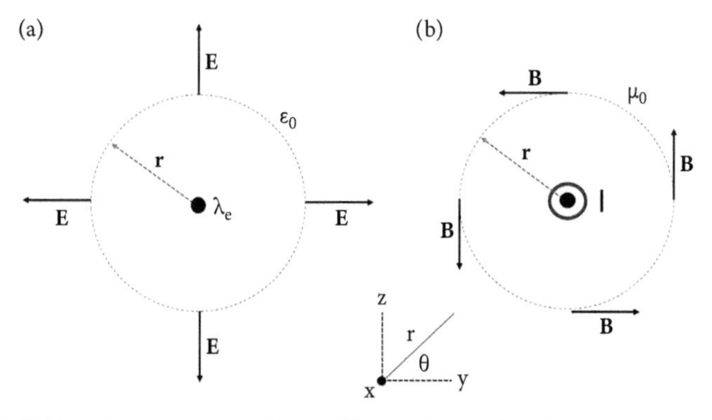

Fig. 4.7 Electric field a distance r away from a linear charge density λ_e on an infinite wire (a) and magnetic field due to an infinitely long current-carrying (I coming out of the plane) wire (b). The infinite wire is along the $\pm x$ direction with (r, θ) describing the (y, z) plane.

$$B(2\pi r) = \oint \mathbf{B} \cdot d\mathbf{l} = \mu_0 I_{enc} \qquad \longrightarrow \qquad B(r) = \frac{\mu_0 I}{2\pi r}, \tag{4.45}$$

while for the electric field, using a Gaussian surface of a cylinder of radius r and length L centered on the wire, we have

$$\begin{aligned} \oint \mathbf{E} \cdot d\mathbf{S} &= \frac{q_{enc}}{\epsilon} \\ E(2\pi r)L &= \frac{1}{\epsilon_0}(\lambda_e L) \\ \Downarrow \quad & \quad \Downarrow \\ E(r) &= \frac{\lambda_e}{2\pi\epsilon_0 r}. \end{aligned} \tag{4.46}$$

We see that DA has correctly captured the dependence on the relevant physical parameters, and the full calculation shows that $C_B = C_E = 1/2\pi$, so that the two geometries have very similar final results. Had we assumed that $1/\epsilon_0$ and μ_0 each bring along with them the standard 4π factors, the values of $C_B = C_E = 2$ would have been appropriate, but again within an order-of-magnitude of unity. DA has nothing to say, however, on the important vector nature of the fields, where we know that

$$\mathbf{E}(\mathbf{r}) = \frac{\lambda_e \hat{\mathbf{r}}}{2\pi\epsilon_0 r} \qquad \text{and} \qquad \mathbf{B}(\mathbf{r}) = \frac{I_e \mu_0 \hat{\boldsymbol{\theta}}}{2\pi r}, \tag{4.47}$$

which are shown in Fig. 4.7. Similar results arise in the problem of the $\mathbf{E}, \mathbf{B}$ fields of a charged/current-carrying **thick** infinite wire (of finite radius a) where we can apply both methods to the case of inside ($r < a$) and outside ($r > a$) the wire, discussed in **P4.13**.

A second (again, highly idealized) geometry corresponds to an infinite (in two dimensions) sheet of charge or current density, as shown in Fig. 4.8. The electric field will depend on the areal charge density or σ_e (with $[\sigma_e] = \mathsf{Q/L^2}$), while the magnetic field requires the linear current density (current per unit length) denoted by j_e (with dimensions $[j_e] = [I/l] = \mathsf{Q/TL}$). The $\mathbf{E}, \mathbf{B}$ fields could presumably depend on the distance away from the plane (z) with directions already suggested in Fig. 4.8, so we might write

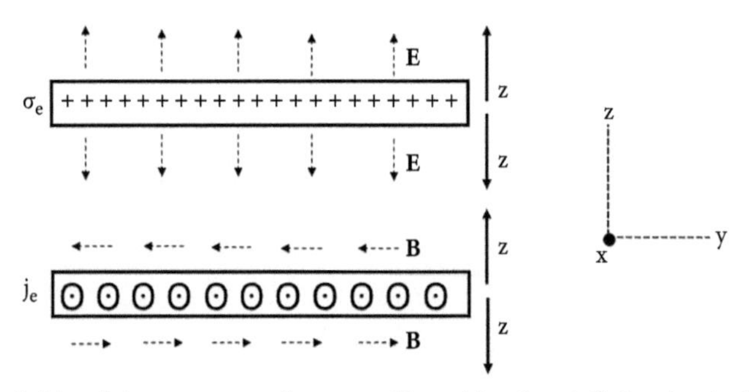

Fig. 4.8 Electric field a distance z away from on either side of an infinite sheet of uniform charge density σ_e (top), and magnetic field due to an infinite sheet of current density (j_e) (bottom).

$$B(z) \propto j_e^{\alpha} \mu_0^{\beta} z^{\gamma} \qquad \text{and} \qquad E(z) \propto \sigma_e^{\overline{\alpha}} \epsilon_0^{\overline{\beta}} z^{\overline{\gamma}} \tag{4.48}$$

and matching dimensions we actually find that $\gamma = \overline{\gamma} = 0$ and

$$B(z) = C_B \mu_0 j_e \qquad \text{and} \qquad E(z) = C_E \frac{\sigma_e}{\epsilon_0} \tag{4.49}$$

with no dependence on z at all. Calculations similar to those above using the integral forms of Ampère's and Gauss's laws find these results with $C_B = C_E = 1/2$, which you can try to verify in **P4.14**, and one can also consider the extension of this planar geometry to "thick" plates as in **P4.16.**

While these geometries may seem very artificial, variations of them are used for two important circuit elements. The structure of a **capacitor** consists of two, oppositely charged, thin charged sheets separated by a small distance (of vacuum or material) with the electric field between the plates getting equal contributions from both plates, giving $E_C = 2(\sigma_e/2\epsilon_0) = \sigma_e/\epsilon_0$, and the fields outside the plates canceling. The parallel (pun intended) situation of two current-carrying sheets, with currents in opposite directions, similarly give $B_{between} = \mu_0 j_e$ and vanishing magnetic field outside the two sheets. The corresponding circuit element making use of these results is a cylindrical **inductor**, where one can imagine wrapping one sheet around a cylindrical form, with a uniform $\boldsymbol{B}$ field pointing along the direction of the axis of symmetry given by $B_L = \mu_0 j_e$. We discuss these two circuit elements in more detail in Example 4.8.

We have considered several very special geometries for several reasons:

- They can make direct/immediate connections between a DA approach and simple implementations of Gauss's and Ampère's law analyses.
- They are connected to technologically relevant device applications (capacitors and inductors, as noted above for the two-dimensional sheet examples) as discussed in many textbooks.
- They naturally lead to discussions of the role of special relativity in Section 4.3.

In advance of that discussion, if we imagine ourselves in a frame of reference moving with speed v with respect to a long, charged wire, we will clearly see a non-zero current related to λ_e and the speed of relative motion v. It therefore might not be surprising that the $\boldsymbol{E}, \boldsymbol{B}$ fields necessarily have the same functional dependence (on r in one case, z in the other) if the relative motion is along the direction of symmetry, with a similar expectation for the two-dimensional planar geometry.

4.3 **Relativity in E&M**

The two cases discussed in Section 4.2.3, the infinite line charge/current and infinite plane charge/current densities, provide valuable examples where DA can correctly identify many of the dimensional dependences of an E&M system. We stressed that their shared geometry gives rise to similar functional forms for the resulting $\boldsymbol{E}, \boldsymbol{B}$ fields, but these "parallel" cases are also excellent examples in which to visualize the effects of relativity on both the current/charge densities of a problem, as well as on the resulting electric and magnetic field configurations.

If, for example, we imagine an observer moving parallel to the infinite wire in Fig. 4.7 in the $-\hat{\boldsymbol{x}}$ direction, in that frame of reference it would appear that there is a positive current in the $+\hat{\boldsymbol{x}}$ direction (assuming that $\lambda_e > 0$ in the first place) present in the wire, giving rise to a magnetic field. In Chapter 1, we quoted the result for the charge and current densities, $(\rho_e, \boldsymbol{J}_e)$, as they appear in different frames of reference, being given by

$$\begin{aligned} c\rho'_e &= \gamma\left(c\rho_e - \frac{v}{c}J_{e,x}\right) \\ J'_{e,x} &= \gamma\left(J_{e,x} - \frac{v}{c}c\rho_e\right) \\ J'_{e,y} &= J_{e,y} \\ J'_{e,z} &= J_{e,z}, \end{aligned} \tag{4.50}$$

and these relations are also valid for two-dimensional charge/current densities $(\sigma_e, \boldsymbol{j}_e)$ and the one-dimensional versions as well, namely (λ_e, I_e).

If we assume that the observer in the $'$ or "prime" frame is moving with velocity $\boldsymbol{v} = (-v, 0, 0)$, and also that $v/c << 1$ (so that $\gamma \approx 1$), then the current seen in that frame is

$$I'_e = -\gamma\left(\frac{-v}{c}(c\lambda_e)\right) \approx +v\lambda_e, \tag{4.51}$$

which is easily seen to make sense dimensionally and physically, and we would then expect the resulting magnetic field to have magnitude

$$B' \approx \frac{\mu_0 I'_e}{2\pi r}. \tag{4.52}$$

To confirm this, we can use the relativistic transformations of the $\boldsymbol{E}, \boldsymbol{B}$ fields themselves, from Chapter 1, namely

$$\begin{aligned} \boldsymbol{E}'_{\parallel} &= \boldsymbol{E}_{\parallel} \\ \boldsymbol{E}'_{\perp} &= \gamma(\boldsymbol{E}_{\perp} + \boldsymbol{v} \times \boldsymbol{B}) \\ \boldsymbol{B}'_{\parallel} &= \boldsymbol{B}_{\parallel} \\ \boldsymbol{B}'_{\perp} &= \gamma\left(\boldsymbol{B}_{\perp} - \frac{1}{c^2}\boldsymbol{v} \times \boldsymbol{E}\right), \end{aligned} \tag{4.53}$$

and we find that the new magnetic field contribution, as viewed in this frame, is given by

$$\boldsymbol{B}'_{\perp} = \gamma\left(-\frac{\boldsymbol{v} \times \boldsymbol{E}}{c^2}\right). \tag{4.54}$$

Using the fact that

$$\boldsymbol{E} = \frac{\lambda_e}{2\pi\epsilon_0 r}\hat{\boldsymbol{r}} \qquad \text{and} \qquad \boldsymbol{v} = -v\hat{\boldsymbol{x}}, \tag{4.55}$$

we find that

$$\mathbf{B}'_{\perp} = -\gamma(-v)\left(\frac{\lambda_e}{2\pi\epsilon_0 r}\right)\frac{\hat{\mathbf{x}}\times\hat{\mathbf{r}}}{c^2} \quad \xrightarrow{\gamma\to 1} \quad \frac{v\lambda_e}{2\pi r(\epsilon_0 c^2)}\hat{\boldsymbol{\theta}} = \frac{I'_e\mu_0}{2\pi r}\hat{\boldsymbol{\theta}}, \tag{4.56}$$

where we assume the non-relativistic limit of $v/c << 1$ so that $\gamma \approx 1$, the fact that $I_e = v\lambda_e$, and the familiar connection $\epsilon_0\mu_0 = 1/c^2$, giving the expected result.

Despite the admittedly somewhat artificial nature of the problem of an infinite length of charged wire, this system does provide not only a great example of DA, but also of the relativistic transformations of $(\rho_e, \mathbf{J}_e)$ and $\mathbf{E}, \mathbf{B}$ fields in a tractable and familiar context.[4] The relativistic transformations of the two-dimensional planar charge/current density example in Section 4.2.3 can also be explored in this same way, as can the **thick** wire and planar cases in **P4.16.**

4.4 **Torques, energies, and forces**

Another example of a symmetry between the structure of $\mathbf{E}$ and $\mathbf{B}$ fields arises in the interactions of electric ($\mathbf{p}$) and magnetic ($\mathbf{m}$) dipole moments with their respective fields. We are all familiar with the behavior of a compass needle as it aligns itself with the local magnetic field of the earth, implying that a magnetic moment will experience a torque in an external $\mathbf{B}$ field. If we imagine that this torque, $\boldsymbol{\tau}_m$, can depend on both $m = |\mathbf{m}|$ and $B = |\mathbf{B}|$, as well as the fundamental magnetic strength μ_0, we would write

$$\tau_m = |\boldsymbol{\tau}_m| = C_\tau m^\alpha B^\beta \mu_0^\gamma \qquad \text{implying that} \qquad \frac{\mathsf{ML}^2}{\mathsf{T}^2} = \left(\frac{\mathsf{QL}^2}{\mathsf{T}}\right)^\alpha \left(\frac{\mathsf{M}}{\mathsf{QT}}\right)^\beta \left(\frac{\mathsf{ML}}{\mathsf{Q}^2}\right)^\gamma, \tag{4.57}$$

which is almost trivially solved to yield $\alpha = \beta = 1$ and $\gamma = 0$, so that $\tau \propto mB$. Given that $\mathbf{m}$ and $\mathbf{B}$ are vectors, the obvious combination (which is not only dimensionally correct, but also geometrically appropriate) is

$$\boldsymbol{\tau}_m = \mathbf{m}\times\mathbf{B}, \tag{4.58}$$

which is the standard result, with the correct order of cross-products, and $C_\tau = 1$. This is indeed consistent with the observation that the magnetic dipole moment will align itself along the direction of the magnetic field.

This behavior also implies that a magnetic dipole will be in a lower energy configuration when aligned with the $\mathbf{B}$ field, suggesting that there is an associated potential energy, $\mathcal{E}_m$, depending on m, B and possibly μ_0 as well, so we again assume $\mathcal{E}_m = C_{\mathcal{E}} m^\alpha B^\beta \mu_0^\gamma$ and repeat the DA in Eqn. (4.57). As emphasized in Section 1.2.1, while torque and energy are conceptually very different, they do share the same dimensions, so that we immediately know that $\mathcal{E}_B \propto mB$. To construct a scalar from two vectors, and to match the observation that the aligned state is of lower energy,

[4] Readers are encouraged to review the dimensional dependences derived in the fascinating exploration of the *Motion of a charged particle in the static fields of an infinite straight wire* by Franklin, Griffiths, and Mann (2022), as well as the connection to the relativistic transformations discussed here.

we might guess that

$$\mathcal{E}_m = -\boldsymbol{m} \cdot \boldsymbol{B}, \tag{4.59}$$

which is the correct expression, with indeed $C_{\mathcal{E}} = -1$. One can use the same techniques (and judicious analogies) to discuss (as in **P4.18**) the torque and energy of an electric dipole moment ($\boldsymbol{p}$) in an electric ($\boldsymbol{E}$) field, with very similar results.

To explore the related concepts of the force experienced by a dipole (either kind) in an external field, let us first consider the case of an external $\boldsymbol{E}$ field acting on an electric dipole moment, $\boldsymbol{p}$, using the same analysis method by writing

$$|\boldsymbol{F}_p| \propto p^{\alpha} E^{\beta} \epsilon_0^{\gamma} \qquad \text{requiring} \qquad \frac{\mathsf{ML}}{\mathsf{T}^2} = (\mathsf{QL})^{\alpha} \left(\frac{\mathsf{ML}}{\mathsf{QT}^2}\right)^{\beta} \left(\frac{\mathsf{Q}^2\mathsf{T}^2}{\mathsf{ML}^3}\right)^{\gamma}. \tag{4.60}$$

This dimensional constraint does indeed have a solution, namely $\alpha = 2/3$, $\beta = -4/3$, and $\gamma = 1/3$, or

$$|\boldsymbol{F}_p| \propto \left(p^2 E^4 \epsilon_0\right)^{1/3}, \tag{4.61}$$

which is dimensionally correct, but not physically reasonable. If nothing else, how would we even match up the vector directions consistent with the 1/3 power?

In fact, if we assume a uniform field acting on the two "ends" of a dipole (charges $+q, -q$) with the same magnitude, we have a total force given by $\boldsymbol{F}_p = q\boldsymbol{E} - q\boldsymbol{E} = \boldsymbol{0}$. This suggests that, for a dipole to experience a net force, there must be different values of $\boldsymbol{E}$ across the physical extent of the dipole, namely that there is a gradient in the electric field, and we can include this effect in the DA approach by writing

$$|\boldsymbol{F}_p| \propto p^{\alpha} E^{\beta} \epsilon_0^{\gamma} \nabla^{\delta} \qquad \text{requiring} \qquad \frac{\mathsf{ML}}{\mathsf{T}^2} = (\mathsf{QL})^{\alpha} \left(\frac{\mathsf{ML}}{\mathsf{QT}^2}\right)^{\beta} \left(\frac{\mathsf{Q}^2\mathsf{T}^2}{\mathsf{ML}^3}\right)^{\gamma} \left(\frac{1}{\mathsf{L}}\right)^{\delta}. \tag{4.62}$$

This can be solved (to the extent possible) to give

$$F_p \propto p^{\alpha} E^{2-\alpha} \epsilon_0^{1-\alpha} \nabla^{3\alpha-2}, \tag{4.63}$$

and given earlier results cited above, we would expect $\boldsymbol{p}$ and $\boldsymbol{E}$ to appear symmetrically, that is, with $\alpha = 2 - \alpha$ or $\alpha = 1$ so that, without yet specifying the appropriate vector combination, we have $\boldsymbol{F}_p \propto \boldsymbol{p}\boldsymbol{E}\boldsymbol{\nabla}$. Given that there are now three vector quantities involved, $(\boldsymbol{p}, \boldsymbol{E}, \boldsymbol{\nabla})$, there are many possible combinations, but the correct one is

$$\boldsymbol{F}_p = (\boldsymbol{p} \cdot \boldsymbol{\nabla})\, \boldsymbol{E}. \tag{4.64}$$

A similar analysis of the force on a magnetic dipole in a non-uniform $\boldsymbol{B}$ field gives an analogous result, $\boldsymbol{F}_m = (\boldsymbol{m} \cdot \boldsymbol{\nabla})\, \boldsymbol{B}$. This expression is perhaps most familiar to physicists as the mechanism used in the famous Gerlach and Stern (1922) experiment to separate and sort silver atoms by their quantized spin angular momentum (which is, of course, related to their magnetic moments.)

Technologically important configurations of charges and currents, such as capacitors and inductors, are designed to store energy in the form of electric and magnetic fields. Given that the energy is distributed in space, the concept of **energy density** (or $U_E(\boldsymbol{r})$ and $U_B(\boldsymbol{r})$ for E/B fields, respectively) is relevant to such devices, and can be considered in parallel (and symmetrical) ways.

For example, for the electric field case, we can initially assume that $U_E = C_E \epsilon_0^\alpha E^\beta \rho_e^\gamma$, where we include the local charge density, $\rho_e(\boldsymbol{r})$ (with $[\rho_e] = \mathsf{Q}/\mathsf{L}^3$) in case it plays some role. Given that energy density has dimensions given by

$$U_E = U_B = \frac{\text{energy}}{\text{volume}} \qquad \text{or} \qquad [U_E] = [U_B] = \left(\frac{\mathsf{ML}^2}{\mathsf{T}^2}\right)\left(\frac{1}{\mathsf{L}^3}\right) = \frac{\mathsf{M}}{\mathsf{LT}^2}, \tag{4.65}$$

we require dimensional consistency by enforcing

$$\frac{\mathsf{M}}{\mathsf{LT}^2} = \left(\frac{\mathsf{Q}^2\mathsf{T}^2}{\mathsf{ML}^3}\right)^\alpha \left(\frac{\mathsf{ML}}{\mathsf{QT}}\right)^\beta \left(\frac{\mathsf{Q}}{\mathsf{L}^3}\right)^\gamma, \tag{4.66}$$

which gives $\alpha = 1$, $\beta = 2$, and $\gamma = 0$. The textbook result is

$$U_E(\boldsymbol{r}) = \frac{1}{2}\epsilon_0 \boldsymbol{E} \cdot \boldsymbol{E} = \frac{1}{2}\epsilon_0 |\boldsymbol{E}|^2, \tag{4.67}$$

with $C_E = 1/2$ and the magnitude squared of $\boldsymbol{E}$ given by the dot product as the only logical scalar combination of two vectors. Thus, even in a charge-free region of space, there is energy stored in the fields themselves. The corresponding expression for the energy density in a magnetic field is

$$U_B(\boldsymbol{r}) = \frac{1}{2\mu_0}|\boldsymbol{B}|^2. \tag{4.68}$$

Both Eqns. (4.67) and (4.68) depend on the modulus-squared of the respective field strengths, and so are invariant under rotations (turning your head) or $\boldsymbol{E}, \boldsymbol{B} \rightarrow -\boldsymbol{E}, -\boldsymbol{B}$ (looking at the problem in a mirror), as we might expect. If we reverse the sign of the charges on a capacitor, or flip the current direction in an inductor, we would not expect any change in the amount of energy stored in the system.

It is easy to imagine configurations where $\boldsymbol{E}$ and $\boldsymbol{B}$ fields are present simultaneously, with both Eqns. (4.67) and (4.68) contributing separately, but is there a mixed contribution dependent on a combination of electric and magnetic fields? We can certainly write

$$U_{EB}(\boldsymbol{r}) \propto \epsilon_0^\alpha \mu_0^\beta E^\gamma B^\delta \tag{4.69}$$

and, after matching dimensions, we find the general result

$$U_{EB} \propto \epsilon_0^\alpha \mu_0^{1-\alpha} E^{2\alpha} B^{2-2\alpha}, \tag{4.70}$$

which reduces to the cases of a pure $\boldsymbol{E}$ or $\boldsymbol{B}$ field when $\alpha = 1$ and $\alpha = 0$, respectively, as it must. If there were going to be an E/B symmetric solution, requiring that $2\alpha = 2 - 2\alpha$, it would be for $\alpha = 1/2$ giving

$$U_{EB} \propto \epsilon^{1/2} \mu_0^{-1/2} E^1 B^1 \quad \longrightarrow \quad U_{EB} \sim \sqrt{\frac{\epsilon_0}{\mu_0}}\, \boldsymbol{E} \cdot \boldsymbol{B}. \tag{4.71}$$

In contrast to the pure U_E, U_B expressions, Eqn. (4.71) does change sign if either $\boldsymbol{E}, \boldsymbol{B}$ do, and so is not a good candidate for standard E&M applications (see Zangwill 2013, section 24.3.3). Not every expression that is dimensionally correct is realized physically!

We note that this term can appear in various (theoretically motivated) extensions of classical electromagnetism, including:

- **High-field quantum electrodynamics (QED)** (Section 7.4) where for very large $\boldsymbol{E}, \boldsymbol{B}$ fields new non-linear terms (including ones involving the square of Eqn. (4.71)) are generated by quantum loop effects, and
- **Axion electrodynamics** where a (still very hypothetical) particle called the **axion** couples to the electromagnetic field via an interaction of the form in Eqn. (4.71), which is one proposed solution to several important problems in theoretical physics.

Another important **volume density** related quantity is given by

$$\mathcal{P} \equiv \boldsymbol{J}_e \cdot \boldsymbol{E} \tag{4.72}$$

with $\boldsymbol{J}_e$ being the current density that gives $\mathcal{P}$ as the **power per unit volume** dissipated by **Joule heating** in a conducting material. We can confirm that this expression is dimensionally correct by noting that

$$[\mathcal{P}] \stackrel{?}{=} [J_e][E]$$
$$\frac{\mathsf{M}}{\mathsf{LT}^3} = \frac{((\mathsf{ML}^2/\mathsf{T}^2)/\mathsf{T})}{\mathsf{L}^3} = \frac{\text{power}}{\text{volume}} \stackrel{!}{=} \left(\frac{\mathsf{Q}}{\mathsf{TL}^2}\right)\left(\frac{\mathsf{ML}}{\mathsf{QT}^2}\right) = \frac{\mathsf{M}}{\mathsf{LT}^3}. \tag{4.73}$$

While seemingly more important for systems where it is clear that there is a transport of energy, for example in the study of E&M waves, any system for which there are simultaneously $\boldsymbol{E}, \boldsymbol{B}$ fields present will have a corresponding **energy flux**, or **energy per unit time per unit area**, or **power per unit area**, or **intensity**, with an associated direction. This quantity is called the **Poynting vector** or **S**, with dimensions given by its definition as $[S] = [energy]/([area][time]) = \mathsf{M}/\mathsf{T}^3$. We can use the same assumptions as in Eqn. (4.69) to write

$$S \propto \epsilon_0^{\alpha}\mu_0^{\beta}E^{\gamma}B^{\delta} \tag{4.74}$$

and matching dimensions we find

$$S \propto \epsilon_0^{\alpha}\mu_0^{\alpha-1}E^{1+2\alpha}B^{1-2\alpha}. \tag{4.75}$$

In this case the only symmetric combination of $\boldsymbol{E}, \boldsymbol{B}$, which also allows us to form a vector quantity, corresponds to $\alpha = 0$, giving

$$\boldsymbol{S} = \frac{1}{\mu_0}\boldsymbol{E} \times \boldsymbol{B}. \tag{4.76}$$

The appropriate statement of energy conservation for such $\boldsymbol{E}, \boldsymbol{B}$ fields is

$$\underbrace{\frac{\partial}{\partial t}(U_E + U_B)}_{\text{rate of energy change}} = -\underbrace{\boldsymbol{J}_e \cdot \boldsymbol{E}}_{\text{heat loss}} - \underbrace{\nabla \cdot \boldsymbol{S}}_{\text{energy flow}}, \tag{4.77}$$

where you can confirm that each term has dimensions of **power per unit volume** or $\mathsf{M}/(\mathsf{LT}^3)$.

One can derive by direct calculation, or explore by DA, expressions for the **radiation pressure** ($\boldsymbol{P}_{rad}$), **E&M momentum density** ($\boldsymbol{p}$), and even **E&M angular momentum density** ($\boldsymbol{l}$) in an $\boldsymbol{E}/\boldsymbol{B}$ field configuration, each given by

$$\boldsymbol{P}_{rad} = \sqrt{\frac{\epsilon_0}{\mu_0}}\boldsymbol{E}\times\boldsymbol{B} = \frac{\boldsymbol{S}}{c} \tag{4.78}$$

$$\boldsymbol{p} = \epsilon_0\boldsymbol{E}\times\boldsymbol{B} = \frac{\boldsymbol{S}}{c^2} \tag{4.79}$$

$$\boldsymbol{l} = \epsilon_0\boldsymbol{r}\times(\boldsymbol{E}\times\boldsymbol{B}) = \boldsymbol{r}\times\boldsymbol{p} = \frac{\boldsymbol{r}\times\boldsymbol{S}}{c^2}, \tag{4.80}$$

where the third expression follows from the generalization of $\boldsymbol{L} = \boldsymbol{r}\times\boldsymbol{p}$. The dimensions of the first two quantities are given by $[\boldsymbol{P}_{rad}] = \mathsf{M}/\mathsf{LT}^2$ and $[\boldsymbol{p}] = \mathsf{M}/\mathsf{L}^2\mathsf{T}$. A very thorough review of many aspects of electromagnetic momentum is given by Griffiths (2012).

4.5 Circuits

We recall from Section 1.3.2 that the application of E&M in circuit analysis (at least in intro- level texts) most often makes use of only three standard components, namely **inductors** (L), **resistors** (R), and **capacitors** (C), with dimensions given in Eqn. (1.67). To explore the behavior of such elements in various circuit configurations, and their DA connections, we recall the definitions of the voltage drop associated with each device as given by

$$V_L = L\frac{dI(t)}{dt}, \qquad V_R = RI(t), \qquad \text{and} \qquad V_C = \frac{q(t)}{C}, \tag{4.81}$$

where of course $I(t) \equiv dq(t)/dt$ and the placement of the capacitance C in the denominator is conventional. We can reproduce the dimensionalities of each component by writing

$$[L] = \left[\frac{V}{dI/dt}\right] = \left(\frac{\mathsf{ML}^2}{\mathsf{QT}^2}\right)\left(\frac{\mathsf{T}^2}{\mathsf{Q}}\right) = \frac{\mathsf{ML}^2}{\mathsf{Q}^2} \tag{4.82}$$

$$[R] = \left[\frac{V}{I}\right] = \left(\frac{\mathsf{ML}^2}{\mathsf{QT}^2}\right)\left(\frac{\mathsf{T}}{\mathsf{Q}}\right) = \frac{\mathsf{ML}^2}{\mathsf{Q}^2\mathsf{T}} \tag{4.83}$$

$$[1/C] = \left[\frac{V}{q}\right] = \left(\frac{\mathsf{ML}^2}{\mathsf{QT}^2}\right)\left(\frac{1}{\mathsf{Q}}\right) = \frac{\mathsf{ML}^2}{\mathsf{Q}^2\mathsf{T}^2} \tag{4.84}$$

to emphasize the similarities between the three expressions, differing only in powers of T. We can also write the ratios of these dimensionalities as

$$[L] : [R] : [1/C] = \mathsf{T} : 1 : (1/\mathsf{T}), \tag{4.85}$$

which can be useful when we examine the time-dependence of various combinations of L, R, C elements in circuit problems; see also **P4.27** where we discuss the generalized notion of **impedance.**

Recall that the **conductance** of a circuit element, $G \equiv 1/R$, is just the inverse of resistance, with $[G] = \mathsf{Q^2T/ML^2}$, and that Ohm's law can be written as $I = GV$ instead of $V = IR$.

Example 4.8 LR circuit

An example of a simple two-component circuit is included as one of the classic Rayleigh *principle of similitude* problems and involves an L, R circuit, described as follows:

> The time-constant (i.e., the time in which the current falls in the ratio $e : 1$) of a linear conducting electric circuit is directly as the inductance and inversely as the resistance ... (Rayleigh 1915).

It is perhaps good to define this problem (actually a closely related one) a bit more precisely by asking about the time-behavior of a pair of L and R circuit elements, connected in series to a battery/voltage source ($\mathcal{E}$) that is suddenly switched on, eventually solving for the time-dependent current $I(t)$ in the circuit. The IR voltage drop across the resistor will eventually balance the applied voltage $\mathcal{E}$, but the inductor (with its $dI(t)/dt$ dependence) provides a generalized inertia[5] to the change in current, so the $\mathcal{E} = IR$ balance is not instantaneous.

If we define the characteristic time scale for the LR circuit to come to equilibrium as τ_{LR}, we can write $\tau_{LR} \propto L^{\alpha}R^{\beta}\mathcal{E}^{\gamma}$ and matching dimensions we easily find that $\alpha = 1, \beta = -1$, and $\gamma = 0$, or $\tau_{LR} = L/R$, which does reproduce the "... directly as the inductance and inversely as the resistance ..." statement about the dependence of the time-constant. We can also see from Eqn. (4.85) that this is an obvious combination.

To then explore the time-dependence of the current, we write $I(t) \propto L^{\alpha}R^{\beta}\mathcal{E}^{\gamma}t^{\delta}$ and match dimensions. If we write the result in terms of δ (thus emphasizing the time-dependence), we find that

$$I(t) \propto L^{-\delta}R^{-1+\delta}\mathcal{E}^{1}t^{\delta} = \frac{\mathcal{E}}{R}\left(\frac{Rt}{L}\right)^{\delta} = \frac{\mathcal{E}}{R}F\left(\Pi = \frac{t}{\tau_{LR}}\right), \tag{4.86}$$

where the dimensionless combination $\Pi = t/(L/R)$ is correctly identified as the scaling ratio, given that we have already identified $\tau_{LR} = L/R$ as the appropriate time scale.

An actual circuit analysis for this problem gives the first-order ordinary differential equation (ODE) $LdI(t)/dt + RI(t) = \mathcal{E}$, with the initial condition $I(0) = 0$, assuming we connect the circuit at time $t = 0$. You can hopefully verify that the solution is given by

$$I(t) = \frac{\mathcal{E}}{R}\left(1 - e^{-t/\tau_{LR}}\right), \tag{4.87}$$

which agrees with Eqn. (4.86), with $F(\Pi) \equiv 1 - e^{-\Pi}$. You're asked to explore the other two pair-wise combinations of circuit elements RC and LC circuit in **P4.24** and **P4.25**.

If we consider a circuit with all three LRC elements, with an added voltage/EMF source, the resulting Kirchhoff's law loop analysis gives the second-order ODE

$$L\frac{dI(t)}{dt} + RI(t) + \frac{1}{C}q(t) = \mathcal{E} \tag{4.88}$$

for $q(t)$, involving its first and second derivative. There is a powerful analogy between this system and that of the damped harmonic oscillator considered in **Example 3.1** (as in Eqn. (3.24)) which,

---

[5] This is a better analogy than you might think, as we'll discuss below.

when an additional external force is included, can be written in the form

$$m\frac{dv(t)}{dt} + bv(t) + kx(t) = \mathcal{F}. \tag{4.89}$$

One can immediately see the similarities between the two systems with identifications between the system variables ($q(t)$ and $x(t)$) and parameters (L, R, C and m, b, k), namely

$$\begin{array}{ccccccc} x(t) & v(t) & a(t) & m & b & k & \mathcal{F} \\ \updownarrow & \updownarrow & \updownarrow & \updownarrow & \updownarrow & \updownarrow & \updownarrow \\ q(t) & I(t) & dI(t)/dt & L & R & 1/C & \mathcal{E} \end{array}$$

which helps justify the description above of the contribution of the voltage of the inductor as being akin to inertia, and $1/C$ playing the role of the spring constant.

These parallels can be immediately used to make connections between other quantities in the mechanical and electromagnetic systems, especially involving energy constructs, such as

$$E_{kin} = \frac{1}{2}mv^2 \quad \longrightarrow \quad E_L = \frac{1}{2}LI^2 \tag{4.90}$$

$$E_{pot} = \frac{1}{2}kx^2 \quad \longrightarrow \quad E_C = \frac{1}{2}\frac{q^2}{C}, \tag{4.91}$$

and a related one involving the power dissipated by mechanical friction (due to $F_{fr} - -bv$) and through Joule heating,

$$P_{diss} = F_{fr}v = bv^2 \quad \longrightarrow \quad P_{Joule} = RI^2. \tag{4.92}$$

You should also be able to use the results of **Example 3.1** to evaluate the oscillation frequencies for an otherwise free (meaning $\mathcal{E} = 0$) *LRC* circuit from Eqn. (3.27).

We have used the mechanical/electrical analogies to derive the energy stored in inductors and capacitors as shown in Eqn. (4.91), but haven't yet focused on the connections of these elements as storage devices for E/B field energies, as suggested by Eqns. (4.67) and (4.68) for $U_E = \epsilon_0\mathbf{E}^2/2$ and $U_B = \mathbf{B}^2/2\mu_0$.

For example, the electric field in a parallel-plate capacitor can be derived from the planar geometry of a single infinite sheet of charge in Eqn. (4.49), where two oppositely charged plates give fields that cancel outside the plates, but add coherently between the plates, giving $E = 2(\sigma_e/2\epsilon_0) = \sigma_e/\epsilon_0$. The surface charge density of two (finite area A) parallel plates is $\sigma_e = q/A$, so that $E = (q/A\epsilon_0)$. If the separation between the plates is d the (electric field) energy E_C stored in the capacitor can be written in terms of the energy density and volume as

$$E_C = U_E V = \left(\frac{\epsilon_0}{2}E^2\right)V = \frac{\epsilon_0}{2}\left(\frac{q}{A\epsilon_0}\right)^2(Ad) = \frac{1}{2}q^2\left(\frac{d}{A\epsilon_0}\right) \equiv \frac{q^2}{2C} \quad \longrightarrow \quad C = \frac{A\epsilon_0}{d}, \tag{4.93}$$

giving a well-known formula for the capacitance in this simple geometry in terms of ϵ_0 and the physical dimensions.

For the standard geometry of a inductor consisting of wires wrapped around a cylindrical form (a **solenoid** geometry), the value of the (uniform) field inside the inductor, pointing along the axis of symmetry, can be derived from Gauss's law (**P4.17**) and can be written in the form

$$B_{ind} = \mu_0 j_e = \mu_0 I \frac{N}{L}, \tag{4.94}$$

where $j_e = I(N/L)$ is the one-dimensional current density, where there are N turns per unit length L along the inductor. This result is reminiscent of Eqn. (4.49), at least dimensionally, and the cylindrical inductor can be thought of as akin to wrapping up the sheet of current into a tube. If we assume a cross-sectional area A of the inductor, the connection between energy density in the magnetic field and the volume reads

$$E_L = U_B V = \left[\frac{1}{2\mu_0}\left(\frac{\mu_0 IN}{L}\right)^2\right](AL) = \frac{1}{2}I^2\left(\frac{\mu_0 N^2 A}{L}\right) = \frac{1}{2}LI^2 \quad \longrightarrow \quad L = \mu_0 N^2 \frac{A}{L}, \tag{4.95}$$

which reproduces a textbook formula for the inductance in this cylindrical geometry, using now μ_0 and the dimensions, along with the dimensionless factor of N.

Both of these examples show how the energy budget for L and C components are directly related to the magnetic/electric field energy densities, the corresponding fundamental interaction strengths (μ_0, ϵ_0) and their geometry. We leave it to **P4.30** to use similar arguments to derive the resistance in terms of the conductance of the material and the dimensions of the resistor, this time using the **power dissipated per unit volume** $\mathcal{P} = \boldsymbol{J}_e \cdot \boldsymbol{E}$ from Eqn. (4.72).

4.6 Maxwell's equations and electromagnetic (EM) radiation

One of the most important applications of Maxwell's equations involves the decoupling of the electric and magnetic fields that appear in the Faraday and Ampère–Maxwell laws to derive the (identical) partial differential equations for both $\boldsymbol{E}(\boldsymbol{r}, t), \boldsymbol{B}(\boldsymbol{r}, t)$ separately describing electromagnetic (EM) waves. (We've often used the notation E&M as a short hand for "electricity and magnetism," as two closely related topics, but we henceforward reserve EM as an abbreviation for "electromagnetic," especially for phenomena connected by Maxwell's equations.) We (briefly) describe that procedure here (focusing more on dimensions than on the detailed vector calculus) as it is presented in almost all textbooks, noting that we extend these results in Section 4.8.2. We initially assume that there are no charges or currents (ρ_e and $\boldsymbol{J}_e$ both vanish, for example as in free-space or vacuum) and focus on Eqns. (4.98) and (4.99), but write all of the Maxwell equations below for completeness.

$$\nabla \cdot \boldsymbol{E}(\boldsymbol{r}, t) = \frac{1}{\epsilon_0}\rho_e(\boldsymbol{r}, t) \qquad \text{Gauss's law} \tag{4.96}$$

$$\nabla \cdot \boldsymbol{B}(\boldsymbol{r}, t) = 0 \qquad \text{No magnetic monopoles} \tag{4.97}$$

$$\nabla \times \boldsymbol{E}(\boldsymbol{r}, t) = -\frac{\partial}{\partial t}\boldsymbol{B}(\boldsymbol{r}, t) \qquad \text{Faraday's law} \tag{4.98}$$

$$\nabla \times \boldsymbol{B}(\boldsymbol{r}, t) = \mu_0 \boldsymbol{J}_e(\boldsymbol{r}, t) + \underbrace{\mu_0 \epsilon_0 \frac{\partial}{\partial t}\boldsymbol{E}(\boldsymbol{r}, t)}_{\text{displacement term}} \qquad \text{Ampère's law (+ Maxwell).} \tag{4.99}$$

Taking the curl of Faraday's law (i.e., applying $\nabla\times$ to both sides), using a vector identity, and interchanging the order of space/time derivatives, we have

$$-\frac{\partial}{\partial t}(\nabla\times\boldsymbol{B}) = \underbrace{\nabla\times\left(-\frac{\partial\boldsymbol{B}}{\partial t}\right) = \nabla\times(\nabla\times\boldsymbol{E})}_{\nabla\times\text{ Faraday's law}} = \nabla\underbrace{(\nabla\cdot\boldsymbol{E})}_{=0} - \nabla^2\boldsymbol{E} = -\nabla^2\boldsymbol{E}, \tag{4.100}$$

where $\nabla\cdot\boldsymbol{E} = \rho_e/\epsilon_0 = 0$ since we assume no free charges are present. Then substituting the Ampère–Maxwell law for $\nabla\times\boldsymbol{B}$, we have

$$-\frac{\partial}{\partial t}\left(\mu_0\epsilon_0\frac{\partial\boldsymbol{E}}{\partial t}\right) = -\nabla^2\boldsymbol{E} \tag{4.101}$$

or

$$\frac{\partial^2\boldsymbol{E}(\boldsymbol{r},t)}{\partial t^2} = \frac{1}{(\mu_0\epsilon_0)}\nabla^2\boldsymbol{E}(\boldsymbol{r},t) = c^2\nabla^2\boldsymbol{E}(\boldsymbol{r},t), \tag{4.102}$$

which is one version of the classical wave equation, with $c \equiv (\mu_0\epsilon_0)^{-1/2}$ being the speed of light. One can also reverse the order of operations (take the curl of the Ampère–Maxwell law and then substitute Faraday's law) and find the same wave equation for $\boldsymbol{B}(\boldsymbol{r},t)$.

A typical **plane wave solution** for either equation can be written in the form

$$\boldsymbol{E}(\boldsymbol{r},t) = \boldsymbol{E}_0 e^{i(\boldsymbol{k}\cdot\boldsymbol{r}-\omega t)} \qquad \text{or} \qquad \boldsymbol{B}(\boldsymbol{r},t) = \boldsymbol{B}_0 e^{i(\boldsymbol{k}\cdot\boldsymbol{r}-\omega t)}, \tag{4.103}$$

where $k = |\boldsymbol{k}| = 2\pi/\lambda$ is the wave-number (with $\hat{\boldsymbol{k}} = \boldsymbol{k}/k$ giving the direction of propagation) and $\omega = 2\pi f$ is the angular frequency, while $\boldsymbol{E}_0, \boldsymbol{B}_0$ are fixed/constant field magnitudes. Substituting Eqn. (4.103) into Eqn. (4.102) we find the usual connections between k, ω (or λ, f), namely

$$\omega^2 = c^2|\boldsymbol{k}|^2, \qquad \omega = ck, \qquad 2\pi f = c\frac{2\pi}{\lambda}, \qquad \text{and} \qquad f\lambda = c. \tag{4.104}$$

Looking ahead to Section 4.8.2, we should recall the quantum mechanical connections encoded in the Einstein and de Broglie relations, and can add one power of $\hbar$ to each side and write

$$E_\gamma = \hbar\omega = (\hbar k)c = p_\gamma c, \tag{4.105}$$

which is the energy-momentum $(E_\gamma, \boldsymbol{p}_\gamma)$ connection for a massless particle, in this case the photon (γ).

We can then "recouple" the equations for the $\boldsymbol{E}, \boldsymbol{B}$ fields by using both Eqns. (4.98) and (4.99) to relate the magnitudes of each field. For example, using Faraday's law, we find

$$\nabla\times\boldsymbol{E} = -\frac{\partial\boldsymbol{B}}{\partial t} \qquad \longrightarrow \qquad \boldsymbol{k}\times\boldsymbol{E}_0 = \omega\boldsymbol{B}_0 \qquad \longrightarrow \qquad |\boldsymbol{E}_0| = \frac{\omega}{k}|\boldsymbol{B}_0| = c|\boldsymbol{B}_0|, \tag{4.106}$$

so that magnitudes of the electric and magnetic field in an EM wave are proportional. The similar result using the Ampère–Maxwell law is

$$\boldsymbol{k}\times\boldsymbol{B}_0 = -\frac{\omega}{c^2}\boldsymbol{E}_0, \tag{4.107}$$

which gives the same connection between the field strengths since

$$k|\mathbf{B}_0| = \frac{\omega}{c^2}|\mathbf{E}_0| \quad \longrightarrow \quad |\mathbf{E}_0| = c^2\left(\frac{k}{\omega}\right)|\mathbf{B}_0| = c|\mathbf{B}_0|. \tag{4.108}$$

In additional to the dimensional/numerical connections, the vector relations require $\mathbf{E}_0, \mathbf{B}_0, \mathbf{k}$ to describe three mutually perpendicular directions, for example, in the $\hat{\mathbf{x}}, \hat{\mathbf{y}}, \hat{\mathbf{z}}$ directions.

While in general the energy densities in the $\mathbf{E}, \mathbf{B}$ fields can be completely unrelated in combined electric/magnetic fields, we find (from Eqns. (4.67) and (4.68)) that

$$U_B = \frac{1}{2\mu_0}|\mathbf{B}_0|^2 = \frac{c^2}{2\mu_0}|\mathbf{E}_0|^2 = \frac{\epsilon_0}{2}|\mathbf{E}_0|^2 = U_E \tag{4.109}$$

and the energy in an EM wave is shared equally. The **Poynting vector** from Eqn. (4.76) can be written in three equivalent forms (all dimensionally consistent) for an EM wave, namely

$$\mathbf{S} = \frac{1}{\mu_0}\mathbf{E}\times\mathbf{B} = \frac{c}{\mu_0}|\mathbf{B}_0|^2\,\hat{\mathbf{k}} = c\epsilon_0|\mathbf{E}_0|^2\,\hat{\mathbf{k}}, \tag{4.110}$$

where $\hat{\mathbf{k}}$ is the unit vector in the direction of propagation.

Example 4.9 Larmor radiation revisited

Returning now in more detail to the topic in **Example 1.2**, the problem of determining the power radiated by an accelerated charge in the form of EM waves is a standard one in many advanced undergraduate texts (see e.g., Good 1999, and Jackson 1999, Zangwill 2013, and Griffiths 2017) and by way of background, we show in Fig. 4.9 the configuration of electric field lines for (a) a stationary charge and (b) a charge moving with constant velocity $\mathbf{v}$, in this case to the right with $v/c = 0.9$. The expression for the electric field for a charge q undergoing uniform motion[6] is

$$\mathbf{E} = \frac{q\hat{\mathbf{R}}}{4\pi\epsilon_0 R^2}\frac{(1 - v^2/c^2)}{[1 - (v^2/c^2)\sin^2(\theta)]^{3/2}} \tag{4.111}$$

where $\mathbf{R}$ points from q to the observation point, and θ is the angle between $\mathbf{v}$ and $\mathbf{R}$, so defined by $\mathbf{v}\cdot\mathbf{R} = vR\cos(\theta)$. Contrast this to that of a particle which has undergone a sudden acceleration, as shown in Fig. 4.10. We note that the "kink" in the field lines corresponds to an EM wave, so we expect that an accelerated particle would emit energy as a result.

To explore the power (P) radiated by such an accelerated charge, we assume that P depends on both ϵ_0 and μ_0 (given the electrodynamic origin of the effect, giving rise to EM radiation), and the magnitude of the charge q and acceleration a, so we posit a dependence $P = C_P q^\alpha a^\beta \epsilon_0^\gamma \mu_0^\delta$ (as we did in **Example 1.2**) and find that

$$P = C_P q^2 a^2 \epsilon_0^{1/2} \mu_0^{3/2}. \tag{4.112}$$

If we use the standard Maxwell unification result relating c, ϵ_0, μ_0, namely $c = 1/\sqrt{\epsilon_0\mu_0}$, we can write this in three equivalent forms, namely

---

[6] See e.g., Griffiths (2017), Jackson (1999), or Zangwill (2013).

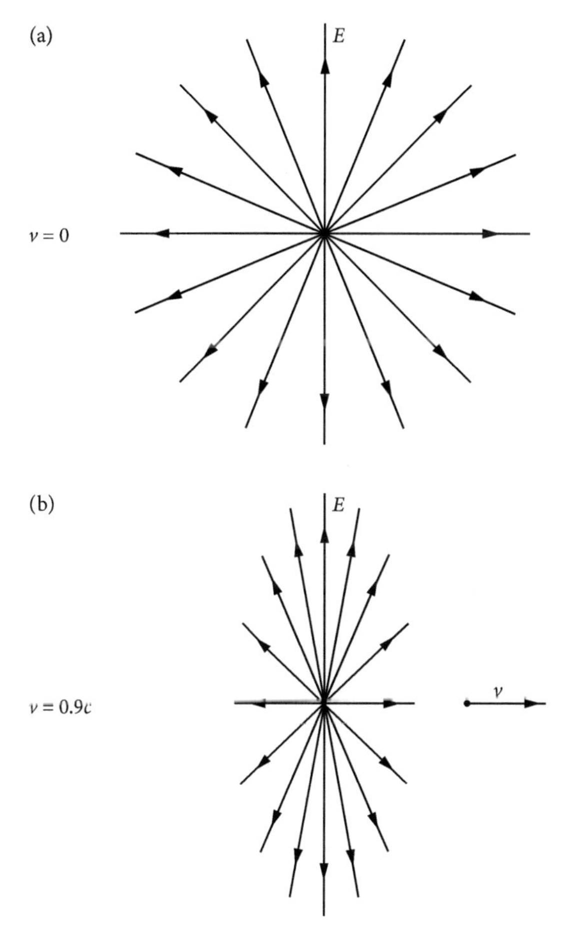

Fig. 4.9 Electric field lines for (a) charge at rest, and (b) moving to the right with $v/c = 0.9$.

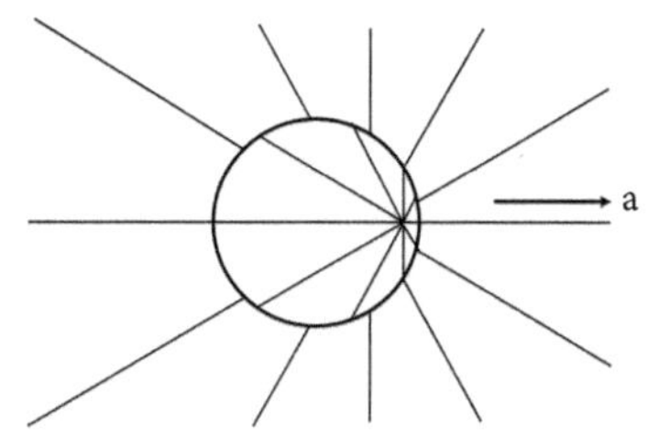

Fig. 4.10 Schematic electric field line configuration from suddenly accelerated charge showing outgoing radiation pattern.

$$P = C_P \frac{\mu_0 q^2 a^2}{c} = C_P \sqrt{\frac{\mu_0}{\epsilon_0}} \frac{q^2 a^2}{c^2} = C_P \frac{q^2 a^2}{\epsilon_0 c^3}. \tag{4.113}$$

With the value of $C_P = 1/6\pi$, this is the standard result, first derived by Larmor (1897), which appears in numerous textbooks. We see that the DA approach has provided several important insights, without the challenging derivation, namely:

- P is independent of the sign of the charge (since it is proportional to q^2), as the only change in Fig. 4.10 for $\pm q$ would be the direction of the arrows on the electric field lines, pointing away from $(+q)$ or towards $(-q)$ the charge.
- P depends on a^2 and given that the acceleration $\boldsymbol{a}$ is, after all, a vector, we associate $a = |\boldsymbol{a}|$ so that the power radiated doesn't depend on the direction of the acceleration, just its magnitude, via $\boldsymbol{a} \cdot \boldsymbol{a} = a^2$.
- It is possible to include that type of "extra information" arising from non-dimensional knowledge (in this case the vector nature of $\boldsymbol{a}$) in a more proactive way. For example, say we had assumed that the radiated power also depended on the mass m of the accelerated charge, via $P = C_P q^\alpha a^\beta \epsilon_0^\gamma \mu_0^\delta m^\rho$. We would have found the relation

$$P \propto q^\alpha a^{1+\alpha/2} \epsilon_0^{-1/2+\alpha/2} \mu_0^{-1/2+\alpha} m^{1-\alpha/2}, \tag{4.114}$$

  and knowing that the power P (a scalar quantity) should depend on $\boldsymbol{a}^2 = a^2$ (or at least powers thereof), it would be natural to assume that $\alpha = 2$, with no dependence then on m.

We note that Larmor (1897) himself makes a comment at the end of his derivation:

> In motion with uniform velocity there is no loss; during uniformly accelerated motion the rate of loss is constant.

Thus, we ask in **P4.31** for the reader to repeat the analysis leading to Eqn. (4.112), but using a constant speed (v) in place of the acceleration (a) to confirm Larmor's first statement; see also **P4.32** and **P4.33** to explore other alternatives.

Note: We could have just as well considered the strengths of the electric and magnetic interactions to be $\epsilon_0 \to 4\pi\epsilon_0$ and $\mu_0 \to \mu_0/4\pi$ (given the prominent roles those dimensionless factors play in Coulomb's law (Eqns. (4.19)) and the Biot–Savart law (Eqn. (4.31)), which would have yielded an extra dimensionless factor of $1/4\pi$ in any of the three formulae in Eqn. (4.112). With that intuition, the truly undetermined value of C_P would have been much closer to unity, namely $C_P = 2/3$. That is, in fact, the value found in the original Larmor (1897) derivation, but also because the author uses a different set of E&M units; see the discussion of CGS versus MKS units in Section 10.6.

When to use (or not use) various factors of 2π (for example, whether $\omega = 2\pi f$ or $\hbar = h/2\pi$) or here 4π benefits from the development of some physical intuition and experience. For examples, problems that more naturally use Eqns. (4.19) or (4.31) directly often retain factors of 4π, while those where Maxwell's equations are used, in either differential or integral form, say Eqns. (4.96)–(4.99), often do not.

It's also important to note that we've implicitly assumed non-relativistic motion, since when the speeds of the charges approach c, then factors of $\gamma = (1 - (v/c)^2)^{-1/2}$ also appear in Eqn. (4.112), so it's not just dimensionless constants that are not determined by DA, but even some dynamical features.

An important case of EM power radiated by an accelerated charge occurs when $\boldsymbol{a}$ is due to an oscillating electric dipole, say of the form $p(t) = qx(t) = p_0 \cos(\omega t)$. In this case the relevant variables are not q, a, but rather p_0, ω, and we can derive the power radiated by such a system by equating $P = C_P p_0^\alpha \omega^\beta \epsilon_0^\gamma \mu_0^\delta$ and finding that

$$P \propto p_0^2 \omega^4 \epsilon_0^{1/2} \mu_0^{3/2} \propto \frac{\mu_0 p_0^2 \omega^4}{c}, \tag{4.115}$$

and a rigorous derivation gives $C_P = 1/12\pi$ (or $C_P = 1/3$, if we'd used $\mu_0/4\pi$). Given that the dipole moment $\boldsymbol{p}_0$ is a vector, the quadratic dependence makes sense. This result also gives the characteristic ω^4 (frequency) or $1/\lambda^4$ (wavelength) dependence related to the scattering of light by molecules (say in the atmosphere) and seen in discussions of why the sky is blue, or as posed in one of the DA problems by Rayleigh (1915):

> The intensity of light scatter in an otherwise uniform medium from a small particle of different refractive index is inversely proportional as the fourth power of the wave-length.

Let us now extend our discussion of EM waves to the case where they propagate in matter, specifically into a conductor which has a finite value of the conductivity σ_e. Because of Ohm's law, the electric field component of the wave can give rise to a current, via $\boldsymbol{J}_e = \sigma_e \boldsymbol{E}$, which, in turn, will appear as a source term in Faraday's law, modifying the EM wave equation. But before discussing the impact this has on the form of the resulting plane wave solution, let us explore the temporal response of a conductor to a sudden change in the charge density, $\rho_e(t)$.

If we add excess charge to a conductor, or otherwise temporarily cause a local change in ρ_e, the charge will rearrange itself on a time scale τ_e. This time might depend on the fundamental electric constant (ϵ_0), the conductivity of the specific material (σ_e), and perhaps on the properties of the free electrons (e, m_e) present in any conductor which are the current carriers. We can write

$$\tau_e \propto \epsilon_0^\alpha \sigma_e^\beta e^\gamma m_e^\delta$$

$$\mathsf{T} = \left(\frac{\mathsf{Q}^2\mathsf{T}^2}{\mathsf{ML}^3}\right)^\alpha \left(\frac{\mathsf{Q}^2\mathsf{T}}{\mathsf{ML}^3}\right)^\beta \mathsf{Q}^\gamma \mathsf{M}^\delta \tag{4.116}$$

$$\downarrow$$

$$\tau_e \propto \frac{\epsilon_0}{\sigma_e}, \tag{4.117}$$

after matching dimensions and finding that that $\alpha = 1$, $\beta = -1$, and $\gamma = \delta = 0$, and this is the exact answer obtained from a more mathematical derivation, as in **P4.36.**[7] On the other hand, the relevant (inverse) time scale for an EM wave is given by the angular frequency, ω, so that a useful dimensionless (Buckingham Pi variable) combination will be $\Pi \equiv \omega\tau_e = \omega\epsilon_0/\sigma_e$, and for fixed ω, very good (very poor) conductors correspond to $\Pi << 1$ ($\Pi >> 1$).

The solutions to the EM wave equation including a σ_e-dependent term are assumed to have the same functional form as before (and that the wave is penetrating a conductor, so the direction of

[7] Where we assume that the values of ϵ, μ in the material are close to their vacuum values, ϵ_0, μ_0 for simplicity.

$\boldsymbol{k}$ is suppressed)

$$\boldsymbol{E}(\boldsymbol{r}, t) = \boldsymbol{E}_0 e^{i(kz-\omega t)}, \tag{4.118}$$

but where $k = \tilde{k} + i\kappa$, with the imaginary part corresponding to exponential decay into the conductor, as $\exp(-\kappa z)$. The inverse of κ corresponds to a **penetration length** or **skin depth** (called $\delta = 1/\kappa$) and we want to explore what DA can tell us about its dependence on the physical parameters of the conductor.

For good conductors, with $\omega\epsilon_0/\sigma_e >> 1$, the skin depth turns out to depend only on ω, μ_0, σ_e and we can write $\delta \propto \sigma_e^\alpha \mu_0^\beta \omega^\gamma$ and matching dimensions we find that

$$\delta = C_\delta (\sigma_e \mu_0 \omega)^{-1/2} = \frac{C_\delta}{\sqrt{\sigma_e \mu_0 \omega}} \tag{4.119}$$

and a detailed analysis gives $C_\delta = \sqrt{2}$ (as explored in **P4.37**.) For poor conductors the skin depth is independent of frequency and depends on $\delta \propto \sigma_e^\alpha \mu_0^\beta \epsilon_0^\gamma$ and dimensional consistency requires that

$$\delta = C_\delta \sqrt{\frac{\epsilon_0}{\mu_0 (\sigma_e)^2}}. \tag{4.120}$$

If we wish to compare to our earlier comments, or the exact textbook result, we can include all four variables and assume $\delta \propto \sigma_e^\alpha \mu_0^\beta \omega^\gamma \epsilon_0^\rho$ and if we solve for the exponents in terms of say α, we find

$$\delta \propto (\sigma_e)^\alpha (\mu_0)^{-1/2} \omega^{-\alpha-1} (\epsilon_0)^{-\alpha-1/2} = \frac{1}{\sqrt{\mu_0 \epsilon_0 \omega^2}} \left(\frac{\sigma_e}{\omega \epsilon_0}\right)^\alpha \longrightarrow \frac{c}{\omega} F\left(\frac{\sigma_e}{\omega \epsilon_0}\right). \tag{4.121}$$

The dimensionless ratio $\sigma_e/(\omega\epsilon_0)$ is just the inverse of the $\Pi = \omega\epsilon_0/\sigma_e$ combination discussed earlier, and so still very much in the spirit of the Buckingham Pi theorem, while the c/ω clearly does set a length scale. A detailed mathematical analysis (see e.g., Griffiths 2017) finds that

$$\frac{1}{\delta} \equiv \kappa = \sqrt{\frac{\epsilon_0 \mu_0 \omega^2}{2}} \left[\sqrt{1 + \left(\frac{\sigma_e}{\epsilon_0 \omega}\right)^2} - 1\right]^{1/2}, \tag{4.122}$$

and you should be able (see **P4.37**) to reproduce both limiting cases in Eqns. (4.117) and (4.118) for $\Pi >> 1$ and $\Pi << 1$.

4.7 Faraday's law, eddy currents, and magnetic braking: A test-bed for dimensional analysis

We have already used two of the Maxwell equations extensively in Section 4.2, namely Gauss's and Ampère's laws, to evaluate electric and magnetic fields, and we consider extensions of the $\nabla \cdot \boldsymbol{B} = 0$ (no magnetic monopoles) law in Section 4.8.1. In this section we focus on Faraday's law, which in integral form is written as

$$\mathcal{E} = \oint \boldsymbol{E} \cdot d\boldsymbol{l} = -\frac{d}{dt} \int \boldsymbol{B} \cdot d\boldsymbol{S} = -\frac{d\Phi_B}{dt}, \tag{4.123}$$

where $\mathcal{E}$ is the induced **electromotive force (EMF)** induced by a changing magnetic flux, $d\Phi_B/dt$. The minus sign can be used as a mnemonic device for **Lenz's law**, namely the statement that any current induced by $\mathcal{E}$ must be in such a direction (or flow pattern) that it opposes the original change in flux.

Since many uses of this law are in applications to technology or devices, including inductors, generators, and transformers, the examples considered here focus on realizations (perhaps simplified) of some such devices. We're also eager to consider how DA can inform quantitative discussions of what otherwise are often considered qualitative examples, namely two compelling lecture demonstrations that illustrate the production of **eddy currents**, especially as applied to **magnetic braking.**

As a starter problem, imagine a circular loop of wire of radius a and resistance R in the x–y plane, with a uniform magnetic field in the z direction of strength B_0, which is turned off on a time scale τ. The resulting EMF will give rise to a current (I) and we can explore how it depends on the quantities involved by writing $I \propto B_0^\alpha a^\beta R^\gamma \tau^\delta$, and solving for the exponents, we find

$$I = C_I \frac{B_0 a^2}{R\tau}. \tag{4.124}$$

To compare to a specific problem, we have to make some assumption about how the field is removed, and one simple model would be to have $B(t) = B_0(1 - t/\tau)$ (at least for $0 < t < \tau$), in which case

$$\Phi_B(t) = B_0\left(1 - \frac{t}{\tau}\right)\pi a^2 \qquad \text{so that} \qquad \mathcal{E} = -\frac{d\Phi_B t}{dt} = \frac{B_0 \pi a^2}{\tau} \tag{4.125}$$

and using $V = \mathcal{E} = IR$ reproduces the dimensions in Eqn. (4.122) with $C_I = \pi$. Most other versions of the time-dependence of $B(t)$, such as being proportional to $e^{-t/\tau}$ or e^{-t^2/τ^2}, do respect this DA result, but add a new Buckingham Pi function depending on the dimensionless combination $\Pi = t/\tau$.

An example where neither the area capturing the flux nor the magnetic field are changing, but only their relative orientation, is to consider the planar loop/coil discussed earlier, in a fixed B field, but which rotates about either the x or y axes at constant angular frequency ω. If we redo the simple DA approach above, replacing $\tau \to \omega$, we can immediately write $I \propto B_0 R^2 \omega / R$, if we ignore the fact that the time-variable t is necessarily present. The Faraday's law problem is straightforward, with

$$\begin{aligned}
\Phi_B(t) &= \boldsymbol{B} \cdot \boldsymbol{A} = B(\pi a^2)\cos(\theta = \omega t) \\
\mathcal{E}(t) &= -\frac{d\Phi_B(t)}{dt} = \pi B a^2 \omega \sin(\omega t) \\
I(t) = \frac{\mathcal{E}}{R} &= \frac{\pi B a^2 \omega}{R} \sin(\omega t),
\end{aligned} \tag{4.126}$$

with a naturally dimensionless Buckingham Pi variable given by $\Pi \equiv \omega t$, as expected.

Example 4.10 EM forces

Consider a rectangular loop of wire of resistance R being pulled at a uniform speed v through a region of uniform magnetic field (B_0), applied perpendicular to its plane, with the length L of wire perpendicular to the direction of motion, as shown in Fig. 4.11. A standard problem in introductory texts is to find the current (I) induced in the loop, the force (F_{ext}) required to maintain the motion at constant speed, and the power dissipated (P_{diss}). Before exploring the basic physics behind this system in an analytical way, we can approach each of these parts using DA, starting by assuming that $I \propto B_0^{\alpha} L^{\beta} v^{\gamma} R^{\delta}$, matching dimensions, and finding that

$$I = C_I \frac{B_0 L v}{R}, \tag{4.127}$$

and, for the simple geometry here, it turns out that $C_I = 1$. For the applied force, we repeat the exercise with the same variables, but different dimensional constraints, and find

$$F = C_F \frac{B_0^2 L^2 v}{R}. \tag{4.128}$$

For the power dissipated, we can use these two results and evaluate the $I^2 R$ loss from circuit theory, or the mechanical power provided by Fv, to obtain

$$P = I^2 R = \left(\frac{B_0 L v}{R}\right)^2 R = \frac{B_0^2 L^2 v^2}{R} \qquad \text{or} \qquad P = Fv = \left(\frac{B^2 L^2 v}{R}\right) v = \frac{B_0^2 L^2 v^2}{R}, \tag{4.129}$$

or redo the question "from scratch" using just DA, always obtaining the same answer.

The actual calculations involved for this idealized problem include only a few steps, namely finding the induced EMF and then current, via

$$\Phi_B(t) = B_0 A = B_0 (L x(t)) \quad \rightarrow \quad \mathcal{E} = \frac{d\Phi_B(t)}{dt} = B_0 L \frac{dx(t)}{dt} = B_0 L v \quad \rightarrow \quad I = \frac{\mathcal{E}}{R} = \frac{B_0 L v}{R}. \tag{4.130}$$

The important question of the direction of the induced current shown in Fig. 4.11 is akin to determining the sign in a problem, about which DA provides no guidance. In this case, it is necessary to invoke Lenz's law, where the standard logic proceeds as follows:

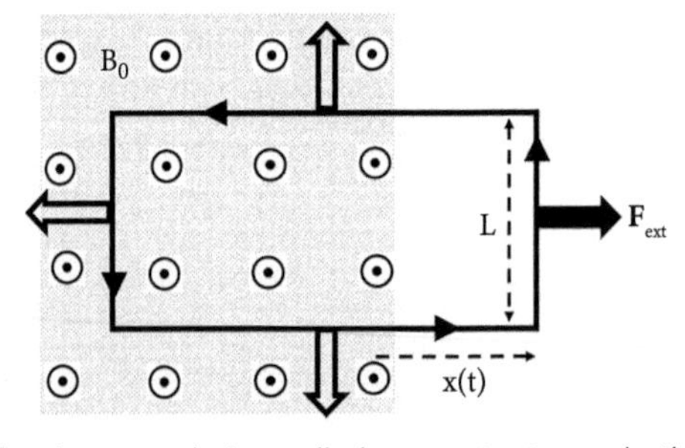

Fig. 4.11 Loop of wire of resistance R being pulled at constant speed v through a magnetic field B_0 by an external force F_{ext}.

- Movement to the right implies less upward flux through the loop'
- Therefore, the induced current should provide more upward flux; and
- The counter-clockwise current (right-hand rule: fingers around the circuit loop, thumb up gives the induced field direction) provides that.

The forces on the current-carrying wires in the magnetic field region are given by $\boldsymbol{F} = I\boldsymbol{l} \times \boldsymbol{B}_0$ and the forces on the upper/lower branches of the loop cancel, leaving a magnetic "drag" force $F_B = ILB_0$ due to the left edge (which is still in the field) and which has to be balanced by the applied F_{ext} to maintain a constant speed, giving

$$F_{ext} = F_B = ILB_0 = B_0 L\left(\frac{B_0 L v}{R}\right) = \frac{B_0^2 L^2 v}{R}, \tag{4.131}$$

which agrees with Eqn. (4.126), determining $C_F = 1$. The fact that $\boldsymbol{F}_{ext}$ is quadratic in the magnetic field is easily conceptualized, since one power of B in the flux gives an induced voltage and hence current, and the current-carrying object then feels the original magnetic field, giving a second power of B. In this relatively simple geometry (wires are, after all, one-dimensional systems), we can calculate everything exactly, but we can also use this as a template for more sophisticated problems where we can still infer dependences, but where a detailed calculation is more challenging.

An extension of this effect to a two-dimensional geometry is visualized in Fig. 4.12, where we imagine a sheet (thickness t) of conducting material being dragged between the poles of a magnet (assume a uniform field B_0 across an effective area $a \times b$) at a constant speed v. The pattern of induced current is now far more complex than the simple wire-constrained, one-dimensional geometry of **Example 4.9**, and given the visualization, the name **eddy current** seems very appropriate. It is still hopefully clear that there is a magnetic drag force (look at small contributions of $d\boldsymbol{F} = Id\boldsymbol{l} \times \boldsymbol{B}_0$ from increments of current that are between the magnetic poles),

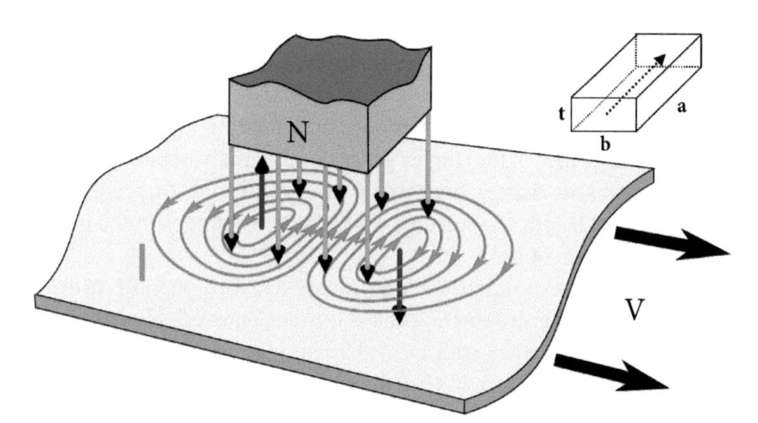

Fig. 4.12 Visualization of current pattern of conducting sheet moving through a magnetic field. Reproduced via Creative Commons License from Wikipedia image authored by Chervorno (2011). Downloaded from https://commons.wikimedia.org/wiki/File:Eddy_currents_due_to_magnet.svg on 9/10/2023. Insert in upper right-hand corner added by R. Robinett showing direction of induced current between the magnetic poles.

but not obviously amenable to a simple, closed form solution. Such problems are now much more mathematically sophisticated (often requiring computer methods) and Lee and Park (2002) note that:

> Since the eddy current problem usually depends on the geometry of the moving conductive sheet and the pole shape, there is no general method to find an analytical solution.

We can still explore the dimensional dependences on the net force, F_{eddy}, by initially assuming general power-law behavior for all of the possible parameters, and balancing dimensions, namely

$$F_{eddy} \propto B_0^{\alpha} \sigma_e^{\beta} v^{\gamma} a^{\delta} b^{\epsilon} t^{\rho}$$

$$\frac{\mathsf{ML}}{\mathsf{T}^2} = \left(\frac{\mathsf{M}}{\mathsf{QT}}\right)^{\alpha} \left(\frac{\mathsf{Q}^2\mathsf{T}}{\mathsf{ML}^3}\right)^{\beta} \left(\frac{\mathsf{L}}{\mathsf{T}}\right)^{\gamma} \mathsf{L}^{\kappa}, \tag{4.132}$$

where, since a, b, d all have dimensions of length, we combine their power-law exponents as $\kappa \equiv \delta + \epsilon + \rho$. Given that we now have four exponents and four constraints, we find that we can uniquely identify the dependences on B, σ_e, v with

$$F \propto \underbrace{B_0^2 \sigma_e v}_{\text{fixed}} \; a^{\delta} b^{\epsilon} t^{3-\delta-\epsilon}. \tag{4.133}$$

Perhaps not surprisingly, we find the same B_0 and v dependence as before, with now the conductivity and shape factors (a, b, d), to be compared to the L^2/R factor in Eqn. (4.130). Given that we expect that any Faraday's law problem using the notion of magnetic flux requires an area, we argue that $F \propto a \times b$, so that $\delta = \epsilon = 1$, which immediately gives $\rho = 3 - \delta - \epsilon = 1$ as well, so that

$$F_{eddy} = B_0^2 \sigma_e v a b t. \tag{4.134}$$

This makes sense since, if either $a, b \to 0$, there is no flux, while if $t \to 0$, there is effectively no current-carrying wire.

We can also connect the two formulae for magnetic braking from Eqns. (4.130) and (4.133) by noting that the effective resistance in the (highly generalized) circuit loop requires the formula for resistance in terms of σ_e and the resistor geometry, namely $R \sim \text{length}/(\sigma_e \times \text{area})$ where we associate (see insert in Fig. 4.12) length $= a$ and area $= bt$, which, when substituted into Eqn. (4.130), agrees dimensionally with Eqn. (4.133).

An application of these ideas to magnetic braking, mentioned early on in **P1.18** as a complicated example, is to consider the torque on a rotating disk due to a magnetic field applied to a small $a \times b$ area a distance R away from the center, with the disk rotating at an angular velocity ω. This next-level problem only requires us to use $\tau = R \times F$ for the torque, and to equate the linear/angular velocities via $v = \omega R$, giving

$$\tau = B_0^2 \sigma_e \omega a b t R^2. \tag{4.135}$$

Magnetic braking of a conducting disk is discussed in textbooks (Zangwill 2013, section 14.12), in pedagogical journals (Wiederick et al. 1987, Heald 1988, and Marcuso et al. 1991), and in applied physics research literature (Lee and Park 2002). The first detailed reference to this problem of which I'm aware is in the historical text by Smythe (first published in 1936, but a 1989 edition is available) where a specific geometry is treated with great mathematical sophistication.

Experimental verification of this behavior requires knowledge of the impact of this magnetic torque on the motion of a disk and we can write Newton's law of rotational motion (with $\mathcal{I}$ being the moment of inertia of the disk)

$$\mathcal{I}\frac{d\omega}{dt} = I\alpha = \tau = -\omega(B_0^2\sigma_e abtR^2) \quad \longrightarrow \quad \frac{d\omega}{dt} = -\frac{\omega}{T_0} \tag{4.136}$$

where

$$\frac{1}{T_0} \equiv \frac{B_0^2\sigma_e abtR^2}{\mathcal{I}}. \tag{4.137}$$

This describes exponential decay of the form $\omega(t) = \omega_0 e^{-t/T_0}$, with $1/T_0 \propto B_0^2$, and Wiederick et al. (1987) confirm both predictions, as shown in Figs. 4.13 and 4.14. You're asked in **P4.41** to determine to what extent the magnetic damping torque, the resulting exponential decay, and dependence of $1/T_0$ on various parameters are uniquely determined by DA.

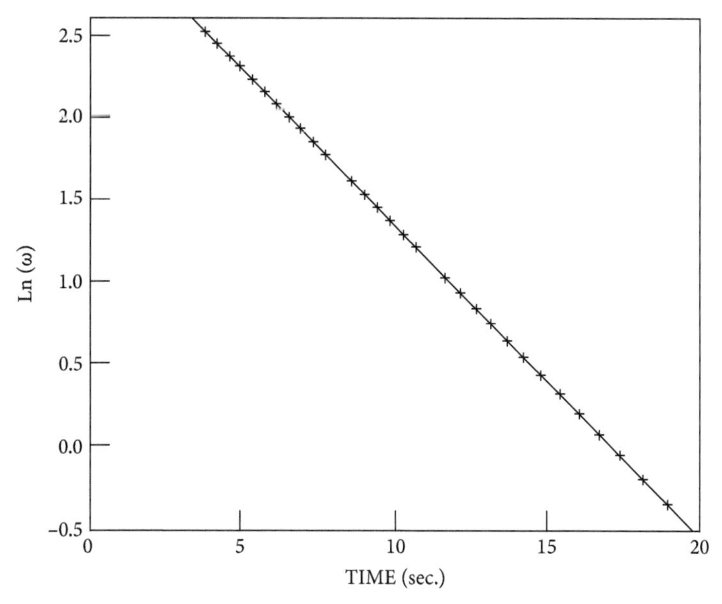

Fig. 4.13 Data on angular velocity ω (log scale) versus time for a magnetic brake showing the expected $\omega_0 \exp(-t/T_0)$ behavior. Reprinted from Wiederick et al. (1987), with the permission of AIP Publishing.

Example 4.11 Eddy currents

A popular lecture demonstration in physics courses that also illustrates the impact of eddy currents, this time on magnets moving near conductors, involves dropping strong (often neodymium) magnets down

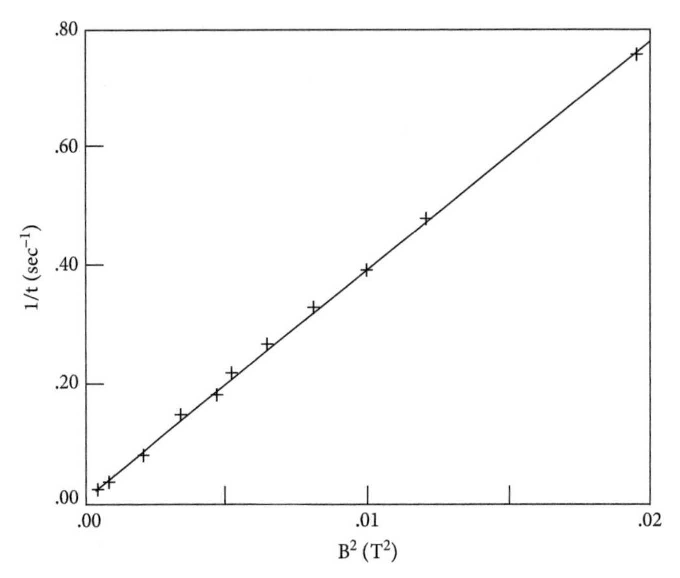

Fig. 4.14 Measured values of $1/T_0$ (sec^{-1}) versus B_0^2 ($Tesla^2$), showing the quadratic behavior predicted by Eqn. (4.136). Reprinted from Wiederick et al. (1987), with the permission of AIP Publishing.

vertical tubes. For non-conducting cylinders, the magnets experience almost free-fall motion (perhaps with a bit of air resistance, or friction from the sides), while for conducting tubes, the eddy current back reaction slows the magnets to a terminal velocity dictated by the geometric and magnetic related variables of the problem, and the magnet's own mg weight.

We can explore the functional dependence of the magnetic retarding force by assuming that the conducting tube has radius R and wall thickness t, conductivity σ_e, and that the falling magnets have magnetic moments given by m, and then write

$$F \propto \mu_0^\alpha m^\beta \sigma_e^\gamma v^\delta t^\epsilon R^\rho, \tag{4.138}$$

where v is the instantaneous speed. Given that there are six quantities, and only four constraints (M, L, T, Q), we can't expect to fully specify the exponents, but we do find that

$$F \propto \mu_0^\alpha m^2 \sigma_e^{\alpha-1} v^{\alpha-1} t^\epsilon R^{\alpha-\epsilon-5}, \tag{4.139}$$

so that at least the dependence on the magnetic moment, m, is determined uniquely to be quadratic, just as it was for the power of B in the magnetic braking example.

If we use the planar magnetic braking example as a metaphor, we might expect the same dependence on σ_e and v (so that a linear exponent for each specifies $\alpha = 2$) and thickness (same linear dependence then determines $\epsilon = 1$) and guess that

$$F \propto \frac{\mu_0^2 m^2 \sigma_e v t}{R^4}, \tag{4.140}$$

which agrees with several theoretical analyses. In fact, we might have tried to actually directly map the planar magnetic braking example onto this problem (imaging wrapping the two-dimensional conductor into a cylinder), where we found that $F = \sigma_e B_{ext}^2 v(ab)t$ and use the analog quantities

$$B_{ext} \rightarrow B_{dipole} \propto \frac{\mu_0 m}{R^3} \qquad \text{and} \qquad ab = A \rightarrow \pi R^2 \tag{4.141}$$

to write

$$F = \sigma_e B_{ext}^2 v(ab)t \quad \longrightarrow \quad F \propto \sigma_e \left(\frac{\mu_0 m}{R^3}\right)^2 v(\pi R^2)t = C_F \frac{\mu_0^2 m^2 \sigma_e v t}{R^4}. \tag{4.142}$$

The problem of *eddy current damping of a magnet moving through a pipe* is a popular one in the pedagogical literature, with analyses by Saslow (1992), MacLatchey et al. (1993), Hahn et al. (1998) (with the title just cited in italics above), Levin et al. (2006), and Donoso et al. (2011) all agreeing on the dimensionality of the result, and most finding a dimensionless prefactor of $C_F = 45/1024$, at least when the same assumptions are made. If we use the diameter of the tube instead of the radius, this coefficient would be $C_F = 45/64$ and much closer to unity. A recent treatment by Lee and Park (2023) includes the impact of the orientation of the magnetic dipole relative to the pipe, and of course DA can provide no guidance on that dimensionless (purely angular) dependence. It's interesting to note that one discussion that uses DA (Pelosko et al. 2005) seemingly leaves out an important factor of pipe thickness (t) and so isn't even dimensionally correct.

4.8 Magnetic monopoles and massive photons: Playgrounds for dimensional analysis

The laws of E&M, with applications across many fields (technology and devices), may be the most "useful" of all of the quadrivium areas we consider. Supplemented by quantum mechanics, E&M, especially in the form of **QED**, is also the best tested of all of the four fundamental forces. Given its continued relevance at the forefront of the standard model of particle physics, it's natural to ask about the possible impact of new physics on this classic topic, and we consider here extensions of E&M, which include magnetic monopoles and the possibility of a non-zero mass for the photon (the carrier of the EM force).

4.8.1 Magnetic monopoles

Of the many foundational laws of physics, perhaps the most famous one which explicitly codifies a null result, is the second of the Maxwell equations in Eqn. (4.97) namely $\nabla \cdot \boldsymbol{B} = 0$, which expresses the fact that, to date, there is no experimental evidence of the existence of isolated magnetic point charges, that is, the separated N/S-pole analogs of +/− point electric charges, at least as fundamental particles.[8]

A number of theoretically motivated models posit such **magnetic monopoles** and there are condensed matter systems that display emergent behavior which is well-described by such constructs (about which more later), so we consider here how E&M might be extended to include such a possibility. This exercise not only extends the pedagogical range of possible DA problems, but

[8] For a series of review articles describing many topics regarding magnetic monopoles, both in theory and experiment, see e.g., Goldhaber and Trower (1990), Milton (2006), or Rajantie (2016).

also can be used to approach modern research-level results. To establish notation, such isolated magnetic charges will be labeled here by $\pm g$, with $+g$ for N-pole and $-g$ for S-poles being a standard convention.

The first obvious extension might well be to copy the electric force component of the Lorentz force law and mirror it for a magnetic monopole, namely

$$\boldsymbol{F}_e = q\boldsymbol{E} \qquad \longrightarrow \qquad \boldsymbol{F}_m = g\boldsymbol{B}, \tag{4.143}$$

from which we can immediately extract the dimensions of a magnetic charge, namely

$$[g] = \frac{[\boldsymbol{F}_B]}{[\boldsymbol{B}]} = \left(\frac{\mathsf{ML}}{\mathsf{T}^2}\right)\left(\frac{\mathsf{M}}{\mathsf{QT}}\right)^{-1} = \frac{\mathsf{QL}}{\mathsf{T}} = \mathsf{Q}\left(\frac{\mathsf{L}}{\mathsf{T}}\right). \tag{4.144}$$

Thus, the dimensions of g are those of ordinary charge (Q) times speed (L/T), which will prove useful later. In fact, given that the only natural speed arising in E&M problems is the speed of light (c), we can define a "baseline" magnetic monopole charge, g_c, one related to the fundamental electric charge e, via

$$g_c \equiv ec = (1.6 \times 10^{-19}\ C)(3 \times 10^8\ m/s) \approx 4.8 \times 10^{-11}\ \frac{C \cdot m}{s}. \tag{4.145}$$

Then, by analogy with Coulomb's law, the $\boldsymbol{B}_g$ field due to a point magnetic charge can be written in the form

$$\boldsymbol{B}_g(r) = \frac{\mu_0}{4\pi}\frac{g\hat{\boldsymbol{r}}}{r^2}, \tag{4.146}$$

which is dimensionally consistent with Eqn. (4.143) since

$$[g] = \frac{[\boldsymbol{B}_g][r^2][4\pi]}{[\mu_0]} = \left(\frac{\mathsf{M}}{\mathsf{QT}}\right)\mathsf{L}^2\left(\frac{\mathsf{Q}^2}{\mathsf{ML}}\right) = \frac{\mathsf{QL}}{\mathsf{T}}. \tag{4.147}$$

The equivalent of the electric potential, and its relation to the monopole $\boldsymbol{B}_g$ field, would be

$$\boldsymbol{B}_g(r) = -\boldsymbol{\nabla} V_g(r) \qquad \text{with the definition} \qquad V_g(r) = \frac{\mu_0}{4\pi}\frac{g}{r}, \tag{4.148}$$

which you can confirm (**P4.45**) are both dimensionally sound, as is the force law between two magnetic monopoles, namely

$$|\boldsymbol{F}_g^{(1,2)}(r)| = \frac{\mu_0}{4\pi}\frac{g_1 g_2}{r^2}, \tag{4.149}$$

all giving $[g] = \mathsf{QL/T}$. Note that the magnetic Coulomb force between two g_c strength monopole satisfies

$$|\boldsymbol{F}_g| = \frac{\mu_0 (g_c)^2}{4\pi r^2} = \mu_0 c^2 \frac{e^2}{4\pi r^2} = \frac{e^2}{4\pi\epsilon_0 r^2} = |\boldsymbol{F}_e|, \tag{4.150}$$

which is the same force between two $\pm e$ point charges, since $c = (\epsilon_0\mu_0)^{-1/2}$.

The analog of the magnetic force on a moving electric charge from the Lorentz law, namely $\boldsymbol{F}_e = q\boldsymbol{v} \times \boldsymbol{B}$, can be guessed (at) by symmetry ideas or approached using relativity arguments (see e.g., Rindler 1989), but let us see how far we can get using DA. In this case, we want to construct a force on a magnetic charge g, with velocity $\boldsymbol{v}$, moving in an electric field $\boldsymbol{E}$ and perhaps we need either/both of the fundamental constants ϵ_0, μ_0. We are thus led to write $F_g^{(E)} = C_F g^\alpha v^\beta E^\gamma \epsilon_0^\delta \mu_0^\rho$, and matching dimensions we find that $\alpha = \gamma = 1$, while $\delta = \rho = (1+\beta)/2$. Thus, the monopole force is constrained to be linear in both g and $\boldsymbol{E}$ and if we assume that it is also linear in $\boldsymbol{v}$ (so $\beta = 1$), we find that the dependence on ϵ_0, μ_0 simplifies to $(\epsilon_0\mu_0)^{(1+\beta)/2} \rightarrow (\epsilon_0\mu_0) = 1/c^2$. Given that we must construct a vector force from $\boldsymbol{v}, \boldsymbol{E}$, the most likely dependence is then

$$\boldsymbol{F}_g^{(E)} = C_F \frac{g\boldsymbol{v} \times \boldsymbol{E}}{c^2}, \tag{4.151}$$

and the correct answer has $C_F = -1$. The complete Lorentz force law is then written as

$$\boldsymbol{F} = q(\boldsymbol{E} + \boldsymbol{v} \times \boldsymbol{B}) + g\left(\boldsymbol{B} - \boldsymbol{v} \times \boldsymbol{E}/c^2\right). \tag{4.152}$$

Just as moving electric charges (currents) can give rise to magnetic fields, we also need the concept of a magnetic current, which in turn can generate electric fields. The analog of one-dimensional currents, two-dimensional (planar) current densities, and their three-dimensional versions have dimensions given by

$$I_m = \frac{dg}{dt} \longrightarrow [I_m] = \left(\frac{\mathsf{QL}}{\mathsf{T}}\right)\left(\frac{1}{\mathsf{T}}\right) = \frac{\mathsf{QL}}{\mathsf{T}^2} \tag{4.153}$$

$$j_m = \frac{I_m}{\text{length}} \longrightarrow [j_m] = \frac{\mathsf{Q}}{\mathsf{T}^2} \tag{4.154}$$

$$J_m = \frac{I_m}{\text{area}} \longrightarrow [J_m] = \frac{\mathsf{Q}}{\mathsf{LT}^2}. \tag{4.155}$$

The analogs of charge densities in three, two, and one dimension(s) (magnetic charge per unit volume, per unit area, or per unit length) are also necessary to extend Maxwell's equations, and have dimensions that satisfy

$$\rho_m = \frac{dg}{dV} \longrightarrow [\rho_m] = \frac{\mathsf{Q}}{\mathsf{L}^2\mathsf{T}} \tag{4.156}$$

$$\sigma_m = \frac{dg}{dA} \longrightarrow [\sigma_m] = \frac{\mathsf{Q}}{\mathsf{LT}} \tag{4.157}$$

$$\lambda_m = \frac{dg}{dl} \longrightarrow [\lambda_m] = \frac{\mathsf{Q}}{\mathsf{T}}. \tag{4.158}$$

We note that ordinary mass density has been consistently labeled ρ_m, but we trust that context clues will avoid confusion with this magnetic charge density having the same symbol, at least in this section.

The familiar Maxwell equations, in differential form, including the possibility of magnetic monopoles, can then be written as

$$\nabla \cdot \boldsymbol{E}(\boldsymbol{r},t) = \frac{1}{\epsilon_0}\rho_e(\boldsymbol{r},t) \qquad \text{Gauss's law} \tag{4.159}$$

$$\nabla \cdot \boldsymbol{B}(\boldsymbol{r},t) = \underbrace{\mu_0\rho_m(\boldsymbol{r},t)}_{\text{monopole density}} \qquad \text{Including magnetic charge density} \tag{4.160}$$

$$\nabla \times \boldsymbol{E}(\boldsymbol{r},t) = -\frac{\partial}{\partial t}\boldsymbol{B}(\boldsymbol{r},t) \underbrace{-\mu_0\boldsymbol{J}_m(\boldsymbol{r},t)}_{\text{monopole current}} \qquad \text{Faraday's law plus monopole magnetic current} \tag{4.161}$$

$$\nabla \times \boldsymbol{B}(\boldsymbol{r},t) = \mu_0\boldsymbol{J}_e(\boldsymbol{r},t) + \mu_0\epsilon_0\frac{\partial}{\partial t}\boldsymbol{E}(\boldsymbol{r},t) \qquad \text{Ampère's law (+ Maxwell),} \tag{4.162}$$

where we note the difference in sign between the two $\boldsymbol{J}_{e,m}$ electric/magnetic current terms. One can also write the new monopole current term with a coefficient given by $1/(\epsilon_0 c^2)$ to make it a bit more symmetric.

The analogous results for the Maxwell equations in integral form are then

$$\oint \boldsymbol{E}\cdot d\boldsymbol{A} = \frac{q_{enc}}{\epsilon_0} \qquad \text{Gauss's law} \tag{4.163}$$

$$\oint \boldsymbol{B}\cdot d\boldsymbol{A} = \underbrace{\mu_0 g_{enc}}_{\text{monopole term}} \qquad \text{Now with magnetic monopoles} \tag{4.164}$$

$$\oint \boldsymbol{E}\cdot d\boldsymbol{l} = -\frac{d}{dt}\int \boldsymbol{B}\cdot d\boldsymbol{A} \underbrace{-\mu_0 I_{m,enc}}_{\text{monopole term}} \qquad \text{Faradays' law} \tag{4.165}$$

$$\oint \boldsymbol{B}\cdot d\boldsymbol{l} = \mu_0 I_{enc} + \underbrace{\mu_0\epsilon_0\frac{d}{dt}\int \boldsymbol{E}\cdot d\boldsymbol{A}}_{\text{displacement term}} \qquad \text{Ampère's law (+ Maxwell)} \tag{4.166}$$

where g_{enc} is the total monopole charge enclosed in the closed surface integral, and $I_{m,enc}$ is the magnetic current "captured" by the line integral.

Example 4.12 Parallel one-dimensional geometries revisited

Using the first example in Section 4.2.3 as a model, we can consider an infinitely long line of magnetic charge density (λ_m) or infinite "wire" of magnetic current (I_m), and use DA to probe the dependence of the resulting $\boldsymbol{B}, \boldsymbol{E}$ fields. Following the example of Eqn. (4.42), we can start by writing

$$B \propto \lambda_m^\alpha r^\beta \epsilon_0^\gamma \mu_0^\delta \qquad \text{and} \qquad E \propto I_m^\alpha r^\beta \epsilon_0^\gamma \mu_0^\delta \tag{4.167}$$

and match dimensions (using the "spelling" of λ_m and I_m from Eqns. (4.158) and (4.153)) and for both cases we find $\alpha = 1$, $\beta = -1$, $\gamma = 0$, and $\delta = 1$, so that

$$B = C_B\frac{\lambda_m\mu_0}{r} \qquad \text{and} \qquad E = C_E\frac{I_m\mu_0}{r}. \tag{4.168}$$

You can use the integral forms of the "magnetic Gauss's law" from Eqn. (4.164) and the new monopole version of Faraday's law from Eqn. (4.165) to find that $C_B = -C_E = 1/2\pi$. These results, and many others considered in Section 4.2.3, show that the same methods used in standard E&M textbooks can be used to approach similar problems involving magnetic charge, taking into account some changes in constants and signs. Many of the comments and connections made in Section 4.3 related to the application of special relativity to E&M can also be carried over directly, including the relativistic transformations of the $\boldsymbol{E}, \boldsymbol{B}$ fields.

Example 4.13 Bound electric and magnetic charge

Consider the system of a point electric charge q and point magnetic charge g, each giving rise to their own $\boldsymbol{E}, \boldsymbol{B}$ Coulomb (or magnetic Coulomb) fields. Even if the two objects are not in motion, the fact that there are simultaneous electric and magnetic fields gives rise to an angular momentum density ($\boldsymbol{l}$ from Eqn. (4.80)) which, when integrated over all space, will produce a net angular momentum $\boldsymbol{L}$. The integration of Eqn. (4.80) for this configuration of q, g fields can be done (see, e.g., Jackson 1999 section 6.13), but here we explore the constraints on q, g, ϵ_0, μ_0 and the separation r imposed solely by dimensional matching. We thus write $L \propto q^\alpha g^\beta \epsilon_0^\gamma \mu_0^\delta r^\rho$, and noting that $[L] = (\mathsf{ML}^2)/\mathsf{T}$, we find that

$$L \propto q^{1-2\gamma} g^{1+2\gamma} \epsilon_0^\gamma \mu_0^{1+\gamma} r^0, \tag{4.169}$$

and a DA approach, by itself, finds that the total field angular momentum is actually independent of the separation between the q, g charges. If we then assume that the electric/magnetic charges contribute in the same way (identical powers), then $\alpha = \beta$, which implies that $\gamma = 0$ and we have $L = C_L q g \mu_0$. A direct calculation[9] gives $C_L = 1/4\pi$, which might have "come along for the ride" anyway, given our intuition. Given that the only direction left in the problem is the vector pointing from q to g (or vice versa), we must also have $\boldsymbol{L} \propto \hat{\boldsymbol{r}}_{qg}$ with only the **unit vector,** $\hat{\boldsymbol{r}} = \boldsymbol{r}/|\boldsymbol{r}|$, being used, since $\rho = 0$ giving no dependence on the actual magnitude of $\boldsymbol{r}_{qg}$.

While there have been (very) occasional reports of the observation of magnetic monopoles (perhaps most famously by Cabrera (1982)), ongoing experimental searches in cosmic-ray/particle astrophysics contexts (Detrixhe et al. 2011, and Abbasi et al. 2022) or in terrestrial accelerators (Acharya et al. 2022) provide increasingly stringent limits on the abundance of such particles as fundamental objects.

However, there are a number of condensed matter/atomic physics systems where emergent configurations of normal matter under extreme conditions are well-described as magnetic monopoles. Perhaps the most famous of these results occurs in a "... class of exotic magnets ..."[10] where one experimental group has made a measurement of the charge and current of magnetic monopoles in spin ice. See Bramwell et al. (2009) and Giblin et al. (2011) for examples of the experimental results and we explore this in **P4.47.** Monopole-like configurations have also been observed in the *synthetic magnetic field* of a Bose–Einstein condensate (BEC) by Ray et al. (2014).

4.8.2 **Massive photons**

One doesn't often see the word **epistemology** in a physics book, but it can sometimes be useful to recall one of its simple definitions:

---

[9] See e.g., Brownstein (1989)

[10] See Castelnovo, Moessner, and Sondhi (2008) for the theoretical prediction.

> The theory of knowledge, especially with regard to its methods, validity, and scope, and the distinction between justified belief and opinion.[11]

This helps us remember to ask ourselves, at least occasionally, questions like "How do we know what we think we know?" For example, one of the first E&M results that's quoted in every textbook is Coulomb's law, given by

$$\boldsymbol{E}(\boldsymbol{r}) = \frac{q\hat{\boldsymbol{r}}}{4\pi\epsilon_0 r^2}, \tag{4.170}$$

with its famous **inverse square law** distance dependence (much like Newton's law of gravitation). Scientists since the time of Cavendish, Coulomb, and Maxwell have experimentally probed to what extent the exponent in Eqn. (4.170) is really $n = -2$, analyzing their experiments in terms of a modified field (or force) law of the form

$$|\boldsymbol{E}_\delta(r)| = \frac{q}{4\pi\epsilon_0 r^{2+\delta}} \quad \text{with the corresponding electric potential} \quad V_\delta(r) = \frac{q}{(1+\delta)4\pi\epsilon_0 r^{1+\delta}}. \tag{4.171}$$

Many of the experiments involved in such determinations have involved the use of two concentric spherical shells, evaluating the difference between the inner and outer shells in electric potential, which would vanish for an inverse square law. The theoretical analysis needed to interpret such experiments is discussed by Fulcher and Telljohann (1975) at an accessible pedagogical level and experimental tests of Coulomb's law and the photon rest mass are extensively reviewed by Tu and Luo (2004), who also outline the long history of experimental limits on δ. Probing this question, Williams, Faller, and Hill (1971) obtained a limit they quote as $\delta = (2.7 \pm 3.1) \times 10^{-16}$, while in a re-analysis of their data, Fulcher 1986 claims an even lower value of $\delta = (1.0 \pm 1.2) \times 10^{-16}$, both consistent with zero.

It thus seems that we're safe to assume that Eqn. (4.170) indeed involves an integral power of r, but this discussion does perhaps beg the question about whether such "alternative" versions of famous results, especially ones which involved the assumption of non-integer power-law dependences (even if infinitesimally close to an integral value) respect the assumptions of DA. For example, if Coulomb's law really involved a $1/r^{2+\delta}$ dependence, would some or all the other dimensionful quantities in the problem $(q, \epsilon_0, \boldsymbol{F}, \boldsymbol{E})$ also be modified to almost integral (up to δ terms) powers of L, T, M, Q?

This is not exactly the fanciful question that it may seem, as some problems in the fully quantum mechanical and relativistically correct version of E&M, namely QED, invoke the notion of **dimensional regularization**, which uses the notation of "almost integral dimensions" as a mathematical tool to analyze field theoretic problems. For some background, albeit at a rather sophisticated level, see e.g., Itzykson and Zuber (2006) and especially Schwartz (2014).

A number of integrals in such theories are formally divergent in $D = 3 + 1 = 4$ space-time dimensions, since relativistic four-vectors like $(ct, \boldsymbol{x})$ exist in a $D =$ four-dimensional space, but can be evaluated in $D - \epsilon$ dimensions[12] with divergences classified in terms of powers of $1/\epsilon$. In such an approach, the electric charge (in the units used in field theory) does acquire a new dimensional dependence as $e \rightarrow \mu^{-\epsilon} e$ where μ is an appropriate dimensional (mass or length) scale in the problem. While this observation may seem to be at a level far above that of the undergraduate curriculum, Olness and Scalise (2011) considered an example where dimensional regularization

[11] Taken from the Oxford English Dictionary.

[12] Note that ϵ here is just a (very) small number, not ϵ_0.

meets freshman E&M, using just such tools to evaluate the electric potential (not the field) of an infinite line of charge, a model system we have discussed in several other contexts here.

Another extension of the standard Coulomb's law (but which is much better motivated theoretically, about which more later) is to consider a variation on the Coulomb potential, by adding an exponential term, namely

$$\phi_\kappa(\boldsymbol{r}) = \frac{Q}{4\pi\epsilon_0}\frac{e^{-\kappa r}}{r}, \tag{4.172}$$

where $[\kappa] = 1/\mathsf{L}$ and such a form is clearly dimensionally consistent. The corresponding electric field is then given by

$$\boldsymbol{E}_\kappa(r) = -\boldsymbol{\nabla}\phi_\kappa(r) = \frac{q\hat{\boldsymbol{r}}}{4\pi\epsilon_0 r^2}e^{-\kappa r}(1+\kappa r) \quad \overset{\kappa\to 0}{\longrightarrow} \quad \frac{q}{4\pi\epsilon_0 r^2}\hat{\boldsymbol{r}}, \tag{4.173}$$

and so reproduces the standard results for $\kappa \to 0$, but also for $r \to 0$. We note that this expression does not satisfy the standard Gauss's law, since the surface integral for this point charge over a spherical surface of radius R is

$$\oint \boldsymbol{E}_\kappa \cdot d\boldsymbol{S} = \left(\frac{q}{4\pi\epsilon_0 R^2}\right)(4\pi R^2)\,e^{-\kappa R}(1+\kappa R) = \frac{q}{\epsilon_0}\left[e^{-\kappa R}(1+\kappa R)\right] \neq \frac{q}{\epsilon_0} \tag{4.174}$$

and only reduces to q/ϵ_0 in the limit that $\kappa \to 0$.

This is one of several examples that demonstrate that the Maxwell equations (two of them in any case) must be extended to be mathematically consistent when $\kappa \neq 0$. We state without proof that the correct versions in integral form are given by

$$\oint \boldsymbol{E}\cdot d\boldsymbol{S} = \frac{q_{enc}}{\epsilon_0} \underbrace{- \kappa^2 \int \phi_\kappa \, dV}_{\kappa\neq 0} \qquad \text{Gauss's law} \tag{4.175}$$

$$\oint \boldsymbol{B}\cdot d\boldsymbol{S} = 0 \qquad \text{No magnetic monopoles} \tag{4.176}$$

$$\oint \boldsymbol{E}\cdot d\boldsymbol{l} = -\frac{d}{dt}\int \boldsymbol{B}\cdot d\boldsymbol{S} \qquad \text{Faraday's law} \tag{4.177}$$

$$\oint \boldsymbol{B}\cdot d\boldsymbol{l} = \mu_0 I_{enc} + \mu_0\epsilon_0\frac{d}{dt}\int \boldsymbol{E}\cdot d\boldsymbol{S} \underbrace{- \kappa^2 \oint \boldsymbol{A}\cdot d\boldsymbol{S}}_{\kappa\neq 0} \qquad \text{Ampère's law (+ Maxwell),} \tag{4.178}$$

where $\boldsymbol{A}$ is the ordinary vector potential, the one which gives $\boldsymbol{B} = \boldsymbol{\nabla}\times\boldsymbol{A}$, and we write the infinitesimal surface area as $d\boldsymbol{S}$ to avoid notational confusion.

The corresponding versions in differential form (which we'll use to explore the wave equation in this version of reality) are

$$\nabla \cdot \boldsymbol{E}(\boldsymbol{r},t) = \frac{1}{\epsilon_0}\rho_e(\boldsymbol{r},t) - \underbrace{\kappa^2\phi_\kappa(\mathbf{r},t)}_{\kappa\neq 0} \qquad \text{Gauss's law} \qquad (4.179)$$

$$\nabla \cdot \boldsymbol{B}(\boldsymbol{r},t) = 0 \qquad \text{No magnetic monopoles} \qquad (4.180)$$

$$\nabla \times \boldsymbol{E}(\boldsymbol{r},t) = -\frac{\partial}{\partial t}\boldsymbol{B}(\boldsymbol{r},t) \qquad \text{Faraday's law} \qquad (4.181)$$

$$\nabla \times \boldsymbol{B}(\boldsymbol{r},t) = \mu_0\boldsymbol{J}_e(\boldsymbol{r},t) + \mu_0\epsilon_0\frac{\partial}{\partial t}\boldsymbol{E}(\boldsymbol{r},t) - \underbrace{\kappa^2\boldsymbol{A}(\boldsymbol{r},t)}_{\kappa\neq 0} \qquad \text{Ampère's law (+ Maxwell).} \qquad (4.182)$$

We note that the inclusion of magnetic monopoles changes two of the Maxwell equations, namely the "no-monopole" one and Faraday's law, while the assumption of massive photons impacts Gauss's and Ampère's laws.[13]

Making use of the investment we've made in documenting (let's not say deriving) the classical wave equations in Eqns. (4.100)–(4.104), we can repeat that exercise, now including any κ-specific terms, starting with

$$\nabla \times (\nabla \times \boldsymbol{E}) = \nabla(\nabla \cdot \boldsymbol{E}) - \nabla^2\boldsymbol{E}$$

Using Faraday's law $\qquad$ Using the new Gauss's law

$$\nabla \times \left(-\frac{\partial \boldsymbol{B}}{\partial t}\right) = \nabla(-\kappa^2\phi_\kappa) - \nabla^2\boldsymbol{E}$$

Using Ampère's law $\qquad$ and moving κ^2 around

$$-\frac{\partial}{\partial t}\left(\mu_0\epsilon_0\frac{\partial \boldsymbol{E}}{\partial t} - \kappa^2\boldsymbol{A}\right) = -\kappa^2\nabla\phi_\kappa - \nabla^2\boldsymbol{E} \qquad (4.183)$$

or

$$-\mu_0\epsilon_0\frac{\partial^2\boldsymbol{E}}{\partial t^2} = -\kappa^2\left(\nabla\phi_\kappa + \frac{\partial \boldsymbol{A}}{\partial t}\right) - \nabla^2\boldsymbol{E}$$

$$-\frac{1}{c^2}\frac{\partial^2\boldsymbol{E}}{\partial t^2} = +\kappa^2\boldsymbol{E} - \nabla^2\boldsymbol{E} \qquad (4.184)$$

since $\boldsymbol{E} = -\nabla\phi_\kappa + \partial\boldsymbol{A}/\partial t$, or

$$\frac{\partial^2\boldsymbol{E}}{\partial t^2} = c^2\left(\nabla^2\boldsymbol{E} - \kappa^2\boldsymbol{E}\right). \qquad (4.185)$$

A plane wave solution of the form in Eqn. (4.103) gives the connection between $\boldsymbol{k}, \omega$

$$\omega^2 = c^2(|\boldsymbol{k}|^2 + \kappa^2) \qquad (4.186)$$

[13] As a young theorist, I was often told to "... only consider one extension of the 'standard model' at a time ..." and it seems that the inclusion of both new effects considered here, magnetic monopoles **and** massive photons, might not be logically consistent; see Ignatiev and Joshi (1996).

which is a different dispersion relation for classical EM waves than the standard linear one, $\omega = |\boldsymbol{k}|c$.

So far, we have not yet even mentioned the phrase "massive photons," despite that being the title of this subsection. To see how this connection arises, let us multiply both sides of Eqn. (4.186) by two powers of Planck's constant, $\hbar$, to give

$$(\hbar\omega)^2 = (\hbar|\boldsymbol{k}|c)^2 + (\hbar\kappa c)^2 \quad \stackrel{?}{\longrightarrow} \quad (E_\gamma)^2 = (|\boldsymbol{p}_\gamma|c)^2 + (m_\gamma c^2)^2. \tag{4.187}$$

This is reminiscent of the relativistic connection between energy and momentum for a massive particle, hence the $m_\gamma c^2$, and we associate

$$\hbar\kappa c = m_\gamma c^2 \quad \longrightarrow \quad \frac{1}{\lambdabar} \equiv \kappa = \frac{m_\gamma c}{\hbar} \quad \longrightarrow \quad \lambdabar = \frac{\hbar}{m_\gamma c}, \tag{4.188}$$

where $\lambdabar$ is the **reduced** de Broglie wavelength (using $\hbar$ versus h) associated with a massive particle.

The fact that the exchange of a massive particle can give rise to a potential energy with an exponential suppression with distance, as in Eqn. (4.172), is familiar from nuclear and particle physics. The form of the so-called **Yukawa potential** describing the effective interaction energy between nucleons (neutrons/protons) via the exchange of pi mesons (pions), is given by

$$V_{Yukawa}(\boldsymbol{r}) \equiv V_Y(\boldsymbol{r}) = g_\pi \frac{e^{-r/\lambdabar_\pi}}{r} \quad \text{where} \quad \lambdabar_\pi = \frac{\hbar}{m_\pi c}. \tag{4.189}$$

Given this connection, experimental limits on the size of $\kappa \equiv 1/\lambdabar$ can be translated into bounds on a possible photon mass, via Eqn. (4.188), and this is explored in **P4.55**. Historical reviews of the limits on m_γ include Goldhaber and Nieto (1971, 2010), Tu, Luo, and Gillies (2005), and there is a nice discussion in (Jackson 1999, section 1.2).

Besides the Maxwell equations, many of the other quantities in standard E&M have extensions, including the Poynting vector and electric and magnetic field energy densities, now depending on the scalar and vector potentials $\phi_\kappa, \boldsymbol{A}$ from Eqn. (1.65), as well as $\boldsymbol{E}, \boldsymbol{B}$, namely

$$\boldsymbol{S} = \frac{1}{\mu_0}\left(\boldsymbol{E} \times \boldsymbol{B} + \underbrace{\kappa^2 \phi_\kappa \boldsymbol{A}}\right) \tag{4.190}$$

and

$$U = \frac{\epsilon_0}{2}\left(|\boldsymbol{E}|^2 + \underbrace{\kappa^2 \phi_\kappa^2}\right) + \frac{1}{2\mu_0}\left(|\boldsymbol{B}|^2 + \underbrace{\kappa^2 |\boldsymbol{A}|^2}\right). \tag{4.191}$$

Experiments that probe modifications of Coulomb's law in either of the forms in Eqns. (4.171) or (4.173) might be expected to gives bounds on either $\delta = n - 2$ or κ and it is natural to ask how such limits might be connected, given that one quantity is dimensionless, while the other has $[\kappa] = 1/\mathsf{L}$. An exact translation between the two parameters can depend on the details of the experiment, but for the classical geometry of two concentric spheres of radius $R_2 > R_2$, one can analyze the results using either approach to give

$$M(R_2, R_1)\,\delta = \frac{V(R_2) - V(R_1)}{V(R_2)} = -\frac{1}{6}\kappa^2\left(R_2^2 - R_1^2\right), \tag{4.192}$$

where $M(R_2, R_1)$ is dimensionless function of order unity. This explicit connection, namely that $\delta \sim (\kappa R_{1,2})^2$, is a specific example of a general theorem due to Goldhaber and Nieto (1971), which says

> If field generating apparatus and field detectors are confined to a region of maximum dimensions D, then effects of a finite photon mass μ are of order $\mathcal{O}[(\mu D)^2]$ or smaller ...

where their μ corresponds to κ in our notation. For example, one can expand the expression for electric field in Eqn. (4.173) for small values of κr and see that the deviation from Coulomb's law is proportional to $(\kappa r)^2$.

4.9 **Problems for Chapter 4**

Q4.1 Reality check: (a) What was the single most important, interesting, engaging, and/or useful thing you learned from this chapter?
(b) Have you ever used DA in approaching any problem in E&M?

P4.1 Ponderomotive force and potential: (a) Assume that the ponderomotive force considered in **Example 4.1** is already known to be derivable from a potential energy function, $\boldsymbol{F}_p = -\nabla\Phi_p$, and use DA with q, m, E_0, ω as inputs to constraint Φ_p. Does your result agree with Eqn. (4.9)?
(b) The ponderomotive force is an especially useful concept in **plasma physics**, where an ionized gas of electrons (so charge e and mass m_e and number density n_e are relevant parameters) interacts with laser fields. Show that the result in Eqn. (4.9) can be written in the form

$$n_e \boldsymbol{F}_p = -\frac{\omega_P^2}{\omega^2}\nabla U_E, \tag{4.193}$$

where $U_E = \epsilon_0 \boldsymbol{E}^2/2$ is the energy density stored in the electric field and ω_P is the **plasma frequency** discussed in Section 2.2 and Eqn. (2.13). Confirm that the dimensions of both sides of Eqn. (4.193) match.

P4.2 Optical tweezers or "single-beam gradient force optical trap": Laser beams with an intensity that varies in space can be used to move/hold objects ranging in sizes from atoms to biological cells. The force $\boldsymbol{F}_T$ of such an **optical tweezer** can depend on the size of the object (assume a spherical shape of radius a), the speed of light (c), the intensity ($I(\boldsymbol{r})$ or power per unit area), and to encode information on the spatial variability, some power of the gradient ∇ operator.

(a) Write $\boldsymbol{F}_T = C_F a^\alpha c^\beta \nabla^\gamma (I(\boldsymbol{r}))^\delta$, match dimensions, and find the exponents to the extent you can.

(b) Given that $\boldsymbol{F}_T$ must be a vector, does this help you determine γ and any other dimensions?

Notes: There are actually two other dimensionless parameters in the problem, namely the indices of refraction of the particle (n_0) and that of the surrounding medium (n_1). It turns out that the dimensionless constant is given by

$$C_F = 2\pi n_0 \left[\frac{(n_0/n_1)^2 - 1}{(n_0/n_2)^2 + 1} \right]. \tag{4.194}$$

This method for the acceleration and trapping of particles by radiation pressurewas invented by Ashkin (1970), for which he shared the 2018 Nobel prize in Physics "... for the optical tweezers and their application to biological systems ... ." He also predicted (Ashkin et al. 1986) with amazing foresight that such methods would "... open a new size regime to optical trapping encompassing macromolecules, colloids, small aerosols, and possibly biological particles."

P4.3 Chasing the charge in the "mirror": A charge q placed a distance D in front of an infinite conducting plane will be attracted to it, due to the induced charge density on the surface. This problem is solved elegantly in most E&M textbooks by the **method of images**, which shows that the boundary value problem is equivalent to assuming that a (fictitious) charge $-q$ is located at an equal distance ($-D$) on the other side of the plane.

(a) Use DA to discuss the time it takes for such a charge, starting from rest, to hit the surface, by assuming $\tau = C_T q^\alpha \epsilon_0^\beta m^\gamma D^\delta$ and solving for $\alpha, \beta, \gamma, \delta$. Given that this is a basically a Coulomb's law problem, do you think we should use ϵ_0 or $4\pi\epsilon_0$?

(b) Assuming that the "mirror charge" is always seen as being a distance $2z$ away (where z is the real distance to the conducting plane), Newton's law would be

$$m\frac{dz^2}{dt^2} = ma = F = -\frac{q^2}{4\pi\epsilon_0(2z)^2}. \tag{4.195}$$

Use non-dimensionalization methods to write $t = \tau y$ and $z = Dw$, where y, w are dimensionless, to find the dependence of τ on D and the other parameters in the problem, and compare to the result of part (a).

(c)* Use the energy methods discussed for the harmonic oscillator in Section 3.1 to write

$$\frac{1}{2}m\left(\frac{dz}{dt}\right)^2 - \frac{q^2}{4\pi\epsilon_0(2z)} = E_0 = -\frac{q^2}{4\pi\epsilon_0(2D)}$$

(where we again use the distance between the charge and its mirror partner as $2z$) to find the connection between dz and dt, noting an important sign. Integrate this respectively over the ranges $(D, 0)$ and $(0, \tau)$. Compare the resulting dimensions with part (a) and evaluate the integral to obtain C_T.

(d) Compare this problem to that discussed in **Example 4.2**, which explored the oscillation frequency ω of a periodic system. This example is more of a "one-shot" experiment, but are there any similarities?

(e) Repeat part (a) for the case where a point dipole ($\boldsymbol{p}$) is released from rest a distance D away from the conducting plane.

P4.4 Cyclotron frequency and Larmor radius: Use Newton's third law and the Lorentz magnetic force equations, $\boldsymbol{F} = m\boldsymbol{a}$ and $\boldsymbol{F}_B = q\boldsymbol{v} \times \boldsymbol{B}$, to derive both parts of Eqn. (4.13).

P4.5 Magnetic focusing: Beams of electrons (so, charge $-e$ and mass m_e) can be focused using magnetic deflection. A simplified version of this problem (Lorrain 2000 pp.320–21), imagines that a beam of electrons has been accelerated through a voltage V and then enters a long solenoid (length D and magnetic field B). Electrons that are "on-axis" (only a $v_{\|}$ component) are not deflected, while those that have $v_{\perp} \neq 0$ undergo Larmor precession, but can return to the axis at the end of the solenoid, provided the magnitude of the magnetic field is chosen appropriately.

(a) Find the dimensional dependence of the required B field by writing $B = C_B m_e^{\alpha} e^{\beta} V^{\gamma} D^{\delta}$ and matching dimensions to find $\alpha, \beta, \gamma, \delta$.

(b) Use the cyclotron and Larmor formulae in Eqn. (4.13), and conservation of energy $eV = mv^2/2$, to solve the problem from first principles and show that $C_B = 2^{3/2}\pi$.

P4.6 Mass spectrometer: Particles of charge q and mass M are accelerated through a potential difference (voltage) V and enter a region of magnetic field (B). They are deflected in the usual circular trajectories and exit the field region after undergoing a half-circular orbit, a distance D away from their starting point. To "weigh" the particles, we want to determine their mass and so write $M = C_M q^{\alpha} B^{\beta} D^{\gamma} V^{\delta}$, match dimensions, and solve for the exponents. Can you then use the equations of motion in Eqn. (4.13) to derive the result, including the constant C_M?

Note: Short and readable histories of mass spectrometry have been given by Griffiths (2008) and Sharma (2013), who include references to important early papers by researchers such as Dempster (1918) and Aston (1919).

P4.7 Hall effect—Theory and experiment: (a) Try to reproduce the expression for the Hall resistance by combining the following equations as suggested by the discussion in Section 4.1.2, namely

$$E_H = vB \text{ (forces balanced)} \tag{4.196}$$

$$E_H = \frac{V_H}{w} \text{(field/potential connection)} \tag{4.197}$$

$$n_e e v = J_e = \frac{I}{A} \text{ (current density definition)} \tag{4.198}$$

$$A = wt \text{ (geometry)} \tag{4.199}$$

to relate $V_H = IR_H$ and solve for R_H.

(b) Armentrout (1990) presents experimental data on the Hall effect (in copper) using "... apparatus ordinarily found in an elementary laboratory ..." and we show in Fig. 4.15 a plot of V_H (in μV, so note the prefix) versus I (*Ampere*) for several values of the applied magnetic field, illustrating that $V_H \propto I$. Using say the values for $B = 0.405$ *Tesla* (denoted by the + signs), find the Hall resistance R_H (slope of the plot) for that value of applied field. Using the relation $R_H = B/(en_e t)$, and the fact that $t = 0.07$ *mm* in the experiment, use your data to find an estimate of the electron carrier density, n_e (in m^{-3}), and compare to the value cited in the paper as $n_e = 12.2 \times 10^{28}\ m^{-3}$.

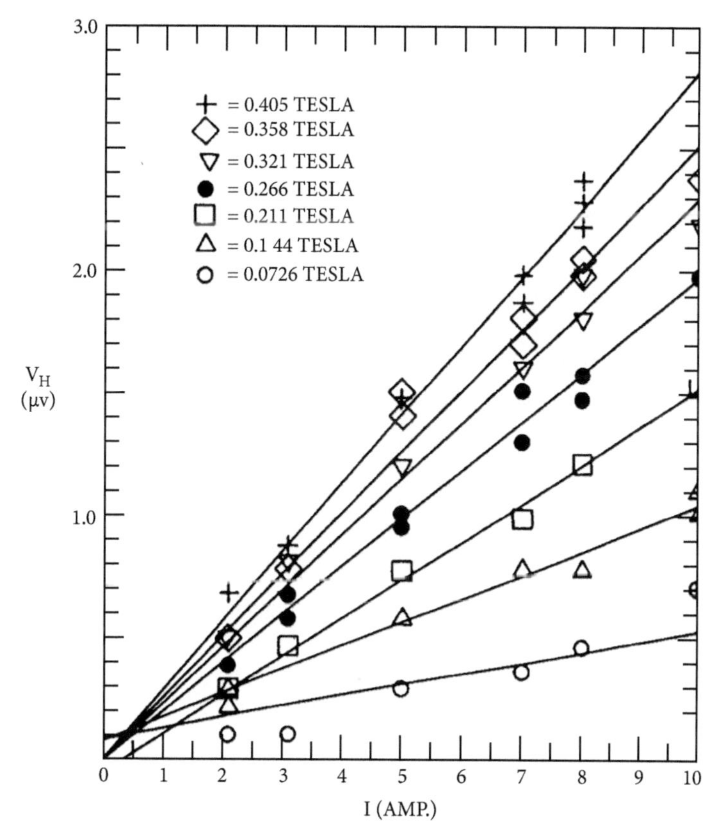

Fig. 4.15 Data on the Hall effect in copper, plots of V_H (in μV) versus I (Ampère) for various values of the applied magnetic field (in *Tesla*). The thickness of the copper strip is cited as $t = 0.07\ mm$. Reprinted from Armentrout (1990) with the permission of AIP publishing. For use with **P4.6**.

P4.8 Hillas criteria for maximum cosmic ray energies: (a) Ultra-high energy ($E > 10^{19}\ eV$) cosmic rays can be produced in many astrophysical situations, and can be confined by magnetic fields until their energy reaches a value for which their Larmor radius (Eqn. (4.13)) exceeds the size of the acceleration region (Hillas 1984). The maximum energy possible for cosmic rays exiting such a region (E_{max}) could then depend on e (or perhaps Ze for charged nuclei), the size of the confining region (R), the magnetic field (B_0) there, and because we expect these charged particles to be ultra-relativistic, the only relevant speed is c.

(a) Assume that $E_{max} \propto e^{\alpha} B_0^{\beta} c^{\gamma} R^{\delta}$ and match dimensions to solve for $\alpha, \beta, \gamma, \delta$.

(b) Using your result, evaluate the maximum possible energy (in eV) for the astrophysical accelerators in Table 4.1.

(c) A standard way to represent the correlation between the size of the acceleration region (R) and the magnetic field (B_0) present there, for a given value of the maximum energy (say

$E_{max} = 10^{20}\ eV$) able to be contained within, is the so-called **Hillas plot**, as it was first discussed systematically by Hillas (1984). A more recent/updated example of a plot of B_0 versus R is an elegant infographic by Kotera and Olinto (2011), shown in Fig. 4.16, listing possible astrophysical sources. Based on your DA result from part (a), why does the plot look the

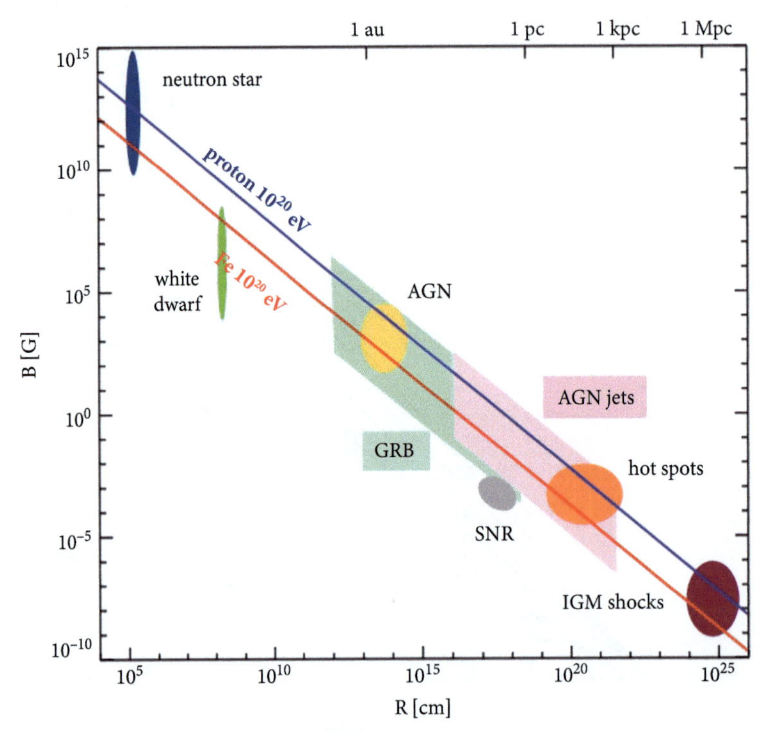

Fig. 4.16 Theoretical plot of magnetic field (in *Gauss*) versus size of region of astrophysical acceleration processes ($R\,(cm)$, for cosmic rays with a fixed value of $E_{max} = 10^{20}\ eV$ for both protons ($Z = 1$) and iron (Fe) nuclei ($Z = 26$)), as discussed in **P4.8**. Used with permission of *Annual Reviews*, from Kotera and Olinto (2011); permission conveyed by Copyright Clearance Center, Inc. Various possible sources such as AGN (Active Galactic Nuclei), Gamma Ray Bursts (GRB), and Supernova Remnants (SNR) are indicated.

Table 4.1 Astrophysical parameters of various cosmic ray acceleration regions, with size (R with various appropriate length scales cited) and magnetic field (B where G stands for Gauss). Note that AU is astronomical unit and pc is parsec (see Section 10.4 for conversion factors) and $10^4\ G = 1\ T$, which is the MKSA unit of magnetic field.

Source	R	B_0	$E_{max}\ (eV)$
Heliosphere	$100\ AU$	$10\ \mu G$	
Interstellar medium	$100\ pc$	$5\ \mu G$	
Neutron star	$10\ km$	$10^{12}\ G$	
AGN (active galactic nuclei)	$10^{12}\ m$	$10^4\ G$	

way it does and can you reproduce the straight line labeled "*proton* 10^{20} *eV*"? Why is the line labeled "*Fe* 10^{20} *eV*" parallel to the proton one, and just how far below it does it lie?

P4.9 "Circular motion of a charged particle in an electric dipole field": The title of this problem is taken from a nice pedagogical article by Jones (1995), who analyses the motion of a charge q and mass m in the field of an electric dipole $\mathbf{p}$. The (fixed) dipole moment is located at the origin, so $(x, y, z) = (0, 0, 0)$, pointing in the $+z$ direction, and the charge is released (from rest) at a point in the (x, y) plane (i.e., so with $z = 0$) a distance R away from the dipole. It's found that the particle "... swings back and forth in a semi-circular arc, as though it were a pendulum supported at the origin" (Griffiths 2017 p. 159).[14]

(a) Use DA to explore the dependence of the oscillation period on the dimensional parameters by writing $\tau = C_T p^{\alpha} q^{\beta} \epsilon_0^{\gamma} m^{\delta} R^{\sigma}$ and solve for $\alpha, \beta, \gamma, \delta, \sigma$ to the extent you can.

(b) Can you simplify this by assuming how p and q must appear in the final expression, similarly to **Example 4.2**?

P4.10 Spherical charge distributions: (a) Approach the problem of the electric field inside a uniformly charged sphere (charge density ρ_e) from **Example 4.4** by assuming that the field for $r < a$ can be written in the form $E \propto \rho_e^{\alpha} \epsilon_0^{\beta} r^{\gamma}$ (so we assume that ρ_e is known, with information on q, a already "baked in") and solve for α, β, γ and compare to the result in Eqn. (4.24).
(b) Assume that the charge density in the spherical region varies with radius, say as $\rho_e^{(n)}(r) = C_n r^n$, which determines the dimensions of C_n. Assume $E \propto C_n^{\alpha} \epsilon_0^{\beta} r^{\gamma}$ and solve for α, β, γ. Does your result reproduce the case of constant charge density for $n = 0$?
(c)* If the variable charge density from part (b) is appropriate in a spherical region of radius a, and gives a total charge q, evaluate C_n in terms of a, q by integrating the charge density over the spherical volume. Do the dimensions of C_n match those from part (b)?
(d)* Use the Gauss's law method from **Example 4.4** to evaluate the electric field inside the sphere in terms of q, a, r using this more general n-dependent charge density and determine the Buckingham Pi theorem function $F(\Pi = a/r)$.

P4.11 * Far-field dependence of arrays of charges: The array of charges shown in Fig. 4.17 correspond to examples of (a) *monopole*, (b) *dipole*, (c) *quadrupole*, (d) *octopole*, etc., electric field configurations at large distances (for $r >> a$). The first corresponds to Coulomb's law, while the second to the electric dipole discussed in **Example 4.4**.

(a) Show that the on-axis electric field for the third configuration is given by

$$E_x = +\frac{q}{4\pi\epsilon_0(r+a)^2} - \frac{2q}{4\pi\epsilon_0 r^2} + \frac{q}{4\pi\epsilon_0(r-a)^2} = \frac{q}{4\pi\epsilon_0 r^2}\left\{\frac{1}{(1+a/r)^2} - 2 + \frac{1}{(1-a/r)^2}\right\}. \quad (4.200)$$

Expand this for $a/r << 1$ and show that the first non-vanishing term goes like a^2/r^4. Feel free to use symbolic math tools such as Mathematica© to expedite your calculation. Can you see how this configuration looks like a pair of opposite sign dipoles, hence the (partial) cancellation?

[14] It's interesting to note that Jones (1995) discovered this "... interesting result while running a numerical simulation ..." and this can be explored in more general cases using computer algebra, as done by McGuire (2003).

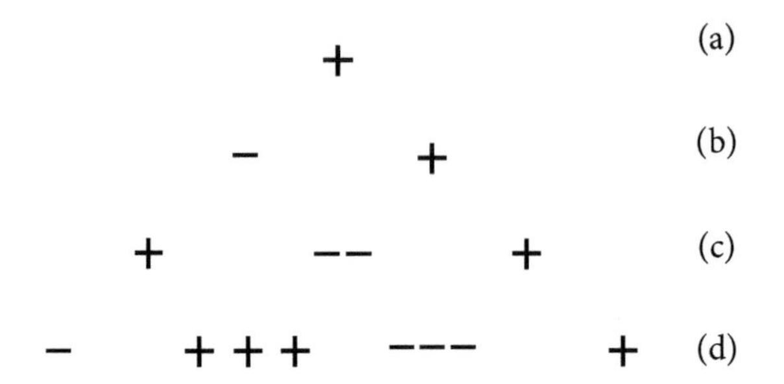

Fig. 4.17 Series of point charge distributions corresponding to (a) point charge (monopole), (b) dipole, (c) quadrupole, and (d) octopole dependences in the far-field electric field. For use with **P4.11** and **P4.12**.

(b) Repeat for the fourth configuration, writing down the exact electric field on-axis, then expanding and show that this field goes like a^3/r^5.

(c) Can you extend the diagram in a logical way several more times to get even higher-pole field configurations? Does this pattern look familiar? Do you recall **Pascal's triangle** from the study of the **binomial expansion**? This example is discussed in nice detail in the context of E&M by Purcell and Morin (2011).

P4.12 * **Far-field dependence of arrays of parallel currents:** Repeat the analysis of **P4.11**, but now assume that the ± signs in Fig. 4.17 refer to infinitely long current-carrying wires, with + corresponding to currents coming out of the paper, and – to current going into the paper, and find the far-field limits for the magnetic fields in each case.

P4.13 ***B* and *E* fields of thick wires:** (a)* Generalize the Ampère's law and Gauss's law results in Eqns. (4.45) and (4.46) to find the $\boldsymbol{B}(r)$ and $\boldsymbol{E}(r)$ fields as a function of r for an infinitely long wire, one with finite radius a, both inside ($r < a$) and outside ($r > a$) the wire. In this case the (standard) three-dimensional volume charge density ρ_e and current density J_e are the relevant quantities.
(b) Show that your results from part (a) are (i) continuous at $r = a$ and (ii) consistent with those from Eqns. (4.45) and (4.46).
(c) Apply DA methods to find the ***B*** and ***E*** fields inside the wires by assuming $B(r) \propto J_e^{\alpha} \epsilon_0^{\beta} \mu_0^{\gamma} r^{\delta}$ and $E(r) \propto \rho_e^{\overline{\alpha}} \epsilon_0^{\overline{\beta}} \mu_0^{\overline{\gamma}} r^{\overline{\delta}}$, matching dimensions, and determining the exponents. Do your results agree with those from part (a)?

P4.14 * ***B* and *E* fields for infinite <u>thin</u> charged and conducting plates:** Use Ampère's law and Gauss's law (and the ***B***, ***E*** field directions shown in Fig. 4.8) to reproduce the DA results in Eqn. (4.49) and confirm that $C_B = C_E = 1/2$.

P4.15 * **Long, but finite-length charge/current segments:** One can evaluate by direct integration (i.e., not using Gauss's and Ampère's laws, but rather Coulomb's law and the Biot–Savart formula) the electric and magnetic fields due to a long, but finite length segment of linear charge density (λ_e) or current (I) as shown in Fig. 4.18, at least at a special point of symmetry, along the perpendicular bisector. Since both systems share a similar geometry, we can use $d = \sqrt{r^2 + x^2}$ and $\sin(\phi) = r/d = \cos(\theta)$ and the line segment is of length $2L$. We're interested to see how

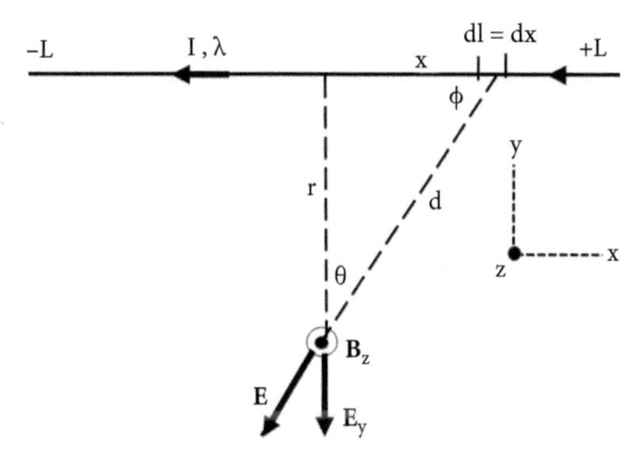

Fig. 4.18 Finite length (L) of linear charge density (λ) or current (I) with resulting electric and magnetic field a distance r on the axis-off symmetry. For use with **P4.15**.

we reproduce the results for the infinitely long segment and especially how the second length dimension, L, "disappears" from the problem.

(a) Show that the contribution to the E_y from a short segment of linear charge density, with charge $dq = \lambda dx$, on the perpendicular bisector, is

$$dE_y = \frac{\lambda_e dz}{4\pi\epsilon_0 d^2}\cos(\theta) = \frac{\lambda_e r}{4\pi\epsilon_0 (r^2 + x^2)^{3/2}}\, dx. \tag{4.201}$$

Integrate these contributions over the length of the segment to obtain

$$E_y = \int_{-L}^{+L} dE_y = \frac{\lambda_e r}{4\pi\epsilon_0}\int_{-L}^{+L}\frac{dx}{(r^2 + x^2)^{3/2}} = \frac{\lambda_e}{2\pi\epsilon_0 r}F(L, r) \tag{4.202}$$

and discuss the behavior of $F(L, r)$ as $L\rightarrow\infty$. Does this general expression have the form of a Buckingham Π function?

(b) Use the same type of analysis to find the magnetic field along the bisector by (i) showing that

$$dB_z = \frac{\mu_0 I dx}{4\pi d^2}\sin(\phi) = \frac{\mu_0 I r}{4\pi (r^2 + x^2)^{3/2}}\, dx \tag{4.203}$$

and then (ii) integrating dB_z over the range $(-L, +L)$. Compare your result to that in part (a), especially showing how the $L\rightarrow\infty$ limit gives the standard result for an infinitely long wire.

P4.16 * ***B* and *E* fields for finite <u>thick</u> charged and conducting plates:** Use Ampère's law and Gauss's law (and the ***B***, ***E*** field directions shown in Fig. 4.8) for a **finite width** plate (thickness a) as a function of z, both inside and outside the plate. For convenience, you might want to measure z from the center of the thick plates. Here again the relevant three-dimensional charge and current densities are ρ_e and J_e.

P4.17 Magnetic field along the axis of a solenoid: Imagine a sheet of current rolled around a circular form (radius a) forming an infinitely long cylinder of current, with the two-dimensional current density j_e circulating around the cylinder. In this (idealized) geometry, there is a uniform field inside the device, along the direction of symmetry, and zero field outside.

(a) Use Ampère's law, with a rectangular integration loop with one long side of length l inside the cylinder and the other outside, to evaluate B. Show that this gives $B_{axis} = \mu j_e$, which is the result for the $\boldsymbol{B}$ field between two parallel sheets of current, with j_e in opposite directions.

(b) A finite length version of this is a model for a **solenoid**, where the current density j_e arises from a single current-carrying wire (I) wrapped around the circular form, so that with N turns along a length l, the current density is $j_e = NI/L$, so that $B_{solenoid} \approx \mu_0 IN/L$, which is dimensionally correct and does not involve the cross-section area (no a) dependence, but does introduce the dimensionless N. Compare this result to that of a single loop of current at the center discussed in **Example 4.7** where the field does involve the radius a.

(c) One can also find the on-axis $\boldsymbol{B}$ field of a long (infinite) solenoid by integrating the magnetic field from individual current loops, as derived in Eqn. (4.33), by writing

$$dB_z = \frac{\mu_0\, dIa^2}{2(a^2 + z^2)^{3/2}} \tag{4.204}$$

writing $dI = j_e\, dz$ and the integrating over all $z \in (-\infty, +\infty)$. Follow this "recipe" and compare to the Gauss's law result from part (a).

P4.18 Torque and energy of electric dipole in electric field: (a) Use the same method used to derive Eqns. (4.58) and (4.59) to explore the torque and energy of an electric dipole $\boldsymbol{p}$ in an external electric field $\boldsymbol{E}$. In both cases, you can start by writing $\tau_E, \mathcal{E}_E \propto p^\alpha E^\beta \epsilon_0^\gamma$ and match dimensions.

P4.19 Torque on a current loop and resulting motion: (a) Consider a planar loop of wire of area a carrying current i in a region of uniform magnetic field B, with the field perpendicular to the loop. If one tips the loop slightly, there will be a torque, $\mathcal{T}$, tending to restore the loop to its original position. Assume that $\mathcal{T} \sim i^\alpha B_0^\beta a^\gamma \mu_0^\delta$ to find the dependence on these quantities.
(b) If the loop has a mechanical moment of inertia $\mathcal{I}$, what is the angular frequency (and period) for small oscillations about equilibrium? Assume a dependence like $\omega \propto i^\alpha B_0^\beta \mathcal{I}^\gamma A^\delta$ and solve the exponents.
(c) Are your results from parts (a) and (b) consistent with Newton's law for rotational motion, namely $\mathcal{T} = \mathcal{I}\alpha$, where $\alpha = \ddot{\theta}$ is the angular acceleration?

P4.20 Dipole–dipole interaction energy: Two magnetic dipoles, $\boldsymbol{m}_1, \boldsymbol{m}_2$, separated by a distance R, will have an interaction energy (E_{mm}) since one dipole can interact with the magnetic field of the other. Apply DA techniques to explore this effect by writing $E_{mm} \propto \mu_0^\alpha m_1^\beta m_2^\gamma R^\delta$ and solve for $\alpha, \beta, \gamma, \delta$ to the extent possible. Can you use symmetry ideas to relate β, γ and simplify your answer? Is your answer consistent with combining the results of Eqns. (4.37)) and (4.59), at least in terms of dependence on dimensional quantities?

P4.21 Solar constant: On any sunny day, we can certainly qualitatively feel the energy output of the Sun. More quantitatively, the power per unit area received on the Earth's surface, on average, is called the **solar constant** and has a value of about $S_\odot \approx 1.4\, kW/m^2$.

(a) Equating this value with the expression for the Poynting vector, $\mathbf{S}$, we know that there will be a corresponding radiation pressure given by Eqn. (4.78). Evaluate this pressure in SI units, and compare to atmospheric pressure, which is roughly $P_{atm} \approx 10^5\ Pa$. Give a one-letter answer to the question of why we feel the suns energy, but not the corresponding momentum.

(b) Evaluate the average values of the electric and magnetic field corresponding to the solar constant.

(c) As a cross-check, use the total power radiated by the Sun ($P_\odot$) and the Earth–Sun (D_{ES}) distance to confirm the value of $S_\odot$ cited above. You're free to use

$$P_\odot \approx 4 \times 10^{26}\ W \qquad \text{and} \qquad D_{ES} \approx 1.5 \times 10^{11}\ m. \tag{4.205}$$

P4.22 *B* and *E* field magnitudes in a laser: (a) A typical laser pointer might have a power of 3–5 mW, with a spot size about 1 mm in radius. What is the associated intensity I (power per unit area) and corresponding $\mathbf{E}$ and $\mathbf{B}$ field strengths. Compare to the values discussed in **P4.21**.
(b) At some point, the most intense laser in the world was cited as having $I = 10^{23}\ W/cm^2$; see Yoon et al. (2021). What are the corresponding value of the electric and magnetic fields in such a beam?
(c) The **critical electric field** discussed in **Example 1.3** associated with the onset of non-linearity of the laws of E&M, was cited as $E_C = (m_e^2c^3)/(e\hbar) \approx 10^{18}\ V/m$. What is the corresponding intensity in W/cm^2 and compare to the "world record value" cited from part (b). A research problem related to this limit is discussed by Bulanov et al. (2010).

P4.23 Relativistic transformations of *E* and *B* fields for a wire: Extend the analysis of Section 4.3 to discuss the relativistic transformations of charge/current (densities) and $\mathbf{E}, \mathbf{B}$ fields for the one-dimensional infinite wire to that of the infinite plane of charge density/current density in Section 4.2.3. You can also consider the **thick** one-dimensional wires and two-dimensional planar geometries discussed in **P4.16**

P4.24 *RC* circuit: Repeat the analysis done in **Example 4.8** to find the natural time scale for an RC circuit connected to a voltage/EMF source $\mathcal{E}$, relevant for charging a capacitor.

(a) Write $\tau_{RC} \propto R^\alpha C^\alpha \mathcal{E}^\gamma$, match dimensions, and solve for α, β, γ. Is your result consistent with the LRC dimensional ratios shown in Eqn. (4.85)?

(b) Constrain the time-dependent charge by writing $q(t) \propto R^\alpha C^\alpha \mathcal{E}^\gamma t^\delta$, match dimensions, and solve for $\alpha, \beta, \gamma, \delta$ to the extent possible and identify the appropriate dimensionless Π ratio.

(c)* The first-order ODE for this system is given by (noting that $I(t) = dq(t)/dt$)

$$R\frac{dq(t)}{dt} + \frac{q(t)}{C} = \mathcal{E} \tag{4.206}$$

and assuming an initially uncharged capacitor, we have $q(0) = 0$. Solve this differential equation/initial value problem by simply "morphing" the solution in Eqn. (4.87) and compare to the results of part (b).

P4.25 *LC* circuit: Repeat the analysis of **P4.24** for an LC circuit, with initial charge q_0 on the capacitor and vanishing current $I(0) = 0$.

(a) Write $\tau_{LC} \propto L^\alpha C^\alpha q_0{}^\gamma$, match dimensions, and solve for α, β, γ.

(b) Write $q(t) \propto L^{\alpha}C^{\beta}(q_0)^{\gamma}t^{\delta}$, match dimensions, and solve for $\alpha, \beta, \gamma, \delta$ to the extent possible and identify the appropriate dimensionless Π ratio.

(c)* The second-order ODE for this system is given by (noting that $dI(t)/dt = d^2q(t)/dt^2 = \ddot{q}(t)$)

$$L\ddot{q}(t) + \frac{1}{C}q(t) = 0\,. \tag{4.207}$$

Can you solve this using the methods for the harmonic oscillator in Section 3.1? Can you extend this problem to include the case where $q(0)$ and $I(0)$ have arbitrary values, in the spirit of **P3.5**?

P4.26 *LRC* **circuit:** (a) Use the analogies between the *LRC* circuit and the damped harmonic oscillator to write down the oscillation frequencies, $\omega_{\pm}$, for the general *LRC* circuit, using Eqn. (3.27).
(b) What is the critical frequency boundary between **underdamped** and **overdamped** solutions?
(c) Write $\omega \propto L^{\alpha}C^{\beta}R^{\gamma}$ and solve for α, β, γ to the extent possible, identify the appropriate dimensionless Π ratio, and compare your DA constraint to the result of part (a).

4.27 (Complex) impedances for alternating current (AC) circuits: An *LRC* circuit may be driven by an (AC) external voltage/EMF source of the form $\mathcal{E}(t) = \mathcal{E}_0\cos(\omega t)$ and each of the three circuit elements can be described as having an **impedance** (generalized resistance) given by

$$Z_L = i\omega L\,, \qquad Z_R = R\,, \qquad Z_C = \frac{1}{i\omega C}, \tag{4.208}$$

where $i = \sqrt{-1}$, so that two of these impedances are complex. Show that the dimensions of Z_L, Z_R, Z_C are identical and that they are consistent with the dimensional ratios in Eqn. (4.85).

P4.28 Impedance of free space: An important quantity in E&M wave propagation (with applications in both antenna and waveguide design in electrical engineering) is a quantity related to the magnitude of the electric and magnetic field values, called the **impedance of free space**, Z_0. For example, the radiation resistance of an antenna of length l emitting E&M waves of wavelength λ is given by

$$R_{ant} = \frac{2\pi}{3}Z_0\left(\frac{l}{\lambda}\right)^2 \tag{4.209}$$

so that the power emitted is $P_{ant} = I^2R_{ant}$ if the current is I.

(a) Recalling from **P4.27** that any impedance should have the same dimensions as resistance, R, determine the dependence of Z_0 on fundamental quantities by writing $Z_0 \propto \epsilon_0^{\alpha}\mu_0^{\beta}e^{\gamma}$ and solving for the exponents.

(b) Insert values for all fundamental quantities and show that $Z_0 \approx 120\pi\,\Omega$ (Ohm), which has historically been a useful mnemonic. Confirm that the units that you get are really those of Ohm.

P4.29 Terminal velocity—*LR* circuit analogy: A particle of mass m is subject to a constant external force $\mathcal{F}$ and a velocity-dependent frictional/viscous force, $F_f = -bv$, and starts from rest with

$v(0) = 0$. Use the solution of the LR circuit problem in **Example 4.8** to find the characteristic time, the dimensional dependence of $v(t) \propto m^\alpha g^\beta b^\gamma t^\delta$, and the complete solution for $v(t)$ by taking advantage of the analogies suggested by Eqns. (4.88) and (4.89).

P4.30 Resistance calculations: Consider a cylindrical resistor (length L and cross-sectional area A) made of material of conductivity σ. The *power dissipated per unit volume*, given by $P/V \equiv \mathcal{P} = \mathbf{J}_e \cdot \mathbf{E}$, can be used to evaluate the actual power lost through Joule heating by (i) writing $\mathbf{E} = \mathbf{J}_e/\sigma_e$, (ii) using $P_{diss} = \mathcal{P}V$, and (iii) noting that $J = I/A$. Use these three steps to write $P_{diss} = I^2 R$ and evaluate R in terms of σ_e, A, L.

P4.31 Larmor radiation from charge in uniform motion: Repeat the analysis leading to Eqn. (4.112), but assume the radiated power arises from a charge moving at constant speed v, namely, assume that $P \sim \epsilon_0^\alpha \mu_0^\beta q^\gamma v^\delta$ and solve for $\alpha, \beta, \gamma, \delta$. Does your "answer" make sense?

P4.32 Larmor radiation from a charge with "jerk": Repeat the analysis of **Example 4.9** or **P4.31**, but now assume the radiated power arises due not to an acceleration ($a = \ddot{x}$), but rather due to a **jerk** (the derivative of acceleration), namely $\mathbf{j} \equiv \dot{\mathbf{a}} = \ddot{\mathbf{v}}$. Do this by assuming that $P \sim \epsilon_0^\alpha \mu_0^\beta q^\gamma \mathbf{j}^\delta$ and solving for $\alpha, \beta, \gamma, \delta$. Do you get a solution from DA? If so, does the dependence on the quantities, especially on $\mathbf{j}$, make physical sense? What did we learn about the DA predictions for quantities involving vectors?

P4.33 Abraham–Lorentz "radiation reaction/damping" force: An accelerated charge that radiates energy (and hence momentum) in the form of EM waves will experience a recoil force ($\mathbf{F}_{AL}$) which **does** depend on the "jerk" $\mathbf{j} \equiv \dot{\mathbf{a}} = \ddot{\mathbf{v}} = \dddot{\mathbf{x}}$. Assume that $\mathbf{F}_{AL} = C_F \epsilon_0^\alpha \mu_0^\beta q^\gamma \mathbf{j}^\delta$ and solve for $\alpha, \beta, \gamma, \delta$. Does your answer make sense vectorially?
Note: It turns out that $C_F = 1/6\pi$, just as in the famous Larmor formula in Eqn. (4.112).

P4.34 Power radiated by an oscillating magnetic dipole: Repeat the analysis leading to Eqn. (4.115), but for the power radiated by an oscillating **magnetic dipole moment,** $\mathbf{m}_0$, by assuming $P \sim \epsilon_0^\alpha \mu_0^\beta \mathbf{m}_0^\gamma \omega^\delta$ and compare the ω dependence to the power radiated by an oscillating **electric dipole moment** from Eqn. (4.115).

P4.35 Current from a lightning strike: Geophysical researchers can use measurements of the electric and magnetic fields far from a thunderstorm to estimate the currents generated in a lightning bolt. (See e.g., Uman and McLain (1970) for a theoretical analysis, which uses a simple "transmission line model", and Willett et al. (1989) for experimental data.)

(a) Assume you have a measurement of the electric field strength (E) a distance D away from the thunderstorm. Can you use DA to estimate the current (I) that initiated the electromagnetic fields at the measurement site? Try to do so by equating $I \sim E^\alpha \epsilon_0^\beta D^\gamma c^\delta$ where c is the speed of light.

(b) The actual analysis involves the "speed of the return strike" (let's call it v), which also has dimensions of speed. Now assume that $I = C_I E^\alpha \epsilon_0^\beta D^\gamma c^\delta v^\epsilon$, and solve for the exponents in terms of ϵ. It turns out that the model solution is that you answer with $C_I = 2\pi$ and $\epsilon = -1$. It also turns out that the speed of the return strike is about one third that of the speed of light

(c) Repeat this analysis assuming you're using measurements of the magnetic field B, and assume $I \sim B^\alpha \mu_0^\beta D^\gamma c^\delta$. Can you rework your solution (here ignoring the dependence on v) to look like something familiar, like the field from a long wire?

P4.36 Time scale for excess charge on a conductor to disperse—Numbers and theory: The time scale for excess/variations in charge density to dissipate was discussed in Section 4.6, where it was found to be $\tau_e = \epsilon_0/\sigma_e$.

(a) Evaluate τ_e for one of the best conductors, namely silver, where $\sigma_e(Ag) = 6.3 \times 10^7\ S/m$ where $S = siemens$ is an SI unit (see Table 10.2 if needed) and use known values for ϵ_0, e, and m_e.

(b)* Try to solve for the exact time-dependence of an excess charge density $\rho_e(\boldsymbol{r}, t)$, if you can, by combining the following basic equations:

$$\frac{\partial \rho_e(\boldsymbol{r}, t)}{\partial t} + \nabla \cdot \boldsymbol{J}_e = 0 \qquad \text{Conservation of charge} \qquad (4.210)$$

$$\boldsymbol{J}_e = \sigma_e \boldsymbol{E} \qquad \text{Ohm's law} \qquad (4.211)$$

$$\nabla \cdot \boldsymbol{E} = \frac{\rho_e(\boldsymbol{r}, t)}{\epsilon_0} \qquad \text{Gauss's law} \qquad (4.212)$$

to find a (hopefully very familiar, and easily soluble) differential equation that determines $\rho_e(t)$.

P4.37 * EM wave skin depth—Limiting cases: (a) If we include a $\boldsymbol{J}_e = \sigma_e \boldsymbol{E}$ term in the derivation of the EM wave equation, in the limit where $\Pi \equiv \epsilon_0 \omega/\sigma_e << 1$, Maxwell's equation reduces to

$$\nabla^2 \boldsymbol{E}(\boldsymbol{r}, t) = \mu_0 \sigma_e \frac{\partial \boldsymbol{E}(\boldsymbol{r}, t)}{\partial t} \qquad (4.213)$$

with an identical equation for $\boldsymbol{B}(\boldsymbol{r}, t)$. Try a solution of the form $\boldsymbol{E} = \boldsymbol{E}_0 \exp(-z/\delta) \sin(\omega t - z/\delta)$, insert into Eqn. (4.213), match terms, and find $\delta(\omega)$ and confirm that $C_\delta = \sqrt{2}$.
(b) Use the exact expression for the skin depth as a function of ω, σ_e from Eqn. (4.122) to find the limits corresponding to $\Pi \equiv \omega\epsilon_0/\sigma_e >> 1$ and $\Pi << 1$, and compare to Eqns. (4.119) and (4.120).
Note: We see that Eqn. (4.213) has the form of the **classical heat equation** and **diffusion equation,** as in Eqns. (5.18) and (5.24), and Krosney et al. (2021) gives a nice discussion of "magnetic diffusion."

P4.38 Surface resistance: A problem from Zangwill's (2013) popular E&M book discusses a property called **surface resistance,** which is described as "... the ohmic resistance to the induced current density $\boldsymbol{j}$ offered by a rectangular slab adjacent to the conductor surface ... ." Assume that this resistance (R_s) depends on e and μ_0, the angular frequency of the applied oscillating fields (ω), and the conductivity of the material (σ_e) via $R_s \propto e^\alpha \mu_0^\beta \omega^\gamma \sigma_e^\delta$, and solve for $\alpha, \beta, \gamma, \delta$.

P4.39 Child–Langmuir relation for space-charge limited current flow: Electrons (charge and mass e, m_e) can be emitted from one side of a vacuum diode in a parallel-plate geometry, accelerated across a distance d through voltage V, forming a current density J_e, which we want to calculate.

(a) Including the fundamental permittivity ϵ_0 as the final dimensional ingredient, write

$$J_e = C_J \epsilon_0^\alpha e^\beta m_e^\gamma V^\delta d^\rho$$

$$\frac{\mathrm{Q}}{\mathrm{TL}^2} = \left(\frac{\mathrm{Q}^2\mathrm{T}^2}{\mathrm{ML}^3}\right)^\alpha \mathrm{Q}^\beta \mathrm{M}^\gamma \left(\frac{\mathrm{ML}^2}{\mathrm{QT}^2}\right)^\delta \mathrm{L}^\rho, \tag{4.214}$$

match dimensions, and evaluate all exponents in terms of δ. Are any of the exponents uniquely determined?

(b) Conservation of energy suggests that the kinetic energy of the electrons ($m_e v^2/2$) comes from the gain in electrostatic potential energy eV, so that e and m_e would appear in the ratio e/m (or m/e) in any equation using them both. Does that connection then completely constrain all of the exponents? (This problem is treated analytically in many E&M books, such as Griffiths (2017), where C_J is found to be 4/9. This is an example of a voltage versus current relation, which not linear, i.e., unlike **Ohm's law.**)

P4.40 Mott–Gurney current-voltage relation: A non-linear relation between I and V that is different in physical origin than Ohm's law, one which is applicable in semi-conductors, relates the current density J_e to the voltage V across a sample of width d and permittivity ϵ, but also invokes the **electron mobility** μ_m. This quantity is defined as the ratio of the **drift velocity** to the **applied electric field**, namely $\mu_m \equiv v_{dr}/E$, which then determines its dimensions. Write $J_e = C_J \mu_m^\alpha \epsilon^\beta V^\gamma d^\delta$, match dimensions, and show that all exponents are determined. A derivation from first principles (Mott and Gurney 1950) finds that $C_J = 9/8$.

P4.41 Magnetic braking of a rotating disk: Starting "from scratch," see to what extent you can predict the magnetic torque on a rotating disk (i.e., Eqn. (4.135)) by writing

$$\tau \propto B_0^\alpha \sigma_e^\beta \omega^\gamma a^\epsilon b^\rho t^\nu R^\mu, \tag{4.215}$$

where a, b, d, R all have dimensions of length. Match dimensions and show that the exponents for B_0, σ, ω_e are all determined. Given the Newton's law connection $\mathcal{I} d\omega/dt = \tau$, show that this gives exponential decay which we can write as $\omega(t) = \omega_0 \exp(-t/T_0)$, and confirm that $1/T_0 \propto B_0^2$.

P4.42 EM rail gun: A problem related to **Example 4.9** assumes a conducting bar of length L, mass m, and resistance R sliding along parallel conducting tracks in a magnetic field B_0, with a similar geometry to Fig. 4.11, but with the addition of a voltage source $\mathcal{E}_{app}$ which provides a current. The bar is accelerated by the $F = ILB$ force, with the current I due to the applied EMF as well as the induced Faraday's law current, and that current induces a velocity-dependent magnetic frictional force. We're interested in finding the terminal velocity (v_T) of the projectile (it is, after all, a rail-gun) and the time scale (τ) to attain it, starting with a DA approach.

(a) Using the five dimensional quantities in the problem, write $v_T \propto m^\alpha R^\beta B_0^\gamma L^\delta \mathcal{E}_{app}^\rho$, match dimensions, and find all exponents to the extent you can. If the terminal speed turns out not to depend on the mass of the bar, does this fix all exponents?

(b) Repeat part (a) to constrain τ as much as possible. If τ doesn't depend on the applied voltage, does this fix all exponents?

(c)* This problem is very similar to that of a falling body subject to $-bv$ frictional force. The equation of motion in this case is determined by the Kirchhoff's loop law for the circuit (which determines the current I) and then Newton's law using the magnetic force on the bar, namely,

$$IR = \mathcal{E}_{tot} = \mathcal{E}_{app} - B_0 L v \qquad \text{and} \qquad m\frac{dv(t)}{dt} = ma = F_{tot} = IB_0 L. \quad (4.216)$$

This is a fairly simple, first-order ODE for $v(t)$ (where we can assume that $v(0) = 0$) which you may be able to solve for $v(t)$. If so, can you check whether the values of v_T and τ discussed above do appear in the exact solution? **Hint:** You can try a solution of the form $v(t) = A + B\exp(at)$ and use the differential equation to fix a and relate A, B, and then apply the initial condition $v(0) = 0$ to finish things off.

P4.43 Watt or Kibble balance: A sensitive electro-mechanical device that can reproducibly measure masses to precisions of something like 10^{-8} balances an object's weight (mg) with the force on a conducting wire of length L carrying a current I in a magnetic field B.

(a) Write $mg \propto B^{\alpha} L^{\beta} I^{\gamma}$ and solve for α, β, γ to see how the balance works.

(b) A problem with using the data in this form is that it's challenging to precisely measure either B or L, so the same conducting wire is moved at constant speed (v) through the same system and the resulting voltage (potential difference) $\mathcal{V}$ is measured. The mass then depends on $m \propto \mathcal{V}^{\alpha} I^{\beta} g^{\gamma} v^{\delta}$, which can be solved for all exponents. Based on your answer, why do you think this is called a Watt balance?

Note: This device was invented by Kibble (1979) (and now named in his honor) and modern versions have become sensitive enough that they have been cited in important aspects of measurement science, including the redefinition of the kilogram; see Section 10.10. Large examples of this device are found in every international metrology lab (including NIST (US), BIPM (FR), and NPL (UK) with nice reviews by Stock (2011) or Robinson and Schlamminger (2016)), while the construction of more low-tech versions have been described for educational labs in the pedagogical literature; see Quinn et al. (2013) or Chao et al. (2015).

P4.44 Gouy balance: A long cylindrical tube (cross-sectional area A) is partially filled with magnetic material and placed between the poles of a magnet (where the field is B). The entire apparatus is hung from a balance, so that the weight (W) of the tube and its contents be measured. Measurements are taken with the field at the location of the magnetic material, and then higher up where the tube is empty, giving two readings, $W' - W = \Delta W$, with the difference arising due to the interaction of the material with the field.

(a) Let us begin by assuming that the apparent change in weight is depends on the area, magnetic field, and the fundamental magnetic constant, μ_0, via $\Delta W = C_W A^{\alpha} B^{\beta} \mu_0^{\gamma}$. Match dimensions and solve for α, β, γ. The constant turns out to be given by $C_W = \chi/2$, where χ is the (dimensionless) **magnetic susceptibility**, which is actually the material property which this device, called a **Gouy balance**, is designed to measure.

Notes: The experimental method was first discussed by Gouy (1889) and can be used in physical chemistry laboratories, as described by Brubacher and Stafford (1962). Values of χ can be either positive (**paramagnetic**) or negative (**diamagnetic**). The description of this device should be considered as very schematic, as the details of actual commercially apparatus vary. The so-called **Evans balance** uses another magnetic geometry to act as a **magnetic susceptibility balance**, and is more seemingly used more frequently in modern laboratories; see Regan (2023).

(b) To explore this problem in more theoretical detail, let us assume that we've learned some of the following relations.

(i) In a magnetic material, we make the replacement $\mu_0 \rightarrow \mu = \mu_0(1+\chi)$, where $|\chi| << 1$ and is dimensionless.

(ii) The change in the energy density due to the magnetic field, between the initial state (material in the B field) and final state (only air, so let us assume vacuum) is given by

$$\Delta U_B = U'_B - U_B = \frac{B^2}{2\mu_0} - \frac{B^2}{2\mu} = \frac{B^2}{2\mu_0}\left(1 - \frac{1}{1+\chi}\right) \approx \chi\frac{B^2}{2\mu_0},$$

since $|\chi| << 1$.

(iii) The change in energy density corresponds to a change in magnetic pressure, $\Delta P_B = \Delta U_B$.

(iv) And finally, the change in magnetic pressure balances the change in apparent pressure due to the measured weight difference, namely, $\Delta P_W = \Delta W/a$.

Combine these steps to find ΔW and show that $C_W = \chi/2$.

P4.45 Checking dimensions for magnetic monopoles: Check that all of the new terms related to magnetic monopoles in the relevant equations in Section 4.8.1 are dimensionally consistent, namely:

(a) Those in Eqns. (4.148) and (4.149).

(b) Those in the revised Maxwell equations in Eqns. (4.159)–(4.162) and (4.163)–(4.266).

P4.46 Dirac monopoles: (a) We've shown that the angular momentum stored in the combined $\boldsymbol{E}/\boldsymbol{B}$ fields of a q, g point charge/monopole pair is $|\boldsymbol{L}| = \mu_0 qg/4\pi$. In quantum mechanics, we're used to seeing angular momenta quantized in units of $\hbar/2$ (the spin of the electron/proton having this value, for example). If we assume that the field angular momentum is constrained to have similar discrete values, we might write (assuming that one of the particles in an electron or proton)

$$\frac{\mu_0 eg}{4\pi} = |\boldsymbol{L}_{em}| = n\frac{\hbar}{2} \tag{4.217}$$

and this would suggest why electric charges are quantized. Given that values of $e, \hbar, \mu_0$ are known, this determines the value of a basic magnetic charge (assuming that $n = 1$), and that value is often written as g_D for the **Dirac monopole.** Show that this monopole charge can be written in the form

$$g_D = \frac{g_c}{2\alpha_{FS}} \qquad \text{where} \qquad \alpha_{FS} \equiv \frac{e^2}{4\pi\epsilon_0\hbar c} \approx \frac{1}{137} \tag{4.218}$$

is the **fine-structure constant** and $g_c \equiv ec$ as defined in Eqn. (1.145). This implies that $g_D \approx 70g_c$ and so a Dirac monopole would have a large magnetic interaction!
(b) If we define a "magnetic fine-structure constant" as $\alpha_m \equiv \mu_0(g_D)^2/(4\pi\hbar c)$, show that (i) α_m is dimensionless, and (ii) that the Dirac quantization condition can be written in the form

$$(\alpha_m)(\alpha_{FS}) = \frac{n^2}{4} \qquad \text{or} \qquad \sqrt{(\alpha_m)(\alpha_{FS})} = \frac{n}{2}. \tag{4.219}$$

(c) To see how a magnetic charge might be constructed from the "building blocks" of E&M and quantum mechanics, write $g \propto \hbar^\alpha e^\beta \epsilon_0^\gamma \mu_0^\delta$, match dimensions, and solve for the exponents to the extent you can. Show that your general result can be written in the form $g \propto (ec)(\alpha_{FS})^n$, and so is consistent with both g_c (for $n = 0$) and g_D (for $n = -1$).

P4.47 Magnetic dipole moment with monopoles: We've learned that the **magnetic dipole moment** for a current (I_e) carrying (planar) closed-circuit element (area A) is given by $|\boldsymbol{m}_e| = I_e A$ with the direction of $\boldsymbol{m}$ determined by the normal to the plane. If magnetic monopoles do exist, the analog of a magnetic dipole could also simply consist of $\pm g$ charges separated by a distance $\boldsymbol{d}$, with $\boldsymbol{m}_g = g\boldsymbol{d}$.

(a) Show that the two definitions have the same dimensions (as they must).
(b) Some of the spin-ice experiments discussed at the end of Section 4.8.1 determine the effective magnetic charge of their "emergent monopoles," with one result citing a value of $g = 4.6\mu_B\text{Å}^{-1}$ where μ_B is the **Bohr magneton** and Å denotes the Ångström unit of length. Look up (in this book or elsewhere) the magnitudes of μ_B and Å, determine g numerically in SI units, and compare to the values of $g_c = ec$ in Eqn. (4.144) and $g_D = g_c/2\alpha_{FS}$ discussed in **P4.46.**
(c) Theoretical expressions for the **Bohr radius** and **Bohr magneton** are discussed in Chapter 6, where they are given by

$$a_0 = \frac{4\pi\epsilon_0\hbar^2}{e^2 m_e} \qquad \text{and} \qquad \mu_B = \frac{e\hbar}{2m_e}, \tag{4.220}$$

where $\hbar$ is Planck's constant and m_e is the electron mass. These values then give the typical atomic physics magnitudes for length scales and magnetic moments, and their dependence on fundamental quantities. Using the results of part (a), the corresponding size of a typical monopole magnetic charge could be associated with $g_H = \mu_B/a_0$ (where H can be a notation for either hydrogen or Hartree, as discussed in Section 6.5.1). Show that

$$g_H \equiv \frac{\mu_B}{a_0} = ec\left(\frac{\alpha_{FS}}{2}\right), \tag{4.221}$$

where α_{FS} is the fine-structure constant, and therefore consistent with the results of **P4.46**(c).

P4.48 Diffusion of magnetic monopoles: In their study of the emergent magnetic monopoles in *spin ice* systems, Bramwell et al. (2009) make use of a dimension<u>**less**</u> quantity (we'll call it b_m) that depends on the effective magnetic charge (g), the applied magnetic field (B), the fundamental magnetic strength (μ_0) and, since the system is at a finite temperature, it also depends on Boltzmann's constant and the temperature (k_B and T).

(a) Write $b_m = C_b \mu_0^\alpha g^\beta B^\gamma k_b^\delta T^\epsilon$, match dimensions, and solve for the exponents to the extent you can. The definition they use is such that b_m is actually linear in the magnetic field strength, so use that fact to see if you can further specify any exponents.

(b) The same authors cite an early work by Onsager (1934) that treats *Deviations from Ohm's Law in Weak Electrolytes*, where they take an important formula and use the electric–magnetic duality we've discussed to derive a magnetic version (the b_m considered here) of the dimensionless b_e used in that paper. Using the result from part (a), map the magnetic variables g, B, μ_0 onto their electric counterparts $e, E, 1/\epsilon_0$ to find the dependences of b_e on those quantities (keeping the k_B, T values the same) and confirm that b_e is also dimensionless.

P4.49 Biot–Savart law for magnetic monopoles: Many textbooks show how the Biot–Savart law from Eqn. (4.31) is directly related to Ampère's law, namely the connection

$$d\boldsymbol{B} = \frac{\mu_0}{4\pi}\frac{I_e d\boldsymbol{l} \times \hat{\boldsymbol{r}}}{r^2} \qquad \longrightarrow \qquad \boldsymbol{\nabla} \times \boldsymbol{B} = \mu_0 \boldsymbol{J}_e. \tag{4.222}$$

Using the differential forms for Faraday's law, now including magnetic monopoles, from Eqn. (4.160), invert the process to go from $\boldsymbol{\nabla} \times \boldsymbol{E} = -\mu_0 \boldsymbol{J}_m$ to find the expression for $d\boldsymbol{E}$ at distance r away from a short ($d\boldsymbol{l}$) line element of magnetic current I_m.

P4.50 "First results from a superconducting detector for moving magnetic monopoles": The title of this problem is taken from the paper by Cabrera (1982), who claimed to have found one candidate event of a magnetic monopole. The experiment monitored the current in a superconducting loop which would have been generated if a monopole had passed through it (anywhere in the loop, at any angle). Assume that the current produced could depend on the monopole charge g and μ_0, as well as the circuit parameters, namely its inductance (L) and resistance (R), via $I \propto \mu_0^\alpha g^\beta L^\gamma R^\delta$ and solve for $\alpha, \beta, \gamma, \delta$.

P4.51 EM radiation from an accelerated magnetic monopole: (a) Repeat the analysis of Section 4.6 to find how the power radiated by an accelerated magnetic monopole would depend on g, a, ϵ_0, μ_0 and discuss your result, comparing it to Eqn. (4.112).
(b) Repeat the analysis of **P4.29** and substitute a constant speed v for the acceleration a to evaluate the radiated power.

P4.52 Checking dimensions for massive photons: Check that all of the new terms related to magnetic monopoles in the revised Maxwell equations in Section 4.8.2 including massive photons are dimensionally consistent, specifically Eqns. (4.175)–(4.178) and (4.179)–(4.182).

P4.53 Energy density contributions from massive photons: Consider the additional terms in the EM energy density in Eqn. (4.191) and explore to what extent they are constrained by DA.

(a) Assume that $U_\kappa \propto \epsilon_0^\alpha \mu_0^\beta \kappa^\rho \phi_\kappa^\delta$, match dimensions to determine $\alpha, \beta, \rho, \delta$.

(b) Repeat part (a), but replace $\phi_\kappa \rightarrow \boldsymbol{A}$.

(c) Assume that $U_\kappa \propto \epsilon_0^\alpha \mu_0^\beta \kappa^\rho \phi_\kappa^\delta \boldsymbol{A}^\sigma$ and see what constraints do you find. Does the fact that U_κ and ϕ_κ are scalars, while $\boldsymbol{A}$ is a vector, provide any further information?

P4.54 Poynting vector contribution from massive photons: Consider the additional term in the Poynting vector in Eqn. (4.190) due to a massive photon (call it $\boldsymbol{S}_\kappa$) and let us see to what extent we can reproduce it with DA. Assume that $S_\kappa = C_S \phi_\kappa^\alpha A^\beta \kappa^\sigma \epsilon_0^\rho \mu_0^\delta$, match dimensions, and solve for all exponents in terms of α.

(a) Do you get a unique answer for any of the powers involved?

(b) For the normal Poynting vector, $\boldsymbol{E}, \boldsymbol{B}$ appear with the same power, so if you assume something similar for $\phi, \boldsymbol{A}$, does that uniquely specify all of the powers?

(c) If you use the fact that $\boldsymbol{S}_\kappa$ is a vector, does that also provide information, say which determines β?

P4.55 Experimental bounds on the photon mass: As we've seen, many standard results in E&M will be changed if one includes a putative photon mass. The expressions for static magnetic fields (like the dipole fields of planets) would then be modified to include terms like $\exp(-\kappa r) = \exp(-r/\lambda\!\!\!^{-})$. Schrödinger (1943/1944) obtained an early bound on a possible photon mass by considering the permanent magnetic field of the Earth. If there are no major deviations in the dipole field of the Earth, we expect that the $\exp(-r/\lambda\!\!\!^{-})$ term is close to unity, or that $R/\lambda\!\!\!^{-} < \mathcal{O}(1)$ (where R is some typical scale for the magnetic field) so that $m_\gamma < \hbar/Rc$.

(a) If we use $R \approx 10R_{earth}$ (as Schrödinger did), what limit does this set on the photon mass?

(b) Calculate the even more restrictive bound on m_γ that has been obtained from *Pioneer-10 observations of Jupiter's magnetic field* by Davis, Goldhaber, and Nieto (1975) using $R \approx 10R_{Jupiter}$. (Feel free to look up that radius!)

(c) The most restrictive limit included in Workman et al. (2022) cites Ryutov (2007), who uses a "... set of magnetohydronamic equations (assuming a finite photon mass) ... to analyze properties of the solar wind at Pluto's orbit" and finds that $\lambda\!\!\!^{-} > 3\,Mkm$ (where here $M = mega = million$). Translate this into a lower bound on m_γ.

References for Chapter 4

Abbasi, R., et al. (IceCube Collaboration). (2022). "Search for Relativistic Magnetic Monopoles with Eight Years of IceCube Data," *Physical Review Letters* 128, 051101.

Acharya, B., et al. (2022). "Search for Magnetic Monopoles via the Schwinger Mechanism," *Nature* 602, 62.

Armentrout, C. (1990). "The Hall Effect in Copper: An Undergraduate Experiment," *American Journal of Physics* 58, 758–62.

Ashkin, A. (1970). "Acceleration and Trapping of Particles by Radiation Pressure," *Physical Review Letters* 24, 156–9.

Ashkin, A., et al. (1986). "Observation of a Single-Beam Gradient Force Optical Trap for Dielectric Particles," *Optics Letters* 11, 288–90.

Aston, F. W. (1919). "LXXIV. A Positive Ray Spectrograph," *The London, Edinburgh, and Dublin Philosophical Magazine and Journal of Science* 38, 707–714.

Bramwell, S. T., et al. (2009). "Measurement of the Charge and Current of Magnetic Monopoles in Spin Ice," *Nature* 461, 956–9.

Brownstein, K. R. (1989). "Angular Momentum of a Charge–Monopole Pair," *American Journal of Physics* 57, 420–1.

Brubacher, L. J., and Stafford, F. E. (1962) "Magnetic Susceptibility—A Physical Chemistry Laboratory Experiment," *Journal of Chemical Education* 39, 574–5.

Bulanov, S. S., et al. (2010). "Schwinger Limit Attainability with Extreme Power Lasers," *Physical Review Letters* 105, 2200407.

Cabrera, B. (1982). "First Results from a Superconducting Detector for Moving Magnetic Monopoles," *Physical Review Letters* 48, 1378–81.

Castelnovo, C., Moessner, R., and Sondhi, S. L. (2007). "Magnetic Monopoles in Spin Ice," *Nature* 451 42–45.

Chao, L. S., et al. (2015). "A LEGO Watt Balance: An Apparatus to Determine a Mass Based on the New SI," *American Journal of Physics* 83, 913–22.

Chen, F. (2018). *Introduction to Plasma Physics and Controlled Fusion*, 3rd ed. (Heidelberg: Springer).

Chien, C. L., and Westgate, C. R., eds. (1980). *The Hall Effect and Its Applications* (New York: Plenum Press).

Dempster, A. J. (1918). "A New Method of Positive Ray Analysis," *Physical Review* 11, 316–25.

Detrixhe, M., et al. (ANITA Collaboration). (2011). "Ultrarelativistic Magnetic Monopole Search with the ANITA-II Balloon-Borne Radio Interferometer," *Physical Review D* 83, 023513.

Donoso, G., Ladera, C. L., and Martin, P. (2011). "Damped Fall of Magnets inside a Conducting Pipe," *American Journal of Physics* 79, 193–200.

Franklin, J., Griffiths, D., and Mann, N. (2022). "Motion of a Charged Particle in the Static Fields of an Infinite Straight Wire," *American Journal of Physics* 90, 513–519.

Fulcher, L. P. (1986). "Improved Result for the Accuracy of Coulomb's Law: A Review of the Williams, Faller, and Hill Experiment," *Physical Review A* 33, 759–61.

Fulcher, L. P., and Telljohann, M. A. (1976). "On the Interpretation of Indirect Tests of Coulomb's Law: Maxwell's Derivation Revisited," *American Journal of Physics* 44, 366–9.

Gerlach, W., and Stern, O. (1922). "Der experimentelle Nachweis der Richtungsquantelung im Magnetfeld (The Experimental Evidence of Directional Quantization in the Magnetic Field)," *Zeitschrift für Physik* 9, 349–52.

Giblin, S. R., et al. (2011). "Creation and Measurement of Long-Lived Magnetic Monopole Currents in Spin Ice," *Nature Physics* 7, 252–8.

Goldhaber, A., and Nieto, M. M. (1971). "Terrestrial and Extraterrestrial Limits on the Photon Mass," *Reviews of Modern Physics* 43, 277–96.

Goldhaber, A., and Nieto, M. M. (2010). "Photon and Graviton Mass Limits," *Reviews of Modern Physics* 82, 939–79.

Goldhaber, A., and Trower, W. P. (1990). "Resource Letter MM-1: Magnetic Monopoles," *American Journal of Physics* 58, 429–39.

Good, R. H. (1999). *Classical Electromagnetism* (Forth Worth, TX: Saunders).

Gouy, L-G. (1889). "Sur l' énergie potentielle magnétique et la mesure des coefficients d'aimantation (On Magnetic Potential Energy and the Measurement of Magnetization Coefficients)," *Comptes rendus de l'Académie des Sciences* T109, 935–9

Griffiths, D. (2012). "Resource Letter EM-1: Electromagnetic Momentum," *American Journal of Physics* 80, 7–18.

Griffiths, D. (2017). *Introduction to Electrodynamics*, 4th ed. (Cambridge: Cambridge University Press).

Griffiths, J. (2008) "A Brief History of Mass Spectroscopy," *Analytical Chemistry* 80, 5678–83.

Hahn, K. D., Johnson E M, Brokken A, and Baldwin S (1998) "Eddy Current Damping of a Magnet Moving through a Pipe," *American Journal of Physics* 66, 1066–1076.

Hall, E. H. (1879). "On a New Action of the Magnet on Electric Currents," *American Journal of Mathematics* 2, 287–92.

Heald, M. A. (1988). "Magnetic Braking: Improved Theory," *American Journal of Physics* 56, 521–2.

Hillas, A. M. (1984). "The Origin of Ultra-High-Energy Cosmic Rays," *Annual Review of Astronomy and Astrophysics* 22, 425–44.

Ignatiev, A. Yu., and Joshi, G. C. (1996). "Massive Electrodynamics and the Magnetic Monopoles," *Physical Review D* 53, 984–92.

Itzykson, C., and Zuber, J-B. (2006). *Quantum Field Theory* (Dover Books in Physics) (Mineola: Dover).

Jackson, J. D. (1999). *Classical Electrodynamics*, 3rd ed. (New York: Wiley and Sons).

Jones, R. S. (1995). "Circular Motion of Charged Particle in an Electric Dipole Field," *American Journal of Physics* 63, 1042–1043.

Kibble, B. P., and Hunt, G. J. (1979). "A Measurement of the Gyromagnetic Ratio of the Proton in a Strong Magnetic Field," *Metrologia* 15, 5–30.

Kotera, K., and Olinto, A. (2011). "The Astrophysics of Ultrahigh Energy Cosmic Rays," *Annual Review of Astronomy and Astrophysics* 49, 119–53.

Krosney, A. E., et al. (2021). "Magnetic Diffusion, Inductive Shielding, and the Laplace Transform", *American Journal of Physics* 89, 490–9.

Larmor, J. (1897). "On the Theory of the Magnetic Influence on Spectra; and on the Radiation from Moving Ions," *Philosophical Magazine* Series 5: 44(271), 503–512.

Lee, C. H., and Park, B-Y. (2023). "A Magnet Falling inside a Conducting Pipe: Dependence of the Drag Force on the Magnet Orientation," *American Journal of Physics* 91, 440–8.

Lee, K., and Park, K. (2002). "Analysis of an Eddy-Current Brake Considering Finite Radius and Induced Magnetic Flux," *Journal of Applied Physics* 92, 5532–8.

Levin, Y., da Silveira, F. L., and Rizzato, F. B. (2006). "Electromagnetic Braking: A Simple Quantitative Model," *American Journal of Physics* 74, 815–17.

(Lord) Rayleigh, (William Strutt). (1915). "The Principle of Similitude," *Nature* 95, 66–68.

Lorrain, P., Corson, D. R., and Lorrain, F. (2000). *Fundamentals of Electromagnetic Phenomena* (New York: Freeman and Company).

MacLatchey, C. S., Backman, P., and Bogan, L. (1993). "A Quantitative Magnetic Braking Experiment," *American Journal of Physics* 61, 1096–1101.

Marcuso, M., Gass, R., Jones, D., and Rowlett, C. (1991). "Magnetic Drag in the Quasi-Static Limit: A Computational Method", *American Journal of Physics* 59, 1118–1123.

McGuire, G. C. (2003). "Using Computer Algebra to Investigate the Motion of an Electric Charge in Magnetic and Electric Dipole Fields," *American Journal of Physics* 71, 809–812.

Milton, K. (2006). "Theoretical and Experimental Status of Magnetic Monopoles," *Reports on Progress in Physics* 69, 1637–1711.

Mott, N. F., and Gurney, R. W. (1950). *Electronic Processes in Ionic Crystals*, 2nd ed. (Oxford: Oxford University Press).

Olness, F., and Scalise, R. (2011). "Regularization, Renormalization, and Dimensional Analysis: Dimensional Regularization Meets Freshman E&M," *American Journal of Physics* 79, 306–312.

Pelosko J A, Cesky M, and Huertas S (2005) "Lenz's Law and Dimensional Analysis," *American Journal of Physics* 73, 37–39.

Plies, E., Marianowski, K., and Ohnweiler, T. (2011). "The Wien Filter: History, Fundamentals, and Modern Applications," *Nuclear Instruments and Methods in Physics Research Section A: Accelerators, Spectrometers, Detectors and Associated Equipment* 645, 7–11.

Quinn, T., Quinn, T., and Davis, R. (2013). "A Simple Watt Balance for the Absolute Measurement of Mass," *Physics Education* 48, 601–606.

Rajantie, A. (2016). "The Search for Magnetic Monopoles," *Physics Today* 69(10), 40–46.

Ray, M. W., et al. (2014). "Observation of Dirac Monopoles in a Synthetic Magnetic Field," *Nature* 505, 657–60.

Regan, A., O'Donoghue, J., Poree, C., and Dunne, P. W. (2023). "Introducing Materials Science: Experimenting with Magnetic Nanomaterials in the Undergraduate Chemistry Laboratory," *Journal of Chemical Education* 100, 2387–93.

Rindler, W. (1989). "Relativity and Electromagnetism: The Force on a Magnetic Monopole," *American Journal of Physics* 57, 993–4.

Robinson, I. A., and Schlamminger, S. (2016). "The Watt or Kibble Balance: A Technique for Implementing the New SI Definition of the Unit of Mass," *Metrologica* 54, A46–A74.

Ryutov, D. D. (2007) "Using Plasma Physics to Weigh the Photon," *Plasma Physics and Controlled Fusion* 49, B429–38.

Saslow, W. M. (1992). "Maxwell's Theory of Eddy Currents in Thin Conducting Sheets, and Applications to Electromagnetic Shielding and MAGLEV," *American Journal of Physics* 60, 693–711.

Schrödinger, E. (1943/1944). "The Earth's and the Sun's Permanent Magnetic Fields in the Unitary Field Theory," *Proceedings of the Royal Irish Academy Section A: Mathematical and Physical Sciences* 49, 135–48.

Schwartz, M. D. (2014). *Quantum Field Theory and the Standard Model* (Cambridge: Cambridge University Press).

Sharma, K. S. (2013). "Mass Spectrometry—The Early Years," *International Journal of Mass Spectrometry* 349–350, 3–8.

Smythe, W. R. (1989). *Static and Dynamic Electricity*, 3rd ed., Revised Printing (New York: Hemisphere). This is an updated version of the original 1936 edition.

Stock, M. (2011). "The Watt Balance: Determination of the Planck Constant and the Redefinition of the Kilogram," *Philosophical Transactions of the Royal Society A: Mathematical, Physical and Engineering Sciences* 369, 3936–53.

Tu, L-C., and Luo, J. (2004). "Experimental Tests of Coulomb's Law and the Photon Rest Mass," *Metrologia* 41, S136–46.

Tu, L-C., Luo, J., and Gillies, G. (2005). "The Mass of the Photon," *Reports on Progress in Physics* 68, 77–130.

Uman, M. A., and McLain, D. K. (1970). "Lightning Return Stroke Current from Magnetic and Radiation Field Measurements," *Journal of Geophysical Research* 75, 5143–7.

Wiederick, H. D., Gauthier, N., Campbell, D. A., and Rochon, P. (1987). "Magnetic Braking: Simple Theory and Experiment," *American Journal of Physics* 55, 500–503.

Willett, J. C., et al. (1989). "Submicrosecond Intercomparison of Radiation Fields and Currents in Triggered Lightning Return Strokes Based on the Transmission Line Model," *Journal of Geophysical Research* 94, 13275–86.

Williams, E. R., Faller, J. E., and Hill, H. A. (1971). "New Experimental Test of Coulomb's Law: A Laboratory upper Limit on the Photon Rest Mass," *Physical Review Letters* 26, 721–4.

Workman, R. L., et al. (Particle Data Group). (2022). "Review of Particle Physics," *Progress of Theoretical and Experimental Physics* 2022, 083C01.

Yoon, J. W., et al. (2021). "Realization of Laser Intensity over 10^{23} W/cm^2," *Optica* 8, 630–5.

Zangwill, A. (2013). *Modern Electrodynamics* (Cambridge: Cambridge University Press).

Chapter 5

Thermal physics

5.1 Transport phenomena and allometric scaling

We start our discussion of classical thermodynamics by reviewing some of the important thermal properties of materials, perhaps imagining the "heat-temperature trajectory" of a sample, starting at absolute zero. A schematic plot of the temperature (K) versus thermal energy added (Q) for water is shown in Fig. 5.1, illustrating the rise in temperature during the solid (S), liquid (L), and vapor (V) phases, along with the "plateaus" (at constant T) corresponding to the solid–liquid (freezing/melting) and liquid–vapor (boiling/condensing) phase transitions.[1]

The relevant connections between heat (added or removed), sample mass (m), and temperature as we move "up" (or "down" or "sideways") along the thermal history of the sample are

$$Q = C_S m\Delta T, \qquad Q = L_F m, \qquad Q = C_L m\Delta T, \qquad Q = L_V m, \qquad \text{and} \qquad Q = C_V m\Delta T \quad (5.1)$$

where C_S, C_L, C_V are the specific heats (per unit mass) in the solid, liquid, and vapor phases respectively, and L_F, L_V are the latent heats of fusion ($S \leftrightarrow L$ transition) and vaporization ($L \leftrightarrow V$ transition). From Chapter 1, or Eqn. (5.1) directly, we recall that their dimensions are given by

$$[C_S] = [C_L] = [C_V] = \frac{\mathsf{L}^2}{\mathsf{T}^2\Theta} \qquad \text{and} \qquad [L_F] = [L_V] = \frac{\mathsf{L}^2}{\mathsf{T}^2}. \quad (5.2)$$

Depending on the application or research area being explored, the relevant specific heat may be instead defined *per unit volume* or *per mole*, via the expressions

$$Q = VC_{vol}\Delta T \qquad \text{or} \qquad Q = nC_{mol}\Delta T, \quad (5.3)$$

where V is the sample volume and n is the number of moles of substance. The appropriate connections are

$$C_{vol} = \left(\frac{m}{V}\right)C = \rho_m C \qquad \text{and} \qquad C_{mol} = \left(\frac{m}{n}\right)C = \mu C, \quad (5.4)$$

where ρ_m is the familiar mass density and μ is the mass per mole, more commonly called the **molecular weight**. Recall, for example, that the molecular weights for hydrogen and oxygen are roughly 1 and 16 grams per mole or **amu** (for atomic mass units), respectively.

5.1.1 Transport phenomena

Besides the important static material properties governing the possible thermal history of matter, there is an important **transport relation** that connects the rate of thermal energy movement

[1] We have used approximate values for water (H_20) in this plot.

Dimensional Analysis Across the Landscape of Physics. Richard W. Robinett, Oxford University Press. © Richard W. Robinett (2024).
DOI: 10.1093/oso/9780192867551.003.0005

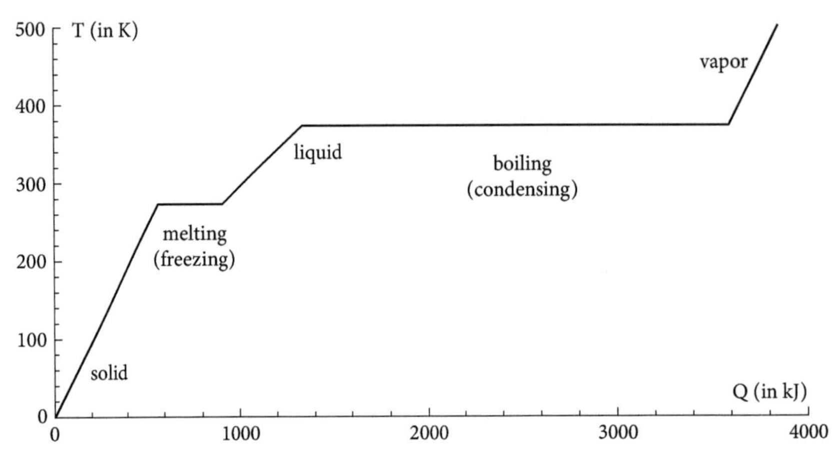

Fig. 5.1 Temperature (*K*) versus heat added (*kJ*) illustrating the phase history of water as it's heated from a solid (ice) at absolute zero to its vapor phase (steam), past two phase transitions (melting and boiling).

(dQ/dt) through a thickness Δx of material with cross-sectional area A, due to a temperature difference ΔT across the sample, namely

$$\frac{dQ}{dt} = \kappa A \frac{\Delta T}{\Delta x}, \tag{5.5}$$

where κ is the **thermal conductivity.** As noted in Chapter 1, this relation has obvious similarities to the relation for the transport of charge in Eqn. (1.63). With these familiar relations in hand, we can already approach some canonical thermodynamics problems using dimensional analysis techniques.

Example 5.1. Growth of ice—Planar geometry

An idealized model, with a simple geometry, of the growth of ice assumes a flat surface with a (constant) temperature difference ΔT across an existing thickness, $x(t)$, of ice. The sheet grows as water freezes at the bottom layer, with the resulting heat flow to the colder surface above. Given that this process depends on thermal transport (hence κ) and a liquid–solid phase transition (so L_F, but also needing the ice mass, hence the mass density ρ_m), we consider all variables and write

$$\begin{aligned} x(t) &= C_x t^{\alpha} \kappa^{\beta} (\Delta T)^{\gamma} L_F^{\delta} \rho_m^{\epsilon} \\ \mathsf{L} &= \mathsf{T}^{\alpha} \left(\frac{\mathsf{ML}}{\mathsf{T}^3 \Theta}\right)^{\beta} \Theta^{\gamma} \left(\frac{\mathsf{L}^2}{\mathsf{T}^2}\right)^{\delta} \left(\frac{\mathsf{M}}{\mathsf{L}^3}\right)^{\epsilon}, \end{aligned} \tag{5.6}$$

and matching $\mathsf{M}, \mathsf{L}, \mathsf{T}, \Theta$ dimensions gives values for all exponents in terms of say α as

$$x(t) \propto t^{\alpha} \kappa^{1-\alpha} (\Delta T)^{1-\alpha} (L_F)^{-3/2+2\alpha} \rho_m^{\alpha-1}. \tag{5.7}$$

We discuss later how to use ones intuition to constrain the exponent α, but in the spirit of the Buckingham Pi theorem, let us first identify any scaling variables. We can certainly write Eqn. (5.7) in the form

$$x(t) \propto \left(\frac{\kappa \Delta T}{\rho_m L_F^{3/2}}\right)\left(\frac{t L_F^2 \rho_m}{\kappa \Delta T}\right)^{\alpha} \longrightarrow l_0 \left(\frac{t}{t_0}\right)^{\alpha} \longrightarrow l_0\, F\left(\frac{t}{t_0}\right), \tag{5.8}$$

where l_0, t_0 are defined as

$$l_0 \equiv \frac{\kappa \Delta T}{\rho_m L_F^{3/2}} \quad \text{and} \quad t_0 \equiv \frac{\kappa \Delta T}{\rho_m L_F^2}, \tag{5.9}$$

which seem to be picked out as natural length and time scales. The two expressions in Eqn. (5.9) look amazingly similar due to the fact that $[L_F] = \mathsf{L}^2/\mathsf{T}^2$ (recall that it's heat/energy per unit mass) so that $[L_F^{1/2}] = \mathsf{L}/\mathsf{T}$ looks like a speed.

Using known handbook values for the material properties of water (κ, ρ_m, L_F), and some nominal value for the temperature difference (which is, of course, the same in both C, K units), say $\Delta T = 10\,^{\circ}C = 10\,K$, we find that $l_0 \approx 10^{-10}\,m$ and $t_0 \approx 2 \times 10^{-13}\,s$. That length scale is a typical atomic size (of order Ångströms), and along with the time scale, are similar to those found in **molecular dynamics** (or MD) simulations of chemistry/materials processes. For example, Matsumoto, Saito, and Ohmine (2002) model the freezing of ice by sampling their (computer) model system at 10 pico-sec ($10^{-11}\,s$) intervals. Whether or not the dimensional analysis approach has singled out these scales as making any connection to atomic/molecular processes, it certainly does not address the original problem of the (very) macroscopic problem of freezing of ice at surfaces in a natural way.

Let us instead focus on using our intuition on how we expect $x(t)$ to depend on the relevant parameters in Eqn. (5.7) to try to place bounds on the power-law exponent α. Some observations that might help include:

- Presumably the ice grows with time, so we expect $\alpha > 0$.
- The thicker the ice, the greater the thermal barrier to heat flow, so the rate of growth, that is, $\dot{x}(t) \propto t^{\alpha-1}$, would slow with time, suggesting that $\alpha - 1 < 0$ or $\alpha < 1$.
- The better the thermal conduction (the bigger is κ), the more efficient the heat flow, and the faster the growth rate, so $\beta = 1 - \alpha > 0$ or $\alpha < 1$ is found again.
- The harder it is to freeze, that is, the bigger L_F, the slower the growth rate, so $\delta = -3/2 + 2\alpha < 0$ or $\alpha < 3/4$,

which, when all combined, require that $0 < \alpha < 3/4$. This might suggest that $\alpha = 1/2$ is a natural choice, and one more physical connection can confirm this. In the freezing process, governed by $Q = mL_F$, we expect the mass (and hence mass density ρ_m) and latent heat parameter (L_F) to appear paired together, so with the same power, in this context implying that $\delta = \epsilon$ or $-3/2 + 2\alpha = \alpha - 1$ or $\alpha = 1/2$. This gives the compact form

$$x(t) = C_x \sqrt{\frac{\kappa \Delta T}{L_F \rho_m} t}, \tag{5.10}$$

and a straightforward calculation (see **P5.1**) gives $C_x = \sqrt{2}$. The classic "first principles" treatment of this problem is due to Stefan (1891) (in a study of the formation of polar ice), whose name is also associated with the famous Stefan–Boltzmann law discussed in Chapter 6.

If the temperature difference is not uniform in time, one can substitute

$$\Delta T \int_0^t dt \quad \longrightarrow \quad \int_0^t \Delta T(t)\, dt \quad \approx \quad \sum_{i=days} \Delta T_i \times (1 \text{ day}). \tag{5.11}$$

The combination of time and temperature (difference) under the square root in Eqn. (5.10) is often measured in **freezing degree days** (or FDD), as that combination is a natural one to keep track of in field studies. In those units, the simple Stefan model prediction can be phrased as

The ice thickness is 1 – 3 centimeters times the square root of the number of freezing degree days (5.12)

and you're asked in **P5.2** to confirm that statement using Eqn. (5.10) and values of the ice parameters.

Data on the thickness of ice as a function of time from Ashton (1989) is shown in Fig. 5.2 compared to this simple model, which seems to give a reasonable dependence for long times. But as with any simple model, this treatment ignores important effects such as snow layers, convection, albedo, and many other phenomena and we explore other options in **P5.3.** A more recent treatment (Desch et al. 2016) cites a best fit for the ice thickness of $x(t) = (0.0133\, m)\, (FDD)^{0.588}$, with an exponent and prefactor close to the classic Stefan result. (We note that when data is fitted, it may well also be scaled to dimensionless values, and that any dimensional dependences can be far from obvious—as here, where non-rational numbers are used as exponents.) While such calculations have been used in the past to model the growth of ice sheets (and Desch et al. (2016) even propose methods of replenishing ice flows), similar methods can be employed for the opposite effect, very relevant to issues related to climate change, global warming, and melting.

A quantity similar to FDD that is used in agriculture is the **growing degree day** (or GDD or even **heat summation**) as some plant and animal development rates can be correlated with such a metric, including the maturity date of crops, the blossoming of flowers, the emergence of insect pests, and even in classifying wine production districts, via the so-called Winkler Index—see Amerine and Winkler (1944) or Winkler et al. (1974).

The spatial and temporal dependence of the temperature, $T(x, t)$ in one dimension, or $T(\mathbf{r}, t)$ in three dimensions, in a sample can also be studied using the relations in Eqns. (5.1) and (5.5), which clearly involve quantities such as ρ_m, C_S (assuming thermal evolution in a solid) and κ. We can then explore the connections between x, t in this context by writing

$$\begin{aligned} x(t) &\propto \kappa^\alpha \rho_m^\beta C_S^\gamma t^\delta \\ \mathsf{L} &= \left(\frac{\mathsf{ML}}{\mathsf{T}^3\Theta}\right)^\alpha \left(\frac{\mathsf{M}}{\mathsf{L}^3}\right)^\beta \left(\frac{\mathsf{L}^2}{\mathsf{T}^2\Theta}\right)^\gamma \Theta^\delta. \end{aligned} \tag{5.13}$$

Matching dimensions then gives

$$x(t) \propto \left(\frac{\kappa}{\rho_m C_S} t\right)^{1/2} = (D_{th} t)^{1/2} \qquad \text{where} \qquad D_{th} \equiv \frac{\kappa}{\rho_m C_S} \tag{5.14}$$

and D_{th} is often called the **thermal diffusion constant** or **thermal diffusivity**, and has dimensions $[D_{th}] = \mathsf{L}^2/\mathsf{T}$.

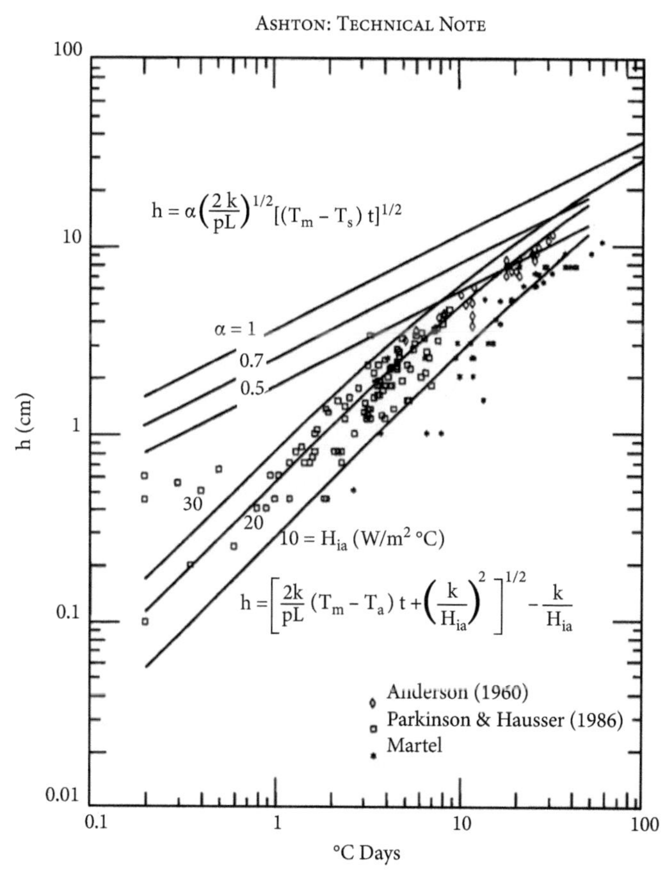

Fig. 5.2 Experimental data on ice thickness (in *cm*) versus *degree-days* compared to two models of ice growth. Reprinted from Ashton (1989) with permission from John Wiley and Sons.

Just as with the Stefan problem from **Example 5.1**, this is a case where the temporal-dependence of a distance scale in a thermal process varies with the square root of the time. Another realization of this type of behavior is seen in the cooling of lava, and we show in Fig. 5.3 data of the ". . . thicknesses of the solidifying crusts on the lava lakes in three pit craters . . ." (Turcotte and Schubert 2002), along with a theoretical curve with of the form $x(t) \propto \sqrt{t}$. Data from Wright et al. (1976) are shown in Fig. 5.4, effectively on a log-log plot, and you should be able to probe the $\sqrt{t}$ dependence and extract an estimate of D_{th} from either of those two figures; see **P5.4**.

Given these two examples, it's natural to explore if the $x^2 \propto t$ dependence is even more general, and to that end, we combine Eqns. (5.1) and (5.5) in a slightly more (but not completely) rigorous manner, mostly keeping track of dimensions. Imagining the standard planar geometry of heat flow across a distance Δx of a sample of area A, we can connect

$$\Delta Q = mC_S\Delta T - (\rho_m V)C_S\Delta T = (\rho_m A\Delta x)\, C_S\Delta T \tag{5.15}$$

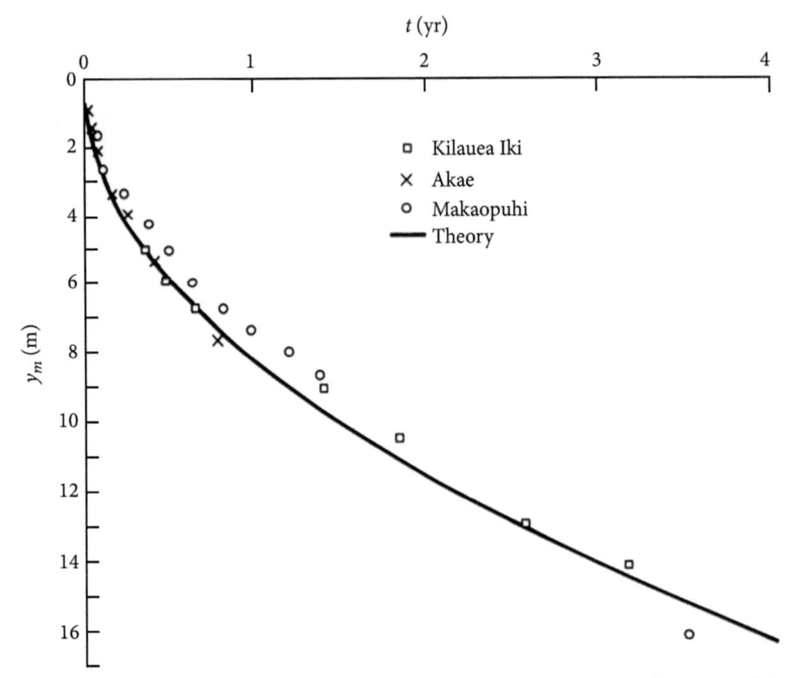

Fig. 5.3 The thicknesses (in *m*) of the solidifying crusts on lava lakes as a function of time (in *yr*). Reprinted from Turcotte and Schubert (2002) with permission of Cambridge University Press.

as the relevant quantity of heat in the thermal conduction equation, allowing us to write

$$(\rho_m A C_S)\,\Delta x \frac{\Delta T}{\Delta t} \sim \frac{\Delta Q}{\Delta t} = \kappa A \frac{\Delta T}{\Delta x}, \tag{5.16}$$

and, cancelling the geometric specific A factors, we have

$$\frac{\Delta T}{\Delta t} = \left(\frac{\kappa}{\rho_m C_S}\right) \frac{\Delta^2 T}{\Delta x^2} = D_{th} \frac{\Delta^2 T}{\Delta x^2} \tag{5.17}$$

and we see that the thermal diffusion constant continues to appear in a natural way.

This "derivation" may well be more at the level of a mnemonic device, but it does capture the physics behind the **classical heat equation** in one dimension, namely

$$\frac{\partial T(x,t)}{\partial t} = D_{th} \frac{\partial^2 T(x,t)}{\partial x^2} \qquad \text{or in 3 dimensions} \qquad \frac{\partial T(\mathbf{r},t)}{\partial t} = D_{th} \nabla^2 T(\mathbf{r},t) \tag{5.18}$$

and does illustrate (dimensionally, at least) the connection

$$\frac{1}{t} \propto D_{th} \frac{1}{x^2} \qquad \text{or} \qquad x^2 \propto D_{th} t \tag{5.19}$$

as being quite general. The famous result in Eqn. (5.18) is attributed to Fourier (1878), and his original work (at least in translation) is eminently readable, including a short discussion

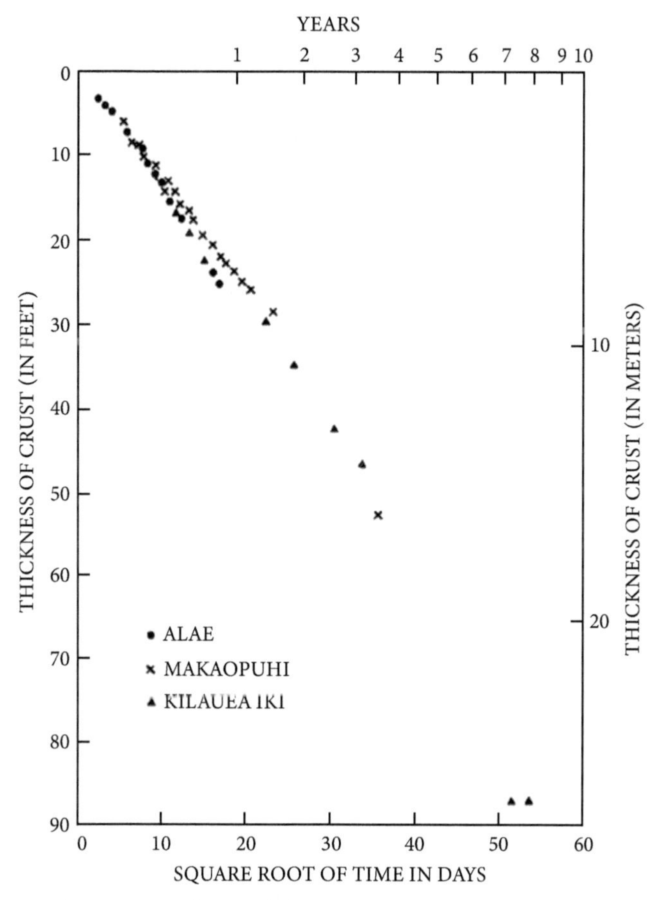

Fig. 5.4 Thickness of lava crusts (in *ft* on left scale, in *m* on right) from three Hawaiian lava lakes as a function of time: note the horizontal scales on top versus bottom! Reprinted from Wright, Peck, and Shaw (1976) with permission of John Wiley and Sons.

of dimensional analysis (his section IX, pp. 126–30), and a wealth of mathematical examples, including use of the Fourier series expansion methods also attributed to him.[2]

We are (or should be) very familiar with experimental data representing physical quantities being displayed in order to be compared to the mathematical solutions of equations in physics, often differential ones, as in Eqn. (5.18) above. It's less frequent that we encounter data representing the quantities on either side of the basic differential equation itself being plotted directly against each other as an analysis tool. One such example (taken from geophysics) is shown in Fig. 5.5 where measured values of $T(z,t)$ have been used to estimate both $\partial T(z,t)/\partial t$ and $\partial^2 T(z,t)/\partial z^2$. The convincing linear relation shown there is a nice example of the classical heat equation and we analyze it in more detail in **P5.5**.

In contrast to the most familiar, and relatively simple, sine/cosine solutions of the classical wave equations in Eqns. (1.33), (3.53), (3.54), or (4.103) that represent wave propagation, a "basic"

[2] In his *Preliminary Discourse*, he says "Profound study of nature is the most fertile source of mathematical discoveries."

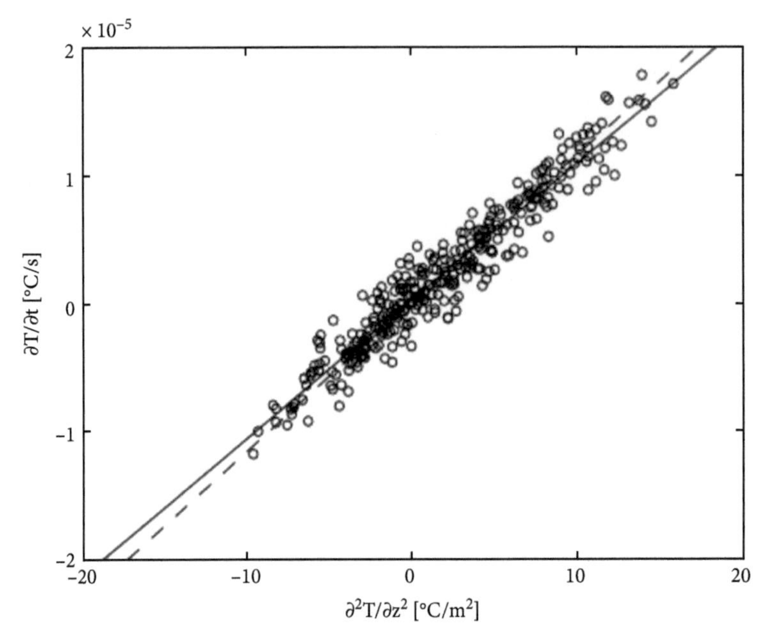

Fig. 5.5 Experimental measurements of $\partial T(z,t)/\partial t$ versus $\partial^2 T(z,t)/\partial z^2$) illustrating the heat equation. Reprinted from Pringle et al. (2003) with permission of John Wiley and Sons. You're asked in **P5.5** to use this data to evaluate the thermal diffusion constant D_{th} and compare to a theoretical prediction.

solution (in one dimension at least) of Eqn. (5.18) is given by

$$T(x,t) \propto \frac{1}{\sqrt{t}} e^{-x^2/4D_{th}t}, \tag{5.20}$$

where the dimensionless Buckingham Pi theorem combination $\Pi = x^2/D_{th}t$ appears naturally. Experimental tests probing the solution of the heat equation are more common in two dimensional systems, with many examples in both pedagogical (Gfroerer et al. 2015) and research (Cernuschi et al. 2001) publications. The relevant functional dependence in a circular geometry with (r,θ) variables is

$$T(r,t) \propto e^{-2r^2/(r_0^2+8D_{th}t)}, \tag{5.21}$$

and we note that in the literature the use of the notation α instead of D_{th} for the thermal diffusivity is common. We show in Fig. 5.6 examples of measurements of $b^2 \equiv r_0^2 + 8D_{th}t$ versus t where the best-fit "slopes" can be used to extract the values of D_{th}; see **P5.7**. Similar experiments can be used to compare to $T(r,t)$ directly to confirm the Gaussian (or $\exp(-x^2/4D_{th}t)$) functional form; Sullivan *et al.* (2008)

The term thermal diffusivity begs the question about the relation of such thermal phenomena to other diffusive processes, such as those in chemistry and related fields. It's typical to describe the behavior of such systems by their **number density**, often labeled by $n(\mathbf{r},t) = N/V$ or sometimes denoted as the **concentration** (with $c(\mathbf{r},t)$ as the variable), with both having dimensions

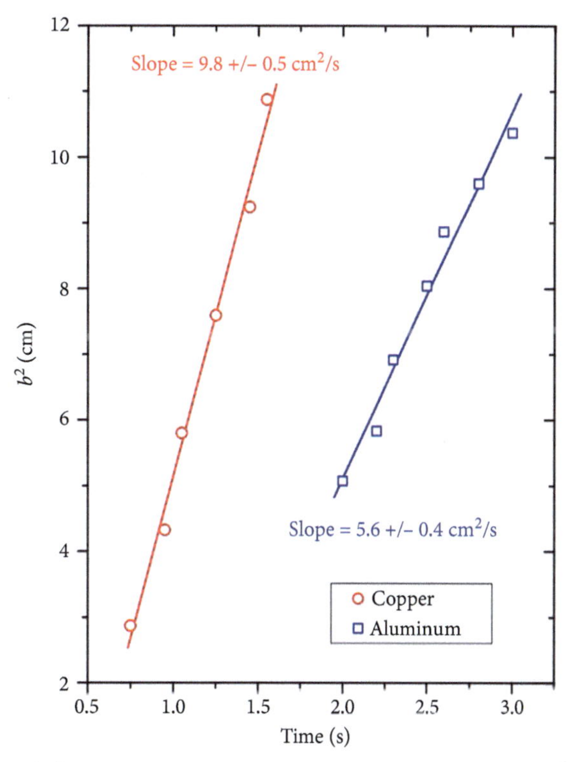

Fig. 5.6 Fitted value of $b^2(t)$ versus t in the Gaussian fit of the form $\exp(-2r^2/b^2)$ from a *thermal diffusivity imaging* experiment. Reprinted from Gfroerer et al. (2015), with the permission of AIP Publishing. The fitted values of $b^2 = b_0^2 + 8D_{th}t$ can be used to extract the values of the thermal diffusivity; see **P5.7**.

$[n] = [c] = \mathrm{L}^{-3}$. The related **diffusive flux** is the *rate of change of number (not number density or concentration) per unit area* or (in 1D) $J_n = (dN/dt)/A$, or $[J_n] = 1/(\mathrm{L}^2\mathrm{T})$.

The relation between this flux and the spatial variation of the density (described by the gradient operator) is called **Fick's law** and given by (this time in three dimensions)

$$\mathbf{J}_n(\mathbf{r}, t) = -D\nabla n(\mathbf{r}, t), \tag{5.22}$$

where the constant of proportionality is the **diffusion constant**, D. Particles will then diffuse from regions of higher to lower concentration.

Combining this expression with the conservation law that assumes that the material in question can't be created or destroyed (only moved around), namely

$$\frac{\partial n(\mathbf{r}, t)}{\partial t} + \nabla \cdot \mathbf{J}_n(\mathbf{r}, t) = 0, \tag{5.23}$$

gives the **diffusion equation**

$$\frac{\partial n(\mathbf{r}, t)}{\partial t} = D\nabla^2 n(\mathbf{r}, t) \qquad \text{or in 1 dimension} \qquad \frac{\partial n(x, t)}{\partial t} = D\frac{\partial^2 n(x, t)}{\partial x^2}. \tag{5.24}$$

One example of the diffusion equation covered in most modern physics books is in the study of **Brownian motion** or the **random walk** problem of thermally agitated particles moving in a liquid medium. In this case, the relevant physical variables might be the properties of the (small) particles, namely their size (say radius R) and the viscosity (η) of the liquid in which they're immersed, along with the thermal variables, k_B and T. We can write the relevant diffusivity (labeled suggestively in this case as D_{Br} for Brownian) as

$$D_{Br} = C_D k_B^{\alpha} T^{\beta} \eta^{\gamma} R^{\delta}$$
$$\frac{\mathsf{L}^2}{\mathsf{T}} = \left(\frac{\mathsf{ML}^2}{\Theta \mathsf{T}^2}\right)^{\alpha} \Theta^{\beta} \left(\frac{\mathsf{M}}{\mathsf{LT}}\right)^{\gamma} \mathsf{L}^{\delta}, \tag{5.25}$$

and then matching dimensions gives $\alpha = \beta = 1$ and $\gamma = \delta = -1$, so that $D_{Br} = C_D(k_B T)/(\eta R)$.

The combination ηR should be familiar from **Stokes' law** (from Chapter 3) for the viscous damping force $F_S = 6\pi v \eta R$ and it indeed turns out that $C_D = 1/6\pi$. This result (attributed to Einstein and Smoluchowski) can also be extended to other situations where a constant external force gives rise to a terminal (or limiting or drift) velocity/speed. This famous relation is our first example where the combination $k_B T$ naturally appears naturally twinned together, and we explore many more such connections in Section 5.2.1.

5.1.2 Allometric scaling

We have already encountered several examples of transport phenomena that are governed by similar physical principles or underlying mechanisms, and not surprisingly, have analogous equations. The two most obvious examples are the flow of charge (q) and thermal energy (Q), encoded via

$$\frac{dq}{dt} = A\sigma_e \frac{dT}{dx} \qquad \text{and} \qquad \frac{dQ}{dt} = A\kappa \frac{dV}{dx}, \tag{5.26}$$

which are connected by microscopic theories of conduction by electrons, linked perhaps most prominently by the Wiedemann–Franz law from Section 2.4. A seemingly obvious feature of both expressions is their dependence on the area in the geometrical realization of the phenomena. Other examples of expressions involving *flux per unit area* of some quantity where such a dependence is implicitly assumed are the *power per unit area* emitted in black-body radiation (discussed in Section 6.1) or the rate of absorption or evaporation from a surface.

For example, the **Hertz–Knudsen**[3] relation gives the number of atoms leaving a surface per unit time as

$$\frac{dN}{dt} = A\left(\frac{(p_v - p_P)}{\sqrt{2\pi m k_B T}}\right), \tag{5.27}$$

where p_v, p_P are the relevant pressures and m the mass of the atomic species involved, with an explicit factor of A. Since such a surface area dependence might seem obvious, it's not necessarily trivial to find examples of experimental verifications of this relationship, but some introductory-level physical chemistry educational modules[4] do describe simple demonstrations which provide

[3] A descriptor that sometimes also includes the names of Langmuir and/or Schrage.

[4] See, e.g., Collection of Physics Experiments, "Dependence of Evaporation Rate of Liquid on Liquid Surface Area." Last modified August 3, 2022. http://physicsexperiments.eu/1774/dependence-of-evaporation-rate-of-liquid-on-liquid-surface-area.

quantitative confirmation of the $dN/dt \propto A$ connection. Other examples from physical chemistry include the rate of catalysis or adsorption and its dependence on the **specific surface area** of a material (defined as the *surface area per unit mass*), which has also been demonstrated to have a linear dependence on area.[5]

Given this "obvious" dependence, it would seem natural to extend these types of area/surface law scaling concepts to biological phenomena as well. For example:

> Allometry, in its broadest sense, describes how the characteristics of living creatures change with size. The term originally referred to the scaling relationship between the size of a body part and the size of the body as a whole, as both grow during development. However, more recently the meaning of the term allometry has been modified and expanded to refer to biological scaling relationships in general, be it for morphological traits (e.g., the relationship between brain size and body size among adult humans), physiological traits (e.g., the relationship between metabolic rate and body size among mammal species) or ecological traits (e.g., the relationship between wing size and flight performance in birds). Indeed, allometric relationships can be described for almost any co-varying biological measurements, resulting in broad usage of the term. However, a unifying theme is that allometry describes how traits or processes scale with one another (Shingleton 2010).

Very often such scaling relationships are described by power-law dependences ($y = Ax^b$) of the form we frequently consider, and much attention is paid to the scaling factor/exponent b.

One extensively explored example (initially for applied agroeconomic motivations) is the variation of **basal metabolic rate (BMR)** with body mass. Values of BMR could be specified in MKS units, as say *Watt* or *Joule/second*, but appear more often in the applied literature as *Calories/day* (where *Calorie* means $kcal = 10^3\ cal$ for food calorie compared to thermal calorie.) Given that animals (at least warm-blooded ones) lose heat through their surfaces, one might expect that the rate at which they expend thermal energy would vary with their surface area, and adopting (for the first, and not the last, time) the "spherical chicken" metaphor for an idealized biological individual, the mass M and volume V would be connected via $M = \rho_m V = \rho_m(4\pi R^3/3)$. Then assuming a surface area $A = 4\pi R^2$, this would immediately suggest that

$$BMR \propto A \propto R^2 \propto V^{2/3} \propto M^{2/3}, \tag{5.28}$$

or a scaling exponent of $b = 2/3$.

In one of the first important studies of the relationship between *body size and metabolism*, Kleiber (1932) collected data on what he called *metabolism* or M (measured in Calories in one day) versus *body weight* or W (in kg) for animals ranging in size from doves (0.15 kg) to steers (679 kg). To probe a power-law dependence, he plotted the data with both M, W shown on logarithmic scales, and we reproduce his results in Fig. 5.7. Instead of the expected 2/3 power-law, he cites that "... the 3/4 power of the body weight was the best-fitting unit ..." and this relationship has come to be known as **Kleiber's law**. (You have the opportunity in Section 10.8 to model his original data using Mathematica© as an example of data fitting to try to confirm his conclusion that $b \approx 3/4$.) Subsequent researchers (including Brody et al. 1934, and Gano 1938) extended the data set to include larger (elephants) and smaller (mice) examples and it's now been said that "Kleiber's law holds across 18 orders of magnitude from microbes to whales" (Smil 2000).

---

[5] See, e.g., Bernard et al. (2021).

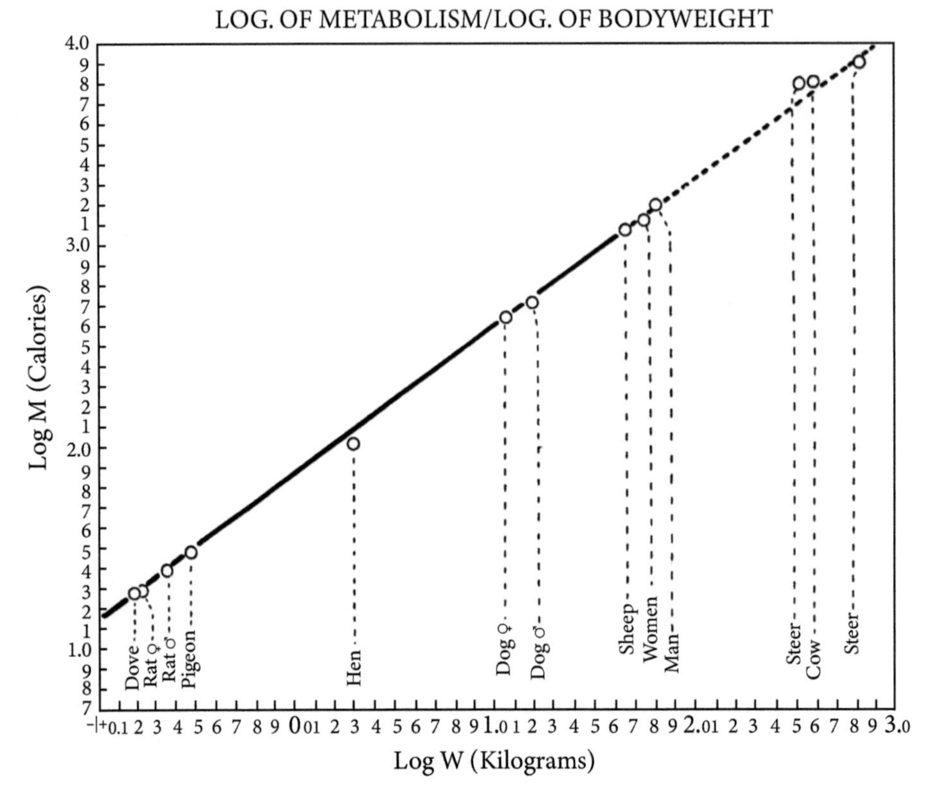

Fig. 5.7 Log–log plot of animal metabolism (M in calories per day) versus mass (W in kg) from Klieber (1932), used by permission of UC Agriculture and Natural Resources.

Many groups have tried to understand such **quarter-power scaling**[6] including West et al. (1997, 1999, 2002), Banavar et al. (1999, 2002, 2010), and others, with many innovative ideas/mechanisms discussed. This even includes the proposal to use fractal geometry[7] so that such transport is effectively occurring in $d = 4$ dimensions,[8] so the relation between volume and area is generalized to $A_d \propto R^{d-1}$ and $V_d \propto R^d$, so that $A_d \propto V^{(d-1)/d}$ giving a $(4-1)/4 = 3/4$ connection. Despite such progress, the underlying physical mechanisms are still unclear, and reviews of the status the field (ca. early 2000s) by Savage (2004), Agutter and Wheatley (2004), White and Seymour (2005), and Savage et al. (2008) all provide more discussion and references. For a broader background, see Kleiber (1961), who surveyed the field of *animal energetics* (including his own work) or more recent popular/public science accounts by McMahon and Bonner (1983) and especially Whitfield (2006), both of which provide a rich historical overview of the field.

In the context of this book, it's important to note that there can certainly be problems involving the contributions of many different physical phenomena for which a handful of dimensionful quantities do **not** suffice to determine the scaling dependence, even if a power-law behavior is clearly present—dimensional analysis isn't always the appropriate tool. For example, the energy

6 It turns out that heart and respiratory rates scale as $M^{-1/4}$; see **P5.9**.
7 See West et al. (1999).
8 Thus giving a different meaning to dimensional analysis!

spectrum of cosmic rays[9] is roughly a power law proportional to $E^{-\alpha}$ with $\alpha \approx 2.7-3.1$ over roughly ten orders of magnitude, but with many possible contributions, from an array of physical sources, leading to the observed behavior. In contrast, familiar thermal diffusivity models can work well to describe many physical systems, giving a 2/3–type dependence in some cases; see **P5.7** and **P5.8** for examples.

5.2 Thermal energy, k_BT physics, and statistical mechanics

5.2.1 The pairing of k_B and T in thermal physics

It's not surprising that a large number of important physical systems are characterized by the appearance of both the Boltzmann constant k_B and the temperature T, paired together in the combination k_BT, or the related twinning of RT, where $R = N_A k_B$ is the **gas constant** with N_A being Avogadro's number. Perhaps the most obvious example is the **ideal gas law**

$$Nk_BT = PV = nRT, \tag{5.29}$$

where N, n are the number of molecules/moles, respectively and all three terms have the dimensions of work/energy

$$[k_BT] = [E] = [W] = [PV] = [RT] = \frac{\mathsf{ML}^2}{\mathsf{T}^2}. \tag{5.30}$$

This connection means that almost any thermal physics problem with another energy scale $\mathcal{E}$ present can automatically include a dimensionless Buckingham Pi ratio $\Pi \equiv \mathcal{E}/k_BT$ or $\Pi = \mathcal{E}/RT$, with the choice between the two depending on context, as the first example corresponds to individual atomic systems, while the second to macroscopic variables.

One example from chemistry of this type of relation describes the temperature dependence of rates of chemical reactions through the exponential **Arrhenius factor.** Very often the **rate constant** k (with $[k] = 1/\mathsf{T}$, definitely not k_B!) that describes the reaction rate for a process is modeled with a form

$$k = Ae^{-E_a/RT} = Ae^{-U/k_BT}, \tag{5.31}$$

where A is, to first approximation, considered T-independent. The **activation energy** for the process can be described at the level of individual particles (atoms, molecules) by U or at the molar level by $E_a = N_A U$. This type of temperature dependence can be probed by appropriate analysis of $k(T)$ versus T data by taking the log of both sides of Eqn. (5.31),

$$\ln(k) = \ln(A) - \frac{E_a}{R}\frac{1}{T} \qquad \text{or} \qquad E_a = R\left(\frac{\partial \ln(k)}{\partial(1/T)}\right), \tag{5.32}$$

so that the activation energy is related to the slope of a log–linear data plot of k versus $1/T$, so long as the "log" is really the natural log or $\ln = \log_e$. (See Section 10.9 for reminders about curve-fitting and extracting power-law and exponential fits from data.) An example of data from a particular chemical reaction (ammonia decomposition) is shown in Fig. 5.8 and you're asked (in **P5.10**) to (carefully) analyze the temperature dependent data presented there to extract a value of E_a.

[9] See, e.g., Verzi et al. (2017) for a review.

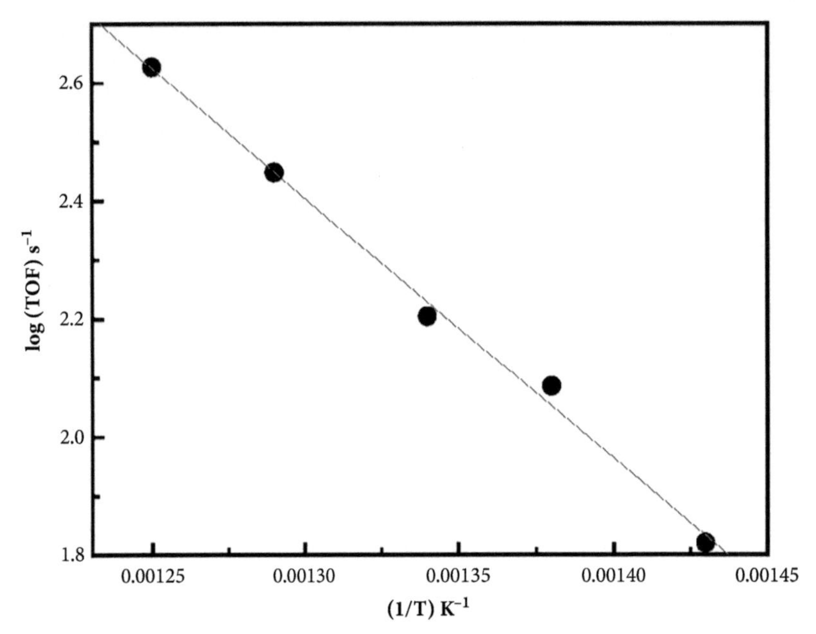

Fig. 5.8 Log–linear data plot of decomposition rate ($1/s$) versus $1/T$ (K^{-1}) showing the Arrhenius law behavior in Eqns. (5.32). Reprinted from Choudhary et al. (2001) with the permission of Springer Nature. You're asked in **P5.10** to evaluate the activation energy E_a from the slope, but note that the authors use the base-10 logarithm on the vertical axis.

See also **Example 10.1** for detailed analysis of a similar log–linear data plot. Laidler (1984) provides an accessible review of the "... development of the Arrhenius equation".

Another example of such an effective rate process, with the Arrhenius factor included, but for which dimensional analysis can be applied in a more proactive way to the important prefactor, occurs in the study of the relaxation of molecules when subjected to external (often oscillating) electric fields. Given that the inverse of the **relaxation time**, τ, gives a rate, one sees relations such as

$$\frac{1}{\tau} = \frac{1}{\tau_0} e^{-\Delta H/RT} = \frac{1}{\tau_0} e^{-\Delta U/k_B T}, \tag{5.33}$$

which is similar to Eqn. (5.31) with differing notation for the macro- or micro-level activation energies. The prefactor, τ_0, has non-trivial dimensions that depend on the thermal and material properties of the surrounding medium namely, k_B, T, and the viscosity η, as well as the typical size scale of the molecules, R_0.

We can explore the dependence of τ_0 on these quantities by writing

$$\begin{aligned} \tau_0 &\propto k_B^{\alpha} T^{\beta} R_0^{\gamma} \eta^{\delta} \\ \mathsf{T} &= \left(\frac{\mathsf{ML}^2}{\Theta \mathsf{T}^2}\right)^{\alpha} \Theta^{\beta} \mathsf{L}^{\gamma} \left(\frac{\mathsf{M}}{\mathsf{LT}}\right)^{\delta} \end{aligned} \tag{5.34}$$

and matching dimensions gives

$$\tau_0 = C'_\tau \frac{R_0^3 \eta}{k_B T} = C_\tau \frac{V\eta}{k_B T}, \tag{5.35}$$

where V is the volume of the molecule (via the scaling with R_0^3), with dimensionless constants C_τ, C'_τ related by geometrical factors. The dimensionless constant in the second form can be written as $C_\tau = f_{stick} C$ where

> "...f_{stick} is the Perrin stick factor dependent upon molecular shape, and C is a measure of the coupling between the rotating molecule and its surroundings" Dote et al. (1981).

The result in Eqn. (5.35) is variously called the **Debye relaxation time** (see Debye 1929 for the original derivation) or **molecular relaxation time**, and has been applied to materials as diverse as water (Grant 1957) and asphalt (Zhang and Greenfield 2007), many showing the $\tau_0 \propto \eta$ proportionality found here. It turns out that the exponential dependences in Eqn. (5.31) are quite general, with many physical properties having $e^{-E/k_B T}$ factors, an important example being the distribution of thermal speeds in a gas, which we explore in the next section. Not every such relation, however, focuses on the temperature dependence of the exponential, but instead on the relevant energy scale.

Example 5.2. The Poole–Frenkel effect

In some materials (dielectrics and semiconductors), the electrical conductivity σ_e has an exponentially sensitive dependence on an external electric field E of the form

$$\sigma_e(E) = \sigma_{e,0} e^{f(E)/k_B T}. \tag{5.36}$$

Assuming that $f(E)$ depends on the applied field, the electron charge and mass (e and m_e) and the dielectric constant in the material, so ϵ and not ϵ_0, we can constrain the form of $f(E)$ by noting that $[k_B T] = [energy]$ and assuming that

$$\begin{aligned} f(E) &\propto e^\alpha E^\beta \epsilon^\gamma m_e^\delta \\ \frac{\mathsf{ML}^2}{\mathsf{T}^2} &= \mathsf{Q}^\alpha \left(\frac{\mathsf{ML}}{\mathsf{QT}^2}\right)^\beta \left(\frac{\mathsf{Q}^2\mathsf{T}^2}{\mathsf{ML}^3}\right)^\gamma \mathsf{M}^\delta, \end{aligned} \tag{5.37}$$

which can be solved to find $\alpha = 3/2$, $\beta = 1/2$, $\gamma = -1/2$, and $\delta = 0$, giving

$$f(E) = C_f \sqrt{\frac{e^3 E}{\epsilon}}, \tag{5.38}$$

and a detailed calculation finds that $C_f = \pi^{-1/2}$. The derivation of this result was first performed by Frenkel (1938) in a less-than-one-page *Letter to the Editor*. Note that this is a different effect than that of cold- or field-emission as discussed in Eqn. (6.170) and **P6.16**, which is a quantum mechanical phenomenon and so includes a factor of $\hbar$.

In such systems, the scaling of σ_e with temperature is less important than that with electric field, so measurements of the induced current, $J(E) = \sigma_e(E)E$, can be used in the form $\ln(\sigma_e(E))/\sqrt{E}$ to extract information on the material under study.

5.2.2 Statistical mechanics

Classical mechanics, of the type taught in the undergraduate curriculum, supplemented by some coding experience, can be used to predict the motion of a small number of point particles, given their starting points ($\boldsymbol{x}_i(0)$ and $\dot{\boldsymbol{x}}_i(0)$), barring any "sensitive dependence on initial conditions" of the kind present in chaotic systems. Conceptually similar methods are used in large-scale, research-level computational studies of galaxy formation that are run on supercomputers, often with over a billion particles tracked (see, e.g., Kuhlen et al. 2012) and make use of an array of sophisticated programming methods (see, e.g., Vogelsberger et al. 2020). It's impossible, however, to extend those techniques to describe the behavior of even just one mole of gas (10^{23} particles), but luckily the methods of **statistical mechanics** can be brought to bear on such problems to reliably predict the average properties of such systems.

In contrast to trying to track the spatial coordinates of every particle, it's easier to focus on the distribution of speeds (v_x in one dimension) or velocities ($\boldsymbol{v}$ in three dimensions) and explore to what extent information on their average values can be extracted, and especially how they're related to the temperature of the gas. Given that the distribution of speeds in a hot gas can be measured, this approach allows one to make more direct contact with experiment, and confirmation of what we'll call the **Maxwell–Boltzmann distribution** has a long history, including results by Eldridge (1927), Zartman (1931), Cohen and Ellet (1937), and perhaps most notably (and most often cited) Estermann, Simpson, and Stern (1947).

As a first step towards such a probabilistic description, we begin in one dimension and define

$$F(v_x)\, dv_x \equiv \mathrm{dProb}[v_x \in (v_x, v_x + dv_x)] \tag{5.39}$$

as the (small) probability that a measurement of the velocity somewhere in the range $(-\infty, +\infty)$ would find it in the narrow "bin" between v_x and $v_x + dv_x$. We can also write this as

$$F(v_x) = \frac{\mathrm{dProb}(v_x)}{dv_x}, \tag{5.40}$$

where $F(v_x)$ is a **probability density** or **probability per unit velocity interval.** Either definition implies that $[F(v_x)] = 1/[v_x] = \mathsf{T}/\mathsf{L}$. Since the probability of finding the particle with some velocity is unity, we must have

$$\int_{-\infty}^{+\infty} F(v_x)\, dv_x = \int \mathrm{dProb} = 1, \tag{5.41}$$

which ensures that the probability distribution $F(v_x)$ is properly normalized. This is similar to the normalization requirement for quantum mechanical wave functions ($\psi(x)$) and their corresponding probability densities ($P(x) = |\psi(x)|^2$) as discussed in Chapter 6, so it's good to gain familiarity with such methods here.

To explore the dependence of $F(v_x)$ on the properties of the particles making up the gas, namely m and v_x, and the thermal parameters, k_B and T, we can write

$$\begin{aligned} F(v_x) &\propto m^\alpha v_x^\beta k_B^\gamma T^\delta \\ \frac{\mathsf{T}}{\mathsf{L}} &= \mathsf{M}^\alpha \left(\frac{\mathsf{L}}{\mathsf{T}}\right)^\beta \left(\frac{\mathsf{ML}^2}{\Theta \mathsf{T}^2}\right)^\gamma \Theta^\delta, \end{aligned} \tag{5.42}$$

and if solved in terms of β we find

$$F(v_x) \propto m^{1/2+\beta} v_x^{\beta} (k_BT)^{-1/2-\beta} = \sqrt{\frac{m}{k_BT}} \left(\frac{mv_x^2}{k_BT}\right)^{\beta/2} \longrightarrow \sqrt{\frac{m}{k_BT}}\, G\left(\Pi = \frac{mv_x^2}{k_BT}\right). \quad (5.43)$$

If we assume that the velocity distribution is proportional to the **Boltzmann factor** of the form e^{-E/k_BT}, namely

$$F(v_x) = Ae^{-mv_x^2/2k_BT}, \quad (5.44)$$

we then determine A by the normalization condition in Eqn. (5.41) by insisting that

$$1 = \int_{-\infty}^{+\infty} F(v_x)\, dv_x = A \int_{-\infty}^{+\infty} e^{-mv_x^2/2k_BT}\, dv_x = A\sqrt{\frac{2k_BT}{m}} \left(\int_{-\infty}^{+\infty} e^{-y^2}\, dy\right) = A\sqrt{\frac{2k_BT}{m}}\sqrt{\pi}, \quad (5.45)$$

so that $A = (m/2\pi k_BT)^{1/2}$, which is consistent with the dimensional analysis (DA) result in Eqn. (5.43), with a dimensionless constant $1/\sqrt{2\pi}$.

With the normalization fixed, we can extract the **thermal averages** or **expectation values** of quantities related to the one-dimensional speed, namely

$$\langle v_x \rangle_{th} \equiv \int_{-\infty}^{+\infty} v_x\, F(v_x)\, dv_x = 0 \qquad \text{and} \qquad \langle v_x^2 \rangle_{th} = \int_{-\infty}^{+\infty} v_x^2\, F(v_x)\, dv_x = \frac{k_BT}{m}, \quad (5.46)$$

where the first result is obvious from symmetry considerations (odd integrand over an even interval), while the second requires a standard Gaussian integral (see Section 10.7 or use any symbolic math program.) The result for $\langle v_x^2 \rangle_{th}$ implies that the thermal average of the kinetic energy is directly related to k_BT by

$$\left\langle \frac{1}{2} m v_x^2 \right\rangle_{th} = \left(\frac{m}{2}\right) \frac{k_BT}{m} = \frac{1}{2} k_BT. \quad (5.47)$$

These results hold for each of the translational degrees of freedom possible for a particle in three dimensions, so that the thermal average of the total kinetic energy is

$$\left\langle \frac{1}{2} m v^2 \right\rangle_{th} = \left\langle \frac{1}{2} m (v_x^2 + v_y^2 + v_z^2) \right\rangle_{th} = 3\left(\frac{m}{2}\right) \frac{k_BT}{m} = \frac{3}{2} k_BT. \quad (5.48)$$

Given that the v_x, v_y, v_z components are independent, we can write for the three-dimensional velocity distribution

$$P(\mathbf{v}) = P(v_x, v_y, v_z) = F(v_x)F(v_y)F(v_z) = \left(\frac{m}{2\pi k_BT}\right)^{3/2} e^{-m(v_x^2+v_y^2+v_z^2)/2k_BT}, \quad (5.49)$$

or we're interested in just the **speed** or the magnitude of the three-dimensional velocity, $v = |\mathbf{v}|$, we have

$$P(v) = 4\pi v^2 \left(\frac{m}{2\pi k_BT}\right)^{3/2} e^{-mv^2/2k_BT}, \quad (5.50)$$

where we use the factor that $dv_x dv_y dv_z = 4\pi v^2 dv$, just as with ordinary Cartesian-spherical coordinates in position space.

The result in Eqn. (5.48) has immediate applications in the prediction of the specific heat of monatomic gases (ones which have only these three translational degrees of freedom, such as noble gases). The **molar specific heat** is given by

$$C_{mol} = N_A \frac{\partial U(T)}{\partial T} = N_A \left(\frac{3k_B}{2}\right) = \frac{3}{2}R, \tag{5.51}$$

where N_A is Avogadro's number and $R = N_A k_B \approx 8.31\, J/(mole \cdot K)$ is again the **gas constant.** We collect in Table 5.1 values of volume specific heats (C_{vol}), molecular weights (μ), and molar specific heats (C_{mol}) for some monatomic and diatomic gases at room temperature, and their ratios to the gas constant, where we have used $C_{mol} = C_{vol}\mu$, and you're encouraged to complete the blank entries. For monatomic gases, the comparison to the simple prediction of $3R/2$ is encouraging.

The result in Eqn. (5.47) that the expectation or average value of the kinetic energy in one demension is equal to $k_B T/2$ is called the **equipartition theorem** and is more generally true when an energy (in this case kinetic) depends quadratically on a variable (in this case v^2).

For diatomic molecules, as shown in Fig. 5.9, which are free to rotate non-trivially around two directions perpendicular to the axis of symmetry, the rotational degrees of freedom have energies given by $E_{rot} = I_x\omega^2/2 = I_y\omega^2/2$, where $I_x = I_y$ is the moment of inertia, and given their quadratic dependence on the variable ω, we also have

$$\left\langle \frac{1}{2} I_x \omega^2 \right\rangle_{th} = \left\langle \frac{1}{2} I_y \omega^2 \right\rangle_{th} = \frac{1}{2} k_B T. \tag{5.52}$$

This suggests that for diatomic molecules, there are two additional contributions, each of $R/2$, to the molar specific heat, giving $C_{diatomic} = 3R/2 + 2(R/2) = 5R/2 \approx 20.8\, J/(mole \cdot K)$, and the (selectively chosen) examples in Table 5.1 seem to bear this out.

Another famous example where the equipartition theorem matches experimental data in a systematic way is the so-called **Dulong–Petit law** for the specific heat of solids. For atoms bound

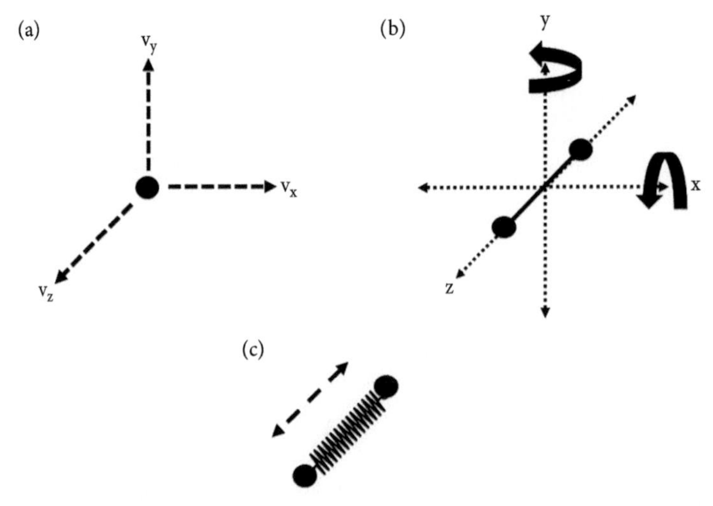

Fig. 5.9 Visualization of the degrees of freedom for a diatomic (and so linear) molecule, illustrating (a) three translation modes, (b) two rotational modes, and (c) two vibrational modes.

Table 5.1: Data on specific heats and molecular weights for several monatomic and diatomic gases at room temperature. You're encouraged (in **P5.20**) to check the filled-in entries in the last two columns, and to complete the rest yourself.

Element	$C_V(kJ/(kg \cdot K))$	$\mu(gr/mol)$	$C_{mol}(J/(mol \cdot K)$	C_{mol}/R
He	3.12	4	12.48	1.5
Ne	0.618	20.1	12.42	1.49
Ar	0.312	39.98		
Kr	0.151	83.8		
Xe	0.097	131.3		
H_2	10.16	2	20.32	2.45
N_2	0.743	28	20.8	2.50
CO	0.72	28		
NO	0.69	30		
O_2	0.659	32		
HCl	0.57	36.4	20.7	2.49

in a solid, we can model the inter-atomic bonds as springs (with spring constant K), in each of the three x, y, z directions. The potential energy in each case is then $E = Kx^2/2$, which is another example of a quadratic energy relation, suggesting that

$$\left\langle \frac{1}{2}Kx^2 \right\rangle_{th} = \left\langle \frac{1}{2}Ky^2 \right\rangle_{th} = \left\langle \frac{1}{2}Kz^2 \right\rangle_{th} = \frac{1}{2}k_BT. \tag{5.53}$$

Then, along with the three degrees of translational freedom, we expect that the molar specific heat would be $C_{mol} = 3R/2 + 3R/2 = 3R \approx 25\,J/(mol \cdot K)$. We plot in Fig. 5.10 the specific heat (at constant pressure) for many solids at room temperature versus their molecular weight and we see generally good agreement, with the experimental values clustering around the horizontal dashed line corresponding to $3R$, with some notable exceptions, like various forms of carbon.

Before congratulating ourselves, we note that for diatomic molecules we would also expect (see Fig. 5.9) vibrational modes to contribute to the total energy, with kinetic and potential energies in the along-axis (or radial) direction,

$$\left\langle \frac{1}{2}\mu \dot{r}^2 \right\rangle_{th} = \left\langle \frac{1}{2}Kr^2 \right\rangle_{th} = \frac{1}{2}k_BT, \tag{5.54}$$

where $\mu = m_1/m_2/(m_1 + m_2)$ is the reduced mass of the diatomic molecule. This would then contribute another $2(k_BT/2)$ to the energy, and another factor R to the specific heat, implying that $C_{vol} = 7R/2$. This seeming discrepancy is one of the first signs that a fully quantum mechanical treatment is required, and we discuss this in Section 5.4.

For the examples considered so far, we've been able to assume that the particles as effectively non-interacting (save for randomizing collisions). We next explore two types of thermal systems where electricity and magnetism (E&M) interactions are present and play an important role, but in very different physical contexts, that is, plasma physics and biophysics.

Example 5.3. Thermal length scales in plasmas

There are several relevant length scales (l_{th}) in a system of ionized gas or **plasma**, many of which can be captured by a quite general dimensional analysis approach. Using the same variables as in Section 2.3, where we discussed the **plasma frequency**, we assume the fundamental electrical (e, ϵ_0) and inertial (m_e, n_e, the electron mass and number density) properties, but now supplemented by k_B and T. We then write

$$l_{th} \propto k_B^\alpha T^\beta e^\gamma \epsilon_0^\delta n_e^\rho m_e^\sigma$$

$$\mathsf{L} = \left(\frac{\mathsf{ML}^2}{\Theta\mathsf{T}^2}\right)^\alpha \Theta^\beta \mathsf{Q}^\gamma \left(\frac{\mathsf{Q}^2\mathsf{T}^2}{\mathsf{ML}^3}\right)^\delta \left(\frac{1}{\mathsf{L}^3}\right)^\rho \mathsf{M}^\sigma \tag{5.55}$$

and match dimensions. The constraints from the Θ and T dimensions immediately specify that $\beta = \alpha = \delta$, while the Q equation gives $\gamma = -2\delta = -2\alpha$. The M equation then requires $\sigma = -\alpha + \delta = 0$ (so no dependence on m_e at all!), and finally the L constraint gives $\rho = -(1+\alpha)/3$ with a final result of

$$l_{th} \propto \left(\frac{k_B T \epsilon_0}{e^2}\right)^\alpha n_e^{-(1+\alpha)/3} m_e^0 . \tag{5.56}$$

Given that one of the basic dimensional ingredients, the number density n_e, depends solely on L, it's not surprising that there is a trivial solution with $\alpha = 0$ (no thermal physics or E&M at all) giving

$$l_{th} = n_e^{-1/3} = l_{ad} \tag{5.57}$$

as basically the average distance (hence the *ad*) between particles. At the other extreme, if $\alpha = -1$, the thermal length scale is independent of n_e, and is given by

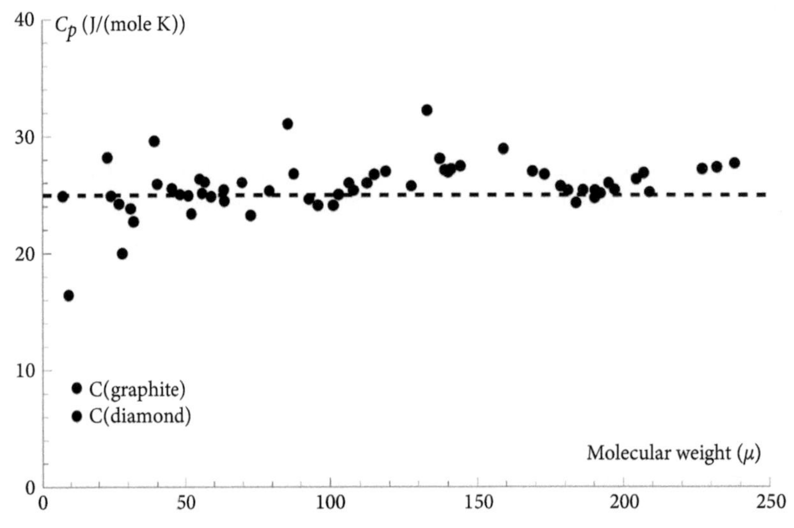

Fig. 5.10 Specific heat at constant pressure (C_p) versus molecular weight (μ) for many solids. The dashed horizontal line corresponds to the Dulong–Petit result of $C_p = 3R \approx 24.9\, J/(mol \cdot K)$. Data taken from "Physical Constants of Organic Compounds", in *CRC Handbook of Chemistry and Physics*, 103rd ed. *(Internet Version 2022)*, edited by J. R. Rumble, Boca Raton, FL CRC Press.

$$l_{th} \propto \frac{e^2}{\epsilon_0(k_BT)} = l_L = l_{ca}, \tag{5.58}$$

which is sometimes called the **Landau distance.** This length scale is associated with a distance of closest approach (hence *ca*) determined by identifying the thermal and Coulomb energies via

$$k_BT \sim E_{th} \sim E_C \sim \frac{e^2}{4\pi\epsilon_0 l_{th}}, \tag{5.59}$$

and therefore very relevant for collisions involving charged particles.

Perhaps the most symmetric solution (in terms of similar exponents) that involves all quantities corresponds to $\alpha = 1/2$, and gives the **Debye length** or sometimes the **Debye shielding distance**, is

$$l_D = \left(\frac{k_BT\epsilon_0}{e^2 n_e}\right)^{1/2}. \tag{5.60}$$

One way to understand the lack of m_e dependence in this relation is to note that one can combine results for the plasma frequency (ω_P) and thermal speed (v_{th}) as

$$\omega_P = \sqrt{\frac{n_e e^2}{m_e \epsilon_0}} \quad \text{and} \quad v_{th} = \sqrt{\frac{3k_BT}{m_e}}$$

$$\text{giving} \qquad l_D \sim \frac{v_{th}}{\omega_P} \sim \sqrt{\frac{k_BT}{m_e}}\sqrt{\frac{m_e\epsilon_0}{n_e e^2}} \sim \sqrt{\frac{k_BT\epsilon_0}{e^2 n_e}}. \tag{5.61}$$

The E&M problem of finding the electric potential due to an isolated point charge q placed in a plasma is discussed in some standard textbooks (e.g., Reitz et al. 1993 or Jackson 1999), and in almost all references on plasma physics (Chen 2018), where one derives the result

$$V_q(r) = \frac{q}{4\pi\epsilon_0 r} e^{-r/l_D}, \tag{5.62}$$

which has the form of the **Yukawa potential** in Eqn. (4.189). The physical implication of this form is that the electrons (and to a lesser extent, ions) in the plasma can easily rearrange themselves to screen out the external charge, and l_D sets the scale for this shielding. In advance of more extensive discussions of plasma physics in Section 9.1, we note that one source qualifies the definition of plasma by noting

> An ionized gas is called a plasma if the Debye length, h, is small compared with other physical dimensions of interest (Reitz et al. 1993, p. 272).

Example 5.4. Length scales in electrolyte solutions near charged membranes—An extended essay

A very different system in which there is a concentration of freely moving charges is an **electrolyte**, described as:

> . . .a medium containing ions that is electrically conducting through the movement of those ions, but not conducting electrons. This includes most soluble salts, acids, and bases dissolved in a polar solvent, such as water.[10]

[10] Wikipedia, s.v. "Electrolyte," https://en.wikipedia.org/wiki/Electrolyte.

Compared to the dimensional quantities characterizing a plasma, some differences (notational and physical) include use of the symbol c for concentration (as opposed to n for number density, but with the same dimensions, $[c] = [n] = \mathsf{L}^{-3}$), and the use of the electric permittivity ϵ of the medium, instead of that for vacuum (but again, with the same dimensions, $[\epsilon] = [\epsilon_0] = \mathsf{Q}^2\mathsf{T}^2/\mathsf{ML}^3$.) Namely, we substitute $\epsilon_0 \rightarrow \epsilon = K\epsilon_0$, where the dimensionless K satisfies $K > 1$, and sometimes $K >> 1$, as in the important case of water. In such systems, the **thermal distance** related to the closest approach, akin to that defined in Eqn. (5.58), is most often denoted by

$$l_{Bjerrum} = l_B \equiv \frac{e^2}{4\pi\epsilon(k_B T)} \tag{5.63}$$

or the **Bjerrum length.**[11]

One important example of such a system is an electrolyte near a charged boundary, here assumed planar for simplicity, with surface charge density σ_e. The relevant length scale in this case could be written in terms of

$$l \propto k_B^\alpha T^\beta e^\gamma \epsilon^\delta \sigma_e^\rho$$

$$\mathsf{L} = \left(\frac{\mathsf{ML}^2}{\Theta\mathsf{T}^2}\right)^\alpha \Theta^\beta \mathsf{Q}^\gamma \left(\frac{\mathsf{Q}^2\mathsf{T}^2}{\mathsf{ML}^3}\right)^\delta \left(\frac{\mathsf{Q}}{\mathsf{L}^2}\right)^\rho, \tag{5.64}$$

and when dimensions are matched we have

$$l \propto (k_B T)^\alpha e^{(1-3\alpha)/2} \epsilon^\alpha \sigma_e^{-(1+\alpha)/2}. \tag{5.65}$$

When $\alpha = -1$, we have $l \propto e^2/(\epsilon(k_B T) \propto l_B$ and we recover the Bjerrum length, while for $\alpha = +1$ we have a new quantity, given by

$$l_{GC} \equiv 2\frac{\epsilon(k_B T)}{e\sigma_e}. \tag{5.66}$$

With the conventional dimensionless factor of 2, this length scale is called the **Gouy–Chapman length,** after a model (to be explored in more detail below) first proposed by Gouy (1910, 1917) and Chapman (1913), which is discussed in many texts on biophysics (see, e.g., Nelson 2004), used in the research literature (see, e.g., Markovich et al. 2021), and even has its own encyclopedia entries (see, e.g., Sposito 2016).

Just as $l_{th} = l_B$ have physical descriptions involving the balance between thermal and electrostatic energies ($\mathcal{E}_{th}$ and $\mathcal{E}_E$), so can l_{GC} can be derived by balancing $k_B T$ with the energy of a charge e in the (constant) electric field E near a surface charge density (σ_e), namely

$$k_B T = \mathcal{E}_{th} = \mathcal{E}_E = (eE)l_{GC} = e\left(\frac{\sigma_e}{\epsilon}\right) l_{GC} \qquad \longrightarrow \qquad l_{GC} = \frac{\epsilon(k_B T)}{e\sigma_e}. \tag{5.67}$$

This length is then a natural one that describes the spatial extent of the concentration of ions near such a charged surface, and sets the scale for the so-called **diffuse layer** near such a boundary.

The one-dimensional problem of self-consistently determining the potential $V(z)$ and concentration $c(z)$ near a two-dimensional charge boundary lends itself both to dimensional analysis as well as an exact treatment. We include both approaches to (i) show how the several length scales relevant to the

[11] After the Danish chemist.

problem interact with each other, and (ii) to provide a tractable one-dimensional problem in electrostatics, compared to the more mathematically sophisticated three-dimensional problems usually covered in standard texts on E&M, and in a more novel area of application.

We assume a planar $(x-y)$ surface of charge density σ_e, and posit that the desired quantities depend on $k_B, T, e, \sigma_e, \epsilon$ and explore the possible dimensional dependences, before citing the exact solution (which can be confirmed by the interested reader). We start with the dimensional magnitude of the electrostatic potential (not including any z-dependence) being defined by combinations of

$$\begin{aligned} V &\propto k_B^\alpha T^\beta e^\gamma \sigma_e^\delta \epsilon^\rho \\ \frac{\mathsf{ML}^2}{\mathsf{QT}^2} &= \left(\frac{\mathsf{ML}^2}{\Theta\mathsf{T}^2}\right)^\alpha \Theta^\beta \mathsf{Q}^\gamma \left(\frac{\mathsf{Q}}{\mathsf{L}^2}\right)^\delta \left(\frac{\mathsf{Q}^2\mathsf{T}^2}{\mathsf{ML}^3}\right)^\rho . \end{aligned} \tag{5.68}$$

As an example of where one's intuition can be profitably used, we write down the five equations constraining the dimensions as

$$\begin{aligned} \mathsf{M}: &\quad 1 = \alpha \qquad\qquad - \rho \\ \mathsf{L}: &\quad 2 = 2\alpha \qquad - 2\delta - 3\rho \\ \mathsf{T}: &\quad -2 = -2\alpha \qquad + 2\rho \\ \mathsf{Q}: &\quad 1 = \qquad \gamma + \delta + 2\rho \\ \Theta: &\quad 0 = -\alpha + \beta. \end{aligned} \tag{5.69}$$

It's hard to miss the fact that the first three equations could easily suggest (before considering any other variables) that $\alpha = 1$, so we will certainly consider that as a possible option. The most general solution, however, after matching dimensions is

$$V \propto (k_B T)^\alpha e^{(1-3\alpha)/2} \sigma_e^{(1-\alpha)/2} \epsilon^{\alpha-1} \tag{5.70}$$

and the expected combination $k_B T$ does indeed appear naturally. A solution corresponding to $\alpha = -1$ gives

$$V \propto \frac{1}{k_B T} \frac{e^2 \sigma_e}{\epsilon} = l_B \left(\frac{\sigma_e}{\epsilon}\right) = l_B E_{surface}, \tag{5.71}$$

which is (of course, it has to be) dimensionally correct, but doesn't reflect the importance of the diffuse boundary layer. The value of $\alpha = 1$, as suggested by the first three lines in Eqn. (5.69), gives a very simple result, namely

$$V \propto \frac{k_B T}{e}, \tag{5.72}$$

which also is consistent with a balance of electrostatic and thermal/kinetic energy, with $eV \sim k_B T$ playing off each other, as well as suggesting a natural dimensionless Buckingham Pi ratio $\Pi = eV/k_B T$ for use in a Boltzmann factor.

For the concentration, c, we write (in a slightly different order)

$$c \propto \sigma_e^\alpha \epsilon^\beta k_B^\gamma T^\delta e^\rho$$

$$\frac{1}{\mathsf{L}^3} = \left(\frac{\mathsf{Q}}{\mathsf{L}^2}\right)^\alpha \left(\frac{\mathsf{Q}^2\mathsf{T}^2}{\mathsf{ML}^3}\right)^\beta \left(\frac{\mathsf{ML}^2}{\Theta\mathsf{T}^2}\right)^\gamma \Theta^\delta \mathsf{Q}^\rho, \tag{5.73}$$

which requires that

$$c \propto \sigma_e^\alpha (\epsilon k_B T)^{3-2\alpha} e^{3\alpha-6}. \tag{5.74}$$

We can consider three special cases very easily, namely

$$\alpha = 0 \quad \longrightarrow \quad c \propto (\epsilon k_B T)^3 e^{-6} = \left(\frac{\epsilon k_B T}{e^2}\right)^3 \sim l_B^{-3} \tag{5.75}$$

$$\alpha = 1 \quad \longrightarrow \quad c \propto \sigma_e (\epsilon k_B T) e^{-3} = \frac{\sigma_e}{e\, l_B} \tag{5.76}$$

$$\alpha = 2 \quad \longrightarrow \quad c \propto \sigma_e^2 (\epsilon k_B T)^{-1} = \frac{\sigma_e}{e\, l_{GC}}. \tag{5.77}$$

Of these options, the first is just the density corresponding to particles separated by a distance l_B, with no connection to the charged membrane. The second and third represents a concentration with a two-dimensional or **areal charge density** of σ_e/e related to the surface, with a thickness of l_B or l_{GC}, respectively. The last case corresponds to a diffuse layer of scale l_{GC}, and this realization is indeed the correct combination of dimensions.

The problem of determining the electrostatic potential given a charge distribution can be couched in the language of the **Poisson equation** (as first mentioned in Eqn. (3.47)), which in three dimensions reads $\nabla^2 V(\mathbf{r}) = -\rho_e(\mathbf{r})/\epsilon$, where we again use the general permittivity ϵ relevant to a given medium. Since we are working in one dimension, the charge density is related to the concentration via $\rho_e(z) = ec(z)$, and we also assume a **Boltzmann** form for the spatial dependence of $c(z)$ related to the potential, via $c(z) = c_0 \exp(-eV(z)/k_B T)$. Combining these two gives the **Poisson–Boltzmann** equation, namely

$$\frac{d^2V(z)}{dz^2} = -\frac{\rho_e(z)}{\epsilon} = -\frac{e}{\epsilon}c(z) = -\frac{ec_0}{\epsilon}e^{-eV(z)/k_BT}, \tag{5.78}$$

which is a highly non-linear relation, which must be solved self-consistently for both $V(z)$ and the constant c_0.

As with any second-order spatial differential equation, we must specify boundary conditions, one of which is that the electric field near the boundary (membrane) must reproduce the $\boldsymbol{E}$ field due to a surface charge density (in a medium), namely

$$E(z=0) \equiv \left.\frac{dV(z)}{dz}\right|_{z=0} = \frac{\sigma_e}{\epsilon}. \tag{5.79}$$

The second boundary condition is somewhat arbitrary, but it's often assumed that $V(z=0) = 0$.

The solution of this system is discussed in textbooks on biophysics (such as Nelson 2004) and is given by

$$V(z) = \frac{2k_BT}{e}\ln\left(1 + \frac{z}{l_{GC}}\right) \qquad \text{and} \qquad c_0 = \frac{\sigma_e}{e\, l_{GC}}, \tag{5.80}$$

both of which are consistent with our purely DA approach above. It's clear that $V(z=0) = 0$, and you're asked (in **P5.18**) to confirm that the relations in Eqn. (5.80) also satisfy Eqns. (5.78) and (5.79) and to calculate related quantities. It's interesting to note the expression for $c(z)$ simplifies, since

$$\frac{eV(z)}{k_BT} = \frac{e}{k_BT}\left\{\left(\frac{2k_BT}{e}\right)\ln\left(1+\frac{z}{l_{GC}}\right)\right\} = 2\ln\left(1+\frac{z}{l_{GC}}\right) = \ln\left(1+\frac{z}{l_{GC}}\right)^2 \tag{5.81}$$

so that

$$\exp\left(-\frac{eV(z)}{k_BT}\right) = \frac{1}{(1+z/l_{GC})^2}. \tag{5.82}$$

This means that the spatial dependence of the concentration is given by

$$c(z) = c_0 e^{-eV(z)/k_BT} = \frac{\sigma}{e\,l_{GC}}\frac{1}{(1+z/l_{GC})^2} = \frac{\sigma l_{GC}}{e}\frac{1}{(z+l_{GC})^2} = \frac{1}{2\pi l_B(z+l_{GC})^2} \tag{5.83}$$

so that another combination of l_B and l_{GC} (and now z) appears.

5.3 **Kinetic theory**

The description of gases in terms of kinetic theory can be used to estimate many of their physical properties, all of which can also be explored using the methods of dimensional analysis. Many standard textbooks at the undergraduate level (such as Kittel and Kroemer 1980, and Reif 1965) make use of methods that Hirschfelder, Curtiss, and Bird (1967) famously describe as "... an ultra-simplified kinetic theory of dilute gases ..." and we will compare our DA expressions to such results. For those interested in more rigorous calculations, one can consult Chapman and Cowling's (1970) *The Mathematical Theory of Non-Uniform Gases: An Account of the Kinetic Theory of Viscosity, Thermal Conduction, and Diffusion in Gases*, which already in its title cites several of the important material properties that can be described by statistical mechanics, and which we consider in what follows.

We first collect some of the relevant physical quantities for a dilute gas, namely

$$\overline{v} = \text{average/mean speed, with } [\overline{v}] = \mathsf{L/T} \tag{5.84}$$

$$\tau = \text{average/mean time between collisions, with } [\tau] = \mathsf{T} \tag{5.85}$$

$$l = \text{mean free path (distance) between collisions, with } [l] = \mathsf{L} \tag{5.86}$$

$$\sigma = \text{scattering cross-section governing collisions, with } [\sigma] = \mathsf{L}^2 \tag{5.87}$$

$$n = \text{number density of the gas, with } [n] = \mathsf{L}^{-3}, \tag{5.88}$$

all of which depend only on the length and time dimensions L, T.

The rate of collisions Γ (the inverse of the collision time) is determined by σ and n, and the average/mean **relative** speed, $\overline{V} = \sqrt{2}\,\overline{v}$, with $\overline{V}, \overline{v} \propto \sqrt{k_BT/m}$ given by kinetic theory, only differing by a dimensionless factor. This connection is given by

$$\Gamma \equiv \frac{1}{\tau} = \overline{V}\sigma n \qquad \text{with} \qquad \frac{1}{\mathsf{T}} \stackrel{!}{=} \left(\frac{\mathsf{L}}{\mathsf{T}}\right)(\mathsf{L}^2)\left(\frac{1}{\mathsf{L}^3}\right), \tag{5.89}$$

which is clearly dimensionally correct. The mean free path is then related to $\tau, \overline{v}$ via

$$l = \tau\overline{v} = \frac{\overline{v}}{\overline{V}}\frac{1}{\sigma n} = \frac{1}{\sqrt{2}\sigma n}. \tag{5.90}$$

With the five dimensional quantities in Eqns. (5.84)–(5.88), and the two relations connecting them in Eqns. (5.89) and (5.90), we will often consider as independent dimensional quantities $n, \overline{v}, l$.

One of the processes of a gas that can be described using such quantities is diffusion, and the relevant physical property is the diffusion constant, D, as in Eqn. (5.24). We can write this as $D = C_D n^\alpha \overline{v}^\beta l^\gamma$ and matching dimensions, we find

$$\frac{\mathsf{L}^2}{\mathsf{T}} = \left(\frac{1}{\mathsf{L}^3}\right)^\alpha \left(\frac{\mathsf{L}}{\mathsf{T}}\right)^\beta \mathsf{L}^\gamma \qquad \longrightarrow \qquad D \propto n^\alpha\, \overline{v}\, l^{1+3\alpha}. \tag{5.91}$$

Given that D itself appears in a differential equation that is supposed to actually determine the number density $n(\mathbf{r}, t)$, we would not expect any dependence on n in the expression above, and so assume that $\alpha = 0$. With the dimensionless prefactor determined by kinetic theory analysis as $C_D = 1/3$ (directly related to the fact that we're in three dimensions!), one finds

$$D = \frac{1}{3}\overline{v}\, l = \frac{1}{3}\frac{l^2}{\tau}, \tag{5.92}$$

which clearly demonstrates the expected dimensional dependence on L, T.

An important transport equation, which defines the thermal conductivity, is given by Eqn. (5.5), namely

$$\frac{dQ}{dt} = \kappa A \frac{\Delta T}{\Delta z}. \tag{5.93}$$

To also make contact with diffusive behavior, we note that we can write

$$J_n = -D\nabla n \qquad \text{or} \qquad \frac{dn}{dt} = -DA\frac{\Delta n}{\Delta z}, \tag{5.94}$$

with D and κ playing parallel roles. A third, seemingly unrelated, expression is one for the viscosity (η) of a fluid (liquid or gas), first introduced in Eqn. (2.28) (with a geometry defined in Fig. 2.8) as

$$\frac{F}{A} = \eta\frac{\Delta v_y}{\Delta z}, \qquad \text{which we can rewrite as} \qquad \frac{dp}{dt} = \eta A \frac{\Delta v_y}{\Delta z}, \tag{5.95}$$

where the force is related to the rate of change of momentum via $F = dp/dt$. This form emphasizes the similarities to Eqns. (5.93) and (5.94), and to emphasize that what is being transported here is momentum.

In contrast to the DA approach for the diffusion constant, where the only quantities involved are $n, \overline{v}, l$, all having only L, T dimensions, for these two transport phenomena, there is another dimensional quantity in play in each case. For the viscosity equation, the **mass per molecule**, m, of the particles is required, while for thermal transport, the **specific heat per molecule**, $c \equiv \partial\epsilon/\partial T$, is needed. We can approach both problems in a parallel manner by writing (and then solving)

$$\begin{aligned} \eta &= C_\eta n^\alpha \bar{v}^\beta m^\gamma l^\delta & \kappa &= C_\kappa n^\alpha \bar{v}^\beta c^\gamma l^\delta \\ \frac{\mathsf{M}}{\mathsf{LT}} &= \left(\frac{1}{\mathsf{L}^3}\right)^\alpha \left(\frac{\mathsf{L}}{\mathsf{T}}\right)^\beta \mathsf{M}^\gamma \mathsf{L}^\delta & \frac{\mathsf{ML}}{\mathsf{T}^2\Theta} &= \left(\frac{1}{\mathsf{L}^3}\right)^\alpha \left(\frac{\mathsf{L}}{\mathsf{T}}\right)^\beta \left(\frac{\mathsf{ML}^2}{\mathsf{T}^2\Theta}\right)^\gamma \mathsf{L}^\delta \\ \eta &= C_\eta n^\alpha \bar{v}\, m\, l^{3\alpha-2} & \kappa &= C_\kappa n^\alpha \bar{v}\, c\, l^{3\alpha-2}. \end{aligned} \tag{5.96}$$

The similarities between the two expressions are striking, and one (perhaps natural) simplifying assumption is obtained by noting that using the combination of mass density, $\rho_m = mn$, might suggest that $\alpha = 1$, while the **volume-specific heat** relation $C_{vol} = cn$ would suggest the same. In fact, the standard kinetic theory arguments leading to the diffusion constant also give

$$\eta = \frac{1}{3}\rho_m \bar{v} l \qquad \text{and} \qquad \kappa = \frac{1}{3} C_{vol} \bar{v} l, \tag{5.97}$$

with $C_D = C_\eta = C_\kappa = 1/3$. These textbook (or "... ultra-simplified kinetic theory ...") results can be combined to predict that

$$\frac{\kappa}{\eta} = \frac{c}{m} = \frac{C_{vol}/n}{\rho_m/n} = \frac{C_{vol}}{\rho_m} = \frac{C_{mol}}{\mu} \quad \text{or the combination } f \text{ satisfies} \quad f \equiv \frac{\kappa}{\eta}\frac{\mu}{C_{mol}} = 1, \tag{5.98}$$

where C_{mol} and μ are the **molar-specific heat** and **molecular mass.** Experiments show that dilute gases are characterized by values of f in the range 1.3–2.5 and one author (rightly) applauds this level of success by saying

> In view of the very simplified nature of the arguments leading to these expressions for η and κ, there is greater justification for being pleased by the extent of agreement with experiment than there is cause for surprise at the discrepancy (Reif 1965, p. 481).

We applaud this sentiment as being very much in the spirit of appreciating any and all steps towards the understanding of physical phenomena, at any level of rigor, including the use of dimensional analysis methods. Kittel and Kroemer (1980, pp. 403–404) note that

> Improved calculations of the kinetic coefficients κ, D, η take account of minor, but difficult effects we have neglected ...

and we explore in **P5.20** just such modifications, as they are nicely characterized by the inclusion of the additional internal degrees of freedom for real molecules.

A final transport relation, for the conduction of electric charge, can also be explored using DA. We can write Ohm's law as

$$\mathbf{J}_e = \sigma_e \mathbf{E} \qquad \text{or} \qquad \frac{dI}{dt} = \sigma_e A \frac{\Delta V}{\Delta x}. \tag{5.99}$$

In this case we identify the collision time τ along with the properties of the conductor, namely n, e, m_e, as the relevant dimensional quantities, and write

$$\sigma_e \propto n^\alpha e^\beta \tau^\gamma m_e^\delta$$
$$\frac{\mathsf{Q}^2\mathsf{T}}{\mathsf{ML}^3} = \left(\frac{1}{\mathsf{L}^3}\right)^\alpha \mathsf{Q}^\beta \mathsf{T}^\gamma \mathsf{M}^\delta \tag{5.100}$$

and find perhaps one of the simplest set of DA constraint equations we'll ever encounter, being able to immediately read off the solutions as $\alpha = 1, \beta = 2, \gamma = 1$, and $\delta = -1$, giving

$$\sigma_e = \frac{ne^2\tau}{m}. \tag{5.101}$$

Despite being derived in the context of dilute gases, this expression continues to be useful in electrical conductivity in solids, and we explore in **P5.22** how the dimensional dependences in the Wiedemann–Franz law (discussed in Section 2.4) can be constrained from Eqns. (5.97) and (5.101)

5.4 Glimmers of quantum mechanics

From the (admittedly, very carefully curated) data on specific heats of (selected) gases presented in Table 5.1, all at room temperature, it would seem that a description based on counting of degrees of freedom (3 translational ones for monatomic molecules, with 2 more rotational ones for diatomic ones) does fairly well in describing the measured values of C_{mol}/R. One obvious failure, not shown there, is that at lower temperatures, below about 100 K, the specific heat of H_2 gets closer to $C_{mol}/R = 3/2$, like that of a monatomic gas, as if the rotational degrees of freedom are "frozen out" and not contributing.

This is one of the first hints that quantum mechanics (with its signature feature of quantized energy levels) is playing a role. The rotational kinetic energy of a system can be written in terms of the moment of inertia (I) and either the angular frequency (ω) or angular momentum (L), using the connection that $L = I\omega$, namely

$$E_{rot} = \frac{1}{2}I\omega^2 = \frac{L^2}{2I}. \tag{5.102}$$

For a diatomic molecule (with two possibly different mass atoms), the moment of inertia will be given by $I = \mu r_0^2$, where r_0 is the separation of the two atoms and $\mu = (m_1m_2)/(m_1 + m_2)$ is the **reduced mass** of the two-body system, and for homo-nuclear (same mass) molecules one has $\mu = m/2$. For the other required variable in Eqn. (5.102), in classical mechanics, L can take on any value, while in quantum mechanics the angular momentum is quantized in units of Planck's constant, $\hbar$. In this case, where L^2 appears quadratically, the corresponding expression for the quantum rotational kinetic energy is

$$E_{rot}^{(J)} = \frac{L^2}{2I} = \frac{J(J+1)\hbar^2}{2I} = J(J+1)\frac{\hbar^2}{mr_0^2} \equiv J(J+1)\mathcal{E}_0, \tag{5.103}$$

where J is the rotational quantum number ($J = 0, 1, 2, ...$) and we assume equal mass atoms for simplicity.

Using nitrogen (N_2) as an example system, we have $m(N) \approx 14(1.67 \times 10^{-27}\,kg)$, $r_0 \approx 1\text{Å} \approx 10^{-10}\,m$, and of course $\hbar \approx 10^{-34}\,J \cdot s$, giving $\mathcal{E}_0 \approx 4 \times 10^{-23}\,J$. These discrete values of $E_{rot}^{(J)}$ require that changes in the rotational energy (from one neighboring quantum state to the next) are given by $\Delta E^J = E^{(J+1)} - E^{(J)} = 2(J+1)\mathcal{E}_0$ and so the minimum energy needed to excite a vibrational mode from $J = 0 \rightarrow J = 1$ would be $\Delta E = 2\mathcal{E}_0 \approx 8 \times 10^{-23}\,J$, and this would correspond to a temperature of roughly $T \sim \Delta E/k_B \approx (8 \times 10^{-23}\,J)/(1.4 \times 10^{-23}\,J/K) \approx 6\,K$. At any temperatures much higher than this value, the rotational degrees of freedom are therefore "in play" and contribute to the accessible degrees of freedom. On the other hand, for H_2, we can scale the energies by a factor of $\mathcal{E}_0(H)/\mathcal{E}_0(N) = (m_N/m_H) = 14$, and the temperature required to excite hydrogen rotational levels would be roughly $14 \times 6\,K \approx 80\,K$, thereby explaining the "freeze-out" at temperatures much lower than this.

We noted earlier that diatomic molecules also have vibrational states (a cartoon version is shown in Fig. 5.9) with an effective radial potential energy given by $V(r) = Kr^2/2$. The data in Table 5.1 do not suggest that the two degrees of freedom associated with this internal motion are "in play," at least at room tempreature, and so we can explore whether the quantization of the vibrational energy levels explains this fact. If we assume that the quantum mechanical E_{vib} energies depend on $\hbar, K, \mu$ (where we again use the reduced mass, μ, of the two-body system), we write $E \propto K^\alpha \mu^\beta \hbar^\gamma$ and match dimensions via

$$\frac{\mathsf{ML}^2}{\mathsf{T}^2} = \left(\frac{\mathsf{M}}{\mathsf{T}^2}\right)^\alpha \mathsf{M}^\beta \left(\frac{\mathsf{ML}^2}{\mathsf{T}}\right)^\gamma \qquad \text{giving} \qquad E \propto K^{1/2}\mu^{-1/2}\hbar^1 \propto \hbar\sqrt{\frac{K}{\mu}}, \tag{5.104}$$

and the corresponding textbook result is $E_{vib}^{(n)} = (n + 1/2)\hbar\sqrt{K/\mu}$ where $n = 0, 1, 2, \ldots$ is another integral quantum number. To see to what extent such states can contribute to the specific heat, we can use spectroscopic data (say again for nitrogen, or N_2) to extract a value of the effective spring constant, $K = 2290\,N/m$, and with $\mu = m(N)/2$, we find $\Delta E_{vib} \approx 5 \times 10^{-20}\,J$ as the typical excitation energy, corresponding to a temperature of $T_{vib} = \Delta E_{vib}/k_B \approx 3800\,K$. This is consistent with observation, where the specific heat of diatomic molecules does approach $7R/2$ at higher temperatures when these degrees of freedom can contribute.

In the last two examples, we have used one of the "hallmark" features of quantum theory, namely that some quantities (in this case both angular momentum and energy) are quantized, depending in various ways on Planck's constant, $\hbar$. But a perhaps even more fundamental property is that particles exhibit wave-like properties and we explore one "corner" of classical thermal physics where this connection appears.

In the limit of high temperatures and low densities, gases can be treated in a classical way via thermodynamics and statistical mechanics, without needing to invoke quantum mechanics. In the opposite limit, of low temperatures and high densities, one needs to include quantum mechanical effects, and especially to treat fermions (half-integral spin particles) differently from bosons (integral spins), as required by the **Pauli principle.**

Let us assume a class of particles of mass m, with the usual k_B, T dimensions, but also include $\hbar$ to see if there is a natural scale for a **quantum (number) density**, n_Q, where $[n_Q] = \mathsf{L}^{-3}$. We do so by writing $n_Q \propto m^\alpha k_B^\beta T^\gamma \hbar^\delta$ and match dimensions to obtain

$$\frac{1}{\mathsf{L}^3} = \mathsf{M}^\alpha \left(\frac{\mathsf{ML}^2}{\Theta \mathsf{L}^2}\right)^\beta (\Theta)^\gamma \left(\frac{\mathsf{ML}^2}{\mathsf{T}}\right)^\delta \quad \text{which gives} \quad n_Q \propto m^{3/2}(k_B T)^{3/2}\hbar^{-3} = \left(\frac{mk_B T}{\hbar^2}\right)^{3/2}. \tag{5.105}$$

To explore to what extent this result naturally encodes information on the wave-nature of particles, we note that if we associate $k_B T \leftrightarrow mv^2$, then we can write $mk_B T \leftrightarrow m(mv^2) \leftrightarrow p^2$, and the quantum density can be written as

$$n_Q \sim \left(\frac{p^2}{\hbar^2}\right)^{3/2} \sim \left(\frac{p}{\hbar}\right)^3 \sim \left(\frac{1}{\lambda_D}\right)^3 \sim \lambda_D^{-3}, \tag{5.106}$$

where $\lambda_D = h/p = 2\pi\hbar/p$ is the **de Broglie wavelength.** Thus, the natural quantum number density corresponds to the case where the interparticle separation is the characteristic quantum length determined by its typical momentum scale and $\hbar$. You can either invert this expression, or "start from scratch" to determine the typical temperature scale at which quantum effects are important for a given number density, as in **P5.23.**

We have not yet explored one of the most important concepts in thermal physics, namely the concept of **entropy,**[12] which allows for a seamless connection between the micro-states of particles as described by quantum theory, and the macroscopic behavior of collections of particles as encoded by thermodynamics. Given that the fundamental constant of quantum mechanics first appeared in the context of discussions of entropy, we choose to begin Chapter 6 with that historical connection.

5.5 Problems for Chapter 5

Q5.1 Reality check: (a) What was the single most important, interesting, engaging, and/or useful thing that you learned from this part of the book?
(b) Have you ever used DA in approaching any problem in thermal physics?

P5.1 Stefan ice thickness—Theory calculation: We can solve the Stefan's ice thickness problem from **Example 5.1** by assuming that the heat flow through an existing thickness of ice, $x(t)$, is related to that released from the freezing process of a new thin (dx) layer of area A. We can then write

$$\frac{dQ}{dt} = \frac{\kappa A \Delta T}{x(t)} \quad \text{and} \quad dQ = L_F\, dm = L_F\left[dV\rho_m\right] = L_F\left[(Adx)\rho_m\right]. \tag{5.107}$$

Combine these two equations to get a relation between dx and dt, which you can integrate (over the ranges $(0, x(t))$ and $(0, t)$, respectively) to find $x(t)$ and confirm that $C_x = \sqrt{2}$.

P5.2 Stefan ice thickness—Numerical evaluation: Confirm the "folk-theorem" description in Eqn. (5.12) of Stefan's model of ice thickness in Eqn. (5.10) by scaling ΔT to $1\,K$ and t to 1 day, respectively, and use values of the ice parameters given by

$$\kappa \approx 2\,\frac{kg \cdot m}{s^3\,K}, \qquad L_F \approx 3 \times 10^5\,\frac{m^2}{s^2}, \qquad \text{and} \qquad \rho_m \approx 10^3\,\frac{kg}{m^3}. \tag{5.108}$$

P5.3 Modified ice growth model: Ashton (1989) extends the simple Stefan model of ice growth to include a term describing the "... bulk heat transfer ... between the top surface temperature of the ice and the air temperature above the ice and the air temperature above the ice ..." given by

$$\frac{dQ}{dt} = H_{ia} A \Delta T, \tag{5.109}$$

where *ia* stands for ice–air interface.

[12] For a nice discussion of *ten steps to entropy*, starting from traditional thermodynamic concepts, see Reed (2023).

(a) What are the M, L, T, Θ dimensions of H_{ia}?

(b) If we ignore the thermal conductivity contribution, we can write $x(t) \propto t^{\alpha} H_{ia}^{\beta} (\Delta T)^{\gamma} L_F^{\delta} \rho_m^{\epsilon}$ for the time-dependence of the ice growth. Match dimensions, constrain the exponents as much as you can, and see if this determines the t exponent. If you use the same connection as in **Example 5.1**, namely that ρ_m, L_F must appear with the same power/exponent, does this simplify things further?

(c) Adding the H_{ia} term to the original problem, and solving the heat flow equations (similarly to **P5.2**), which are now more mathematically involved, Ashton (1989) finds the expression

$$x(t) = \left[\frac{2\kappa \Delta T}{\rho_m L_F} t + \left(\frac{\kappa}{H_{ia}} \right)^2 \right]^{1/2} - \frac{\kappa}{H_{ia}}. \tag{5.110}$$

Show that all terms in this expression are dimensionally correct.

(d)* For short times (and you should be able to define short), show that $x(t)$ in Eqn. (5.110) is linear in time, while for long times (again, where is the cutoff between long and short in terms of these parameters?) it reproduces the original Stefan calculation. You'll have to expand a function of the form $(1+y)^{1/2}$ for small values of y. You can compare the results of this model to real data in Fig. 5.2.

(d) You presumably found that the small parameter used in the expansion in part (c) was

$$y = \frac{2(H_{ia})^2 (\Delta T\, t)}{\kappa \rho_m L_F}, \tag{5.111}$$

so that $y << 1$ and $y >> 1$ correspond to short and long times, and the crossover from one behavior to the other then corresponds to $y \approx 1$. Given that the quantity of interest is actually $\Delta T\, t$ (degree-days, as in Fig. 5.2), find the value of that product corresponding to $y \approx 1$. You'll need to assume a value for H_{ia}, and Ashton (1989) used $H_{ia} = (10\text{–}30)\ W/(m^2 \cdot K)$.

P5.4 Lava cooling data: A complete mathematical solution to the lava cooling problem finds that $x(t) = 2\lambda(D_{th} t)^{1/2}$ where $D_{th} = \kappa/(\rho_m C_S) = \kappa/C_{vol}$ (as we've discussed) and λ is a dimensionless constant of order unity (determined by $L_F, \Delta T$ and detailed solutions of the heat equation).

(a)† From Fig. 5.3 we can extract a few points (from the theoretical curve, which seems to fit the experimental data fairly well), which are included in the table below.

Thickness (m) - $x(t)$	time (year) - t
4.4	0.25
6.1	0.5
8.4	1.0
11.9	2.0
14.2	3.0

Use whatever "data fitting" methods with which you might be familiar, and try to match this data with a functional form $x(t) = At^n$ and see how close n is to the expected value of 1/2. (One method is to code the data-set in Mathematica©, take the log of the data, and use the

Fit[ ] command as discussed in Section 10.9, but you're free to use any method you can find.)

(b) We can also use the data to estimate the thermal diffusion constant, D_{th}, of the cooling magma by writing the heat equation in the form $D_{th} \sim (x(t))^2/(4\lambda^2 t)$. Assuming that $\lambda = \mathcal{O}(1)$, evaluate D_{th} using any of the data points above. Many geophysics books cite values of $D_{th} \approx 1\, mm^2/s$, where *millimeter* is the length unit used.

(c)† Similar data on the thickness of cooling lava as a function of time is shown in Fig. 5.4, but plotted in ways designed to demonstrate the $x(t) \propto t^{1/2}$ dependence more directly. The scale on the bottom is "square root of time in days," while the top is basically a logarithmic one in "years." Use whatever fitting methods you're comfortable with to confirm the $\sqrt{t}$ time dependence and try to extract an estimate for D_{th} following the methods of part (b).

P5.5 Heat equation applied to arctic permafrost: The data from Fig. 5.5 are taken from the paper by Pringle et al. (2003) and show a respectable linear relation between $\partial T(z,t)/\partial t$ and $\partial^2 T(z,t)/\partial z^2$.

(a) Use the slope of that data to estimate the thermal diffusion constant, D_{th}, in m^2/s.

(b) The material property values for the volume specific heat (C_{vol}) and thermal conductivity (κ) that the authors use to analyze their results to compare to theory are cited as

$$C_{vol} = (1.7\text{–}1.8)\,\frac{MJ}{m^3\,{}^\circ C} \qquad \text{and} \qquad \kappa \approx (2.5\text{–}3.0)\frac{W}{m\,{}^\circ C}. \tag{5.112}$$

Note that *MJ* stands for *mega-Joule*. Use these values and the relation $D_{th} = \kappa/(\rho_m C_S) = \kappa/C_{vol}$ to compare this prediction to the slope found in part (a).

P5.6 Heat equation: (a)* Show that the solution in Eqn. (5.20) does indeed solve the one-dimensional classical heat equation in Eqn. (5.18).

(b) In *An Experiment on the Dynamics of Thermal Diffusion*, Sullivan *et al.* (2008) take an (effectively) infinitely long metal rod of circular cross-section (area A) and add a heat pulse Q at the center (at $z = 0$) and measure the temperature profile, $T(z,t)$, along the rod (on either side) as a function of distance and time, hence the *dynamics* of the title. The material has thermal conductivity κ and **volume specific heat,** $C_{Vol} = \rho_m C_S$. They model the solution of the heat equation as

$$T(z,t) = \frac{C_T}{\sqrt{t}} e^{-z^2/4D_{th}t} \qquad \text{where} \qquad D_{th} = \frac{\kappa}{C_{vol}}. \tag{5.113}$$

The constant C_T is determined by the geometry and material properties to depend on $C_T \propto Q^\alpha A^\beta \kappa^\gamma C_{vol}^\delta$. Match dimensions, then solve for $\alpha, \beta, \gamma, \delta$. How would one determine both κ and C_{vol} from the data?

P5.7† Thermal diffusivity experiment analysis: Gfroerer *et al.* (2015) present measurements (see Fig. 5.6) of b^2 versus t that can be used to extract values of the thermal diffusivity constant D_{th} via the relation $b^2 = r_0^2 + 8D_{th}t$, based on the solution in Eqn. (5.21).

(a) Confirm that the slopes cited in the figure are as advertised, and use them to find values of D_{th} for copper and aluminum (which just means dividing by 8).

(b) Compare those experiment results to the theoretical predictions of $D_{th} = \kappa/(\rho_m C_S)$. You can use the values in the following table:

material	κ $(W/(m \cdot K))$	ρ_m (kg/m^3)	C_S $(J/(kg \cdot K)$
copper	398	8.96×10^3	385
aluminum	167	2.17×10^3	892

P5.8 Cooking a turkey: A familiar and practical application of thermal diffusivity relates to the time it takes to cook foods immersed in a heat bath (oven, boiling water, etc.) and especially how the cook-time (t_C) scales with the size of the item, let's say its mass m_F. Assuming the various material properties, including density (ρ_m), thermal conductivity (κ), and the specific heat (c) are involved, write $t_C \propto m_F^{\alpha} \rho_m^{\beta} \kappa^{\gamma} c^{\delta}$ and solve for all exponents. Does the dependence you get on m_F and ρ_m make sense in the light of the $x(t) \sim t^{1/2}$ examples we've discussed? To explore that, can you use the definition of **diffusivity** as $D \equiv \kappa/(\rho_m c)$ to simplify your results at all, using $m_F = \rho_m V_F$ where V_F is the food sample volume?
Note: In the spirit of the famous "spherical chicken" approximation, experiments on thermal diffusivity in spherical samples have been nicely analyzed in terms of the standard Fourier theory of heat conduction using polymers (rubber) as the sample material; see Unsworth and Duarte (1979). The same methods have been applied to spherical samples of various food items by Wang, Jin, and Wang (2022), and the same authors (in a different order) explore the applicability of those methods to the annual problem (at least in the United States) of how long it takes to cook a turkey; see Jin, Wang, and Wang (2021).

P5.9† Allometric scaling analysis of mammalian heart rate data and engine RPMs: Scaling analyses can be applied equally to biological and technological systems to explore any possible allometric (scaling with size/mass) behavior. Two examples that can both be broadly defined as **combustion** are explored below.

(a) Values of the heart rate (or HR, in beats per minute) versus 'weight' (W in kg) for a range of mammals spanning 7.5 orders of magnitude are shown below, taken from a paper by Günther and Morgado (2005). Fit the data to a power-law form $HR = aW^b$, using whatever tools you have, and see if you get a good match, and if so, what value of b you estimate.

Mammal	$W(kg)$	HR (min^{-1})
Blue whale	100,000	11.4
Elephant	3,000	27.3
Human	70	70
Shrew	0.003	862

(b) Analogous data[13] on a class of mechanical systems (motors, i.e., internal combustion engines) is shown below, where the revolutions per minute (RPM) of variously sized engines are given as a function of their mass M (in kg). Once again, fit this data to a power-law and extract the most likely exponent b.

[13] Taken from McMahon and Bonner (1983).

$M(kg)$	RPM (rev per minute)
0.135	22,000
0.67	12,000
2.45	8,000
75	2,550
229	2,000
1,775	900
2,000	1,100
12,860	400

Note: Marsden and Allen (2002) discuss the universal performance characteristics of motors (mechanical and biological) in the context of allometric scaling.

P5.10† Arrhenius factor: Data from on ammonia decomposition taken from Choudhary et al. (2001) are shown in Fig. 5.8, which plots the decomposition rate ($\log(TOF, s^{-1})$) versus $1/T K^{-1}$ which is indeed seemingly well-fit by a straight line on this log–linear plot. The data are said to span a range in temperature ($700\,K$, $800\,K$) which are the two end points. Use the relation in Eqn. (5.32) to evaluate the activation energy E_a and compare to the value of $E_a = 84\,kJ/mol$ obtained by the authors.
Note: The authors are plotting the base-10 log and **not** the natural log (ln), so you'll have to rework Eqn. (5.32) a bit. This is a good reminder about sometimes needing to translate "dialects" from one discipline (chemistry) to another (physics).

P5.11 Thermal skin depth: Variations in the surface temperature are not reflected in the below-ground temperature profile immediately, due to thermal diffusion effects. Say that the ambient surface temperature is T_0, but also has a time-dependent component $\Delta T \sin(\omega t)$, where $\omega = 2\pi/\tau$ and τ is the time scale over which the temperature changes on hourly, daily, monthly, or annual cycles.

(a) Let $\delta(\omega)$ be the typical depth over which such temperature variations extend, write $\delta(\omega) = C_\omega \kappa^\alpha \rho_m^\beta C_S^\gamma \omega^\sigma$ (where C_ω is the usual dimensionless, and therefore undetermined, prefactor), match dimensions, and solve for the exponents. Is your result consistent with the assumption that longer cycles (smaller values of ω) presumably allow ΔT variations to penetrate to deeper profiles.
Hint: Can you use the results from Eqns. (5.13)–(5.14) and save yourself lots of work?

(b) Without specifying values for the soil parameters (ρ_m, κ, C_S), what can you say about the ratio δ(year) to δ(day), namely how different the thermal skin depth is due to annual versus daily variations?

(c)* Use a trial solution of the form $T(z,t) = T_0 + \Delta T \exp(-z/\delta) \sin(\omega t - z/\delta)$ in the one-dimensional version of the heat equation in Eqn. (5.18), match terms to find an expression for $\delta(\omega)$, show that it agrees with your DA result from part (a), and that it gives $C_\omega = \sqrt{2}$.

(d) Do you see any similarities between this problem and the **viscous skin depth** discussion in **P3.31** or the **EM skin depth** example in Section 4.6?

Note: This type of problem is discussed in many texts on environmental physics, including Monteith and Unsworth (2008) (their section 15.3), Hillel (1982) (section 9.F), and Campbell and

Norman (1998) (section. 8.1). The most compelling comparison to real field (quite literally, from a field) data of which I'm aware is by West (1952), but a pedagogically accessible discussion of the *thermal properties of soil* is given by McIntosh and Sharrat (2001).

P5.12 Einstein–Smoluchowski diffusion constant: The Stokes' law motivated diffusion constant in Eqn. (5.25) is a specific example of a more general relation. If we assume that the diffusive particles are subject to an external force (F_{ext}) and acquire a terminal or drift speed (v_{dr}) we can write $D = k_B^\alpha T^\beta (v_{dr})^\gamma (F_{ext})^\delta$ and you can solve for the exponents and confirm that the Einstein–Stokes result is a special case of your answer here. The ratio of drift speed to applied force, $\mu \equiv v_{dr}/F_{ext}$, is sometimes called the **mobility**, but that term is more often reserved for the quantity $\mu_m = v_{dr}/E_{ext}$, for an external electric field, and so has different dimensions. The electron mobility plays an important role in condensed matter and solid state physics; see, for example, **P4.40.**

P5.13 Adiabatic lapse rate: The rate of change (it's a decrease) of the temperature in the atmosphere as a function of vertical distance, $dT(z)/dz$, is called the **(adiabatic) lapse rate.** It might depend on the average properties of the air (say the density ρ_m and specific heat c_P), the acceleration of gravity (g), and perhaps k_B as well. Write $-dT/dz \equiv \Gamma_d \propto \rho_m^\alpha c_P^\beta g^\gamma k_B^\delta$, match dimensions, and solve for the exponents. Do the various dependences make sense to you, even if you're not an expert in meteorology? Using back-of-the-envelope values for $\rho_{air} \sim 1\, kg/m^3$, $c_P \sim 10^3\, J/(kg \cdot K)$, $g \sim 10\, m/s^2$, and the handbook value of k_B, estimate Γ_d in K/km.
Note: We have not been careful to specify the precise nature of the specific heat used in a detailed analysis, but the notation c_P is a "nod" to the fact that it's the specific heat at constant pressure. For a derivation, see any almost any book on meteorology, such as Holton (2004).

P5.14 Johnson–Nyquist noise: Two papers, published back-to-back, by Johnson (1928) and Nyquist (1928), presented experimental evidence for, and a theoretical explanation of, what the second author called thermal agitation of electric charge in conductors. The thermal average/expectation value of the voltage, $\langle V \rangle_{th}$, in a resistance (R) due to the random motion of the charge carriers vanishes (equally likely to be moving left or right, after all), but the thermal average of V^2 does not, giving rise to power dissipation at any finite temperature.

(a) Assume that $\langle V^2 \rangle_{th} \propto k_B^\alpha T^\beta R^\gamma (\Delta f)^\delta$, where Δf is the frequency bandwidth of the measurements. Match dimensions and solve for $\alpha, \beta, \gamma, \delta$. With four unknowns and five constraints, we hopefully get a solution.

(b) What if we add the the mass (m_e) of the charge carriers (electrons) as another possible dimension. What constraints do we get on the exponent of m_e? If we add the basic electric charge e as another dimensional parameter, what constraints do we get on the exponent of e, and the other parameters?

P5.15 Surface evaporation I—Distance scale: When material evaporates from a surface, there is a region where the gas "relaxes" from the outflow velocity profile to a standard Maxwellian equilibrium distribution—this is called the **Knudsen layer**, with a characteristic length scale designated as l_K. Assume that this distance depends on thermal properties such as k_B and T, the local pressure P, and a measure of the size of the gas molecules, say their diameter d. Write $l_K \propto k_B^\alpha T^\beta d^\gamma P^\delta$, match dimensions, and evaluate the exponents to the extent possible. Compare your result to the mean free path defined in Eqn. (5.90), and see if that expression suggests the values of any underdetermined exponents.

P5.16 Surface evaporation II—Rate: To see to what extent the **Hertz–Knudsen** equation for evaporation rate from Eqn. (5.27) is constrained by dimensional analysis, write

$$\frac{dN}{dt} \propto A^{\alpha}(\Delta P)^{\beta} m^{\gamma} k_B^{\delta} T^{\sigma}, \tag{5.114}$$

where dN/dt is the rate (number of molecules per time), A the surface area, ΔP the pressure difference across the surface, m the mass of the molecules, and k_B, T have their usual meaning. Match dimensions and see to what extent any of the parameter dependences are specified.

P5.17 Maxwell–Boltzmann distribution for rotation and vibration: Let us consider the probability distributions for angular velocity (ω) or position (x) for objects undergoing rotational or vibrational motion, respectively, where the relevant energies are given by $I\omega^2/2$ and $kx^2/2$.

(a) First define

$$G(\omega)\, d\omega \equiv \text{dProb}[\omega \in (\omega, \omega + d\omega)] \tag{5.115}$$

as the (small) probability of finding a particle rotating with angular velocity ω in a $d\omega$-size "bin," which gives the dimensions of $[G(\omega)]$. Write $G(\omega) \propto I^{\alpha}\omega^{\beta}k_B^{\gamma}T^{\delta}$, match dimensions, and determine the exponents and any dimensionless ratios.

(b) Writing $G(\omega) = A_{\omega}\exp(-I\omega^2/2k_BT)$, use the fact that $G(\omega)$ must be normalized appropriately (as in Eqn. (5.45)) to evaluate the prefactor A_{ω} and compare to your DA result. Then use your normalized result to evaluate $\langle \omega^2 \rangle_{th}$ and comment.

(b) Repeat for vibrational motion, using

$$H(x)\, dx \equiv \text{dProb}[x \in (x, x + dx)] \tag{5.116}$$

and $H(x) = A_x \exp(-Kx^2/2k_BT)$ to find A_x, first using dimensional matching of the form $H(x) \propto K^{\alpha}x^{\beta}k_B^{\gamma}T^{\delta}$, and then normalization constraints, and finally to evaluate $\langle x^2 \rangle_{th}$.

Note: You've hopefully just confirmed two more cases of the equipartition theorem.

P5.18* Checking the Gouy–Chapman solution: Confirm that the results in Eqn. (5.80) for $V(z)$ and c_0 solve the Poisson–Boltzmann differential equation in Eqn. (5.78) and the boundary condition in Eqn. (5.79).

P5.19 Surface "pressure" (tension) near a charged membrane: Because of the thermal and electrostatic energies near a charged membrane (as considered in **Example 5.3**), there is an energy per unit area ($\mathcal{E}/A$) associated with the surface of such a system.

(a) We can first analyze this by writing $\mathcal{E}/A \propto \sigma_e^{\alpha} e^{\beta} \epsilon^{\gamma} k_B^{\delta} T^{\rho}$, where ϵ is the permittivity of the medium ($\epsilon = K\epsilon_0 > \epsilon$). Match dimensions to see how far you can constrain the result, say writing your answer in terms of the exponent α, and use your intuition to either provide bounds of likely values of α, or one that gives the simplest form.

(b) The surface energy density can also be explored by using results from basic E&M and the Gouy–Chapman model of **Example 5.3.** We can write down the results for the energy of a charged capacitor and the charge on such a device, the definition of capacitance in terms of geometrical properties, and the typical distance scale (d) of a capacitor (in his case the Gouy–Chapman length, $d = l_{GC}$. Use the following relations to find $\mathcal{E}/A$,

$$\mathcal{E} = \frac{Q^2}{2C}, \qquad \sigma_e = \frac{Q}{A}, \qquad \text{and} \qquad C = \frac{\epsilon A}{d} = \frac{\epsilon A}{l_{GC}}, \tag{5.117}$$

and express your final answer in terms of $\sigma_e, e, \epsilon, k_b, T$. Is your answer consistent with the general result from part (a) and if so, for what value of α?

(c) The energy ($\mathcal{E}$) in a capacitor can calculated using the *energy density* stored in the electric field ($|E|$), and the volume of the device, using the expressions

$$U_E = \frac{\epsilon}{2}|E|^2, \qquad |E| = \frac{\sigma_e}{\epsilon}, \qquad \text{and} \qquad \mathcal{E} = U_E \times V = U_E \times (A\, l_{GC}), \tag{5.118}$$

from which you can find $\mathcal{E}/A$. Compare to the results of both parts (a) and (b).

(d) In discussing this problem, Nelson (2004) describes the solution that you've found as follows:

> This makes sense: The environment is willing to give up about $k_B T$ of energy per counterion. This energy gets stored in forming the diffuse layer.

Does your answer seem consistent with this description?

P5.20 Specific heats of monatomic and diatomic gases: Check the filled-in values of the molar specific heats in the last two columns of Table 5.1, calculate values for the blank entries, and comment on the extent of agreement with simple "degrees of freedom" counting arguments.

P5.21 Thermal conductivity and viscosity in rigorous models: The "... ultra-simplified kinetic theory ..." prediction for the ratio $f = (\kappa\mu)/(\eta C_{mol}) = 1$ is consistently lower than experimentally observed values. The f values for some select gases (at 0°C) are quoted in one reference[14] as

Gas	f
Helium	2.45
Neon	2.52
Argon	2.48
Krypton	2.54
Xenon	2.58
Hydrogen (H_2)	2.02
Deuterium (D_2)	2.07
Carbon monoxide (CO)	1.91
Nitrogen (N_2)	1.96
Oxygen (O_2)	1.94
Nitric oxide (NO)	1.90

A more rigorous approach known as **Chapman–Enskog theory**

> ... provides a framework in which equations of hydrodynamics for a gas can be derived from the Boltzmann equation. The technique justifies the otherwise phenomenological constitutive relations

[14] Chapman and Cowling (1970).

appearing in hydrodynamical descriptions such as the Navier–Stokes equations. In doing so, expressions for various transport coefficients such as thermal conductivity and viscosity are obtained in terms of molecular parameters.[15]

For spherical molecules, with varying types of interatomic forces assumed, a robust prediction of the theory is that $f \approx 2.5$, which seems to be roughly correct for the monatomic (noble) gases in the table above. For more complex molecules, ones with internal degrees of freedom, a prediction by Eucken (1913) suggests that this structure can be taken into account by a dependence on the ratio of specific heats, $\gamma = C_P/C_V$ (constant pressure to constant volume) that is seen in the $P - V$ relations of ideal gases, and in the prediction for sound speeds. His prediction can be written in the form

$$f_{spherical} = 2.5 \quad \longrightarrow \quad f_{Eucken} = \frac{1}{4}(9\gamma - 5). \tag{5.119}$$

Recalling that $C_P = C_V + R$ (where R is the gas constant), evaluate γ for monatomic ($C_V = 3R/2$) and diatomic gases ($C_V = 5R/2$) and compare the results in the table above to the Eucken prediction.

P5.22 Wiedemann–Franz law from kinetic theory. Using the results of Eqns. (5.97) and (5.101), write down the ratio κ/σ_e. Then use the fact that the molecular specific heat per particle is $c = \partial\epsilon/\partial T = 3k_B/2$, the relation $\overline{v} = l/\tau$, and the connection $m\overline{v}^2/2 \leftrightarrow 3k_BT/2$ to find a simple expression for the Lorenz factor $\mathcal{L}_0 \equiv \kappa/(\sigma_e T)$ and compare to the discussion in Section 2.4.

P5.23 Quantum gases: (a) The result for the quantum number density n_Q in Eqn. (5.105) has a specific dependence on $\hbar$. One often hears that classical physics can be recovered from quantum mechanics in the limit that $\hbar \to 0$. Is that equation consistent with that statement?

(b) If we instead fix the gas density (n), we can find a critical quantum temperature by writing $T_Q \propto \hbar^\alpha n^\beta m^\gamma k_B^\delta$ and matching dimensions. Does your answer scale with $\hbar \to 0$ as you expect?

(c) Evaluate the quantum number density in Eqn. (5.105) for the electrons in a metal at room temperature, so using m_e and $T = 300$ K. Compare your value to that used in free electron models of such systems where $n_e \sim 10^{28.5}\ m^{-3}$ to see if quantum mechanics is indeed relevant.

References for Chapter 5

Agutter, P. S., and Wheatley, D. N. (2004). "Metabolic Scaling: Consensus or Controversy?," *Theoretical Biology and Medical Modelling* 1, 1–13.

Amerine, M. A., and Winkler, A. T. (1944). "Composition and Quality of Musts and Wines of California Grapes," *Hilgardia* 15, 493–673.

Ashton, G. D. (1989). "Thin Ice Growth," *Water Resources Research* 25, 564–6.

Banavar, J. R., et al. (1999). "Size and Form in Efficient Transportation Networks," *Nature* 399, 130–2.

Banavar, J. R. et al. (2002). "Supply-Demand Balance and Metabolic Scaling," *Proceedings of the National Academy of Sciences of the United States of America* 99, 10506–509.

Banavar, J. R., et al. (2010). "A General Basis for Quarter-Power Scaling in Animals," *Proceedings of the National Academy of Sciences of the United States of America* 107, 15816–20.

Benedict, F. G. (1938). *Vital Energetics: A Study in Comparative Basal Metabolism* (Washington, DC: Carnegie Institute).

[15] Wikipedia, s.v. "Chapman-Enskog," https://en.wikipedia.org/wiki/Chapman%E2%80%93Enskog_theory.

Bernard, P., et al. (2021). “Demonstration of the Influence of Specific Surface Area on Reaction Rate in Heterogeneous Catalysis,” *Journal of Chemical Education* 98, 935–40.

Brody, S., Procter, R. C., and Ashworth, U. S. (1934). “Basal Metabolism, Endogenous Nitrogen, Creatinine and Neutral Sulphur Excretions as Functions of Body Weight,” *University of Missouri Agricultural Experiment Station Research Bulletin* 220, 1–40.

Campbell, G. S., and Norman, J. M. (1998). *An Introduction to Environmental Biophysics*, 2nd ed. (New York: Springer).

Cernuschi, F., Russo, A., Lorenzoni, L., and Figari, A. (2001). “In-Plane Thermal Diffusivity Evaluation by Infrared Thermography,” *Review of Scientific Instruments* 72, 3988–95.

Chapman, D. L. (1913). “A Contribution to the Theory of Electrocapillarity,” *Philosophical Magazine* 25, 475–81.

Chapman, S., and Cowling, T. G. (1970). *The Mathematical Theory of Non-Uniform Gases: An Account of the Kinetic Theory of Viscosity, Thermal Conduction, and Diffusion in Gases* (Cambridge: Cambridge University Press).

Chen, F. (2018). *Introduction to Plasma Physics and Controlled Fusion*, 3rd ed. (Heidelberg: Springer).

Choudhary, T. V., et al. (2001). “Ammonia Decomposition on Ir(100): From Ultrahigh Vacuum to Elevated Pressures,” *Catalysis Letters* 77, 1–5.

Cohen, V. W., and Ellet, A. (1937). “Velocity Analysis by Means of the Stern–Gerlach Effect,” *Physical Review* 52, 502–508.

Debye, P. (1929). *Polar Molecules* (New York: Chemical Catalog Company).

Desch, S. J., et al. (2016). “Arctic Ice Management,” *Earth's Future* 5, 1–21.

Dote, J. L., Kivelson, D., and Schwartz, R. N. (1981). “A Molecular Quasi-Hydrodynamic Free-Space Model for Molecular Rotational Relaxion in Liquids,” *Journal of Physical Chemistry A* 85, 2169–80.

Eldridge, J. A. (1927). “Experimental Test of Maxwell's Distribution Law,” *Physical Review* 30, 931–5.

Estermann, I., Simpson, O. C., and Stern, O. (1947). “The Free Fall of Atoms and the Measurement of the Velocity Distribution in a Molecular Beam of Cesium Atoms,” *Physical Review* 71, 238–49.

Eucken, A. (1913). “Über das Wärmeleitvermögen, die spezifische Wärme und die innere Reibung der Gase (On the Thermal Conductivity, Specific Heat, and Viscosity of Gases),” *Physikalische Zeitschrift* 14, 324–32.

Fourier, J. (1878). *The Analytical Theory of Heat*. Translated by A. Freeman (London: Cambridge University Press).

Frenkel, J. (1938). “On Pre-Breakdown Phenomena in Insulators and Electronic Semi-Conductors,” *Physical Review* 54, 647–8.

Gfroerer, T., Phillips, R., and Rossi, P. (2015). “Thermal Diffusivity Imaging,” *American Journal of Physics* 83, 923–7.

Günther, B., and Morgado, E. (2005). “Allometric Scaling of Biological Rhythms in Mammals,” *Biological Research* 38, 207–212.

Gouy, M. (1910). “Sur la constitution de la charge électrique a la surface d'un électrolyte (On the Constitution of the Electric Charge on the Surface of an Electrolyte),” *Journal of Physics: Theories and Applications* 9, 457–68.

Gouy, M. (1917). “Sur la fonction électrocapillaire (On the Electrocapillary Function),” *Annales de Physique* 9, 129–84.

Grant, E. H. (1957). “Relationship between Relaxation Time and Viscosity in Water,” *Journal of Chemical Physics* 26, 1575–7.

Günther, B., and Morgado, E. (2005). “Allometric Scaling of Biological Rhythms in Mammals,” *Biological Research* 38, 207–212.

Hillel, D. (1982). *Introduction to Soil Physics* (New York: Academic Press).

Hirshfelder, J. O., Curtiss, C. F., and Bird, R. B. (1954). *Molecular Theory of Gases and Liquids* (New York: Wiley).

Holton, J. R. (2004). *An Introduction to Dynamic Meteorology*, 4th ed. (Amsterdam: Elsevier Academic Press).

Jackson, J. D. (1999). *Classical Electrodynamics*, 3rd ed. (New York: Wiley and Sons).

Jin, Y. J., Wang, L. R., and Wang, J. J. (2021). "Physics in Turkey Cooking: Revisit the Panofsky Formula," *AIP Advances* 11, 115316.

Johnson, J. G. (1928). "Thermal Agitation of Electricity in Conductors," *Physical Review* 32, 97–109.

Kittel, C., and Kroemer, H. (1980). *Thermal Physics*, 2nd ed. (New York: W. H. Freeman).

Kleiber, M. (1932). "Body Size and Metabolism," *Hilgardia* 6, 315–53.

Kleiber, M. (1961). *The Fire of Life: An Introduction to Animal Energetics* (New York: Wiley and Sons).

Kuhlen, M., Vogelsberger, M., and Angulo, R. (2012). "Numerical Simulations of the Dark Universe: State of the Art and the Next Decade," *Dark Universe* 1, 50–93.

Laidler, K. J. (1984). "The Development of the Arrhenius Equation," *Journal of Chemical Education* 61, 494–8.

Leventhal, I., Janmay, P. A., and Cēbers, A. (2008). "Electrostatic Contribution to the Surface Pressure of Charged Monolayers Containing Polyphosphoinositides," *Biophysical Journal* 95, 1199–1205.

Markovich, T., Andelman, D., and Podgornik, R. (2021). "Charged Membranes: Poisson–Boltzmann Theory, DLVO Paradigm and Beyond." In *Handbook of Lipid Membranes*, edited by C. R. Safinya and J. Radler, Chapter 6 (Boca Raton: CRC Press).

Marden, J. H., and Allen, L. R. (2002). "Molecules, Muscles, and Machines: Universal Performance Characteristics of Motors," *Proceedings of the National Academy of Sciences of the United States of America* 99, 4161–6.

Matsumoto, M., Saito, S., and Ohmine, I. (2002). "Molecular Dynamics Simulation of the Ice Nucleation and Growth Process Leading to Water Freezing," *Nature* 416, 409–413.

McIntosh, G., and Sharrat, B. S. (2001). "Thermal Properties of Soils," *Physics Teacher* 39, 458–60.

McMahon, T. A., and Bonner, J. T. (1983). *On Size and Life* (New York: Scientific American).

Monteith, J., and Unsworth, M. (2008). *Principles of Environmental Physics*, 3rd ed. (Amsterdam: Elsevier).

Nelson, P. (2004). *Biological Physics* (Philadelphia, PA: W.H. Freeman).

Nyquist, H. (1928). "Thermal Agitation of Electric Charge in Conductors," *Physical Review* 32, 110–113.

Pringle, D. J., Dickinson, W. W., Trodahl, H. J., and Pyne, A. R. (2003). "Depth and Seasonal Variations in the Thermal Properties of Antarctic Dry Valley Permafrost from Temperature Time Series Analysis," *Journal of Geophysical Research* 108(B10), 2474.

Reed, B. C. (2023). "Ten Steps to Entropy," *European Journal of Physics* 44, 045102.

Reif, F. (1965). *Fundamentals of Statistical and Thermal Physics* (New York: McGraw-Hill).

Reitz, J. R., Milford, F. J., and Christy, R. W. (1993). *Foundations of Electromagnetic Theory*, 4th ed. (Reading: Addison-Wesley).

Savage, V. M., et al. (2004). "The Predominance of Quarter-Power Scaling in Biology," *Functional Ecology* 18, 257–82.

Savage, V. M., Deeds, E. J., and Fontana, W. (2008). "Sizing up Allometric Scaling Theory," *PLoS Computational Biology* 4, e1000171.

Shingleton, A. (2010). "Allometry: The Study of Biological Scaling," *Nature Education Knowledge* 3(10), 2.

Smil, V. (2000). "Laying down the Law," *Nature* 403, 597.

Sposito, G. (2016). "Gouy–Chapman Theory." In *Encyclopedia of Geochemistry*, edited by W. M. White. Encyclopedia of Earth Sciences Series (Cham: Springer Nature).

Stefan, J. (1891). "Ueber die Theorie der eisbildung, insbesondere über die eisbildung im polarmeere (The Theory of Ice Formation, Especially Regarding Ice Formation in the Polar Sea)," *Annalen der Physik* 42, 269–86. Note: This is the same author whose name appears in the Stefan–Boltzmann law.

Sullivan, M. C., Thompson, B. G., and Williamson, A. P. (2008). "An Experiment on the Dynamics of Thermal Diffusion," *American Journal of Physics* 76, 637–42.

Turcotte, D. L., and Schubert, G. (2002). *Geodynamics*, 2nd ed. (Cambridge: Cambridge University Press).

Unsworth, J., and Duarte, F. J. (1979). "Heat Diffusion in a Solid Sphere and Fourier Theory: An Elementary Practical Example", *American Journal of Physics* 47, 981–3.

Verzi, V., Ivanov D., and Tsunesada, Y. (2017). "Measurement of Energy Spectrum of Ultra-High Energy Cosmic Rays," *Progress of Theoretical and Experimental Physics* 12A, 103.

Vogelsberger, M., Marinacci, F., Torrey, P., and Puchwein, E. (2020). "Cosmological Simulations of Galaxy Formation," *Nature Reviews Physics* 2, 42–66.

Wang, L. R., Jin, Y. J., and Wang, J. J. (2022). "A Simple and Low-Cost Experimental Method to Determine the Thermal Diffusivity of Various Types of Foods," *American Journal of Physics* 90, 568–72.

West, E. W. (1952). "A Study of the Annual Soil Temperature Wave," *Australian Journal of Scientific Research* 5, 303–14.

West, G. B., et al. (1997). "A General Model for the Origin of Allometric Scaling Laws in Biology," *Science* 276, 122–6.

West, G. B., et al. (1999). "The Fourth Dimension of Life: Fractal Geometry and Allometric Scaling of Organisms," *Science* 284, 1677–9.

West, G. B., et al. (2002). "Allometric Scaling of Metabolic Rate from Molecules and Mitochondria to Cells and Mammals," *Proceedings of the National Academy of Sciences of the United States of America* 99, 2473–8.

White, C. R., and Seymour, R. S. (2005). "Review: Allometric Scaling of Mammalian Metabolism," *Journal of Experimental Biology* 208, 1611–1619.

Whitfield, J. (2006). *In the Beat of a Heart: Life, Energy, and the Unity of Nature* (Washington, DC: Joseph Henry Press).

Winkler, A. J., Cook, J. A., Kliewer, W. M., and Lider, L. A. (1974). *General Viticulture* (Berkeley: University of California Press).

Wright, T. L., Peck, D. L., and Shaw, H. R. (1976). "Kilauea Lava Lakes: Natural Laboratories for Study of Cooling, Crystallization, and Differentiation of Basaltic Magma." In *Geophysics of the Pacific Ocean Basin and its Margins*, edited by G. H. Sutton, M. H. Manghnani, and R. Moberly, 375–90. (Washington, DC: American Geophysical Union).

Zartman, I. F. (1931). "A direct measurement of molecular velocities," *Physical Review* 37, 383–91.

Zhang, L., and Greenfield, M. L. (2007). "Relaxation time, diffusion, and viscosity analysis of model asphalt systems using molecular simulation," *Journal of Chemical Physics* 127, 194502

Chapter 6

Quantum mechanics

Quantum mechanics has historically been (and continues to be) applied to a bewilderingly broad array of physical phenomena, ranging from elementary particles and nuclear physics at the smallest experimentally accessible scales to molecular/atomic physics, chemistry, and solid state physics reaching to mesoscopic/macroscopic lengths, all the way to the impact of quantum effects in astrophysics, cosmology (so the largest length scales observed), and the early universe. In addition to its areas of application, it can also be conceptualized, described mathematically, and taught in a wide variety of very different ways. It is very likely then a huge understatement that there are just *nine formulations of quantum mechanics*,[1] as textbooks increasingly include presentations extending beyond wave or matrix mechanics approaches to include connections to modern quantum information science.

We've seen in Chapters 3–5 that there is a fundamental dimension**ful** physical constant (or constants) underpinning the physics of each of those areas of the *quadrivium*, including G (Newton's constant) for mechanics, e, ϵ_0, μ_0 for electricity and magnetism (E&M), and k_B for thermal physics. Whichever approach/application is being studied in quantum mechanics, the corresponding quantity that sets the scale for quantum effects is **Planck's constant**, either h or $\hbar = h/2\pi$, and we start our review of a dimensional analysis (DA) approach to the subject with a discussion of its introduction into the pantheon of fundamental constants.

We note, however, that there are many crucial/critical aspects of quantum theory that are not directly addressed in the language of DA, on which a DA approach is mute. These include the symmetry properties of the quantum wave functions of collections of fermions or bosons, the first implying the Pauli exclusion principle, which is responsible for many features of nuclear and atomic physics (and therefore chemistry and condensed matter physics) or quantum entanglement, and the second leading to very different phenomena, such as Bose–Einstein condensation or lasers.

6.1 Historical developments and the introduction of Planck's constant

6.1.1 Blackbody radiation: Classical thermodynamics and the need for quanta

From the use of infrared sensors to estimate your body temperature remotely, to the measurement of the Cosmic Microwave Background (CMB), which fills the universe with $3\,K$ photons, blackbody radiation is an important concept in thermodynamics that can be addressed using the methods of statistical mechanics. It was also the first system in which what we now call

[1] The title of a pedagogical article by Styer et al. (2002).

Dimensional Analysis Across the Landscape of Physics. Richard W. Robinett, Oxford University Press. © Richard W. Robinett (2024).
DOI: 10.1093/oso/9780192867551.003.0006

Planck's constant was conceptualized, and we will explore that connection, not so much in a dedicated historical retrospective, but rather using DA, to see how $h/\hbar$ made its first appearance in the quantitative understanding of thermal emission of electromagnetic (EM) radiation from hot bodies.

We begin by defining several relevant physical, and experimentally related, quantities that are often used as the touchstones of blackbody radiation.

- The **power per unit area** emitted in thermal radiation by a hot object will be denoted here as S_T, to be notationally consistent with quantities such as the **Poynting vector** in E&M (which has the same dimensions), with the subscript T standing here for thermal. This observable has dimensions

$$[S_T] = \frac{[power]}{[area]} = \left(\frac{\mathsf{ML}^2}{\mathsf{T}^3}\right)\left(\frac{1}{\mathsf{L}^2}\right) = \frac{\mathsf{M}}{\mathsf{T}^3}. \tag{6.1}$$

It was known experimentally that S_T is proportional to T^4, which is known as **Stefan's law**, with a coefficient of proportionality given by $S_T = \sigma T^4$, where σ is called the **Stefan–Boltzmann constant.** (For an accessible brief history of the T^4 law, see Crepeau (2009), or almost any modern physics textbook.) The value of σ has been experimentally determined and is known to be

$$\sigma = 5.67 \times 10^{-8} \frac{W}{m^2 \cdot K^4}. \tag{6.2}$$

- The **power per unit area** **per unit frequency interval**, $R_T(f)$, is related to S_T via

$$dS_T = R_T(f)\, df \quad \text{with the integral of } R_T(f) \text{ over all frequencies giving} \quad S_T = \int_0^\infty R_T(f)\, df \tag{6.3}$$

and its dimensions are given by

$$[R_T(f)] = \frac{[dS_T]}{[df]} = \frac{\mathsf{M/T}^3}{1/\mathsf{T}} = \frac{\mathsf{M}}{\mathsf{T}^2}. \tag{6.4}$$

- Finally, the **energy per unit volume** **per unit frequency interval** (say inside the hot oven radiating the thermal energy, or in the microwave background all around us), labeled $\rho_T(f)$, has dimensions

$$[\rho_T(f)] = \frac{[energy]}{[volume][frequency]} = \frac{(\mathsf{ML}^2/\mathsf{T}^2)}{(\mathsf{L}^3)(1/\mathsf{T})} = \frac{\mathsf{M}}{\mathsf{LT}}. \tag{6.5}$$

The last two quantities are related by

$$R_T(f) = \frac{c}{4}\rho_T(f) \tag{6.6}$$

(where c is the speed of light), which is clearly dimensionally consistent, and is much like the relationship between the Poynting vector and the energy densities (U_E and U_B) in the E&M fields.

The (dimensionless) factor of 1/4 arises from counting of polarizations and geometric effects when averaging over the angles (actually solid angle) of emission. The final connection between S_T and $\rho_T(f)$ we will use is then

$$S_T = \int_0^\infty R_T(f)\,df = \frac{c}{4}\int_0^\infty \rho_T(f)\,df. \tag{6.7}$$

We first explore how S_T depends on the fundamental constants of E&M (ϵ_0, μ_0) and thermal physics (k_B, T), and since we already know that this analysis was done at the birth of quantum mechanics (about which more below) we also assume Planck's constant, h, is in play—we choose h instead of $\hbar = h/2\pi$ for historical purposes. We then write

$$S_T \propto h^\alpha \epsilon_0^\beta \mu_0^\gamma k_B^\delta T^\sigma \tag{6.8}$$

and matching all five M, L, T, Q, Θ dimensions, we find that $\alpha = -3, \beta = 1, \gamma = 1, \delta = \sigma = 4$, or

$$S_T = C_S h^{-3}(\epsilon_0\mu_0)^1(k_BT)^4 = \left(C_S\frac{k_B^4}{h^3c^2}\right)T^4 = \sigma T^4, \tag{6.9}$$

since $c = 1/\sqrt{\epsilon_0\mu_0}$. A detailed calculation (which we'll discuss below) gives the dimensionless constant $C_S = 2\pi^5/15 \approx 41$. Thus, the **Stefan–Boltzmann constant** is given by

$$\sigma = \frac{2\pi^5}{15}\frac{k_B^2}{c^2h^3}, \tag{6.10}$$

which agrees well (check for yourself!) with the experimentally observed value from Eqn. (6.2).

Now that we've confirmed that ϵ_0 and μ_0 naturally appear only as the obvious combination, giving powers of c, we can turn to the frequency-dependent energy density and write

$$\rho_T(f) \propto h^\alpha c^\beta k_B^\gamma T^\delta f^\sigma. \tag{6.11}$$

Matching dimensions (we now only have M, L, T, Θ constraints), we find that $\beta = -3$, $\gamma = \delta = 1 - \alpha$, and $\sigma = 2 + \alpha$, or

$$\rho_T(f) \propto h^\alpha c^{-3}(k_BT)^{1-\alpha} f^{2+\alpha} = \frac{f^2}{c^3}(k_BT)\left[\frac{hf}{k_BT}\right]^\alpha \quad\longrightarrow\quad \rho_T(f) \propto \frac{f^2}{c^3}(k_BT)F\left[\Pi = \frac{hf}{k_bT}\right]. \tag{6.12}$$

Given the dimensionless $\Pi = hf/k_BT$ ratio we've obtained, this could equally well be written as

$$\rho_T(f) \propto \frac{f^2}{c^3}(hf)\,G\left[\Pi = \frac{hf}{k_bT}\right], \tag{6.13}$$

and both forms are consistent with the actual result, derived by Planck (1900), namely

$$\rho_T(f) = \frac{8\pi f^2}{c^3}\frac{hf}{e^{hf/k_BT} - 1}. \tag{6.14}$$

We can make contact with Stefan's law by using the relation in Eqn. (6.7) to write

$$
\begin{aligned}
S_T &= \frac{c}{4}\int_0^\infty \rho_T(f)\,df \\
&= \left(\frac{c}{4}\right)\left(\frac{8\pi h}{c^3}\right)\int_0^\infty \frac{f^3}{e^{hf/k_BT}-1}\,df \\
&= \left(\frac{2\pi h}{c^2}\right)\left(\frac{k_BT}{h}\right)^4\left(\int_0^\infty \frac{x^3}{e^x-1}\,dx\right) \\
&= 2\pi\left(\frac{k_B^4T^4}{h^3c^2}\right)\left(\frac{\pi^4}{15}\right) = \left(\frac{2\pi^5}{15}\frac{k_B^4}{h^2c^2}\right)T^4 \\
S_T &= \sigma T^4,
\end{aligned}
\tag{6.15}
$$

which reproduces the value for the Stefan–Boltzmann constant cited above. (We note that the integral can be done using many automated mathematics programs, such as Wolfram Alpha or Mathematica©.) The energy density for such a photon gas is then

$$
U_T = \frac{4}{c}S_T = \frac{4\sigma}{c}T^4 = aT^4, \tag{6.16}
$$

where $a \equiv 4\sigma/c$ is called the **radiation constant** and is often used in astrophysical calculations (such as in Section 9.2).

The frequency-dependent intensity spectrum from the CMB has been measured in many experiments, both ground-based and satellite, and we present in Fig. 6.1 data compiled from several of them (including from the COBE, FIRA, and ARCADE collaborations) and note the excellent agreement, over five orders of magnitude of intensity.

Noting the change of variable used in the integral in Eqn. (6.15), namely $x = hf/k_BT$, we can also write

$$
\rho_T(x) = \frac{8\pi(k_BT)^3}{c^3h^2}\left(\frac{x^3}{e^x-1}\right) \equiv \frac{8\pi(k_BT)^3}{c^3h^2}H(x). \tag{6.17}
$$

The function $H(x)$ of the dimension<u>**less**</u> combination $x \equiv hf/k_BT$ is another example of the Buckingham Π theorem (and was extensively discussed by Buckingham (1912) himself in exactly this context) and describes the frequency dependence of any blackbody emission. The maximum of $H(x)$ is easily found (again, please feel free to use any symbolic math program) to occur at $x_{max} \approx 2.82$, so that blackbody spectra for two different temperatures will have peaks at frequencies satisfying

$$
\frac{f_{max,1}}{T_1} = \frac{f_{max,2}}{T_2} \qquad \text{since} \qquad \frac{hf_{max}}{k_BT} = 2.82. \tag{6.18}
$$

This relation (and a more familiar one using wavelength, see **P6.2**) is called the **Wien's displacement law** and was one of the other important inputs into Planck's derivation. You should be able to use the data plotted in Fig. 6.1 to find the value of f_{max} at the peak of the spectrum, and use Eqn. (6.18) to confirm that the CMB temperature is consistent with $T_{CMB} \approx 2.7$ K.

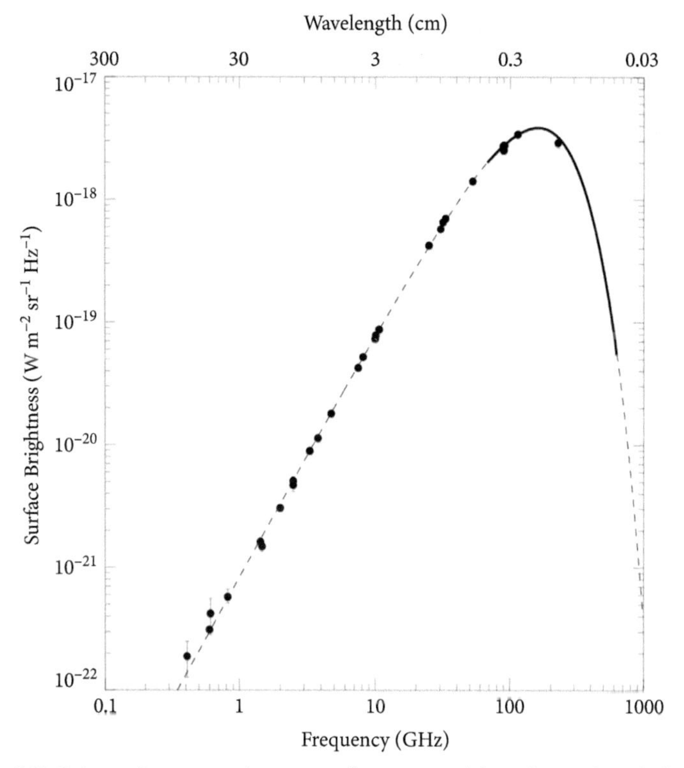

Fig. 6.1 Plot of CMB intensity versus frequency (bottom scale) and wavelength (top scale). The quantity plotted is **surface brightness** (Eqn. (6.218)), which is proportional to $R_T(f)$, but more relevant for a diffuse source such as the CMB, compared to values relevant for blackbody radiation emitted by a small source. Figure courtesy of A. J. Kogut (NASA).

One can also express all of these ideas citing the dependence on wavelength instead of frequency, with the result

$$\rho_T(\lambda) = \frac{8\pi hc}{\lambda^5}\frac{1}{e^{hc/\lambda k_B T} - 1} \tag{6.19}$$

being the **energy per unit volume <u>per unit wavelength</u> interval** and this can also be explored using DA methods, or simply rewriting Eqns. (6.14), as discussed in **P6.2.**

In order to give some flavor of the original derivation of Planck's constant (which we have assumed *ab initio* as being known), we note that the "default" DA result in the prefactor in Eqn. (6.12), namely that $\rho_T(f) \propto (f^2/c^3)(k_B T)$, is actually a purely classical prediction based on the number of EM standing wave modes in a cavity, each having a thermal energy $k_B T$. Very roughly, for standing waves of wavelength λ confined in a cubical volume $V = L^3$, where half-wavelengths would fit into each side of length L, we might have the number of allowed modes scaling (we don't pretend to get the numerical coefficient correct) as

$$N(f) = \frac{V}{(\lambda/2)^3} = \frac{8Vf^3}{c^3}, \tag{6.20}$$

so that the number of such modes in the frequency range $(f, f + df)$ would be

$$dN(f) = N(f + df) - N(f) = \frac{8V}{c^3}\left[(f + df)^3 - f^3\right] \sim \frac{24Vf^2}{c^3}\,df. \tag{6.21}$$

An actual rigorous enumeration of the number of modes (as presented in many modern physics textbooks, see e.g., Eisberg and Resnick 1974), gives the number of modes as

$$dN(f) = \frac{8\pi V f^2}{c^3}\,df, \tag{6.22}$$

with the correct numerical factor of 8π instead of 24, but our hand-waving approach does give the correct dimensions and a numerical factor, which is surprisingly close. This implies that

$$\rho_T(f) = (\text{number of modes per unit volume per df}) \cdot (\text{thermal energy})$$

$$\rho_T(f) = \left(\frac{dN(f)}{df}\frac{1}{V}\right)(k_B T) \propto \frac{8\pi f^2}{c^3} k_B T, \tag{6.23}$$

and this is indeed the low-frequency, or $hf << k_B T$, limit of Eqn. (6.14).

However, it's immediately clear that the integral over all frequencies required in Eqn. (6.7) would diverge if this was the complete f-dependence. Planck hypothesized that the energies in the standing waves of the EM field inside the blackbody did not have continuous values, but rather were quantized. This is one of the most famous **origin stories** about the discovery of quantum phenomena, but there will be no attempt here to reproduce the original Planck (1900) derivation in detail, given that there are many well-researched and densely referenced (and therefore more trustworthy) history of science presentations on the topic; see the especially accessible one by Nauenberg (2016) or the recent book-length treatment by Lemons, Shanahan, and Buchholtz (2022). We focus instead on a DA commentary, since pedagogical clarity can often be inconsistent with actual historical analysis (see e.g., Perrson 2018). We do want to note, however, that Planck made use of the notion of entropy (which we explore further in Section 6.1.2) and so can profitably begin a discussion of that topic, not covered in Chapter 5, citing some of his results.

Planck gave up the assumption of continuous field variables and posited instead that there were discrete quanta of energy (ϵ) in the EM field. With this assumption, he was able to sum the modes and derive an expression for the entropy as

$$S = k_B\left[\left(1 + \frac{\mathcal{U}}{\epsilon}\right)\ln\left(1 + \frac{\mathcal{U}}{\epsilon}\right) + \frac{\mathcal{U}}{\epsilon}\ln\left(\frac{\mathcal{U}}{\epsilon}\right)\right], \tag{6.24}$$

where $\mathcal{U}$ is internal energy. Then using the second law of thermodynamics, he found

$$\frac{1}{T} = \frac{dS}{d\mathcal{U}} = k_B\left[\frac{1}{\epsilon}\ln\left(1 + \frac{\mathcal{U}}{\epsilon}\right) + \frac{1}{\epsilon} - \frac{1}{\epsilon}\ln\left(\frac{\mathcal{U}}{\epsilon}\right) - \frac{1}{\epsilon}\right] = \frac{k_B}{\epsilon}\ln\left(\frac{1 + \mathcal{U}/\epsilon}{\mathcal{U}/\epsilon}\right) \tag{6.25}$$

or

$$\frac{\epsilon}{k_B T} = \ln\left(\frac{1 + \mathcal{U}/\epsilon}{\mathcal{U}/\epsilon}\right) \quad \text{giving} \quad e^{\epsilon/k_B T} = \frac{1 + \mathcal{U}/\epsilon}{\mathcal{U}/\epsilon}, \tag{6.26}$$

which can be solved for

$$\mathcal{U} = \frac{\epsilon}{e^{\epsilon/k_B T} - 1}. \tag{6.27}$$

Planck (1900) then argued that the the entropy should have a simple dependence on $\mathcal{U}$ and frequency, namely $S = F(\mathcal{U}/f)$, so that comparing this to Eqn. (6.24), he said

> If we apply Wien's displacement law . . . for the entropy S, we then find that the energy element $\mathcal{E}$ must be proportional to the frequency . . .

and wrote this in the form $\mathcal{E} = hf$ (in our notation), which is the first appearance of his constant in the literature. Then, with this identification, using the number of normal modes per unit volume per unit frequency from Eqns. (6.22) and (6.27), gives

$$\rho_T(f) = \frac{8\pi f^2}{c^3} \frac{hf}{e^{hf/k_B T} - 1}, \tag{6.28}$$

which is the famous expression. Using the data available at the time, Planck found the value $h = 6.55 \times 10^{-27}\ erg \cdot s$ (note the units!) compared to the currently recognized value of $h = 6.62607015 \times 10^{-34} J \cdot s$ taken from the most recent CODATA collection (Mohr et al. 2016) of values of fundamental constants. We note that $h/\hbar$ is no longer a value measured directly, but is now a defined quantity in the new SI system of units (about which more in Section 10.10.)

6.1.2 The Sackur–Tetrode equation: Entropy meets quantum mechanics

The definition of entropy in Eqn. (1.78) in terms of the logarithm of the number of accessible states is deceptively simple, but in principle can be a challenge to evaluate for specific cases, even in advanced statistical mechanics courses. Perhaps the most trivial example is the system of a single spin-1/2 particle, which has only two possible states (spin-up and spin-down), so that

$$S_{1/2} = k_B \ln(\Omega_{1/2}) = k_B \ln(2). \tag{6.29}$$

This quantity has actually been measured by Hartman et al. (2018), who

> . . . directly measure the entropy of just a few electrons . . . using the well-understood spin statistics of the first, second, and third electron ground states in a GaAs quantum dot. The precision of this technique, quantifying the entropy of a single spin-1/2 to within 5% of the expected value of $k_B \ln(2)$, shows its potential for probing more exotic systems.

A more standardly discussed example, which nicely covers the (historical and intellectual) overlap between classical thermodynamics, statistical mechanics, and quantum mechanics,[2] and also makes use of Planck's constant in a fundamental way, is the calculation of the entropy, S_{IG}, of an ideal (so ignore interactions) monatomic gas (hence the IG). Such a system will have only translation degrees of freedom (three of them) contributing to the entropy, but we consider later (text and problems) the analogous results for the entropy from rotational and vibrational degrees of freedom of diatomic molecules, using similar methods, but with very different dimensional dependences.

---

[2] See the aptly named popular science article titled *The Sackur–Tetrode Equation: How Entropy Met Quantum Mechanics* by Williams (2009).

The textbook result[3] for the entropy of an ideal gas consisting of N particles, each of mass m, confined to a volume V, maintained at a temperature T, is often seen written as

$$S_{Ideal\ Gas} \equiv S_{IG} = Nk_B \left\{ \ln\left(\frac{V}{N}\right) + \frac{3}{2}\ln\left(\frac{2\pi m k_B T}{h^2}\right) + \frac{5}{2} \right\}, \tag{6.30}$$

where h is Planck's constant (again using h and not $\hbar$, for historical consistency/accuracy, just as for blackbody radiation and the Einstein hypothesis of $E_\gamma = hf$ discussed below) and is called the **Sackur–Tetrode equation.** For a retrospective review of its historical development and verification, see Williams (2009) and Grimus (2013a), and especially Grimus (2013b) and Paños and Pérez (2015) for details of how this relation can be tested experimentally.

At first blush, Eqn. (6.30) looks incorrect as the arguments of the two logarithms clearly have dimensions, but we can easily recombine terms to give

$$\frac{S_{IG}}{Nk_B} = \ln\left[\left(\frac{V}{N}\right)\left(\frac{2\pi m k_B T}{h^2}\right)^{3/2}\right] + \frac{5}{2}. \tag{6.31}$$

The dimensions of the argument of the log are then (since we can use $[k_B T] = [E] = \mathsf{ML}^2/\mathsf{T}^2$)

$$\left[\left(\frac{V}{N}\right)\left(\frac{2\pi m k_B T}{h^2}\right)^{3/2}\right] = \mathsf{L}^3\left(\frac{\mathsf{M}(\mathsf{ML}^2/\mathsf{T}^2)}{(\mathsf{ML}^2/\mathsf{T})^2}\right)^{3/2} = \mathsf{L}^3(\mathsf{L}^{-2})^{3/2} = 1, \tag{6.32}$$

or dimension**<u>less</u>** as expected. The result that S_{IG} is directly proportional to the number of particles, N, is not unexpected (even though DA cannot provide any guidance about how it might appear), but the factors of 2π and $5/2$ are the result of a detailed calculation.

We can at least confirm more proactively that the result in Eqn. (6.31) is consistent with DA by writing the constraint $S_{IG} \propto k_B^\alpha T^\beta m^\gamma V^\delta h^\epsilon$ and matching dimensions to find $S_{IG} \propto k_B^{1+\beta} T^\beta m^\beta V^{2\beta/3} h^{2\beta}$ or

$$S_{IG} \propto k_B\left(\frac{mk_B T}{h^2}V^{2/3}\right)^\beta \quad \longrightarrow \quad k_B F\left[\frac{mk_B T}{h^2}V^{2/3}\right] \quad \text{or} \quad k_B G\left[\left(\frac{mk_B T}{h^2}\right)^{3/2} V\right], \tag{6.33}$$

with some arbitrary function $F[\Pi]$ or $G[\Pi]$, but of course we gain no information on the N dependence, either "outside" or "inside" the logarithm.

To see if we can do any better using a more first principles calculation, but keeping to the spirit of DA, and not claiming a rigorous derivation, let us first note that for a classical ID ideal gas, if we could keep track of the position of every particle, we'd apply Newton's $F = ma$ law, with initial conditions, to evaluate $x(t)$ and $\dot{x}(t) = v(t)$ for each atom. We can, of course, equally well identify the momentum of the particle as the relevant quantity and use instead $p(t) = mv(t)$ The notion of plotting the "trajectory" of a classical object in the two-dimensional $x - p$ or **phase space** is one often used in more advanced treatments of classical mechanics.

The initial phase space location of a given particle in one dimension would then (classically) be an infinitesimal point in $x - p$ space and so, in principle, could be able to be placed in an arbitrarily large number of accessible states. Quantum mechanics, however, puts restrictions on our ability to

[3] See, e.g., Kittel and Kroemer (1980), Pathria (1996), Thorne and Blandford (2017), and many other advanced undergraduate or graduate texts on statistical mechanics.

simultaneously specify x and p, with the uncertainty principle limiting our ability to do so, since the measurement ambiguities or uncertainties in each must satisfy $\Delta x \cdot \Delta p \geq \hbar/2$, and we attempt to visualize this in Fig. 6.2.

Using that "cartoon" as guidance, we assume that there is a minimal "cell" or element of phase space of size h (and not $\hbar$, again for historical reasons) in the enumeration of possible states. Note that in this approach the dimensions of h (and therefore $\hbar$) are given by an "area" in phase space, so that $[h] = [\hbar] = [x \cdot p] = \mathsf{L}(\mathsf{ML}/\mathsf{T}) = \mathsf{ML}^2/\mathsf{T}$ as before, but realized in a very different way than by assuming quantized energies as in $\mathcal{E}/f = h$, which also gives $[h] = (\mathsf{ML}^2/\mathsf{T}^2)(1/(1/\mathsf{T})) = \mathsf{ML}^2/\mathsf{T}$.

For a single particle in one dimension, the spatial extent might be L_x, while the momentum can be related to the temperature by using the **equipartition theorem** and the identification

$$\frac{1}{2}mv^2 = \frac{p_x^2}{2m} = \frac{1}{2}k_B T \qquad \text{implying} \qquad p_x = \sqrt{mk_B T}. \tag{6.34}$$

We can then approximate the number of accessible states, Ω_x, by dividing the available phase space by the elemental phase space area (sometimes called the **quantum of action**, for reasons we explore in **P6.15**) to write

$$\underbrace{\Omega_x}_{\text{1D}} = \frac{L_x p_x}{h} = \frac{L_x (mk_B T)^{1/2}}{h}, \tag{6.35}$$

which is dimension<u>**less**</u>, as it must. Considering that the particle can actually move in three dimensions, the total number of states accessible is then the product

$$\Omega = \underbrace{\Omega_x \Omega_y \Omega_z}_{\text{3D}} = \frac{(L_x L_y L_z)(mk_B T)^{3/2}}{h^3} = \frac{V(mk_B T)^{3/2}}{h^3}. \tag{6.36}$$

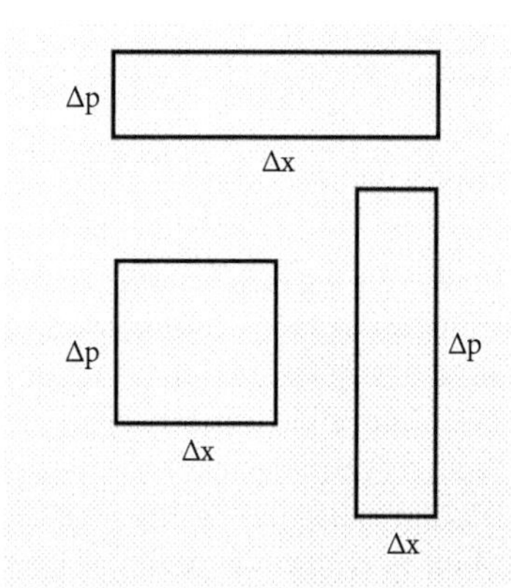

Fig. 6.2 Visualization of $x - p$ phase space showing minimum size (quantum of action) as determined by the Heisenberg uncertainty principle $\Delta x \cdot \Delta p \geq \hbar/2$.

Then, since there are N particles (let us label them $1, 2, \ldots$), all assumed, for the moment, to be distinguishable (or DIS for future reference), the total number of possible states of the system is

$$\Omega_{DIS}^{(N)} = \underbrace{\Omega_1 \Omega_2 \ldots \Omega_N}_{\text{N particles in 3D}} = \left[V\left(\frac{mk_BT}{h^2}\right)^{3/2}\right]^N. \tag{6.37}$$

This gives for the entropy

$$\begin{aligned} S_{IG} = k_b \ln\left(\Omega_{DIS}^{(N)}\right) &= k_B \ln\left[V\left(\frac{mk_BT}{h^2}\right)^{3/2}\right]^N \\ &= Nk_B\left[\ln(V) + \frac{3}{2}\ln\left(\frac{mk_BT}{h^2}\right)\right], \end{aligned} \tag{6.38}$$

which reproduces the dimensionful terms in Eqn. (6.30), and also explains the "outside the log" factor of N.

This treatment misses the important "inside the log" factor of N, the absence of which gave rise to the so-called **Gibbs paradox** in the treatment of the mixing of two gases. Gibbs attempted to address this problem by assuming that the molecules of gas were, in fact, not distinguishable (which we now know to be true) and that we were therefore overcounting by a large factor.

By way of background, for a set of $N = 3$ objects, if they are distinguishable there are $3! = 6$ distinct orderings, namely $(1, 2, 3), (1, 3, 2), (2, 1, 3), (2, 3, 1), (3, 1, 2), (3, 2, 1)$, all of which must be included in the enumeration of the accessible states. But if one can't tell them apart (say there are no numbers on billiard balls, for example, just three identical red ones), there is only one distinct state, since all of the orderings yield the same triplet. Thus, for N **indistinguishable** (or *IND*) particles, we can only count those states that are truly different. We correct for this overcounting by dividing by $N!$ to give

$$\Omega_{IND}^{(N)} = \frac{\Omega_{DIS}^{(N)}}{N!}, \tag{6.39}$$

and using the first approximation for the **factorial function** $N!$, namely that $\ln(N!) \approx N\ln(N)$ (named after **Stirling**; see Section 10.7), we find that

$$\begin{aligned} S_{IG} &= Nk_B\left[\ln(V) + \frac{3}{2}\ln\left(\frac{mk_BT}{h^2}\right) - \ln(N)\right] \\ &= Nk_B\left[\ln\left(\frac{V}{N}\right) + \frac{3}{2}\ln\left(\frac{mk_BT}{h^2}\right)\right], \end{aligned} \tag{6.40}$$

which solves the Gibbs dilemma by accounting for the *inside the log* factor of N. So, besides making fundamental use of Planck's constant as being related to the quantum of phase space, this derivation also introduces one of the first applications of the notion of indistinguishable particles, thereby connecting to one of the most important aspects of particle physics.

As noted in our discussion in Section 5.2, molecular systems have more accessible states than just translational ones, with polyatomic molecules having rotational and vibrational degrees of freedom, both of which can contribute to entropy. Recalling Fig. 5.9, we first explore the entropy contribution from the rotational motion of a diatomic molecule which is given by (see e.g., Glasstone 1947, Fowler and Guggenheim 1956, or Hill 1960)

$$\frac{S_{rot}}{Nk_B} = \ln\left[\frac{8\pi^2 Ik_BT}{\sigma h^2}\right] + 1, \tag{6.41}$$

where I is the **moment of inertia** of the molecule. The dimensionless quantity σ is another result of the proper counting due to indistinguishability, with $\sigma = 1$ for **heteronuclear** molecules (ones where the two atoms are of different types, such as $NaCL$ or OH) and $\sigma = 2$ for **homonuclear** molecules (with identical atoms, such as O_2 or N_2). We can also apply some of the same heuristic phase space arguments used earlier to see how close we can come to reproducing that result, at least in terms of dimensions.

For a two-atom molecule, rotations about either axis perpendicular to the line of symmetry are possible, so there are two degrees of angular freedom, each with a contribution to the kinetic energy given by

$$\frac{1}{2}I\omega^2 = \frac{\mathcal{L}^2}{2I} = \frac{1}{2}k_BT \qquad \text{implying} \qquad \mathcal{L} = (Ik_BT)^{1/2} \tag{6.42}$$

where $\mathcal{L}$ is the angular momentum (using the notation $\mathcal{L}$ to avoid confusion with L_x, just used for a spatial dimension) and we've used again the equipartition theorem. In this case, the relevant two-dimensional phase space is characterized not by $x - p$ (position and momentum), but rather by $\theta - \mathcal{L}$ (angle and angular momentum), and a related uncertainty principle of the form $\Delta\theta \cdot \Delta\mathcal{L} \geq \hbar/2$ also gives a similar quantum of action for this pair of variables. Noting that θ is dimensionless, we might use 2π as its angular extent, and including that factor gives the dimensional dependence of the number of accessible states for one rotational degree of freedom as

$$\Omega_x = \frac{(2\pi)\mathcal{L}_x}{h} = \frac{2\pi\sqrt{Ik_BT}}{h}, \tag{6.43}$$

with a similar expression for Ω_y. The number of accessible states for N such diatomic molecules would be

$$\Omega^{(N)} = \left(\Omega_x\Omega_y\right)^N \qquad \text{giving for the entropy} \qquad S_{rot} = Nk_B \ln\left(\frac{4\pi^2 Ik_BT}{h^2}\right), \tag{6.44}$$

which does capture the dimensionality of Eqn. (6.41).

The contribution to the molecular entropy of the **vibrational degree of freedom** of a diatomic molecule, at least for high temperatures (see, e.g., Fultz 2010), has a similar form to those above, namely

$$\frac{S_{vib}}{Nk_B} = \ln\left[\frac{k_BT}{h\nu}\right] + 1. \tag{6.45}$$

Here we imagine a two-body system with reduced mass $\mu = m_1m_2/(m + 1 + m_2)$, with an effective spring constant K, oscillating with frequency ν (and not f to avoid confusion with photon frequency in blackbody radiation) and a position space amplitude A satisfying

$$\frac{1}{2}KA^2 - \frac{1}{2}k_BT \qquad \text{giving} \qquad A = \sqrt{\frac{k_BT}{K}}. \tag{6.46}$$

For such a vibrating system, we can use the standard result for the harmonic oscillator that $\omega = 2\pi\nu = \sqrt{K/\mu}$ as the relevant (angular) frequency of oscillation. Then along with the expression for momentum from Eqn. (6.34), namely $p = (\mu k_B T)^{1/2}$, we have the phase space enumeration of states to be

$$\Omega_{vib} = \frac{Ap}{h} = \sqrt{\frac{k_B T}{K}}\frac{\sqrt{\mu k_B T}}{h} = \frac{k_B T}{h\sqrt{K/\mu}} = \frac{k_B T}{h(2\pi\nu)}. \tag{6.47}$$

Then with $\Omega_{vib}^{(N)} = [\Omega_{vib}]^N$, we do reproduce the dimension**ful** parts of Eqn. (6.45). This discussion can actually be connected to the Planck blackbody expression for entropy in Eqn. (6.24), where the frequency in question was that of the EM modes (so f reappears), while here is that of the vibrational states, but can be written in identical form; we discuss this in **P6.3**.

6.1.3 Entropy and quantum mechanics still connected

While we have focused here on discussions of entropy in atomic, molecular, and chemical physics applications, most of them from the historical literature, or at the textbook level, the notion of entropy continues to be a very important one in modern research. Examples include:

- The **entropy of a black hole**, first discussed by Bekenstein (1972) and Hawking (1975), connects thermodynamics to quantum mechanics, relativity, gravitational physics, and cosmology in fundamental ways. The entropy associated with a black hole is given by (check dimensions in **P6.5**)

$$S_{BH} = k_B\left(\frac{Ac^3}{4G\hbar}\right), \tag{6.48}$$

where A is the area of the so-called **event horizon** of the black hole and G is Newton's constant. A natural (and seemingly still open) question is how to relate this expression to the standard Boltzmann formula, $S = k_B\ln(\Omega_{BH})$, and what the nature[4] is of the accessible microstates, Ω_{BH}, for such an object. Recent theoretical analyses related to such phenomena include *Testing the Black-Hole Area Law with GW150914*, where Isi et al. (2021) probe the **second law of black hole thermodynamics** (that the total horizon area of classical black holes cannot decrease in time) and find agreement with experimental data from gravitational wave observations, pre- and post-merger.

- Connections of classical thermodynamics and statistical mechanics notions of entropy to **information theory** abound in the modern research literature. The familiar expression for entropy in terms of the number of accessible states, via $S = k_B\ln(\Omega)$, can be rewritten in terms of the individual probabilities, P_r, of finding a system in a particular micro-state, namely

$$P_r \equiv \frac{e^{-E_r/k_B T}}{\sum_r e^{-E_r/k_B T}} \equiv \frac{e^{-E_r/k_B T}}{Z}, \tag{6.49}$$

which also defines the **partition function**, Z. In terms of these probabilities, the entropy can be written as

$$S = -k_B\sum_r P_r\ln(P_r) \qquad \text{such that} \qquad \sum_r P_r = 1. \tag{6.50}$$

This formulation is important as it's identical (save for the physical factor of k_B and possibly the choice of base of the logarithm) to the **Shannon entropy** or (S_{Sh}) formula (Shannon 1948, Shannon and Weaver 1949) for the information content of messages as discussed in

[4] For a discussion, see, e.g., Carlip (2014).

communication theory. For example, with N possible messages, each with probability p_i of being sent, the information content (or entropy) is

$$S_{Sh} = -\sum_i p_i \log(p_i) \qquad \text{with} \qquad \sum_i p_1 = 1, \tag{6.51}$$

where, in information theory at least, it may be natural to use $\log_2$ instead of ln.

- Classical thermodynamic definitions of entropy involving probability were extended by von Neumann to quantum theory (with several new symbols and mathematical constructs introduced below) as

$$S_{QM} = -Tr[\rho \ln(\rho)], \tag{6.52}$$

where ρ is the quantum mechanical **density matrix** (built from the quantum wave function) and Tr stands for the **trace** of the matrices, and there is at least some "visual overlap" with the symbols in Eqn. (6.51). If the density matrix is written in its simplest form, involving the energy eigenstates ψ_j weighted by dimensionless constants c_i, we have

$$\rho = \sum_j c_j |\psi_j\rangle\langle\psi_j| \qquad \text{which gives a quantum entropy} \qquad S_{QM} = \sum_j c_j \ln(c_j). \tag{6.53}$$

For our purposes, there is no need to continue with the detailed definitions, or mathematical manipulations, but suffice it to say that this definition then allows for the discussion of the **entropy of quantum entanglement;** see Horodecki et al. (2009) for a review. **Note:** It's often the case that neither S_{Sh} nor S_{QM} include the factor of k_B.

- To conclude this section, we note that there is another powerful connection of the notion of entropy to information theory and computation. In *Irreversibility and Heat Generation in the Computing Process*, Landauer (1961) found that there is a minimum amount of energy required to erase one bit of information (or for any other non-reversible manipulation of information), namely

$$E_{min} \geq k_B T \ln(2), \tag{6.54}$$

and subsequent experiments (see e.g., Bérut et al. 2012) have verified what has come to be known as **Landauer's principle.** For a nice review of this basic tenet of information processing, see Bennett (2003), and for a very accessible, pedagogical level discussion of entropy, information, and computation, see Machta (1999).

6.1.4 Wave mechanics: Whither h versus $\hbar$

Planck's educated (but directly informed by experiments) suggestion that the energy stored in the EM field modes of a blackbody was quantized and proportional to frequency (f) was formalized by Einstein (1905), who posited that individual photon energies were related by

$$E_\gamma = hf. \tag{6.55}$$

That result helped explain the **photoelectric effect** involving the emission of electrons from surfaces subject to EM radiation, namely that the kinetic energy of the emitted electrons obeyed

$$\frac{1}{2}mv^2 = hf - W = \frac{hc}{\lambda} - W = eV_0, \tag{6.56}$$

where W is the **work function** of the material, and V_0 (eV_0) is the electric potential (potential energy) needed to bring the photoelectrons to rest. Millikan (1916) determined h directly via the photoelectric effect, by measuring V_0 versus f, and you're asked to use his data to reproduce

Table 6.1: Collection of double-slit diffraction experiments probing the wave nature of matter, ranging from point particles (such as the electron) to large bio-molecules.

Particle	**Mass (amu)**	**Size**	λ_{dB}	**Reference**
	(atomic units)	(Å)	(Å)	
electron (e^-)	5×10^{-4}	$< 10^{-10}$	$1 - 2$	Davisson and Germer (1927)*
electron (e^-)	5×10^{-4}	$< 10^{-10}$	0.1	Thomson and Reid (1927)*
electron (e^-)	5×10^{-4}	$< 10^{-10}$	0.05	Jönnson (1961, 1974)
neutron (n)	1	$\sim 10^{-5}$	$15 - 30$	Zeilinger et al. (1988)
helium (He)	4	~ 1	$0.5 - 3$	Carnal et al. (1991)
sodium (Na)	23	~ 4	0.16	Keith et al. (1991)
helium (He_2)	$2 \cdot 4 = 8$	~ 50	$1 - 2$	Schollkopf et al. (1994)
sodium (Na_2)	$2 \cdot 23 = 46$	~ 8	0.1	Chapman et al. (1995)
buckyballs (C_{60})	$60 \cdot 12 = 720$	~ 7	0.025	Arndt et al. (1999)
$C_{48}H_{26}F_{24}N_8O_8$	1298		0.05	Juffman et al. (2012)
$C_{284}H_{190}F_{320}N_4S_{12}$	10123		0.005	Ellenberger et al. (2013)
"functionalized oligoporphryins"	26777†		0.005	Fein et al. (2019)

* The seminal Davisson/ and Germer and Thomson/ and Reid experiments relied on diffraction from crystals andthin films (respectively) and not on two-slit geometries as did those of Jönnson.
† The mass of this complex biomolecule was cited as *beyond* **25 kDa** where **Da** (for the Dalton) is another mass unit often used by chemists.

that result in **P6.6.** Einstein and Millikan both won Nobel prizes citing their contributions to the understanding of the photoelectric effect (and not, in Einstein's case, for his work on either special or general relativity!).

Both the blackbody radiation and photoelectric effect "origin stories"appear in every modern physics text and provide early examples of the particle nature of EM waves. The corresponding statement that particles exhibit wave properties is due to de Broglie (1923), with the wavelength associated with the momentum of a "material particle" given by

$$\lambda_{dB} = \frac{h}{p} = \frac{2\pi\hbar}{p}, \tag{6.57}$$

where we include definitions using both h and $\hbar$.

The wave nature of matter has been repeatedly demonstrated with "particles" of increasing mass, size, and structure, ranging from (seemingly point-like) electrons (Davisson and Germer 1927, and Thomson and Reid 1927) to C_{60} molecules or buckyballs (Nairz, Arndt, and Zelinger 2003), which has something like $60(6 + 6 + 6) = 1080$ particles if one includes protons and neutrons in the nucleus, and electrons. We reproduce above (adapted and updated from Robinett 2006) examples of results from diffraction experiments on particles showing the applicability of Eqn. (6.57) for objects of increasing complexity.

The Einstein and de Broglie relations can be rewritten in ways that emphasize their expression in the language of wave physics (as in Chapter 1), namely

$$E = hf = \frac{h}{2\pi}(2\pi f) = \hbar\omega \qquad \text{and} \qquad p = \frac{h}{\lambda_{dB}} = \frac{2\pi\hbar}{\lambda_{db}} = \hbar k, \tag{6.58}$$

where ω and k are the **angular frequency** and **wave-number.** The functional forms for standing wave or traveling wave solutions of the wave equation are often written using k, ω as

$$\cos(kx - \omega t) \qquad \text{or} \qquad e^{i(kx-\omega t)} \tag{6.59}$$

instead of in terms of λ, f. If we follow that example when approaching quantum mechanics from the wave perspective, via the Schrödinger equation, it's then appropriate to write

$$\cos\left(\frac{px - Et}{\hbar}\right) \qquad \text{or} \qquad e^{i(px-Et)/\hbar}, \tag{6.60}$$

and we can see that it's natural in many situations to use the reduced Plancks constant, $\hbar$, when citing results that are derived via the wave equation. We explore many cases in Section 6.4 where a wave function has a dimension**<u>less</u>** argument, with some dimension**<u>ful</u>** quantity divided by (scaled by) $\hbar$.

A final example of another foundational appearance of Planck's constant arises in the quantization of angular momentum (á la Bohr) where we write $\mathcal{L}_z = m\hbar$ where m is a quantum number (and not a particle mass) and it's easy to remember that the dimensions of $\hbar$ are also those of angular momentum. We will see that again in our discussion of the hydrogen atom in Section 6.3.4.

6.2 Formalism of quantum mechanics

In many treatments of quantum mechanics beyond the modern physics level, in the advanced undergraduate or graduate curriculum, it can seem like the mathematics is actually the "star of the show," with applications of differential equations, eigenvalue and boundary value problems, special functions (some familiar, some arcane), Fourier series and transforms, matrix methods, singular distributions (such as the Dirac δ-function) and even infinite-dimensional vector spaces prominently introduced and extensively utilized. Even when the use of these formalisms is beyond the current level of one's mathematical fluency, all of these tools, however abstract, still have to respect the basic rules of DA, and a DA approach (and related ones) can continue to be useful, even at an early stage of a student's quantum mechanical initiation.

In its simplest (most short-hand, or zen) form, the **time-dependent Schrödinger equation** can be written as[5]

$$\hat{H}\psi = \hat{E}\psi, \tag{6.61}$$

where the **Hamiltonian operator** ($\hat{H}$) and **energy operator** ($\hat{E}$) act on a **position-space wave function** (ψ). (We discuss the corresponding **momentum-space wave function** ϕ below.) Expanding on this abbreviated notation, we often see

$$\left(\frac{\hat{p}^2}{2m} + V(x,t)\right)\psi(x,t) = \hat{E}\psi(x,t) \tag{6.62}$$

or

$$-\frac{\hbar^2}{2m}\frac{\partial^2\psi(x,t)}{\partial x^2} + V(x,t)\psi(x,t) = i\hbar\frac{\partial\psi(x,t)}{\partial t}, \tag{6.63}$$

[5] This was the author's first tattoo.

where $\hat{p}, \hat{E}$ are the **momentum** and **energy operators** (first seen in **P1.17**), namely

$$\hat{p} = \frac{\hbar}{i}\frac{\partial}{\partial x} \qquad \text{and} \qquad \hat{E} = i\hbar\frac{\partial}{\partial t}, \tag{6.64}$$

where $i = \sqrt{-1}$ is the **imaginary unit** (and so dimension**less**) and $V(x,t)$ is the **potential energy** function relevant to the problem at hand.

We can see that Eqn. (6.63) is dimensionally consistent, regardless of the (still-to-be-determined) dimensions of $[\psi(x,t)] = [\psi]$, since

$$\left[-\frac{\hbar^2}{2m}\frac{\partial^2\psi(x,t)}{\partial x^2}\right] = \left(\frac{\mathsf{ML}^2}{\mathsf{T}}\right)^2\left(\frac{1}{\mathsf{M}}\right)\left(\frac{1}{\mathsf{L}^2}\right)[\psi] = \frac{\mathsf{ML}^2}{\mathsf{T}^2}[\psi]$$
$$[V(x,t)\psi(x,t)] = [V(x,t)][\psi] = \frac{\mathsf{ML}^2}{\mathsf{T}^2}[\psi]$$
$$\left[i\hbar\frac{\partial\psi(x,t)}{\partial t}\right] = \left(\frac{\mathsf{ML}^2}{\mathsf{T}}\right)\left(\frac{1}{\mathsf{T}}\right)[\psi] = \frac{\mathsf{ML}^2}{\mathsf{T}^2}[\psi]. \tag{6.65}$$

Many texts actually focus much of their attention on the **time-independent Schrödinger equation** (since it's especially relevant for the study of bound state systems) by assuming a specific form for the time-dependence of $\psi(x,t)$, namely

$$\psi(x,t) = \psi_n(x)\, e^{-iE_n t/\hbar} \tag{6.66}$$

to obtain

$$-\frac{\hbar^2}{2m}\frac{d^2\psi_n(x)}{dx^2} + V(x)\psi_n(x) = E_n\psi_n(x). \tag{6.67}$$

This is the version on which many textbooks spend the majority of their time/space/pages, exploring many (sometimes all) of the handful (almost literally, about five familiar and tractable examples) of the exactly soluble problems in quantum mechanics. In Section 6.4.3, we discuss in more detail the roles of $\psi_n(x)$ and E_n as the energy eigenfunctions and energy eigenvalues of bound state systems.

Anticipating our discussion of the physical interpretation (and dimensional spelling) of $\psi(x,t)$, we first provide a (very) brief review of **complex numbers.** A complex number can be written in the form $z = a + ib$ where $i \equiv \sqrt{-1}$ is the **imaginary unit** and a, b are themselves real numbers, which are referred to as the **real** and **imaginary** parts of z, respectively. This can be correspondingly expressed as

$$Re(z) = a \qquad \text{and} \qquad Im(z) = b, \tag{6.68}$$

and a related quantity is the **complex conjugate** $z^* \equiv a - ib$ (obtained by just flipping the sign of the imaginary part). The **modulus** of a complex number is given by

$$|z|^2 \equiv zz^* = (a+ib)(a-ib) = a^2 + b^2 \qquad \text{yielding} \qquad |z| = \sqrt{a^2+b^2}, \tag{6.69}$$

and $|z|$ can be thought of as the generalized length of z, in the two-dimensional (real/imaginary) complex plane.

Making contact with DA, in much the same way that vector components (at least when expressed in Cartesian coordinates) must all have the same dimensions, we must have for the dimensional content of complex numbers that

$$[z] = [a] = [b] = [\,|z|\,]. \tag{6.70}$$

A complex number can also be expressed in terms of its **modulus** and **phase** as

$$z = |z|e^{i\theta} = |z|\,\{\cos(\theta) + i\sin(\theta)\} = a + ib \qquad \text{implying} \qquad \tan(\theta) = \frac{b}{a}, \tag{6.71}$$

where θ is clearly dimensionless.

Returning to matters of interpretation, the **position-space Schrödinger wave function**, $\psi(x,t)$, is a (complex) **probability amplitude** with the corresponding square (more precisely, modulus squared) given by

$$P(x,t) \equiv |\psi(x,t)|^2 = \psi(x,t)^* \psi(x,t) \tag{6.72}$$

defining a **probability density**

$$P(x,t)\,dx = |\psi(x,t)|^2\,dx = \mathrm{dProb}\,(x \in [x, x+dx])\,. \tag{6.73}$$

This statement is to be interpreted as saying that the (real-valued) quantity $P(x,t)$ gives the (small) probability that a measurement of the position of the particle (at time t) will find it to be in the (narrow) range $[x, x+dx]$.[6] In this sense we can write

$$P(x,t) = |\psi(x,t)|^2 = \frac{\mathrm{dProb}}{dx}, \tag{6.74}$$

and since probabilities are dimensionless, this gives the dimensionality of $\psi(x,t)$ as

$$[\,|\psi(x,t)|\,]^2 = \left[\frac{\mathrm{dProb}}{dx}\right] = \frac{1}{\mathsf{L}} \quad \text{or} \quad [\psi(x,t)] = [\,|\psi(x,t)|\,] = \mathsf{L}^{-1/2} = [Re(\psi(x,t))] = [Im(\psi(x,t))]. \tag{6.75}$$

We note that while it has no real impact on a DA discussion of quantum mechanics, the fact that $\psi(x,t)$ is fundamentally a complex quantity is a subject of experimental verification, with recent results having confirmed that purely real theories of quantum mechanics are excluded at the level of many standard deviations; see Li et al. (2022), Chen et al. (2022), or Wu et al. (2022).

Given the statistical nature of the quantum wave function, $\psi(x,t)$, the notion of a measurement requires the concept of **average value** or **expectation value**, where for any physical quantity represented by a quantum mechanical variable or operator, $\hat{O}$, we write

$$\langle \hat{O} \rangle = \int_{-\infty}^{+\infty} \psi^*(x,t)\,\hat{O}\,\psi(x,t)\,dx, \tag{6.76}$$

[6] This probability interpretation is attributed to Born (1927).

which is dimensionally consistent for any $\hat{O}$, since

$$\left[\langle\hat{O}\rangle\right] = \left(\frac{1}{\sqrt{\mathsf{L}}}\right)[\hat{O}]\left(\frac{1}{\sqrt{\mathsf{L}}}\right)(\mathsf{L}) = [\hat{O}]. \tag{6.77}$$

One of the new mathematical constructs that finds repeated use in the formalism of quantum mechanics is the **Dirac delta function**, or $\delta(s)$, where s can be any argument, and may well have dimensions.[7] This function (often called a **distribution** because of its singular nature) is defined as

$$\delta(s) = \begin{cases} \infty & \text{if s} = 0 \\ 0 & \text{if s} \neq 0 \end{cases} \tag{6.78}$$

or, if the singular δ-function is centered at a different point, we can similarly write

$$\delta(s - s_0) = \begin{cases} \infty & \text{if s } = s_0 \\ 0 & \text{if s } \neq s_0 \end{cases}. \tag{6.79}$$

The nature of the singularity is such that the total area (generalized "area under the curve," in the sense of integration) is normalized to unity (1), so that it can then "pick out" the value of a function at the singular point, such that

$$\int_{-\infty}^{+\infty} \delta(s - s_0)\, ds = 1 \qquad \text{implying that} \qquad \int_{-\infty}^{+\infty} \delta(s - s_0)\, f(s)\, ds = f(s_0). \tag{6.80}$$

The first of these relations determines the dimensionality of the $\delta(s)$ function via

$$\left[\int_{-\infty}^{+\infty} \delta(s)\, ds\right] = [\delta(s)]\,[ds] = 1 = \textit{dimensionless} \quad \text{implying} \quad [\delta(s)] = \frac{1}{[s]}, \tag{6.81}$$

whatever the dimensions of s might be. Examples of these dimensionalities for δ-functions routinely used in quantum mechanics, namely for position, time, momentum, and energy, include

$$[\delta(x - x_0)] = \frac{1}{\mathsf{L}}, \quad [\delta(t - t_0)] = \frac{1}{\mathsf{T}}, \quad [\delta[p - p_0)] = \frac{\mathsf{T}}{\mathsf{ML}}, \quad \text{and} \quad [\delta[E - E_0)] = \frac{\mathsf{T}^2}{\mathsf{ML}^2}. \tag{6.82}$$

Example 6.1 Bound state for a single, attractive δ-function potential

While the Dirac δ-function clearly is a mathematical abstraction, it can be used to model a highly localized attractive potential in one dimension by writing

$$V_\delta(x) \equiv -K\delta(x), \tag{6.83}$$

[7] For more mathematical details, see almost any advanced undergraduate textbook on quantum mechanics or the chapter-length discussion (with problems) in Kusse and Westwig (1998). See also Blennow (2018), which is one of the only treatments I've encountered which emphasizes the dimensionality of the δ-function, as done here, and uses that idea in discussing topics such as coordinate transformations.

where K measures the strength of the attractive force. The dimensionality of K is determined by

$$[K] = \left[\frac{V_\delta(x)}{\delta(x)}\right] = \left(\frac{\mathsf{ML}^2}{\mathsf{T}^2}\right)\left(\frac{1}{\mathsf{L}}\right)^{-1} = \frac{\mathsf{ML}^3}{\mathsf{T}^2}. \tag{6.84}$$

It turns out that this potential admits exactly one bound state, and to explore the length (ρ) and energy (E) scales associated with that solution (without having to solve the Schrödinger equation completely), we write

$$\rho \propto m^\alpha K^\beta \hbar^\gamma \qquad \text{and} \qquad E \propto m^{\overline{\alpha}} K^{\overline{\beta}} \hbar^{\overline{\gamma}} \qquad \text{giving} \qquad \rho, E \propto M^\alpha \left(\frac{ML^3}{T^2}\right)^\beta \left(\frac{ML^2}{T}\right)^\gamma. \tag{6.85}$$

Matching dimensions, we find $\alpha = -1, \beta = -1, \gamma = +2$ and $\overline{\alpha} = +1, \overline{\beta} = +2, \overline{\gamma} = -2$, or

$$\rho = c_\rho \frac{\hbar^2}{mK} \qquad \text{and} \qquad E = c_E \frac{mK^2}{\hbar^2}. \tag{6.86}$$

We imagine that since $E < 0$, given the attractive nature of the bound state, we'd have $c_E < 0$, so we at least can infer the sign, if not the magnitude, of the dimensionless constant. You should be able to also find dimensional dependences for the typical values of momentum (p), speed (v), and characteristic time (τ) for this problem in terms of $\hbar$, K, and m; see **P6.13**(a).

This problem is a staple of the undergraduate quantum mechanics curriculum, and is therefore covered in many textbooks, where it's noted that for the single bound state, the length scale, energy, and position-space wave function corresponding to the unique $E < 0$ solution are given by

$$\rho = \frac{\hbar^2}{mK}, \qquad E_\delta = -\frac{1}{2}\frac{mK^2}{\hbar^2}, \qquad \text{and} \qquad \psi(x) = \frac{1}{\sqrt{\rho}} e^{-|x|/\rho}, \tag{6.87}$$

and once again DA has provided us with the correct dimensions, with a dimensionless coefficient close to (or exactly) unity in both cases. Even if we have not encountered this problem in a more advanced context, the dimensional dependences found this way in Eqn. (6.86) can be used to hone one's intuition about the nature of quantum mechanical bound states, even before proceeding to more well-known examples in Section 6.3.

We also note, for future reference, that despite the seemingly very artificial nature of this (zero-range) potential, there are many similarities between these results and those of the (very long-range, $V(r) \propto 1/r$) quantum Coulomb problem, that is, the first approximation to the hydrogen atom, discussed in Section 6.3.4.

Extending the range of quantum mechanics to cover more realistic three-dimensional problems, we next write the (differential) momentum operator as a vector, using the gradient, so that

$$\hat{p} = \frac{\hbar}{i}\frac{\partial}{\partial x} \qquad \longrightarrow \qquad \hat{\boldsymbol{p}} = \frac{\hbar}{i}\boldsymbol{\nabla}, \tag{6.88}$$

and then the wave function as $\psi(\boldsymbol{r}, t)$, giving the three-dimensional Schrödinger equation

$$-\frac{\hbar^2}{2m}\boldsymbol{\nabla}^2\psi(\boldsymbol{r}, t) + V(\boldsymbol{r}, t)\psi(\boldsymbol{r}, t) = i\hbar\frac{\partial\psi(\boldsymbol{r}, t)}{\partial t}. \tag{6.89}$$

The Schrödinger wave function still has a probability interpretation, but now as

$$|\psi(\boldsymbol{r}, t|^2\, dV = \text{dProb(in a small region dV near } \boldsymbol{r}) \qquad \text{or} \qquad |\psi(\boldsymbol{r}, t)|^2 = \frac{\text{dProb}}{dV}, \tag{6.90}$$

which gives the probability of finding a particle in a small spatial region dV, near the point $\boldsymbol{r}$, at time t, and importantly for us, giving the dimensions of a three-dimensional wave function to be

$$[\,|\psi(\boldsymbol{r}, t)|\,]^2 = \frac{1}{[dV]} = \frac{1}{\mathsf{L}^3} \quad \longrightarrow \quad [\psi(\boldsymbol{r}, t)] = L^{-3/2}. \tag{6.91}$$

The corresponding Dirac δ-function (for position) in three-dimensional space can be written as

$$\delta(\boldsymbol{r} - \boldsymbol{r}_0) = \begin{cases} \infty & \text{if } \ \boldsymbol{r} = \boldsymbol{r}_0 \\ 0 & \text{if } \ \boldsymbol{r} \neq \boldsymbol{r}_0 \end{cases}, \tag{6.92}$$

with the integrals corresponding to those in Eqn. (6.80) being

$$\int \delta(\boldsymbol{r} - \boldsymbol{r}_0)\, dV = 1 \qquad \text{and} \qquad \int \delta(\boldsymbol{r} - \boldsymbol{r}_0)\, F(\boldsymbol{r})\, dV = F(\boldsymbol{r}_0) \tag{6.93}$$

implying that

$$[\delta(\boldsymbol{r} - \boldsymbol{r}_0] = \frac{1}{[dV]} = \frac{1}{\mathsf{L}^3}. \tag{6.94}$$

A familiar example of the use of $\delta(\boldsymbol{r})$ from electrodynamics is the description of a point-like charged particle, associated with a singular charge density, given by $\rho_e(\boldsymbol{r}) = q\delta(\boldsymbol{r} - \boldsymbol{r}_0)$, but which has a finite total charge

$$Q_{tot} = \int \rho_e(\boldsymbol{r})\, dV = q \int \delta(\boldsymbol{r} - \boldsymbol{r}_0)\, dV = q. \tag{6.95}$$

Thus the notion of the Dirac δ-function is useful far beyond the pages of quantum mechanics texts.

Example 6.2 Model of short-range boson–boson interactions

In some atomic physics systems, the interactions of identical bosons can be modeled by a repulsive Dirac δ-function potential, implying that the particles have very short-range forces. One such effective potential for a two-body interaction (for particles i, j) is given by

$$V_{bb}(\boldsymbol{r}_i - \boldsymbol{r}_j) = \frac{4\pi\hbar^2 a_S}{m}\delta(\boldsymbol{r}_i - \boldsymbol{r}_j) \equiv K_S\delta(\boldsymbol{r}_i - \boldsymbol{r}_j), \tag{6.96}$$

where m is the particle mass (same for all particles), $\hbar$ is Planck's constant, and a_S is described as the **S-wave scattering length** for the interactions of the two particles. Even if we are not "fluent" in the language of particle scattering, this description alone implies that the dimensions of a_S are given by $[a_S] = \mathsf{L}$, and we can at the very least confirm that the dimensions in Eqn. (6.96) match, since

$$\begin{aligned}
[V_{bb}] &\stackrel{?}{=} \left[\frac{4\pi\hbar^2 a_S}{m}\delta(\boldsymbol{r}_i - \boldsymbol{r}_j)\right] \\
\frac{\mathsf{ML}^2}{\mathsf{T}^2} &\stackrel{?}{=} \left(\frac{\mathsf{ML}^2}{\mathsf{T}}\right)^2 \frac{\mathsf{L}}{\mathsf{M}}\left(\frac{1}{\mathsf{L}^3}\right) \\
\frac{\mathsf{ML}^2}{\mathsf{T}^2} &\stackrel{!}{=} \frac{\mathsf{ML}^2}{\mathsf{T}^2}\,.
\end{aligned} \tag{6.97}$$

This type of term leads to a **non-linear Schrödinger equation** of a type first studied by Gross (1961) and Pitaevskii (1961), which is used to model the ground state of a system of interacting identical bosons; see **P6.9**.

Perhaps because of our own (human) evolutionary history, we are conditioned to focus on the descriptions of objects in position-space, but in quantum mechanics, there is a duality between $x \leftrightarrow p$ and one can consider all quantities in **momentum-space**. In that language, $\phi(p, t)$ is a **momentum-space probability amplitude**, and similarly to Eqns. (6.73) and (6.74), we associate

$$|\phi(p,t)|^2\, dp = \mathrm{dProb}\,(p \in [p, p + dp]) \tag{6.98}$$

as the (small) probability that a measurement of the momentum of the particle (at time t) will find it to be in the (narrow) range $[p, p + dp]$. We can thus write can write

$$|\phi(p,t)|^2 = \frac{\mathrm{dProb}}{dp} \tag{6.99}$$

and since probabilities are still dimensionless, this gives

$$[\,|\phi(p,t)|\,]^2 = \left[\frac{\mathrm{dProb}}{dp}\right] = \frac{1}{[p]} = \frac{\mathsf{T}}{\mathsf{ML}} \qquad \text{or} \qquad [\phi(p,t)] = [\,|\phi(p,t)|\,] = \left(\frac{\mathsf{T}}{\mathsf{ML}}\right)^{1/2}. \tag{6.100}$$

In a manner perhaps familiar from classical wave physics, where there is a connection between frequency and time (f or ω and t), the two Schrödinger wave functions, $\psi(x, t)$ and $\phi(p, t)$, are related by **Fourier transforms**. Going from one to the other in one direction (from $x \to p$), we have the relation below, which we can also check to be dimensionally consistent via

$$\begin{aligned}
\phi(p,t) &= \frac{1}{\sqrt{2\pi\hbar}} \int_{-\infty}^{+\infty} \psi(x,t)\, e^{-ipx/\hbar}\, dx \\
\left(\frac{\mathsf{T}}{\mathsf{ML}}\right)^{1/2} &\stackrel{?}{=} \left(\frac{\mathsf{ML}^2}{\mathsf{T}}\right)^{-1/2} (\mathsf{L}^{-1/2})\,(1)\,(\mathsf{L}) \\
\left(\frac{\mathsf{T}}{\mathsf{ML}}\right)^{1/2} &\stackrel{!}{=} \left(\frac{\mathsf{T}}{\mathsf{ML}}\right)^{1/2}.
\end{aligned} \tag{6.101}$$

For realistic physical systems, we almost always encounter bound states (atoms, molecules, solids, etc.) of two or more ($N \geq 2$) particles, and so for such multi-particle states, the Schrödinger wave function has as many arguments as constituent particles. In one dimension this implies

$$\psi(x,t) \quad \longrightarrow \quad \psi(x_1,x_2,\ldots,x_N;t), \tag{6.102}$$

with an associated probability

$$|\psi(x_1,\ldots,x_N;t)|^2 dx_1 dx_2 \ldots dx_N = \text{dProb}(x_1 \in (x_1,x_1+dx_1),\ldots,x_N \in (x_N,x_N+dx_N)) \tag{6.103}$$

so that

$$|\psi(x_1,\ldots,x_N;t)|^2 = \frac{\text{dProb}}{dx_1\ldots dx_N} \quad \text{implying} \quad [\psi(x_1,\ldots,x_N;t)] = \left(\frac{1}{\sqrt{\mathsf{L}}}\right)^{\mathsf{N}} = \mathsf{L}^{-\mathsf{N}/2}, \tag{6.104}$$

and for an N-particle three-dimensional wave function,

$$|\psi(\boldsymbol{r}_1,\ldots,\boldsymbol{r}_N;t)|^2 = \frac{\text{dProb}}{dV_1\ldots dV_N} \quad \text{implying} \quad [\psi(\boldsymbol{r}_1,\ldots,\boldsymbol{r}_N;t)] = \left(\frac{1}{\sqrt{\mathsf{L}^3}}\right)^{\mathsf{N}} = \mathsf{L}^{-3\mathsf{N}/2}. \tag{6.105}$$

6.3 **Bound states**

Wigner (1960) celebrated the fact that

> The miracle of the appropriateness of the language of mathematics for the formulation of the laws of physics is a wonderful gift which we neither understand nor deserve. We should be grateful for it and hope that it will remain valid in future research and that it will extend, for better or for worse, to our pleasure, even though perhaps also to our bafflement, to wide branches of learning...

and we can be especially thankful for its application in the study of bound state systems in quantum mechanics. Four important model systems, which can be solved in terms of known mathematical special functions (some as simple as sines and cosines), have varying degrees of relevance to experimentally realizable physical systems, namely

- The **infinite well**, as a model of highly localized one-dimensional, two-dimensional, or three-dimensional bound systems, such as quantum dots, but even for modeling of white dwarf or neutron stars (as in Section 9.3);
- The **harmonic oscillator**, the "best first guess" (recall Section 3.1) for many systems characterized by stable equilibrium, including the vibrational states of molecules;
- The **linear potential**, as a model for the bound neutron system discussed in Section 2.5 when $V(z) = mgz$, or to describe external constant forces, such as electric fields when $V(x) = Fx = qEx$; and
- The **quantum Coulomb problem**, as the zero-th order approximation to the hydrogen atom.

For all of these systems (and others) there are a number of approaches, all involving the use of dimensionality, which can be used to obtain information about the nature of the quantized energies (E_n) and typical scales for quantities such as length (ρ), momentum (p), and periodicity (τ), and how they depend on the appropriate dimensionful physical variables. We will repeatedly consider four different strategies, **S1–S4**, either explicitly in the text, or as suggested problems, namely:

- **S1—Pure DA:** This is the most familiar approach and can be used proactively in a number of bound state problems.
- **S2—Uncertainty principle arguments:** The relationship between the kinetic and potential energies in quantum theory is far more constrained than in classical mechanics, as information about the position and momentum of a particle is highly correlated by the **Heisenberg uncertainty relation**, $\Delta x \cdot \Delta p \geq \hbar/2$. This limitation is responsible for the zero-point energy of bound state systems, since we cannot simultaneously have x perfectly localized at the minimum of a potential **and** be at rest with $p = 0$. This uncertainty principle constraint can be used derive the dimensional dependence of the ground state energy (and so all quantized energies) and, by inference, many other quantities.
- **S3—Non-dimensionalization methods:** We have discussed this in the context of other differential equation approaches to physical systems in Chapter 1, and this method is often the starting point of textbook discussions of bound states in quantum mechanics, where changes in variables are employed to turn the Schrödinger equation into a dimensionless "pure" mathematical physics problem, sadly, often not replacing the dimensions back at the end to compare to physical observables.
- **S4—"Fitting de Broglie waves" and more mathematically sophisticated extensions:** Instead of solving the Schrödinger equation directly, one can use the analog of waves on a one-dimensional string, with the **de Broglie wavelength** as the relevant length scale, and insist that appropriate boundary conditions be met. In this simplest of forms, it works only for the first case we consider below, but more rigorous derivations involving quantum mechanical wave functions arise in the **Wentzel–Kramers–Brillouin** (WKB) approximation method, which is valid in the semi-classical limit of large quantum numbers.

Noting the comments in Section 6.1.4, we will use the reduced Planck's constant, $\hbar$, as the fundamental dimensionful constant in quantum mechanics, given that many of our results come from discussions based on wave mechanics and the Schrödinger equation.

6.3.1 Infinite well

The simplest problem involving bound states, treated at every level of quantum mechanics pedagogy, is that of the **infinite square well** or *ISW*, which is relevant for an otherwise free particle, but one confined in one dimension to a region of length L, described by a potential energy function given by

$$V_{ISW}(x) = \begin{cases} \infty & \text{for } x < 0 \\ 0 & \text{for } 0 < x < L \\ \infty & \text{for } L < x \end{cases} . \tag{6.106}$$

If we start with the most familiar **S1** approach, we would identify $\hbar$, m, and L as the relevant dimensionful quantities, and write for the energy scale

$$E \propto \hbar^{\alpha} m^{\beta} L^{\gamma} \qquad \text{requiring that} \qquad \frac{\mathsf{ML}^2}{\mathsf{T}^2} = \left(\frac{\mathsf{ML}^2}{\mathsf{T}}\right)^{\alpha} \mathsf{M}^{\beta}\, \mathsf{L}^{\gamma}, \tag{6.107}$$

and matching dimensions we find $\alpha = 2, \beta = -1, \gamma = -2$ or $E \propto \hbar^2/mL^2$. The exact answer for the quantized energy levels, including the dimensionless quantum number n (a result that we reproduce in a semi-classical way below) is cited in even introductory level texts as

$$E_n = \frac{n^2\pi^2\hbar^2}{2mL^2} = n^2\left(\frac{\pi^2}{2}\right)\frac{\hbar^2}{mL^2} \tag{6.108}$$

and is consistent with the DA result with a dimensionless $C_E = \pi^2/2$, but with the new feature of a quantum number ($n = 1, 2, 3, \ldots$) about which DA, by itself, is mute. We will see, however, that combined with other semi-classical results, we can sometimes gain information on the n-dependence of some quantum systems where n can appear "partnered" with $\hbar$.

As a trivial example of the **S2** strategy, we note that the classical kinetic energy of a particle in one dimension can be written as

$$E_{KE} = \frac{1}{2}mv^2 = \frac{p^2}{2m} \tag{6.109}$$

and if we use the spirit (if not the precise mathematical rigor) of the uncertainty principle to write $\Delta x \cdot \Delta p \sim xp \sim \hbar$, and use the well size L as a measure of Δx, we have (at least dimensionally)

$$E_{KE} \sim \frac{(\hbar/L)^2}{2m} \sim \frac{\hbar^2}{2mL^2}, \tag{6.110}$$

which agrees with the DA result. In this case, we seem to be using this as a reactive or a posteriori check, but in other cases we find that it can be used in a more powerful and predictive way.

We've already explored the **non-dimensionalization** approach of **S3** in several examples, and in this case the natural change of variables involving length would be to write $x = \rho y = Ly$ since L is already singled out as the only relevant scale. With this substitution, the time-independent Schrödinger equation then reads

$$-\frac{\hbar^2}{2mL^2}\frac{d^2\psi_n(y)}{dy^2} = E_n\psi_n(y), \tag{6.111}$$

and since y is dimensionless, we have that $E_n \sim \hbar^2/2mL^2$ is the natural connection between the quantized energy and the length scale.

Finally, given the trivial nature of the potential energy function within the well, we can use the **de Broglie wavelength** as the analogous quantity in a "waves on a string" argument to write

$$\lambda_{dB} = \frac{h}{p} = \frac{2\pi\hbar}{p} \tag{6.112}$$

and demand that an integral number of half-wavelengths "fit" into the box (the way we would for a string fixed at both ends) to satisfy the quantum mechanical boundary conditions that $\psi(0) = \psi(L) = 0$. This implies that

$$n\left(\frac{\pi\hbar}{p}\right) = n\left(\frac{\lambda_{dB}}{2}\right) = L \quad\longrightarrow\quad p_n = \frac{n\pi\hbar}{L} \quad\longrightarrow\quad E_n = \frac{p_n^2}{2m} = \frac{n^2\pi^2\hbar^2}{2mL^2}, \tag{6.113}$$

which is the exact answer in this most simplified of all potential energy functions.

6.3.2 Harmonic oscillator

The harmonic oscillator is perhaps the **most soluble** problem in all of mechanics (classical, quantum mechanical, and even in field theory), due to its high degree of symmetry, with

- kinetic energy given by $p^2/2m$, and
- potential energy given by $kx^2/2$,

both being quadratic in their variables, and so truly unique. We discussed it in detail in Section 3.1 in its classical version, and even earlier on in Chapter 1, using non-dimensionalization and pure DA, for the quantum case. Using what we now have labeled the **S1** and **S3** strategies, we solved for the typical length and energy scales by writing $\rho, E \propto \hbar^{\alpha} m^{\beta} k^{\gamma}$ (with k, m the spring constant and particle mass, respectively) and found (Section 1.3 and **P1.16**)

$$\rho \propto \left(\frac{\hbar^2}{mk}\right)^{1/4} \quad \text{and} \quad E \propto \hbar \left(\frac{k}{m}\right)^{1/2}, \tag{6.114}$$

where the actual quantized energy eigenvalues are given by

$$E_n = (n + 1/2)\hbar\sqrt{\frac{k}{m}}, \tag{6.115}$$

where $n = 1, 2, 3, ..$ is the relevant quantum number. The ground state wave function is given by

$$\psi_{GS}(x) = \frac{1}{\sqrt{\rho\pi}} e^{-x^2/2\rho^2}, \tag{6.116}$$

showing that ρ is indeed the correct length scale, both in the functional form (with the dimensionless ratio $\Pi = x/\rho$ singled out), and in the overall normalization factor.

As an illustration of the power of the **S2** strategy, we again simplify the rigorously defined uncertainty principle relation $\Delta x \cdot \Delta p \geq \hbar/2$ to write $x \cdot p \sim \hbar/2$ to model the effect of the position-momentum correlations in quantum mechanics to write the classical energy as

$$E = \frac{p^2}{2m} + \frac{1}{2}kx^2 \sim \frac{\hbar^2}{8mx^2} + \frac{1}{2}kx^2. \tag{6.117}$$

If we associate $E = E(x)$ and try to minimize the total energy with respect to x, to model the ground state energy of the quantum system, we write

$$0 = \frac{dE(x)}{dx} = -\frac{\hbar^2}{4mx^3} + kx \quad \longrightarrow \quad (x_{min})^4 = \frac{\hbar^2}{4mk} \quad \text{or} \quad x_{min} = \left(\frac{\hbar^2}{4mk}\right)^{1/4}. \tag{6.118}$$

This approach does reproduce the appropriate length scale in Eqn. (6.114), and if we use this result in Eqn. (6.117) we find the minimum energy in this classical–quantum hybrid approach to be

$$E(x_{min}) = \frac{\hbar^2}{8mx_{min}^2} + \frac{1}{2}kx_{min}^2 \quad \longrightarrow \quad E = \frac{1}{4}\hbar\sqrt{\frac{k}{m}} + \frac{1}{4}\hbar\sqrt{\frac{k}{m}} = \frac{1}{2}\hbar\sqrt{\frac{k}{m}}, \tag{6.119}$$

reproducing the exact zero-point energy in the complete quantum mechanical solution. Just as the "fitting de Broglie waves" method was especially successful in the ISW situation, this approach is also suited to this problem with unique symmetry.

To see to what extent the strategy **S4** can be used in problems more general than the ISW, let us first rewrite Eqn. (6.113) in the form

$$n\pi\hbar = p_n L = \sqrt{2mE_n}L \qquad \text{since in the ISW we have} \qquad E_n = \frac{p_n^2}{2m}. \tag{6.120}$$

The more sophisticated version of this constraint in a general 1D potential, $V(x)$, is given by the **WKB** semi-classical approximation where we generalize

$$\begin{array}{ccc} n\pi\hbar & = & \sqrt{2mE_n}L \\ \Downarrow & & \Downarrow \end{array}$$
$$(n+1/2)\pi\hbar = \int_a^b \sqrt{2m(E_n - V(x))}\,dx. \tag{6.121}$$

The values of a, b are the classical turning points in the potential energy, defined by $V(a) = E = V(b)$. The inclusion of the additional factor of 1/2 in the "counting half-wavelengths" term arises from the fact that the quantum mechanical wave functions at those turning points need not vanish identically, as they must at a truly infinite wall/barrier, as they can tunnel into the classically disallowed region, "spilling over" by about a quarter-wavelength at each boundary. The fact that $p_n(x) = \sqrt{2m(E_n - V(x)}$ varies with position is taken into account by summing contributions of $p_n(x)\,dx$ over the classically allowed region. This approach is strictly valid only for large values of the quantum number, n, but for the case of the uniquely symmetric harmonic oscillator, also happens to give the correct results.

We see this by noting that the turning points are given by $\pm a = \sqrt{2E_n/k}$ and the WKB quantization condition reads

$$\begin{aligned} (n+1/2)\pi\hbar &= \int_{-a}^{+a} \sqrt{2m(E_n - kx^2/2)}\,dx \\ &= a\sqrt{mk}\int_{-a}^{+a} \sqrt{1-(x/a)^2}\,dx \\ &= a^2\sqrt{mk}\int_{-1}^{+1} \sqrt{1-y^2}\,dy \\ (n+1/2)\pi\hbar &= \left(\frac{2E_n}{k}\right)\sqrt{mk}\left(\frac{\pi}{2}\right) \end{aligned} \tag{6.122}$$

or $E_n = (n+1/2)\hbar\sqrt{k/m}$, once again reproducing the exact result.

One important observation about this WKB approach, is that for large quantum numbers, the combination $n\hbar$ appears on the left-hand side, so we expect that whenever DA is used, whether in any of the **S1–S3** approaches, that we may expect to see n appear with the same power as $\hbar$. This was certainly the case for the ISW where we found that $E_n \propto n^2\hbar^2$. Thus in some situations we can "piggyback" on any DA approach and also obtain information on the quantum number dependence, even when DA would otherwise be blind to its functional form.

6.3.3 **Linear potential and general power-law case**

Another quantum mechanical problem that has a physical motivation is the (prosaically named) **linear potential on the half-plane**, which we can describe by the potential energy function

$$V(x) = \begin{cases} \infty & \text{for } x<0 \\ Fx & \text{for } x \geq 0 \end{cases}. \tag{6.123}$$

We have already seen that this system is useful in modeling the gravitational bound states of neutrons (in Section 2.5), where we specialized to $F = mg$.

Any of the four strategies **S1–S4** can be profitably employed in this case, with the simplest DA approach giving results for the typical length and energies as

$$\rho \propto \left(\frac{\hbar^2}{mF}\right)^{1/3} \qquad \text{and} \qquad E \propto F\rho \propto \left(\frac{\hbar^2 F^2}{m}\right)^{1/3}, \tag{6.124}$$

which reduce to the results in Section 2.5 if we associate $F = mg$. The WKB approach also gives the quantum number (n) dependence as $E_n \propto n^{2/3}$, again for $n >> 1$. Explicit calculations using the **S1–S4** strategies are left to the homework in **P6.12.**

We now have three examples of the dimensional (and quantum number) dependence of the energy eigenvalues for three different one-dimensional confining potentials, namely

$$E_n^{(ISW)} \sim \frac{n^2\hbar^2}{mL^2}, \qquad E_n^{(HO)} \sim n\hbar\left(\frac{k}{m}\right)^{1/2}, \qquad \text{and} \qquad E_n^{(linear)} \sim \left(\frac{n^2\hbar^2F^2}{m}\right)^{1/3}, \tag{6.125}$$

with dimensions strictly determined, and quantum number dependence valid for $n >> 1$. We immediately notice some similarities involving the dependence on $\hbar$ and m, namely

- In all cases, the quantized energies (and hence the zero-point energy) are proportional to a (positive) power of Planck's constant, so that in the limit that $\hbar \to 0$ (presumably when quantum mechanics is not applicable) then $E_n \to 0$ as well, consistent with our **S2** uncertainty principle strategy.
- Each of the energies are proportional to a (negative) power of m, so that the larger the mass (inertia) the smaller the zero-point energy, presumably consistent with a $E_{kin} = p^2/2m$ relation.

Similar patterns can be seen in the values of the length scales, namely

$$\rho^{(ISW)} = L, \qquad \rho^{(HO)}) \sim \left(\frac{\hbar^2}{mk}\right)^{1/4}, \qquad \text{and} \qquad \rho^{(linear)} \sim \left(\frac{\hbar^2}{mF}\right)^{1/3}, \tag{6.126}$$

where the ISW is a special case which we discuss below.

One can approach all three of these problems (and, as we'll see, even the quantum Coulomb problem in three dimensions!) as special cases of a (symmetric) power-law potential of the form

$$V^{(P)}(x) = K_P|x|^P. \tag{6.127}$$

For $P = 2$, the similarity to the harmonic oscillator is obvious, assuming the association $K_2 = k/2$ with the normal spring constant. For the linear potential, we can easily extend the definition of the linear potential on the half-plane to the **symmetric linear potential**

$$V^{(1)}(x) = K_1|x| \tag{6.128}$$

where we associate K_1 with F, with none of the dimensional arguments changed, and the **S4** WKB approach simply requiring a integral over the range $(-a, +a)$ instead of just $(0, +a)$.

The infinite square well of length L can be easily reformulated as a potential of width $2L$ (with $E_n \to E_n/4$, using the L dependence we've found) and then shifted so that it's symmetrically placed about the origin (which can't, of course, change the physics in way), namely

$$\tilde{V}_{ISW}(x) = \begin{cases} \infty & \text{for} \quad x < -L \\ 0 & \text{for} \quad -L < x < +L \\ \infty & +L < x \end{cases}. \tag{6.129}$$

To connect this version to the general power-law potential in Eqn. (6.127), we note that for $|x/L| < 1$ (i.e., inside the well), we have $|x/L|^P \to 0$ when $P \to \infty$, while for $|x/L| > 1$, we have $|x/L| \to \infty$ in the same limit. For this particular limit, it's best to write the power-law potential as

$$V^{(P)}(x) = V_0 \left|\frac{x}{L}\right|^P \qquad \text{so that} \qquad K_P = \frac{V_0}{L^P} \tag{6.130}$$

to make the *ISW* geometry more obvious. We can then include $\tilde{V}^{(ISW)}(x)$ in the family of power-law potentials as the $P \to \infty$ limit of Eqn. (6.130).

We can now apply any of our **S1–S4** strategies to evaluate the characteristic length and energy scales for this entire class of potentials. We only need to know that

$$V^{(P)}(x) = K_P|x|^P \qquad \text{implies that} \qquad [K_P] = \left[\frac{V(x)}{|x|^P}\right] = \left(\frac{\mathsf{ML}^2}{\mathsf{T}^2}\right)\frac{1}{\mathsf{L}^\mathsf{P}} = \frac{\mathsf{ML}^{2-\mathsf{P}}}{\mathsf{T}^2}. \tag{6.131}$$

Proceeding as usual with a DA approach, we write $E \propto \hbar^\alpha m^\beta K_P^\gamma$ and match dimensions as

$$\frac{\mathsf{ML}^2}{\mathsf{T}^2} = \left(\frac{\mathsf{ML}^2}{\mathsf{T}}\right)^\alpha \mathsf{M}^\beta \left(\frac{\mathsf{ML}^{(2-P)}}{\mathsf{T}^2}\right)^\gamma, \tag{6.132}$$

which is easily solved for this general case to obtain

$$\alpha = \frac{2P}{2+P}, \qquad \beta = -\frac{P}{2+P}, \qquad \text{and} \qquad \gamma = \frac{2}{2+P} \tag{6.133}$$

or

$$E^{(P)} \sim \hbar^{2P/(2+P)}\, m^{-P/(2+P)}\, K_P^{2/(2+P)} \propto \left(\frac{\hbar^{2P}K_P^2}{m^P}\right)^{1/(2+P)}. \tag{6.134}$$

For the oscillator case, we have $P = 2$ and $K_2 = k/2$ giving

$$E^{(P=2)} \sim \hbar^{4/4} m^{-2/4}\, k^{2/4} \propto \hbar\sqrt{k/m} \tag{6.135}$$

as expected, while for the linear potential, where $P = 1$ and $K_1 = F$, we find

$$E^{(P=1)} \sim \hbar^{2/3} m^{-1/3} F^{2/3} \sim \left(\frac{\hbar^2 F^2}{m}\right)^{1/3}, \tag{6.136}$$

consistent with Eqn. (6.124).

For the case of the infinite square well, we use the form in Eqn. (6.130) and take the limit of Eqn. (6.134) when $P \rightarrow \infty$ to obtain

$$E^{(P\rightarrow\infty)} \sim \hbar^{2\infty/\infty} m^{-\infty/\infty} (V_0)^{2/\infty} L^{-2\infty/\infty} \longrightarrow \hbar^2 m^{-1} (V_0)^0 L^{-2} \propto \frac{\hbar^2}{mL^2} \tag{6.137}$$

also as expected. We note that the actual value of V_0 in Eqn. (6.130) does not play a role in the $P \rightarrow \infty$ limit and is just a placeholder to ensure dimensional consistency.

Using the same methods, you should be able to find the natural length scale for the general P case to be

$$\rho^{(P)} \propto \hbar^{2/(2+P)} m^{-1/(2+P)} K_P^{-1/(2+P)} \propto \left(\frac{\hbar^2}{mK_P}\right)^{1/(2+P)} \tag{6.138}$$

and confirm that this expression reproduce the results collected in Eqn. (6.125); see **P6.13.** For example, for the ISW, using $K_P = V_0/L^P$, we have

$$\rho^{(P\rightarrow\infty)} \sim \hbar^{2/\infty} m^{-1/\infty} (V_0)^{-1/\infty} L^{\infty/\infty} = \hbar^0 m^0 V_0^0 L^1 = L, \tag{6.139}$$

which is the only logical length scale. Similar expressions for characteristic values of the momentum (p), speed (v), and period (τ) are all derivable using the basic DA methodology; see **P6.13.** We note that we were able to approach the same class of power-law potentials in **P3.4** to extract information on the classical period as a function of the particle mass (m), amplitude (A), and exponent (P). You can also confirm that a non-dimensionalization **S3** approach using the general power-law potential Eqn. (6.131) also gives the P-dependent length scale in Eqn. (6.138).

If we use the **S4** WKB strategy, we can then derive the quantum number dependence (at least in the large n limit) as being the same as that for $\hbar$ in the energy derivation, namely

$$E_n^{(P)} \sim n^{2P/(2+P)}, \tag{6.140}$$

which reproduces the oscillator ($P = 2$), linear potential ($P = 1$), and ISW ($P \rightarrow \infty$) limits. This then gives for the energy and length dependences the forms

$$E_n^{(P)} \propto \left(\frac{n^{2P} \hbar^{2P} K_P^2}{m^P}\right)^{1/(2+P)} \quad \text{and} \quad \rho^{(P)} \propto \left(\frac{n^2 \hbar^2}{mK_P}\right)^{1/(2+P)}. \tag{6.141}$$

The same "trick" can be used to extract the n-dependence of all other dimensional quantities, such as the length scale ρ, as well as that for p, v, τ (momentum, velocity, and period) mentioned above. This strategy was nicely discussed at a pedagogical level by Sukhatme (1973). Quigg and Rosner (1979) provide many other examples of scaling laws (dependences on m, n, P, etc.) in power-law (and other) potentials in the context of quark–antiquark bound states, including providing the next-level correction to the n dependence for three-dimensional problems involving angular momentum (via the l quantum number) where $n \rightarrow n - \gamma_l(P)$ in Eqn. (6.140).

6.3.4 Quantum Coulomb problem: First glimpse of the hydrogen atom

Counting (yet again) our blessings, in the spirit of Wigner (1960), another of the small number of completely soluble model problems in the mathematical physics literature is of direct relevance to the simplest of all atomic systems, the bound state of a proton and electron, namely the hydrogen atom, this time as a three-dimensional problem. Because of the (unexpectedly) high degree of symmetry of this problem, besides conservation of angular momentum (expected for any central potential), there is a hidden symmetry that gives rise to extra degeneracies in the energy spectrum (via the so-called **Lenz–Runge vector**; see, e.g., Robinett 2006 for a discussion). There are also interesting connections (see, e.g., Grant and Rosner 1994) to the other most famous soluble model system, the harmonic oscillator, and the two problems can be morphed into each other by a simple change of variables. Finally, the quantum Coulomb problem can also be couched in the language of **supersymmetric quantum mechanics** (see, e.g., Cooper, Khare, and Sukhatme 2001), adding another layer of provenance, elegance, and richness to this important physically motivated mathematical physics problem.

However, in the spirit followed in this book, we want to explore, to the extent possible, the physics behind this idealized model, or zeroth-order approximation, of the hydrogen atom system, using the methods of DA. We will explore many other aspects of the H-atom system (in Section 7.2.3) involving more dynamical effects, but we begin with the quantum version of a system of two particles interacting only via an electrostatic interaction.

Given that the proton and electron charges are $Q_p = +e$ and $Q_e = -e$, we can begin by writing the (classical) Coulomb potential energy in the form(s)

$$V_C(r) = \frac{Q_p Q_e}{4\pi\epsilon_0 r} = -\frac{e^2}{4\pi\epsilon_0 r} \equiv -\frac{K_e}{r} \qquad \text{where} \qquad K_e \equiv \frac{e^2}{4\pi\epsilon_0}, \tag{6.142}$$

where, for notational simplicity and compactness, we will often use the combination K_e in our analyses. We immediately note that

$$[K_e] = \left[\frac{e^2}{4\pi\epsilon_0}\right] = \mathsf{Q}^2\left(\frac{\mathsf{Q}^2\mathsf{T}^2}{\mathsf{ML}^3}\right)^{-1} = \frac{\mathsf{ML}^3}{\mathsf{T}^2}. \tag{6.143}$$

Since for the purposes of the Schrödinger equation, only the combination K_e appears, we will not require any dimensional constraints from the Q base dimension, using the results from Eqn. (6.143) and the other three relevant $\mathsf{M}, \mathsf{L}, \mathsf{T}$ dimensions. We note that had we been using CGS/Gaussian units, the only change would be to write $K_e = e^2$, but with the same $\mathsf{M}, \mathsf{L}, \mathsf{T}$ dimensions as in Eqn. (6.143). (See Eqn. (10.4) in Section 10.6 for a discussion.)

Given the fact that $m_e << m_p$, we can also use another zeroth-order approximation, one assumed in introductory level studies of the solar system and Kepler's law (since there one has $M_{planet} << M_{sun}$) and assume that the electron mass (m_e) is the relevant one; we will be more careful in Section 7.2.1, where we properly include the **reduced mass** for this two-body problem.

We immediately note that we have encountered a DA problem almost identical to this in **Example 6.1** for the single attractive δ-function, in Eqn. (6.83), where the $-K\delta(x)$ term has the same dimensionality (see Eqn. (6.84), with $[K] = [K_e] = \mathsf{ML}^3/\mathsf{T}^2$) as the $1/r$ function here. We can then use the **S1** type DA results in Eqn. (6.86) directly to write

$$\rho = C_\rho \frac{\hbar^2}{m_e K_e} \qquad \text{and} \qquad E = C_E \frac{m_e K_e^2}{\hbar^2}, \tag{6.144}$$

and despite being in three dimensions, the DA constraints are identical.

The simple **Bohr atom model** for circular orbits of the hydrogen atom makes use of the assumption that angular momentum is quantized, via

$$rp = |\boldsymbol{r} \times \boldsymbol{p}| = L = n\hbar \tag{6.145}$$

and this relation can be used in a manner similar to the **S2** uncertainty principle approach, but it also has a connection to the **S4** de Broglie method. If we associate $\lambda_{dB} = h/p = 2\pi\hbar/p$ and demand that an integral number of wavelengths fit into a circular orbit (periodic boundary conditions here, not infinite wall boundary conditions), we then have

$$2\pi r = n\lambda_{dB} = \frac{2\pi n\hbar}{p} \qquad \text{or} \qquad n\hbar = rp = L, \tag{6.146}$$

which is the **Bohr quantization condition.** Since we have $p = n\hbar/r$, we can then write

$$E = \frac{p^2}{2m_e} - \frac{K_e}{r} \qquad \longrightarrow \qquad E_n(r) = \frac{n^2\hbar^2}{2m_e r^2} - \frac{K_e}{r} \tag{6.147}$$

which is minimized by

$$0 = \frac{dE_n(r)}{dr} = -\frac{n^2\hbar^2}{m_e r^3} + \frac{K_e}{r^2} \qquad \longrightarrow \qquad r_n = n^2\left(\frac{\hbar^2}{m_e K_e}\right) \equiv n^2 a_0, \tag{6.148}$$

where $a_0 \equiv \hbar^2/mK_e$ is the **Bohr radius,** which sets the length scale for all of atomic physics. We see that our DA approach in Eqn. (6.144) has reproduced this important dimensionality, with $C_p = 1$.

The resulting quantized energies from Eqn. (6.147) are given by

$$\begin{aligned} E_n(r_n = n^2 a_0) &= \frac{n^2\hbar^2}{2m(n^2 a_0)^2} - \frac{K_e}{n^2 a_0} \\ &= \frac{K_e^2 m_e}{2n^2\hbar^2} - \frac{K_e^2 m_e}{n^2\hbar^2} \\ E_n &= -\frac{1}{2n^2}\left(\frac{K_e^2 m_e}{\hbar^2}\right), \end{aligned} \tag{6.149}$$

which turns out to be the exact result for the three-dimensional quantum Coulomb problem.

We can even compare the ground state (GS) wave functions for the one-dimensional attractive δ-function potential in Eqn. (6.87) (there is, after all, only one bound state in that case) and that for the quantum Coulomb problem (H-atom approximation), namely

$$\psi_{GS}(x) = \frac{1}{\sqrt{\rho}}\, e^{-|x|/\rho} (1D\ \delta-function) \quad \text{vs.} \quad \psi_{GS}(\boldsymbol{r}) = \frac{1}{\sqrt{\pi a_0^3}}\, e^{-r/a_0}\ \text{(3D Coulomb problem)} \tag{6.150}$$

with obvious similarities. Note, however, the dimensionality of $[\psi_{GS}(\boldsymbol{r})] = \mathsf{L}^{-3/2}$, as required by a three-dimensional wave function.

We can make numerous comparisons to previous results:

- Once again, the quantum number n and Planck's constant $\hbar$ appear together as the combination $n\hbar$, consistent, of course, with the angular momentum relation in Eqn. (6.146), but also very similar to the WKB condition in Eqn. (6.121), at least for large n.
- We note that for $\hbar \to 0$, the ground state energy tends to negative infinity, as there is no uncertainty principle constraint, that is, no "bottom floor" from the potential energy function. Similarly, for $m_e \to \infty$ we have $E_n \to -\infty$.
- We see that the dimensional dependences for power-law potentials (in one dimension) found in Eqns. (6.134) and (6.138) on $K, m, \hbar$ are consistent with the results in Eqn. (6.144) if we use $P = -1$. So despite being in three-dimensions, the DA constraints continue to predict the dependence on physical quantities in meaningful ways.
- We can also (rather blindly) apply Eqn. (6.121) directly to this system, assuming that $x \to r$, $E_n = -|E_n|$, $V(x) \to -K_e/r$, and $n + 1/2 \to n$, to write

$$n\hbar\pi = \int_0^b \sqrt{2m\left(-|E_n| + \frac{K_e}{r}\right)}\, dr, \tag{6.151}$$

where $|E_n| = K_e/b$ is the outer turning point. This turns out to reproduce (see **P6.14**) the standard result.

We can also use these methods to extract the dimensional and quantum number dependence on other physical quantities, though in quantum mechanics such expressions may well be derived more formally via expectation values, as in Eqn. (6.76). We can argue that for momentum, speed, and period (say for a putative semi-classical circuit orbit) we'd have

$$p_n = \frac{n\hbar}{r_n} = \frac{n\hbar}{n^2(\hbar^2/K_e m_e)} = \frac{K_e m_e}{n\hbar} \tag{6.152}$$

$$v_n = \frac{p_n}{m_e} = \frac{K_e}{n\hbar} \tag{6.153}$$

$$\tau_n \equiv \frac{2\pi r_n}{v_n} = 2\pi \frac{n^3\hbar^3}{K_e^2 m_e}. \tag{6.154}$$

We can use these expressions to make an analogy between these (clearly) quantum mechanical results and one from classical gravitational physics, namely **Kepler's third law**. For the gravitational system, we have

$$V_G = -\frac{GMm}{r} \quad \longrightarrow \quad a^3 = \left(\frac{GM}{4\pi^2}\right) T^3, \tag{6.155}$$

where a, T are the radius (actually semi-major axis of a elliptical orbit) and period of the planetary orbits, respectively. If we make the connection

$$V_G = -\frac{GMm}{r} \quad \longrightarrow \quad V_C = -\frac{K_e}{r}, \tag{6.156}$$

we can associate $GMm \leftrightarrow K$ or $GM \to K_e/m$ (using the mass of the lighter object) and with a, T replaced by r_n, τ_n to guess that

$$r_n^3 \stackrel{?}{=} \left(\frac{K_e}{4\pi^2 m_e}\right)\tau_n^2 \qquad \text{(quantum Kepler's law?)}$$

$$\left(\frac{n^2\hbar^2}{K_e m_e}\right)^3 \stackrel{?}{=} \left(\frac{K_e}{4\pi^2 m_e}\right)\left(2\pi\frac{n^3\hbar^3}{K_e^2 m_e}\right)^2 = \frac{4\pi^2}{4\pi^2}\left(\frac{K_e}{m_e}\right)\left(\frac{n^6\hbar^6}{K_e^4 m_e^2}\right)$$

$$\frac{n^6\hbar^6}{K_e^3 m_e^3} \stackrel{!}{=} \frac{n^6\hbar^6}{K_e^3 m_e^3}. \tag{6.157}$$

A rotating charge would induce a magnetic moment, classically given by $\mu = IA$ where I, A are the current and area. For this semi-classical description of the hydrogen atom, we associate $I = e/\tau_n$ and $A = \pi r_n^2$, so that using the results of Eqns. (6.148) and (6.154), we'd have

$$\mu = \left[e\left(\frac{K^2 m_e}{2\pi n^3\hbar^3}\right)\right]\left[\pi\left(\frac{n^2\hbar^2}{m_e K}\right)^2\right] = \frac{e}{2m_e}(n\hbar) = \left(\frac{e\hbar}{2m_e}\right)n \equiv \mu_B n, \tag{6.158}$$

where we have associated $\mu_B \equiv e\hbar/2m_e$. This is one example of the more general notation for the **Bohr magneton** for any charged particle (call it X), namely

$$\mu_X = \frac{q_X\hbar}{2m_X}. \tag{6.159}$$

Since we also associate the angular momentum of the electron as $L = n\hbar$, this result also shows that

$$\frac{\mu}{L} = \frac{e}{2m_e}, \tag{6.160}$$

much like the classic gyromagnetic ratio of a spinning object, as in Eqn. (4.41).

We will revisit the "real" hydrogen atom, examining the electron–proton (and related) bound state system(s), including more physical phenomena, in Sections 7.2 and 7.3.

6.4 Dimensionless phases in quantum mechanics

The simplest solution of the one-dimensional Schrödinger equation, for the case of a free-particle (so with $V(x) = 0$), namely

$$\hat{H}\psi(x,t) = \hat{E}\psi(x,t) \qquad \text{or more explicitly} \qquad -\frac{\partial^2\psi(x,t)}{\partial x^2} = -i\hbar\frac{\partial\psi(x,t)}{\partial t}, \tag{6.161}$$

is

$$\psi_{free}(x,t) \propto e^{i(px-Et)/\hbar} \tag{6.162}$$

where $p^2/2m = E$. We might naturally refer to the $(px - Et)/\hbar$ combination as a phase, but we will extend this definition to the dimensionless argument of any quantum wave function, including ones representing tunneling phenomena as well.

An approximate (and we stress that) solution, one which is used in the WKB approximation scheme mentioned earlier for the Schrödinger equation, but now including a non-zero potential energy, so that $p^2/2m = E - V(x)$, is

$$\psi(x) \sim \psi_0 \exp\left(\pm i \frac{\int^x p(x)\, dx}{\hbar}\right)$$
$$\sim \psi_0 \exp\left(\pm i \frac{\sqrt{2m}}{\hbar} \int^x \sqrt{E - V(x)}\, dx\right), \qquad (6.163)$$

which gives an oscillatory solution (so, with what we call a phase in the exponential) in the region where $E > V(x)$, namely in the classically allowed regions. This form captures much of the change in **phase** of the wave function as it traverses regions of changing potential and is the basis of the WKB extension of the "fitting de Broglie waves" argument **S4** discussed in Section 6.3. Taking appropriate combinations of the $\pm i$ solutions gives the real (meaning not complex) "wiggly" waveforms of the type plotted in many quantum mechanics text, where it's easy to see the correlations between the x-location in the classical potential, and the oscillatory behavior of $\psi(x)$, namely

- When $p^2/2m = E - V(x)$ is small, so, near the turning points of the potential, where a classical particle would be slowing down and reversing direction, the momentum (and classical speed) are small and the solutions are less "wiggly," one can also invoke the de Broglie connection of $\lambda_{dB} = h/p$ to see the correlation between small p and large wavelength.
- When $p^2/2m = E - V(x)$ is large, p and v are large, the de Broglie wavelength is smaller, hence more "wiggles."

For the case when $E < V(x)$, where quantum mechanical tunneling can occur, this solution is easily transformed to "ordinary" exponentials, namely

$$\psi(x) \sim \exp\left(\pm \frac{\sqrt{2m}}{\hbar} \int^x \sqrt{V(x) - E}\, dx\right), \qquad (6.164)$$

since we switch from $(E - V(x)) \rightarrow (V(x) - E)$ giving a factor of i. We will also describe the (necessarily) dimensionless argument of the exponential in this case as a generalized phase and therefore consider both oscillatory and tunneling tunneling type behaviors in a similar way. We proceed in Section 6.4.1 by reviewing the second case to extract information on the large $|x|$ behavior of quantum wave functions, which can then be used to describe quantum tunneling rates, then in Section 6.4.2 turn to the observability of dimensionless phases for unbound wave functions.

We are interested in comparing well-known, but qualitatively different, theoretical predictions, confirmed by seminal experiments, from both categories, especially emphasizing the role that $\hbar$ plays, given that it always appears in the denominator of the phase. That fact is the source of the "fixed point" dependence of such phases on $\hbar^{-1}$, which can then be used to constrain dimensional analyses.

6.4.1 Large $|x|$ behavior and quantum tunneling

The approximate solution in Eqn. (6.164) can be used, for example, to extract information on the large x (tunneling) behavior of the quantum wave function in the classically disallowed regions, such as for problems like the harmonic oscillator. In that specific case, we have $V(x) = kx^2/2$ and for $V(x) >> E$ (or x far beyond the classical turning points) we write

$$\psi(x) \sim \exp\left(\pm\frac{\sqrt{2m}}{\hbar}\int^{x}\sqrt{kx^2/2}\,dx\right) \sim \exp\left(\pm\frac{\sqrt{mk}}{\hbar}\int^{x}x\,dx\right)$$
$$\sim \exp\left(\pm\frac{\sqrt{mk}}{\hbar}\frac{x^2}{2}\right) \propto \exp\left(-\frac{x^2}{2\rho^2}\right), \tag{6.165}$$

where we have used our definition of $\rho = (\hbar^2/mk)^{1/4}$ for the harmonic oscillator, and chosen the physically appropriate (exponentially decaying and not growing) minus sign for the solution.

We can approach this problem in a different way, noting that for a dimensionless ratio involving x to appear as the argument of a functional form (exponential or not) in the tunneling regime, we will naturally find the ratio $\Pi = x/\rho$ (or powers thereof) where ρ is the natural length scale of the bound state problem. But if we also require, as the solutions in Eqn. (6.163) or (6.164) imply, that the argument Π must have a single power of $\hbar$ in the denominator, then we are more constrained, and require

$$\textit{dimensionless} = \Pi = \frac{m^{\alpha}k^{\beta}x^{\gamma}}{\hbar} \qquad \text{implying that} \qquad [\hbar] = [m]^{\alpha}[k]^{\beta}[x]^{\gamma}, \tag{6.166}$$

or $\alpha = \beta = 1/2$ and $\gamma = 2$, consistent with the functional dependence of Eqn. (6.165). You should be able to determine the tunneling behavior of the solution of the linear potential on the half-plane, defined by the potential in Eqn. (6.122), using both the WKB tunneling solution approach and this DA method, to show that for large x one has

$$\psi(x) \sim \exp\left[-\frac{2}{3}\left(\frac{x}{\rho}\right)^{3/2}\right], \tag{6.167}$$

with the appropriate combination of $\hbar, F$, and m giving the length scale ρ in Eqn. (6.124). For more details on the properties of the linear potential problem, and the mathematics of the solutions thereof, see Belloni and Robinett (2014) and Valleé and Soares (2010), respectively.

One can, in fact, use both approaches to explore the tunneling dependence in a general power-law potential of the form in Eqn. (6.130) to find that $\psi(x) \sim \exp[-c(x/\rho_P)^{(2+P)/2}]$. It's important to note that the exponential decay of the probability density scales like the relevant value of ρ for the potential, and that in general $\rho^{(P)} \propto \hbar^{2/(2+P)}$, which vanishes as $\hbar \to 0$ in the limit of classical physics, hence no quantum mechanical tunneling in that limit.

For bound state problems such as the harmonic oscillator, no measurement is taking place to probe whether the wave function extends beyond the classical turning points, but for several tractable geometries, one can determine the **tunneling probability**[8] for the escape from/emission

[8] For a review, see Merzbacher (2002).

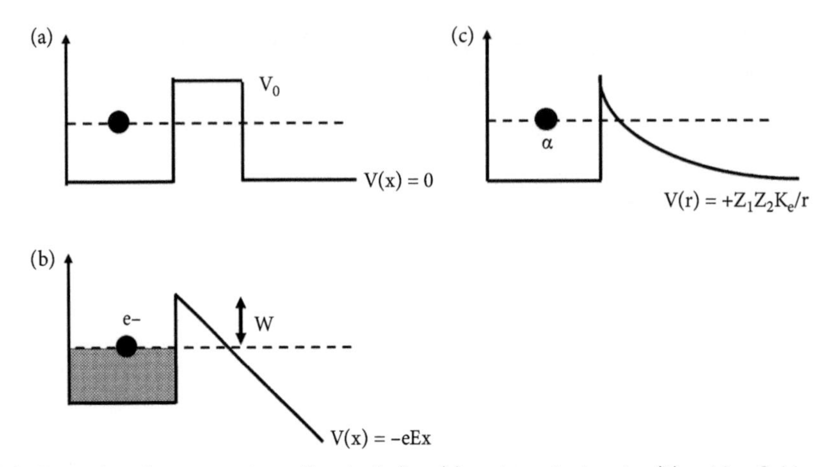

Fig. 6.3 Examples of quantum tunneling, including (a) rectangular barrier, (b) cold or field emission of electrons from a metal due to an external electric field, and (c) schematic representation of α-particle decay from a heavy nucleus.

of a particle from the system, as shown in Fig. 6.3. For example, in Fig. 6.3(a), the Schrödinger equation for the case when $V_0 > E > 0$ is

$$\frac{d^2\psi(x)}{dx^2} = +\frac{2m(V_0 - E)}{\hbar^2}\psi(x) = +\kappa^2\psi(x) \tag{6.168}$$

with solutions $\psi(x) \propto \exp(\pm\kappa x)$ and the corresponding length scale for tunneling given by $\rho = 1/\kappa = \hbar/\sqrt{2m(V_0 - E)}$. One can understand this connection dimensionally from an uncertainty principle argument which proceeds along the following lines:

- The uncertainty in energy during the tunneling is $\Delta E \sim (V_0 - E)$ and the corresponding uncertainty in time, during which conservation of energy is seemingly violated, is $\Delta t \sim \hbar/\Delta E$.
- If we associate a speed with $mv^2/2 \sim \Delta E$, we have $v \sim \sqrt{2\Delta E/m}$, suggesting that the particle can "travel" a distance

$$\Delta x = v \cdot \Delta t = \left(\sqrt{\frac{2\Delta E}{m}}\right)\left(\frac{\hbar}{\Delta E}\right) \sim \frac{\hbar}{\sqrt{m\Delta E}}, \tag{6.169}$$

which has the right dimensions.

If the potential barrier is of size D, then the wave function suppression factor is $e^{-\kappa D} \equiv e^{-G}$ and the corresponding reduction in probability would be $e^{-2\kappa D} = e^{-2G}$. The notation G is suggested by a case to be considered below (tunneling of α-particles from heavy nuclei, through a Coulomb barrier) first considered by Gamow (1928) and Gurney and Condon (1929).

A more realistic example from the early days of quantum mechanics came from the study of **field emission** (*FE*) of electrons from metals subject to intense (static) electric fields. A cartoon in Fig. 6.3(b) shows a filled "electron sea" (shaded area), with an energy W (the work function, described in Eqn. (6.56)), needed to initiate the photoelectric effect. An external electric field (***E***) can "turn over" the potential and provide a finite (triangular-shaped) barrier given by $V(x) = -eEx$, through which the electrons can tunnel. The integral required in the tunneling wave function is

$$G_{FE} = \frac{\sqrt{2m_e}}{\hbar}\int_0^{a=W/eE}\sqrt{W - eEx}\,dx$$

$$G_{FE} = \frac{\sqrt{2m_e W}}{\hbar}\left(\frac{W}{eE}\right)\int_0^1 \sqrt{1-y}\,dy = \frac{2}{3}\frac{\sqrt{2m_e W^3}}{eE\hbar} \equiv \frac{E_0}{E}, \tag{6.170}$$

where $E_0 \propto \sqrt{mW^3}/e\hbar$ sets the scale the electric field required to see an effect. This type of analysis was first performed by Fowler and Nordheim (1928) (again, in the early days of quantum mechanics) to help explain data obtained by Millikan and Eyring (1926), who plotted the electron current as a function of the applied E field and found that it could be fit well by $\ln(I) = a - b/E$, suggesting the $I(E) = I_0 e^{-E_0/E}$ form.

Returning to a DA approach, we can ask what combinations of m, W, e, E would scale correctly with $\hbar$ to give

$$dimensionless = G_{FE} \propto \frac{m_e^{\alpha}W^{\beta}e^{\gamma}E^{\delta}}{\hbar} \qquad \text{requiring} \qquad [\hbar] = [m_e]^{\alpha}[W]^{\beta}[e]^{\gamma}[E]^{\delta} \tag{6.171}$$

and matching M, L, T, Q dimensions, gives $\alpha = 1/2, \beta = 3/2, \gamma = \delta = -1$, consistent with the theoretical derivation. One could also explore the dimensional spelling of E_0 in a similar way (**P6.16**).

Another early success of quantum mechanics was the quantitative prediction of the decay rates for heavy nuclei via α emission. A heavy nucleus with atomic number A and charge Z can decay by the emission of a ${}^4_2He$ nucleus (or α-particle) via the process ${}^A_ZX \rightarrow {}^{A-4}_{Z-2}Y + {}^4_2He$. A simple model (again, see Gamow 1928) has the α particle inside a sphere of radius R_A (the nuclear radius), which has to the tunnel through the Coulomb barrier (or CB), $V(r) = +Z_1Z_2K_e/r$, as shown in Fig. 6.3(c). For α decay we'd have $Z_1 = Z - 2$ and $Z_2 = +2$ and use the mass of the alpha particle, m_α, but this model has been applied to decays involving single protons (so $Z_2 = +1$) and heavier decay products ($Z_2 > +2$) as well. The appropriate G factor is then

$$G_{CB} = \frac{\sqrt{2m_\alpha}}{\hbar}\int_{R_A}^{b=Z_1Z_2K/E_\alpha}\sqrt{\frac{Z_1Z_2K_e}{r} - E_\alpha}$$

$$= \pi\sqrt{\frac{2m_\alpha}{E_\alpha}}\left(\frac{Z_1Z_2K_e}{\hbar}\right), \tag{6.172}$$

and the tunneling probability $e^{-2G_{CB}}$ is sometimes written as $e^{-\sqrt{E_G/E_\alpha}}$ where E_G is the **Gamow energy**, which sets the scale for the probability of the decay. The observed decay rates, R, (which are proportional to this factor) of many nuclei can be well described by an expression of the form $\ln(R) = -a/\sqrt{E_\alpha} + b$, just as suggested by this relation, which was first observed by Geiger and Nuttall (1911). This exponentially sensitive tunneling factor is also of importance in the inverse process of nuclear fusion, especially in stellar calculations. You're asked to explore the structure of G_{CB} in more depth in **P6.17**

6.4.2 Wave function phases demonstrated by interference experiments

Returning to the oscillatory WKB solutions in Eqn. (6.163), we explore next how theory and experiment can probe the nature of the phases that appear there, given that formally the probability density determined by $|\psi(x)|^2$ would have factors of $\psi(x)\psi^*(x) \propto e^{\pm i\,phase}e^{\mp i\,phase} = 1$, with the phase information seemingly disappearing.

Historically, one of the most important methods of probing wave-like behavior is to make use of **phase differences** between two sources, a technique pioneered by Thomas Young (1807) in his double-slit experiments, which provided evidence for the wave nature of light. In this section, we discuss four such experiments, illustrated (in cartoon fashion) in Fig. 6.4, where S and O stand for source and observer, while P and Q denote two different paths, along which the particles acquire differing amount of phase according to Eqn. (6.163). The common thread in many such experiments is that the quantum wave function at the position of an observer O finds contributions from the two paths, given by

$$\begin{aligned}\psi_O(x) &\sim \psi_S^1 e^{i\theta_1(x)} + \psi_S^2 e^{i\theta_2(x)}\\ &\sim \left(\psi_S^1 e^{i(\theta_1(x)-\theta_2(x))} + \psi_S^2\right) e^{i\theta_2(x)}\\ &\sim \left(\psi_S^1 \underbrace{e^{i\Theta(x)}} + \psi_S^2\right) e^{i\theta_2(x)},\end{aligned} \tag{6.173}$$

where $\Theta(x) \equiv \theta_1(x) - \theta_2(x)$ is the phase difference. For values of $\Theta(x) = 2n\pi$ ($(2n+1)\pi$) the complex exponential of the phase difference gives a factor of $+1$ (-1) corresponding to constructive (destructive) interference.

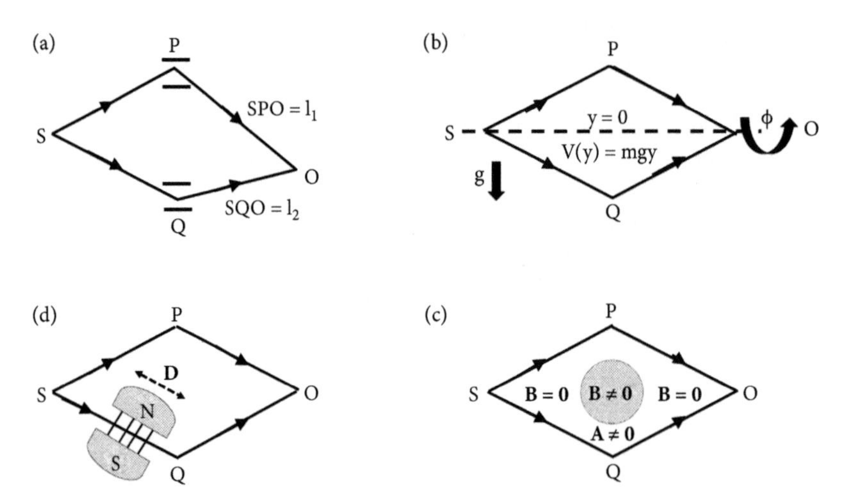

Fig. 6.4 Interference experiments measuring phase differences induced by (a) geometrical path lengths, (b) gravitational effect, (c) magnetic moment interactions with $\boldsymbol{B}$ fields, and (d) the Aharonov–Bohm effect.

We first discuss an interference experiment where a beam of particles (see Table 6.1 for examples of systems studied) illustrated in Fig. 6.4(a) impinges on the two slits labeled P and Q and recombines at the observer O. For a free particle, the phase difference picked up over a distance l will be

$$phase = \frac{1}{\hbar}\int^{x} p(x)\,dx = \frac{pl}{\hbar} = kl \tag{6.174}$$

where $k = p/\hbar = 2\pi/\lambda$ is the wave-number. The wave function observed at O will then be

$$\psi_O = \psi_S^1 e^{ikl_1} + \psi_S^1 e^{ikl_2} = \left(\psi_1^S e^{i\Theta} + \psi_2^S\right) e^{ikl_2}, \tag{6.175}$$

where $\Theta \equiv k(l_1 - l_2) = 2\pi\Delta l/\lambda$ is the (dimensionless) phase difference, so that path differences equal to a half-integral (integral) number of wavelengths will give rise to destructive (constructive) interference. Two-slit (and multiple-slit!) experiments (see, e.g., Jönnson 1961, 1974) observing such diffraction effects for electrons confirmed their wave properties and have been described by many authors in a variety of very laudatory ways:

- Beyond the technical *tour de force*, however, lies the conceptual simplicity of a real, pedagogically clean, fundamental experiment whose description and study can now enrich and simplify the learning of quantum physics (French and Taylor 1974).
- Beyond its pedagogic value in illuminating the wave–particle duality in modern physics, this article is a classic in the way that it reports the "experimental vise" that all investigators face (Jönnson 1974).
- As the winner of a "most beautiful experiment" survey, see Crease (2002), who also says "But uniquely among the top 10, the most beautiful experiment in physics . . . does not have a name of an individual associated with it".

Perhaps even more fundamentally, more advanced versions of these experiments also provided direct evidence for the statistical nature of the observation process, with the observed interference patterns being "built up", electron by electron (see e.g., Merli, Missroli, and Pozzi 1976, and Tonomura et al. 1989). Two-slit diffraction experiments involving increasingly large and complex molecules continue to be performed, with the largest object so far seen demonstrating such effects being a biomolecule with atomic mass $>25,000\,amu$ (Fein et al. 2019).

A second example, visualized in Fig. 6.4(b), is based on work by Colella, Overhauser, and Werner (1975) in a paper entitled "*Observation of Gravitationally Induced Quantum Interference*". We are to imagine that a beam of neutrons is split and traverses two paths in which the particles experience differing gravitational potential energy ($V(x) = mgy(x)$), hence acquiring non-trivial quantum phases (not through geometrical path length differences) and then recombine. For simplicity, we assume that there are similar upper and lower physical path lengths, and that the SO axis defines $V(y) = mgy = 0$.

The phase suggested by Eqn. (6.163) would be

$$
\begin{aligned}
upper\,phase = \theta_1(x) &= \frac{1}{\hbar}\int^x p(x)\,dx \\
&= \frac{\sqrt{2m}}{\hbar}\int^x \sqrt{E - V(x)}\,dx \\
&= \frac{\sqrt{2mE}}{\hbar}\int^x \left(1 - \frac{mgy(x)}{E}\right)^{1/2} dx \\
&= \frac{\sqrt{2mE}}{\hbar}\left(l_1 - \frac{mg}{2E}\underbrace{\int^x y(x)\,dx}_{\text{upper area}}\right) \\
\theta_1(x) &= kl_1 + \frac{\sqrt{2mE}}{\hbar}\left(\frac{mg}{2E}\right)a_{SPO},
\end{aligned}
\tag{6.176}
$$

while the corresponding phase for the lower path is

$$
lower\,phase = \theta_2(x) = kl_2 - \frac{\sqrt{2mE}}{\hbar}\left(\frac{mg}{2E}\right)a_{SQO} \tag{6.177}
$$

since we've assumed that the upper/lower paths correspond to $y > 0$ and $y < 0$, respectively. Combining these two results we have for the component of the *phase difference* depending on gravity as

$$
\Theta_g \sim \frac{\sqrt{2mE}}{\hbar}\frac{mg}{2E}(a_{SPO} + a_{SQO}) \sim \frac{m^2 g a_{total}}{\hbar p} = \frac{mg a_{total}}{\hbar v}, \tag{6.178}
$$

where $mv = p = \sqrt{2mE}$ is the momentum. You should be able to confirm that this combination is dimensionless.

Since the upper and lower path lengths may not be exactly equal, giving a $\Theta = k(l_1 - l_2)$ term as well, to avoid systematic errors in trying to observe such a phase difference, the experimenters (Colella et al. 1975) rotated the apparatus about the *SO*-axis as shown, so that Θ_g not only had the magnitude determined by the dimensionless combination noted above, but also a $\cos(\phi)$ dependence, so the interference pattern could be observed to change periodically with ϕ and demonstrate the effect described here. That angular dependence can be incorporated more directly into Eqn. (6.178) by expressing the total area as a vector $\boldsymbol{a}_{total} = a_{total}\hat{\boldsymbol{n}}$, with the associated direction being the normal to the plane, and generalizing $g a_{total}\cos(\phi) \rightarrow \boldsymbol{g}\cdot\boldsymbol{a}_{total}$.

Another seminal interference experiment employing neutrons was suggested by Bernstein (1967) and Aharonov and Susskind (1967) and performed by Werner et al. (1975) and Rauch et al. (1975).[9] In this system, the relevant interaction potential is no longer $V(x) = mgy(x)$, but rather the magnetic interaction $V(\boldsymbol{r}) = -\boldsymbol{\mu}\cdot\boldsymbol{B}$ of the neutron's intrinsic magnetic moment (here labeled

[9] See Rauch and Werner (2015) for a comprehensive and authoritative reference.

$\boldsymbol{\mu}$ to avoid confusion with the particle mass m), and an external magnetic field ($\boldsymbol{B}$). Because of the vector (so-called **spinor**) description of the spin-1/2 neutrons quantum wave function, the expected interference pattern was predicted to correspond to a phase difference of 4π, instead of the typical 2π associated with pure rotations. The experiment is visualized (again, in cartoon fashion) in Fig. 6.4(c), with the phase difference arising due to the $\boldsymbol{\mu} \cdot \boldsymbol{B}$ interaction acting over a distance D. You should be able to reproduce the arguments leading to Eqn. (6.178) or using analogy and substituting $mgy \longrightarrow \mu B$, or even pure DA, to show that the phase difference (Θ_s with s for spinor) goes like

$$\Theta_s = \frac{\mu BD}{\hbar v}, \tag{6.179}$$

where v is again the neutron speed.

Figure 6.4(d) illustrates a fourth and final (at least until the problems) example of a historically important interference experiment involving electrons again. We imagine a beam of electrons split at point S and taking different paths to the observer O around a region of magnetic field B encircled by the $SPOQ$ area. The magnetic field is carefully constrained to be in the cylindrical region (in one early experiment, confined in a magnetic whisker) so that the electrons are never actually in a region of non-zero B field at any time. Despite that, an interference effect is still observed and one can again try to use analogy (and constraints from DA) to predict the phase difference, Θ_{AB}.

Using the result in Eqn. (6.178), imagining that this result must also depend on the enclosed area, we might take, if only on dimensional grounds the substitution $mg \longrightarrow qvB$ to guess that

$$\Theta_g = \frac{mga_{tot}}{\hbar v} \qquad \longrightarrow \qquad \Theta_{AB} = \frac{(qvB)a_{tot}}{\hbar v} = \frac{q(BA)}{\hbar} = \frac{q\Phi_B}{\hbar}, \tag{6.180}$$

where Φ_B is the magnetic flux encircled by the $SPOQ$ area, completely contained, by construction, in the cylindrical volume.

It is interesting to note that this is actually the right answer, despite the use of a very questionable analogy—there is, after all, absolutely no $q\boldsymbol{v} \times \boldsymbol{B}$ force present as the particle never enters the region of non-zero magnetic field. We include this "red herring" argument as we think it's an example of where the very real constraints of DA limit the possible outcomes of the dependence of physical observables on dimensional quantities, with the limited alphabet of M, L, T, Q, Θ and internal consistency sometimes leading us along an otherwise dark path to the proverbial "light at the end of the tunnel" (wrong approach or physics, but right answer, since dimensions demand it.)

The actual physics involved in the real derivation is nicely suggested by the title of a seminal paper by Aharonov and Bohm (1959), namely *Significance of Electromagnetic Potentials in the Quantum Theory*. The relevant phase difference experienced by a particle of charge q in the presence of the EM vector potential $\boldsymbol{A}(\boldsymbol{r})$ is given by $\theta(x) = (q/\hbar)\int^x \boldsymbol{A} \cdot d\boldsymbol{r}$, so that even when $\boldsymbol{B} = 0$ a phase contribution can be built-up in a region of non-zero $\boldsymbol{A}$.

The phase difference between the upper and lower path is then given by

$$\begin{aligned}
\Theta_{AB} &= \frac{q}{\hbar}\int_{SPO} \boldsymbol{A}\cdot d\boldsymbol{r} - \frac{q}{\hbar}\int_{SQO} \boldsymbol{A}\cdot d\boldsymbol{r} = \frac{q}{\hbar}\int_{SPO} \boldsymbol{A}\cdot d\boldsymbol{r} + \frac{q}{\hbar}\int_{OQS} \boldsymbol{A}\cdot d\boldsymbol{r} \\
&= \frac{q}{\hbar}\oint_{SQOPS} \boldsymbol{A}\cdot d\boldsymbol{r} = \frac{q}{\hbar}\oint_{SQOPS} (\boldsymbol{\nabla}\times\boldsymbol{A})\cdot d\boldsymbol{S} \\
&= \frac{q}{\hbar}\oint_{SQOPS} \boldsymbol{B}\cdot d\boldsymbol{S} \\
\Theta_{AB} &= \frac{q\Phi_B}{\hbar}.
\end{aligned} \tag{6.181}$$

In deriving this relation, we have made use of a non-trivial vector calculus relation (that the line integral about a closed loop of a vector function can be expressed as the integral of its curl over the enclosed area, or the **Kelvin–Stokes theorem**) and the relation between the vector potential and the magnetic field from Eqn. (1.65), namely that $\boldsymbol{\nabla}\times\boldsymbol{A} = \boldsymbol{B}$. But perhaps more importantly, we stress that the limited number of appropriate dimensional combinations of the basic experimental quantities $(q, B, a_{tot}, \hbar)$ constrain the final result in Eqn. (6.181), perhaps so much that we can intuit the correct dependence, even having followed the wrong path to do so. This effect was first observed experimentally by Chambers (1960) and perhaps most convincingly by Tonomura et al. (1986).

We encourage you to explore in the problems a number of other measurable phase difference effects (for neutrons and electrons), but we choose to close our discussion of this topic with the **Aharonov–Bohm** (hence the *AB* subscript) effect just considered.[10]

6.4.3 Time-development of quantum wave functions as a dimensionless phase

The discussions of quantum mechanical phases in the last two sections have focused on the spatial dependence of wave functions, but equally important dimensionless phases appear in the temporal development of solutions of the Schrödinger equation (or SE). The most compact statement of the space- and time-dependence of the quantum wave function $\psi(x,t)$ (the time-dependent Schrödinger equation) in one dimension can be written as

$$\hat{H}\psi(x,t) = \hat{E}\psi(x,t) \quad \Longrightarrow \quad \left(-\frac{\hbar^2}{2m}\frac{\partial^2}{\partial x^2} + V(x)\right)\psi(x,t) = i\hbar\frac{\partial}{\partial t}\psi(x,t), \tag{6.182}$$

and if we assume the simple time-dependence

$$\psi(x,t) = \psi_n(x)e^{-iE_nt/\hbar} \quad \Longrightarrow \quad \left(-\frac{\hbar^2}{2m}\frac{\partial^2}{\partial x^2} + V(x)\right)\psi_n(x) = E_n\psi_n(x), \tag{6.183}$$

[10] We are happy to include this example, not only because of its theoretical impact, but partly to acknowledge and honor the fact that David Bohm, who made many contributions to fields as diverse as quantum theory, plasma physics, neuroscience, and even philosophy, was a BS graduate (and ΣΠΣ inductee) of Penn State University (when it was still called a College) in 1939, an institution where this author has happily spent his entire faculty career.

we obtain the time-independent Schrödinger equation. The $\psi_n(x)$ are the energy-eigenfunction solutions, with quantized energies E_n, determined by the form of the potential energy function used in the differential equation, characterized by the quantum number n. The probability densities corresponding to these special solutions (recall Eqn.(6.66)) satisfy

$$P_n(x,t) = |\psi_n(x,t)|^2 = \left[\psi_n^*(x)e^{+iE_nt/\hbar}\right]\left[\psi_n(x)e^{-iE_nt/\hbar}\right] = |\psi_n(x)|^2 = P_n(x), \tag{6.184}$$

which are actually independent of time, hence the name **stationary state.**

Much like the examples in the last section, the importance of the exponential time-dependence of the solutions is most easily seen in "interference" behavior. Since the Schrödinger equation is a linear equation, the sum (or linear combination) of two (or more) solutions is also a solution, and the simplest such generalization would represent a two-state (or 2S) system, with quantized energies $E_2 > E_1$, for example,

$$\psi^{(2S)}(x,t) = \frac{1}{\sqrt{2}}\psi_1(x)e^{-iE_1t/\hbar} + \frac{1}{\sqrt{2}}\psi_2(x)e^{-iE_2t/\hbar}, \tag{6.185}$$

where the factor of $1/\sqrt{2}$ is to ensure that the wave function is properly normalized, and is an example of a more general solution considered below. The corresponding probability density (assuming that the $\psi_{1,2}(x)$ are themselves real) can be written as

$$P^{(2S)}(x,t) = |\psi^{(2S)}(x,t)|^2 = \frac{1}{2}\left[\psi_1(x)^2 + \psi_2(x)^2 + 2\psi_1(x)\psi_2(x)\cos\left(\frac{\Delta Et}{\hbar}\right)\right], \tag{6.186}$$

where $\Delta E = E_2 - E_1$, which does exhibit non-trivial time-dependence. In this case, we have an interference effect in time, not in space, akin to Eqn. (6.173).

The most general time-dependent solution to a one-dimensional problem would consist of a linear combination of all energy-eigenstates, namely

$$\psi(x,t) = \sum_{n=1}^{\infty} a_n\psi_n(x)e^{-iE_nt/\hbar}, \tag{6.187}$$

where we formally require that $\sum_n |a_n|^2 = 1$ for normalization purposes (which was clearly satisfied for the example in Eqn. (6.185).) While such an expression may well seem like pure formalism, it's possible to excite wave packets (e.g., via ultra-short laser pulses) for real physical atomic and molecular systems with a relatively narrow spread of quantum numbers, centered around an $n_0 >> 1$ value. In such cases, it makes sense to expand the energy eigenvalues about the peak of the a_n distribution, namely

$$E(n) = E(n_0) + \frac{dE}{dn}(n_0)(n-n_0) + \frac{1}{2}\frac{d^2E}{dn^2}(n_0)(n-n_0)^2 + \cdots. \tag{6.188}$$

In that limit, the wave function can be written in the form

$$\begin{aligned}\psi(x,t) &\approx \sum_{n=1}^{\infty} a_n\psi_n(x)e^{-iE(n_0)t/\hbar}\, e^{-iE'(n_0)(n-n_0)t/\hbar}\, e^{-iE''(n_0)(n-n_0)^2t/2/\hbar} \\ &= e^{-iE(n_0)t/\hbar}\sum_{n=1}^{\infty} a_n\psi_n(x)e^{-2\pi i(n-n_0)t/T_{cl}}\, e^{-2\pi i(n-n_0)^2t/T_{rev}},\end{aligned} \tag{6.189}$$

where we have defined two important time scales, the **quantum period** T_{qp} (which we'll soon associate with the classical period) and the **quantum revival time** T_{rev} via

$$T_{qp} = \frac{2\pi\hbar}{|E'_n(n_0)|} \qquad \text{and} \qquad T_{rev} = \frac{4\pi\hbar}{|E''_n(n_0)|}. \tag{6.190}$$

The phase in the prefactor due to the $E(n_0)$ term plays no essential role, as it's common to all terms and factors out of the sum, while the term linear in $n - n_0$ (if it was the only part of the temporal phase) would mean that the wave function would satisfy

$$\begin{aligned}\psi(x, t + T_{qp}) &= e^{-iE(n_0)(t+T_{qp})/\hbar} \sum_{n=1}^{\infty} a_n \psi_n(x) e^{-2\pi i(n-n_0)(t+T_{qp})/T_{qp}} \\ &= e^{-iE(n_0)T_{qp}/\hbar} \left[e^{-iE(n_0)t/\hbar} \sum_{n=1}^{\infty} a_n \psi_n(x) \underbrace{\left(e^{-2\pi i(n-n_0)}\right)}_{=1} e^{-2\pi i(n-n_0)t/T_{qp}} \right] \\ &= e^{-iE(n_0)T_{qp}/\hbar}\, \psi(x,t)\end{aligned} \tag{6.191}$$

(since $e^{-2\pi i(n-n_0)} = 1$), so that the probability density, $P(x, t + T_{qp}) = |\psi(x, t + T_{qp})|^2 = |\psi(x,t)|^2 = P(x,t)$ is periodic with period T_{qp}.

To further justify the association of T_{qp} with the classical periodicity, let us consider the case of the harmonic oscillator, where the quantized energies are given by

$$E_n = (n + 1/2)\hbar\sqrt{\frac{k}{m}} = (n + 1/2)\hbar\omega, \tag{6.192}$$

where m, k are the mass and spring constant of the system, and $\omega = \sqrt{k/m}$ is the classical vibrational frequency. Using the definition from Eqn. (6.190), we have

$$\frac{dE}{dn} = E'(n) = \hbar\omega \qquad \text{giving} \qquad T_{qp} = \frac{2\pi\hbar}{|E'_n|} = \frac{2\pi\hbar}{\hbar\omega} = \frac{2\pi}{\omega} = T_{cl}, \tag{6.193}$$

which is indeed the classical period. You're asked to explore similar connections to other famous textbook-level quantum systems in the problems, namely the infinite square well in **P6.22**, the quantum-Coulomb system in **P6.23** (also comparing to experiment), and the general power-law potential in **P6.24**.

The less-familiar T_{rev} term refers to the existence of **quantum wave packet revivals**, where initially localized solutions evolve in time, at first dominated by the T_{qp} or classical periodicity time scale, but then dephasing due to the higher-order terms in Eqn. (6.188). But because of the $(n - n_0)^2$ dependence, $\psi(x,t)$ can still reform (approximately) at a later time, $T_{rev} >> T_{cl}$. The simplest example of such behavior is the infinite well (or ISW) of width L, where

$$E(n) = \frac{\hbar^2\pi^2 n^2}{2mL^2}, \qquad E''(n) = \frac{\hbar^2\pi^2}{mL^2}, \qquad \text{and} \qquad T_{rev} = \frac{4mL^2}{\hbar\pi^2}. \tag{6.194}$$

One of the first discussions of this behavior appeared in a pedagogical article by Segre and Sullivan (1976) in the context of just this (ISW) model system and so may seem of only academic interest. But experimental realizations of quantum wave packet revivals have been demonstrated repeatedly, in numerous atomic and molecular systems, exhibiting a rich structure, including the existence of fractional revivals where multiple "mini-copies" of the original wave packet are temporarily formed. The resulting highly regular pattern formation in quantum wave packet evolution (Kaplan et al. 2000) in the probability density $P(x, t) = |\psi(x, t)|^2$ has been labeled as a **quantum carpet.** (For a review of the theory and laboratory realizations[11] of these phenomena, see Robinett 2004.)

Examples of the short-term periodicity and medium-term dephasing of a wave packet in the infinite square well are shown in Fig. 6.5. The subsequent longer-term revival (and fractional revival) behavior is illustrated in Fig. 6.6, where we show both the position-space ($|\psi(x, t)|^2$) and momentum-space ($|\phi(p, t)|^2$) probability densities for various times. At $t = 0$ (and a very short time after) we see the initial wave packet moving to the right with momentum centered about $+p_0$, while for $t = T_{rev}$ we see it fully reformed, back at the same location in space, and again with positive momentum. For $t = T_{rev}/2$ the packet has revived, but is moving to the left (so a half period out of phase) with $p = -p_0$, while for $t = T_{rev}/4$ there are temporarily two "half-copies" of the initial packet, moving out of phase with each other, in opposite directions, hence the two smaller peaks at $\pm p_0$ in momentum-space. You're asked to compare these model predictions to real experimental data in **P6.26.**

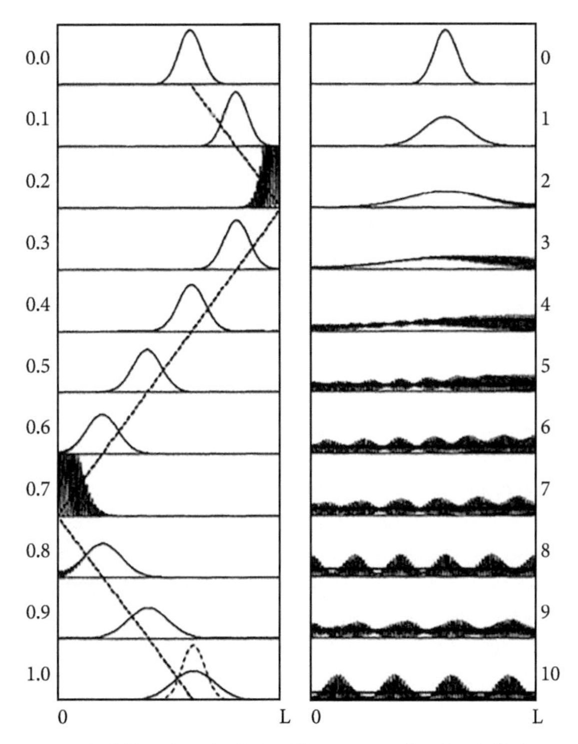

Fig. 6.5 Position-space probability density $|\psi(x, t)|^2$ versus x for the infinite square well for various times over one classical period or T_{qp} (left), and over ten classical (T_{qp}) cycles (right), showing the initial periodicity, and dephasing. Note that even after one period the wave packet can be seen to have spread/dispersed.

[11] For examples of revivals of wave packets in Rydberg atoms, see Yeazell, Mallalieu, and Stroud (1990), Yeazell and Stroud (1991), and Meacher et al. (1991).

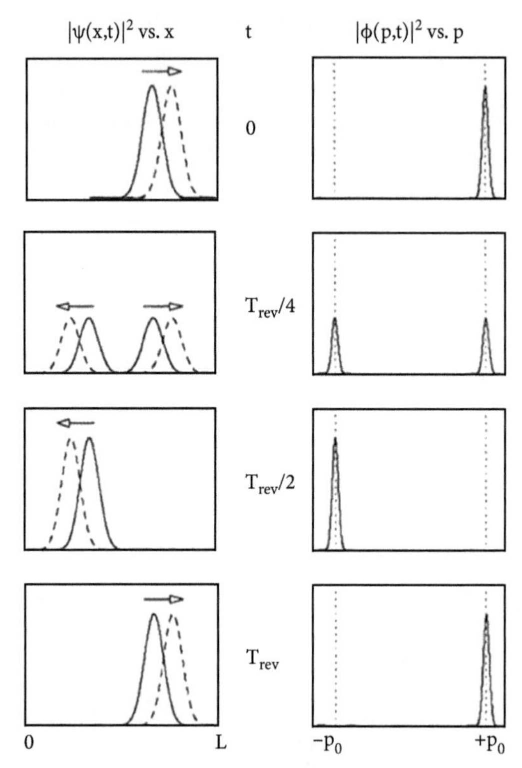

Fig. 6.6 Position-space probability density $|\psi(x,t)|^2$ versus x (left), and momentum-space probability density $|\phi(p,t)|^2$ versus p (right), for a time-dependent wave packet in the infinite square well. The times indicated are $t = 0, T_{rev}/4, T_{rev}/2$ and T_{rev} corresponding to the initial state, a *fractional revival* (with two copies of the wave packet, moving in opposite directions), a *half-revival* (full packet revived, but moving out of phase), and a full-revival.

One reason to focus on such behavior is to note that the dimensional dependence of all the time scales discussed here, T_{cl} and T_{rev}, and any higher-order terms in the expansion in Eqn. (6.188), will be determined by same set of physical parameters in the system, with the hierarchy of temporal evolution (the fact that $T_{rev} >> T_{cl}$) being dictated by functions of the dimension**less** quantum number n. While that connection may not seem to be directly related to a DA approach, we recall that the n-dependence of the quantized energies in power-law potentials was tied directly to the factors of $\hbar$ present, with the combination $n\hbar$ often appearing together. For example in any of the power-law potentials discussed so far, we have

$$E_n^{(P)} \approx \mathcal{E}_0^{(P)} n^{2P/(2+P)} = \mathcal{E}_0^{(P)} n^{k_P} \tag{6.195}$$

so that

$$T_{cl} \propto \frac{1}{|E'(n_0)|} \propto n_0^{1-k_P}, \qquad T_{rev} \propto \frac{1}{|E''(n_0)|} \propto n_0^{2-k_P}, \qquad \text{and} \qquad \frac{T_{rev}}{T_{cl}} \propto n_0 >> 1, \tag{6.196}$$

at least for large quantum number wave packet realizations. Chapter 7 explores other cases where there are also hierarchies of time (and length) scales dictated by dimensionless constants, but in that case it will be the value of the (dimension**less**) fine-structure constant, $\alpha_{FS} \equiv e^2/(4\pi\epsilon_0\hbar c) \approx 1/137$, that plays that role.

While we have focused on complex phases governing the time-dependence of quantum wave functions, we can also make contact between the spatial dependence of $\psi(x)$ in quantum tunneling (namely exponential suppression) from Section 6.4.1 and that of the familiar time-dependence from the study of exponential decay, often seen in say a modern physics course.

If we assume that the energy of a quantum state has an imaginary part, namely that $E_n \to E_n - i\Gamma/2$, then the wave function and corresponding probability density satisfy

$$\psi(x,t) = \psi(x)e^{-iE_n t/\hbar}e^{-\Gamma t/2\hbar} \quad \Longrightarrow \quad P(x,t) = |\psi(x,t)|^2 \propto e^{-\Gamma t/\hbar} = e^{-\lambda t} = e^{-t/\tau}, \tag{6.197}$$

where

$$\frac{\Gamma}{\hbar} = \lambda = \frac{1}{\tau} \tag{6.198}$$

and λ and τ are the **decay rate** and **(mean) lifetime**, respectively. These connections are consistent with the energy–time uncertainty principle, since if we associate $\Delta E \sim \Gamma$ and $\Delta t \sim \tau$, we have $\Delta E \cdot \Delta t \sim \Gamma\tau \sim \hbar$. In many cases, the "decay rate" of a process may actually be cited in terms of an energy, implicitly assuming the connection in Eqn. (6.198) providing (or requiring) the relevant factor of $\hbar$; we discuss other examples of what are often called "$\hbar = c = 1$" type conversions in Section 7.3.

6.5 **Natural or universal units: Hartree versus Planckian quantities**

While our main focus has been on the understanding and functional use of dimensional reasoning in exploring physical problems, we have also advocated for acquaintance with (if not necessarily fluency in) the different systems of units in which physics is spoken (or written, or calculated with, or thought about, or taught, etc.) on a daily basis, in different disciplinary specialties. For example, astronomers or planetary scientists may well use as their base units for mass, length, and time the solar mass ($M_\odot$), the mean Earth–Sun distance (AU being short for **astronomical unit**), and the orbital period of the Earth around the Sun (the year), namely

$$\begin{aligned} 1\,M_\odot &= 1.99 \times 10^{30}\,kg \\ 1\,AU &= 1.496 \times 10^{9}\,m \\ 1\,yr &= 3.15 \times 10^{7}\,s \approx \pi \times 10^{7}\,s, \end{aligned}$$

where we have included a popular mnemonic approximation for how many seconds are in a year. Astrophysicists who must consider distances on cosmological scales may use other (much larger) distance scales, such as the **light-year** (ly) or **parsec** (pc), given by

$$\begin{aligned} 1\,ly &= 9.46 \times 10^{15}\,m \\ 1\,pc &= 3.09 \times 10^{16}\,m \approx 3.26\,ly, \end{aligned}$$

which, while defined in very different ways, share the same order of magnitude.

The use of English units (foot, second, pound) or SI (meter, second, kilogram) are, of course, understandable for everyday scales, but for scientists (physicists or chemists) studying atomic- or molecular-level phenomena, the relevant sizes are well-captured by the dimensions of the hydrogen atom, determined by combinations of $\hbar$, m_e, and K_e, as discussed in Section 6.3.4. At the ultimate extreme of Nature, where quantum mechanics, relativity, and gravitation all come into play, the dimensional scale for **relativistic quantum gravity** has already been mentioned in Section 1.3 and is determined by $\hbar$, c, and G and appears in studies of the early universe. To gain some experience with the use of both scales (and as always, including a focus on DA), the next two sections explore system of units that may seem "natural" in their appropriate fields of study.

6.5.1 **Hartree or atomic units**

DA singled out length and energy scales for the hydrogen atom (and by implication, for all of atomic and molecular physics for which atoms are the constituents) in Eqns. (6.148) and (6.149). Substituting numerical values for the fundamental constants K_e, m_e, and $\hbar$, we find

$$a_0 = 5.29 \times 10^{-11}\, m = 0.53\,\text{Å} \tag{6.199}$$

$$E_n = -\frac{2.2 \times 10^{-18}\, J}{n^2} \sim -\frac{13.6\, eV}{n^2}, \tag{6.200}$$

and we often use the **Angstrom** (Å) and **electron-volt** (eV) as relevant units, as they are the appropriate length and energy scales for atomic/molecular physics. (Recall that there is an extensive dictionary of conversion factors in Section 10.4.)

If we try to systematize the expressions for physical quantities relevant for the hydrogen atom, using just K_e, m_e, $\hbar$ and keeping dimensionless constants in place, we can write

$$r_n = \left(\frac{\hbar^2}{m_e K_e}\right)\left(n^2\right) = n^2 a_H \quad \text{where} \quad a_H \equiv \left(\frac{\hbar^2}{m_e K_e}\right) = a_0 \tag{6.201}$$

$$E_n = -\frac{1}{2}\left(\frac{K_e^2 m_e}{\hbar^2}\right)\left(\frac{1}{n^2}\right) = -\frac{E_H}{2n^2} \quad \text{where} \quad E_H \equiv \left(\frac{K_e^2 m_e}{\hbar^2}\right) = \frac{K_e}{a_H} \tag{6.202}$$

$$p_n = \left(\frac{K_e m_e}{\hbar}\right)\left(\frac{1}{n}\right) = \frac{p_H}{n} \quad \text{where} \quad p_H \equiv \left(\frac{K_e m_e}{\hbar}\right) = \frac{\hbar}{a_H} \tag{6.203}$$

$$v_n = \left(\frac{K_e}{\hbar}\right)\left(\frac{1}{n}\right) = \frac{v_H}{n} \quad \text{where} \quad v_H \equiv \left(\frac{K_e}{\hbar}\right)\left(\frac{1}{n}\right) = \frac{p_H}{m_e} \tag{6.204}$$

$$\tau_n \equiv \frac{2\pi r_n}{v_n} = 2\pi\left(\frac{\hbar^3}{K_e^2 m_e}\right)\left(n^3\right) = 2\pi\tau_H n^3 \quad \text{where} \quad \tau_H \equiv \left(\frac{\hbar^3}{K_e^2 m_e}\right) = \frac{a_H}{v_H} = \frac{\hbar}{E_H}, \tag{6.205}$$

and the subscript H might well suggest hydrogen. The order-of-magnitude values for some dimensionful quantities are given by

$$v_H = 2.2 \times 10^6\, m/s \tag{6.206}$$

$$\tau_H = 2.4 \times 10^{-17}\, s \tag{6.207}$$

$$p_H = 2.0 \times 10^{-24}\, kg \cdot m/s. \tag{6.208}$$

In Section 1.2.4 we noted that the speed associated with the ground state of the electron in the H-atom can be written in the form

$$v_H = \frac{K_e}{\hbar} = \frac{e^2}{4\pi\epsilon_0\hbar} = \left(\frac{e^2}{4\pi\epsilon_0\hbar c}\right)c \equiv \alpha_{FS}\, c, \tag{6.209}$$

where $\alpha_{FS} \approx 1/137 << 1$ is the fine-structure constant, and its magnitude implies that the H-atom problem is, to first-approximation, a non-relativistic one.

The suggestion to formalize the use of such combinations of dimensionful quantities was first popularized by Hartree (1927),[12] who says:

> Both in order to eliminate various universal constants from the equations and also to avoid high powers of 10 in numerical work, it is convenient to express quantities in terms of units, which may be called "atomic units" defined as follows:
>
> Unit of length, $a_H = h^2/4\pi^2 m\, e^2$, on the orbital mechanics the radius of the 1-quantum circular orbit of the H-atom with fixed nucleus.
>
> Unit of charge, e, the magnitude of the charge on the electron.
>
> Unit of mass, m, the mass of the electron.

The unit of length he described as a_H (so the H subscript can now also stand for Hartree) is said to be

$$a_H = \frac{h^2}{4\pi^2 m e^2} = \frac{\hbar^2}{m\, e^2}, \tag{6.210}$$

which is missing the conventional SI/MKS(C) factor of $4\pi\epsilon_0$, but only since he uses different (CGS/Gaussian) type E&M units, but has the same numerical value as the Bohr radius. The E_H unit of energy is indeed called the **hartree** (or **Hartree energy**) and is given by $E_H = 27.2\ eV$.

Other "order-of-magnitude" quantities that can be approximated using such units include the current due to the "rotating" electron (assuming a semi-classical picture of a circulating electron). On purely dimensional grounds, we might associate $I_H = e/\tau_H$, or perhaps more reasonably we can use the orbital period, $2\pi\tau_H$, to write

$$I = \frac{e}{2\pi\tau_H} = \frac{1.6 \times 10^{-19}\, C}{2\pi(2.4 \times 10^{-17}\, s)} \approx 1 \times 10^{-3} A \left(\text{or } \frac{C}{s}\right), \tag{6.211}$$

from which one can estimate the magnetic field at the center of the atom (see **P6.30**). Other important materials properties such as the **atomic polarizability** and **conductivity** can be expressed in Hartree-inspired units and we explore those in **P6.28–P6.31**, and later in **P8.1** and **P8.2**. Section 7.7 briefly explores a different set of length and time units, based on other physical aspects of the hydrogen atom.

6.5.2 **Planckian units**

Very early in the history of quantum mechanics, Planck himself recognized the fundamental nature of what has come to be known as his constant (which he associated with thermal physics, given its provenance from the study of blackbody radiation) alongside those of gravitation (G), special relativity (c), and thermodynamics (k_B). He discussed (Planck 1900) what he described as "natürlichen Einheiten" (or natural units) and cited the foundational nature of such a new set of base units by saying:

> On the other hand, it should not be without interest to note that ... it is possible to set up units for length, mass, time and temperature, which, independently of special bodies or substances retain their

[12] And encouraged by Shull and Hall (1959).

meaning for all times and for all cultures, including extraterrestrial and extra-human ones, and which can therefore be designated as "natural units" of measure (Planck 1900).

We have already seen one of these in Chapter 1, the so-called **Planck length** (or L_P) and the other three base units (M_P for mass, T_P for time, and Θ_P for temperature) are given by

$$L_P = \sqrt{\frac{\hbar G}{c^3}} \approx 1.6 \times 10^{-35}\, m \tag{6.212}$$

$$M_P = \sqrt{\frac{\hbar c}{G}} \approx 2.2 \times 10^{-8}\, kg \tag{6.213}$$

$$T_P = \sqrt{\frac{\hbar G}{c^5}} \approx 5.4 \times 10^{-44}\, s \tag{6.214}$$

$$\Theta_P = \sqrt{\frac{\hbar c^5}{G k_B^2}} \approx 1.4 \times 10^{32}\, K, \tag{6.215}$$

all of which you should confirm by writing $\chi_P \propto \hbar^\alpha G^\beta c^\gamma$ and balancing dimensions, for each of the $\chi_P = L_P, M_P, T_P, \Theta_P$ units. Given the extreme values for L_P, T_P, Θ_P, they are most used in studies of the early universe (i.e., cosmology) or black-hole physics where such length, time, mass, and temperature scales might be relevant. Since one imagines that many (all?) intelligent civilizations would have understood these four basic components of nature, and learned the "spelling words," $\hbar, G, c, k_B$, Planck (1900) posited these quantities

> ... must, from the most diverse Intelligence measured by a variety of methods, arise again and again as the named ones....

You should be able to "spell" many other fundamental quantities in Planckian units as outlined in **P6.32** and evaluate them numerically.

One might imagine a similar set of natural dimensional quantities based on purely classical quantities. For example, if one assumes that there is a fundamental electric charge (e), then a combination of the speed of light (c), and the basic "strengths" of the gravitational and electrostatic forces (G and ϵ_0) might form a basis for the space of the $\mathsf{L}, \mathsf{T}, \mathsf{M}, \mathsf{Q}$ dimensions. Given that e would naturally be the relevant fundamental Q dimension, the other three could only depend on a combination of e and ϵ_0 involving $\mathsf{L}, \mathsf{T}, \mathsf{M}$, namely in the ratio of e^2/ϵ_0, or in our language, $K_e \equiv e^2/(4\pi\epsilon_0)$.

This approach is essentially the same as one advocated by G. J. Stoney (1881),[13] even before the rationalization of E&M units and the discovery of the electron. In our notation, we would then use combinations like $K_e^\alpha G^\beta c^\gamma$ to extract mass, length, and time units, finding

$$M_{St} = \sqrt{\frac{K_e}{G}}, \qquad L_{St} = \sqrt{\frac{K_e G}{c^4}}, \qquad T_{St} = \sqrt{\frac{K_e G}{c^6}}, \qquad \text{and} \qquad Q_{St} = e, \tag{6.216}$$

where, of course, the subscript is St for Stoney. Instead of evaluating each of these quantities numerically, we simply note that the ratios of these quantities to relevant Planckian ones are all given by

[13] Also cited as having introduced the term *electron* (originally *electrine*) and suggested the symbol E for the basic physical charge.

$$\frac{M_{St}}{M_{Pl}} = \frac{L_{St}}{L_{Pl}} = \frac{T_{St}}{T_{Pl}} = \sqrt{\frac{K_e}{\hbar c}} = \sqrt{\alpha_{FS}} \approx \frac{1}{\sqrt{137}} \sim \frac{1}{10} \tag{6.217}$$

and are thus the same order of magnitude, even closer if we'd not chosen to add the typical 4π factor to K).

Some of the history of this early approach to "natural units" is nicely discussed by Duff, Okun, and Veneziano (2002) (who also delightfully argue about just how many fundamental constants there are!) and especially in the popular *The Constants of Nature* by Barrow (2002). Other interesting discussions of Planckian units are contained in Adler (2010) and Wilczek (2005, 2006a, 2006b). Lapidus (1981) has even speculated on fundamental units and dimensionless constants in a universe with one, two, and four space dimensions.

Besides being a (perhaps) interesting historical connection, we find it appropriate to end this chapter with another nod to the importance of the fine-structure constant, α_{FS}. We continue to explore the relevance of this dimensionless constant as we start Chapter 7 as a pathway to more advanced quantum mechanics topics, including field theory, but focus on quantum electrodynamics as an example.

6.6 **Problems for Chapter 6**

Q6.1 Reality check: (a) What was the single most important, interesting, engaging, and/or useful thing you learned from this chapter?
(b) Have you ever used DA in approaching any problem in quantum mechanics?

Q6.2 "What if" scenarios for Planck's constant: In a popular book by the physicist George Gamow (1965), a ". . . bank clerk interested in modern science . . ." named C. G. H. Tompkins explores many aspects of modern (c. 1940) physics, including relativity, gravitational physics, and quantum mechanics.[14] One chapter describes a world in which $\hbar$ is macroscopically big, say $\hbar = 1\, kg \cdot m^2/s$. Imagining a less-drastic scenario, what would happen to atomic and molecular physics, and chemistry and materials science, if $\hbar$ were 1% bigger or smaller? 50% Fifty percent bigger or smaller?

P6.1 Testing the CMB spectrum: (a) Using the plot of the CMB data in Fig. 6.1, confirm that the frequency (f_{max}) corresponding to the peak of the intensity curve gives $T_{CMB} \approx 2.7\, K$, using the relation in Eqn. (6.18).
Note: The data shown in Fig. 6.1 actually corresponds to **surface brightness** which, for a diffuse source like the CMB, is proportional to the expression for $R_T(f) = c\rho_T(f)/4$, just not integrated over solid angles, specifically

$$SB(f) = \frac{\pi f^3}{2c^2} \frac{h}{e^{hf/k_B T} - 1}, \tag{6.218}$$

hence the extra unit of sr^{-1} or *per steradian* on the vertical axis.
(b) Show that for $f << f_{max}$ the surface brightness goes like

$$SB(f) = \frac{\pi}{2} \frac{(k_B T) f^2}{c^2}, \tag{6.219}$$

which is the analog of the classical result in Eqn. (6.23).

[14] Please note the protagonist's initials!

(c)† Use the data on the plot to show that for $f << f_{max}$ (the low-frequency end of the spectrum, corresponding to $hf/k_BT << 1$) that the spectrum is well described by $SB(f) \propto f^2$, as in Eqn. (6.219). Do this by fitting the straight-line log–log plot to a power-law of the form $y = ax^b$, in whatever you know how, and see what value of b you obtain. Can you also extract a value of a to compare to the expected value of the prefactor of $(\pi/2)(k_BT/c^2)$?
(d) Confirm that Eqn. (6.218) also gives the correct value of $SB(f_{max})$ at the peak.

P6.2 Blackbody spectrum—Wavelength dependence: The wavelength dependent energy spectrum, $\rho_T(\lambda)$, given by Eqn. (6.19) corresponds to the **energy per unit volume per unit wavelength interval.**

(a) Using that definition, evaluate the dimensions of $\rho_T(\lambda)$, and confirm that the expression in Eqn. (6.19) is dimensionally consistent with that.

(b) Show that Eqn. (6.28) can be derived from the wavelength form in Eqn. (6.19) by the variable relation $\lambda = c/f$, and noting that (up to a sign)

$$\rho_T(\lambda)\, d\lambda = \rho_T(f)\, df. \tag{6.220}$$

(c) Rewrite Eqn. (6.19) in terms of the dimensional variable $y = hc/\lambda k_BT$ and see if you can show that the maximum value of the intensity occurs when $y_{max} \approx 4.96$. (Do this in any way you wish, including plotting the dimensionless function analogous to $H(x)$, or using symbolic math programs to find a more precise answer.) Using values of h, c, k_B, show that this corresponds to the statement that

$$\lambda_{max} T = b = \text{ Wien's displacement constant } \approx 2.9 \times 10^{-3}\, m \cdot K \approx 2900\, \mu m \cdot K, \tag{6.221}$$

which is the more familiar version of Wien's displacement law.

(d) Fig. 6.7 shows historical data (Coblentz 1917) on the blackbody spectrum as a function of wavelength. Show that the peak wavelength ($\lambda_{max} \approx 1.82\, \mu m$) and stated temperature ($T \approx 1323$ C, so you'll have to change this to K) satisfy Eqn. (6.221).

(e) Finally, write $\rho_T(\lambda) \propto h^\alpha c^\beta k_B^\gamma T^\delta \lambda^\sigma$, solve for the exponents to the extent you can, and show that your result is consistent with Eqn. (6.19).

P6.3 Vibrational entropy—Full expression: The complete expression for the entropy due to vibrational motion for a diatomic molecule (with oscillation frequency given by $2\pi\nu = \sqrt{K/\mu}$) is

$$\frac{S_{vib}}{Nk_B} = \frac{x}{e^x - 1} - \ln(1 - e^{-x}), \tag{6.222}$$

where $x \equiv h\nu/k_BT$.

(a)* In the high-temperature limit, we'd approximate this by assuming that $x << 1$, so use appropriate series expansions (included in Section 10.7) to show that this expression reduces to Eqn. (6.45) as advertised.

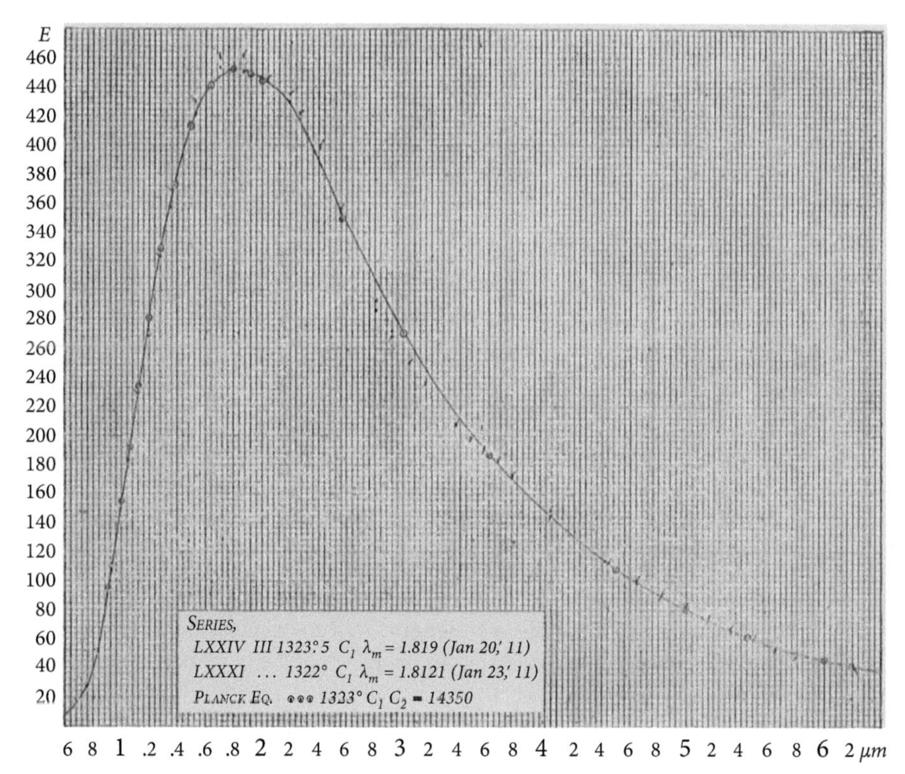

Fig. 6.7 Blackbody spectrum from Coblentz (1917) showing intensity versus wavelength (μm).

(b) Following the notation of the Planckian blackbody derivation using entropy ideas, we know that

$$n \equiv \frac{\mathcal{U}}{\epsilon} = \frac{1}{e^x - 1}. \tag{6.223}$$

Show that

$$e^x = \frac{n+1}{n}, \qquad x = \ln\left(\frac{n+1}{n}\right), \qquad \text{and} \qquad 1 - e^{-x} = \frac{1}{n+1}. \tag{6.224}$$

(c) Use these results to show that Eqn. (6.222) can be written as

$$\frac{S_{vib}}{Nk_B} = (n+1)\ \ln(n+1) - n\ \ln(n) \tag{6.225}$$

and since $n \equiv \mathcal{U}/\epsilon$, this reproduces Eqn. (6.24).

P6.4 Rotational entropy for polyatomic molecules: Consider a molecule with ≥ 3 atoms (such as H_2O, NH_3 or C_2H_4), which can rotate non-trivially about any of three axes, with moments of inertia about each labeled I_A, I_B, I_C. Using the same methods that were used to derive Eqn. (6.44), see if you can capture the dimensional dependence (if not all of the constants) of the textbook result for this system (see, e.g., Glasstone 1947, or Hill 1960) shown below as

$$\frac{S_{rot}}{Nk_B} = \ln\left[\frac{8\pi^2(8\pi^3 I_A I_B I_C)^{3/2}(k_B T)^{3/2}}{\sigma h^3}\right] + \frac{3}{2}. \tag{6.226}$$

Also confirm that this relation is dimensionally consistent, with the argument inside the log being dimensionless.

P6.5 Black hole thermodynamics: (a) Confirm that the dimensions of Eqn. (6.48) are consistent. (b) Can you rewrite Eqn. (6.48) in terms of the **Planck length** from Eqn. (1.129) or Section 6.5.2? (c) Associated with the entropy of the black hole, there is an effective temperature, T_H (H for Hawking) that depends on the mass of the black hole, M_{BH}. Explore how this temperature depends on the fundamental dimensionful constants and M_{BH}, by writing $T_H \propto \hbar^\alpha c^\beta G^\gamma M_{BH}^\delta k_B^\epsilon$ and solving to the extent possible. Given that there are more exponents (5) than constraints (4) you won't be able to determine them all, but can you use intuition (say the fact that G and M_{BH} might appear together) to pin things down further?

P6.6 Millikan's determination of Planck's constant: The data below, taken from Millikan (1916), shows the stopping potential V_0 for various **wavelengths** (not frequencies) in Angströms or Å units. The wavelengths are taken from Table 1 of his paper, while the potentials are read off his Figure 6.

λ (Å)	V_0 (*Volts*)
5461	−2.05
4339	−1.48
4047	−1.30
3651	−0.92
3125	−0.38

Convert the wavelengths to frequencies and plot V_0 versus f, find the slope and use the relation in Eqn. (6.56) to evaluate h and compare to the value found by Millikan, namely $h = 6.57 \times 10^{-27}\ erg \cdot sec$ (noting the units he used).

P6.7 Davisson–Germer experiment on electron diffraction: (a) The de Broglie wavelength associated with a particle of charge Ze and mass m that has been accelerated through a voltage V can be explored with DA by writing $\lambda_{dB} = C_\lambda h^\alpha m^\beta (Ze)^\gamma V^\delta$ (where Z is, of course, dimensionless, but necessarily tags along with e.) Match dimensions to solve for the exponents.
(b) Compare this to the exact answer obtained by equating the charged particles kinetic energy with that provided by the voltage, namely

$$\frac{1}{2}mv^2 = \frac{p^2}{2m} = eV, \tag{6.227}$$

to solve for p and then use the de Broglie definition to also determine C_λ.
(c) Davisson and Germer (1927) in their paper on electron diffraction say that:

> Rewriting the de Broglie formula for λ in terms of the kinetic energy V of the electrons expressed in equivalent volts, we have:
>
> $$\lambda \text{ (in Angstrom units)} = (150/V)^{1/2}$$

Use your result from part (b) to confirm that this is correct, once all of the dimensionful and numerical factors are put in.
(d) Thomson and Reid (1927) accelerated electrons to higher energies than Davisson and Germer, in one run to 13,800 V, which they quote as giving $\lambda = 1.0 \times 10^{-9}\ cm$. Confirm that by using the formula from part (c).

P6.8 Dirac δ-functions in two dimensions and three dimensions: (a) The three-dimensional δ-function can be written in Cartesian coordinates as $\delta(\boldsymbol{r} - \boldsymbol{r}_0) = \delta(x - x_0)\delta(y - y_0)\delta(z - z_0)$ and the dimensions obviously satisfy Eqn. (6.94). In a spherical coordinate system, where the labels are given by $\boldsymbol{r} = (r, \theta, \phi)$, the corresponding expression is (Blennow 2018)

$$\delta(\boldsymbol{r} - \boldsymbol{r}_0) = \frac{1}{r^2 \sin(\theta)} \delta(r - r_0)\delta(\theta - \theta_0)\delta(\phi - \phi_0). \tag{6.228}$$

Show that this equation is dimensionally correct. Does it matter if we use r, θ or r_0, θ_0 in the denominator? Can you multiply the expression on the right by the three-dimensional spherical volume element $dV = \sin(\theta)\, d\theta\, d\phi\, r^2\, dr$ to see how the δ-functions work out to give unit area?
(b) Can you then guess what the two-dimensional δ-function, $\delta(x - x_0)\delta(y - y_0)$, in **polar coordinates** (i.e., just (r, θ)) would look like? (For the answer, see, e.g., Morse and Feshbach 1953, p. 825.) Can you guess what the three-dimensional δ-function might look like for a general coordinate transformation $(x, y, z) \rightarrow (u, v, w)$? Do you remember what a Jacobian is from multivariable calculus? If not, it's totally fine!

P6.9 Short-range boson–boson interactions and the Gross–Pitaeveskii equation: (a) Assume that the δ-function boson–boson interaction is written in the form

$$V_{bb}(\boldsymbol{r}_i - \boldsymbol{r}_j) = K_S \delta(\boldsymbol{r}_i - \boldsymbol{r}_j) \qquad \text{and write} \qquad K_S \propto \hbar^{\alpha} a_S^{\beta} m^{\delta} \tag{6.229}$$

and solve for α, β, γ by matching dimensions.
(b) An approximate equation (first explored by Gross 1961 and Pitaevskii 1961) for the quantum mechanical ground state of identical bosons, using the V_{bb} interaction in Eqn. (6.96) (after several simplifying assumptions) is written as

$$-\frac{\hbar^2}{2m}\nabla^2 \psi(\boldsymbol{r}) + V(\boldsymbol{r})\psi(\boldsymbol{r}) + \underbrace{\frac{4\pi\hbar^2 a_S}{m} |\psi(\boldsymbol{r})|^2 \psi(\boldsymbol{r})}_{\text{non-linear interaction}} = E\psi(\boldsymbol{r}). \tag{6.230}$$

Show that the new (non-linear) interaction term has the same dimensionality as the rest of the terms in the original three-dimensional Schrödinger equation

P6.10 Three-dimensional momentum-space wave function: (a) What are the dimensions of a momentum-space wave function in three dimensions, namely what is $[\,\phi(\mathbf{p}, t)\,]$?
(b) How about the dimensions of a multi-particle momentum-space wave function $[\,\phi(p_1, p_2, \ldots, p_n, t)\,]$ in one dimension?
(c) The momentum-space probability distribution for the ground state of the hydrogen atom has actually been measured experimentally (McCarthy and Weigold 1983) and the theoretical formula from many textbooks is given as

$$|\phi_{(1,0,0)}(\mathbf{p})|^2 = \frac{8}{\pi^2} \frac{p_0^5}{(p^2 + p_0^2)^4}, \tag{6.231}$$

where $p_0 = \hbar/a_0$ and a_0 is the Bohr radius. Show that this expression is dimensionfully correct. McCarthy and Weigold (1983) write this in the form

$$|\phi_{1S}(p)|^2 = (2^3/\pi^2)(1+p^2)^{-4}. \qquad (6.232)$$

Show that this is consistent with the Hartree/atomic units discussed in Section 6.5.1.

P6.11 Fourier transforms and Wigner distributions: The **inverse Fourier transform** of Eqn. (6.101), which takes $\phi(p,t)$ to $\psi(x,t)$, is given by

$$\psi(x,t) = \frac{1}{\sqrt{2\pi\hbar}} \int_{-\infty}^{+\infty} \phi(p,t)\, e^{+ipx/\hbar}\, dp. \qquad (6.233)$$

Show that this is dimensionally consistent.

(b) The closest thing in quantum mechanics to what one might call a **phase-space** description is given by the **Wigner quasi-probability distribution** or sometimes **quasi-phase space distribution**, given by two equivalent definitions

$$\begin{aligned} W(x,p;t) &\equiv \frac{1}{\pi\hbar} \int_{-\infty}^{+\infty} \psi^*(x+y,t)\, \psi(x-y,t)\, e^{2ipy/\hbar}\, dy \\ &= \frac{1}{\pi\hbar} \int_{-\infty}^{+\infty} \phi^*(p+q,t)\, \phi(p-q,t)\, e^{-2ixq/\hbar}\, dq. \end{aligned} \qquad (6.234)$$

Show that the two definitions are dimensionally consistent with each other, and that the dimensions of $W(x,p;t)$ are inverse phase-space.

P6.12 Length and energy scales for the linear potential: Use the four **S1—S4** strategies to determine the dimensional (and quantum number) dependences for the linear potential in Eqn. (6.122) to evaluate how ρ and E depend on $\hbar, m, F$, comparing the results to each other. Can you also confirm that $E_n \propto n^{2/3}$, at least for large n where the WKB method is valid? And that this is consistent with the general power-law prediction in Eqn. (6.141)?

P6.13 Momentum, velocity, and time scales for general power-law potential: (a) Extend the results of Eqn. (6.86) for E and ρ for the single δ-function potential to determine how momentum, velocity, and time depend on $\hbar, m, K$.

(b) Extend the results for $E_n^{(P)}$ and $\rho_n^{(P)}$ from Eqn. (6.141) for the general power-law potential to find expressions for momentum (p), velocity (v), and a characteristic time (τ) in terms of $m, K_P, \hbar$ and quantum number n.

(c) Compare the results for the single δ-function attractive potential in Eqn. (6.86) to the general formulae in expressions used in part (a). Are the dimensional dependences consistent with any power of P? Compare the results from parts (a) and (b) for the dependences of p, v, τ.

(d) Discuss the quantum mechanical tunneling behavior in the classically disallowed region for the general power-law potential $V_P(x) = K_P|x|^P$ especially in the $P\to\infty$ limit where there is an infinite wall boundary. Is your general result consistent with the boundary condition that $\psi(L) = 0$ there, with no tunneling beyond?

P6.14 WKB-like analysis of the quantum Coulomb problem: Complete the WKB analysis suggested by Eqn. (6.151) by changing to dimensionless variables and doing the resulting "pure" integral to solve for $|E_n|$. If you can't find the integral using a handbook or symbolic math program, feel free to use the result that

$$\int_0^1 \sqrt{\frac{1}{y} - 1}\, dy = \int_0^1 \sqrt{\frac{1-y}{y}}\, dy = \frac{\pi}{2}. \tag{6.235}$$

P6.15 Planck's constant as the "elementary quantum of action":[15] In classical mechanics, the **action** is a so-called **functional** (function of a function), which takes a trajectory or path, $x(t)$, and its desired end points, and associates with it a number (the action). It makes use of the **Lagrangian** for the system, defined by

$$L(x(t), \dot{x}(t)) \equiv \frac{1}{2} m\dot{x}(t)^2 - V(x(t)), \tag{6.236}$$

and so is a function of position and velocity. The **action** is the defined as

$$S[x(t)] = \int_{t_1}^{t_2} \left\{ \frac{m}{2} \left(\frac{dx(t)}{dt} \right)^2 - V(x(t) \right\} dt = \int_{t_1}^{t_2} L(x(t), \dot{x}(t))\, dt. \tag{6.237}$$

The classical path, $x(t)$, that solves Newton's equation is the one that satisfies the boundary conditions at the start/end of the path, namely $x(t_1) = x_1$ and $x(t_2) = x_2$, **and** which minimizes $S[x(t)]$, hence the description as the **principle of least action.**

(a) Show that the dimensions of the Lagrangian are the same as those for energy, even though this is not E: note the minus sign of the potential energy.

(b) Show that the dimensions of action are the same as those of $\hbar$.

(c) In **P6.13** we developed expressions for the energy ($E_n^{(P)}$) and time ($\tau_n^{(P)}$) scales for an arbitrary power-law potential as a function of $\hbar$, m, K_P and quantum number n. Using Eqn. (6.237) as a guide, assume that the dimensions of action for a quantized state are given by $[S_n] = [E_n][\tau_n]$ and show how this quantum version of action scales with $\hbar$, m, K_P, and n. Does this help explain the name "quantum of action"?

(d) The time-development of a quantum wave function can be written in terms of a **propagator** $K(x, t; x', 0)$ in the form

$$\psi(x, t) = \int_{-\infty}^{+\infty} \psi(x', 0)\, K(x, t; x', 0)\, dx'. \tag{6.238}$$

What are the dimensions of $K(x, t; x', 0)$?

(e)* The propagator can only be calculated in closed form for a small number of tractable problems, including the free particle and the harmonic oscillator, where one finds

$$\text{free particle:} \quad K(x, t; x', 0) = \sqrt{\frac{m}{2\pi i\hbar t}}\, e^{-m(x-x')^2/2i\hbar t} \tag{6.239}$$

$$\text{oscillator:} \quad K(x, t; x', 0) = \sqrt{\frac{m\omega}{2\pi i\hbar \sin(\omega t)}}\, e^{-(m\omega((x^2+(x')^2)\cos(\omega t)-2xx')/2i\hbar \sin(\omega t)}. \tag{6.240}$$

---

[15] Or "elemenarten wirklungsquantum".

Show that the two expressions are dimensionally correct, both in their prefactors and in their (dimensionless) phases. If one takes the limit that where the spring constant vanishes, one has $\sqrt{k/m} = \omega \to 0$, and the oscillator problem should reduce to the free-particle. Can you confirm this for the propagators?

Note: The propagator for a general problem can be written (here described very schematically) as the sum of the phases $e^{iS[x(t)]/\hbar}$ over all possible classical paths, hence the term **path integral** formulation. The path with the **least action,** S_{min}, determines the classical trajectory, but those close to it in value, say with $\Delta S = |S - S_{min}| \leq \hbar$, will have similar phases and so contribute significantly to the propagator. On the other hand, those with $\Delta S >> \hbar$ will have phase factors that oscillate wildly and, on average, cancel. In this sense, in the limit where $\hbar \to 0$, the trajectory with the least action is singled out and classical mechanics is recovered.
Comment: A long way to go, perhaps, to explain the "elementary quantum of action," but it's important to see what part $\hbar$ plays in this formulation, its dimensional connection to $S[x(t)]$, and another foundational role that dimensionless phases can play in quantum theory, along those discussed in Sec. 6.4.2. The original presentation of *Quantum Mechanics and Path Integrals* by Feynman and Hibbs (1965) is still the standard introduction to the subject.

P6.16 Cold/field emission: (a) Complete the DA suggested by Eqn. (6.171) to solve for the dependence of the cold/field emission G factor on m_e, W, e, E.
(b) Noting that the experimental data are consistent with a function of $F(E_0/E)$, write $E_0 \propto m_e^{\alpha} W^{\beta} e^{\gamma} \hbar^{\delta}$ and solve the exponents.
(c) For metals, a typical value of the work function is $W \sim 3 - 5\ eV$. To what values of $\mathcal{E}_0$ do these correspond?

P6.17 Coulomb barrier: (a) Use DA alone to explore the dependence of the (dimensionless) tunneling factor (phase) G_{CB} in Eqn. (6.172) by writing

$$G_{CB} = \frac{m_\alpha^{\rho} E_\alpha^{\sigma} (Z_1 Z_2 K_e)^{\delta}}{\hbar} \tag{6.241}$$

and solve for ρ, σ, δ. We have used different Greek letters to avoid confusion with the α-particle subscripts.
(b) Evaluate the integral in Eqn. (6.172) to reproduce the Gamow factor for α-decay. For simplicity, you can assume that $R_A << b \equiv K_e Z_1 Z_2 / E_\alpha$ and use Eqn. (6.235) from **P6.14.** But also try to justify using that integral, by assuming sample values like $Z_1 = 50$, $Z_\alpha = 2$, $E_\alpha \sim 2\ MeV$, and $R_A \sim 5\ F$ to estimate the lower limit of the integral, namely $y = R_A/b << 1$.
(c) Show that the **Gamow energy**, as defined just below Eqn. (6.172) by the relation $\exp(-2G_{CB}) = \exp(-\sqrt{E_G/E_\alpha})$, is given by

$$E_G = 2\left(\frac{\pi Z_1 Z_2 K_e}{\hbar}\right)^2 m_\alpha. \tag{6.242}$$

Evaluate E_G for a typical case of α-decay, say with $Z_1 = 50$ and $Z_2 = 2$. Perhaps the simplest way to represent your result is to find a dimensionless value of the ratio $E_G/m_\alpha c^2$, perhaps recalling that $\alpha_{FS} = K_e/\hbar c$.

P6.18 "Sparking the vacuum" as barrier penetration: Instead of having to tunnel through a barrier defined by the work function W, as in the Fowler–Nordheim effect in Eqns. (6.170) and (6.171), imagine instead that the energy barrier was given by $2m_ec^2$. What form would the natural electric field $\mathcal{E}_0$ and tunneling probability take? Compare this to the expression in the earlier discussion in Section 1.3 (**Example 1.4** and Eqn. (1.132)) for the **Schwinger effect** and the discussion in **Example 7.5.**

P6.19 WKB wave function—Phase and amplitude information: The approximate WKB solution in Eqn. (6.163) can be refined to include not only information on the phase of the quantum wave function, but also its amplitude. Several standard references write down this more detailed solution in the form

$$\psi(x) \sim \frac{C_\pm \sqrt{\hbar}}{\sqrt[4]{2m(E - V(x))}} \exp\left(\pm i \frac{\sqrt{2m}}{\hbar} \int^x \sqrt{E - V(x)}\, dx\right), \tag{6.243}$$

where $C_\pm$ is a constant to be determined by the fact that the probability density has to be properly normalized, namely that $\int_{-\infty}^{+\infty} |\psi(x)|^2\, dx = 1$.

(a) We have most often used C to represent a dimensionless constant, but in this standardly used notation, that is not the case. What dimensions must $[C]$ have in order for this expression to be dimensionally correct?

(b) For the bound state problems discussed in Section 6.3, what dimensionful quantities do you think determine C? For example, for the harmonic oscillator problem, how do you think that C depends on $m, k, \hbar$?

(c) With this "new and improved" WKB expression, including more information on the magnitude of $\psi(x)$, what correlations are there between the "wiggliness" of the wave function (local de Broglie wavelength) and the magnitude of $|\psi(x)|$, and hence the local probability density, $|\psi(x)|^2$? Classically, how would the probability of finding a particle in a bound state depend on the local speed $v(x) = \sqrt{2(E - V(x))/m}$?

P6.20 More neutron interference experiments: (a) Using similar techniques to Colella, Overhauser, and Werner (1975), Werner, Staudenmann, and Colella (1979) "... observed the effect of the Earth's rotation on the phase of the neutron wave function ..." or the **Coriolis effect.** Assuming that the phase difference depends on the neutron mass (m), the angular rotation frequency of the earth ($\mathbf{\Omega}_E$, a vector, since it has a rotation axis), the area enclosed by the interference apparatus ($\boldsymbol{a}_{tot}$, with the vector direction along the normal to the area), and $\hbar$, try writing

$$\Theta_{Cor} \propto \frac{m^\alpha \Omega_E^\beta a_{tot}^\gamma}{\hbar} \tag{6.244}$$

and solve for α, β, γ. How do you think the two vector quantities combine?
(b) *The effect of the Earth's rotation on the velocity of light* (the name of two papers by Michelson and Gale (1925), with Part II (1925) adding Pearson as a co-author) is the EM/optical equivalent of the experiment discussed in part (a). For this purely classical experiment, there is no power of

$\hbar$ and the phase shift depends on $\Theta_{cor} \propto \boldsymbol{\Omega}_e^{\alpha} \boldsymbol{a}_{tot}^{\beta} \lambda^{\gamma} c^{\delta}$, where λ is the wavelength of light used and c is the speed of light. Match dimensions to constrain the exponents to the extent possible, noting that only L, T are involved, but then see if the results from part (a) can provide a useful analogy to specify them more completely. Can you see other analogies between the optical and particle results by making the identifications $\lambda \rightarrow \lambda_{dB} = h/p = h/mv$ and $c \rightarrow v$?

(c) One of the first examples of neutron interferometry by Rauch, Treimer, and Bonse (1974) probed "... phase shifting material inserted in the beams ...", which resulted in a "... marked intensity modulation behind the interferometer." In this case, the effective neutron-matter potential is similar to that in Eqn. (6.96), in this case

$$V_{nN}(\boldsymbol{r}_n - \boldsymbol{R}_N) = \frac{2\pi\hbar^2 a_{nN}}{m_n}\delta(\boldsymbol{r}_n - \boldsymbol{R}_N) \equiv K_{nN}\delta(\boldsymbol{r}_n - \boldsymbol{R}_N), \tag{6.245}$$

where a_{nN} is the neutron–nucleus[16] **scattering length** and $K_{nN} \equiv 2\pi\hbar^2 a_{nN}/m_n$. Assuming that the phase shifting material has a (number, not mass) density n, is of thickness D, and the neutrons have speed v, write the phase difference due to nuclear interactions in the form

$$\Phi_{nN} \propto \frac{K_{nN}^{\alpha} n^{\beta} (D/v)^{\gamma}}{\hbar}, \tag{6.246}$$

where $[n] = \mathsf{L}^{-3}$ as appropriate for a number density, and solve for α, β, γ. This phase difference is often cited in the form

$$\Phi_{nN} = n a_{nN} \lambda D, \tag{6.247}$$

where $\lambda = \lambda_{dB}$ is the de Broglie wavelength. Substitute the definition of K_{nN} from Eqn. (6.245) into your answer above and see if you can reproduce Eqn. (6.247); confirm that this expression for Φ_{nN} is indeed dimensionless.

(d) Aharonov and Casher (1984) predicted that there would be an Aharonov–Bohm type phase shift for neutrons with magnetic moment μ_n moving in the electric field of a long, straight wire, one with linear charge density λ_e. Explore how this phase shift can depend on the relevant electric/magnetic dimensional parameters (ϵ_0, μ_0) by writing

$$\Phi_{AC} \propto \frac{\epsilon_0^{\alpha} \mu_0^{\beta} \lambda_e^{\gamma} \mu_n^{\delta}}{\hbar} \tag{6.248}$$

and solve for $\alpha, \beta, \gamma, \delta$. This effect was first observed experimentally by Cimmino et al. (1989). **Note**: Both the theory and experimental references cited here use CGS units, and not SI ones as we have, so you may have to translate using the information in Section 10.6.

P6.21 Variation on the Aharonov–Bohm experiment: In the original Aharonov–Bohm (1959) paper, they also discussed an interference effect for electrons where an electric potential $\phi(\boldsymbol{r}, t)$ was turned on for a short time τ, but in such a way that the electrons were never in a region of electric field $\boldsymbol{E}$, so analogous to the magnetic AB effect. How might the phase difference Φ_{ϕ} depend on $e, \epsilon_0, \phi, \tau, \hbar$, assuming that $\hbar$ is always "on the bottom"? This effect has yet to be observed due to experimental challenges; for a discussion, see Vavagiakis, Bachlechner, and Klegan (2021).

P6.22 Classical and quantum periodicity in the infinite well: Using the energy eigenvalues from Eqn. (6.194), and the expression for the wave packet periodicity from Eqn. (6.190), evaluate T_{qp}

[16] Note the difference of a factor of 2 between this expression and that for identical bosons due to symmetry effects.

for the ISW. The classical period for such a system would be $T_{cl} = \tau = 2L/v$ and we would associate $v = v_n$ with $E_n = mv_n^2/2$. Confirm that this approach gives the same answer.

P6.23 Classical and quantum periodicity for the hydrogen atom: The Bohr atom results for the quantized energies, radii, and speeds for the H-atom are

$$E_n = -\frac{1}{2n^2}\left(\frac{K_e^2 m_e}{\hbar^2}\right), \qquad r_n = n^2\left(\frac{\hbar^2}{m_e K_e}\right), \quad \text{and} \quad v_n = \frac{1}{n}\left(\frac{K_e}{\hbar^2}\right), \tag{6.249}$$

where $K_e \equiv e^2/(4\pi\epsilon_0)$.

(a) Use the relation $\tau_n = 2\pi r_n/v_n$ to evaluate the analog of the classical period, $T_{cl} = \tau_n$.

(b) Use $T_{qp} = 2\pi\hbar/|E'(n)|$ from Eqn. (6.190) to find the quantum periodicity and compare to the results of part (a).

(c) An experiment by Yeazell et al. (1989) produced signals of the type shown in Fig. 6.8, where Rydberg wave packets were excited with values of $n_0 \approx 89$. Use the results of parts (a) or (b) to evaluate τ_n or T_{qp} for this value of n_0 and compare to the experimental results, where the periodicity is approximately 107 *ps*.

P6.24 Quantum periodicity in general power-law potentials: Use the scaling of $E_n^{(P)}$ for the potential energy function $V_P(x) = K_P|x|^P$, from Eqn. (6.141) and **P6.13** (including the dependences on $n, \hbar, K_P$, and m) and Eqn. (6.190) to extract $T_{qp}^{(P)}$. Compare this to the classical periodicity derived from **P6.13** and show that they match, including the dependence on the quantum number.

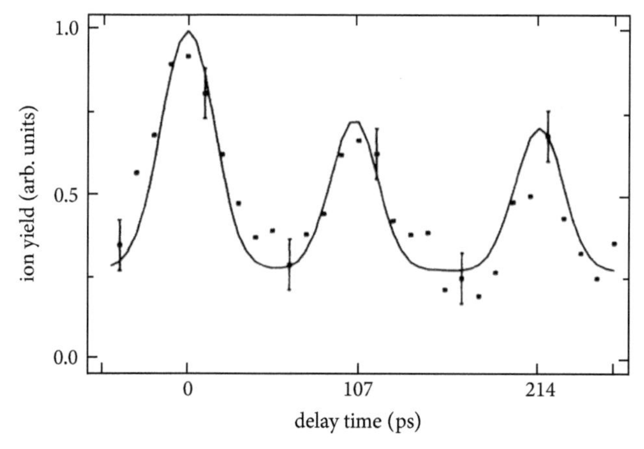

Fig. 6.8 Experimental measurement of the classical periodicity of a large quantum number ($n = 89$) Rydberg atom. Reprinted with permission from Yeazell et al. Yeazell (1989). Copyright (1989) by the American Physical Society. You're asked in **P6.23** to compare the formula for the classical H-atom period with this data.

P6.25 Quantum periodicity in the Morse potential: A model of the potential energy function of a diatomic molecule is the **Morse potential**, given by the function

$$V(r) = D\left(1 - e^{-(r-r_0)/a}\right)^2, \tag{6.250}$$

where r_0 is the equilibrium separation of the atoms and a, D are measures of the width and depth of the potential well. The quantized energy eigenvalues for this potential are known to be

$$E_n = (n + 1/2)\hbar\omega - \frac{[(n + 1/2)\hbar\omega]^2}{4D}, \tag{6.251}$$

where $\omega = \sqrt{2D/(ma^2)}$. What can you say about the quantum periodicity T_{qp} of this system as a function of n?

P6.26 Wave packet revivals in Rydberg atoms: Experimental data on the time-dependence of a Rydberg atom wave packet state (with $n_0 = 62$) from Meacher et al. (1991) is shown in Fig. 6.9.

(a) Find the short-term classical periodicity from the plot and compare to T_{qp} from Eqn. (6.190).

(b) The first large revival with the same classical periodicity, centered about 700 *ps*, corresponds to $T_{rev}/2$, while the structure at about 350 *ps* is smaller in magnitude and has twice the period, both of which are nicely consistent with the ISW model visualization in Fig. 6.6. Evaluate T_{rev} from Eqn. (6.190) and compare $T_{rev}/4$ and $T_{rev}/2$ to the data.

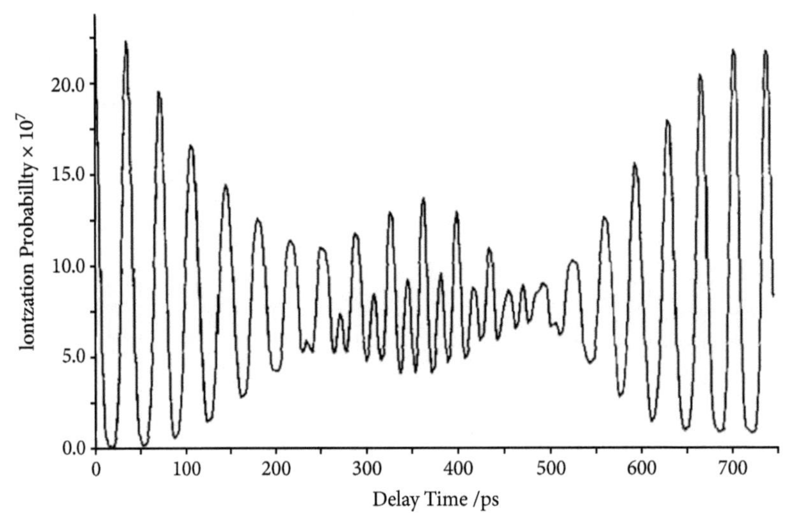

Fig. 6.9 Experimental measurement of a fractional- and half-revival for a Rydberg atom wave packet ($n = 62$). Reprinted from Meacher et al. (1991). ©IOP Publishing. Reproduced with permission. All rights reserved. Note: You're asked in **P6.26** to compare predictions for the quantum period and revival times, T_{qp} and T_{rev}, to this data.

P6.27 Lifetime versus decay rates: (a) The decay "rate" for the process of muon decay is often cited as $\Gamma(\mu^- \to e^- \overline{\nu}_e \nu_\mu) = 3 \times 10^{-10}\, eV$ (so in eV units). Convert this to a lifetime in micro-seconds (μs).

(b) The lifetime for the decay of the neutral pion to two photons is cited as $\tau(\pi^0 \to \gamma\gamma) = 85\, as$ (which is to be read as 85 attoseconds). Look up the prefix if needed (in Section 10.2), convert this into a "decay rate" Γ (in eV), and compare to the result of part (a).
Note: We explore both of these processes in a bit more detail in **P7.7** and **P7.20**, as they provide important tests of the weak and E&M/strong (QCD) interactions respectively.

P6.28 Hartree units I—Hartree electric and magnetic field: The Hartree units in Eqns. (6.201)–(6.205) all describe "mechanical" quantities (such as energy, length, etc.), and so only use M, L, T dimensions. They are described by the dimensional parameters $m_e, \hbar$, and the combination $K_e = e^2/(4\pi\epsilon_0)$. For E&M quantities, we also need to include either e or ϵ_0, in order to obtain the relevant Q dimensionality, and the simplest choice is often to use e.

(a) For a Hartree electric field, write $\mathcal{E}_H = e^\alpha m_e^\beta K_e^\gamma \hbar^\delta$, match dimensions, and solve for the exponents. Evaluate $\mathcal{E}_H$ in MKSA/SI units, $Volt/m$ or N/C.

(b) Show that the Hartree electric field can be written in the form $\mathcal{E}_H = E_H/(ea_0)$, where E_H is the hartree energy unit.

(c) For the Hartree magnetic field, write $\mathcal{B}_H \propto e^\alpha m_e^\beta K_e^\gamma \hbar^\delta$, match dimensions, and solve for the exponents. Evaluate $\mathcal{B}_H$ in units of $Tesla$.

(d) Show that a particle with a magnetic moment given by the Bohr magneton, $\mu_B = e\hbar/2m_e$, in a field of strength $\mathcal{B}_H$, has an energy of roughly E_H (a Hartree), namely that $E_H \sim \mu_B \mathcal{B}_H$.

P6.29 Hartree units II—Magnetic field of the hydrogen atom: (a) Assume a classical formula for the magnetic field at the center of a current loop of radius R, namely $B_{center} = \mu_0 I/2R$ to estimate the $\boldsymbol{B}$ field at the center of an H-atom. Use $R = a_0$, and I from Eqn. (6.211) as the approximate electron current, and give your answer in $Tesla$. Compare your numerical answer to the result of **P6.28**(c).
(b) Compare the dimensional dependences of your answer for B_{center} to that found for $\mathcal{B}_H$ in **P6.28.** In order to do so, rewrite $\mu_0 = 1/(c^2\epsilon_0)$ and use the fact that $\alpha_{FS} \equiv K_e/\hbar c$ is the dimensionless fine-structure constant to show that $B_{center} = (\alpha_{FS})^2 \mathcal{B}_H$.

P6.30 Hartree units III—Atomic polarizabilities: (a) The **polarizability** of a system (either classically or quantum mechanically) is defined by the relation $\boldsymbol{p} = \alpha_p \boldsymbol{E}$, where $\boldsymbol{p}, \boldsymbol{E}$ are the dipole moment of an individual atom/molecule and the external/applied electric field respectively.

(a) Find the M, L, T, Q dimensions of α_p.

(b) Polarizability can also be defined by the energy shift of an atom/molecule ($\Delta\mathcal{E}$) in an external electric field ($\boldsymbol{E}$) via $\Delta\mathcal{E} = -\alpha_p|\boldsymbol{E}|^2/2$. Show that this definition gives the same dimensions for α_p as in part (a).

(c) Instead of using e, K_e as relevant E&M quantities, we can also use DA to determine how α_p depends on fundamental atomic physics parameters by writing $\alpha_p = C_\alpha e^\alpha \epsilon_0^\beta m_e^\gamma \hbar^\delta$ and solving for $\alpha, \beta, \gamma, \delta$. Show that your result can be written in the form $C_\alpha 4\pi\epsilon_0 a_0^3$, where a_0 is the Bohr radius.
Note: For the hydrogen atom, an explicit calculation of α_p can be done yielding a result $C_\alpha = 9/2$. See Bowers (1986) for both a classical and quantum mechanical analysis.

(d) Given the simple form of α_p in terms of ϵ_0 and a_0, can you obtain the same result by writing $\boldsymbol{p}$ in terms of e and a_0, and $\boldsymbol{E}$ in terms of e, ϵ_0 and a_0 separately, and then taking the ratio?

(e) Some sources cite the hartree units of atomic polarizability as $e^2 a_0^2/E_H$. Show that this expression is consistent with what you found in part (c).

P6.31 Hartree units IV—Energies in different "dialects": (a) A statement from a NIST (National Institute of Standards and Technology) website[17] from their online *Computational Chemistry Comparison and Benchmark DataBase* says:

> A hartree is a unit of energy used in molecular orbital calculations. A hartree is equal to 2, 625.5 *kJ/mol*, 627.5 *kcal/mol*, 27.211 *eV*, and 219, 474.6 cm^{-1}

and we have already noted the value of the hartree in *eV* agrees with this. Confirm that the other three "definitions" of the hartree (seemingly very different) are, in fact, all consistent. Recall that one mole (or *mol*) is 6×10^{23} particles for the first two, and that $E = hf = hc/\lambda$ is a way to translate between energy and (inverse) wavelength.
(b) The definition of the **Rydberg constant** is $R = (e^4 m_e)/(8\epsilon_0^2 h^3 c)$. How is the Rydberg constant related to the Hartree energy?

P6.32 Planckian units: Use DA to find expressions in terms of $\hbar, G, c$ for the Planck (a) momentum, (b) energy, (c) mass density, (d) acceleration, (e) force or tension, (f) luminosity (i.e., power), and (g) pressure. Do any of these **not** depend on $\hbar$ (meaning they don't involve quantum mechanics), and if so, why?

P6.33 Stoney units: Starting with e, ϵ_0, G, c as fundamental quantities, confirm that you obtain the results in Eqn. (6.217) for M_{St}, L_{St}, T_{St}. Then repeat **P6.32** to find the dimensional combinations giving momentum, energy, force/tension, density, and luminosity. Are they all related to the corresponding Planckian results by the same factor of $\sqrt{\alpha_{FS}}$?

References for Chapter 6

Adler, R. J. (2010). "Six Easy Roads to the Planck Scale," *American Journal of Physics* **78**, 925–32.

Aharonov, Y., and Bohm, D. (1959). "Significance of Electromagnetic Potentials in the Quantum Theory", *Physical Review* **115**, 485–91.

Aharonov, Y., and Casher, A. (1984). "Topological Quantum Effects for Neutral Particles," *Physical Review Letters* 53, 319–21.

Aharonov, Y., and Susskind, L. (1967). "Observability of the Sign Change of Spinors under 2π Rotations," *Physical Review* **158**, 1237–8.

Aronstein, D. L., and Stroud, Jr. C. R. (1997). "Fractional Wave-Function Revivals in the Infinite Square Well," *Physical Review A* **55**, 4526–37.

Barrow, J. D. (2002). *The Constants of Nature: From Alpha to Omega—The Numbers that Encode the Deepest Secrets of the Universe* (New York: Pantheon).

Bekenstein, A. (1972). "Black Holes and the Second Law," *Lettere al Nuovo Cimento* 4, 99–104.

Belloni, M., and Robinett, R. (2014). "The Infinite Well and Dirac Delta Function Potentials as Pedagogical, Mathematical and Physical Models in Quantum Mechanics," *Physics Reports* 540, 25–122.

Bennett, C. H. (2003). "Notes on Landauer's Principle Reversible Computation, and Maxwell's Demon," *Studies in the History and Philosophy of Modern Physics* 34, 501–510.

[17] See Computational Chemistry Comparison and Benchmark DataBase. "VII. What's a Hartree?" July 30, 2023. https://cccbdb.nist.gov/hartree.asp.

Bernstein, H. J. (1967). "Spin Precession during Interferometry of Fermions and the Phase Factor Associated with Rotations through 2π Radians," *Physical Review Letters* 18, 1102–1103.

Bérut, A. et al. (2012). "Experimental Verification of Landauer's Principle Linking Information and Thermodynamics," *Nature* 483, 187–189.

Blennow, M. (2018). *Mathematical Methods for Physics and Engineering* (Boca Raton, FL: CRC Press, Taylor and Francis, Boca Raton).

Born,M. (1926). "Zur Quantenmechanik der Stoßvorgänge (On the Quantum Mechanics of Collisions)," *Zeitschrift für Physik* 37, 863–7.

Bowers, W. (1986). "The Classical Polarizability of the Hydrogen Atom," *American Journal of Physics* 54, 347–350.

Buckingham, E. (1912). "On the Deduction of Wien's Displacement Law," *Journal of the Washington Academy of Sciences* 2, 180–182.

Carlip, S. (2014). "Black Hole Thermodynamics," *International Journal of Modern Physics D* 23, 1430023.

Chambers, R. G. (1960). "Shift of an Electron Interference Pattern by Enclosed Magnetic Flux," *Physical Review Letters* 5, 3–5.

Chen, M-C., et al. (2022). "Ruling out Real-Valued Standard Formalism of Quantum Theory," *Physical Review Letters* 128, 040403.

Cimmino, A., et al. (1989). "Observation of the Topological Aharonov-Casher Phase Shift by Neutron Interferometry," *Physical Review Letters* 63, 380–383.

Coblentz, W. W. (1917). "Constants of Spectral Radiation of a Uniformly Heated Inclosure or So-Called Black Body II," *Bulletin of the Bureau of Standards* 13, 459–477.

Colella, R., Overhauser, A. W., and Werner, S. A. (1975). "Observation of Gravitationally Induced Quantum Interference," *Physical Review Letters* 34, 1471–1474. The original suggestion and theoretical analysis underpinning this experiment was done by Overhauser and Colella (1974).

Cooper, F., Khare, A., and Sukhatme, U. (2001). *Supersymmetry in Quantum Mechanics* (River Edge, NJ: World Scientific).

Crease, R. P. (2002). "The most beautiful experiment," *Physics World* 15, 19–20.

Crepeau, J. (2009). "A Brief History of the T^4 law," *Proceedings of the 2009 ASME Summer Heat Transfer Conference*, 59–65.

Davisson, C., and Germer, L. H. (1927). "Diffraction of Electrons by a Crystal of Nickel," *Physical Review* 30, 705–740.

de Broglie, L. (1923). "Waves and quanta," *Nature* 112, 540.

Duff, M., Okun, L. B., and Veneziano, G. (2002). "Trialogue on the Number of Fundamental Constants," *Journal of High Energy Physics* 2002, 023.

Einstein, A. (1905). "Über einen die Erzeugung und Verwandlung des Lichtes betreffenden heuristischen Gesichtspunkt (On a Heuristic Point of View Concerning the Production of and Transformation of Light)," *Annalen der Physik* 17, 132–148.

Eisberg, R., and Resnick, R. (1974). *Quantum Physics of Atoms, Molecules, Solids, Nuclei, and Particles* (New York: Wiley and Sons).

Ellenberger, S., et al. (2013). "Matter-Wave Interference of Particles Selected from a Molecular Library with Masses Exceeding 10000 amu," *Physical Chemistry Chemical Physics* 15, 14696–700.

Fein, Y. Y., et al. (2019). "Quantum Superposition of Molecules beyond 25 kDA," *Nature Physics* 15, 1242–5.

Feynman, R. P., and Hibbs, A. R. (1965). *Quantum Mechanics and Path Integrals* (New York: McGraw-Hill). There is a 2010 emended edition from Dover Publications, with many typos corrected by D. F. Styer in 2005.

Fowler, R. H., and Nordheim, L. (1928). "Electron Emission in Intense Electric Fields," *Proceedings of the Royal Society of London. Series A, Mathematical and Physical Sciences* 119, 173–81.

Fowler, R., and Guggenheim, E. A. (1956). *Statistical Thermodynamics* (Cambridge: Cambridge University Press).

French, A. P., and Taylor, E. F. (1974). "Editorial: The Pedagogically Clean, Fundamental Experiment," *American Journal of Physics* 42, 3.

Fultz, B. (2010). "Vibrational Thermodynamics of Materials," Prog. Mat. Sci. 55, 247–352.

Gamow, G. (1928). "Zur Quantentheorie des Atomkernes (Quantum Theory of the Atomic Nucleus)," *Zeitschrift für Physik* 51, 204–12.

Gamow, G. (1965). *Mr. Tompkins in Paperback* (Cambridge: Cambridge University Press). This edition comprises *Mr. Tompkins in Wonderland* (1940) and *Mr. Tompkins Explores the Atom* (1946), both by Cambridge University Press.

Geiger, H., and Nuttall, J. M. (1911). "LVII: The Ranges of the Particles from Various Radioactive Substances and a Relation between Range and Period of Transformation," *The London, Edinburgh, and Dublin Philosophical Magazine and Journal of Science* 22, 613–21.

Glasstone, S. (1947). *Thermodynamics for Chemists* (New York: Van Nostrand).

Grant, A. K., and Rosner, J. L. (1994). "Classical Orbits in Power-Law Potentials," *American Journal of Physics* 62, 310–15.

Grimus, W. (2013a). "On the 100th Anniversary of the Sackur–Tetrude Equation," Report no. UWTHPH02-11-34, arXiv:1112.3748v2 [physics.hist-ph]. This expanded preprint version of Grimus (2013b) has comparisons to experiments and determines a value of h from data.

Grimus, W. (2013b). "100th Anniversary of the Sackur–Tetrode Equation," *Annalen der Physik* 525, A32–A35.

Gross, E. P. (1961). "Structure of a Quantized Vortex in Boson Systems," *Il Nuovo Cimento* 20, 454–7.

Gurney, R. W., and Condon, E. U. (1929). "Quantum Mechanics and Radioactive Disintegration," *Physical Review* 33, 127–40.

Hartman, N., et al. (2018). "Direct Entropy Measurement in a Mesoscopic Quantum System," *Nature Physics* 14, 1083–1086.

Hartree, D. R. (1927). "The Wave Mechanics of an Atom with a Non-Coulomb Central Field Part I," *Mathematical Proceedings of the Cambridge Philosophical Society* 24, 89–110.

Hawking, S. (1975). "Pair Creation by Black Holes," *Communications in Mathematical Physics* 43, 199–220.

Hill, T. L. (1960). *Introduction to Statistical Thermodynamics* (New York: Addison-Wesley).

Horodecki R, Horodecki P, Horodecki M, and Horodecki K (2009) "Quantum Entanglement," *Reviews of Modern Physics* 81, 865.

Isi, M., et al. (2021). "Testing the Black-Hole Area Law with GW150914," *Physical Review Letters* 127, 011102.

Jönnson, C. (1961). "Elektroneninterferenzen an mehreren ktinstlich hergestellten Feinspalten (Electron Interference at Several Artificially Produced Fine Slits)," *Zeitschrift für Physik* 161, 454–474.

Jönsson, C. (1974). "Electron Diffraction at Multiple Slits," *American Journal of Physics* 42, 4–11. This is an abridged and translated version of Jönnson (1961). The quotation from this work appearing in Section 6.4.1 is actually from the translators' foreword by D. Brandt and S. Hirschi.

Kaplan, A. E., et al . (2000). "Multimode Interference: Highly Regular Pattern Formation in Quantum Wave-Packet Evolution," *Physical Review A* 61, 032101.

Kittel, C., and Kroemer, H. (1980). *Thermal Physics* (San Francisco: W. H. Freeman and Company).

Kusse, B., and Westwig, E. (1998). *Mathematical Physics: Applied Mathematics for Scientists and Engineers* (New York: Wiley and Sons).

Landauer, R. (1961). "Irreversibility and Heat Generation in the Computing Process," *IBM Journal of Research and Development* 5, 183–91.

Lapidus, I. R. (1981). "Fundamental Units and Dimensionless Constants in a Universe with One, Two, or Four Space Dimensions," *American Journal of Physics* 49, 890–1.

Lemons D S, Shanahan W R, and Buchholtz L J (2022) *On the Trail of Blackbody Radiation: Max Planck and the Physics of his Era* (Cambridge, MA: MIT Press).

Li, Z-D., et al. (2022). "Testing Real Quantum Theory in an Optical Quantum Network," *Physical Review* 128, 040402.

Machta, J. (1999). "Entropy, Information, and Computation," *American Journal of Physics* 67, 1074–1077.

McCarthy, I. E., and Weigold, E. (1983). "A Real 'Thought' Experiment for the Hydrogen Atom," *American Journal of Physics* 51, 152–5. See also Lohmann, B., and Weigold, E. (1981). "Direct Measurement of the Electron Momentum Probability Distribution in Atomic Hydrogen," *Physics Letters* 86A, 139–41.

McKagan, S. B., et al. (2009). "A Research-Based Curriculum for Teaching the Photoelectric Effect," *American Journal of Physics* 77, 87–94.

Meacher, D. R., et al. (1991). "Observation of the collapse and fractional revival of a Rydberg wavepacket in atomic rubidium," *Journal of Physics B: Atomic, Molecular and Optical Physics* 24, L63–L69.

Merli, P, Missiroli, G., and Pozzi, G. (1976). "On the statistical aspect of electron interference phenomena," *American Journal of Physics* 44, 306–307.

Merzbacher, E. (2002). "The early history of quantum tunneling," *Physics Today* 55(4), 44–49.

Michelson, A. A., and Gale, H. G. (1925). "The effect of the Earth's rotation on the velocity of light," *Nature* 115, 566.

Michelson, A. A., Gale, H. G., and Pearson, F. (1925). "The effect of the Earth's rotation on the velocity of light Part II," *Astrophysical Journal* 61, 140–5.

Millikan, R. A. (1916). "A direct photoelectric determination of Planck's 'h'," *Physical Review* 7, 355–88.

Millikan, R. A., and Eyring, C. F. (1926). "Law governing the pulling of electrons out of metals by intense electrical fields," *Physical Review* 27, 51–67.

Mohr P, Newell D, and Taylor B (2016) "CODATA recommended values of the fundamental physical constants: 2014," Rev. Mod. Phys. **88** 035009 (73 pages). See also the NIST (National Institute of Science and Technology site at https://physics.nist.gov/cuu/Constants/index.html for regular updates.

Morse, P., and Feshbach, H. (1953). *Methods of Mathematical Physics*, Part I: Chapters 1 to 8 (New York: McGraw-Hill).

Nairz, O., Arndt, M., and Zeilinger, A. (2003). "Quantum Interference Experiments with Large Molecules," *American Journal of Physics* 71, 319–25.

Nauenberg, M. (2016). "Max Planck and the Birth of the Quantum Hypothesis," *American Journal of Physics* 84, 709–20.

Overhauser, A. W., and Colella, R. (1974). "Experimental Test of Gravitationally Induced Quantum Interference," *Physical Review Letters* 33, 1237–9.

Paños F. J., and Pérez, E. (2015). "Sackur–Tetrode Equation in the Lab," *European Journal of Physics* 36, 055033. doi:10.1088/0143-0807/36/5/055033

Pathria, R. K. (1996). *Statistical Mechanics*, 2nd ed. (Amsterdam: Elsevier).

Perrson, J. R. (2018). "Evolution of Quasi-History of the Planck Blackbody Radiation Equation in a Physics Textbook," *American Journal of Physics* 86, 887–92.

Pitaevskii, L. P. (1961). "Vortex Lines in an Imperfect Bose Gas," *Journal of Experimental and Theoretical Physics* 13, 451–4.

Planck, M. (1900). "Über irreversible Strahlungsvorgänge (About Irreversible Radiation Processes)," *Sitzungsberichte der Königlich Preußischen Akademie der Wissenschaften zu Berlin* 5, 440–80. See also Planck, M. (1901). "Ueber das Gesetz der Energieverteilung im Normalspectrum (On the Law of Distribution of Energy in the Normal Spectrum)," *Annalen der Physik* 4, 553–63.

Quigg, C., and Rosner, J. L. (1979). "Quantum Mechanics with Applications to Quarkonium," *Physics Reports* 56, 167–235.

Rauch H., et al. (1975). "Verification of Coherent Spinor Rotation of Fermions," *Physics Letters A* 54, 425–7.

Rauch, H., Treimer, W., and Bonse, U. (1974). "Test of a Single Crystal Neutron Interferometer," *Physics Letters A* 47, 369–71.

Rauch, H., and Werner, S. A. (2015). *Neutron Interferometry: Lessons in Experimental Quantum Mechanics, Wave-Particle Duality, and Entanglement*, 2nd ed. (Oxford: Oxford University Press).

Robinett, R. W. (2004). "Quantum Wave Packet Revivals," *Physics Reports* 392, 1–119.

Robinett, R. W. (2006). *Quantum Mechanics: Classical Results, Modern Systems, and Visualized Examples*, 2nd ed. (Oxford: Oxford University Press).

Segre, C., and Sullivan, J. D. (1976). "Bound-State Wave Packets," *American Journal of Physics* 44, 729–32.

Shannon, C. E. (1948). "A Mathematical Theory of Communication," *The Bell System Technical Journal* 27, 379–423, 623–56.

Shannon, C. E., and Weaver, W. (1949). *The Mathematical Theory of Communication* (Chicago: University of Illinois Press).

Shull, H., and Hall, G. G. (1959). "Atomic Units," 1959 *Nature* 184, 1559–60.

Stoney, G. J. (1881). "On the Physical Units of Nature," *The London, Edinburgh, and Dublin Philosophical Magazine and Journal of Science*, 11, 381–90. (Described in print as "From the Scientific Proceedings of the Royal Dublin Society of February 16, 1881, being a paper which had been read before Section A of the British Association at the Belfast Meeting in 1881".)

Styer, D., et al. (2002). "Nine Formulations of Quantum Mechanics," *American Journal of Physics* 70, 288–97.

Sukhatme, U. P. (1973). "WKB Energy Levels for a Class of One-Dimensional Potentials," *American Journal of Physics* 41, 1015–1016.

Thomson, G., and Reid, A. (1927). "Diffraction of Cathode Rays by a Thin Film," *Nature* 119, 890.

Thorne, K., and Blandford, R. (2017). *Modern Classical Physics* (Princeton: Princeton University Press).

Tonomura, A., et al. (1989). "Demonstration of Single-Electron Buildup of an Interference Pattern," *American Journal of Physics* **57**, 117–120.

Tonomura, A. et al. (1986). "Evidence for Aharonov–Bohm Effect with Magnetic Field Completely Shielded from Electron Wave," *Physical Review Letters* 56, 792–5.

Valleé, O., and Soares, M. (2010). *Airy Functions and Applications to Physics*, 2nd ed. (New Jersey: World Scientific).

Vavagiakis, E. M., Bachlechner, T. C., and Klegan, M. (2021). "Is the Electric Potential Physical?," *Physics Today* 74(8), 62–63.

Werner S. A., et al. (1975). "Observation of the Phase Shift of a Neutron Due to Precession in a Magnetic Field," *Physical Review Letters* **35**, 1053–5.

Werner, S. A., Staudenmann, J. -L., and Colella, R. (1979). "Effect of Earth's Rotation on the Quantum Mechanical Phase of the Neutron," *Physical Review Letters* 42, 1103–1106

Wigner, E. (1960). "The Unreasonable Effectiveness of Mathematics in the Natural Sciences (Richard Courant Lecture in Mathematical Sciences)," *Communications on Pure and Applied Mathematics* 13, 1–14.

Wilczek, F. (2005). "On Absolute Units. I: Choices,". *Physics Today* 58(10), 12–13.

Wilczek, F. (2006a). "On Absolute Units. II: Challenges and Responses," *Physics Today* 59(1), 10–11.

Wilczek, F. (2006b). "On Absolute Units. III: Absolutely Not!," *Physics Today* 59(5), 10–11.

Williams, R (2009) "This Month in Physics History: September 1911—The Sackur–Tetrode Equation: How Entropy Met Quantum Mechanics," *American Physical Society* https://www.aps.org/archives/publications/apsnews/200908/physicshistory.cfm

Wu, D., et al. (2022). "Experimental Refutation of Real-Valued Quantum Mechanics under Strict Locality Conditions," *Physical Review Letters* **129**, Article no. 140401.

Yeazell, J. A., Mallalieu, M., and Stroud, Jr. C. R. (1990). "Observation of the Collapse and Revival of a Rydberg Electronic Wave Packet," *Physical Review Letters* **64**, 2007–10.

Yeazell, J. A., et al. (1989). "Classical Periodic Motion of Atomic-Electron Wave Packets", *Physical Review A* **40**, 5040–3.

Yeazell, J. A., and Stroud, Jr. C. R. (1991). "Observation of Fractional Revivals in the Evolution of a Rydberg Atomic Wave Packet", *Physical Review A* **43**,5153–6.

Young T. (1807). *A Course of Lectures on Natural Philosophy and the Mechanical Arts*, Vol. 1, Joseph Johnson. 463–4.

Part III

Exploring more advanced topics

Chapter 7

Advanced quantum mechanics

7.1 The role of the fine-structure constant (α_{FS}) in quantum theory

Of the four fundamental forces of nature, the strong, weak, electromagnetic (EM), and gravitational interactions, Newtonian gravity (and extensions by Einstein) play the most important role in systems ranging in length scales from geophysical to planetary and eventually over cosmological distances, with Newton's constant G being the fundamental dimensional quantity. In contrast, EM interactions dominate the structure of atoms and molecules, which in turn provide the templates for higher-level organizational constructs such as chemistry and biology, or solid-state physics and technology.

For the study of electricity and magnetism (E&M), using either classical or quantum mechanics, the basic electro-static interaction strength can be described by e and ϵ_0, most often occurring in the combination $K_e = e^2/4\pi\epsilon_0$. At length scales where quantum theory provides the appropriate calculational tool, we add $\hbar$ as a natural dimensional quantity, while for relativistic systems, the speed of light c is also relevant. Even though we have yet to discuss the details of the energy level structure of the hydrogen atom, we have already seen (in Eqn. (1.92)) that an important dimension**less** combination of $e, \epsilon_0, \hbar$, and c is

$$\alpha_{FS} \equiv \frac{e^2}{4\pi\epsilon_0\hbar c} \equiv \frac{K_e}{\hbar c} \approx \frac{1}{137}, \tag{7.1}$$

which we have called the **fine-structure constant** in advance of later discussions of the detailed spectroscopy of the H-atom.

The first time one might encounter this combination is in the Bohr model of hydrogen, where a semi-classical analysis finds that the electron speed in the state with principal quantum number n is $v_n = v_0/n$, where

$$v_0 = \left(\frac{e^2}{4\pi\epsilon_0\hbar}\right) = \left(\frac{K_e}{\hbar}\right) = \left(\frac{K_e}{\hbar c}\right)c \equiv (\alpha_{FS})c, \tag{7.2}$$

so that $v_0/c = \alpha_{FS} \approx 1/137 << 1$ and the electron in the hydrogen atom can be assumed (to first-approximation) to be non-relativistic. Even though the quantity c does not explicitly appear in the basic quantum-Coulomb problem, we can use this dimensionless ratio to quantify the effects of relativistic (and, as we'll see, other) corrections in the real H-atom system.

Following that suggestion, we recall that the relativistically correct formulae connecting energy (E), momentum (p), and mass (m) can be expanded for speeds much less than c, in two familiar,

Dimensional Analysis Across the Landscape of Physics. Richard W. Robinett, Oxford University Press.
DOI: 10.1093/oso/9780192867551.003.0007

and equivalent, ways, namely, for $pc << mc^2$ as

$$E = \left[(mc^2)^2 + (pc)^2\right]^{1/2} = mc^2\left[1 + \left(\frac{pc}{mc^2}\right)^2\right]^{1/2}$$
$$= mc^2\left[1 + \frac{1}{2}\left(\frac{p}{mc}\right)^2 - \frac{1}{8}\left(\frac{p}{mc}\right)^4 + \cdots\right]$$
$$= mc^2 + \frac{p^2}{2m} - \frac{1}{8}\frac{p^4}{m^3c^2} + \cdots \tag{7.3}$$

or for $v/c << 1$ as

$$E = \gamma mc^2 = mc^2\left(1 - \frac{v^2}{c^2}\right)^{-1/2} = mc^2\left[1 + \frac{1}{2}\left(\frac{v}{c}\right)^2 + \frac{3}{8}\left(\frac{v}{c}\right)^4 + \cdots\right]$$
$$= mc^2 + \frac{1}{2}mv^2 + \frac{3}{8}\frac{mv^4}{c^2} + \cdots \tag{7.4}$$

and the coefficients of the two expansions are different in second order since relativistically $p = \gamma mv = mv/\sqrt{1 - (v/c)^2}$ and not just $p = mv$. Thus, for the hydrogen atom, the first relativistic corrections are then roughly a factor of $\mathcal{O}(v_0/c)^2 = \mathcal{O}(\alpha_{FS}^2) \approx 10^{-4}$ smaller than any leading-order effects.

Extending the range of the use of α_{FS}, we note that the Bohr atom results for the energy eigenvalues for the quantum Coulomb system model of the hydrogen atom can be written in the form $E_n = -E_H^{(0)}/2n^2$, where

$$E_H^{(0)} \equiv \left(\frac{K_e^2 m_e}{\hbar^2}\right) = \left(\frac{K_e^2}{\hbar^2 c^2}\right)(m_e c^2) = (\alpha_{FS})^2(m_e c^2)$$
$$= (\alpha_{FS})^2(0.511\,MeV) \approx 27.2\,eV. \tag{7.5}$$

Once again, while there is no requirement to write the energy eigenvalues including factors of c, we will see that it's very useful to classify various contributions to the detailed energy level structure of the H-atom[1] in the form $(\alpha_{FS})^n(m_e c^2)$ with various powers of $n \geq 2$.

As an example, we know that the kinetic and potential energy terms for the H-atom problem (say from Eqn. (6.147)) will be certainly of the same dimension, but will also be of the same order of magnitude[2], so that $E_H^{(0)} \sim p^2/2m \sim mv_0^2/2$, where the superscript $^{(0)}$ denotes the (non-relativistic) Bohr result. We then have from Eqn. (7.4) that the first relativistic corrections to the H-atom spectra will scale as

$$E_H^{(rel)} \propto \frac{mv_0^2}{2}\left(\frac{v_0^2}{c^2}\right) \propto E_H^{(0)}\left(\frac{v_0}{c}\right)^2 \propto E_H^{(0)}(\alpha_{FS})^2 \propto \alpha_{FS}^4(m_e c^2)$$
$$\sim 10^{-4}E_H^{(0)}. \tag{7.6}$$

[1] A pedagogical approach used by Griffiths (2004).

[2] For a general discussion of the sharing of kinetic/potential energy in bound state systems, see almost any book on classical or quantum mechanics, looking in the index under the entry for **virial theorem**.

The actual values of such corrections depend on the details of the quantum state (and their quantum numbers n, l), and although small, can be easily identified in the detailed spectroscopy of atomic lines, giving rise to the fine structure alluded to in the subscript of α_{FS}.

In addition to speeds and energies, we can make similar use of factors of α_{FS} when discussing time scales for the H-atom problem. For example, the classical/quantum periodicity from Eqn. (6.154) can be written as $\tau_n = 2\pi n^3 \tau_0$, where

$$\tau_0 \equiv \frac{\hbar^3}{K_e^2 m_e} = \left(\frac{\hbar^2 c^2}{K_e^2}\right)\frac{\hbar}{m_e c^2} = \frac{\hbar}{(\alpha_{FS})^2(m_e c^2)} = \frac{\hbar}{E_H^{(0)}}, \tag{7.7}$$

which is another dimensionally natural connection. To see the utility of writing the H-atom time scale in this form, we note that the decay rate λ (and related lifetime, τ) for the transition from the first-excited ($2P$) to ground state ($1S$) can be calculated[3] to be

$$\frac{1}{\tau(2P \to 1S)} = \lambda(2P \to 1S) = \left(\frac{2}{3}\right)^8 \frac{(\alpha_{FS})^5(m_e c^2)}{\hbar}. \tag{7.8}$$

Comparing this decay time to the period in the $n = 2$ state (namely $\tau(n = 2) = 2\pi(2)^3\tau_0$), we find the ratio

$$\frac{\tau(2P \to 1S)}{\tau(n = 2)} = \frac{1}{16\pi}\left(\frac{3}{2}\right)^8 \frac{1}{(\alpha_{FS})^3} \approx 10^6, \tag{7.9}$$

so that the first excited state "lives" for many (classical) rotational periods/cycles before decaying to the ground state. For comparison, if we assume that an average human life time is 80 *years*, this same ratio would correspond to a "personal period" of about an hour. You're asked in **P7.1** to also compare these times to the **classical free-fall time** for the H-atom, when treated without quantum mechanics, but couched in the same "power counting" language of factors of α_{FS}.

For length scales, one of the first dimensional quantities we discussed in the context of the Bohr atom model was the Bohr radius, obtained in the semi-classical energy minimization in Eqn. (6.148), namely

$$a_0 \equiv \frac{\hbar^2}{m_e K_e} \approx 5.3 \times 10^{-11}\, m \approx 0.53\,\text{Å}, \tag{7.10}$$

which sets the scale for all atomic and molecular interaction lengths as Ångström size. We also note that a_0, τ_0 are related via

$$\frac{a_0}{\tau_0} = \left(\frac{\hbar^2}{m_e K_e}\right)\left(\frac{\hbar^3}{m_e K_e^2}\right)^{-1} = \frac{K_e}{\hbar} = \left(\frac{K_e}{\hbar c}\right)c = (\alpha_{FS})c = v_0 \tag{7.11}$$

as expected.

A second natural length scale arises from the radiative transitions between quantized energy states, namely the wavelength of the emitted photons. The connection between the photon energies (E_γ) and wavelengths (λ_γ) are given by

[3] See e.g., Bethe and Salpeter (1957).

$$
\begin{aligned}
E_\gamma &= \Delta E_{n_1,n_2} = E_{n_1} - E_{n_2} \\
\frac{2\pi\hbar c}{\lambda} &= hf = \frac{1}{2}\left(\frac{K_e^2 m_e}{\hbar^2}\right)\left(\frac{1}{n_2^2} - \frac{1}{n_1^2}\right) \\
\frac{1}{\lambda_{n_1,n_2}} &= \frac{1}{4\pi}\left(\frac{K_e^2 m_e}{\hbar^3 c}\right)\left(\frac{1}{n_2^2} - \frac{1}{n_1^2}\right) \\
(\lambda_{n_1,n_2})^{-1} &= Ry\left(\frac{1}{n_2^2} - \frac{1}{n_1^2}\right),
\end{aligned} \tag{7.12}
$$

where the **Rydberg constant** (Ry) is defined by

$$
Ry \equiv \frac{1}{4\pi}\frac{K_e^2 m_e}{\hbar^3 c} \approx 1.1 \times 10^5 \, cm^{-1}. \tag{7.13}
$$

The typical photon wavelengths then scale as

$$
\begin{aligned}
\lambda_\gamma \propto \frac{1}{Ry} &= 4\pi\left(\frac{\hbar^3 c}{K_e^2 m_e}\right) = 4\pi\left(\frac{\hbar c}{K_e}\right)\left(\frac{\hbar^2}{K_e m_e}\right) \\
&= \left(\frac{4\pi}{\alpha_{FS}}\right) a_0 \approx 900\,\text{Å} >> a_0.
\end{aligned} \tag{7.14}
$$

This hierarchy implies that the emitted radiation has a wavelength that is much larger than the **antenna size** (in this case, the physical dimension of the atom/molecule), which implies (from classical EM theory) that such systems are relatively poor (inefficient) radiators. While such a statement might seem like an off-hand allusion to engineering design lore, there are important implications of this α_{FS} hierarchy in the quantum mechanical description of atomic transitions.

For example, the amplitude for a radiative transition between an initial atomic/molecular state, $\psi_i(\boldsymbol{r})$, and a final configuration, $\psi_f(\boldsymbol{r})$, accompanied by a photon (represented by a plane-wave of the form $e^{+i\boldsymbol{k}_\gamma\cdot\boldsymbol{r}}$) can be written (somewhat schematically) as an overlap integral of the form

$$
\begin{aligned}
\text{Amplitude}(i \to f + \gamma) &\sim \int \underbrace{\psi_f(\boldsymbol{r})\; e^{+i\boldsymbol{k}_\gamma\cdot\boldsymbol{r}}}_{\text{final state}} \; \underbrace{\psi_i(\boldsymbol{r})}_{\text{initial state}} \; dV \\
&\sim \int \psi_f(\boldsymbol{r})\left(\underbrace{\;1\;}_{=0} + i\boldsymbol{k}_\gamma\cdot\boldsymbol{r} - \frac{1}{2}(\boldsymbol{k}_\gamma\cdot\boldsymbol{r})^2 + \cdots\right)\psi_i(\boldsymbol{r})\, dV,
\end{aligned} \tag{7.15}
$$

where we've expanded the plane-wave exponential in powers of $\boldsymbol{k}_\gamma \cdot \boldsymbol{r}$. This expansion makes sense, given that

$$
k_\gamma = \frac{2\pi}{\lambda_\gamma} \quad \text{and} \quad |\boldsymbol{r}| \sim a_0 \qquad \text{so that} \qquad \boldsymbol{k}_\gamma\cdot\boldsymbol{r} \sim \frac{2\pi a_0}{\lambda_\gamma} \sim \frac{\alpha_{FS}}{2} << 1. \tag{7.16}
$$

The first term in the expansion (from the leading $\mathcal{O}(1)$ term) always vanishes, because the overlap between the initial and final quantum states is zero (they are so-called **orthogonal** solutions of the Schrödinger equation), while the subsequent terms are each of order α_{FS} smaller. The systematic evaluation of such decay amplitudes (and their squares, which give the actual decay rates) can lead to selection rules depending on the symmetries of the $\psi_{i,f}(\boldsymbol{r})$ states (via their quantum numbers) and so radiative lifetimes can be characterized by various powers of $(\alpha_{FS})^2$.

One of the new length scales introduced in a modern physics course, often as the first application of relativistic kinematics in a collision process, is related to **Compton scattering**, namely photon–electron scattering or the process $\gamma + e^- \rightarrow \gamma + e^-$. The application of conservation of energy-momentum to this process gives

$$\lambda'_\gamma - \lambda_\gamma = \frac{h}{m_e c}\left[1 - \cos(\theta_\gamma)\right], \tag{7.17}$$

where $\lambda_\gamma, \lambda'_\gamma$ are the initial/final wavelengths of the incoming/outgoing photons and θ_γ is the scattering angle. The length-scale on the right-hand side is

$$\lambda_C^{(e)} \equiv \frac{h}{m_e c} \qquad \text{or the related combination involving } \hbar,$$

$$\lambda\!\!\!^{-}{}_C^{(e)} \equiv \frac{\hbar}{m_e c} = \frac{\lambda_C^{(e)}}{2\pi} \tag{7.18}$$

for the **Compton wavelength** or the **reduced Compton wavelength** of a particle, respectively, in this case the electron. More generally, we have $\lambda\!\!\!^{-}{}_C^{(X)} \equiv \hbar/m_X c$ for any particle labeled X.

Despite this being an EM scattering process (about which more in **Example 7.1**), this length scale does <u>**not**</u> include factors of e or ϵ_0, but <u>**is**</u> related to the two earlier fundamental lengths via

$$\lambda\!\!\!^{-}{}_C^{(e)} = \frac{\hbar}{m_e c} = \left(\frac{K_e}{\hbar c}\right)\left(\frac{\hbar^2}{m_e K_e}\right) = (\alpha_{FS})a_0 \approx 0.0039\,\text{Å} = 390\,F, \tag{7.19}$$

which is about "half-way" (in a logarithmic sense) between nuclear (F) and atomic (Å) length scales.

A final distance scale, which appears in both classical and quantum mechanical applications, is the **classical electron radius**, r_e. This is obtained by assuming that the electro-static potential energy of an electron is equivalent to its rest mass, thus a combination of classical E&M and relativity, so that

$$\frac{e^2}{4\pi\epsilon_0 r_e} = \frac{K_e}{r_e} = m_e c^2 \qquad \longrightarrow \qquad r_e \equiv \frac{K_e}{m_e c^2} = \left(\frac{\hbar}{m_e c}\right)\left(\frac{K_e}{\hbar c}\right)$$
$$= (\alpha_{FS})\lambda\!\!\!^{-}{}_C^{(e)} \approx 2.8\,F, \tag{7.20}$$

which is a typical nuclear length scale. In this case, r_e has no dependence on $\hbar$ and so can be confidently described as classical, at least in not being (directly) related to quantum mechanical effects.

We have now found four very distinct length scales involved in the EM interactions of photons (or the EM field) and electrons, for bound states or scattering processes, each separated by single powers of the fine-structure constant, namely

$$r_e : \lambda\!\!\!^{-}{}_C : a_0 : (Ry)^{-1} \qquad \longleftrightarrow \qquad \alpha_{FS}^2 : \alpha_{FS} : 1 : \alpha_{FS}^{-1} \tag{7.21}$$

spanning a dynamic range of $\alpha_{FS}^{-3} \approx 10^6$, from nuclear to atomic/molecular distance scales.

We note that despite the description of r_e as the classical electron radius, all data from experiments probing the structure of the electron indicate that it's actually a point-like particle, with limits on its possible size of $R_e \lesssim 2 \times 10^{-5}\,F = 2 \times 10^{-20}\,m$, while the proton and neutron are known to be composite objects (made of quarks) with sizes of roughly $R_{p,n} \sim 1\,F$.

Example 7.1 Thomson scattering

The scattering cross-section (σ_T) of EM radiation from a free charged particle can be described by classical methods (as discussed below) as the low-energy limit ($E_\gamma << m_e c^2$) of Compton scattering and this process can be explored using the methods of dimensional analysis (DA).

We can write σ_T in terms of the quantities e, ϵ_0, m_e, c, and recalling that the dimensions of a cross-section are $[\sigma] = \mathsf{L}^2$, we have

$$\sigma_T \propto e^\alpha \epsilon_0^\beta m_e^\gamma c^\delta$$

$$\mathsf{L}^2 = \mathsf{Q}^\alpha \left(\frac{\mathsf{Q}^2\mathsf{T}^2}{\mathsf{ML}^3}\right)^\beta \mathsf{M}^\gamma \left(\frac{\mathsf{L}}{\mathsf{T}}\right), \tag{7.22}$$

which can be solved giving $\alpha = 4$, $\beta = \gamma = -2$, and $\delta = -4$, or (adding the usual factors of 4π alongside the ϵ_0)

$$\sigma_T = C_T\left(\frac{e^2}{4\pi\epsilon_0 m_e c^2}\right)^2 = C_T\left(\frac{K_e}{m_e c^2}\right)^2 = C_T r_e^2, \tag{7.23}$$

where r_e is the classical electron radius, and a detailed calculation (next paragraph) gives $C_T = 8\pi/3$. So, while the familiar kinematic connection in Eqn. (7.17) contains $\lambda\!\!\!^{-}{}_C^{(e)}$, with no e/ϵ_0 dependence, the scattering cross-section σ_T does, via the r_e scale.

We can use classical arguments to solve this problem in more detail, and a shorthand version of such a "proof" starts by using the Larmor formula from Eqn. (4.113) for the radiation from an accelerated charge, namely

$$P_{rad} = \frac{\mu_0}{6\pi}\frac{e^2 a^2}{c} = \frac{1}{6\pi}\frac{e^2 a^2}{\epsilon_0 c^3} = \frac{2}{3}\left(\frac{e^2}{4\pi\epsilon_0}\right)\frac{a^2}{c^3} = \frac{2}{3}\frac{K_e a^2}{c^3}. \tag{7.24}$$

The relevant acceleration (a) is due to the external electric force/field (E_0) via

$$m_e a = F_E = eE_0 \qquad \text{or}$$

$$a = \frac{eE_0}{m_e} \qquad \longrightarrow \qquad P_{rad} = \frac{2}{3}\frac{K_e e^2 E_0^2}{m_e^2 c^3}. \tag{7.25}$$

The power incident on the target is the energy flux provided by the EM field, in the form of the Poynting vector,

$$\mathbf{S} = \epsilon_0\, \mathbf{E} \times \mathbf{B} \qquad \text{and for an EM wave} \qquad |\mathbf{S}| = \epsilon_0 c E_0^2. \tag{7.26}$$

In terms of dimensions, the two quantities $\mathbf{S}$ and P_{rad} can be described as

$$[\mathbf{S}] = \left[\frac{\textit{incident power}}{\textit{area}}\right] \quad \text{and} \quad [P_{rad}] = \left[\frac{\textit{energy radiated}}{\textit{time}}\right] \tag{7.27}$$

$$\text{so that} \qquad \left[\frac{P_{rad}}{|\mathbf{S}|}\right] = [\textit{area}], \tag{7.27}$$

so their ratio gives the cross-section. We then have a final result

$$\sigma_T = \frac{P_{rad}}{S} = \frac{2}{3}\left(\frac{e^2 K_e}{\epsilon_0 m_e^2 c^4}\right) = \frac{8\pi}{3}\left(\frac{e^2}{4\pi\epsilon_0}\right)\left(\frac{K_e}{m_e^2 c^4}\right)$$
$$= \frac{8\pi}{3}\left(\frac{K_e}{m_e c^2}\right)^2 = \frac{8\pi}{3} r_e^2 . \tag{7.28}$$

This result does not depend on $\hbar$ and so can be considered as a purely classical EM process. For future reference, we note that since we can also write $r_e = \alpha_{FS} \lambda\!\!\!^{-}_C$, the Thomson scattering cross-section can be expressed as

$$\sigma_T = \frac{8\pi}{3}(\alpha_{FS})^2 \lambda\!\!\!^{-2}_C, \tag{7.29}$$

where $\hbar$ appears in both α_{FS} and $\lambda\!\!\!^{-}_C$. In much the same way that Eqn. (7.5) was rewritten to express a non-relativistic expression (which didn't depend on c) in terms of α_{FS} and $m_e c^2$ to facilitate comparisons among physical process (see Section 7.2.3), the expression in Eqn. (7.29) can be used when exploring Feynman-diagram motivated discussions of many QED cross-sections.

Finally, for a particle of mass m_X, charge q_X, and with intrinsic angular momentum or spin $\mathbf{S}_X$, its magnetic moment can be written as

$$\boldsymbol{\mu}_X = g_X\left(\frac{q_X}{2m_X}\right)\mathbf{S}_X, \tag{7.30}$$

where g_X is the dimension**less** gyro-magnetic ratio, given in terms of the dimension**ful** combination $q_X/2m_X$, motivated by the classical result from Eqn. (4.41). While the interaction of such a magnetic moment with a $\boldsymbol{B}$ field is indeed an EM one, at this stage the value of α_{FS} seems to play no role.

For the (seemingly) point-like electron (e) and muon (μ), the $g_{e,\mu}$ factors are predicted by the Dirac equation to be exactly $g = 2$, which is indeed very close to the values $g_e \approx g_\mu \approx 2.0023...$ found experimentally. For the proton and neutron, which are known to be composite objects (made of quarks), and of finite size, their g-values are $g_p = 5.585...$ and $g_n = -3.826...$ which are far from their Dirac values. The magnetic moments of such strongly interacting particles (with complex internal struture) can be predicted using numerically intensive and computationally expensive first-principles **quantum chromodynamics (QCD)** calculations (performed on a space-time lattice) with good agreement with such values; you can review the progress of such efforts, from the early results of Martinelli et al. (1982) to more recent ones by Beane et al. (2014) or Parreño et al. (2017).

Much more impressively, the calculations of the electron (and muon) magnetic moments from **quantum electrodynamics (QED)**, and their comparison to experimental results, are considered among the landmark achievements of quantum field theory, with calculations and data agreeing to something like twelve orders of magnitude. For example, a recent measurement of g_e is described (Fan et al. (2023)) as follows:

> The electron magnetic moment, $\mu/\mu_B = g/2 = 1.00115965218059(13)[0.13ppt]$, is determined 2.2 times more accurately than the value that stood for fourteen years. The most precisely determined property of an elementary particle tests the most precise prediction of the standard model (SM) to 1 part in 10^{12}.

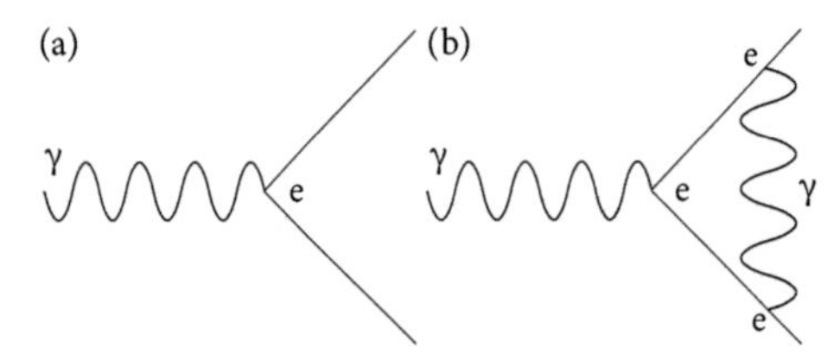

Fig. 7.1 Cartoon pictograph of (a) the interaction of a photon with a charged particle, like the electron or muon, at leading order (which gives the gyromagnetic ratio $g_{e,\mu} = 2$) and (b) the exchange of a virtual photon (γ), which gives rise to the first-order correction, $a_{e,\mu} \equiv (g_{e,\mu} - 2)/2 = \alpha_{FS}/2\pi$.

The leading corrections to both the electron and muon g-factor are due to **(QED)** loop effects arising from the exchange of virtual photons, and a cartoon version of the leading term is shown in Fig. 7.1. The first two terms of such a systematic expansion are often given as contributions to the **deviation** of g_e from its Dirac value of 2 (or a_e for **anomalous magnetic moment**), namely

$$\begin{aligned} a_e &\equiv \frac{g_e - 2}{2} \\ &= \frac{1}{2}\left(\frac{\alpha_{FS}}{\pi}\right) + \left(\frac{\alpha_{FS}}{\pi}\right)^2 \left[\frac{197}{144} + \frac{\pi^2}{12} - \frac{\pi^2}{2}\ln(2) + \frac{3}{4}\zeta(3)\right] + \cdots \end{aligned} \tag{7.31}$$

where the first-order ($\mathcal{O}(\alpha_{FS})$) and second-order ($\mathcal{O}(\alpha_{FS})^2$) corrections were derived by Schwinger (1948) and Sommerfeld (1958), respectively. A recent review by Aoyama, Kinoshita, and Nio (2019) describes the "state-of-the-art" predictions in this field. This is one example of the perturbative calculations for many QED processes, including lifetimes and cross-sections, and most successfully for the e, μ magnetic moments, as a power series in the dimensionless ratio α_{FS}.

7.2 Scaling of the quantum Coulomb problem and the "real" hydrogen atom

The quantum Coulomb problem considered in Section 6.3.4, either approached by semi-classical methods such as the Bohr atom, or via the standard differential equation approaches found in all standard textbooks on quantum theory, is both one of the hallmarks of the undergraduate curriculum as well as one of the canonical results of mathematical physics. Many of the dependences on physical parameters already highlighted in such discussions can be used to explore systems other than the hydrogen-atom, while many new physical effects are present in the "real" H-atom[4] system, beyond the simple Coulomb attraction considered so far, both of which we can explore by DA.

[4] For a fascinating "biography of hydrogen" as the *Essential Element*, and the role its study has played in understanding quantum theory and the properties of atoms, see Rigden (2002). For reviews of the spectrum of atomic hydrogen, see Series (1957), Hänsch (1979), and Karshenboim et al. (2001).

7.2.1 Scaling the Bohr atom to describe new systems

The Bohr atom results for the energy eigenvalues of the H-atom problem of a single proton ($Q_p = +e$) and electron ($Q_e = -e$) interacting via the Coulomb force (assuming that $m_e << m_p$) can be written as

$$E_n = -\frac{1}{2n^2}\left(\frac{K_e^2 m_e}{\hbar^2}\right) = -\frac{(\alpha_{FS})^2}{2n^2}(m_e c^2) = -\frac{E_H^{(0)}}{2n^2} = -\frac{13.6\,eV}{n^2}. \tag{7.32}$$

One of the simplest variations on this problem is to consider multiply ionized atoms, ones where all but one electron is stripped away from a heavier nucleus, changing the appropriate nuclear charge felt by the remaining single e^- from

$$Q_P = +e \quad \longrightarrow \quad Q_Z = +Ze \qquad \text{so that} \qquad K_e = \frac{e^2}{4\pi\epsilon_0} \rightarrow ZK_e. \tag{7.33}$$

Examples of such ions include

$Z = 1$	$Z = 2$	$Z = 3$	$Z = 4$	$\cdots$	$Z = 92$
H	He^+	Li^{++}	Be^{3+}	$\cdots$	U^{91+}

and precise atomic physics experiments have been performed on systems as heavy as "... hydrogenlike uranium measured on cooled, decelerated ion beams ..." as discussed by Stöhlker et al. (2000).

The resulting scaling of the Bohr-atom results for energies, radii, and speeds with nuclear charge Z are then

$$E_n^{(Z)} = Z^2 E_n^{(Z=1)}, \qquad r_n^{(Z)} = \frac{1}{Z} r_n^{(Z=1)}, \qquad \text{and} \qquad v_n^{(Z)} = Z v_n^{(Z=1)}. \tag{7.34}$$

The scaling of v_n with Z means that the lowest-order (non-relativistic) and relativistic corrections to the H-atom spectrum scale differently with nuclear charge and allow for another "knob" to turn in exploring the contributions of non-Coulombic interactions in such systems. For large enough Z values, the system approaches relativistic speeds and so other mathematical approaches may need to be employed. The fact that Bohr orbit radii can be much smaller in such systems with $Z >> 1$, also means that the effect of the nucleus, assumed so far to be point-like, can be more pronounced.

A different realization of what might be called "quasi-single electron" systems comes from the study of alkali atoms, namely those in column 1 of the periodic table. In such systems, closed electron shells (the other $Z - 1$ electrons) shield the single outermost nS^1 electron (think of Gauss's law arguments) giving an effective value of $Z \approx 1$. Examples of such atoms are

H	Li	Na	K	Rb	Cs	Fr
hydrogen	lithium	sodium	potassium	rubidium	cesium	francium
$1S^1$	$2S^1$	$3S^1$	$4S^1$	$5S^1$	$6S^1$	$7S^1$

and because of their similarity to the simplest H-atom system, they are often used as the workhorses of many atomic physics experiments. Recall, for example, the quantum revivals observed by Yeazell et al. (1990) discussed in Section 6.4.3 that used *K* atoms, and similar studies by Meacher et al. (1991), which used *Rb* atoms—other applications include atomic clocks and even the definition of the second relies on *Cs* (cesium) atoms!

In these systems, the scaling of energy levels is determined less from any Z dependence, but rather from a systematic change in the effective quantum number, via the substitution of the Bohr quantum number, $n \to n^* = n - \delta_l$, where δ_l is **quantum defect** depending most on the angular momentum state (value of the l quantum number) of the "outside the closed shells" electron, and less on the n value. For systems with $l \geq 3$, the classical orbits are more and more circular, the electron does not penetrate the inner closed shell(s), and the quantum defect is often close to zero.

Besides the scaling with e, K_e, Z, and n, the dependence on the particle mass(es) is an important one that we should explore more rigorously, as we have assumed that $m_P >> m_e$, ignoring the role of the nuclear mass. For any mutually interacting two-body system, with a potential energy that depends on the separation via $V(x_1, x_2) = V(x_1 - x_2)$, the correct set of degrees of freedom (here shown in one dimension, but easily extendable to a realistic three-dimensional problem by simply using vector notation) are the **relative** and **center-of-mass** coordinates, given by

$$x = x_1 - x_2 \quad \longleftrightarrow \quad \mu \equiv \frac{m_1 m_2}{(m_1 + m_2)} \tag{7.35}$$

(relative or REL coordinate) (reduced mass)

$$X = \frac{m_1 x_1 + m_2 x_2}{m_1 + m + 2} \quad \longleftrightarrow \quad M_{tot} = m_1 + m_2 \tag{7.36}$$

(center-of-mass or COM coordinate) (total mass).

The corresponding expressions for the kinetic energy, written in terms of speeds, which is most relevant for classical Newtonian mechanics, or momentum, which is more important for Schrödinger equation applications in quantum mechanics, are

$$\frac{1}{2}m_1 v_1^2 + \frac{1}{2}m_2 v_2^2 \quad \longrightarrow \quad \underbrace{\frac{1}{2}M_{tot}V^2}_{\text{COM kinetic energy}} + \underbrace{\frac{1}{2}\mu v^2}_{\text{REL kinetic energy}} \tag{7.37}$$

$$\frac{p_1^2}{2m_1} + \frac{p_2^2}{m_2} \quad \longrightarrow \quad \underbrace{\frac{P^2}{2M_{tot}}}_{\text{COM kinetic energy}} + \underbrace{\frac{p^2}{2\mu}}_{\text{REL kinetic energy}} . \tag{7.38}$$

For particles interacting only via their mutual force/potential, the center-of-mass moves freely, while the bound state problem for hydrogen (or H-type atoms or any two-body Coulombic system) is solved by simply extending earlier results with the substitution

$$m_e \longrightarrow \mu = \frac{m_e m_N}{m_e + m_N} = m_e \left(1 + \frac{m_e}{m_N}\right)^{-1} . \tag{7.39}$$

For hydrogen, or more complex atoms with nuclear masses $M_N \geq m_p$, the correction factor is of order $m_e/m_N < 0.05\%$ or smaller.

While such effects may seem small, with the precision methods available in atomic physics experiments, such changes in the energy levels can be easily seen spectroscopically, as shifts in the wavelengths of photons emitted in radiative decays. In fact, this was this method used to discover

deuterium, a bound electron–**deuteron** pair, where $D = {}^2H$ consists of a proton and neutron, so with $M_D \approx 2m_p$; you're asked to explore this in **P7.3.**

7.2.2 **Other two-body quantum Coulomb bound states**

In addition to "swapping out the proton for a heavy nucleus" by considering ionized atoms, it's possible to substitute other particles for the electron in the quantum Coulomb problem. One of the most familiar examples are **muonic atoms**, where the e^- is replaced (or added to in heavier atoms) by another spin-1/2, singly charged, massive particle that interacts by the electro-static interaction, namely the **muon.** While unstable, the muon lifetime is $\tau(\mu^- \to e^- \overline{\nu}_e \nu_\mu) = 2.2\,\mu s$, so they live long enough on atomic scales to allow precision atomic physics experiments probing many new aspects of atomic and even nuclear structure. The main feature that sets them aside from electrons is their larger mass, namely

$$\frac{m(\mu)}{m(e)} = \frac{105.7\,MeV/c^2}{0.511\,MeV/c^2} \approx 207,$$

where we use the MeV/c^2 type mass units common in particle physics, as discussed in Section 7.3. Given the scaling relations we've found (and noting that $m_\mu << M_N$ for most nuclei), we know that energies, radii, speeds, and periods scale as

$$E_n(m_\mu) = E_n \times \left(\frac{m_\mu}{m_e}\right), \quad r_n(m_\mu) = r_n \times \left(\frac{m_e}{m_\mu}\right),$$

$$v_n(m_\mu) = v_n, \quad \text{and} \quad \tau_n(m_\mu) = \tau_n \times \left(\frac{m_e}{m_\mu}\right). \tag{7.40}$$

The spin-1/2 electrons in an atomic system are fermions and so must satisfy the Pauli exclusion principle, leading to restrictions on how many of them can be contained in a given (n, l) quantum state, giving rise to closed shells. An individual muon, however, once captured in a high n orbital, can decay its way down to the ground state with no such constraints,[5] allowing it to be much closer to the atomic nucleus. If this happens in a heavy atom, where the scaling with Z from Eqn. (7.34) also comes into play, the size of the muonic atom can approach that of nuclear scales. To illustrate this, the following table shows values of the muonic Bohr radius for various sets of e, μ masses and Z values.

System	*Radius*	*Radius (m)*
$a_0(H, m_e)$	$a_0 = 0.53\,Å$	5×10^{-11}
$a_0(H, m_\mu)$	$a_0 \times (m_e/m_\mu)$	2.5×10^{-13}
$a_0({}^{A=208}_{Z=82}Pb, m_\mu)$	$a_0(H, m_\mu)/(Z = 82)$	3×10^{-15}
$R_N({}^{A=208}_{Z=82}Pb)$	$A^{1/3} R_0 \sim 7\,F$	7×10^{-15}
R_p	$1\,F$	10^{-15}

[5] Think single rider lines at a theme park, compared to those for couples (S-states with 2 electrons) or families (P-states with 6).

We see that for heavy muonic atoms, the simple scaling with m and Z suggests an effective Bohr radius, which is actually inside the nucleus, and which should only be taken to suggest that, for such heavy nuclei, the EM interaction must be reformulated to consider the finite size of the charged nucleus changing the energy level structure, see **P7.4.** In addition, for a muon that spends more of its time inside the nucleus, its lifetime can increasingly be determined by nuclear capture by a proton (rather than by its weak decay), via the process $\mu^- p \rightarrow \nu_\mu n$, which in turn provides another probe of nuclear structure.

We note that the finite lifetime of the muon still allows it to probe nuclear and atomic (and even material science) properties since on such time scales it lives for quite a long time, namely

$$\tau_{n=1}(Z=1) = \frac{1.5 \times 10^{-16}\, s}{207} \sim 7.2 \times 10^{-19}\, s \qquad \text{so that}$$
$$\frac{\tau(\text{decay})}{\tau(\text{orbit})} \sim 3 \times 10^{12}. \tag{7.41}$$

The experimental status and theoretical background of muonic atoms has been reviewed many times and continues to be a subject of research interest, as documented by Wu and Wilets (1969), Borie and Rinker (1982), Walker (2012), and Knecht, Skawran, and Vogiatzi (2020), and the last authors have noted that ". . . . almost all stable elements have been examined using muons ...". In "Muonic Hydrogen and the Proton Radius Puzzle," Pohl et al. (2013) discuss the utility of precision muonic hydrogen experiments to probe the size of the proton, for which differing values have been found from bound-state versus scattering experiments. For a recent discussion, see Bezginov et al. (2019).

Like the electron, which has its own anti-particle partner (the **positron** or e^+), so too does the muon have a positively charged counterpart, the μ^+, and while the μ^- can be considered as a heavy electron, the μ^+ can act as a light proton. The $\mu^+ e^-$ bound state is called **muonium** and was discovered by Hughes et al. (1960) and has been reviewed several times over the decades; see Hughes (1966, 1992). Besides being of fundamental interest in atomic physics, *Muon and Muonium Chemistry* (Walker 2012) is a rich field of study in materials science, and $\mu^+ e^-$ bound states can preferentially migrate towards sites of interest in crystal lattices and can act as a probe of the magnetic properties of materials. This experimental field of study is often called **muon spin spectroscopy** or **μSR**, similar to the names NMR and ESR, but more often with the R standing for **rotation** and **relaxation.** See Blundell et al. (2022) for an excellent introduction to this field.

The positron can also capture an electron forming an e^+e^- pair called **positronium** (*PS*), which is an even simpler quantum Coulomb bound state, due to the lack of nuclear structure, and can act as another precision probe of QED. In this case, the replacement of the reduced mass gives $m_e \rightarrow \mu = m_e/2$, with simple scaling results, namely

$$E_n(PS) = \frac{1}{2} E_n, \qquad r_n(PS) = 2r_n,$$
$$v_n(PS) = v_n, \qquad \text{and} \qquad \tau_n(PS) = 2\tau_n \tag{7.42}$$

in terms of the H-atom results. Historical reviews of the basic quantum mechanics of *PS* and experimental measurements on this system are very readable, including older accounts by

Stroscio (1975), Berko and Pendleton (1980), and Rich (1981), but also much more recent ones such as Bass et al. (2023), citing applications not only in fundamental research, but also in biomedical applications such as positron emission tomography.

Other two-body quantum Coulomb systems to which such scaling ideas can be extended include:

- Pionic and kaonic atoms, where strongly interacting π^- and K^- mesons replace an electron, also providing probes of nuclear interactions; see Backenstoss (1970), Seki and Wiegand (1975), and Curceanu et al. (2019).
- Antiprotonic-hydrogen atoms, where an antiproton ($\overline{p}$) instead of an e^- is bound to another proton, again probing nuclear structure; see Batty (1989).
- Antihydrogen formed by a $\overline{p}\,e^+$ pair, the study of which has "... long been seen as a route to test some of the fundamental principles of physics" (Bertsche et al. 2015) and continues to be studied experimentally; see, for example, Ahmadi et al. (2022).

Even more speculative investigations of possible models for dark matter have used scaling rules such as those discussed here (and similar ones for molecules) to probe the research frontier of cosmology; see, for example, Ryan (2022a, 2022b) and Gurian (2022) for examples.

7.2.3 Physics of the "real" hydrogen-atom

While the mathematical solution of the quantum Coulomb system is an important first step towards understanding the structure of the H-atom and its energy levels, and therefore making connections to observable quantities via spectroscopy, there are a number of smaller effects that can be studied that contribute to the detailed spectrum of hydrogen (and other atoms). We've already seen that first-order relativistic effects from Eqn. (7.6) contribute energy shifts of order $E^{(rel)} \propto (\alpha_{FS})^4(m_e c^2) \propto (\alpha_{FS})^2 E_H$ to the H-atom spectrum. Another, seemingly unrelated, physical effect is the interaction of the fundamental magnetic moment of the electron (μ_e) with the magnetic field it 'sees' due to the relative motion of the electron/proton, leading to the phrase **spin-orbit coupling** or *SO*.

In a similar way to observers on the Earth thinking that the Sun rotates around us, an electron can experience an effective magnetic field from what it sees as the protons motion. A classical expression for the magnetic field at the center of a current loop (which you could confirm or re-derive using DA if needed) is

$$B = \frac{\mu_0 I}{2R} \qquad \text{and we'll use} \qquad I = \frac{e}{\tau_0} = e\left(\frac{K_e^2 m_e}{\hbar^3}\right) \qquad \text{and} \qquad R \sim a_0 = \frac{\hbar^2}{K_e m_e} \tag{7.43}$$

and eventually we'll swap in $\mu_0 = 1/(\epsilon_0 c^2)$. The magnetic moment of the electron is given by Eqn. (7.30) and with $S = \hbar/2$, we have $\mu_e \sim (e\hbar)/m_e$. The interaction of such a moment in a $\boldsymbol{B}$ field is

$$E_{mag} = \boldsymbol{\mu_e} \cdot \boldsymbol{B}, \tag{7.44}$$

and combining all of these individual dimensional results, we find energy shifts due to such spin-orbit (SO) interactions of order

$$\Delta E_{SO} \sim \mu_e \left(\frac{\mu_0 I}{R}\right) \sim \overbrace{\left(\frac{e\hbar}{m_e}\right)}^{\mu_e} \overbrace{\left(\frac{1}{\epsilon_0 c^2}\right)}^{\mu_0} \overbrace{\left(\frac{eK_e^2 m_e}{\hbar^3}\right)}^{I} \overbrace{\left(\frac{\hbar^2}{m_e K_e}\right)^{-1}}^{1/a_0}$$

$$\sim \left(\frac{e^2}{\epsilon_0}\right)\left(\frac{K_e^3}{\hbar^4 c}\right) m_e$$

$$\sim \left(\frac{e^2}{4\pi\hbar c}\right)\left(\frac{K_e^3}{\hbar^3 c^3}\right)(m_e c^2)$$

$$\Delta E_{SO} \propto (\alpha_{FS})^4 (m_e c^2). \tag{7.45}$$

This contribution is of the same order as the relativistic corrections, $E^{(rel)}$, and together are responsible for the "fine structure" of the H-atom (and more complex systems) spectrum and such corrections have even been confirmed in anti-hydrogen at the few percent level; see Ahmadi et al. (2022).

A secondary magnetic effect in the H-atom is that due to the dipole–dipole interaction of the electron and proton magnetic moments. We can either recall the form of such an interaction from E&M texts, review **P4.20**, or perhaps refresh ourselves by writing such an interaction energy in terms of μ_0, μ_e, μ_p, and their separation R, and matching dimensions, as

$$E_{(\mu_e,\mu_p)} \propto \mu_0^{\alpha} \mu_e^{\beta} \mu_p^{\delta} R^{\delta}$$

$$\frac{\mathsf{ML}^2}{\mathsf{T}^2} = \left(\frac{\mathsf{ML}}{\mathsf{Q}^2}\right)^{\alpha} \left(\frac{\mathsf{QL}^2}{\mathsf{T}}\right)^{\beta} \left(\frac{\mathsf{QL}^2}{\mathsf{T}}\right)^{\gamma} \mathsf{L}^{\delta}. \tag{7.46}$$

This is easily solved for $\alpha = 1$, $\delta = -3$, and $\gamma = 2 - \beta$, and if we assume that the two magnetic moments appear symmetrically, so that $\beta = \gamma$, we then have

$$E_{(\mu_e,\mu_p)} \sim \mu_0 \frac{\mu_e \mu_p}{R^3}. \tag{7.47}$$

Once again substituting the basic magnetic strength in preference for ϵ_0, using the scaling of the two moments, and the separation dictated by Bohr radius type length scale, we identify

$$\mu_0 = \frac{1}{\epsilon_0 c^2}, \qquad \mu_{e,p} \propto g_{e,p} \frac{e\hbar}{m_{e,p}}, \qquad \text{and} \qquad R \sim a_0 = \frac{\hbar^2}{K_e m_e}. \tag{7.48}$$

Substituting these values into Eqn. (7.47), using $g_e = 2$ but retaining g_p, and collecting terms we find the possible energy shifts due to such magnetic dipole–dipole interactions scale as

$$\Delta E_{hyp} \propto \alpha_{FS}^4 (m_e c^2) \left(\frac{m_e}{m_p}\right) g_p. \tag{7.49}$$

We have introduced the subscript hyp as being short for **hyperfine** as these energy shifts are suppressed from those of fine structure by a power of $m_e/m_p \approx 5 \times 10^{-4}$. An energy shift of this magnitude will arise when the electron/proton moments are in an aligned (total spin $S = 1$) configuration versus being in an anti-aligned (total spin $S = 0$) state, with the otherwise spin degenerate ground state being split by this amount, allowing a transition between the two levels.

To see how close we've gotten using some DA and basic physics, we note that the exact expression for the spin triplet to singlet transition energy cited in a famous textbook (Griffiths 2004) is

$$\Delta E_{hyp} = \frac{4g_p\hbar^4}{3m_p m_e^2 c^2 a_0^4} = \alpha_{FS}^4(m_e c^2)\left(\frac{m_e}{m_p}\right)\left(\frac{4g_p}{3}\right). \tag{7.50}$$

We can evaluate this transition energy by using values of the fundamental constants, and $g_p = 5.85$, to find the values of the energy, frequency, period, and wavelength of the emitted photon, namely

$$\Delta E_{hyp} = 5.88 \times 10^{-6}\, eV = E_\gamma \qquad \text{and} \qquad f_\gamma = \frac{E_\gamma}{h} = 1420\, mHz$$

$$\tau_\gamma = \frac{1}{f_\gamma} = 7.1 \times 10^{-8}\, s \qquad \text{and} \qquad \lambda_\gamma = \frac{c}{f_\gamma} = c\tau_\gamma = 21\, cm. \tag{7.51}$$

Section 7.7 discusses examples of the impact that hyperfine interactions have in hydrogen and other atoms.

The final higher-order (in α_{FS}) shift for the hydrogen atom (or other Coulombic bound states) is the **Lamb shift**, which is the energy difference between the $2P_{1/2}$ and $2S_{1/2}$ levels, which would otherwise be degenerate for the quantum Coulomb problem. This effect can be "... explained as a shift in the energy of the atom arising from its interaction with the radiation field ..." as noted by Welton (1948), who provided a heuristic discussion of the effect which, is another manifestation of the exchange of virtual photons as sketched in Fig. 7.1(b). The complete calculation is the stuff of advanced QED texts (we again cite Beresteskii et al. 1982) with the result for hydrogen being cited (in spectroscopic notation) as

$$E_{20(1/2)} - E_{21(1/2)} = 0.41(\alpha_{FS})^5(m_e c^2), \tag{7.52}$$

and you're asked in **P7.5** to evaluate this shift and the corresponding wavelength.

7.3 **Field theory, $\hbar = c = 1$, and Feynman diagrams**

Elementary particle physics has often been at the leading edge of the high-energy and short-distance search for new fundamental phenomena and has helped confirm many aspects of the wildly successful standard model of the strong, weak, and EM interactions. In terms of the diversity of dimensionful quantities, however, there are only a handful that form the focus of the experimental techniques used to probe the basic constituents and forces of nature.

In practice, one often starts with stable charged particles such as electrons (or positrons) or protons (or anti-protons), which can be accelerated by electric/magnetic fields to high energies, scattering/interacting with fixed (at-rest) targets or in colliding beam geometries, to produce new particles whose production rates (i.e., cross-sections), masses, lifetimes, and couplings are the subject of the studies.

Given that the energy gain of a particle of charge q accelerated through a voltage V is $|\Delta E| = |qV|$, where $q = \pm e$, a natural unit of energy for all such processes is

$$\text{one electron-volt} = 1\,eV = (1.6 \times 10^{-19}\,C) \cdot (1\,V) = 1.6 \times 10^{-19}\,J, \tag{7.53}$$

where eV energies are the typical order of magnitude of atomic and molecular processes. At higher energies (probing shorter distance scales), one encounters combinations such as

$$keV = 10^3\,eV, \quad MeV = 10^6\,eV, \quad GeV = 10^9\,eV, \quad TeV = 10^{12}\,eV, \quad \text{and} \quad PeV = 10^{15}\,eV. \tag{7.54}$$

For example, the energy of each proton/anti-proton beam at the CERN Large Hadron Collider (LHC) is 6.8 TeV, so that the total center-of-mass energy is double that, namely $E_{cm} = 13.6\,TeV$. In contrast, the highest-energy cosmic rays ever measured (see Bird et al. 1995, and Abbasi *et al.* 2023 contrast) had $E \approx (2-3) \times 10^{20} eV \sim 40\text{–}50\,J$, meaning a macroscopic amount of energy in a single elementary particle.

Given that such processes are often clearly relativistic, so the speed of light is always assumed involved, the connection between energy, momentum, and mass is often "abbreviated" as

$$E^2 = (pc)^2 + (mc^2)^2 \qquad \Longrightarrow \qquad E^2 = p^2 + m^2, \tag{7.55}$$

with the c being understood, hence the "$c = 1$" language used in particle physics, so that E, p, and m are often all cited in eV units. We can also write (slightly more carefully)

$$\{E\} = eV, \quad \{p\} = \frac{eV}{c}, \qquad \text{and} \quad \{m\} = \frac{eV}{c^2}, \tag{7.56}$$

where the notation $\{\}$ is to remind us to include appropriate powers of c, either by hand, or by simple DA constraints. The rest masses of important fundamental particles are shown in Table 7.1 in eV/c^2 units.

In the production of new (unstable) particles or resonances in nuclear and particle physics, one often encounters peaks in a collision/production cross-section (σ) or probability modeled via the

Table 7.1 Rest masses for many subatomic particles, in MeV/c^2 or GeV/c^2 units.

Particle	Rest mass	Particle	Rest mass
electron (e)	$0.511\,MeV/c^2$	muon (μ)	$105.7\,MeV/c^2$
neutral pion (π^0)	$135.0\,MeV/c^2$	charged pion ($\pi^\pm$)	$159.6\,MeV/c^2$
proton (p)	$928.1\,MeV/c^2$	neutron (n)	$939.4\,MeV/c^2$
charm quark (c)	$1.27\,GeV/c^2$	bottom quark (b)	$4.18\,GeV/c^2$
W-boson ($W^\pm$)	$80.4\,GeV/c^2$	Z-boson (Z^0)	$91.2\,GeV/c^2$
Higgs boson (H)	$125.2\,GeV/c^2$	top quark (t)	$172.8\,GeV/c^2$

Breit–Wigner distribution, namely

$$\sigma(E) \propto \frac{1}{(E^2 - M^2)^2 + (M\Gamma)^2}, \tag{7.57}$$

where E is the center-of-mass energy, M the mass of the new particle, and Γ is the **decay width** of the (unstable) particle, where Γ is expressed in energy units, and is related to the lifetime of the state. To make Eqn. (7.57) dimensionally correct, we should of course use Mc^2 and reintroduce powers of c as needed, while to connect Γ to a decay rate (λ) or lifetime (τ), we have to add appropriate powers of $\hbar$ via

$$\frac{\Gamma}{\hbar} = \lambda = \frac{1}{\tau}, \tag{7.58}$$

which should be familiar from Section 6.4.3 and Eqn. (6.198) and you can practice such translations in **P7.7**. Factors of $\hbar$ are often needed to translate from energy (or momentum or mass) quantities to times.

We have encountered several fundamental length scales that are directly related to Planck's constant, including the de Broglie and Compton wavelengths, namely

$$\lambda_{dB} \equiv \frac{h}{p} = \frac{2\pi\hbar}{p} = \frac{2\pi\hbar c}{pc} \quad \text{and} \quad \bar{\lambda}_e^C \equiv \frac{\hbar}{mc} = \frac{\hbar c}{mc^2}. \tag{7.59}$$

We see that many distance metrics in high energy physics will have very similar forms,

$$\text{length scales} \sim \frac{\hbar c}{E}, \quad \frac{\hbar c}{pc}, \quad \text{or} \quad \frac{\hbar c}{mc^2}, \tag{7.60}$$

which have implications for the form of scattering cross-sections in high-energy experiments as we'll see. For massless photons, we might write $E_\gamma = p_\gamma c = hf = \hbar\omega$, in which case the relevant length scale could be $(\hbar c)/E_\gamma = (\hbar c)/(\hbar\omega) = c/\omega = \bar{\lambda}_\gamma$. In each case, the relevant "translation" factor from energy to length is

$$\hbar c = 1973\, eV \cdot \text{Å} = 197.3\, MeV \cdot F = 0.1973\, GeV \cdot F \tag{7.61}$$

and we may have to insert powers of both $\hbar$ and c to make familiar formulas dimensionally obvious.

Example 7.2 Thomson scattering in field theory language

As an example, let us consider the process of Compton scattering, which is a staple in many field theory texts. In the laboratory system (the geometry in which an incoming photon hits an electron "at rest") we have $\gamma + e^- \rightarrow \gamma + e^-$, where $E_\gamma = hf = \hbar\omega$ and $m_e c^2$ are the two relevant energy scales. For low energies, $\hbar\omega << m_e c^2$, any expression should reduce to the classical Thomson results in Eqns. (7.28) and (7.29). A standard reference on QED processes (Berestetskii et al. 1982) has the following expression for the

Compton cross-section as a function of those two parameters:

$$\sigma_C = \frac{2\pi r_e^2}{x}\left\{\left(1 - \frac{4}{x} - \frac{8}{x^2}\right)\ln(1+x) + \frac{1}{2} + \frac{8}{x} - \frac{1}{2(1+x)^2}\right\}, \tag{7.62}$$

where the authors define the parameter $x \equiv 2\omega/m_e$, which must clearly be dimensionless given the form of the equation. This definition is, however, not dimensionally correct as written, at least in our $\hbar, c \neq 1$ world. We can write $x \sim (\omega/m_e)\hbar^\alpha c^\beta$, and using the fact that $[x] = \mathsf{M}^0\mathsf{L}^0\mathsf{T}^0$, we find $\alpha = 1$ and $c = -2$, but these should already be obvious "by inspection" given our earlier discussions.

The classical Thomson limit corresponds to $x << 1$ and you should be able to expand Eqn. (7.62) in that limit to confirm that you obtain Eqn. (7.28). In the opposite extreme, where $E_\gamma = \hbar\omega >> m_e c^2$ or $x >> 1$, we find

$$\begin{aligned}\sigma_C &= \frac{2\pi r_e^2}{x}\ln(x) = 2\pi\left(\frac{K_e}{m_e c^2}\right)^2\left(\frac{m_e c^2}{2\hbar\omega}\right)\ln\left(\frac{2\hbar\omega}{m_e c^2}\right) \\ &= \pi\left(\frac{K_e^2}{\hbar^2 c^2}\right)^2\left(\frac{\hbar c}{m_e c^2}\right)\left(\frac{\hbar c}{\hbar\omega}\right)\ln\left(\frac{2\hbar\omega}{m_e c^2}\right) \\ \sigma_C &= (\alpha_{FS})^2\left(\pi\lambda_e^C\bar{\lambda}_\gamma\right)\ln\left(\frac{2\hbar\omega}{m_e c^2}\right)\end{aligned} \tag{7.63}$$

and the dimensions of area are due to factors of both λ_e^C and $\bar{\lambda}_\gamma$. In this case, the result is closer in spirit to Eqn. (7.29) where we include the factors of α_{FS} explicitly.

Example 7.3 Glimpses of Feynman diagrams

To gain some intuition about the use of Feynman diagram techniques in the description of fundamental processes, let us introduce some of the ingredients needed in a full field-theoretic evaluation of a physical quantity, in this case the Compton scattering cross-section. We illustrate in Fig. 7.2 one of the two lowest-order (so called **tree-level**) diagrams contributing to this process. For many QED processes, we then need:

- A factor of e at each electron-photon vertex.
- A factor of $S_F(\mathcal{P}) = (\mathcal{P} - m)^{-1}$ for every internal electron line, where $\mathcal{P}$ is the four-momentum (energy-momentum, recall Eqn. (1.94)) of the virtual particle and m is

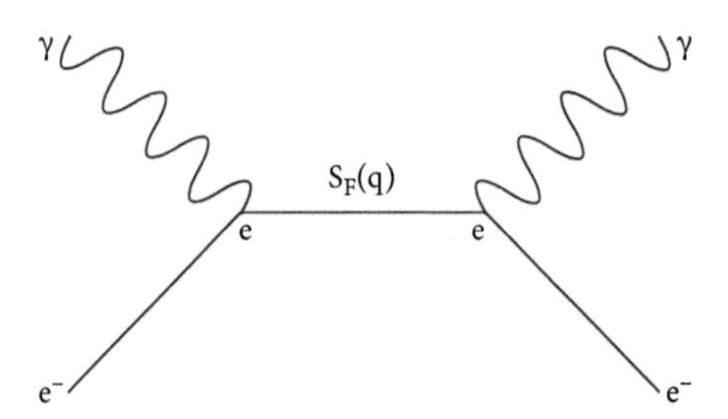

Fig. 7.2 Example Feynman diagram (one of two at tree-level) illustrating the Compton scattering process $\gamma + e^- \to \gamma + e^-$. A factor of electric charge e is associated with each electron–photon vertex, and a propagator $S_F(q)$ with each internal electron line.

its mass. This is a **propagator** for the internal fermion (hence F) line. Notice already that one should insert powers of c to make this dimensionally consistent, namely $\not{P}c$ and mc^2.

- The amplitude for any process (scattering or decay) will be proportional to a product of such factors.
- The cross-section (σ) or decay rate (λ) will be proportional to the amplitude squared!
- For every factor of e^2, one includes a $1/4\pi\epsilon_0$ to ensure the EM dimensions work in our familiar parlance.
- For each factor of $e^2/4\pi\epsilon_0 = K_e$ add enough powers of $\hbar c$ to produce a factor of α_{FS}.
- And finally add enough factors of $\hbar, c$ to ensure that σ or λ have the units of area or inverse time, as needed.

For low energy Compton scattering, $S_F(\not{P})$ is dominated by the m_e (mass) term so we have

$$\text{Amp} \propto (e)\left(\frac{1}{m_e}\right)(e) \rightarrow \left(\frac{e^2}{4\pi\epsilon_0}\right)\left(\frac{1}{m_e}\right) \propto \frac{\alpha_{FS}}{m_e}, \tag{7.64}$$

so that the scattering cross-section is

$$\sigma \propto |\text{Amp}|^2 \propto \frac{(\alpha_{FS})^2}{m_e^2} \quad \xrightarrow{\text{adding } \hbar, c}$$

$$\sigma \propto (\alpha_{FS})^2 \left(\frac{\hbar}{m_e c}\right)^2 \propto (\alpha_{FS})^2 \bar{\lambda}_C^2 \propto (r_e)^2, \tag{7.65}$$

which is consistent with the Thomson scattering limit. We also get some sense of the systematics of how the QED interaction strength (now measured by α_{FS}), and mass/energy scales from Eqn. (7.60), contribute to the physical cross-section.

A process related to Compton scattering by a simple re-drawing of the Feynman diagrams (as in Fig. 7.3) is that of electron–positron annihilation to two photons, namely $e^- + e^+ \rightarrow \gamma + \gamma$. The cross-section for this process is also derived in the same seminal reference as above (Berestetskii et al. 1982) with the expression

$$\sigma_{e^+e^- \rightarrow \gamma\gamma} = \pi r_e^2 \left(\frac{1-\beta^2}{4\beta}\right)\left[\frac{3-\beta^4}{\beta} \ln\left(\frac{1+\beta}{1-\beta}\right) - 2(2-\beta^2)\right], \tag{7.66}$$

where $\beta = v_{rel}/c$ is the relative velocity of the e^+e^- pair. The actual formula as written in the text uses v_{rel} instead of the dimensionless β, so we have already done an obvious translation at this stage.

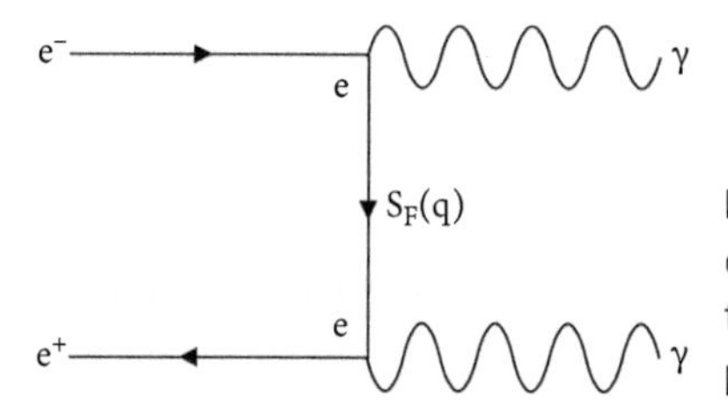

Fig. 7.3 Example Feynman diagram (one of two at tree-level) illustrating the process $e^+ + e^- \rightarrow \gamma + \gamma$.

For small relative speeds, you should be able to show that Eqn. (7.66) reduces to

$$\sigma_{e^+e^-\to\gamma\gamma} = \frac{\pi r_e^2}{2\beta}, \tag{7.67}$$

where the area dimensions are once again determined by $r_e = \alpha_{FS}\lambda_C^{(e)}$. The factor of β^{-1} implies that the cross-section becomes very large as their relative speed goes to zero, since the e^+e^- pair can interact more easily given time to "find each other". At high energies, where $E_e = \gamma m_e c^2 >> m_e^2$, we have

$$\left(\frac{E}{m_e c^2}\right) = \gamma = \frac{1}{\sqrt{1-\beta^2}} \qquad \text{or} \qquad 1-\beta^2 = \left(\frac{m_e c^2}{E}\right)^2, \tag{7.68}$$

so that the relevant cross-section area scales as

$$\begin{aligned}\sigma_{e^+e^-\to\gamma\gamma} \propto (\alpha_{FS})^2 \left(\lambda\!\!\!^{-}{}_C^{e}\right)^2 (1-\beta^2) &= (\alpha_{FS})^2 \left(\frac{\hbar}{m_e c}\right)^2 \left(\frac{m_e c^2}{E}\right)^2 \\ &= (\alpha_{FS})^2 \left(\frac{\hbar c}{E}\right)^2,\end{aligned} \tag{7.69}$$

once again consistent with Eqn. (7.60). Experimental results for the process $e^+ + e^- \to \gamma + \gamma$ (see, e.g., Abdallah et al. 2004) at center-of-mass energies up to 200 *GeV* were described at the time as "... confirming the validity of QED at the highest energies ever attained in electron-positron collisions."

The cross-section formula in Eqn. (7.66) is not only useful in the pair-annihilation of free e^+e^- pairs at ultra-high energies ($E_e >> m_e c^2$), but also in the decay of their bound state, using the low-energy limit in Eqn. (7.67). To make contact with decay rates, we recall that a familiar formula from modern physics texts for the interaction rate (λ) of particles in terms of their velocity, interaction cross-section, and the number density of scatterers is

$$\begin{aligned}\lambda &= (\text{number density}) \cdot (\text{cross-section}) \cdot (\text{velocity}) \\ &= \quad n \quad \cdot \quad \sigma \quad \cdot \quad v \\ \frac{1}{\mathsf{T}} &\stackrel{!}{=} \left(\frac{1}{\mathsf{L}^3}\right)\left(\mathsf{L}^2\right)\left(\frac{\mathsf{L}}{\mathsf{T}}\right),\end{aligned} \tag{7.70}$$

which we confirm is dimensionally correct. For the two-photon decay of the e^+e^- bound state, the cross-section from Eqn. (7.67) diverges as $v_{rel}\to 0$, but in this formula they appear together as $\sigma \cdot v_{rel}$ and so give a finite result in that limit. For a bound state, the relevant number density is actually determined by the *PS* wave function via $|\psi_S(0)|^2 \propto a_0^{-3}$ (basically one e^+e^- pair per *PS* volume), which indeed has dimensions of L^{-3}, giving

$$\lambda(PS\to\gamma\gamma) = |\psi_S(0)|^2 \left(v\sigma_{e^+e^-\to\gamma\gamma}\right)_{v\to 0}. \tag{7.71}$$

The resulting decay rate then scales as

$$\lambda(PS\to\gamma\gamma) \propto \left(\frac{1}{a_0^3}\right)^2 (v) \left(\frac{r_e^2}{v/c}\right) \propto \frac{r_e^2 c}{a_0^3} = C_\lambda \frac{(\alpha_{FS})^5 (m_e c^2)}{\hbar}, \tag{7.72}$$

where a detailed analysis[6] gives $C_\lambda = 1/2$. It's interesting to compare this result to the $2P \rightarrow 1S$ transition rate in hydrogen from Eqn. (7.8) and that for the classical free-fall time considered in **P7.1** as another example of the classification of QED processes in terms of m_e, appropriate factors of $\hbar$ and c, and especially powers of α_{FS}.

Measurements of this decay process (Al-Ramadhan and Gidley 1994) have reached precisions at the 215 *ppm* (parts per million) level which allow for comparison to higher-order radiative corrections to such decays, as reviewed by Czarneck and Karshenboim (1999), though not nearly at the level of comparison to, say $g - 2$ experiments.

Example 7.4 Light-by-light scattering

We've seen in Fig. 7.1(a) that the basic QED interaction can be sketched as a single photon interacting with a charged particle. There is no similar direct coupling of photons to each other (no $\gamma - \gamma$ vertex), which is a reflection of the fact that classical E&M is a linear theory, with $\boldsymbol{E}, \boldsymbol{B}$ fields appearing everywhere, in say Maxwell's equations, singly. But QED does allow for the exchange of virtual particles (as in Fig. 7.1(b)), which can allow for processes not seen in classical electrodynamics. One such process is $\gamma + \gamma \rightarrow \gamma + \gamma$, which can proceed through a **box diagram**, as shown Fig. 7.4.

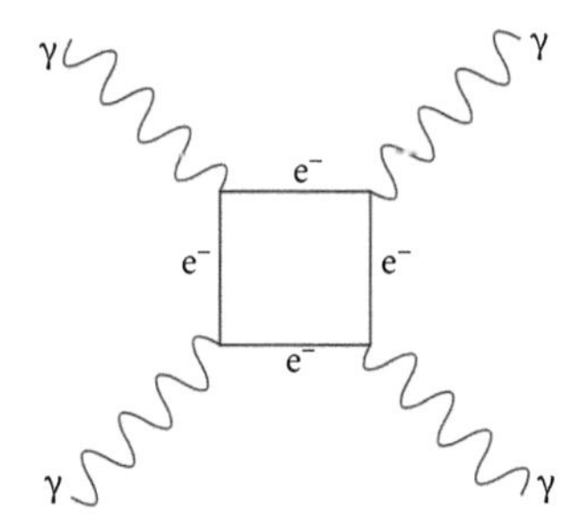

Fig. 7.4 Example Feynman diagram for photon-photon scattering, $\gamma + \gamma \rightarrow \gamma + \gamma$, or light-by-light scattering.

Using what intuition we've gained about the structure of Feynman diagrams, and how to translate those pictographs into calculations, let us assume that for low energies where the S_F fermion propagators are dominated by their m_e mass terms, we guess that the amplitude for this process would scale as

$$\text{amplitude} \propto (e)\left(\frac{1}{m_e}\right)(e)\left(\frac{1}{m_e}\right)(e)\left(\frac{1}{m_e}\right)(e)\left(\frac{1}{m_e}\right) \rightarrow \frac{(\alpha_{FS})^2}{m_e^4} \quad \text{so that}$$

$$\sigma_{\gamma\gamma\rightarrow\gamma\gamma} \propto \frac{(\alpha_{FS})^4}{m_e^8}. \tag{7.73}$$

To turn this into a cross-section with dimensions we understand, we also need to consider not only the ubiquitous factors of $\hbar$ and c, but also the energy dependence of the initial photons, each having an energy $E = \hbar\omega_\gamma$, so assumed to be interacting in a (symmetric) colliding beam geometry. With partial

---

[6] The first calculation of this process was done by Wheeler (1946) in an article with the surprising title *Polyelectrons*.

knowledge of the dependence of σ on α_{FS} and m_e already in place, we can write

$$\sigma_{\gamma\gamma\to\gamma\gamma} \propto \frac{(\alpha_{FS})^4}{m_e^8} \hbar^\alpha c^\beta \omega_\gamma^\delta$$
$$\mathsf{L}^2 = \mathsf{M}^{-8}\left(\frac{\mathsf{ML}^2}{\mathsf{T}}\right)^\alpha \left(\frac{\mathsf{L}}{\mathsf{T}}\right)^\beta \left(\frac{1}{\mathsf{T}}\right)^\delta, \tag{7.74}$$

which is easily solved to obtain $\alpha = 8, \beta = -14$, and perhaps most importantly that $\delta = 6$, giving a strong dependence on the initial photon energy, $E_\gamma = \hbar\omega_\gamma$. The theoretical prediction[7] for low-photon energies ($\hbar\omega_\gamma << m_e c^2$) is

$$\sigma_{\gamma\gamma\to\gamma\gamma} = \frac{973}{10125\pi}(\alpha_{FS})^2 r_e^2 \left(\frac{\hbar\omega}{m_e c^2}\right)^6, \tag{7.75}$$

which you should be able to show (**P7.12**) has the same form as suggested above, when expressed in terms of α_{FS}. This phenomena was predicted in the very early days of quantum mechanics by Euler and Kockel (1935) and it's noteworthy that experiments have seen this light-by-light scattering process, using the γ flux associated with heavy ions (Pb), with enhancements of order Z^2 from the two incoming photons; see Aad et al. (ATLAS Collaboration) (2019).

Example 7.5 The Schwinger effect

Without providing much background, we have already mentioned in **Example 1.4** and **P1.14** that sufficiently large electric fields can produce e^+e^- pairs from the vacuum. In those two earlier discussions we found that the critical electric[8] (and magnetic) fields for this were easy to obtain by DA and we found

$$E_{cr} = \frac{m_e^2 c^3}{e\hbar} \quad \text{and} \quad B_{cr} = \frac{m_e^2 c^2}{e\hbar}. \tag{7.76}$$

We can explore with more physical insight the origin of the critical $\boldsymbol{E}$ field by equating the work done by such a field on a unit electric charge over a distance equal to the (reduced) Compton wavelength $\lambda\!\!\!^{-}{}_C^{(e)}$ of an electron with the energy needed for pair production, namely

$$(eE_{cr})(\lambda\!\!\!^{-}{}_C) = F \cdot d = W = m_e c^2 \quad \longrightarrow$$
$$E_{cr} = \frac{m_e c^2}{e\lambda\!\!\!^{-}{}_C} = \left(\frac{m_e c^2}{e}\right)\left(\frac{m_e c}{\hbar}\right) = \frac{m_e^2 c^3}{e\hbar}. \tag{7.77}$$

To gauge the actual **production rate per unit volume** of such pairs, often denoted by Γ, we can apply DA methods again. Noting that $[\Gamma] = (1/\mathsf{T})/\mathsf{L}^3 = 1/(\mathsf{TL}^3)$, and assuming that Γ could depend on m_e, c, $\hbar$, e

[7] See again the compendium of QED results by Berestetskii et al. (1982).

[8] Called "Kritsche Feldstärke" in the original paper by Heisenberg and Euler (1936).

and the value of the external field E, we can write

$$\Gamma \propto m_e^\alpha c^\beta \hbar^\gamma E^\delta e^\epsilon$$

$$\frac{1}{\mathrm{TL}^3} = \mathrm{M}^\alpha \left(\frac{\mathrm{L}}{\mathrm{T}}\right)^\beta \left(\frac{\mathrm{ML}^2}{\mathrm{T}}\right)^\gamma \left(\frac{\mathrm{ML}}{\mathrm{QT}^2}\right)^\delta \mathrm{Q}^\epsilon$$

$$\Gamma \propto m_e^\alpha c^{-1+3\alpha/2} \hbar^{-2-\alpha/2} E^{2-\alpha/2} e^{2-\alpha/2}$$

$$\propto \frac{e^2 E^2}{\hbar^2 c} \left(\frac{m_e c^{3/2}}{\hbar^{1/2} e^{1/2} E^{1/2}}\right)^\alpha$$

$$\Gamma = C_\Gamma \left(\frac{e^2 E^2}{\hbar^2 c}\right) F\left(\frac{E_{cr}}{E}\right) \tag{7.78}$$

and the dimensionless factor $\Pi \equiv E_{cr}/E$ is naturally found. Two early discussions of this effect were presented by Sauter (1931) and Heisenberg and Euler (1936), and a calculation by Schwinger (1951) finds a very interesting functional form for $F(\Pi)$,

$$C_\Gamma = \frac{1}{4\pi^3} \quad \text{and} \quad F(x) = \sum_{n=1}^{\infty} \frac{1}{n^2} e^{-n\pi x}. \tag{7.79}$$

Despite the E^{+2} term in the prefactor, the exponential dependence shows that this process is highly suppressed and only possible for $E \gtrsim E_{cr}$. Studies of the Schwinger limit attainability with extreme power lasers by Bulanov et al. (2010) discuss many of the experimental and physical constraints/limitations, and recall our own **P4.22** as well. Ruffini et al. (2010) presents an excellent review of e^+e^- pair production across many topics in physics and astrophysics.

7.4 Lagrangians in field theory

The importance of providing *Multiple Representations in Physics Education*[9] has long been understood as being crucial in improving student learning of what can be otherwise challenging topics. The use of formulae, graphs, word problems, comparison with experiment in labs, even dare we say DA, all provide students with more tools to master our discipline. Even when focusing on the purely mathematical description of Nature, having an expanded tool kit can not only facilitate computation, but also can suggest new approaches to problem solving, with the application of coding and computer modeling being obvious examples. Even if approaching problems with purely "pencil-and-paper" mathematical techniques, being able to formulate a problem in different ways can still be useful.

One example from classical mechanics is the extension of Newtonian methods, based on $\mathbf{F} = ma$ techniques, to **Lagrangian** and eventually **Hamiltonian** descriptions. The combination of kinetic and potential energy terms defining the Lagrangian,

$$L(x, \dot{x}) \equiv \frac{1}{2} m\dot{x}^2 - V(x) \tag{7.80}$$

[9] The title of a collection of PER essays on this subject by Treagust, Duit, and Fischer (2017).

might seem counter intuitive, given the relative minus sign, as it is not the total mechanical energy, but is an important construct nonetheless. This expression is dimensionally consistent, since $[L] = [E] = [KE] = [V] = \mathsf{ML}^2/\mathsf{T}^2$. The quantity $L(x, \dot{x})$ is then used to construct the **action,** S, via the integral

$$S[x(t), \dot{x}(t)] = \int L(x, \dot{x})\, dt \tag{7.81}$$

where the $S[x, \dot{x}]$ is a **functional** or function of a function, taking as its argument, any possible path $x(t)$ that connects the appropriate initial and final positions, namely those satisfying $x(t_1) = x_1$ and $x(t_2) = x_2$. (For background, see any undergraduate textbook on classical mechanics, such as Marion and Thornton (2003), or the brief description in **P6.15.**)

The classical path which minimizes this quantity, turns out to be the same one as determined by Newton's law, as the calculus-of-variations problem of finding the minimum $S[x, \dot{x}]$ (or least action) gives the **Euler–Lagrange equations,** namely

$$\underbrace{\frac{\partial L}{\partial x} - \frac{d}{dt}\left(\frac{\partial L}{\partial \dot{x}}\right) = 0}_{\text{Euler-Lagrange equation}} \quad \longrightarrow \quad -\frac{\partial V(x)}{\partial x} - \frac{d}{dt}(m\dot{x}) = 0 \quad \longrightarrow \quad \underbrace{m\ddot{x} = -\frac{\partial V(x)}{\partial x} = F(x)}_{\text{Newton's third law}}. \tag{7.82}$$

The relevance of action and Euler–Lagrange methods is not limited to mechanics, as one can couch classical E&M in similar language as well. Given that the quantities of interest there are E&M fields, spread out over space (and defined not only on specific one-dimensional trajectories) and so are naturally three-dimensional quantities, the concept of a **Lagrangian density** is a natural extension from Eqn (7.80). For $\boldsymbol{E}, \boldsymbol{B}$ fields in source-free regions, the appropriate Lagrangian density is

$$\mathcal{L}_{EM}^{(free)} = \frac{\epsilon_0}{2}|\boldsymbol{E}|^2 - \frac{1}{2\mu_0}|\boldsymbol{B}|^2 = U_E - U_B, \tag{7.83}$$

which is constructed from the free-field electric and magnetic field energy densities from Eqns. (4.67) and (4.68), again with a relative minus sign. The dimensionality of any Lagrangian density will be *energy/volume*, so that

$$[\mathcal{L}] = \frac{[E]}{[V]} = \frac{\mathsf{M}}{\mathsf{LT}^2}. \tag{7.84}$$

To include sources of $\boldsymbol{E}, \boldsymbol{B}$ fields, namely electric charges and currents, one writes for the **interaction Lagrangian density**

$$\mathcal{L}_{EM}^{(int)} = \boldsymbol{J}_e \cdot \boldsymbol{A} - \rho_e \phi, \tag{7.85}$$

where $\boldsymbol{J}_e, \boldsymbol{A}$ are the current density and vector potential, while ρ_e, ϕ are the charge density and scalar potential. This expression can be confirmed to be dimensionally consistent, as we have

$$[\boldsymbol{J} \cdot \boldsymbol{A}] = \left(\frac{\mathsf{Q}}{\mathsf{TL}^2}\right)\left(\frac{\mathsf{ML}}{\mathsf{QT}}\right) \stackrel{!}{=} \frac{\mathsf{M}}{\mathsf{LT}^2} \stackrel{!}{=} \left(\frac{\mathsf{Q}}{\mathsf{L}^3}\right)\left(\frac{\mathsf{ML}^2}{\mathsf{QT}^2}\right) = [\rho\phi]. \tag{7.86}$$

One can then extend the concept of action as an integral of $\mathcal{L}$ not only over time, but now over space as well, giving

$$S[\boldsymbol{E}, \boldsymbol{B}, \boldsymbol{J}_e, \rho_e] = \int d\boldsymbol{x}\, dt \left(\mathcal{L}_{EM}^{(free)} + \mathcal{L}_{EM}^{(int)}\right) \tag{7.87}$$

and the dimensions of action are still $[S] = [dV][dt][E/V] = \mathsf{ML^2/T}$. Most importantly, the corresponding Euler–Lagrange equations obtained by minimizing this action reproduce the Maxwell equations. For such a derivation (which is beyond the scope of this treatment), see either of the excellent textbooks by Jackson (1999) or Zangwill (2013).

This extension applies to classical EM field theory, but such methods continue to be important in relativistic quantum mechanics as well. The Lagrangian density for a single spin-1/2 fermion (such as an electron) described by a relativistic wave function ψ can be written as

$$\mathcal{L}_{Dirac}^{(free)} = \overline{\psi}\left(i\hbar c \not{\partial} - mc^2\right)\psi, \tag{7.88}$$

where we have included the appropriate factors of $\hbar$ and c that most graduate-level textbooks which treat the subject would have taken for granted. The $\not{\partial}$ symbol is constructed from spacetime derivatives, along with 4×4 (dimensionless) γ matrices, which are needed to encode the spin-information of the Dirac particle, but which play no role in the DA, namely

$$\partial \equiv \left(\frac{1}{c}\frac{\partial}{\partial t}, \nabla\right) \quad \text{and} \quad \not{\partial} = \partial_\mu \gamma^\mu \quad \text{where} \quad \gamma^\mu = (\gamma_0, \boldsymbol{\gamma}). \tag{7.89}$$

To confirm that the dimensions of $\mathcal{L}_{Dirac}^{(free)}$ are correct, we note that

$$[\psi] = [\overline{\psi}] = \frac{1}{\mathsf{L}^{3/2}}, \quad \left[mc^2\right] = [E] = \frac{\mathsf{ML}^2}{\mathsf{T}^2}, \quad \text{and}$$
$$[\hbar c \not{\partial}] = \left(\frac{\mathsf{ML}^2}{\mathsf{T}}\right)\left(\frac{\mathsf{L}}{\mathsf{T}}\right)\left(\frac{1}{\mathsf{L}}\right) = \frac{\mathsf{ML}^2}{\mathsf{T}^2} \tag{7.90}$$

where the dimensions of ψ were discussed in Section 6.2. With these identifications, we have

$$\left[\mathcal{L}_{Dirac}^{(free)}\right] = \frac{1}{\mathsf{L}^{3/2}}\left(\frac{\mathsf{ML}^2}{\mathsf{T}^2}\right)\frac{1}{\mathsf{L}^{3/2}} = \frac{\mathsf{M}}{\mathsf{LT}^2} = \frac{energy}{volume} \tag{7.91}$$

as required. Charged fermions can couple to E&M fields, and the QED interaction term of ψ with photons is written not in terms of $\boldsymbol{E}, \boldsymbol{B}$ fields, but rather their E&M potentials, namely

$$\mathcal{L}_{Dirac}^{(int)} = qc\,\overline{\psi}(\not{\mathcal{A}})\psi \quad \text{with} \quad \mathcal{A} = \left(\frac{\phi}{c}, \boldsymbol{A}\right) \tag{7.92}$$

and $\mathcal{A}$ is the EM four-potential (recall Eqn. (1.98)) including the scalar (ϕ) and vector ($\boldsymbol{A}$) terms, and with $\not{\mathcal{A}} = \mathcal{A}_\mu \gamma^\mu$. You can confirm that this form is dimensionally consistent since $[\mathcal{L}_{Dirac}^{(int)}] = \mathsf{Q(L/T)L^{-3/2}(ML/QT)L^{-3/2}} = \mathsf{M/LT^2}$. The Euler–Lagrange equations obtained from the use of $\mathcal{L}_{Dirac}^{(free)} + \mathcal{L}_{Dirac}^{int}$ give the Dirac equation and the interaction term in Eqn. (7.92) is the source for the pictographic Feynman diagram in Fig. 7.1(a) of the basic fermion–photon vertex.

We have already considered two quantum mechanical phenomena related to the box diagram in Fig. 7.4, namely light-by-light scattering and the Schwinger mechanism of e^+e^- pair production. The same types of effects can be encoded in **Euler–Heisenberg Lagrangian density** obtained from the one-loop box diagram giving

$$\begin{aligned}
\mathcal{L}_{EM}^{(1-loop)} &= \frac{2\alpha_{FS}^2\hbar^3\epsilon_0^2}{45m_e^4c^5}\left[(|E|^2 - c^2|B|^2)^2 + 7c^2(E\cdot B)^2\right] \\
&= \frac{8\alpha_{FS}^2\hbar^3}{45m_e^4c^5}\left[\frac{\epsilon_0^2}{4}\left(|E|^2 - c^2|B|^2\right)^2 + \frac{7}{4}\epsilon_0^2c^2(E\cdot B)^2\right] \\
&= \frac{8\alpha_{FS}^2\hbar^3}{45m_e^4c^5}\left[(U_E - U_B)^2 + \frac{7}{4}(U_{EB})^2\right] \\
&= \mathcal{P}\left[(U_E - U_B)^2 + \frac{7}{4}(U_{EB})^2\right], \qquad (7.93)
\end{aligned}$$

where U_E, U_B are the ordinary electric and magnetic field energy densities and $U_{EB} \equiv \sqrt{\epsilon_0/\mu_0}\,E\cdot B$, which was introduced in Section 4.4 (as Eqn. (4.71)). That discussion pointed out that while U_{EB} has the appropriate dimensions for an energy density, it does not have the correct symmetry properties (since it changes sign if either E, B do), but in this context it appears squared and so is allowed.

The prefactor in Eqn. (7.93) can be written in the form of the inverse of an energy density, $\mathcal{P} = 1/U_{QM}$, which is clearly due to quantum effects (note the factors of $\hbar$) and can be explored in the context of DA, informed by some field-theoretic intuition. The 1-loop QED Lagrangian in Eqn. (7.93) arises from the same light-by-light scattering diagram shown in Fig. 7.4 and so we might expect that any relevant prefactor could include amplitude combinations as in Eqn. (7.73), namely proportional to α_{FS}^2/m_e^4. The total prefactor, including dimensional factors of $\hbar$, and c, would then be

$$\mathcal{P} = \frac{\alpha_{FS}^2}{m_e^4}\hbar^\alpha c^\beta \quad \longrightarrow \quad \frac{\mathsf{LT}^2}{\mathsf{M}} = \left[\frac{1}{U_{QM}}\right] = [\mathcal{P}] = \mathsf{M}^{-4}\left(\frac{\mathsf{ML}^2}{\mathsf{T}}\right)^\alpha\left(\frac{\mathsf{L}}{\mathsf{T}}\right)^\beta, \qquad (7.94)$$

which is easily solved to find $\alpha = 3$ and $\beta = -5$, just as in Eqn. (7.93). We note that $\mathcal{L}_{EM}^{(1-loop)}$ can also be written in terms of the critical fields (E_{cr}, B_{cr}) for the Schwinger effect from Eqn. (7.76); see **P7.14.**

7.5 **The Casimir effect**

The final quantum field theoretic effect we will explore, partly by DA, but also using analogies with earlier examples, is perhaps one of the least obvious implications of QED, namely the **Casimir effect.** We begin classically by noting that the total E&M field energy stored in a region of space will be the integral of the electric and magnetic energy densities, namely

$$\begin{aligned}
E_{TOT} &= \int d\mathbf{r}\left\{\frac{\epsilon_0}{2}|\mathbf{E}(\mathbf{r},t)|^2 + \frac{1}{2\mu_0}|\mathbf{B}(\mathbf{r},t)|^2\right\} \\
&= \int d\mathbf{r}\{U_E(\mathbf{r},t) + U_B(\mathbf{r},t)\}. \qquad (7.95)
\end{aligned}$$

We've already seen examples of field theory preferring to express $\boldsymbol{E}, \boldsymbol{B}$ in terms of the corresponding potentials, $\boldsymbol{A}, \phi$, and in situations where $\phi(\boldsymbol{r}, t)$ is not present, we have

$$\boldsymbol{E}(\boldsymbol{r}, t) = -\frac{\partial \boldsymbol{A}(\boldsymbol{r}, t)}{\partial t} \quad \text{and} \quad \boldsymbol{B}(\boldsymbol{r}, t) = -\nabla \times \boldsymbol{A}(\boldsymbol{r}, t), \tag{7.96}$$

which is still a very classical representation. The vector potential $\boldsymbol{A}(\boldsymbol{r}, t)$ in "real" (position) space can be Fourier transformed back and forth to wave-number-space (or in quantum mechanical terms, momentum-space) via

$$\boldsymbol{A}(\boldsymbol{r}, t) = \frac{1}{(2\pi)^{3/2}} \int d\boldsymbol{k}\, \tilde{\boldsymbol{A}}(\boldsymbol{k}, t)\, e^{+i\boldsymbol{k}\cdot\boldsymbol{r}} \tag{7.97}$$

$$\tilde{\boldsymbol{A}}(\boldsymbol{k}, t) = \frac{1}{(2\pi)^{3/2}} \int d\boldsymbol{r}\, \boldsymbol{A}(\boldsymbol{r}, t)\, e^{-i\boldsymbol{k}\cdot\boldsymbol{r}} \tag{7.98}$$

where the dimensions of $\tilde{\boldsymbol{A}}(\boldsymbol{k}, t)$ are

$$[\tilde{\boldsymbol{A}}] = [d\boldsymbol{r}][\boldsymbol{A}][e^{-i\boldsymbol{k}\cdot\boldsymbol{r}}] = (\mathsf{L}^3)\left(\frac{\mathsf{ML}}{\mathsf{QT}}\right)(1) = \frac{\mathsf{ML}^4}{\mathsf{QT}}. \tag{7.99}$$

In this language, the total EM energy from Eqn. (7.95) can be written as

$$E_{TOT} = \int d\boldsymbol{k} \left\{ \frac{\epsilon_0}{2} \left| \frac{\partial \tilde{\boldsymbol{A}}(\boldsymbol{k}, t)}{\partial t} \right|^2 + \frac{k^2}{2\mu_0} |\tilde{\boldsymbol{A}}(\boldsymbol{k}, t)|^2 \right\} \tag{7.100}$$

and it's good practice to confirm that the dimensions of the first term are consistent by writing

$$\begin{aligned} [E_{TOT}] &\stackrel{?}{=} [d\boldsymbol{k}][\epsilon_0]\left[\frac{\partial \tilde{\boldsymbol{A}}}{\partial t}\right]^2 \\ \frac{\mathsf{ML}^2}{\mathsf{T}^2} &\stackrel{?}{=} \left(\frac{1}{\mathsf{L}^3}\right)\left(\frac{\mathsf{Q}^2\mathsf{T}^2}{\mathsf{ML}^3}\right)\left(\frac{1}{\mathsf{T}}\frac{\mathsf{ML}^4}{\mathsf{QT}}\right)^2 \stackrel{!}{=} \frac{\mathsf{ML}^2}{\mathsf{T}^2} \end{aligned} \tag{7.101}$$

and you can confirm the second term yourself.

One striking thing about the form in Eqn. (7.100) is its similarity to classic oscillators, such as the mass/spring or inductor/capacitor systems, where we include their oscillation frequencies, namely

$$E_{HO} = \frac{1}{2} m\dot{x}(t)^2 + \frac{1}{2} kx(t)^2 \quad \text{or} \quad E_{LC} = \frac{1}{2} L\dot{q}(t)^2 + \frac{1}{2C} q(t)^2 \tag{7.102}$$

$$\omega_{HO} = \sqrt{\frac{k/2}{m/2}} = \sqrt{\frac{k}{m}} \quad \text{or} \quad \omega_{LC} = \sqrt{\frac{1/2C}{L/2}} = \frac{1}{\sqrt{LC}}. \tag{7.103}$$

To see if we can extend this analogy to the EM field as a collection of oscillators, we take the coefficients of the $\tilde{\boldsymbol{A}}$ and $\partial\tilde{\boldsymbol{A}}/\partial t$ terms as representing generalized restoring forces and inertias, similar to

k $(1/C)$ and m (L) for the mass/spring and LC circuit examples. The oscillation frequency would then be

$$\omega_k = \sqrt{\frac{(k^2/2\mu_0)}{\epsilon_0/2}} = \frac{1}{\sqrt{\epsilon_0\mu_0}}k = kc \tag{7.104}$$

and we indeed recognize this as the dispersion relation for EM waves, identical to the classical result $2\pi f = c(2\pi/\lambda)$ or $f\lambda = c$. So far, this approach may seem only like a more complicated way to rewrite classical EM results.

If we then apply quantum mechanics to each k mode of the $\tilde{\boldsymbol{A}}(\boldsymbol{k}, t)$ field, we expect quantized energies of the form $E_{HO}^{(k)} = (n + 1/2)\hbar\omega_k$ for each three-dimensional wave-number $\boldsymbol{k}$, and a resulting zero-point energy

$$E_{TOT}^{(0)} = \sum_{\boldsymbol{k}} \frac{1}{2}\hbar\omega_{\boldsymbol{k}} = \frac{\hbar c}{2}\sum_{\boldsymbol{k}} \sqrt{k_x^2 + k_y^2 + k_z^2}. \tag{7.105}$$

Given that this sum should be over all possible $\boldsymbol{k}$ values, it implies that otherwise empty space has an infinite energy due to the zero-point energies of all of these modes. This isn't quite as bad as it sounds, since physical phenomena in classical mechanics do not depend on the definition of the zero of potential energy, only on changes in potential energy, via $F(x) = -dV(x)/dx$, so that $V(x) \rightarrow V(x) + V_0$ has no impact on the motion.

A simple geometry where the impact of Eqn. (7.105) could be explored is that of two (infinite) parallel conducting plates, as in Fig. 7.5. The inclusion of the conductors imposes boundary conditions on the $\boldsymbol{E}$ field modes, requiring them to vanish at the two boundaries. The effect of these plates is to transform an infinite integral (over the $\boldsymbol{k}_x$ wave-number) to an infinite sum, and the difference of those two calculations turns out to be a finite, and as we'll see, measurable **energy per unit area** ($\mathcal{E}$) when the plates are present.

To approach the problem of calculating $\mathcal{E}$ (and the resulting **force per unit area**) via DA, we assume that it could depend on the plate separation (a), the constants of quantum mechanics and relativity ($\hbar, c$), and those of E&M (e, ϵ_0). We would then write

$$\begin{aligned}\frac{\mathcal{E}}{A} &= a^\alpha \hbar^\beta c^\gamma e^{2\delta} \epsilon_0^\sigma \\ \frac{\mathsf{M}}{\mathsf{T}^2} &= \mathsf{L}^\alpha \left(\frac{\mathsf{ML}^2}{\mathsf{T}}\right)^\beta \left(\frac{\mathsf{L}}{\mathsf{T}}\right)^\gamma \mathsf{Q}^{2\delta} \left(\frac{\mathsf{Q}^2\mathsf{T}^2}{\mathsf{ML}^3}\right)^\sigma\end{aligned} \tag{7.106}$$

and matching dimensions, one finds $\alpha = -4$, $\beta = \gamma = 1 - \delta$, and $\sigma = -\delta$, or

$$\frac{\mathcal{E}}{A} \propto a^{-3}(\hbar c)^{1-\delta} e^{2\delta} \epsilon_0^{-\delta} \propto \frac{\hbar c}{a^3}\left(\frac{e^2}{\epsilon_0 \hbar c}\right)^\delta \propto \frac{\hbar c}{a^3}(\alpha_{FS})^\delta. \tag{7.107}$$

With a fair amount of either hindsight or intuition, or taking the cartoon picture of the effect in Fig. 7.5 seriously, we can argue that the only impact of the presence of the conductors is to enforce geometrical boundary conditions on the EM field modes, and does not represent the exchange of virtual photons (as say in Fig. 7.1(b)). In that case, we would expect no dependence on e, ϵ_0, giving an unambiguous prediction for the $\hbar$, c and especially a dependence of the Casimir energy as $\mathcal{E} \propto (\hbar c)/a^3$.

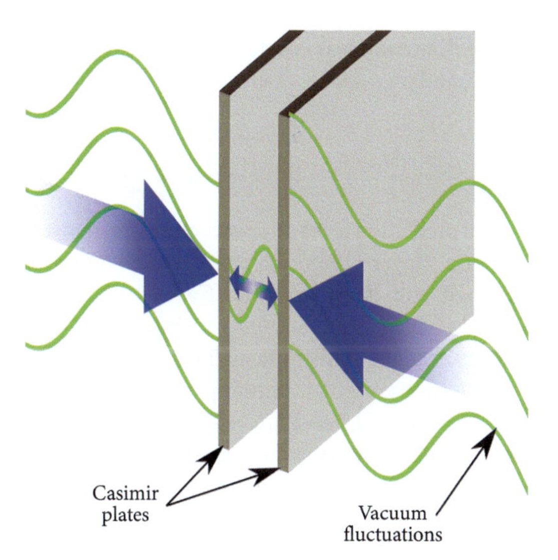

Fig. 7.5 Parallel plate geometry for the Casimir effect, with metallic plates imposing boundary conditions on the quantum/vacuum fluctuations of the EM fields. Reproduced via Creative Commons License from Wikipedia authored by Emok (2008). Downloaded from https://commons.wikimedia.org/wiki/File:Casimir_plates.svg on 20 October, 2023.

The resulting **force per unit area** is obtained by differentiation as

$$\mathcal{F} = \frac{F}{A} \propto -\frac{\partial}{\partial a}\mathcal{E}(a) \propto \frac{\hbar c}{a^4} \qquad \text{or} \qquad \mathcal{F} = C_{\mathcal{F}}\frac{\hbar c}{a^4}. \tag{7.108}$$

While this DA approach gives the correct dependence on all relevant quantities, it is mute on one of the most important questions, namely the **sign** of $C_{\mathcal{F}}$, and whether the force between plates is attractive or repulsive. A detailed calculation by Casimir (1948, 1951) found that $C_{\mathcal{F}} = -\pi^2/240$, giving an attractive force. Spaarnaay (1958) confirmed the effect, albeit with large error bars, and later determinations by Bressi et al. (2002) improved the precision to 15%.

More recent experiments, with a slightly different geometry, by Lamoreaux (1997) and Mohideen and Roy (1998), demonstrated consistency with theory at the level of 5% and 1% respectively; see **P7.16** for an exploration. Given the importance of this effect on the understanding of quantum field theory, it's not a surprise that many reviews of the theoretical background and experimental results of the Casimir effect have appeared; see, for example, Elizalde and Romero (1991), Lamoreaux (1999, 2005), Milton (2011), and Stange et al. (2021), while textbook derivations at the level of Itzykson and Zuber (1980) are very readable.

An intriguing suggestion made by Casimir (1953) himself was that if the electron were to consist of a spherical shell of charge, and **if** the Casimir force for this geometry were to also be attractive, then there could be a natural balance between the electro-static self-repulsion and zero-point energy forces/potentials. In this finite geometry, the Casimir energy would not be "per unit volume," and would depend on $V_{Cas}(r) \propto \hbar^{\alpha}c^{\beta}r^{\gamma}$, which gives (by inspection) $V_{Cas}(r) \propto \hbar c/r$. The balance would come from two terms, namely

$$V_{Coulomb} = +C_{Cou}\frac{e^2}{4\pi\epsilon_0 r} \qquad \text{and} \qquad V_{Casimir} = -C_{Cas}\frac{\hbar c}{r}, \tag{7.109}$$

where C_{Cou} and C_{Cas} would be dimension**less** constants depending only on the geometry of the spherical shell (or other shape), and presumably calculable. Balancing forces/energies would then give

$$+ C_{Cou}\frac{e^2}{4\pi\epsilon_0 r} = C_{Cas}\frac{\hbar c}{r} \qquad \text{or} \qquad \alpha_{FS} \equiv \frac{e^2}{4\pi\epsilon_0 \hbar c} = \frac{C_{Cas}}{C_{Cou}}, \tag{7.110}$$

independent of r, and provide a prediction for the fine-structure constant α_{FS} from first principles.

Motivated by this observation, Boyer (1968) calculated the quantum EM zero-point energy of a conducting spherical shell and found that the Casimir energy in this geometry was $E_{Cas} \approx +0.09\hbar c/2r$, and so of the wrong sign, a result which he noted "... invalidates Casimir's intriguing model ..." Lukosz (1971) did a similar calculation for a rectangular parallelpiped (with sides $L_x \times L_y \times L_z$) that reproduced the original Casimir parallel plate result when $L_{x,y} >> L_z$, and for a cubical shell (where $L_x = L_y = L_z \equiv L$) gave $E_{Cas} = +0.0918\hbar c/L$, which is very similar to the Boyer result if we associate $2r \leftrightarrow L$, presumably as it must be on dimensional grounds. Not all things that are dimensionally consistent are realized in nature, and DA is always silent on the important $\pm$ sign!

7.6 Exploring field theory

This chapter focuses on the application of quantum theory to electromagnetism, starting our discussion with the role of the fine-structure constant (α_{FS}) in bound state, scattering, and decay rate problems in QED. We think that this emphasis is particularly appropriate for a volume such as this, allowing for connections to classical E&M results, while highlighting the role that quantum mechanics and relativity play via $\hbar$ and c. Besides its important role in fundamental physics, we note that two of the components of α_{FS}, namely $h = 2\pi\hbar$ and e^2, form the **quantum of resistance**, $R_K = h/e^2$, discovered by von Klitzing in the quantum Hall effect, which plays an important part in modern metrology, given that it can be measured to $\sim 10^{-8}$ precision.[10] We note that the original paper by von Klitzing et al. (1980) was actually called *New Method for High-Accuracy Determination of the Fine-Structure Constant Based on Quantized Hall Resistance.* We explore the **quantum Hall effect** in **Example 8.3.**

Interested readers are encouraged to continue the exploration of the ideas and methods outlined here for QED to the study of the strong (**QCD**) interactions of quarks and/or the **weak interactions** of quarks and leptons, either in more advanced presentations in textbooks, or even just in the few extended problems (**P7.17–P7.21**) below. In the study of the strong and weak interactions, we find analogs of α_{FS} in both systems, as well as intuitive similarities in the application of Feynman diagram type methods to these two other fundamental forces. A compendium of decay rates and scattering cross-sections for all three types of forces is available in the *Review of Particle Physics* (with the most recent edition being Workman et al. (2022)) against which students can practice their (very brief and incomplete) experiences with Feynman diagrammatics. We focus our attention on the fourth basic force of nature, gravity, in Chapter 9, in astrophysical contexts.

We have already seen examples where quantum effects of the exchange of virtual particles in QED can generate measurable differences in familiar physical quantities (as in the anomalous magnetic moments of the e, μ) or induce new physical processes (such as light-by-light scattering),

[10] See Mohr et al. (2008) for a discussion of the connection between the "... von Klitzing constant R_K and α ..." as well as many detailed QED effects in atomic physics.

classified by powers α_{FS}. While we have consistently described the combination $\alpha_{FS} = K_e/(4\pi\epsilon_0\hbar c)$ as the fine-structure **constant**, loop effects also induce radiative corrections to even Coulomb's law, effectively modifying the value of e as a function of distance (r), or equivalently momentum (p), as would be measured in say in a scattering experiment. The only familiar E&M textbook that mentions these effects (of which we are aware) is by Zangwill (2013), who in turn quotes results from the classic QED compendium by Berestetskii et al. (1982).

Phrased in the language most often cited in field theory sources, the **running coupling constant** for QED can be expressed by a momentum-dependence of α_{FS} (with a reference momentum scale Λ) as

$$\alpha_{FS}(p^2) = \frac{\alpha_{FS}(\Lambda^2)}{1 - \frac{\alpha_{FS}(\Lambda^2)}{3\pi}\ln(p^2/\Lambda^2)} \tag{7.111}$$

so that e effectively becomes stronger (weaker) when measured at larger (smaller) values of momentum or smaller (larger) distances. Berestetskii et al. (1982) note that

> These corrections may be intuitively described as resulting from the polarization of the vacuum round a point charge...

and you can imagine a photon line splitting into a virtual e^+e^- pair for a short time (as allowed by the uncertainty principle) before "returning to itself," during which its interactions with external charges are modified. This effect has been confirmed experimentally several times, including by Odaka et al. (1998) and Abbiendi et al. (OPAL Collaboration) (2006). One group[11] cites a value of $\alpha_{FS}^{-1}(p^2) = 128.5 \pm 1.8(stat) \pm 0.7(syst)$ at $p \approx 58\ GeV/c$, citing both statistical and systematic errors, showing the clear change from the $\alpha_{FS}^{-1} = 137.0$ value at low energies.

7.7 Natural or universal units: The role of hyperfine transitions

We ended Chapter 6 with a discussion of two systems of what might be called natural units, both depending on Planck's constant ($\hbar$), one most used for physics on the small scales relevant for atoms, molecules, and chemistry (Hartree units) and one far more appropriate for the length scales relevant for quantum gravity, cosmology, and the study of the early universe (Planck units).

We think it fitting to close Chapter 7 with a similar mention of applications of a fundamental physics phenomenon and its implications to metrology and beyond, namely the hyperfine energy shifts in atomic systems due to the dipole–dipole (spin–spin) interactions of the electron(s) and the nucleus. While they may seem very small on the scale of the bound-state energies ($\sim 10^{-5}\ eV$ compared to $\sim 10\ eV$), their impact goes far beyond the realm of precision spectroscopy.

For example, the 2019 revision of the SI (or Système International) units included a new definition of the **second**, which is included verbatim on all of the websites for national organizations responsible for *weights and measures*, including the BIPM (Bureau International des Poids et Mesures, France), the NPL (National Physical Lab, UK), and NIST (National Institute of Science and Technology, US). They all cite:

> The second, symbol s, is the SI unit of time. It is defined by taking the fixed numerical value of the cesium frequency, $\Delta\nu_{Cs}$, the unperturbed ground-state hyperfine transition frequency of the cesium 133 atom, to be 9,192,631,770 when expressed in the unit Hz, which is equal to s^{-1}.

[11] Levine et al. TOPAZ Collaboration (1997).

The corresponding definition of the **meter** is then listed as:

> The meter is defined by taking the fixed numerical value of the speed of light in vacuum c to be 299,792,458 when expressed in the unit m/s^{-1}, where the second is defined in terms of $\Delta\nu_{Cs}$.

Despite their ultra-precise "instructions," these definitions retain their approximate human-sized dimensions, and so might not be immediately obvious realizations of T, L units to intelligent life forms with different physical appearances, no matter the sophistication of their scientific knowledge.

In that context, two (seemingly unrelated) questions related to aspects of communication in the SETI (Search for Extraterrestrial Life)[12] endeavor involve (a) the optimal channel (wavelength/frequency) for EM wave transmission/reception (or the *how*) and (b) what messages to include in any such attempts (namely the *what*). The first question was explored in *Searching for Interstellar Communications* by Cocconi and Morrison (1959) who noted that the "... wide radio band from, say, 1 *Mc.* to 10^4 *Mc./s.*, remains as the rational choice ..."[13] after considering the fact that such signals would not be attenuated in their travel through space, nor through the Earth's atmosphere, especially since there is a microwave window in the 1–10 *GHz* range which is fairly transparent. Cocconi and Morrison (1959) then note that

> But, just in the favoured radio region, which must be known to every observer in the universe: the outstanding radio emission line at 1,420 *Mc./s.* ($\lambda = 21\ cm$) of neutral hydrogen ...

singling out the hyperfine transition of hydrogen as a preferred frequency "channel" for communication, at least by remote means.

In a different context, more than fifty years ago, space scientists were noting that the Pioneer 10/11 spacecraft would be the first objects of human construction to leave the solar system, and in their paper *A Message from Earth*, Sagan, Sagan, and Drake (1972) suggested that such a mission should

> ... carry some indication of the locale, epoch, and nature of its builders ...

which led to the design, fabrication, and installation of the famous Pioneer plaque, reproduced in Fig. 7.6.

The "dumbbell" pictograph in the upper left corner represents the two states of the H-atom with electron–proton spins aligned and anti-aligned, with a single "digit" included, intending to suggest that the τ_γ and λ_γ associated with this hyperfine transition be used as the time and distance units in interpreting the rest of the message. A representation of the spacecraft, alongside two human-sized figures, is designed to confirm the intended length scale (since someone intercepting the physical object would be able to measure its dimensions first hand) and the | – –– symbol next to that tableau is shorthand for the binary representation 1000 or 8 length units. That corresponds to $8 \times 21\ cm = 1.68\ m \approx 5\ ft\ 5\ in$ and provides a double check on the length units assumed.

The "starburst" pattern on the left is designed to specify the locations in space (and also in time, as it turns out) of fourteen pulsars. The numerical values associated with each one (again, in binary) are cited to something like 9–10 place accuracy and so can't possibly represent distances, but rather are the pulsar periods, at least circa the 1970s. Long before the direct experimental measurement of gravitational radiation by Abbott et al. (2019), it was well known that pulsar

---

[12] For an early historical review, see Sagan and Drake (1975).
[13] Note the units used are *Mc.* for *mega-cycle* instead of *MHz.*

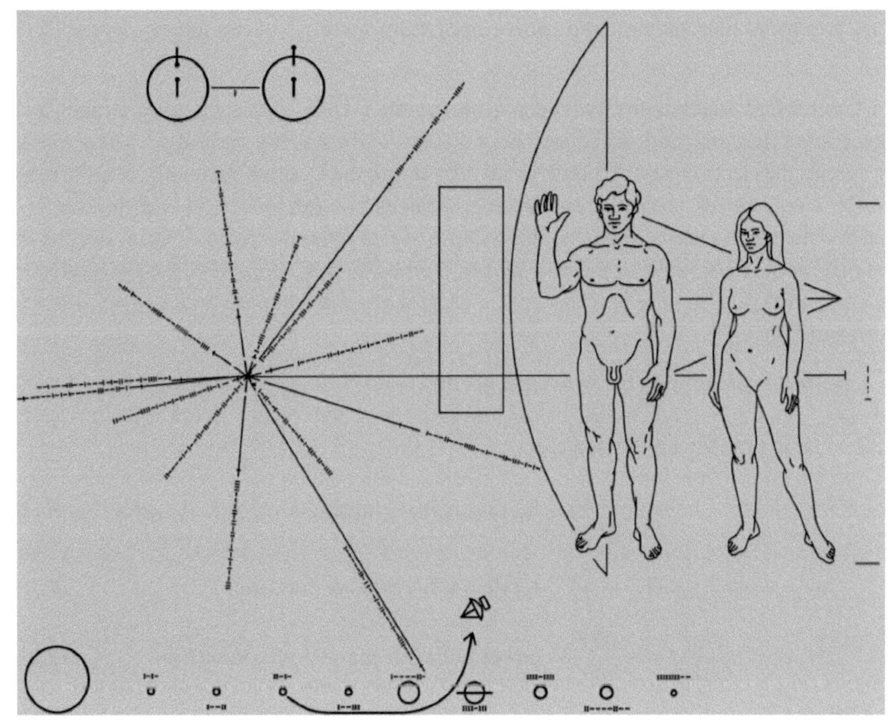

Fig. 7.6 Metal plaque attached to the Pioneer 10 and 11 space craft with *A Message from Earth*. See Sagan, Sagan, and Drake (1972) for a discussion.

periods changed in time[14] in a predictable way due to the loss of energy by gravitational wave emission, and so observers with an extensive enough database of such objects, could triangulate where (from the pictograph), and when (from the present-day period values compared to these "snapshots in time" entries), the object originated.

We note that this artifact provides another valuable object lesson in experimental physics, namely that frequencies/periods are among the most well-measured physical properties in science, since they can often require only the counting of large numbers (N) of cycles of some periodic phenomena, with the resulting "error to signal" accuracy scaling as $\sqrt{N}/N = 1/\sqrt{N}$. We include in **P7.24** two such binary representations of pulsar periods and their experimental values and you're asked (for the first and only time in this text) to translate from binary to base-10 arithmetic to confirm the translation suggested by the plaque.[15]

7.8 **Problems for Chapter 7**

Q7.1 Reality check: (a) What was the most important, interesting, engaging, or novel things you learned about how to use DA in this chapter?

[14] See, e.g., Taylor and Weisberg (1982).

[15] A slightly more extensive discussion of these topics, at the level of a pedagogical journal, is provided by Robinett (2001).

(b) Have you previously encountered (or learned) many (or any) of the topics covered in this chapter?

P7.1 Classical free-fall time for hydrogen ground state I–DA: Classically, an electron in a circular orbit is accelerating, and so radiates away energy via the Larmor formula, $P = e^2a^2/6\pi\epsilon_0 c^3$. Losing energy, the electron would therefore fall deeper into the Coulomb potential and the radius of the orbit would shrink, eventually collapsing. A calculation (an analytical one you can try in the next problem if you wish) finds that the collapse or *free-fall* time for an electron starting in a circular orbit at the Bohr radius, a_0, is $\tau_{ff} = a_0^3/(4r_e^2 c)$ where $r_e = K_e/(m_e c^2)$ is the classical electron radius. Write this collapse time in terms of $m_e c^2$, $\hbar$, and α_{FS} and compare to the $\tau_{2P\to 1S}$ radiative decay time in Eqn. (7.8).

P7.2* Classical free-fall time for hydrogen ground state II—Calculation: Following the logic in **P7.1**, start with the "ingredient list" below, and then follow the "recipe" to calculate the classical radiative collapse time for a hydrogen atom:

$$E(r) = -\frac{K_e}{2r} \quad \text{energy–radius connection for circular orbit} \tag{7.112}$$

$$m\frac{v^2}{r} = ma_c(r) = F(r) = \frac{K_e}{r^2} \quad \text{acceleration–radius connection} \tag{7.113}$$

$$P_{Larmor} = \frac{e^2 a_c^2}{6\pi\epsilon_0 c^3} = -\frac{dE(r)}{dt} \quad \text{power radiated–energy loss connection} \tag{7.114}$$

$$-\frac{K_e}{2}\frac{d}{dt}\left(\frac{1}{r(t)}\right) = -\frac{2}{3}\frac{K_e^3}{m_e^2 c^3}\left(\frac{1}{r(t)}\right)^4 \quad dr(t)/dt\text{–r(t) connection.} \tag{7.115}$$

You now have a relation between dt and dr that you can integrate over the limits $(0, \tau_{ff})$ and $(a_0, 0)$ to solve for τ_{ff}.

Notes: (a) You should be able to derive Eqn. (7.112) by writing the total energy as $E(r) = mv^2/2 - K_e/r$ and then using the acceleration relation in Eqn. (7.113).

(b) This exercise is a purely classical calculation, and the only connection to quantum mechanics ($\hbar$) comes from the initial condition of being in a circular orbit of radius a_0. This is also a non-relativistic approximation, as the particle eventually speeds up and $v \to c$ as $r(t)$ becomes small, but we ignore that effect for this estimation problem. Can you guess at what $r(t)$ value the approximation starts to fail?

P7.3 Reduced mass effects in "heavy hydrogen":[16] The energy levels of hydrogen-like atoms with a proton nucleus (ordinary hydrogen or ^1H) will be slightly different from those with a deuteron (D) nucleus ("heavy hydrogen" or deuterium, ^2H, a neutron/proton bound state, which has roughly twice the mass of a proton) due to reduced mass effects.

(a) Show that the shift in wavelength of a given emission line for ^2H relative to ^1H is roughly

$$\frac{\Delta\lambda}{\lambda} \approx m_e\left(\frac{1}{m_D} - \frac{1}{m_p}\right), \tag{7.116}$$

where $m_D \approx 2m_p$. Evaluate this fractional change numerically.

16 This problem is taken from Robinett (2006).

(b) The discovery of deuterium was made by looking for such shifts in the *visible* atomic Balmer spectra of hydrogen. The original paper[17] says that:

> When with ordinary hydrogen, the times of exposure required to just record the strong ^1H lines were increased 4000 times, very faint lines appeared at the calculated positions for the ^2H lines ... on the short wave-length side and separated from them by between 1 and 2 Å (Urey, Brickewedde, and Murphy 1932b).

Use the results of part (a) to quantitatively explain the observed pattern of wavelength shifts.

P7.4 Hydrogen atom energy shift due to finite proton size: While the spatial extent of the proton ($R_p \approx 1\,F$) is much less than the size of any Bohr orbit ($r_p << a_0$), at least for the H-atom with an electron, it is not a perfectly point-like charge, and so there will be some deviation from Coulomb's law at short distances. A standard expression for the energy shift due to such effects (see Pohl et al. (2013) for a discussion in the context of the "proton radius puzzle", but the first calculation seems to have been done by Karplus, Klein, and Schwinger (1952)) is written as

$$\Delta E = \frac{2\pi}{3}\alpha_{FS}|\psi_S(0)|^2 R_p^2, \tag{7.117}$$

where $|\psi_S(0)|$ is the value of the H-atom wave function at the origin.

(a) As it stands, this expression is not dimensionally correct when $\hbar, c \neq 1$, so first add enough factors of $\hbar$ and/or c to make it consistent.

(b) What is the mostly likely dimensional scale for $|\psi_S(0)|$?

(c) Use your results from parts (a) and (b) and the fact that $r_p \approx 1\,F$ (Fermi) to estimate ΔE in eV units for the H-atom and compare to the Bohr energies, and the fine-structure and hyperfine energies.

(d) How would you expect Eqn. (7.117) to scale with m_e, say for a muonic atom where $m_e \to m_\mu$?

P7.5 Lamb shift calculations: Scully and Zubairy (1997) provide a derivation of the Lamb shift, following the methods of Bethe (1947) and Welton (1948), and derive a formula for the energy shift (even using SI units!) which we reproduce exactly here, namely

$$\langle \Delta V \rangle = \frac{4}{3}\frac{e^2}{4\pi\epsilon_0}\frac{e^2}{4\pi\epsilon_0 \hbar c}\left(\frac{\hbar}{mc}\right)^2 \frac{1}{8\pi a_0^3}\ln\left(\frac{4\epsilon_0 \hbar c}{e^2}\right). \tag{7.118}$$

where $m = m_e$ is the electron mass.

(a) Rewrite this in terms of $m_e c^2$ and powers of α_{FS} and compare to the dependence on those quantities of the Bohr energies, the fine-structure terms, and hyperfine energies considered in this chapter.

(b) Compare your result from part (a) to the more exact formula from Eqn. (7.52) and obtain a numerical value for the energy shift. Compare that to the value of the hyperfine shift for

---

[17] See Urey, Brickwedde, and Murphy (1932a, b) for details on the original experiment, and Brickwedde (1982) for a historical review of how "Chemistry, nuclear physics, spectroscopy, and thermodynamics came together to predict and detect heavy hydrogen before the neutron was known."

hydrogen from Eqn. (7.51) and discuss why they are of similar magnitude despite having different powers of α_{FS}.

(c) The so-called **Bethe logarithm** in Eqn. (7.118) arises from a sum over the energies of the QED vacuum modes (much like those in the Casimir effect) where an integral over k space appears, namely a factor of $\int_{k_{min}}^{k_{max}} dk/k = \ln(k_{max}/k_{min})$. The upper bound is seemingly always chosen to satisfy $\hbar c k_{max} = m_e c^2$ or $k_{max} = m_e c/\hbar$, while Scully and Zubairy (1997) use $k_{min} = \pi/a_0$ as the lower bound. Using these two limits, show that you can you reproduce the logarithm in Eqn. (7.118), perhaps noting the relationship between a_0 and $\lambda\!\!\!^{-}{}_C^e$ discussed in Eqn. (7.19).

(d) Most treatments of the Lamb shift (in addition to Bethe (1947) and Welton (1948), see Berestetskii et al. (1982) or Eides et al. (2000) or other QED reviews) use a lower bound on the dk integral given by $\hbar c k_{min} \sim E_H$ where E_H is the Hartree energy or the "average excitation energy" and we know that $E_H = (\alpha_{FS})^2(m_e c^2)$. Show that this gives a different power of α_{FS} in the Bethe logarithm, and also improves agreement with the experimental value.

P7.6 Photon decays of positronium (PS): (a) One calculation of the $PS \to \gamma\gamma$ decay rate (Crater 1991) describes the result of their evaluation as follows:

Taking this factor into account our decay rate becomes ...

$$\Gamma = \frac{\alpha^5 m_e}{2}$$

where of course $\alpha = \alpha_{FS}$. Add enough factors of $\hbar$ and c to have this reflect a real decay rate with dimensions $[\Gamma] = \mathsf{T}^{-1}$.
(b) Evaluate your result from part (a) and compare to the observed decay rate of $\lambda = 7990.9 \pm 1.7\ \mu s^{-1}$ observed experimentally by Al-Ramadhan and Gidley (1994).
(c) The original calculation of the $PS \to \gamma\gamma$ decay rate by Wheeler (1946) cited the lifetime (for the state with quantum number n) as

$$T = \frac{2\hbar}{m_e}(137n)^3\left(\frac{\hbar}{e^2}\right)^2.$$

In the notation and units used by Wheeler, we know that we should replace $e^2 \to e^2/(4\pi\epsilon_0) = K_e$, and with that change, show that this formula is consistent with Eqn. (7.72) with $C_\lambda = 1/2$, and the expression in part (a) above.
(d) A different state of PS can decay into three (3) photons. How do you think that $\lambda(PS \to \gamma\gamma\gamma)$ depends on m_e, $\hbar$, and especially α_{FS}? It turns out that the dimensionless prefactor[18] for this process is $C_\lambda = 2(\pi^2 - 9)/(9\pi)$ so the decay is also suppressed for more technical phase-space reasons, and the coincidence that $\pi^2 \approx 9$.

P7.7 Decay rates/lifetimes and distance scales: In many processes involving elementary particles, many/most of the relevant quantities are expressed in eV-related units, as in Eqn. (7.56). To translate these into decay rates and lifetimes, often only one power of $\hbar$ is required to convert such an energy dimension to a time (or its inverse) using Eqn. (7.58), while to express them in terms of a length, one usually needs the combination $\hbar c$ as in Eqns. (7.60) and (7.61).

[18] See Ore and Powell (1949).

(a) The lifetime and decay rate for the neutral pion to two photons (so, an EM process) can be calculated in field theory as

$$\frac{1}{\tau(\pi^0 \to \gamma\gamma)} = \Gamma(\pi^0 \to \gamma\gamma) = \frac{(\alpha_{FS})^2 (m_\pi)^3 (N_C)^2}{576\pi^3 F_\pi^2}, \tag{7.119}$$

where in "particle language" the pion mass is $m_\pi = 135\,MeV$ and the pion coupling is $F_\pi = 92\,MeV$. The number of quark colors (from QCD) is $N_C = 3$, while α_{FS} is the familiar fine-structure constant. Calculate $\Gamma(\pi^0 \to \gamma\gamma)$ in eV units and convert to a lifetime and compare to the measured value of $\tau(\pi^0 \to \gamma\gamma) = 8.5 \times 10^{-17}\,s$.

Notes: The fact that this observable is directly related to the number of colors of quarks is a nice test of QCD.[19] If you happen to see research papers where a value of $F_\pi = 130\,MeV$ is cited, it's likely due to a different normalization convention, where an extra factor of $\sqrt{2}$ is used, and indeed $130\,MeV/\sqrt{2} = 92\,MeV$.

(b) The lifetime and decay rate for the **muon** via the weak interaction can be calculated to be

$$\frac{1}{\tau_\mu} = \Gamma_\mu = \frac{G_F^2 m_\mu^5}{192\pi^3}, \tag{7.120}$$

where the muon mass is $m_\mu = 105.6\,MeV$ (again in particle language) and the **Fermi constant** is $G_F = 1.166 \times 10^{-5} eV^{-2}$. Evaluate the lifetime and compare to the experimental value of $\tau_\mu = 2.2\,\mu s$.

(c) The limit on the possible scale of compositeness or finite size of the electron is often cited as an energy scale, with one of the most stringent lower bounds being $\Lambda_e > 10\,TeV$. Translate this into an upper bound on a possible electron size, in m and F(*Fermi*). See Bourilkov (2001) for a derivation of this from e^+e^- scattering experiment data.

P7.8 Magnetic fields in high-energy heavy-ion collisions: In the peripheral (meaning not head-on, but almost grazing) collisions of highly charged heavy ions, very large magnetic fields can be generated by the effective currents of the $Z >> 1$ charges moving past each other at relativistic speeds. In many research articles discussing this subject, one sees statements like "... such a magnetic field is estimated to be of the order of $m_\pi^2 \approx 10^{18}$ *Gauss*" (see, e.g., Zhong et al. 2014). It's easy enough to find the mass of the pion (say from Table 7.1, at least in MeV/c^2 units), and the conversion from *Gauss* to *Tesla* (say from Table 10.8) is straightforward, but in the $\hbar, c \neq 1$ world, this statement is dimensionally challenging, to say the least.

(a) To confirm this association in our language, write

$$m_\pi^2 = C_m B e^\alpha \hbar^\beta c^\gamma, \tag{7.121}$$

match dimensions, and solve for α, β, γ. One detailed calculation (Kharzeev, McLerran, and Warringa 2008) suggests that $C_m = 2$.

(b) Evaluate m_π^2 in kg^2, and your expression for $C_m B e^\alpha \hbar^\beta c^\gamma$ using values for the fundamental constants, $C_m = 2$, and $B = 10^{18}$ *Gauss*, and see how closely they match.

[19] For a review, see Bernstein and Holstein (2013).

P7.9 "Ultrarelativistic electron–positron plasma": In a review article about the topic cited in the title of this problem, Thoma (2009a, 2009b) cites a value of the **shear viscosity** for a hot e^+e^- gas as

$$\eta_e = \frac{55.8T^3}{e^4 \ln(1/e)}, \tag{7.122}$$

where T is the plasma temperature and e is the electric charge (and **NOT** the base of the natural logs as one might first think!) In the E&M system the author uses, he notes that "... In these units $e = 0.3$, corresponding to a fine structure constant $\alpha = e^2/(4\pi) = 1/137$" a statement that is easy enough to verify. Thoma also says that "... we use natural units, i.e., $\hbar = c = k_B = 1$, as usual in quantum field theory..." so in this problem our job will be to replace three fundamental constants into a formula where they are "missing."

(a) Add enough factors of $\hbar, c, k_B$ to Eqn. (7.122) to make it dimensionally correct in the way we use the term. Do this by writing $\eta_e \propto T^3 k_B^\alpha \hbar^\beta c^\gamma$, match dimensions, and solve for the exponents, recalling that $[\eta_e] = \mathsf{M/LT}$ for viscosity.

(b) The author provides a numerical value for Eqn. (7.122) for one case by noting that "At $T = 10\,MeV$ the shear viscosity coefficient is $\eta_e = 7.9 \times 10^{10}\,Pa \cdot s$." Use your results from part (a), this numerical value for the "temperature" (which is given in energy units), his value of e, and the other fundamental constants to check this statement.

(c) In the same paper, the author gives an expression for the **damping rate** γ_e, which has the same dimensions as the angular frequencies (ω) he cites (so $[\gamma_e] = \mathsf{T}^{-1}$) writing

$$\gamma_e = \frac{e^2 T}{4\pi} \ln\left(\frac{1}{e}\right). \tag{7.123}$$

Write $\gamma_e \propto T k_B^\alpha \hbar^\beta c^\gamma$, match dimensions, and solve for the exponents.

(d) The author provides the opportunity for another numerical comparison by saying "For $10\,MeV$ we obtain $\gamma_e = 86\,keV$..." and you should try to confirm that statement, noting that he cites γ_e in energy units and not inverse time.

P7.10 Multiphoton states in positron production: In the first paper to describe observations of the process $\gamma + \gamma \to e^+ + e^-$, Burke et al. (1997) that

> In strong electromagnetic fields the interaction need not be limited to initial states with two photons [3], but rather the number of interacting photons becomes large as the dimensionless invariant parameter η ... approaches or exceeds unity.

In this experiment, one of the two incident photons arises from an high-intensity laser beam, and so the parameter η can depend the mean-square electric field ($\mathcal{E}_{rms}$) and frequency (ω_0) of the laser, the electron charge and mass (e and m_e), and the speed of light (c).

(a) Write $\eta \propto (\mathcal{E}_{rms})^\alpha e^\beta m_e^\gamma \omega_0^\delta c^\delta$ and solve to find a dimensionless ratio. Presumably, the higher the value of $\mathcal{E}_{rms}$, the more photons, which should give you information on the sign of say α. **Note:** most definitions use $\alpha = 1$.

(b) In the experiment, Burke et al. (1997), the beam parameters were $\lambda = 527\,nm$ with a beam intensity $I = 1.3 \times 10^{18}\,W/cm^2$, which they cite as corresponding to a value of $\eta = 0.36$. Use these values to try to exact the power α in the dimensionless ratio you found in part (a).

You'll need to recall the connection between the intensity of a laser beam (the Poynting vector) and the electric field of an E&M wave as $I = E_{rms}B_{rms}/\mu_0 = E_{rms}^2/(\mu_0 c)$ and that $\omega_0 = 2\pi f = 2\pi c/\lambda$.

P7.11 Rutherford scattering—Impact parameter: One of the early estimates of the size of the nucleus was given by Rutherford (1911) in *The Scattering of α and β Particles by Matter and the Structure of the Atom* in terms of the elastic scattering of charged particles via the Coulomb interaction. The geometry involved is sketched in Fig. 7.7 for a particle with charge $Z_1 e$ (and mass m and speed v_0) incident on a stationary target particle of charge $Z_2 e$. The diagram illustrates the classical hyperbolic trajectory obtained from Newton's laws for an inverse square force, specifically for two positively charged particles, such as for α scattering with an ${}^4_2He$ nucleus. (A similar trajectory for β scattering for an incident electron would also give a hyperbolic trajectory in a different direction.) The length scale that eventually determines the size of the scattering cross-section is the **impact parameter** (b), which can be determined from the physical parameters above and the observed scattering angle, θ.

(a) To determine the dimensional dependence of $b(\theta)$ write $b \propto (Z_1 e)^\alpha (Z_2 e)^\beta \epsilon_0^\gamma m^\delta v_0^\sigma$ and evaluate the exponents, using whatever intuition you can bring to bear. Do the dependences you find on m and v_0 make sense?

(b) Does this distance/length scale seem familiar from discussions of length scales in thermal problems such as in **Example 5.2** and **5.3**.

(c) DA is unable to provide any guidance on the angular dependence, but a derivation using only kinematic constraints from classical mechanics gives

$$b(\theta) \propto \sqrt{\frac{1 + \cos(\theta)}{1 - \cos(\theta)}}. \tag{7.124}$$

Confirm that this dependence makes physical sense in the limits when $\theta \to 0$ and $\theta \to \pi$.

(d) Is there a maximum size for the impact parameter where such a Coulomb scattering analysis stops making sense? For example, what about for $b(\theta) >> a_0$, namely much larger than the Bohr radius?

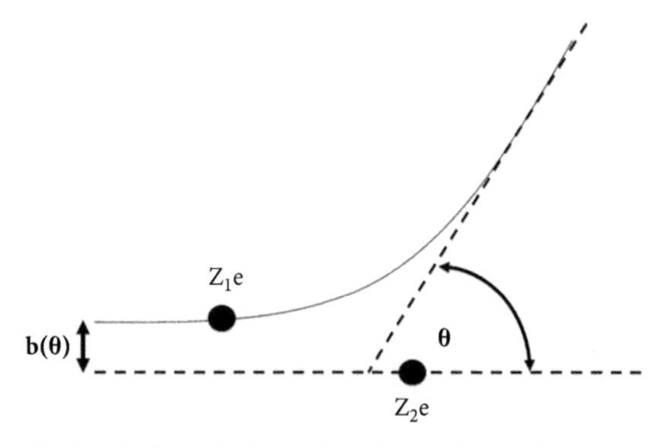

Fig. 7.7 Geometry for Rutherford scattering with a charge $Z_1 e$ and impact parameter b interacting with nuclear charge $Z_2 e$ being deflected through an angle θ.

P7.12 Light-by-light scattering cross-sections: (a) Show that the textbook result for $\sigma_{\gamma\gamma\to\gamma\gamma}$ from Eqn. (7.75) agrees with the DA predictions arising from Eqns. (7.74) in terms of their dependence on $\alpha_{FS}, m_e, \hbar, c$.
(b) The light-by-light scattering cross-sections for high-energy ($\hbar\omega >> m_e c^2$) photons is given by the expression

$$\sigma_{\gamma\gamma\to\gamma\gamma} = 4.7(\alpha_{FS})^4 \left(\frac{c}{\omega}\right)^2. \tag{7.125}$$

Show that this is consistent with our α_{FS} "power-counting" arguments as well as the dimensional discussion from Eqn. (7.60).

P7.13 One-loop QED Lagrangian prefactor of "quantum energy density" U_{QM}: (a) The prefactor in Eqn. (7.93) $\mathcal{P} = 1/U_{QM}$ can be written in terms of an inverse energy-density, which we know to be due to QED loop effects. To see to what extent DA alone can determine the dependence of U_{QM} on $m_e, \hbar, c$, and K_e, write $U_{QM} = m_e^\alpha \hbar^\beta e^\gamma K_e^{2\delta}$ and solve for as many exponents as you can, in terms of δ. Does your result agree with the form of $\mathcal{P} = 1/U_{QM}$ as expected? Do you find a relevant/familiar Buckingham Pi theorem dimensionless ratio?
(b) The two relevant energy (E) and length (l) scales for this problem are the electron rest energy, $m_e c^2$, and the Compton wavelength $\lambda_C^{(e)}$. Is there a logical way to combine these to form an energy per unit volume that might relate to U_{QM}? And does this help better specify the exponents in part (a)? If not, why not?

P7.14 One-loop QED Lagrangian in terms of critical fields: If we take the first term of the one-loop QED Lagrangian, namely

$$\mathcal{L}_{EM}^{(1\text{-}loop)} = \frac{8\alpha_{FS}^2\hbar^3}{45m_e^4c^5}(U_E - U_B)^2, \tag{7.126}$$

the (dimensionless) ratio ($\mathcal{R}$) of this term to the classical $\mathcal{L}_{EM}^{(free)} = U_E - U_B$ term is

$$\mathcal{R} \equiv \frac{\mathcal{L}_{EM}^{(1\text{-}loop)}}{\mathcal{L}_{EM}^{(free)}} = \frac{8\alpha_{FS}^2\hbar^3}{45m_e^4c^5}(U_E - U_B). \tag{7.127}$$

Rewrite this expression in terms of $|\mathbf{E}|/E_{cr}$ and $|\mathbf{B}|/B_{cr}$ where E_{cr}, B_{cr} are the critical Schwinger fields in Eqn. (7.76).

P7.15 Numerical value for the Casimir force: In the first experimental test of the Casimir force between two parallel plates, Sparnaay (1958) cites Casimir (1948) as has having

> ... derived the following expression for the attractive force F, which exists between two metal plates a distance d apart:
>
> $$F = 0.013/d^4 \quad dyne/cm^2 \qquad (1)$$
>
> The distance d, which in this expression measured in microns ...

Show that inserting values of $\hbar$ and c in Eqn. (7.108), and using the value of $C_{\mathcal{F}} = -\pi^2/240$, does reproduce the expression above.

P7.16 Casimir force between an infinite flat surface and a sphere: (a) Experiments on the Casimir force are much more easily done with a spherical object (radius R) close to a flat surface, as there are fewer issues with precise alignment of two parallel plates. In this case there is a finite Casimir force (not force per unit area) on the sphere, and writing

$$F_{sp} = C_F a^\alpha R^\beta \hbar^\gamma c^\delta, \tag{7.128}$$

match dimensions, and see to what extent you can determine the exponents. Given that this result is, in a sense, "half-way" between the parallel plate geometry of Eqn. (7.108) and that for the spherical shell in Eqn. (7.109), what does your intuition say about how to pin down your result? (b) To help specify the exponents more completely for this geometry, we note that Lamoreaux (1997) made use of a *Proximity Force Theorem* (Blocki et al. 1977), which proves that:

> The force between two gently curved surfaces as a function of the separation degree of freedom s is proportional to the interaction potential per unit area, $e(s)$, between two flat surfaces, the proportionality factor being 2π times the reciprocal of the square root of the Gaussian curvature of the gap width function at the point of closest approach (Blocki et al. 1977),

or (somewhat more prosaically, and hence more readably) in the specific context of this experimental geometry,

> ... which in the present case reduces to $F = 2\pi RE$ where R is the radius of curvature of the spherical surface and E is the potential energy per unit surface area which gives rise to the force of attraction between flat plates (Lamoreaux 1997).

First of all, show that the equation $F = 2\pi RE$ is dimensionally correct. Then, use this result, along with Eqn. (7.108), to find the attractive force on the sphere due to the infinite plane, and use it to specify the exponents you found from part (a) and find the dimensionless constant, C_F. **Note:** In the spirit that there is "nothing new under the sun," we note that Derjaguin (1934) derived an even more general result for the attractive force between two curved bodies, namely $F = 2\pi E[R_1 R_2/(R_1 + R_2)]$, where R_1, R_2 are the respective radii of curvature, which reduces to the expression above if one of the surfaces is a flat plane ($R_1 \to \infty$). He did this in the context of research on colloidal matter, not QED!

P7.17 Heavy quark–antiquark potential energy and the strong interactions I: The Cornell model (Eichten et al. 1978) of the interaction potential energy of a heavy[20] quark–antiquark ($Q\overline{Q}$) pair is often written in the form

$$V_{Q\overline{Q}}(r) = -\frac{4}{3}\frac{\alpha_S}{r} + \sigma r. \tag{7.129}$$

The dimensionless constant α_S is the analog of the fine-structure constant of QED, but represents the short-distance **QCD** Coulomb-like potential due to **gluon** (as opposed to photon) exchange and has an approximate value of $\alpha_S \approx 0.3$ (where S is for the strong interaction) and the factor of 4/3 is a group-theoretic dimensionless constant. The term linear in r arises from the confining nature of the long-distance QCD interaction, and has a coefficient often cited as $\sigma \approx 0.3\ GeV^2$. The constant σ is sometimes called the *string tension*.[21] This "strong" force has a larger dimensionless coupling strength $\alpha_S > \alpha_{FS}$, but is also "strong" because of the linear confining potential.

---

[20] In this context, the light quarks are the u, d, s ones, while the c, b, t quarks are considered 'heavy'.

[21] While originally introduced as a 'theory-informed' phenomenological model, this potential can be evaluated from first principles, using the basic QCD equations (solved numerically on a space-time lattice) and the results are very close to this form – see e.g., Bali et al. (2000) for the original research or Bali (2001) for a review.

(a) In our $\hbar, c \neq 1$ language, Eqn. (7.129) is not dimensionally correct, so add enough factors of $\hbar$ and c to each term to make it consistent.

(b) A linear potential corresponds to a constant force (F) via $V(r) = Fr$. After adding factors of $\hbar$ and c, evaluate F in both GeV/F and Newtons.

(c) Evaluate the distance scale r_0 which marks the cross-over between the "color Couloumb" ($\propto 1/r$) short-distance term and the linear ($\propto r$) confining potential at large r, by finding where $V_{Q\overline{Q}}(r_0) = 0$. The estimated radial separation between the quark–antiquark pair in the ground state for the $c\overline{c}$ (charm quark) system is $r_{c\overline{c}} \approx 0.4\,F$, while that for the heavier $b\overline{b}$ (bottom quark) pair is $r_{b\overline{b}} \approx 0.25\,F$. How do these values compare to the cross-over distance, r_0? The radial separations for higher level excitations of the $c\overline{c}$ system are roughly 0.7–0.8 F. Which term in the potential dominates those excited states?

(d) For heavy enough quarks, where the "color Coulomb" part of the potential dominates, we should be able to use Bohr atom ideas to extract information on the system. For example, the "color Bohr radius" should be obtainable via scaling by taking

$$a_0^e = \frac{\lambda_e^C}{\alpha_{FS}} = \frac{\hbar}{\alpha_{FS} m_e c} \tag{7.130}$$

and substituting $\alpha_{FS} \rightarrow \alpha_S$ and $m_e \rightarrow m_Q/2$ (remember reduced mass effects, like for PS), so that

$$a_0^Q = \frac{2\hbar c}{\alpha_S(m_Q c^2)}. \tag{7.131}$$

Use the values $m_c \approx 1.4\,GeV/c^2$ and $m_b \approx 4.2\,GeV/c^2$ to evaluate $a_0^{c,b}$ and compare to the estimated radial separation values above. Do either agree with the experimental values?

P7.18 Heavy quark–antiquark potential energy and the strong interactions II: (a) The values of $a_0^{c,b}$ for the $c\overline{c}$ and $b\overline{b}$ systems seem to fall near the cross-over between a $P = -1$ and $P = +1$ power-law potential, $V^{(P)}(r) \propto r^P$, corresponding to roughly $P \approx 0$ behavior. Assuming that $P = 0$ actually corresponds to a logarithmic potential (and not a constant), write $V_{(log)}(r) = V_0 \log(r/r_0)$, use the value of r_0 (from **P7.15 (c)**) for the Cornell potential in Eqn. (7.129) and part (c), and plot $V_{(log)}(r)$ versus $V_{Q\overline{Q}}(r)$ for several values of V_0 to see how close that phenomenological fit can get. **Hint:** Assume that $V_0 = 0.1–1.0\,GeV$ and see if any values in that range work. Quigg and Rosner (1979) in *Quantum Mechanics with Applications to Quarkonium* make use of many scaling results for power-law potentials, as in Section 6.3.3, including ones for this logarithmic form.
(b) Assuming that $V^{(P=0)} = V_0 \ln(r/r_0)$ is the relevant form of the power-law potential for this logarithmic form, so that $K_{P=0} = V_0$, use the scaling results in Eqn. (6.141) to find how the energy ($E_n^{(P=0)}$) and length ($\rho^{(P=0)}$) scales depend on $n, \hbar, m, K_{P=0}$.
(c) Use your scaling result for $\rho^{(0)}$ from part (b) to estimate the ratio $\rho(b\overline{b})/\rho(c\overline{c})$ using $m_c \approx 1.4\,GeV/c^2$ and $m_b \approx 4.2\,GeV/c^2$ and compare to the experimental values of $r_{b\overline{b}} \sim 0.25\,F$ and $r_{c\overline{c}} \sim 0.4\,F$. If we use the general power-law scaling result, show that $\rho(b\overline{b})/\rho(c\overline{c}) \sim (m_c/m_b)^{1/(2+P)}$ and use the data above to estimate the value of P.

P7.19 The weak interaction I—Preamble: The "weak" interaction is mediated by massive spin-one bosons, either charged ones ($W^{\pm}$), or neutral (Z^0) ones, both with masses in the 80 GeV/c^2 range. A cartoon of the Feynman diagram vertices for some of these interactions is shown in Fig. 7.8 where one can "paste together" two such vertices with a propagator corresponding to the exchange of W/Z bosons. By analogy with earlier Feynman diagram "recipes," the amplitude for such a weak interaction process (either for a 1→3 particle decay or a 2→2 particle scattering cross-section) will have a form like

$$\text{amplitude} = \text{AMP}_W \propto (g_W)\left(\frac{1}{M_W^2}\right)(g_W) \quad \longrightarrow \quad \frac{\alpha_W}{M_W^2} \tag{7.132}$$

where the dimensionless α_W is the analog of the EM fine-structure constant α_{FS} and has a value of approximately $\alpha_W \sim 1/28$, which is actually larger than $\alpha_{FS} \sim 1/137$. The weak interactions are therefore not "weak" due to the strength of the coupling constant, but rather due to the exchange of a (very) heavy intermediate force carrier.

(a) Recalling the discussion about massive photons in Section 4.8.2, write down the likely form for the potential energy function corresponding to W/Z exchange as

$$V_W(r) = \frac{\alpha_W}{r} e^{-M_W r} \tag{7.133}$$

and add factors of $\hbar$ and c as needed to make this expression dimensionally correct.

(b) What is the effective range of the weak force, in F (Fermi) and m?

P7.20 The weak interaction II—Muon decay: The muon decays via the weak interaction, $\mu^- \to e^- + \nu_\mu + \overline{\nu}_e$, with a decay rate involving the amplitude in Eqn. (7.132). The lifetime and decay rate then have factors of $|\text{AMP}_W|^2$ and powers of $\hbar$, c, and the mass of the muon, m_μ. Write

$$\frac{1}{\tau(\mu^- \to e^- \nu_\mu \overline{\nu}_e)} = \lambda(\mu^- \to e^- \nu_\mu \overline{\nu}_e) = C_\lambda \left(\frac{\alpha_W}{M_W^2}\right)^2 m_\mu^\alpha \hbar^\beta c^\gamma. \tag{7.134}$$

(a) Balance dimensions to solve for α, β, γ.

(b) Using values of $\alpha_W = 1/28$, $M_W = 80\ GeV/c^2$, and $m_\mu = 106\ MeV/c^2$, evaluate the muon lifetime in this simple DA-only approach and compare to the experimental value of $\tau_\mu = 2.2\ \mu s$.

(c) It turns out that the dimensionless prefactor is actually $C_\lambda = 1/(384\pi) \approx 10^{-4}$, which is not nearly as close to $\mathcal{O}(1)$ as in many other DA problems. Update your prediction from part (b) using this value of C_λ and see if you're closer to the observed value.

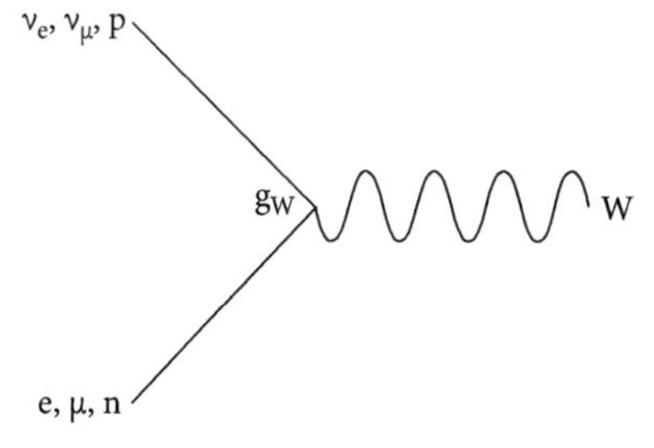

Fig. 7.8 Example weak interaction vertex illustrating (e, ν_e), (μ, ν_μ), and (p, n) coupling to a charged weak boson. For processes with energies below $M_W c^2$, the W propagator will add a factor of $1/M_W^2$ when two such vertices are connected, giving a weak amplitude $\text{AMP}_W \propto g_W^2/M_W^2$.

Notes: One reason that scattering cross-sections and decay rates for more complex processes may have a "smaller than DA would suggest" dimensionless prefactor is that for each additional particle appearing in the final state, there is an associated **phase space** factor of $d^4p/(2\pi)^4$. The additional powers of $(2\pi)^{-4} \approx 1500$ can lead to suppressions, beyond any accounted for by powers of the coupling constants (α_{FS}, α_S or α_W) and other factors. See also **P7.6**(d) for another example of a three-body decay with a small dimensionless constant. This effect is similar to the ambiguity arising from the use of f versus $\omega = 2\pi f$ or h versus $\hbar = h/2\pi$ in any physical problem, but magnified by the number of powers. In addition, the phase-space factor is often written as $d^4p/(2\pi\hbar)^4 = d^4p/h^4$, with factors of Planck's constant in the denominator, much like the expressions in Eqns. (6.35) and (6.36).

P7.21 The weak interaction III—Scattering cross-section: Neutrinos can scatter from electrons or nucleons (protons or neutrons) via the weak interaction. The scattering/interaction cross-section (σ_W) for such a process will be proportional to

$$\sigma_W \propto \left(\frac{\alpha_W}{M_W^2}\right)^2 E_\nu^\alpha M_T^\beta \hbar^\gamma c^\delta, \tag{7.135}$$

where m_T is the mass of the (stationary) target electron or nucleon and E_ν is the energy of the incident neutrino.

(a) Recalling that $[\sigma] = \mathsf{L}^2$, balance dimensions and solve for $\alpha, \beta, \gamma, \delta$ to the extent you can, say in terms of α. Considering that there would be no scattering if either E_μ or m_T were zero, what are the most likely values of α, β.

(b) To set the scale for the size of σ_W, the proton–proton scattering cross-section is roughly $\sigma_{pp} \sim \pi(2 \times R_p)^2$ where $R_p \approx 1\,F$ is the proton radius. Evaluate this in m^2. A recent experimental measurement (Antchev et al. 2013) of the total p–p cross-section at LHC energies finds that $\sigma_{pp} = 98.6 \pm 2.2\,mb$ where $1\,barn = 10^{-24}\,cm^2$. Do these two values roughly agree? **Note:** The unit **barn** alludes to the fact that this scattering cross-section is actually a rather large value in the world of elementary particle collisions, namely "as big as a barn."

(c) Use your best guess for the neutrino cross-section from Eqn. (7.135) for a proton/neutron target (so $m_T \approx 940\,MeV/c^2$) and an $E_\mu = 1\,GeV$ incident neutrino and compare to the *pp* cross-section from part (b).

(d) The **path length** (l_ν) of a neutrino passing through matter will roughly be determined by the condition that the number of collisions/interactions will be at least one, namely $l_\nu n_T \sigma_\nu \sim 1$, where n_T is the number density of the targets. For water, with $\rho_m = 10^3\,kg/m^3$, and a typical nucleon having a mass of $m_{p,n} \sim 1.6 \times 10^{-27}\,kg$, find n_T and use the neutrino-cross-section from part (c) to estimate l_ν.

(e) Such calculations are often used to demonstrate just how weakly interacting neutrinos can be. If this is the case, how do experiments that measure neutrino properties, either at accelerator facilities, or in particle-astrophysics settings (such as the IceCube detector) ever detect such particles?

P7.22 Graviton scattering processes: While we focus most of any remaining discussions of the fourth and final fundamental force, namely gravity, to astrophysical problems (including a bit of general relativity) in Chapter 9, there are some problems involving gravitational physics that can be couched in the language of particle interactions. In the same way that photons (γ) are treated as the quanta of the EM field, and can be analyzed in many scattering processes, we can associate massless **gravitons** (labeled as g, since Newton's constant G is still used as the effective coupling strength) as the particle-like, quantized excitations of the gravitational field.

(a) The process of **graviton photo-production** can be symbolized by the reactions

$$\gamma + s \to g + s \qquad \text{or} \qquad \gamma + f \to g + f, \tag{7.136}$$

where g stands for the graviton particle. For the target particles, s stands for a *scalar* (or $S = 0$) particle of mass m and charge e, while f is for a $S = 1/2$ fermion, assumed to have the same m, e values. Voronov (1973) considered this process for electrons ($f = e^-$), while Holstein (2006) reviewed both processes at a more pedagogical level. The differential cross-section for the scalar process is

$$\frac{d\sigma}{d\Omega} = G\alpha_{FS}\cos^2\left(\frac{\theta}{2}\right)\left(\frac{\omega_f}{\omega_i}\right)^2 \left[\cot^2\left(\frac{\theta}{2}\right)\cos^2\left(\frac{\theta}{2}\right) + \sin^2\left(\frac{\theta}{2}\right)\right], \tag{7.137}$$

where ω_i, ω_f are the angular frequencies of the incoming γ and outgoing graviton (g) so that $E_\gamma = \hbar\omega_i$ and $E_g = \hbar\omega_f$. As it stands, this formula is not dimensionally correct, so add enough factors of $\hbar$ and c to the right-hand side to ensure that $d\sigma/d\Omega$ has the dimensions of L^2. Using what you know about α_{FS} does the final result depend on $\hbar$?

(b) The **gravitational analog of Compton scattering for the graviton** would be

$$g + s \to g + s \qquad \text{or} \qquad g + f \to g + f \tag{7.138}$$

and the differential cross-section for this process for $S = 0$ particles (Holstein 2006) is

$$\frac{d\sigma}{d\Omega} = G^2 m^2 \left(\frac{\omega_f}{\omega_i}\right)^2 \left[\cot^4\left(\frac{\theta}{2}\right)\cos^4\left(\frac{\theta}{2}\right) + \sin^4\left(\frac{\theta}{2}\right)\right]. \tag{7.139}$$

Once again, add enough factors of $\hbar$ and c to make Eqn. (7.139) dimensionally correct.

P7.23 Landau pole: A famous observation about the "running" E&M coupling constant in Eqn. (7.111) is that for sufficiently high energies/momenta (or small enough distance scales), the fine-structure constant would actually diverge. Let us assume that the typical reference momentum scale is $\Lambda = m_e c$, where $\alpha_{FS}(\Lambda^2) = 1/137$. If we set

$$1 - \frac{\alpha_{FS}(\Lambda^2)}{3\pi}\ln\left(\frac{p^2}{\Lambda^2}\right) = 0, \tag{7.140}$$

find the value of $E = pc$ where this divergence would be expected, say in eV. Can you translate that into a length scale? Should we be worried? (If you're still worried, read any popular accounts of **grand unified theories** of the strong (S), weak (W), and E&M interactions which describe how α_S

and α_W also "run" with energy/momentum, and might merge into a single force at a much lower scale than predicted here.)

P7.24 Pulsar periods in terms of "hyperfine units": After reviewing the discussion in Sec. 7.7, use the binary representation of the scaled periods below, and the base unit time period

$$\tau_{hyp} = \frac{1}{f_{hyp}} = \frac{1}{1420\,MHz} = 7.04 \times 10^{-8}\,s, \tag{7.141}$$

corresponding to the hyperfine transition from Eqn. (7.51), to compare to the experimentally measured periods in seconds in the last column. If you want more practice in checking such binary arithmetic, you can see all fourteen of the pulsar periods etched on the Pioneer plaque in Sagan, Sagan, and Drake (1972) shown in this format. See also Robinett (2001) for more worked out examples.

Pulsar	**Scaled period (binary representation)**	**Period (in seconds) 1970/1971 epoch**
1240	100,000,110,110,010,100,001,001,111,000	3.8800×10^{-1}
0531	10,110,011,100,000,101,010,000,010	3.31296×10^{-2}

References for Chapter 7

Aad, G., et al. (ATLAS Collaboration). (2019). "Observation of Light-by-Light Scattering in Ultraperipheral Pb + Pb Collisions with the ATLAS Detector," *Physical Review Letters* 123, 052001.

Abbasi, et al., (Telescope Array Collaboration). (2023). "An Extremely Energetic Cosmic Ray Observed by a Surface Detector," *Science* 382, 903–7.

Abbiendi, G., et al. (OPAL Collaboration). (2006). "Measurement of the Running of the QED Coupling in Small-Angle Bhabha Scattering at LEP," *European Physical Journal C* 45, 1–21.

Abdallah, J., et al. (DELPHI Collaboration). (2004). "Determination of the $e^+e^- \rightarrow \gamma\gamma(\gamma)$ Cross-Section at LEP 2," *European Journal of Physics C* 37, 405–19.

Ahmadi, M., et al. (The ALPHA Collaboration). (2022). "Investigation of the Fine Structure of Antihydrogen," *Nature* 578, 375–80.

Al-Ramadhan, A. H., and Gidley, D. W. (1994). "New Precision Measurement of the Decay Rate of Single Positronium," *Physical Review Letters* 72, 1632–5.

Antchev, G., et al. (TOTEM Collaboration). (2013). "Measurement of Proton–Proton Elastic Scattering and Total Cross-Section at $\sqrt{s} = 7$ *TeV*," *Europhysics Letters* 101, 21002.

Aoyama, T., Kinoshita, T., and Nio, M. (2019). "Theory of the Anomalous Magnetic Moment of the Electron," *Atoms* 7, 28.

Backenstoss, G. (1970). "Pionic Atoms," *Annual Review of Nuclear and Particle Science* 20, 467–508.

Bali, G. S., et al. (SESAM/TCL Collaborations). (2000). "Static Potentials and Glueball Masses from QCD Simulations with Wilson Sea Quarks," *Physical Review D* 62, 054503.

Bali, G. S. (2001). "QCD Forces and Heavy Quark Bound States," *Physics Reports* 343, 1–136.

Bass, S. D., Marizaai, S., Moskal, P., and Stępień, E. (2023). "Colloquium: Positronium Physics and Biomedical Applications," *Reviews of Modern Physics* 95, 0211002.

Batty, C. J. (1989). "Antiprotonic-Hydrogen Atoms," *Reports on Progress in Physics* 52, 1165–1216.

Beane, S. R., et al. (2014). "Magnetic Moments of Light Nuclei from Lattice Quantum Chromodynamics," *Physical Review Letters* 113, 252001.

Berestetskii, V. B., Lifshitz, E. M., and Pitaevskii, L. P. (1982). *Quantum Electrodynamics*, 2nd ed. (Oxford: Butterworth-Heinemann).

Berko, S., and Pendleton, H. N. (1980). "Positronium," *Annual Review of Nuclear and Particle Science* 30, 543–81.

Bernstein, A. M., and Holstein, B. R. (2013). "Neutral Pion Lifetime Measurements and the QCD Chiral Anomaly," *Reviews of Modern Physics* 85, 49–78.

Bertsche, W. A., Butler, E., Charlton, M., and Madsen, N. (2015). "Physics with Antihydrogen," *Journal of Physics B: Atomic, Molecular and Optical Physics* 48, 232001.

Bethe, H. A. (1947). "The Electromagnetic Shift of Energy Levels," *Physical Review* 72, 339–41.

Bethe, H. A., and Salpeter, E. E. (1957). *Quantum Mechanics of One- and Two-Electron Atoms* (Berlin: Springer-Verlag).

Bezginov, N., et al. (2019) "A Measurement of the Atomic Hydrogen Lamb Shift and the Proton Charge Radius," *Science* 365, 1007–1012.

Bird, D. J., et al. (1995). "Detection of a Cosmic Ray with Measured Energy Well beyond the Expected Spectral Cutoff Due to Cosmic Microwave Background," *Astrophysical Journal* 441, 144–50.

Blocki, J., Randrup, J., Świątecki, W. J., and Tsang, C. F. (1977). "Proximity Forces," *Annals of Physics* 105, 427–62.

Blundell, S. J., de Renzi, R., Lancaster, T., and Pratt, F. L. (2022). *Muon Spectroscopy: An Introduction* (Oxford: Oxford University Press).

Borie, E., and Rinker, G. A. (1982). "The Energy Levels of Muonic Atoms," *Reviews of Modern Physics* 54, 67–118.

Bourilkov, D. (2001). "Hint for Axial-Vector Contact Interactions in the Data on $e^+e^- \rightarrow e^+e^-(\gamma)$ at Center-of-Mass Energies 192–208 GeV," *Physical Review D* 64, 071701.

Boyer, T. H. (1968). "Quantum Electromagnetic Zero-Point Energy of a Conducting Spherical Shell and the Casimir Model of a Charged Particle," *Physical Review* 174, 1764–76.

Bressi, G., Carugno, G., Onofrio, R., and Ruoso, G. (2002). "Measurement of the Casimir Force between Parallel Metallic Surfaces," *Physical Review Letters* 88, 041801.

Brickwedde, F. G. (1982). "Harold Urey and the Discovery of Deuterium," *Physics Today* 35(9), 34–39.

Bulanov, S. S., et al. (2010). "Schwinger Limit Attainability with Extreme Power Lasers," *Physical Review Letters* 105, 220407.

Burke, D. L., et al. (1997). "Positron Production in Multiphoton Light-by-Light Scattering," *Physical Review Letters* 79, 1626.

Casimir, H. B. G. (1948). "On the Attraction of Two Perfectly Conducting Plates," *Proceedings of the Section of Sciences, Koninklijke Nederlandsche Akademie van Wetenschappen* 51, 793–5.

Casimir, H. B. G. (1951). "On the Theory of Electromagnetic Waves in Resonant Cavities," *Philips Research Reports* 6, 162–82.

Casimir, H. B. G. (1953). "Introductory Remarks on Quantum Electrodynamics," *Physica* 19, 846–9.

Cocconi, G., and Morrison, P. (1959). "Searching for Interstellar Communication," *Nature* 184, 844–6.

Crater, H. (1991). "Singlet-Positronium Decay Using a Relativistic Wave Function," *Physical Review A* 44, 7065–70.

Curceanu, C., et al. (2019). "The Modern Era of Light Kaonic Atom Experiments," *Reviews of Modern Physics* 91, 25006.

Czarneck, A., and Karshenboim, S. G. (1999). "Decays of Positronium." In *Proceedings of the 14th International Workshop on High Energy Physics and Quantum Field Theory* (QFTHEP99, Moscow 1999), edited by B. B. Levchenko and V. I. Savrin, 538–44. (East Lansing: Michigan State University Press).

Derjaguin, G. (1934). "Untersuchungen über die Reibung und Adhäsion, IV - Theorie des Anhaftens kleiner Teilchen (Studies on Friction and Adhesion, IV: Theory of Adhesion of Small Particles)," *Kolloid-Zeitschrift* 69, 155–64.

Eichten, E., et al. (1978). "Charmonium: The Model," *Physical Review C* 17, 3090–3117.

Eides, M. I., Grotch H., and Shelyuto, V. A. (2000). "Theory of Light Hydrogen-Like Atoms," *Physics Reports* 342, 63–261.

Elizalde, E., and Romero, A. (1991). "Essentials of the Casimir Effect and its Computation," *American Journal of Physics* 59, 711–19.

Euler, H., and Kockel, B. (1935). "Über die Streuung von Licht an Licht nach der Diracschen Theorie (On the Scattering of Light on Light According to Dirac's Theory)," *Naturwissenschaften* 23, 246–7.

Fan, X., Myers, T. G., Sukra, B. A. D., and Gabrielse, G. (2023). "Measurement of the Electron Magnetic Moment," *Physical Review Letters* 130, 071801.

Griffiths, D. J. (2004). *Introduction to Quantum Mechanics*, 2nd ed. (Upper Saddle River, NJ: Pearson/ Prentice-Hall).

Gurian, J., Jeong, D., Ryan, M., and Shandera, S. (2022). "Molecular Chemistry for Dark Matter. II. Recombination, Molecule Formation, and Halo Mass Function in Atomic Dark Matter," *Astrophysical Journal* 934, 121.

Hänsch, T. W., Schawlow, A. L., and Series, G. W. (1979). "The Spectrum of Atomic Hydrogen," *Scientific American* 240, 94–111.

Heisenberg, W., and Euler, H. (1936). "Folgerungen aus der Diracschen Theorie des Positrons (Consequences of Dirac's Theory of the Positron)," *Zeitschrift für Physik* 98, 714–32.

Holstein, B. R. (2006). "Graviton Physics," *American Journal of Physics* 74, 1002–1011.

Hughes, V. W., McColm, D. W., and Ziock, K. (1960). "Formation of Muonium and Observation of its Larmor Precession", *Physical Review Letters* 5, 63–65.

Hughes, V. W. (1966). "Muonium," *Annual Review of Nuclear and Particle Science* 16, 445–70.

Hughes, V. W. (1992). "Muonium," *Zeitschrift für Physik C: Particles and Fields* 56, S35–S43. This paper was part of a supplemental issue from Vol. 56 of Z. Phys. C., March 1992, dedicated to muon physics, with contributions from leading experts in both experiment and theory as of the time.

Itzykson, C., and Zuber, J -B. (1980). *Quantum Field Theory* (New York: McGraw-Hill).

Jackson, J. D. (1999). *Classical Electromagnetism*, 3rd ed. (New York: Wiley).

Karplus, R., Klein, A., and Schwinger, J. (1952). "Electrodynamic Displacement of Atomic Energy Levels. II. Lamb Shift," *Physical Review* 86, 288–301.

Karshenboim, S. G., et al., eds. (2001). *The Hydrogen Atom: Precision Physics of Simple Atomic Systems* (Berlin: Springer-Verlag).

Kharzeev, D. E., McLerran, L. D., and Warringa, H. J. (2008). "The Effects of Topological Charge Change in Heavy Ion Collisions: 'Event by Event P and CP Violation'," *Nuclear Physics A* 803, 227–53.

Knecht, A., Skawran, A., and Vogiatzi, S. M. (2020). "Study of Nuclear Properties with Muonic Atoms," *European Physical Journal - Plus* 135, 777.

Lamoreaux, S. K. (1997). "Demonstration of the Casimir Force in the 0.6 to 6-μm Range," *Physical Review Letters* 78, 5–8.

Lamoreaux, S. K. (1998). "Erratum: Demonstration of the Casimir Force in the 0.6 to 6-μm Range [Phys. Rev. Lett. 78, 5 (1997)]," *Physical Review Letters* 81, 5475.

Lamoreaux, S. K. (1999). "Resource Letter CF-1: Casimir Force," *American Journal of Physics* 67, 850–61.

Lamoreaux, S. K. (2005). "The Casimir Force: Background, Experiments, and Applications," *Reports on Progress in Physics* 68, 201–36.

Levine, I., et al. (TOPAZ Collaboration). (1997). "Measurement of the Electromagnetic Coupling at Large Momentum Transfer," *Physical Review Letters* 78, 424–7.

Lukosz, W. (1971). "Electromagnetic Zero-Point Energy and Radiation Pressure for a Rectangular Cavity," *Physica* 56, 109–20.

Marion, J. B., and Thornton, S. T. (2003). *Classical Dynamics of Particles and Systems*, 5th ed. (Boston: Cengage).

Martinelli, G., Parisi, G., Petronzio, R., and Rapuano, F. (1982). "The Proton and Neutron Magnetic Moments in Lattice QCD," *Physics Letters B* 116, 434–6.

Meacher, D. R., et al. (1991). "Observation of the Collapse and Fractional Revival of a Rydberg Wavepacket in Atomic Rubidium," *Journal of Physics B: Atomic, Molecular and Optical Physics* 24, L63–L69.

Milton, K. (2011). "Resource Letter VWCPF-1: Van der Waals and Casimir–Polder Forces," *American Journal of Physics* 79, 697–711.

Mohideen, U., and Roy, A. (1998). "Precision Measurement of the Casimir Force from to 0.1 to 0.9 μm," *Physical Review Letters* 81 4549–52.

Mohr, P. J., Taylor, B. N., and Newell, D. B. (2008). "CODATA Recommended Values of the Fundamental Physical Constants: 2006*," *Reviews of Modern Physics* 80, 633–730.

Odaka, S., et al. (VENUS Collaboration). "Measurement of the Running of the Effective QED Coupling at Large Momentum Transfer in the Space-Like Region," *Physical Review Letters* 81, 2428–31.

Ore, A., and Powell, J. L. (1949). "Three-Photon Annihilation of an Electron-Positron Pair," *Physical Review* 75, 1696–9.

Parreño, A., et al. (2017). "Octet Baryon Magnetic Moments from Lattice QCD: Approaching Experiment from a Three-Flavor Symmetric Point," *Physical Review D* 95, 114513.

Pohl, R., Gilman, R., Miller, G. A., and Pachucki, K. (2013). "Muonic Hydrogen and the Proton Radius Puzzle," *Annual Review of Nuclear and Particle Science* 63, 175–204.

Quigg, C., and Rosner, J. L. (1979). "Quantum Mechanics with Applications to Quarkonium," *Physics Reports* 56, 167–235.

Rich, A. (1981). "Recent Experimental Advances in Positronium Research," *Reviews of Modern Physics* 53, 127–65.

Rigden, J. S. (2002). *Hydrogen: The Essential Element* (Cambridge, MA: Harvard University Press).

Robinett, R. W. (2001). "Spacecraft artifacts as a physics teaching resource," *Physics Teacher* 39, 476–9.

Robinett, R. W. (2006). *Quantum Mechanics: Classical Results, Modern Systems, and Visualized Examples*, 2nd ed. (Oxford: Oxford University Press).

Ruffini, R., Vereshchagina, G., and Xue, S-S. (2010). "Electron–Positron Pairs in Physics and Astrophysics: From Heavy Nuclei to Black Holes," *Physics Reports* 487, 1–140.

Rutherford, E. (1911). "LXXIX: The Scattering of α and β Particles by Matter and the Structure of the Atom," *Philosophical Magazine* Series 6, 21, 125, 669–88.

Ryan, M., Gurian, J., Shandera, S., and Jeong, D. (2022a). "Molecular Chemistry for Dark Matter," *Astrophysics Journal* 934, 120.

Ryan, M., Shandera, S., Gurian, J., and Jeong, D. (2022b). "Molecular Chemistry for Dark Matter. III. DarkKROME," *Astrophysics Journal* 934, 122.

Sagan, C., Sagan, L. S., and Drake, F. (1972). "A Message from Earth," *Science* 175, 881–4.

Sagan, C., and Drake, F. (1975). "The Search for Extra-Terrestrial Intelligence," *Scientific American* 232, 80–89.

Sauter, F. (1931). "Über das Verhalten eines Elektrons im homogenen elektrischen Feld nach der relativistischen Theorie Diracs (On the Behavior of an Electron in a Homogeneous Electric Field According to Dirac's Relativistic Theory)," *Zeitschrift für Physik* 69, 742–64.

Schwinger, J. (1948). "On Quantum-Electrodynamics and the Magnetic Moment of the Electron," *Physical Review* 73, 416–17.

Schwinger, J. (1951). "On Gauge Invariance and Vacuum Polarization," *Physical Review* 82, 664–79.

Scully, M. O., and Zubairy, M. S. (1997). *Quantum Optics* (Cambridge: Cambridge University Press).

Seki, R., and Wiegand, C. (1975). "Kaonic and Other Exotic Atoms," *Annual Review of Nuclear and Particle Science* 25, 241–81.

Series, G. W. (1957). *The Spectrum of Atomic Hydrogen* (Oxford: Oxford University Press).

Sommerfeld, C. M. (1958). "The Magnetic Moment of the Electron," *Annals of Physics* 5, 26–57.

Sparnaay, M. J. (1958). "Measurement of Attractive Forces between Flat Plates," *Physica* 24 751–64.

Stange, A., Campbell, D. K., and Bishop, D. J. (2021). "Science and Technology of the Casimir Effect," *Physics Today* 74(1), 42–48.

Stöhlker, Th., et al. (2000), "1S Lamb Shift in Hydrogenlike Uranium Measured on Cooled, Decelerated Ion Beams," *Physical Review Letters* 85, 3109.

Stroscio, M. A. (1975). "Positronium: A Review of the Theory," *Physics Reports* 22, 215–77.

Taylor, J. H., and Weisberg, J. M. (1982). "A New Test of General Relativity: Gravitational Radiation the Binary Pulsar PSR 1913+16," *Astrophysical Journal* 253, 908–20.

Thoma, M. H. (2009a). "Ultrarelativistic Electron–Positron Plasma," *European Physical Journal D: Atomic, Molecular, Optical and Plasma Physics* 55, 271–8.

Thoma, M. H. (2009b). "Colloquium: Field Theoretic Description of Ultrarelativistic Electron-Positron Plasmas," *Reviews of Modern Physics* 81, 959–68.

Treagust, D. F., Duit, R., and Fischer, H. E., eds. (2017). *Multiple Representations in Physics Education* (Cham: Springer Nature).

Urey, H. C., Brickwedde, F. G., and Murphy, G. M. (1932a). "A Hydrogen Isotope of Mass 2," *Physical Review* 39, 164–5.

Urey, H. C., Brickwedde, F. G., and Murphy, G. M. (1932b). "A Hydrogen Isotope of Mass 2 and its Concentration," *Physical Review* 40, 1–15.

von Klitzing, K., Dorda, G., and Pepper, M. (1980). "New Method for High-Accuracy Determination of the Fine-Structure Constant Based on Quantized Hall Resistance," *Physical Review Letters* 45, 494–7.

Voronov, N. A. (1973). "Gravitational Compton Effect and Photoproduction of Gravitons by Electrons," *Journal of Experimental and Theoretical Physics* 37, 953–8.

Walker, D. C. (2012). *Muon and Muonium Chemistry* (Cambridge: Cambridge University Press).

Welton, T. A. (1948). "Some Observable Effects of the Quantum-Mechanical Fluctuations of the Electromagnetic Field," *Physical Review* 74, 1157–67.

Wheeler, J. A. (1946). "Polyelectrons," *Annals of the New York Academy of Sciences* 46, 219–38.

Workman, R. L. (2022). "Review of Particle Physics," *Progress of Theoretical and Experimental Physics* 2022, 083C01.

Wu, C. S., and Wilets, L. (1969). "Muonic Atoms and Nuclear Structure," *Annual Review of Nuclear and Particle Science* 19, 527–606.

Yeazell, J. A., Mallalieu, M., and Stroud, Jr., C. R. (1990). "Observation of the Collapse and Revival of a Rydberg Electronic Wave Packet," *Physical Review Letters* 64, 2007–10.

Zangwill, A. (2013). *Modern Electrodynamics* (Cambridge University Press, Cambridge).

Zhong, Y., et al. (2014). "A Systematic Study of Magnetic Field in Relativistic Heavy-Ion Collisions in the RHIC and LHC Energy Regions," *Advances in High Energy Physics* 2014, 193039.

Chapter 8

Condensed Matter Physics

More advanced study of the topics covered in the traditional "quadrivium" of undergraduate (and graduate) physics coursework can focus on continuing to explore the "landscape of physics" at the research frontiers, or to utilize them in engineering or technological applications. For example, the basics of classical mechanics for point objects can be extended to continuous systems (gases, liquids, solids, granular materials) with use in civil, mechanical, aerospace, geosciences, or other engineering disciplines, **or** can be applied (especially when extended to include special and general relativity) to confront "forefront"[1] problems in astrophysics and cosmology. The fundamental laws of electricity and magnetism (E&M) as encoded in Maxwell's equations can equally well be brought to bear to describe device performance or electromagnetic (EM) wave propagation, but also can provide a starting point for fundamental (gauge) theories such as quantum electrodynamics (QED) and as a template for the study of the weak and strong forces relevant at the elementary particle level.

The continued study of quantum mechanics beyond simple two-body (or few-body) systems like atoms and molecules to describe the behavior of matter in its condensed form (especially in the **solid state**) can be argued to have produced some of the most important recent technological applications of all. Examples of such **many-body** phenomena include superconductivity (for accelerator magnets to magnetic resonance imaging machines to quantum computer qubits), semiconductors, transistors and other magnetic storage media (some based on giant magneto-resistance effects), liquid crystal displays, and by extension, much of what underlies the information storage, transfer, and retrieval infrastructure on which modern society relies.[2]

In thinking about the relationship between what is sometimes described as **intensive** versus **extensive** research, Weisskopf (1965) suggests:

> In short: intensive research goes for the fundamental laws, extensive research goes for the explanation of phenomena in terms of known fundamental laws.

In a famous article in the journal *Science* entitled *More is Different*, P. W. Anderson lucidly discussed some of the issues with scientists taking a **reductionist implying constructionist** mindset too literally, noting that:

> The ability to reduce everything to simple fundamental laws does not imply the ability to start from those laws and reconstruct the universe . . .
>
> The behavior of large and complex aggregates of elementary particles, it turns out, is not to be understood in terms of a simple extrapolation of the properties of a few particles. Instead, at each level of complexity entirely new properties appear, and the understanding of the new behaviors requires research which I think is as fundamental in its nature as any other . . .

[1] Note the quotation marks, in light of the statements in the next paragraph.
[2] "I spend my money on the regular miracles, just like you, like me, like everybody else." *New Politics* (2013).

Dimensional Analysis Across the Landscape of Physics. Richard W. Robinett, Oxford University Press. © Richard W. Robinett (2024).
DOI: 10.1093/oso/9780192867551.003.0008

> At each stage entirely new laws, concepts, and generalizations are necessary, requiring inspiration and creativity to just as great a degree as in the previous one. Psychology is not just applied biology, nor is biology applied chemistry (Anderson (1972)).

Condensed matter physics is home to some of the most unexpected and important emergent phenomena of the twentieth (and now twenty-first) century, and we consider in this chapter to what extent some of the most famous of them can be explored using dimensional analysis (DA), focusing on the "research side" of things, but noting especially their impact on the field of metrology. Excellent references for many topics in solid state physics for further study include classics by Kittell (1971) or Ashcroft and Mermin (1976), and more recently those by Marder (2000), Snoke (2009), Cohen and Louie (2016), or Girvin and Yang (2019).

8.1 Atoms to molecules to solids

One of the early successes of quantum mechanics was to explain why as simple a system as the hydrogen atom is stable, given that classically it would radiate away energy and collapse, as discussed in **P7.1** and **P7.2**. It's perhaps less obvious that more complex multi-electron atoms are themselves stable, or that collections of atoms in bulk are either.

An early acknowledgment of the physical principle most responsible for the *stability of matter*[3] was made by Ehrenfest (1959), who said:

> We take a piece of metal. Or a stone. When we think about it, we are astonished that this quantity of matter should occupy so large a volume. Admittedly, the molecules are packed tightly together, and likewise the atoms within each molecule. But why are the atoms themselves so big?
>
> Consider for example the Bohr model of an atom of lead. Why do so few of the 82 electrons run in the orbits close to the nucleus? The attraction of the 82 positive charges in the nucleus is so strong. Many more of the 82 electrons could be concentrated into the inner orbits, before their mutual repulsion would become too large. What prevents the atom from collapsing in this way?
>
> Answer: only the Pauli principle, "No two electrons in the same state." That is why atoms are so unnecessarily big, and why metal and stone are so bulky.

So far, the only effects for which we've cited the role of the electron spin is the spin-orbit interaction in the hydrogen atom (in Section 7.2.3), and then (including the proton spin) the hyperfine or magnetic dipole–dipole interaction, where the particle's spin-1/2 nature determines their relevant magnetic moments. One reason for this lack of coverage of a critical part of quantum theory is that DA is blind to the crucial impact of the Pauli principle and the details of the (anti)symmetry properties of the electrons in an atomic wave function.[4] While we will return to a more detailed discussion of the role of electron (and neutron) spin in the context of astrophysics in Section 9.3, we will assume that stable atoms exist and will focus on what DA can tell us about the energy states of molecules, and then focus on condensed matter systems.

---

[3] The title of a review article by Lieb (1976), but see also the mathematical proofs by Dyson (1967) and Dyson and Lenard (1967, 1968).

[4] But, like all other aspects of science, it is subject to experimental tests. Ramberg and Snow (1990) have presented an experimental limit on a small violation of the Pauli principle of $\sim 2 \times 10^{-26}$, Bartalucci et al. (2006) find a bound of $\sim 5 \times 10^{-28}$, while Kekez et al. (1990) claim an upper limit of $\sim 7 \times 10^{-34}$.

We've argued that the length and energy scales appropriate for atomic physics are determined by the electron mass, m_e, via

$$a_0 \sim \frac{\hbar^2}{m_e K_e} \sim 0.5\,\text{Å} \qquad \text{and} \qquad E_H \sim \frac{K_e}{a_0} \sim \frac{K_e^2 m_e}{\hbar^2} \sim 27\,eV, \tag{8.1}$$

and we will continue to use those relations in scaling arguments. For back-of-the-envelope calculations, it's often appropriate to consider length and energy scales in the ranges (2–4) Å and (2–4) eV. For molecular motion, we need to include a second mass scale M to encode information on the presence of the atomic nucleus, to which end we can write (with a non-standard exponent written as $-\delta$ for future simplicity) for the energy scale

$$E \propto K_e^\alpha \hbar^\beta m_e^\gamma M^{-\delta}$$

$$\frac{\mathrm{ML}^2}{\mathrm{T}^2} = \left(\frac{\mathrm{ML}^3}{\mathrm{T}^2}\right)^\alpha \left(\frac{\mathrm{ML}^2}{\mathrm{T}}\right)^\beta \mathrm{M}^\gamma \mathrm{M}^{-\delta}, \tag{8.2}$$

which gives $\alpha = 2$ and $\beta = -2$ and $\gamma = 1 + \delta$, or

$$E \propto \frac{K_e^2}{\hbar^2} m_e \left(\frac{m_e}{M}\right)^\delta \quad \longrightarrow \quad E \propto \frac{K_e^2 m_e}{\hbar^2} F\left(\frac{m_e}{M}\right) = E_H F\left(\frac{m_e}{M}\right), \tag{8.3}$$

with a (hopefully unsurprising) dimensionless ratio of $\Pi = m_e/M$. This result should be obvious from the appearance of two mass scales, and we already know from Chapter 7 one analytic/exact example for any two-body system, namely that the introduction of the **reduced mass** $\mu = m_e M/(m_e + M) = m_e(1 + m_e/M)^{-1}$, which determines the form of $F(m_e/M)$ for that simple case.

To model the next level of molecular complexity, it's helpful to have an explicit form for the atom–atom interaction. A common approximation for the inter-atomic potential is the **Lennard-Jones** or **LJ** or **6-12** functional form given by

$$V_{LJ}(R) = \frac{\epsilon}{4}\left[\left(\frac{\sigma}{R}\right)^{12} - \left(\frac{\sigma}{R}\right)^6\right], \tag{8.4}$$

where R is the inter-atomic distance and ϵ, σ have dimensions of energy and length, respectively. The long-distance attractive term proportional to R^{-6} can be motivated physically (see **P8.3** for how), while the form of the short-distance repulsive term is designed to encode the energy cost dictated by the physics of the overlapping electron shells (and the exclusion principle) to the extent possible, while making "pencil-and-paper" calculations easier.

As with almost all potentials with a local minimum, the "best first guess" of the behavior near the bottom is a harmonic oscillator form, so we model the inter-atomic potential near equilibrium with a form

$$V(R) = \frac{1}{2} k_M (R - R_{min})^2 \qquad \text{where} \qquad k_M \sim \frac{\text{energy}}{\text{length}^2} \sim \frac{eV}{\text{Å}^2}. \tag{8.5}$$

We can certainly obtain a value for k_M in terms of ϵ, σ from expanding the LJ potential in Eqn. (8.4) about the minimum, but we are more interested in understanding the dimensional dependences and scaling behavior of such a model in terms of fundamental constants. Assuming that the *energy* ~ E_H and *length* ~ a_0 scales above arise from the details of the atomic physics, we can write the spring constant in terms of fundamental parameters as

$$k_M \sim \frac{E_H}{a_0^2} \sim \left(\frac{m_e K_e^2}{\hbar^2}\right)\left(\frac{\hbar^2}{m_e K_e}\right)^{-2} \sim \frac{m_e^3 K_e^4}{\hbar^6}. \tag{8.6}$$

Given this effective k_M value, we can analyze the energy spectra of molecular vibrational states where we assume a typical nuclear mass scale M as the relevant one (actually the reduced mass $\mu = M_1 M_2/(M_1 + M_2)$ of say a diatomic molecule would be even more appropriate) instead of the electron scale m_e. The vibrational energies from the quantum harmonic oscillator then scale as

$$E_{vib} \sim \hbar\sqrt{\frac{k_M}{M}} \sim \hbar\sqrt{\frac{m_e^3 K_e^4}{\hbar^6}\frac{1}{M}} \sim \hbar\left(\frac{m_e^2 K_e^4}{\hbar^6}\right)^{1/2}\sqrt{\frac{m_e}{M}} \sim \left(\frac{m_e K_e^2}{\hbar^2}\right)\sqrt{\frac{m_e}{M}} \sim E_H\left(\frac{m_e}{M}\right)^{1/2}, \tag{8.7}$$

which shows that the vibrational energy levels are roughly $\sqrt{m_e/M} \sim 10^{-2}$ smaller than the *eV* electron energies.

A similar scaling argument can be used for the rotational states of molecules, using the quantum version of a rigid rotor as a model, and estimating the moment of inertia of a diatomic molecule as $I \sim M a_0^2$. Knowing that angular momentum is quantized in units of $L \sim \hbar$, the rotational energies scale as

$$E_{rot} = \frac{1}{2}I\omega^2 = \frac{L^2}{2I} \sim \frac{\hbar^2}{M a_0^2} \sim \frac{\hbar^2}{M}\left(\frac{\hbar^2}{m_e K_e}\right)^{-2} \sim \left(\frac{m_e K_e^2}{\hbar^2}\right)\left(\frac{m_e}{M}\right) \sim E_H\left(\frac{m_e}{M}\right), \tag{8.8}$$

which is about 10^{-4} smaller than those on the *eV* scale. We thus have the (very approximate) scaling of atomic, vibrational, and rotational energy levels as

$$E_H : E_{vib} : E_{rot} \quad \longleftrightarrow \quad 1 : \sqrt{\frac{m_e}{M}} : \frac{m_e}{M} \quad \longleftrightarrow \quad 1 : 10^{-2} : 10^{-4}, \tag{8.9}$$

which can be a useful mnemonic if nothing else.

One might well ask why such DA arguments are needed, given that AMO (atomic, molecular, and optical) physics is such a mature and well-developed field, and the fundamental mass scales (m_e, M) and interactions are well known. One example of being able to scale well-known results to study new phenomena was provided by Ryan et al. (2022) who studied *molecular chemistry for dark matter* to constrain speculative models in astrophysics and cosmology. The authors considered **dark hydrogen**, with new versions of the electron ($\tilde{m}$), proton ($\tilde{M}$), photon ($\tilde{m}_\gamma$ assumed massless) and a new fine-structure constant $\tilde{\alpha}_{FS}$, and are able to scale many molecular bound state and scattering processes in terms of $\tilde{m}, \tilde{M}$, and $\tilde{\alpha}_{FS}$.

Moving on from these few-body systems, let us now extend our DA exploration of systems interacting via E&M interactions, but now in a condensed state, characterized by a number density n_e. We then have for any general quantum energy the constraints, and connections,

$$E_Q \propto n_e^\alpha \hbar^\beta m_e^\gamma e^\delta \epsilon_0^\sigma$$

$$\frac{\mathsf{ML}^2}{\mathsf{T}^2} = \left(\frac{1}{\mathsf{L}^3}\right)^\alpha \left(\frac{\mathsf{ML}^2}{\mathsf{T}}\right)^\beta \mathsf{M}^\gamma \mathsf{Q}^\delta \left(\frac{\mathsf{Q}^2\mathsf{T}^2}{\mathsf{ML}^3}\right)^\sigma$$

$$E_Q \propto n_e^\alpha \hbar^{-2+6\alpha} m_e^{1-3\alpha} e^{4-6\alpha} \epsilon_0^{-2+\alpha}$$

$$\propto n_e^\alpha \left(\frac{m_e}{\hbar^2}\right)^{1-3\alpha} \left(\frac{e^2}{4\pi\epsilon_0}\right)^{2-3\alpha}$$

$$\propto \left(\frac{m_e K_e^2}{\hbar^2}\right)\left(n_e\left(\frac{\hbar^2}{m_e K_e}\right)^3\right)^\alpha$$

$$E_Q \propto E_H \left(n_e a_0^3\right)^\alpha, \tag{8.10}$$

where E_H and a_0 are the Hartree energy and Bohr radius, with an obvious dimensionless combination appearing.

It's interesting to note that Eqn. (8.10) includes (or at least suggests) many special cases of quantum mechanical bound states that we've already encountered. For example, when $\alpha = 0$ (i.e., no dependence on n_e, so not really a multi-particle system at all in that case) we have $E_Q \propto K_e^2 m_e/\hbar^2 \propto E_H$, which gives the (single-particle) hydrogen atom (or Hartree) energy scale. If we associate the number density with a typical inter-particle distance d_e, via $d_e = (n_e)^{-1/3}$, then the case of $0 = \beta = \gamma = 1 - 3\alpha$ or $\alpha = 1/3$ gives $E_Q \propto K_e/d_e$ or the classical Coulomb energy of two particles separated by d_e, independent of quantum mechanics. The last obvious case is $0 = \gamma = \delta = 2 - 3\alpha$ giving $\alpha = 2/3$ and $E_Q \propto \hbar^2/(m_e d_e^2)$, which is reminiscent of the infinite square well of width d_e.

The first and third examples correspond to the extreme values of the power-law potentials $V^{(P)}(x) = V_0|x/a|^P$ considered in Section 6.3.3, where we found $E^{(P)} \propto \hbar^{(2P/(2+P)}$. That gives dependences on Planck's constant of $E_Q \propto \hbar^2$ for $P = \infty$ (the infinite well) and $E_Q \propto \hbar^{-2}$ for $P = -1$ (the quantum Coulomb problem), respectively and we want to explore an important intermediate case next.

Example 8.1 Plasmons or quantized plasma oscillations

That last observation begs the question of whether there is a physically relevant system in this family of quantized energies that is analogous to the quantum harmonic oscillator, that is, one with power-law exponent $P = 2$, giving $E_Q \propto \hbar^1$. That option corresponds to $1 = \beta = -2 + 6\alpha$ or $\alpha = 1/2$ giving

$$E_Q \propto n_e^{1/2} \hbar\, m_e^{-1/2} K_e^{1/2} = \hbar \sqrt{\frac{n_e K_e}{m_e}} \propto \hbar\omega_P \equiv E_P, \tag{8.11}$$

where ω_P is the plasma frequency introduced in Section 2.3, and to be discussed more extensively in Section 9.1. Such states were first discussed by Bohm and Pines (1951, 1952) who described them as the "... organized behavior produced by the interactions in an electron gas of high density ..." and called them **plasma oscillations**, and given their quantized nature they are also called **plasmons**. The existence of such states has been confirmed by the measurement of *collective energy losses in solids*[5] where

[5] The title of a review article by Pines (1956).

the electron energy loss spectrum exhibited peaks at quantized values given by $E_{loss} = \hbar\omega_P \sim 10-15\ eV$; see e.g. Powell and Swan (1959a, 1959b) for the experiments.

While quantized plasmon energies can presumably be found for any value of n_e, the experiments above were done using the free electrons in condensed matter systems, and one finds experimentally that $E_P \sim E_H$. In such systems we might expect n_e to scale with the typical atomic length scale, via $n_e \sim a_0^{-3} = (m_eK_e/\hbar^2)^3$. In that case we'd have

$$E_P = \hbar\sqrt{\frac{n_eK_e}{m_e}} \sim \hbar\left[\left(\frac{m_eK_e}{\hbar^2}\right)^3\frac{K_e}{m_e}\right]^{1/2} = \hbar\left(\frac{m_eK_e^2}{\hbar^3}\right) = \frac{m_eK_e^2}{\hbar^2} = E_H. \tag{8.12}$$

Despite arising from two very different physical phenomena (quantized collective oscillations of an electron gas versus the two-body quantum Coulomb problem), in this limit, where the atomic physics sets the scale for the number density, the two energies scale in exactly the same way.

We have focused so far on the nature of discrete energy states in AMO or condensed matter systems, but the world of solid state physics is rich with many other quantized phenomena that are amenable to the methods of DA.

8.2 Paths to quantization

We often first hear of quantization phenomena in physics starting with the story of blackbody radiation and the $E_\gamma = hf_\gamma = \hbar\omega_\gamma$ association made by Einstein, using Planck's constant very directly. We then see it applied in bound state problems where the de Broglie wavelength of the particle is invoked to describe its wave behavior and where we find that angular momenta (L_n) and then energies (E_n) are quantized. For example, if we write for a particle in a circular obit

$$\lambda_{dB} = \frac{h}{p} = \frac{2\pi\hbar}{p} \qquad \text{giving} \qquad n\left(\frac{2\pi\hbar}{p}\right) = n\lambda_{dB} = 2\pi R \qquad \text{we find} \qquad n\hbar = pR = L_n, \tag{8.13}$$

or for a particle-in-a-box of width d, we write

$$n\left(\frac{\lambda_{dB}}{2}\right) = d \qquad \text{giving} \qquad p_n = \frac{n\pi\hbar}{d} \qquad \text{requiring} \qquad E_n = \frac{p_n^2}{2m} = n^2\frac{\hbar^2\pi^2}{2md^2}. \tag{8.14}$$

The Wentzel–Kramers–Brillouin (WKB) approximation discussed in Section 6.3.2 applies such "fitting de Broglie waves" arguments to the Schrödinger equation as an approximate method of solution. And ultimately, while solutions of the SE are possible for any energy value, only those for which the wave function satisfies $\psi(x)\to 0$ as $x\to\pm\infty$ are acceptable, as those allow for the interpretation of $|\psi(x)|^2$ as a probability density, which generalizes the particle-in-a-box argument.

In some cases, the source of the quantization constraint is not as obvious as it is in some familiar bound state problems, and we next explore one such example which is important in condensed matter physics.

Example 8.2 Landau levels—Quantized motion in a magnetic field

Consider the problem of a particle of charge q and mass m moving under the influence of a constant magnetic field $\boldsymbol{B}_0$. The classical solution (discussed in Section 4.1.2) consists of circular orbits (in the plane perpendicular to $\boldsymbol{B}_0$) and free-particle motion in the direction along the field, which suggests that the behavior of the particle is "constrained", at least in the plane, but that's not exactly what we associate with quantum mechanical bound states, and certainly not quantization. The quantum version (so we'll include $\hbar$) of this problem does turn out to lead to quantized energies and we can explore how E depends on the four relevant quantities. We first write

$$E \propto \hbar^{\alpha} q^{\beta} B_0^{\gamma} m^{\delta}$$

$$\frac{\mathsf{ML}^2}{\mathsf{T}^2} = \left(\frac{\mathsf{ML}^2}{\mathsf{T}}\right)^{\alpha} \mathsf{Q}^{\beta} \left(\frac{\mathsf{M}}{\mathsf{QT}}\right)^{\gamma} \mathsf{M}^{\delta}, \tag{8.15}$$

which is easily solved to find $\alpha = \beta = \gamma = +1$ and $\delta = -1$ or

$$E \propto \hbar \left(\frac{qB_0}{m}\right) \equiv \hbar\omega_C, \tag{8.16}$$

where ω_C is the **cyclotron frequency** discussed in Eqn. (4.13), which determines the period of the particle moving in a classical circular orbit via $\omega_C = 2\pi/\tau_C$.

The dependence of E on **one** power of Planck's constant, and the appearance of an angular frequency alongside it, clearly suggests a relation with the quantum harmonic oscillator, where $E \sim \hbar\sqrt{k/m} = \hbar\omega$ (as we also found for plasmons), but it's perhaps not clear at the moment how such an association might arise. If this is the case, however, we'd also expect that $E_n \propto n$, at least for large quantum numbers, given the $n, \hbar$ scaling behaviors discussed in Eqn. (6.141).

We can first make contact with the classical circular motion by noting that if the magnetic field is in the $\hat{\boldsymbol{z}} = \hat{\boldsymbol{k}}$ direction, with the circular motion restricted to the $x - y$ plane, the planar trajectory is given in polar, and then Cartesian coordinates, by

$$r(t) = R \text{ and } \theta(t) = \omega_C t \quad \text{or} \quad x(t) = R\cos(\omega_C t) \text{ and } y(t) = R\sin(\omega_C t). \tag{8.17}$$

In this notation, there is indeed simple harmonic (sinusoidal) motion in both the $x(t)$ and $y(t)$ coordinates, with the same angular frequency ω_C, much like the harmonic oscillator, but that observation still doesn't yet make direct connection with a Schrödinger equation approach.

To do so, we explore the formalism of treating the effects of EM fields in quantum mechanics, not by directly including the $\boldsymbol{E}, \boldsymbol{B}$ fields, but rather their EM scalar and vector potentials, ϕ and $\boldsymbol{A}$, which from Eqn. (1.65) give

$$\boldsymbol{E}(\boldsymbol{r}, t) = -\boldsymbol{\nabla}\phi(\boldsymbol{r}, t) - \frac{\partial}{\partial t}\boldsymbol{A}(\boldsymbol{r}, t) \qquad \text{and} \qquad \boldsymbol{B}(\boldsymbol{r}, t) = \boldsymbol{\nabla} \times \boldsymbol{A}(\boldsymbol{r}, t). \tag{8.18}$$

The canonical procedure in quantum mechanics to generalize the free-particle Hamiltonian to include E&M fields is to write

$$\hat{H} = \frac{1}{2m}\hat{\boldsymbol{p}}^2 \qquad \longrightarrow \qquad \hat{H} = \frac{1}{2m}\left[\hat{\boldsymbol{p}} - q\boldsymbol{A}(\boldsymbol{r}, t)\right]^2 + q\phi(\boldsymbol{r}, t), \tag{8.19}$$

where we recall (see Eqn. (1.99)) that $q\boldsymbol{A}$ and $q\phi$ do have the dimensions of momentum and energy, respectively.

Because of the so-called **gauge freedom** of EM, there are any number of choices of the vector potential $\boldsymbol{A}$, which can reproduce a constant magnetic field in the $\hat{z}$ direction, one of which is

$$\boldsymbol{A}(\boldsymbol{r}) = \frac{B_0}{2}(-y, +x, 0) \qquad \text{since} \qquad \nabla \times \boldsymbol{A} = \begin{pmatrix} \hat{x} & \hat{y} & \hat{z} \\ \partial/\partial x & \partial/\partial y & \partial/\partial z \\ -B_0 y/2 & B_0 x/2 & 0 \end{pmatrix} = 2\left(\frac{B_0}{2}\right)\hat{z} = B_0\hat{z}. \quad (8.20)$$

This choice of vector potential is called the **symmetric gauge**, but others such as the **Landau gauge**, where $\boldsymbol{A} = (0, B_0 x, 0)$, give the same physical results.

Substituting Eqn. (8.20) into Eqn. (8.19) gives the Hamiltonian

$$\hat{H} = \frac{\hat{\boldsymbol{p}}^2}{2m} - \frac{q}{2m}(\hat{\boldsymbol{p}} \cdot \boldsymbol{A} + \boldsymbol{A} \cdot \hat{\boldsymbol{p}}) + \frac{1}{2m}\left(\frac{qB_0}{2}\right)^2 (x^2 + y^2), \quad (8.21)$$

where the kinetic energy term can be written in terms of $\hat{p}_x, \hat{p}_y, \hat{p}_z$ operator components as

$$\frac{1}{2m}\hat{\boldsymbol{p}}^2 = \frac{1}{2m}\left(\hat{p}_x^2 + \hat{p}_y^2\right) + \frac{1}{2m}\hat{p}_z^2. \quad (8.22)$$

The last kinetic energy term corresponds to free-particle motion in the $\hat{z}$ direction, leaving the other contributions to describe the x–y planar motion in quantum mechanical terms. The $\hat{p}_x^2 + \hat{p}_y^2$ and $x^2 + y^2$ terms clearly describe a two-dimensional harmonic oscillator problem, with a two-dimensional potential of the form

$$V(x, y) = \frac{m}{2}\left(\frac{qB_0}{2m}\right)^2 (x^2 + y^2) \equiv \frac{m}{2}\omega_L^2(x^2 + y^2) = \frac{1}{2}\left(\frac{q^2 B_0^2}{4m}\right)(x^2 + y^2) \equiv \frac{1}{2}k(x^2 + y^2). \quad (8.23)$$

We can rightly describe this as a two-dimensional **isotropic** oscillator, since it has the same spring constant in both the x and y directions. We have also written this using the **Larmor frequency** $\omega_L = (qB_0)/2m = \omega_C/2$, which is half the cyclotron frequency, but still proportional to it, dimensionally. From our knowledge of the bound state solutions of the harmonic oscillator, this approach now makes the quantized $E_n \propto \hbar\omega_L \propto \hbar\omega_C$ relation more obvious in a wave-mechanics language, and in the context of a familiar confining potential.

The middle term in Eqn. (8.21) looks formally rather different, but can be shown to reduce to

$$-\frac{q}{2m}(\hat{\boldsymbol{p}} \cdot \boldsymbol{A} + \boldsymbol{A} \cdot \hat{\boldsymbol{p}}) = -\frac{qB_0}{2m}\hat{L}_z \quad (8.24)$$

in terms of the angular momentum operator for the relevant $\hat{z}$ direction. Given that the eigenvalues of $\hat{L}_z$ are quantized in units of $L_z \sim \hbar$, this term also produces contributions to the quantized energies of the form $E \propto (qB_0/m)\hbar \propto \hbar\omega_{L,C}$ as well, so everything is indeed dimensionally consistent.

In the language of two-dimensional quantum mechanics, where there are two quantum numbers, n_r being the one that counts nodes in the radial wave function, and m the angular momentum eigenvalue of $\hat{L}_z$ (not to be confused with the mass), the actual quantized energies[6] in this gauge are given by

$$E(n_r, m) = \hbar\omega_L\,(2n_r + |m| - m + 1) = \hbar\omega_C(n_L + 1/2) = E(n_L). \tag{8.25}$$

We have associated $\omega_L = \omega_C/2$ and $n_L = n_r + (|m| - m)/2$ is an integer, since n_r, m both are, and we do indeed recover the energy spectrum for the one-dimensional harmonic oscillator.

While the DA in Eqns. (8.15)–(8.16) strongly suggested the correct final form (including the $(n + 1/2)$ quantum number dependence), a complete analysis is always reassuring. Many treatments of this problem do use the Landau gauge and we encourage readers to explore other presentations they'll find in the literature, to see how they differ in mathematical detail "along the way," but ultimately give the same physical result, as they must.

Given the analogies with the quantum oscillator, we can also identify the natural **magnetic length scale**, l_B, associated with this problem in two ways, namely (i) using the spring constant k in Eqn. (8.23) and the relation in Eqn. (6.114),

$$l_B \sim \left(\frac{\hbar^2}{mk}\right)^{1/4} = \left(\frac{\hbar^2}{m}\frac{4m}{q^2B_0^2}\right)^{1/4} \sim \sqrt{\frac{\hbar}{qB_0}} \tag{8.26}$$

or (ii) "from scratch" by using DA to equate

$$\begin{aligned} l_B &\propto \hbar^\alpha q^\beta B_0^\gamma \\ \mathsf{L} &= \left(\frac{\mathsf{ML}^2}{\mathsf{T}}\right)^\alpha \mathsf{Q}^\beta \left(\frac{\mathsf{M}}{\mathsf{QT}}\right)^\gamma \end{aligned} \tag{8.27}$$

giving $\alpha = 1/2$ and $\beta = \gamma = -1/2$, or $l_B \sim \sqrt{\hbar/qB_0}$.

These quantized states are especially relevant for two-dimensional systems with applied fields, such as two-dimensional electron gas (2DEG) materials, and are called **Landau levels** (Landau 1930) as he is credited with their introduction, though he did use the symmetric gauge we've cited here instead of the choice that carries his name. We will see such ideas used again in the study of the quantum Hall effect (QHE).

Two of the most dramatic emergent behaviors of condensed matter systems at low temperature are **superfluidity** and **superconductivity** which are characterized by macroscopic systems with vanishing viscosity (no resistance to fluid/mass flow) and zero electrical conductivity (no resistance to charge flow), respectively. For each system there is a measurable quantity found to be quantized (and predicted in advance to be so), namely the **circulation** in a superfluid and the **magnetic flux** in a superconductor. We show in Figs. 8.1 and 8.2 examples of experimental data illustrating the discrete nature of these two variables. We can initially approach the dimensionality of each variable, and then explore their "path" to quantization a bit more formally to confirm.

We begin with the circulation, C, a classical hydrodynamical quantity defined via a line-integral of the velocity field, $\boldsymbol{v}(\boldsymbol{r})$, over any closed path. It is widely used in atmospheric physics and

[6] Not counting the free-particle motion in the $\hat{\boldsymbol{z}}$ direction, corresponding to uniform translation along that axis, where $E_z = \hbar^2k_z^2/2m$.

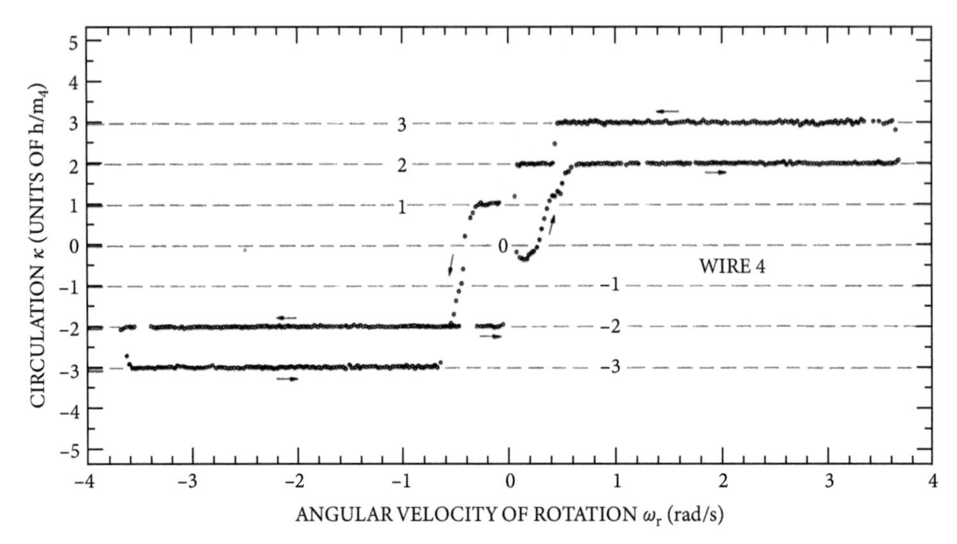

Fig. 8.1 Plot of the circulation (in units of h/m_4) as a function of angular velocity (rad/s) in a sample of superfluid ^{4}He. Reprinted with permission from Karn, Starks, and Zimmerman (1980). Copyright (1980) by the American Physical Society.

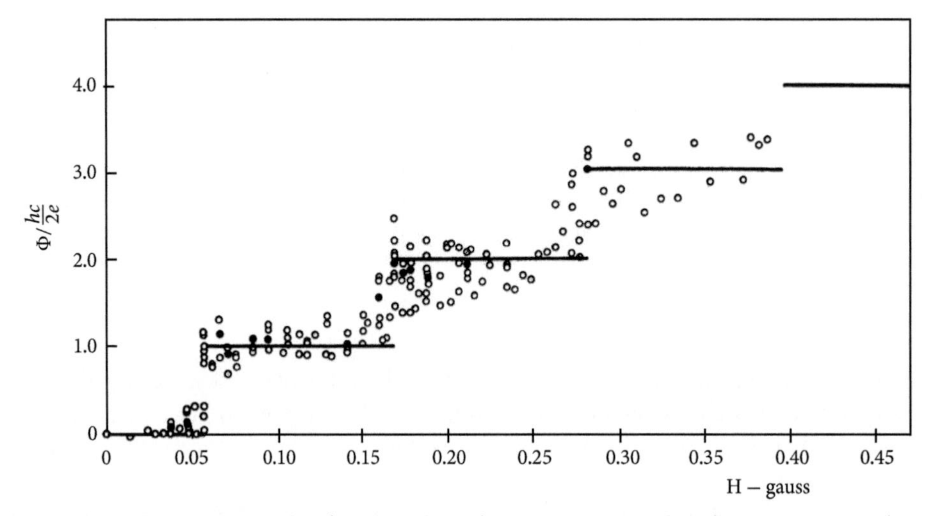

Fig. 8.2 Plot of the magnetic flux (in units of $hc/2e$) versus magnetic field (labeled H in $Gauss$) in a superconductor. Reprinted with permission from Deaver and Fairbank (1961). Copyright (1961) by the American Physical Society. The value for magnetic flux quoted here is in CGS-Gaussian units and corresponds in our SI notation to $\Phi_0 = h/2e$. See Section 10.6 for a discussion of how to convert from one convention to another.

meteorology and, under some assumptions, is a conserved quantity in fluid dynamics, giving the so-called **Kelvin's theorem.** To constrain C in terms of the fundamental quantities relevant to condensed matter systems, we assume dependences on $\hbar$, m, and the number density n, and write

$$C \equiv \oint \mathbf{v} \cdot d\mathbf{l} = C_C \hbar^\alpha m^\beta n^\gamma$$

$$\left(\frac{\mathsf{L}}{\mathsf{T}}\right)\mathsf{L} = \frac{\mathsf{L}^2}{\mathsf{T}} = \left(\frac{\mathsf{ML}^2}{\mathsf{T}}\right)^\alpha \mathsf{M}^\beta \left(\frac{1}{\mathsf{L}^3}\right)^\gamma, \tag{8.28}$$

which is easily solved for $\alpha = +1$, $\beta = -1$, and $\gamma = 0$, or $C \propto \hbar/m$. We note that Fig. 8.1 plots C in terms of h/m_4, with a mass appropriate for superfluid ^{4}He, and h instead of $\hbar$, and we clearly see the quantized values in those units.

For the magnetic flux, we include appropriate E&M quantities and similarly write

$$\int \mathbf{B} \cdot d\mathbf{A} = \Phi_0 = C_\Phi \hbar^\alpha q^\beta \epsilon_0^\gamma n^\delta$$

$$\left(\frac{\mathsf{M}}{\mathsf{QT}}\right)\mathsf{L}^2 = \left(\frac{\mathsf{ML}^2}{\mathsf{T}}\right)^\alpha \mathsf{Q}^\beta \left(\frac{\mathsf{Q}^2\mathsf{T}^2}{\mathsf{ML}^3}\right)^\gamma \left(\frac{1}{\mathsf{L}^3}\right)^\delta \tag{8.29}$$

with $\alpha = +1$, $\beta = -1$, and $\gamma = \delta = 0$, giving $\Phi_0 \propto \hbar/q$. Results from experiments showing the quantized nature of Φ_0 are shown in Fig. 8.2. The basic flux quantum is cited there as $hc/2e$ because for the (electron–electron) Cooper pairs leading to superconductivity, one has $q = -2e$, and the "extra" factor of c is because the results are cited in CGS-Gaussian units! It turns out that the dimensionless constants in both cases are the same, namely $C_c = C_\Phi = 2\pi i$, where $i = 0, 1, 2, \ldots$ giving

$$C^{(i)} = (2\pi i)\frac{\hbar}{m} = i\frac{h}{m} = iC_0 \qquad \text{and} \qquad \Phi^{(i)} = (2\pi i)\frac{\hbar}{q} = i\frac{h}{q} = i\Phi_0, \tag{8.30}$$

which might suggest a common origin of the quantization condition.

Instead of describing either phenomenon by a many-particle quantum wave function $\Psi(\mathbf{r}_1, \ldots . \mathbf{r}_N)$, the phenomenological **Ginzburg–Landau model** of the quantum state of the system makes use of a complex order-parameter field involving the number density $(n(\mathbf{r}))$, given by

$$\psi = |\psi| e^{i\theta} \qquad \text{and} \qquad \psi^* = |\psi| e^{-i\theta} \qquad \text{where} \qquad |\psi| = \sqrt{n}. \tag{8.31}$$

Since the magnitude of n is approximately constant, the most important part of ψ is actually its phase, θ, and that dependence is the origin of both quantum constraints.

Using this type of quantum vocabulary, we first make contact with the circulation by associating

$$\mathbf{v} \sim \frac{1}{m}\hat{\mathbf{p}} \sim \frac{1}{m}\frac{\hbar}{i}\nabla \qquad \longrightarrow \qquad \mathbf{v} = \frac{1}{m}\left(\frac{\hbar}{i}\right) i\nabla\theta = \frac{\hbar}{m}\nabla\theta, \tag{8.32}$$

where here $i = \sqrt{-1}$, and not the quantum number. The line-integral of $\mathbf{v}$ is then

$$C_n = \oint \mathbf{v} \cdot d\mathbf{l} = \frac{\hbar}{m}\oint \nabla\theta \cdot d\mathbf{l} = \frac{\hbar}{m}\Delta\theta = \frac{\hbar}{m}(2\pi i) = i\frac{h}{m} \qquad \text{where } i = 1, 2, \ldots \tag{8.33}$$

since the $\Delta\theta$ must be an integral number of factors of 2π in order for the ψ field to be well defined.

To discuss the quantization of the magnetic flux, we need to add information on the E&M fields, again via the extension $\hat{\boldsymbol{p}} \to \hat{\boldsymbol{p}} - q\boldsymbol{A}$ where $\boldsymbol{A}$ is the vector potential which, in turn, gives the magnetic field via $\nabla \times \boldsymbol{A} = \boldsymbol{B}$. Using this identification, the current density $\boldsymbol{J}_e$ is associated with

$$\boldsymbol{J}_e \sim nq\boldsymbol{v} \sim q\psi^* \boldsymbol{v}\psi = \frac{nq}{m} \underbrace{(\hbar\nabla\theta - q\boldsymbol{A})}, \tag{8.34}$$

which connects the field phase factor to the vector potential. Since the current ($\boldsymbol{J}_e$) inside a superconductor vanishes, we can relate the $\nabla\theta$ and $\boldsymbol{A}$ terms via

$$\hbar(2\pi i) = \underbrace{\hbar\oint \nabla\theta \cdot d\boldsymbol{l} = q\oint \boldsymbol{A}\cdot d\boldsymbol{l}}_{\text{since } \boldsymbol{J}_e=0} = q\int \nabla\times\boldsymbol{A}\cdot d\boldsymbol{S} = q\int \boldsymbol{B}\cdot d\boldsymbol{S} = q\Phi_B \quad \text{or} \quad \Phi_B = i\frac{h}{q}, \tag{8.35}$$

where we use Stokes' theorem to rewrite a line integral in terms of a surface integral. We see that flux quantization also depends on the same type of constraint on a (macroscopic) quantum wave function being well-defined, via the phase θ.

Zieve (2023)[7] has rightly described the initial direct observation of quantized circulation by Vinen (1961) as being

> ... the first demonstration of quantization on a macroscopic scale, followed soon after by observations of quantized flux in superconductors.

Other important advances in the experimental observation of quantized circulation were made by Rayfield and Reif (1964), Whitmore and Zimmerman (1968), Karn et al. (1980), and Davis et al. (1991), and research on quantum vortices continues to be an active research area in Bose–Einstein condensates, such as in Matthews et al. (1999). The first detection of magnetic flux quantization is attributed to Doll and Näbauer (1961) and Deaver and Fairbank (1961).

Example 8.3 The quantum Hall effect (QHE)

The classical Hall effect (Hall 1879) was one of the first "condensed matter" experiments to probe the nature of the fundamental charge carriers, their sign and three-dimensional number density (n_e) in matter, long before the actual discovery of the electron (by Thomson in 1897.) Later generations of such experiments, under more extreme laboratory conditions, namely much lower temperatures, and at much higher magnetic fields, found dramatic evidence of **quantized resistance**, giving the **QHE**. We show an example of such data in Fig. 8.3, where the single most striking features are the flat plateaus, characterized by discrete values of resistance.

Measurements on devices consisting of 2DEG geometries (with the charge carrying layer having thicknesses as small as $t \sim 10\,nm$) do show the expected linear relation between the Hall resistance and the applied field (at least for low B field values), namely the classical Hall effect,

$$R_H = \frac{B}{etn_e} = \frac{B}{en_{2D}}, \tag{8.36}$$

where the effective two-dimensional carrier density is related to the three-dimensional n_e by $n_{2D} = tn_e$. You're asked in **P8.14** to use the data in Fig. 8.3 to confirm this for $B \lesssim 20\,kG$.

[7] Zieve (2023) nicely reviews "Sixty Years of Quantized Circulation".

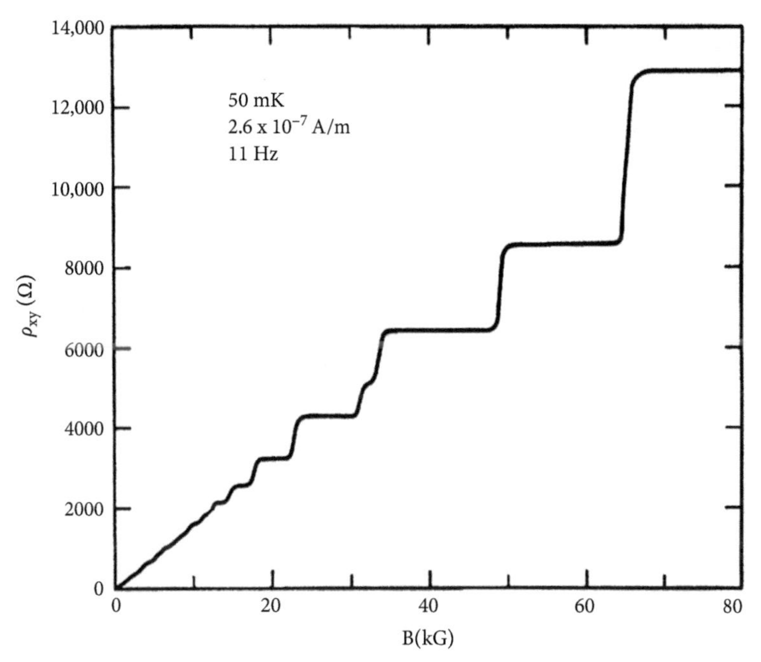

Fig. 8.3 Plot of the Hall resistance (in Ω), labeled here as ρ_{xy}, versus applied magnetic field (in *kG*) illustrating the QHE. Reprinted with permission from Paalanen, Tsui, and Gossard (1982). Copyright (1982) by the American Physical Society.

On the other hand, the quantized values of resistance on the "plateaus" do **not** seem to depend experimentally on the detailed nature of the device, or the presence of impurities, and are remarkably uniform across the center of each step, being "flat" to something like one part in 10^{10}, and reproducible from device to device, and lab to lab; see Delahaye et al. (2000) for examples of this uniformity.

We want to explore to what extent we can elucidate the dependence of these quantized resistance values on fundamental constants, and start by writing (using h instead of $\hbar$, motivated by earlier results)

$$R_H = C_R e^{\alpha} h^{\beta} m_e^{\gamma} (n_{2D})^{\delta}$$

$$\frac{\mathsf{ML}^2}{\mathsf{Q}^2\mathsf{T}} = \mathsf{Q}^{\alpha}\left(\frac{\mathsf{ML}^2}{\mathsf{T}}\right)^{\beta}\mathsf{M}^{\gamma}\left(\frac{1}{\mathsf{L}^2}\right)^{\delta}, \tag{8.37}$$

which has solutions $\alpha = -2$, $\beta = 1$, and $\gamma = \delta = 0$ or $R_H \propto h/e^2$, and experiment and theory both find $C_R = 1$. This gives the **quantum of resistance**, and the associated plateau values seen by experiment, as

$$R_H = \frac{h}{e^2} \equiv R_K \approx 25.8\,k\Omega \qquad \text{and} \qquad R_H^{(i)} = \frac{R_K}{i}, \tag{8.38}$$

where $i = 1, 2, 3...$ and the notation R_K for the **von Klitzing constant** is in honor of von Klitzing et al. (1980), the discoverer of the effect.

The incredible uniformity of the quantized R_K values has led to its use in metrology, with the latest (2019) revisions of the SI systems of units making use of this effect in a fundamental way, which we discuss (very briefly) in Section 10.10. We note that von Klitzing (2017, 2019) himself has provided two accessible "public science" reviews of the impact of the QHE on metrology, while his own Nobel prize lecture on the physics background and experimental realization of the *Quantized Hall Effect* (1986) is very readable.

A "derivation" of the physics leading to the quantization of R_K is not much beyond the level we've described for the quanta of circulation or magnetic flux. It relies on both the formalism of Landau levels, with the addition of an applied electric field to Eqn. (8.21) (via a term of the form $V(y) = e\phi = eE_0 y$), the quantization of magnetic flux, and gauge invariance. Interested readers can read the original argument by Laughlin (1981) (which is in CGS-Gaussian units though) or consult any of the excellent reviews of many aspects of the QHE(s)[8] such as Yoshioka (2002) or Ezawa (2013).

Instead of pursuing such an argument in more depth, we focus instead on another (perhaps less well-known) **quantum of transport**, this one related to thermal conduction. If the quantization suggested by Eqn. (8.38) seems different than that of $C^{(i)}$ or $\Phi^{(i)}$ in Eqn. (8.30), with the quantum number appearing in the denominator, a trivial rewriting of Ohm's law from $V = IR \rightarrow I = G\Delta V$ where $G \equiv 1/R$ gives the **quantum of electrical conductance**, written as $G_H^{(i)} = i(e^2/h)$.

That form suggests again (we've noted this connection before) the similarity in form between the equations for electrical/charge and thermal transport. In this sense, we can write

$$\frac{dq}{dt} = I = \sigma A \frac{dV}{dx} \quad \text{and} \quad \frac{dQ}{dt} = \kappa A \frac{dT}{dx} \quad \text{(conductivities, } \sigma, \kappa) \tag{8.39}$$

$$\downarrow \qquad\qquad \downarrow$$

$$\frac{dq}{dt} = I = G_H \Delta V \quad \text{and} \quad \frac{dQ}{dt} = K_H \Delta T \quad (\textit{conductances}, \mathrm{G_H}, \mathrm{K_H}) \tag{8.40}$$

in terms of the charge and thermal **conductances**, and we note that $[K_H] = \mathsf{ML^2/T^3\Theta}$. We might then suspect a quantized value of K_H, akin to that of $G_H^{(i)}$, and to explore its dimensions we write

$$K_H \propto k_B^\alpha T^\beta \hbar^\gamma m_e^\delta$$

$$\frac{\mathsf{ML^2}}{\mathsf{T^3\Theta}} = \left(\frac{\mathsf{ML^2}}{\mathsf{T^2\Theta}}\right)^\alpha \Theta^\beta \left(\frac{\mathsf{ML^2}}{\mathsf{T}}\right)^\gamma \mathsf{M}^\delta, \tag{8.41}$$

giving $\alpha = 2$, $\beta = 1$, $\gamma = -1$, and $\delta = 0$. The final result, including the dimensionless constant (and the quantization, with $i = 1, 2, 3, \ldots$) is

$$K_H^{(i)} = i\left(\frac{\pi}{6}\right)\frac{k_B^2 T}{\hbar} = i\left(\frac{\pi^2}{3}\right)\frac{k_B^2 T}{h}, \quad \text{which gives the ratio} \quad \frac{K_H^{(i)}}{G_H^{(i)} T} = \frac{\pi^3}{3}\left(\frac{k_B}{e}\right)^2, \tag{8.42}$$

and the final equation is identical to the Wiedemann–Franz law for the ratio of thermal to electrical conductivities discussed in Section 2.4.

The value of the **quantum of thermal conductance** $K_H^{(i)}$ in Eqn. (8.42) was predicted by Rego and Kirczenow (1998)[9] and experiments probing this have been reported by Schwab et al. (2000), Meschke

[8] Effects plural, since there are **many** realizations of related phenomena, such as the fractional and anomalous QHE.

[9] Pendry (1983) obtained a similar bound on K_H by studying the *quantum limits to the flow of information and entropy* even earlier.

et al. (2006), Jezouin et al. (2013), Banerjee et al. (2017), and Banerjee et al. (2018) in systems involving a number of different mechanisms of thermal transport (via phonons, photons, electrons, and even "anyons", particles with fractional particle statistics) showing the wide applicability of this prediction for both bosons and fermions. One analysis rightly notes:

> While dimensional analysis would lead us to expect the same factor $k_B^2 T/\hbar$ to occur independently of the statistics, there is no a priori reason to expect the same numerical factor $\pi/6$ as well . . . which have qualitatively different forms for particles obeying different statistics. This remarkable property is unique to the thermal conductance: all other single-channel transport coefficients depend on the particle statistics. Blencowe and Vitelli (2000)

One final quantum effect with both technological and metrological implications is the tunneling of Cooper (electron–electron or e^-e^-) pairs through an insulating barrier (junction) from one superconductor to another, one of several possible new effects in superconducting tunneling predicted by Josephson (1962, 1965). Two important realizations of such tunneling are:

- **DC Josephson effect:** The application of a DC (constant) voltage V across the junction gives rise to current oscillations of the form $I(t) = I_0 \cos(\omega_J t)$. The first observation of this effect was by Anderson and Rowell (1963).
- **Inverse Josephson effect:** The application of microwave photons of frequency ω_J can give rise to quantized voltages V. For the original measurements of this effect, see Langenberg et al. (1966).

If we apply only DA constraints on ω_J and V, using e and h (since we've see many examples where h is the relevant quantum constant), and either m_e or n_e we write

$$\begin{aligned} \omega_J &= C_J V^\alpha e^\beta h^\gamma m_e^\delta \\ \frac{1}{\mathsf{T}} &= \left(\frac{\mathsf{ML}^2}{\mathsf{T}^2\mathsf{Q}}\right)^\alpha \mathsf{Q}^\beta \left(\frac{\mathsf{ML}^2}{\mathsf{T}}\right)^\gamma \mathsf{M}^\delta, \end{aligned} \tag{8.43}$$

which give $\alpha = \beta = +1$, $\gamma = -1$, and $\delta = 0$ (with the same results if we instead use n_e). With the correct dimensional constant of $C_J = 2$ (from electron Cooper pairs with $q = -2e$), this gives the relation

$$\omega_J = \frac{2eV}{h} = \left(\frac{2e}{h}\right) V = K_J V = \frac{V}{\Phi_0}, \tag{8.44}$$

where $K_J \equiv 2e/h = 1/\Phi_0$ is the **Josephson constant**, which is just the inverse of the **quantum of magnetic flux**. Because of their relation to Φ_0, Josephson junction devices can be used to measure very small magnetic fields; look up SQUID, or <u>S</u>uperconducting <u>Qu</u>antum <u>I</u>nformation <u>D</u>evice for details.

The fact that frequencies can be measured very accurately (simply integrating a signal over many cycles can produce a large statistical data set) led to one of the early precision measurements of e/h, long before the quantized Hall effect using the first relation. For a review of the measurement of $2e/h$ *using the ac Josephson effect, . . .* see Taylor, Parker, and Langenberg (1969) and many references therein.

Such devices can also be used in the second mode as **superconducting qubits** for quantum computers. One large-scale IBM quantum computer uses such elements that are described by saying[10]

> By firing microwave photons at these qubits, we can control their behavior and get them to hold, change, and read out individual units of quantum information.

You should be able to estimate the frequencies ($f = \omega_J/2\pi$) that are needed to obtain voltages in the 10–20 μV range seen in experiments (Langenberg et al. 1966) and confirm that they are in the (broad) microwave range of 0.3–300 GHz.

In just two decades, the use of Josephson junctions in quantum computing has gone from proposals for quantum-state engineering with Josephson-junction devices (Makhlin, Schön, and Shnirman 2001) to recent experimental demonstrations of > 100 qubit devices (Kim et al. 2023), with discussions of the need for methods for the efficient and cost-effective fabrication (Osman et al. 2021) of the individual qubit elements. For an honest appraisal of the status of, and prospects for, quantum computing in the NISQ era and beyond, see Preskill (2018).

We have focused so far on systems of particles where quantum mechanics plays the predominant role, and where the explicit dependence on $\hbar$ can be identified in every result. But there are any number of condensed matter systems that can be treated essentially classically, once the constituents and their interactions are known, namely the array of materials collectively known as soft matter.

8.3 **Soft matter physics**

We have already considered some topics related to solids, gases, liquids, granular materials, and even plasmas (with even more on that in Chapter 9). Many of the classic texts (e.g., Peierls 1955, Kittell 1971, or Ashcroft and Mermin 1976), which deal with the topics covered so far in this chapter include the phrase **solid state physics** in their titles. More modern treatments (e.g., Cohen and Louie 2016, and Girvin and Yang 2019) of the subject advertise a somewhat expanded coverage by instead citing **condensed matter physics**,[11] perhaps partly to acknowledge such topics as superfluidity, but very often still covering a very similar set of canonical topics.

The study of matter in its condensed state, however, includes many (MANY!) other areas and types of materials. As an example, one of the newer research journals in the *Physical Review* "family" of publications, namely *Physical Review E*, cites its topical areas as

> ... covering statistical, nonlinear, biological, and soft matter physics ...

(emphasis ours), and among other subjects mentions (i) colloids, complex fluids, and active matter, (ii) polymers, (iii) liquid crystals, (iv) films and interfaces, and (v) fluid dynamics.

Familiar examples of materials that might be considered as soft matter are often cited in the context of human usage, including many food products (gelatins, mayonnaise, whipped cream), cosmetics[12] (toothpaste, shampoo, soaps), many artificial substances with industrial applications

[10] Taken from https://www.ibm.com/topics/quantum-computing. Accessed September 5, 2023.

[11] A rubric often attributed to P. W. Anderson.

[12] A former undergraduate advisee of mine who got his PhD studying "... light-based microscopy of colloidal and lyotropic liquid crystal systems ..." now works for Johnson & Johnson.

(pigmented ink, polymers, plastic 3D-printer material) and, of course, biomaterials, including the skin we live in. In fact, one textbook author (Jones 2002) reminds us that

> We are ourselves **soft machines,**[13] in William Burroughs' apt phrase, and the material we are made of is soft matter.

The term soft matter was made popular by Pierre-Gilles de Gennes as the title of his Nobel acceptance lecture (de Gennes 1992), though he himself gives credit (de Gennes 2005) for the name (*matière molle* in French) to a colleague, M. Veyssie, evidently initially intended as an inside joke. While there is no single all encompassing definition of soft matter, de Gennes (2005) described it as follows:

> ... soft matter appeared as a significant concept: comprising all physicochemical systems which have *large response functions.*

Given the wide range of fields covered under the umbrella of soft matter physics, textbooks and monographs on the subject tend to cover a much more diverse set of topics than do standard books on condensed matter. Examples include Daoud and Van Damme (1999), Jones (2002), Kleman and Lavrentovich (2003), Poon and Andelman (2006), Doi (2013), and Terentjev and Weitz (2015), along with review articles by Frenkel (2002) and Nagel (2017), as well the more public science level presentation by McLeish (2020).

Most treatments of the subject do agree that soft matter is characterized by weaker bond strengths (energies) and longer typical distance and time scales than "hard matter," and several of these properties contribute to their "squishiness" in ways that can be seen by scaling arguments. For example, many sources on soft matter physics cite binding energies for Coulomb (atomic, hard, C) versus van der Waals (molecular, soft, vdW) systems as being of order

$$E_C \sim 250\, kJ/mole \sim 4 \cdot 10^{-19}\, J \sim 2.5\, eV \tag{8.45}$$

$$E_{vdW} \sim 1\, kJ/mole \sim 2 \quad \text{and} \quad 10^{-21}\, J \sim 10^{-2}\, eV, \tag{8.46}$$

respectively, which we see are consistent with our dimensional scaling arguments in Eqn. (8.9). These can be compared to typical thermal energies at room temperature (T_r), namely

$$E_{th} = k_B T_r \sim 2.4\, kJ/mole \sim 4 \times 10^{-21}\, J \sim \frac{1}{40}\, eV \sim 2.5 \times 10^{-2}\, eV, \tag{8.47}$$

so that thermal fluctuations can easily play an important role in the structure of such systems. It's perhaps not surprising that materials which have evolved, or are synthesized/produced, to be used on an everyday basis in our homes, and on/in our own bodies, would exhibit such thermal "matching." This also suggests that modest mechanical stresses can also exhibit large strains, hence the "soft" or "... large response functions ..." descriptions.

To explore that connection, we can use atomic dimensions to estimate the various elastic moduli of a "hard" material, given that the units of $E \sim Y \sim B \sim G$ are all pressure, or energy per

---

[13] The title of the 1961 novel by Burroughs.

unit volume[14] and are all roughly the same in magnitude. We'd then expect that say the shear stress (measuring deformation related to slippage or indentation) of a normal solid to be roughly given by

$$G \sim \frac{(2-4)\,eV}{(2-4\,\text{Å})^3} \sim (0.5-0.03)\frac{(2\times 10^{-19}\,J)}{(10^{-10}\,m)^3} \sim (1-0.06)\times 10^{11}\,\frac{N}{m^2} \sim (100-6)\,GPa, \qquad (8.48)$$

where GPa is *giga-Pascal.* This is in the right range, given that diamond and tin have values of 530 and 16 GPa, respectively.

In contrast, a material such as gelatin would have $E_{vdW} \sim 10^{-2}\,eV$ and length scales more like 100 Å, giving values of $E \sim 10^3 N/m^2 \sim 1\,kPa$. Given that this particular class of materials is of use in the medical physics field (and not just for dessert), it's not surprising that measurements of their values are important, and they can made using a variety of methods, including mechanical indentation. The resulting values can also be compared to the speed of shear waves (recall Eqn. (3.60)) and values in the $\mathcal{O}(1\text{–}10)\,kPa$ range are indeed found, with sound speeds *in situ* corresponding to $v_s = \sqrt{G/\rho_m} \sim \mathcal{O}\left(\sqrt{(10^3-10^4)\,Pa/(10^3 kg/m^3)}\right) \sim \mathcal{O}(1\text{–}3\,m/s)$; see, for example Amador et al. (2011). We note that the range of densities of typical solids span roughly only one order of magnitude, $\rho_m \sim 1-10\,gr/cm^3$, while the density of, say aerogels[15] (a synthetic solid that is roughly 90–99% air) can be $1-10\,gr/m^3$, or about 10^{-6} times smaller.

Given the diversity of possible topics related to soft matter, we can focus on only a few examples (and problems), all related to the fact that Brownian motion (random thermal agitation of particles) is important for many such systems, including connections to the mathematical problem of random walks and especially diffusion processes. We have already cited many examples of the classical diffusion equation, or related diffusion-like processes, including viscous penetration depth (**P3.31**), the Ekman spiral (**P3.32**), E&M skin depth (**P4.36**), diffusion of magnetic monopoles (**P4.46**), the one-dimensional/three-dimensional heat and diffusion equations (Eqns. (5.18) and (5.24) and **P5.5**), and thermal skin depth (**P5.10**). Including several of the most well-known effects of similar physics here then extends the discussion of that concept across the "landscape" even further.

We also feel that such connections are very much in the spirit of Feynman et al. (1964) (Section 12-1) who said:

> Finally, there is a most remarkable coincidence: *The equations for many different physical situations have exactly the same appearance.* Of course, the symbols may be different—one letter is substituted for another—but the mathematical form of the equations is the same. This means that having studied one subject, we immediately have a great deal of direct and precise knowledge about the solutions of the equations of another.

Example 8.4 Gravitational density gradient

One of the many demonstrations of the molecular nature of matter is provided by the exponential distribution of number density of dispersed particles in a column of fluid under the influence of gravity, as predicted by kinetic theory. The density of particles suspended in a vertical column of liquid is known to be described by $n(z) = n_0 e^{-z/z_0}$ (see **P8.18** for more details of how this arises) and we assume

---

[14] This approach is suggested by Frenkel (2002) as a way of obtaining scaling estimates of shear moduli and to explain "Why Colloidal Materials are Soft."

[15] The IUPAC definition is "Gel comprised of a microporous solid in which the dispersed phase is a gas."

that z_0 depends on the thermal environment and we will henceforth always use the combination $k_B T$ paired together as the relevant variable. Including a dependence on the mass m of the particles and the (perhaps "effective," as we'll discuss) acceleration of gravity g, we then write

$$z_0 \sim (k_B T)^\alpha m^\beta g^\gamma$$
$$\mathsf{L} = \left(\frac{\mathsf{ML}^2}{\mathsf{T}^2}\right)^\alpha \mathsf{M}^\beta \left(\frac{\mathsf{L}}{\mathsf{T}^2}\right)^\gamma, \tag{8.49}$$

giving $\alpha = +1$ and $\beta = \gamma = -1$, or $z_0 \sim k_B T/mg$. The resulting dimensionless combination of $z/z_0 = z/(k_B T/(mg)) = mgz/k_B T$ is, of course, consistent with many other $E/k_B T$ ratios encountered already.

As a reality check, we apply this result to the molecules in the atmosphere, which has a roughly 80/20 mix of nitrogen (N_2) and oxygen (O_2), with atomic masses of 28/32, respectively. Assuming then that $m \sim 30m_p$, we find the rough estimate (arbitrarily using room temperature for $k_B T$)

$$z_0 \approx \frac{4 \times 10^{-21}\,J}{(30 \times 1.7 \cdot 10^{-27}\,kg)(10\,m/s^2)} \sim 8 \times 10^3\,m \approx 8\,km. \tag{8.50}$$

The troposphere, where most of the earth's weather occurs, and where about 75% of the atmosphere resides, is illustrated in the infographic in Fig 2.9, and is cited there as being about 12 km deep, so this is a reasonable first approximation.

To see how this density gradient distance scales with particle mass, we note that in terrestrial lab conditions, the masses of polymer molecules can easily reach 300,000 amu, which would imply that $z_0 \sim 1\,m$. The mass of the largest stable synthetically produced molecule, PG5, (Zhang 2011) is given as 200 $MDa \approx 2 \times 10^8\,gr/mole$, while the mass of a DNA strand is sometimes cited in units of $\approx 6\,pg \sim 1 \times 10^{12}\,gr/mole$. The reader (and the author) may need to recall that $Da = amu$ is the Dalton unit in chemistry, and pg is *pico-gram*.

Besides the dependence on mass, it's possible to effectively change g by using centrifuges, where $g^* = \omega^2 R$ provides two tweakable parameters, ω and R. Readily available, off-the-shelf versions of such devices are cited as producing RCF (relative centrifugal force) values of $g^* = 3 \times 10^3 - 3 \times 10^5\,g$, while those used for uranium separation for nuclear fuel production are guessed to be closer to $g^* \sim 10^6\,g$.

One can also write the expression for z_0 in terms not of the individual particles mass, but in terms of their molecular weight, μ, namely

$$z_0 = \frac{k_B T}{mg} = \frac{RT}{\mu g} \tag{8.51}$$

where $\mu = N_A m$. A measurement of the density profile giving z_0 then provides information on Avogadro's number and Perrin (1910) used this technique to measure N_A. In fact, his Nobel prize citation reads "... for his work on the discontinuous structure of matter, and especially for his discovery of sedimentation equilibrium" (emphasis again ours).

To extend the range of systems considered under the broad rubric of soft matter, we note that a useful definition of a colloidal system is from Wikipedia,[16] namely

> A **colloid** is a mixture in which one substance consisting of microscopically dispersed insoluble particles is suspended throughout another substance. Some definitions specify that the particles

[16] From https://en.wikipedia.org/wiki/Colloid. Accessed 10/9/2023.

must be dispersed in a liquid, while others extend the definition to include substances like aerosols and gels.

A slightly more specific, practitioners level, definition has been given by Poon (2015), who notes:

> In modern scientific usage, a colloid is a dispersion of particles or droplets in a liquid that, despite the density difference between the dispersed phase and the dispersing medium, is stable against sedimentation or creaming.

So, while ball bearings will fall to the bottom of a glass of water, the small particles of pigment in watercolor paint will remain suspended, which suggests that there is a maximum size for such particles, and this description also helps us focus on more carefully enunciating the parameters defining a colloid.

The statement above also reminds us to use the effective density of any suspended material, namely $\rho_m^* \equiv \rho_{material} - \rho_{medium}$ in any calculation involving the balance of thermal energies and gravitational energies. A number of experiments using suspended particles intentionally density match[17] the material and medium densities to limit any sedimentation effects, or even to vary the "effective" value of g in a way different than centrifugation; see, for example, Costantino et al. (2011).

Example 8.5 Size and time scales for colloids

To explore the size/time domain relevant for colloidal physics, let us assume that the maximum size of such a particle is given by

$$d_{max} \sim (k_B T)^\alpha (\rho_m^*)^\beta g^\gamma$$

$$\mathsf{L} = \left(\frac{\mathsf{ML}^2}{\mathsf{T}^2}\right)^\alpha \left(\frac{\mathsf{M}}{\mathsf{L}^3}\right)^\beta \left(\frac{\mathsf{L}}{\mathsf{T}^2}\right)^\gamma, \tag{8.52}$$

requiring $\alpha = 1/4$ and $\beta = \gamma = -1/4$, or $d_{max} \sim (k_B T/g\rho_m^*)^{1/4}$. Inserting values corresponding to room temperature, a "normal" value of g, and $\rho_m^* \sim \rho_{H_20} \sim 1\,gr/cm^3$ as an appropriate order-of-magnitude value, we find $d_{max} \sim 1\,\mu m$.

To set the scale, the sizes of sand, silt, and clay particles are in the ranges $(0.05-2)\,mm$, $(0.002-0.005)\,mm = 2-50\,\mu m$, and $< 2\,\mu m$, respectively, so sand will eventually sediment, while fine silt or clay may remain suspended—think the Amazon river.[18] Human blood cells have diameters/thicknesses of $\sim 8\,\mu m/2\,\mu m$, while soot particles in the air have sizes $30-50\,nm$. Since we also want to treat such colloidal systems using classical methods,[19] we want them to be larger than atomic scales, so it's usually assumed that $d_{min} \geq 10-100$ Å $= 1-10\,nm$, which gives the typical lower limit of lengths in such systems.

In contrast to the *ps* (*pico-second*) time scales for atomic systems, diffusion processes often dictate the pace of temporal development of colloidal particles. The **Brownian time**, τ_{Br}, is one such measure and is roughly the time it takes a particle to diffuse across its own spatial dimension, say its radius a. To see the scale for this, let us assume that the particles are in a medium of viscosity η, and then write

---

[17] See, e.g., Wiederseiner et al. (2011).
[18] See, e.g., Nittrouer et al. (2021).
[19] You'll notice no further mention of $\hbar$ in this section!

$$\tau_{Br} \propto \eta^{\alpha} a^{\beta} (k_B T)^{\gamma}$$

$$\mathsf{T} = \left(\frac{\mathsf{M}}{\mathsf{LT}}\right)^{\alpha} \mathsf{L}^{\beta} \left(\frac{\mathsf{ML}^2}{\mathsf{T}^2}\right)^{\gamma}, \tag{8.53}$$

which gives $\alpha = 1, \beta = 3$, and $\gamma = -1$ or $\tau_{Br} = C_{\tau}\eta a^3/(k_B T)$ and some treatments (see, e.g., Terentjev and Weitz 2015) specify $C_{\tau} = \pi$. (We note that we have seen this combination already, in the study of the Debye relaxation time in Eqn. (5.35).)

Assuming a value of $\eta = 10^{-3}$ $Pa \cdot s$ for water, and a particle at the upper end of the colloidal length scale, $a = 1\,\mu m$, we find $\tau_{Br} \sim 1\,sec$. Such time scales allowed Perrin to make "real-time" observations of the random-walk motions of his particles, as we discuss next.

In addition to sedimentation equilibrium, one of Perrin's important discoveries was the confirmation of the prediction of Einstein (and others) of the statistical nature of Brownian motion as being similar to a random walk process. The problem of a particle diffusing (let us consider only in one dimension to start) in a viscous medium (hence, with a frictional force proportional to $-v$) subject to a random external thermal force $\mathcal{F}(t)$ (due to collisions with the ambient molecules), can be written as

$$ma = m\frac{dv}{dt} = F = -bv + \mathcal{F}(t). \tag{8.54}$$

The time-average of the thermal force function due to collisions with molecules vanishes, so $\langle \mathcal{F}(t) \rangle = 0$. Given that it plays no role dimensionally, we're left with the thermal parameters $k_B T$, the particle mass m, and the "viscous coefficient" b as system parameters, along with t, to describe any time-development. Knowing that the **average** position of the particle is $\langle x \rangle = 0$ (equally likely, after all, to move left/right in this one-dimensional model), we want to constrain the temporal behavior of $\langle x^2 \rangle_t$, by writing

$$\langle x^2 \rangle_t = C_x (k_B T)^{\alpha} m^{\beta} b^{\gamma} t^{\delta}$$

$$\mathsf{L}^2 = \left(\frac{\mathsf{ML}^2}{\mathsf{T}^2}\right)^{\alpha} \mathsf{M}^{\beta} \left(\frac{\mathsf{M}}{\mathsf{T}}\right)^{\gamma} \mathsf{T}^{\delta}$$

$$\downarrow$$

$$\langle x^2 \rangle_t \propto (k_B T) m^{1-\delta} b^{-2+\delta} t^{\delta} = \frac{m}{b^2}(k_B T)\left(\frac{tb}{m}\right)^{\delta}, \tag{8.55}$$

and the natural time scale $\tau = m/b$ has been singled out. (You might want to review the results of **P3.7**.)

Two obvious values of δ are physically relevant (and also arise from the limiting cases of the exact solution, as in **P8.17**), namely

(short times or $t << \tau = m/b$) (long times or $t >> \tau = m/b$)

$$\delta = 2 \Rightarrow \langle x^2 \rangle_t = (k_B T)\frac{t^2}{m} \quad \text{and} \quad \delta = 1 \Rightarrow \langle x^2 \rangle_t = (k_B T)\frac{t}{b}. \tag{8.56}$$

The long-time solution with $\langle x^2 \rangle_t \propto t$ corresponds to diffusive behavior, familiar from discussions of the heat equation in Eqns. (5.18) and (5.19), and is the one relevant for the random-walk explanation of Brownian motion. For that case, we then use Stokes's law for the viscous force on a spherical body of radius a, namely $b = 6\pi a\eta$, and cite $C_x = 2$ in this limit, giving

$$\langle x^2 \rangle_t = \frac{k_B T}{3\pi\eta a} t = 2D_{Br} t \qquad \text{with} \qquad D_{Br} \equiv \frac{k_B T}{6\pi\eta a}, \tag{8.57}$$

where D_{Br} is diffusion constant appropriate for Brownian motion.

Perrin (1910) tracked the motion of individual particles of a gum extract (gamboge) of size $\sim 0.5\,\mu m$ suspended in water as a function of time, using a *camera lucida* apparatus attached to a microscope to project their motion onto an enlarged grid. He was able to confirm this familiar diffusion-driven $\langle x^2 \rangle_t \propto t$ result, and to extract a value of k_B, thereby providing another measure of Avogadro's number via $N_A = R/k_B$ using the known value of the gas constant; you can analyze some of his data in **P8.18.** The size and time scales used here are consistent with the values discussed in **Example 8.5.**

This mathematical approach is credited to Langevin, but Einstein arrived at this same result for the diffusion constant D_{Br} a few years earlier (as did Sutherland and Smoluchowski) through different means and a nice comparison between the two derivations, as well as a presentation of real data using modern charge-coupled device imaging methods, is presented in Newburgh et al. (2006).

The problem of a random walk in one dimension or two dimensions (or in three dimensions, where it's sometimes called random flight, at least in the chemistry literature) is an important one in statistical physics and applied mathematics. Phrased as a discrete process, with integral "steps" of specified size a (in one dimension) or $\boldsymbol{a}$ (in two/three dimensions), there is little DA can say about the problem, other than the obvious length scale being $a = |\boldsymbol{a}|$. But given that a random-walk analysis can be also applied to the problem of the length of a polymer, one of the canonical soft matter systems, with some resulting (dimension**ful**) physical modeling included, we discuss next some of the basic math involved, leading to another diffusion-like result.

Example 8.6 Polymers as (self-avoiding) random walks

Let us consider the problem of an object moving in three dimensions, with each "step" taken along the direction $\boldsymbol{a}_i$, with fixed distance ($|\boldsymbol{a}_i| = a$) between movements, but in an arbitrary direction each time. We can also use this identical visualization to model the construction of a long polymer, with each molecular subunit (or monomer) being added to the end of the chain as the analogous "step". For example, polystyrene has a structure described as $(C_6H_5CH{=}CH_2)_n$ or $(C_8H_8)_n$ and n can be of order a few thousand.

The total vector distance (or polymer length) is then given by

$$\boldsymbol{R}_N = \sum_{i=1}^{N} \boldsymbol{a}_i, \tag{8.58}$$

and since on average each of the $\boldsymbol{a}_i$ can point in a different random direction, the average $\langle \boldsymbol{R}_N \rangle = 0$.

In contrast, the distance squared is given by

$$(\boldsymbol{R}_N)^2 = \left(\sum_{i=1}^{N} \boldsymbol{a}_i\right)\left(\sum_{j=1}^{N} \boldsymbol{a}_j\right) = \sum_{i=1}^{N} |\boldsymbol{a}_i|^2 + \sum_{i \neq j}^{N} \boldsymbol{a}_i \cdot \boldsymbol{a}_j = Na^2 + \sum_{i \neq j}^{N} \boldsymbol{a}_i \cdot \boldsymbol{a}_j. \tag{8.59}$$

If we average over a large number ($N >> 1$) of steps/monomers, the random orientations of the $\boldsymbol{a}_i \cdot \boldsymbol{a}_j = a^2 \cos(\theta_{i,j})$ dot products will also average to zero, leaving

$$\langle(\boldsymbol{R}_N)^2\rangle = Na^2 \quad \text{or} \quad R_0 \equiv \sqrt{\langle \boldsymbol{R}_N^2 \rangle} = aN^{1/2}. \tag{8.60}$$

For many random walk problems, such relations are written as $R_0 \propto N^\nu$ where ν is a scaling exponent, which in this case is predicted to be $\nu = 1/2 = 0.5$. DA here correctly gives the scaling of $R_0 \sim a$ (after all, what else is there?), but is otherwise mute on the dimensionless function $F(N) = N^{1/2}$, which we know could be present. Elementary notions of counting and probability are also clearly useful skills, especially in statistical mechanics applications, so this is perhaps a useful reminder.

We can make a stronger connection to basic physical principles by noting that the potential energy corresponding to such a polymer chain is often written as

$$V_{el}(R) = V_0 + \frac{3k_BT}{2}\frac{R^2}{a^2N} = V_0 + \frac{1}{2}\left(\frac{3K_BT}{Na^2}\right)R^2 = V_0 + \frac{1}{2}K_{poly}R^2, \tag{8.61}$$

where $K_{poly} \equiv 3k_BT/(Na^2)$ is the effective spring constant of the polymer. We note that the factor of N in the denominator (from the random-walk analysis) is reminiscent of the result for a number of springs connected in <u>series</u> (i.e., connected end to end), where one has for equal strength springs ($K_1 = \ldots = K_N \equiv K_0$)

$$\frac{1}{K_{eff}} = \frac{1}{K_1} + \ldots + \frac{1}{K_n} = \frac{N}{K_0} \quad \text{or} \quad K_{eff} = \frac{K_0}{N}. \tag{8.62}$$

The exponent of $\nu = 1/2$ predicted by this simple model is not observed for real polymers, where values in the range 0.55–0.6 are seen instead (de Gennes 1979), and one reason is that this model does not include the fact that real chains cannot overlap with themselves, and so should be modeled as **self-avoiding** random walks.

Given that there is some region, v_0, around any monomer/subunit which must be avoided, an **excluded volume** interaction term that models the energy "cost" of such effects can be written as

$$V_{int}(r) = (k_BT)v_0\left(\frac{N^2}{2R^3}\right), \tag{8.63}$$

which is dimensionally correct since $[V_{int}] = [k_BT]\mathsf{L}^3\mathsf{L}^{-3}$ or $[E] = [E]$. (We discuss the motivation for this form, including the important N dependence, in **P8.20.**)

The total energy, now including this term, can then be minimized with respect to R to find

$$V(R) = V_0 + \frac{3k_BT}{2}\frac{R^2}{Na^2} + (k_BT)v_0\left(\frac{N^2}{2R^3}\right)$$

$$\Downarrow \qquad\qquad \Downarrow$$

$$0 = \frac{\partial V(R)}{\partial R} = 3k_BT\frac{R}{Na^2} - 3(k_BT)v_0\frac{N^2}{2R^4}, \tag{8.64}$$

giving

$$R^5 = v_0a^2N^3 \quad \text{or} \quad R(N) = (v_0a^2)^{1/5}N^{3/5}. \tag{8.65}$$

This result has, of course, the correct dimensions since $[R] = \mathsf{L} = ((\mathsf{L}^3)(\mathsf{L}^2))^{1/5} = [(v_0a^2)]^{1/5}$, but now with a different dependence than the trivial $R \propto a$ scaling as there is a second length scale in the problem via the v_0 term. Given that presumably $(v_0)^{1/3}$ is probably of the same physical size as a, numerically it doesn't likely make much difference.

More interestingly, however, this result corresponds to a scaling exponent of $\nu = 3/5 = 0.6$, much closer to that seen in experiments. This approach was originally discussed by Florry (1949), using such a phenomenological physical model, and later confirmed in a more mathematical way using statistical mechanical methods by Edwards (1965), who found that

$$\langle R^2 \rangle(N) = \left(\frac{5}{3}\right)^{6/5} \left(\frac{v_0 l^2}{3\pi}\right)^{2/5} N^{6/5}, \tag{8.66}$$

which is very similar to Eqn. (8.65), even down to the numerical prefactor.

This value of the scaling exponent turns out to be very close to the mathematically rigorous result of $\nu = 0.588$ obtained by renormalization group methods (Le Guillou and Zinn-Justin 1977, or Des Cloiseaux et al. 1985) or Monte Carlo calculations (Li et al. 1995). For a recent review of self-avoiding walks, see Slade (2019), but for a more complete review of such scaling concepts in polymer physics, see the very readable treatment by de Gennes (1979).

An extension of such ideas to two-dimensional surfaces (instead of one-dimensional chains) can be made using similar Flory "excluded volume" ideas (Kantor et al. 1986) to describe, for example, the fractal geometry in crumpled paper balls, where one predicts that the mass of such objects scale with their (compacted) size R as $M \sim R^{(d+2)/2} = R^{5/2}$, a result observed experimentally by Gomes (1987) in our $d = 3$ world; see **P8.22.**

As a final example of the diversity of systems that fall under the rubric of soft matter, let us consider an example of a complex fluid, where many competing material properties are in play, and as another example of where dimensionless Buckingham Pi ratios can be identified and used to quantify complex relationships.

Example 8.7 "Steady-state thickness of liquid–gas foams"

Pilon et al. (2001) start a paper (with the title of this Example) by stating that "Pneumatic foams are produced by a continuous stream of gas bubbles rising to the surface of a foaming liquid" and attempt to model how the equilibrium surface foam thickness, H_∞, depends on the properties of the material.

One obvious parameter is the average bubble size, r_0, which of course has the same dimensions as H_∞, automatically giving one dimensionless parameter, H_∞/r_0. The liquid density (ρ_m), viscosity (μ), and surface tension (σ) are all considered, and gravitational effects taken into account by including g. The rate at which gas is bubbled through the material is characterized by a quantity $\Delta j = j - j_m$, where j is the "superficial gas velocity" defined as the gas flow rate Q (so $[Q] = \mathsf{L}^3/\mathsf{T}$) divided by the cross-sectional area A (with $[A] = \mathsf{L}^2$) of the nozzle, giving $[j] = [Q/A] = \mathsf{L}/\mathsf{T}$ or just the dimensions of speed. The threshold value required to produce bubbles at all is j_m, hence m for minimum, and for considering Δj.

To explore how H_∞ might depend on all of these values, we begin by assuming a form for the dimensionless ratio

$$\frac{H_\infty}{r_0} \propto \rho_m^\alpha g^\beta \sigma^\gamma r_0^\delta \mu^\epsilon (\Delta j)^\tau. \tag{8.67}$$

Matching dimensions we find that $\delta = \alpha + \beta$, $\epsilon = -\alpha - \gamma$, and $\tau = \alpha - 2\beta - \gamma$, which gives

$$\frac{H_\infty}{r_0} \propto \left(\frac{\rho_m r_0(\Delta j)}{\mu}\right)^\alpha \left(\frac{g r_0}{(\Delta j)^2}\right)^\beta \left(\frac{\sigma}{\mu(\Delta j)}\right)^\gamma. \tag{8.68}$$

Three dimensionless quantities have been singled out, all of which are known to play important roles in the dynamics of complex fluids, namely

$$\frac{\text{inertia}}{\text{viscosity}} = \text{Reynolds number} = \mathcal{R}e \equiv \frac{\rho_m v L}{\mu} \tag{8.69}$$

$$\frac{\text{inertia}}{\text{gravity}} = \text{Froude number} = \mathcal{F}r \equiv \frac{v^2}{gL} \tag{8.70}$$

$$\frac{\text{viscosity}}{\text{surface tension}} = \text{Capillary number} = \mathcal{C}a \equiv \frac{\mu v}{\sigma}, \tag{8.71}$$

where v and L are characteristic speeds and length for the problem, here associated with Δj and r_0, respectively. (We note that in some treatments, the Froude number is given as $\mathcal{F}r = v/\sqrt{gL}$, but still dimensionless.) In just the same way that the Reynolds number was singled out as an important dimensionless combination in **Example 3.4**, we see here how other well-known (at least to those interested in fluid mechanics) named ratios can naturally appear.

The constraints that we have from just matching M, L, T dimensions can then be written as

$$\frac{H_\infty}{r_0} = (\mathcal{R}e)^\alpha (\mathcal{F}r)^{-\beta} (\mathcal{C}a)^{-\gamma}. \tag{8.72}$$

Given that the maximum foam height is likely due to a balance of the foam weight, we expect that the combination $\rho_m g$ will appear paired with the same powers, hence $\alpha = \beta$, giving

$$(\mathcal{C}a)^\gamma \left(\frac{H_\infty}{r_0}\right) \propto \left(\frac{\mathcal{R}e}{\mathcal{F}r}\right)^\alpha, \tag{8.73}$$

and Pilon et al. (2001) use non-dimensionalization methods to argue that $\gamma = 1$.

So, using DA and information encoded in the differential equations, and of course professional intuition and experience, they proposed a relation $\Pi_2 = K\Pi_1^n$ where $\Pi_2 = \mathcal{C}a(H_\infty/r_0)$ and $\Pi_1 = \mathcal{R}e/\mathcal{F}r$ are actually the product/ratio of two already dimensionless quantities. Using data from ten different experiments, they find a power-law fit with $K \approx 2900$ and $n \approx -1.8$, with a correlation coefficient $R^2 = 0.95$, a value (close to 1), which indicates a good fit; we reproduce the plot of their results in Fig. 8.4, illustrating the relatively good agreement.

While far from being an *ab initio* calculation of H_∞ starting from first principles of hydrodynamics, this simple relation does seem to correctly describe much of the complexity encoded in the physics of the system, matching data over three to four orders of magnitude, from multiple systems, and so is clearly a useful approach.

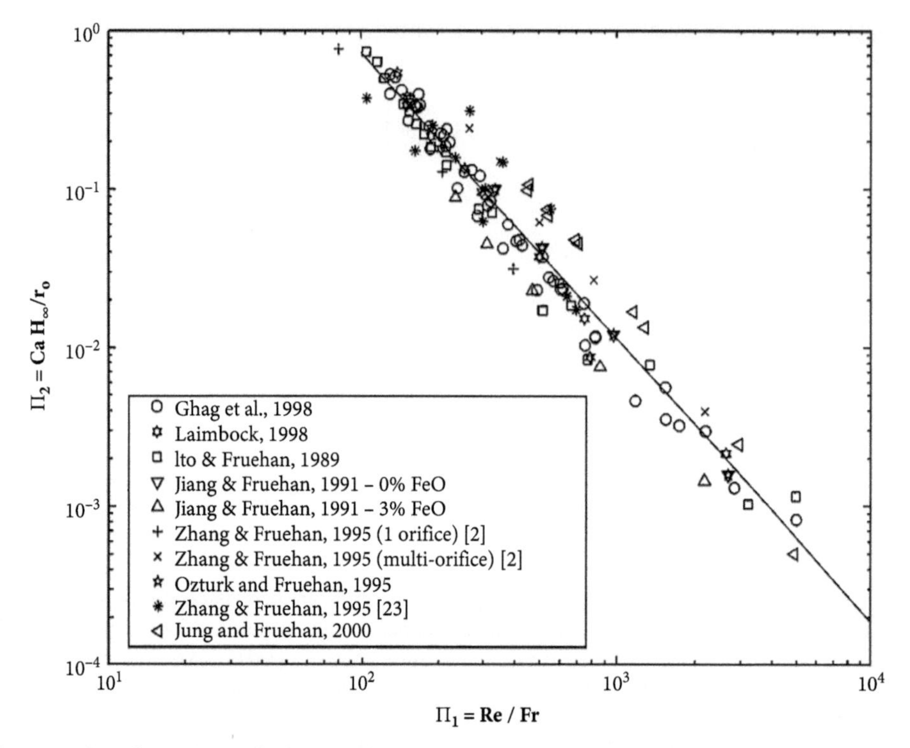

Fig. 8.4 Plot of two dimensionless Buckingham Pi ratios, $\Pi_2 = CaH_\infty/r_0$ versus $\Pi_1 = \mathcal{R}e/\mathcal{F}r$, data from various experiments, and power-law fit (straight line). Reprinted from Pilon, Fedorov, and Viskanta (2001) with permission of Elsevier.

8.4 **Problems for Chapter 8**

Q8.1 Reality check: (a) What was the most important, interesting, engaging, or novel thing you learned about how to use DA, or about anything else, in this chapter?
(b) Have you previously encountered (or learned) many (or any) of the topics covered in this chapter?

P8.1 Mass densities in terms of fundamental constants: Let's explore how the mass densities of solids depend on basic quantities in atomic physics by writing $\rho_m \propto \hbar^\alpha K_e^\beta m_e^\gamma M^\delta$ where m_e, M are the electron and nuclear masses.

(a) Match dimensions and solve to the extent you can. Are any of the exponents uniquely determined?

(b) A reasonable assumption is that density of solids, $\rho_m = M/V$, is determined by the nuclear masses ($M >> m_e$), but by the volume occupied by the atoms (which is determined by the electrons), so that $\rho_m \sim M/(2a_0)^3$ is one option. Is that consistent with your dimensional constraint from part (a)

(c) Use your result from part (b) to estimate a typical density for a solid and compare to the range $1-10\,gr/cm^3$.

P8.2 Electrical conductivity in terms of fundamental constants: (a) Use DA to explore how electrical conductivity (σ_e) might depend on atomic physics parameters by writing $\sigma_e \propto e^\alpha \epsilon_0^\beta \hbar^\gamma m_e^\delta$ and solving for $\alpha, \beta, \gamma, \delta$. In examples in the text, or in other problems, where we evaluate energies, densities, and viscosities, all M, L, T (i.e., 'just mechanics') related quantities, we used only the $K_e = e^2/4\pi\epsilon_0$ combination. In this case, where conductivity is explicitly an E&M quantity, we consider the e and ϵ_0 (or perhaps $4\pi\epsilon_0$) dependences separately. Can you write your final answer in terms of e, $\hbar$, and a_0?

(b) Evaluate your answer numerically (in SI/MKS(C) units) and compare to values for familiar metals, where σ_e ranges from 10^6 to *few* $\times$ 10^7 *siemens/meter*. Use the collection of units in Sec. 10.1 if needed.

(c) The relation between the resistance of a conductor of length l and area A and the conductivity of the material is $R = l/\sigma_e A$. If the magnitudes of both l, A are given in terms of atomic dimensions (via a_0), what does this say about the typical atomic resistance in terms of fundamental constants. Is this relation familiar from the QHE?

P8.3 Long-range van der Waals potential: The R^{-6} term in the Lennard-Jones potential in Eqn. (8.4) is often called the **van der Waals term** and is physically motivated as resulting from the electric dipole-dipole interactions of the two atoms, with dipole moments $\boldsymbol{p}_1, \boldsymbol{p}_2$. In this problem you're asked to put together several ingredients of obtain a crude (but hopefully dimensionally correct) idea of how that particular functional form arises. You can use the facts that:

- The electric dipole fields (E_1, E_2) and interaction energy (U_{dip}) are given by:

$$E_1 \sim \frac{p_1}{4\pi\epsilon_0 R^3}, \quad E_2 \sim \frac{p_2}{4\pi\epsilon_0 R^3}, \qquad \text{and} \qquad U_{dip} = -p_2 E_1 = -p_1 E_2. \tag{8.74}$$

- One dipole can be induced by the electric field of the other via polarization effects, described by

$$p_2 = \alpha_P E_1 \qquad \text{or} \qquad p_1 = \alpha_P E_2 \tag{8.75}$$

  where α_P is the **electric polarizability**.
- The value of α_P, and a rough estimate of an atomic dipole moment, are

$$\alpha_P \sim (4\pi\epsilon_0)a_0^3 \qquad \text{and} \qquad p_{1,2} \sim ea_0. \tag{8.76}$$

Put these pieces together to get a long-distance potential of the form $U_{dip} \sim E_H(a_0/R)^6$, where E_H is the Hartree energy or hartree unit, defined in Eqn. (6.202).

Note: Even though each (instantaneously fluctuating) vector electric dipole $\boldsymbol{p}_{1,2}$ will average to zero, this approach gives a result proportional to $\boldsymbol{p}^2 \sim (ea_0)^2$, which has a non-vanishing expectation value.

P8.4 Quantum viscosities in terms of fundamental constants: A recent research paper by Trachenko and Brazhkin (2020) has an abstract which begins with the words

> Viscosity of fluids is strongly system dependent, varies across many orders of magnitude, and depends on molecular interactions and structure in a complex way not amenable to first-principle theories.

They then go on to argue[20] that there is a **minimal kinematic viscosity of fluids**, ν_m, given in terms of fundamental constants.

(a) Write $\nu_m \propto \hbar^\alpha K_e^\beta m_e^\gamma M^\delta$, match dimensions, and find the exponents to the extent you can. You will, of course, find the dimensionless ratio m_e/M (or its inverse) since there are two masses, but are there any exponents which are uniquely determined? Use the Table 10.1 if needed to find $[\nu_m] = \mathsf{L}^2/\mathsf{T}$.

(b) A scaling argument to help fix the remaining powers has some of the following ingredients:
   (a). The kinematic viscosity can be written as $\nu = \eta/\rho \sim (\rho v L)/\rho \sim vL$ where v, L are the typical speed and length scales in the problem.
   (b). $L \sim a_0$ is a reasonable length scale for atomic/molecular processes.
   (c). $v \sim a_0 f_{vib}$ is a reasonable velocity, where $E_{vib} = \hbar\omega_{vib} = \hbar(2\pi f_{vib})$ are the molecular vibrational state quantities.
   (d). And finally, $E_{vib} \sim E_H\sqrt{m_e/M}$ from Eqn. (8.7).

Combine these to help completely determine the exponents of ν_m and see if they're consistent with your answer from part (a).

P8.5 Nernst effect in terms of fundamental constants: In some materials, when a magnetic field (B_z) is applied to a conducting sample along with a temperature gradient (dT/dx), there is a resulting electric field (E_y), in a perpendicular to both, which is called the **Nernst effect.** Two measures of this effect are given by coefficients ν or e_N defined by

$$\nu = \frac{E_y}{B_z}\frac{1}{dT/dx} = \frac{e_N}{B_z}. \tag{8.77}$$

Theory papers have sought to find relations for e_N and other thermo-electric transport coefficients in terms of fundamental constants (similar to the discussions in Sec. 2.6). To see what parameters are most important, write $e_N \propto k_B^\alpha e^\beta \epsilon_0^\gamma \hbar^\delta m^\sigma$, match dimensions (using Eqn. (8.77) to find the dimensions $[e_N]$), and extract as many exponents as you can. (For an example of a theoretical analysis, see Hartnoll et al. (2007)).

P8.6 Bose-Einstein condensate (BEC) temperature: Below a certain temperature, T_C, systems of bosons can condense, with all of the particles in the ground state.

(a) Estimate this temperature for a system of particles of mass m and number density n by writing $T_C = C_T\hbar^\alpha n^\beta m^\gamma k_b^\delta$, match dimensions, and determine the exponents. Real calculations exhibit the dimensional dependences you'll find, with the dimensionless constant given by $C_T = 2\pi/\zeta(3/2) \approx 3.31$, where $\zeta(s)$ is the Riemann zeta function.

(b) One oft-stated criteria for producing a BEC is that the **thermal de Broglie wavelength** (λ_{th}) and number density satisfy the relation $\rho_{ps} = n(\lambda_{th})^3 > 2.612$ where ρ_{ps} is the dimension<u>**less**</u> phase-space density. The thermal wavelength λ_{th} is determined by the average particle momentum which, in turn, depends on the temperature. Try to compare this description to your result in part (a) to see if they're consistent.

P8.7 Debye temperature: In the Debye model of the heat capacity of solids, lattice vibrations (phonons) are treated in much the same way as photons in blackbody radiation (normal modes of

[20] See also Trachenko and Brazhkin (2021).

waves in a box) in a way that reproduces much of the low- and high-temperature behavior of C_V. The temperature corresponding to the highest-frequency phonon is called the **Debye temperature** and could depend on variables in the problem as $T_D = C_T \hbar^\alpha v_s^\beta k_B^\gamma n^\delta$, where v_s is the velocity of sound in the material and n the density. Match dimensions and solve for the exponents. The textbook answer is consistent with your result and gives $C_T = (6\pi^2)^{1/3} \approx 3.2$ as the undetermined dimensionless constant.

P8.8 Low-temperature specific heat: At low temperatures ($T << T_D$), the specific heat per unit volume, c_V, of solids is observed to scale as T^3, meaning that c_V/T^3 approaches a constant as $T \to 0$. Assume that this constant depends on fundamental parameters, write $c_V/T^3 = C_c k_B^\alpha \hbar^\beta c^\gamma m^\delta$, match dimensions, and determine the exponents. The textbook answer is consistent your result and gives $C_c = 2\pi^5/5$. Given the array of specific heat definitions used in condensed matter physics, we specify here that $c_V = (dU/dT)/V$ giving $[c_V] = ((\mathsf{ML}^2/\mathsf{T}^2)/\Theta)/(\mathsf{L}^3) = \mathsf{M}/(\mathsf{LT}^2\Theta)$.

P8.9 London penetration depth: Magnetic fields in superconductors are expelled (the so-called **Meissner effect**) and are only allowed as exponentially damped solutions at the interface with an external field. The magnetic field satisfies an equation of the form $\nabla^2 \mathbf{B} = \mathbf{B}/\lambda_L^2$ where λ_L is the **London penetration depth.**

(a) Since it appears squared, let's explore how $(\lambda_L)^2$ can depend on m, q, n, μ_0 by writing $(\lambda_L)^2 = m^\alpha q^\beta n^\delta \mu_0^\gamma$. Match dimensions and see to what extent you can determine the exponents. You'd expect with four quantities and four constraints from the M, L, T, Q dimensions that you could get a unique answer, but you don't (why?). Do you have any intuition about what might be the best choice for the exponents?

(b)* To explore this problem in a more detailed way, still focusing on dimensions and not trying to be too rigorous, take the following 'ingredient' list and try to put them together to see which combination you've found in part (a) is most likely to give $(\lambda_L)^2$:

$\mathbf{p} = m\mathbf{v} + q\mathbf{A}$	momentum, including E & M effects, via $\mathbf{A}$	(8.78)
$\mathbf{J}_e = nq\mathbf{v}$	current density	(8.79)
$\nabla \times \mathbf{B} = \mu_0 \mathbf{J}_e$	Ampere's law in differential form	(8.80)
$\mathbf{B} = \nabla \times \mathbf{A}$	relation between magnetic field and $\mathbf{A}$	(8.81)
$\nabla \times (\nabla \times \mathbf{F}) = \nabla(\nabla \cdot \mathbf{F}) - \nabla^2 \mathbf{F}$	general vector identity for any $\mathbf{F}$,	(8.82)

where we make use of some of the connections between the vector potential $\mathbf{A}$ cited in Eqns. (8.18) and (8.19) and a general vector identity involving the ∇ gradient operator. You can also remember the "no magnetic monopole" Maxwell equation $\nabla \cdot \mathbf{B} = 0$ if you wish, but again focus on keeping track of the dimensions more than anything else.

P8.10 Degeneracy of Landau levels: Given that the radii of the classical orbits as encoded in the magnetic length in Eqn. (8.27) are typically much smaller than the average size of a 2D Hall effect sample, there can be many orbits which 'fit' into a $L_x \times L_y$ size device, hence a large degeneracy per Landau level, with $N_{deg} >> 1$ per n_L value in Eqn. (8.25).

(a) Explore how N_{deg} depends on the parameters in the problem by writing $N_{deg} = C_N q^\alpha B_0^\beta h^\gamma (L_x \times L_y)$ where we use h instead of $\hbar$ given our experience with other quantized systems we've discussed, and we assume a linear dependence on the product of the two lengths or the area. Match dimensions and determine the exponents. The dimensionless coefficient C_N turns out to be unity with this choice of variables.

(b) Can you write your answer for N_{deg} in terms of the product $L_x \times Ly$ and the magnetic length l_B?

(c) Can you also write your answer to part (a) in terms of the total flux through the sample and the flux quantum in Eqn. (8.30)?

(d) The 2D number density of states will be given by $n_{2D} = N_{deg}/(L_x \times L_y)$ and it turns out that when this value is proportional to an integer i, one obtains the quantized values of $R_H^{(i)}$. Show this by using $R = B/(en_{2D})$, and this condition, to show how this reproduces the quantum of resistance R_H in Eqn. (8.38).

Note: This bound on the maximum quantum number is derived more rigorously by Peierls (1955) (Sec. 7.2 on *Diamagnetism of free electrons (Landau)*).

P8.11 de Haas – van Alphen effect: A important quantized variable related to the motion of electrons in an applied magnetic field (B) is given by $S_l = S_0(l + 1/2)$, and one text (Kittell (1971), Chap. 10) cites S_l as being

> ... the area of the orbit l in $\mathbf{k}$ space ... Thus in a magnetic field the area of the orbit in $\mathbf{k}$ space is quantized.

and l is an integer quantum number. The notion of **wave-number** or $\mathbf{k}$ space, with $k_{x,y,z} = 2\pi/\lambda_{x,y,z}$, is an important one in condensed matter systems, especially for crystalline solids, where there are lattice structures in (x, y, z) or position space, which translate into preferred directions in the space of propagating waves. The most important dimensional part of the statement above, however, is that the area of an orbit in $\mathbf{k}$ space will be a 2D product of the form $S_0 \propto k_x k_y$, and therefore have dimensions $[S_0] = [k_x k_y] = \mathsf{L}^{-2}$.

(a) Assume that $S_0 = C_S e^\alpha B^\beta \hbar^\gamma m^\delta$, match dimensions and extract the exponents. A complete derivation finds that $C_S = 2\pi$, along with the quantum number factor $(l + 1/2)$ which appears in the full expression for S_l.
Note: Mapping out values of S_l for various orientations of the applied $\mathbf{B}$ field can give important information on the shape of the **Fermi surface**, an abstract 3D geometric object which is determined by the condition that $E_F(k_x, k_y, k_z) = constant$. For a free electron gas, where particles can move equally well in any direction, and $E_F \propto (k_x^2 + k_y^2 + k_z^2)$, the shape of the Fermi surface is a sphere. The concept of the Fermi surface is an extremely important one in condensed matter physics and an accessible review is *Life on the edge: a beginner's guide to the Fermi surface* by Dugdale (2016) and interested readers should search (online or otherwise) for the striking images of Fermi surfaces obtained by such techniques.

(b) With that **Note** in place, for a spherical Fermi surface, the 2D 'area' of any $\mathbf{k}$ orbit is simply $S = \pi k_F^2$ and the corresponding quantized energies are $E_F = \hbar^2 k_F^2/2m$. Use this result and your answer from part (a) to reproduce the expression for the Landau level energies in Eqn. (8.25).

P8.12 Classical hydrodynamics for vortex rings: One of the early papers to cite evidence for *Quantized vortex rings in superfluid helium*, by Rayfield and Reif (1964), used classical hydrodynamical relations[21] to connect the speed v of the rings with their density (ρ_m), circulation (C),

[21] Citing the famous *Hydrodynamics* text by Lamb (1945).

and energy (E). They then used that connection to extract values of C from data of v versus E. Write $v = C_v \rho_m^\alpha C^\beta E^\gamma$ and extract the exponents to the extent you can. The actual result has a $C_v = 1/8\pi$ factor, but also includes a non-trivial function of a dimensionless ratio of the form $\eta = \ln(8R/a)$, where R, a are the outer and core radii of the ring, factors which they then had to model.

P8.13 Quantum numbers for macroscopic systems: Some intro-level quantum texts ask readers to estimate the quantum numbers (n) of 'everyday' or macroscopic systems for quantities such as angular momentum or energy, for example, of the harmonic oscillator, infinite square well, or even the bouncing ball (gravity) examples studied in Sec. 6.3.

(a) Let's do this first for $rp = L_z = n\hbar$, and then the three quantized energies

$$E_n^{(HO)} = \hbar\sqrt{\frac{k}{m}}(n + 1/2), \qquad E_n^{(ISW)} = n^2\frac{\hbar^2\pi^2}{2mL^2}, \qquad \text{and} \quad E_n^{(gravity)} \sim n^{2/3}\left(mg^2\hbar^2\right)^{1/3} \tag{8.83}$$

harmonic oscillator, particle in a box, and bouncing ball.

Assume 'unit amounts' of classical quantities or variables in SI units, namely $l \sim 1\,m$, $v \sim 1\,m/s$, $m \sim 1\,kg$, $g \sim 10\,m/s^2$, and $k = F/x \sim 1\,N/m$ for any length, speed, mass, g-value, spring constant or other quantities you need. Find n for each of the macroscopic limits of these quantized variables. Are all of the n values roughly the same? Do they have to be?

(b) Now do this for macroscopic examples of the circulation, so say for water swirling down a drain with macroscopic values of v, l, m, using $C^{(i)} = ih/m$, and find the quantum number i.

(c) Finally, estimate how many flux quanta are required if a $B = 1$ *Tesla* field pierces a $1\,m \times 1\,m$ area.

P8.14 Numerics of the quantum Hall effect (QHE): Given that there are many energy, length, and other scales for the QHE, in this problem you'll be asked to explore some of them with real data, that shown in Fig. 8.3, along with other values also taken from Paalanen et al. (1982).

(a) For magnetic fields below about 15–20 kG, the Hall resistance is seen to be proportional to magnetic field. Use the relation $R_H = B/en_{2D}$ to extract the value of the two-dimensional electron density and compare to the cited value of $n_{2D} \sim 4 \times 10^{15}\,m^{-2}$.

(b) Using the relation $R_H^{(i)} = R_K/i$, identify as many of the plateaus in Fig. 8.3 by their quantum number i.

(c) The thickness of the current carrying strip in such experiments is roughly $t \sim 10\,nm$. An electron confined to such a dimension will have quantized energies of order $E_z = n^2\hbar^2\pi^2/(2m_et^2)$. Estimate the size of the zero-point ($n = 1$) energy state in meV (where $m = milli$).

(d) The experiment was done at a temperature of 50 mK. Evaluate the thermal energy (k_BT) associated with this, again in meV. Is the temperature cool enough that it's unlikely that any of the 1D infinite well higher energy states ($n > 1$) will be excited?

(e) What is the energy of the lowest ($n_L = 0$) Landau level from Eqn. (8.25) for $B = 40\,kG$, again in meV? Compare this to the results of parts (c) and (d).

(f) What is the magnetic length $l_B = \sqrt{\hbar/eB}$ for $B = 40\,kG$?

P8.15 Hall viscosity: Besides the quantized Hall electric and thermal conductances, some quantum Hall fluid systems can also exhibit a **Hall viscosity**. Explore the dimensional dependence of this property by writing $\eta = C_\eta n_e^\alpha \hbar^\beta m_e^\gamma e^\delta$ where n, m_e, e are the number density, mass, and charge of the carriers. Match dimensions and determine the exponents to the extent you can. The dimensionless prefactor turns out to be $C_\eta = S/4$ where S is an integer-valued filling factor.

P8.16 Molecular scale of surface tension: Values of the surface tension of liquids can range from $S = 10\, dyne/cm$ for liquid oxygen to $S = 75\, dyne/cm$ for water with its strong hydrogen bonds. Convert these two values to units of $eV/\text{Å}^2$ (where Å is Angström). Is your result consistent with the magnitude of the van der Walls (molecular) interactions between atoms in such materials, namely 4–40 meV per bond, and atomic distance scales?

P8.17* Langevin solution for Brownian motion: The solution of the mathematical problem described in Eqn. (8.54) for the time-development of the average value of x^2 is cited as

$$\langle x^2 \rangle_t = \frac{2k_BT}{b}\left[t - \tau\left(1 - e^{-t/\tau}\right)\right], \tag{8.84}$$

where $\tau \equiv m/b$. Expand this answer for $t/\tau << 1$ and $t/\tau >> 1$ and compare to the dependences found by DA in Eqn. (8.56).

P8.18 Sedimentation equilibrium equation: From the long term diffusive behavior result for $\langle x^2 \rangle_t$ in Eqn. (8.56), we found that $D_{br} = k_BT/b$ for an arbitrary viscous force proportional to $-v$. An increased concentration of particles due to gravity will lead to a concentration gradient and a resulting diffusive particle flux given by **Fick's law**, namely $J_{diff} = -Ddn(z)/dz$, from regions of high concentration to low, so upwards. In equilibrium, this must be balanced by the rate of sedimentation given by $J_{sed} = n(z)v_{sed}$, where $mg = bv_{sed}$ gives the sedimentation velocity. Balancing these two fluxes, show that you get a differential equation which has the solution $n(z) = n_0 e^{-z/z_0}$, and confirm that your expression for z_0 matches that in **Example 8.4.**

P8.19† Perrin data on Brownian motion: In a popular account of his own work on Brownian motion, Perrin (1910) provides data (his *second series*) on observations of the mean displacement versus time for his 'granules', as well as the fitted value of Avogadro's number, namely

Time in seconds	Mean horizontal displacement (in μm)	$N \times 10^{-22}$
30	8.4	68
60	11.6	70
90	14.8	71
120	17.5	62

(a) Find the mean of Avogadro's number from the four data points listed and compare to the modern value.

(b) Fit the values of $\overline{x} \equiv \sqrt{\langle x^2 \rangle}$ in the second column versus t to a power-law of the form $\overline{x} = at^n$, using whatever tools you know, to find the "best fit" power n and compare to the random-walk prediction of 1/2.

P8.20 "Excluded volume" interaction term: The "excluded volume" interaction term from Eqn. (8.63) can be derived from a more basic, subunit–subunit (monomer/monomer), potential energy function of the form[22]

$$V_{ij}(\boldsymbol{r}_i, \boldsymbol{r}_j) = v_0(k_B T)\delta(\boldsymbol{r}_i - \boldsymbol{r}_j), \tag{8.85}$$

where the Dirac δ-function in 3D was defined in Eqn. (6.92) and has $[\delta(\boldsymbol{r})] = \mathsf{L}^{-3}$. The δ-function term is clearly responsible for the R^{-3} factor in Eqn. (8.63) with the dimensions being consistent, leaving only the $N^2/2$ dependence to be explained as being due to number of possible subunit-subunit overlaps. Look up, or derive for yourself, how many distinct pairs can be made from a total of N items and see if that explains this factor. Once again, DA plays no role in this, while simple "counting" does.

P8.21 Florry scaling for polymers in d-dimensions: The analysis leading to the $R(N) \propto N^{3/5}$ dependence from Eqn. (8.64) can be easily extended to d-dimensions by generalizing the excluded volume interaction term to be

$$V_{int}^{(d)}(R) = (k_B T)v_0^{(d)} \frac{N^2}{2R^d}, \tag{8.86}$$

where $[v_0^{(d)}] = \mathsf{L}^{\mathsf{d}}$ is an excluded length for $d = 1$, or area for $d = 2$, and so forth. Minimize the total energy to find $R^{(d)}(N) \sim N^{\nu^{(d)}}$ and extract the scaling exponent $\nu^{(d)}$. Does your answer make sense in the restrictive geometry of one dimension? Renormalization group arguments (Nienhuis 1982) and Monte Carlo calculations (Li et al. 1995) have confirmed the value you find here for $d = 2$ is also correct. For $d = 3$ we noted that the prediction of $\nu = 3/5 = 0.6$ is very close to the exact answer of 0.588.

P8.22 Florry scaling of crumpled paper: Kantor et al. (1986) used excluded volume ideas to write the energy for a "... surface of internal size L, occupying a region R_G in d-dimensional space ..." and obtain

$$V(R_G) = \frac{1}{2}KR_G^2 + \frac{\nu L^4}{(R_G)^d}. \tag{8.87}$$

Write down the dimensions of each parameter (especially K and ν) to confirm that everything is dimensionally correct.

(a) Minimize this to show that $R_G \propto L^{4/(d+2)}$.

(b) The mass of a two-dimensional sheet scales as its area, namely $M \sim L^2$. Combine this with the result from part (a) to obtain the fractal scaling relation $M \sim R_G^{(d+2)/2}$. For the case of $d = 3$ (crumpling actual paper in the real world), what does this predict for the scaling exponent? See Gomes (1987) and Daoud and Van Damme (1999) for experimental data presented in two different ways, which are consistent with this result.

P8.23 Power-law decays of correlation functions of colloidal particles: One measure of the (complicated) dynamics of colloidal particles in suspension is the **velocity autocorrelation function**, defined as $R_v(t) \equiv \langle v_x(t)v_x(0)\rangle_{th}$, where we take the thermal average over many diffusive particle paths. $R_v(t)$ gives information about how knowledge of the initial motion determines the speed at later times, and how that connection decays in time as the particle diffuses away. For long times, this quantity turns out to not depend on the details of the particles properties (mass, moment of inertia, etc.), but rather on the thermal combination $k_B T$, properties of the medium (the density ρ_m and kinematic viscosity ν), and, of course, the time t. Experiments can often probe the long-time dependence of $R_v(t)$, so this limit can be the most relevant.

---

[22] See, e.g., Doi and Edwards (1986).

(a) Write $R_v(t) \propto (k_B T)^\alpha \rho_m^\beta \nu^\gamma t^\delta$, match dimensions, and find the exponents to the extent you can.

(b) The distance a particle diffuses in a time t is given by $\sqrt{2\nu t}$, so that these two variables often appear paired as the combination νt. Does this association help specify all of the exponents? **Notes:** A simple argument leading to the complete dimensional dependence you've found using parts (a) and (b) is given by Frenkel (2002). Measurements of $R_v(t)$ by Zhu et al. (1992) find a universal time dependence consistent with this same result.

(c) For a particle (let us assume spherical, as always) undergoing Brownian rotational motion, the relevant autocorrelation function is $R_\omega \equiv \langle \omega_z(t)\omega_z(0)\rangle$ using the angular velocity $\omega_z(t)$. Write $R_\omega(t)$ in terms of the same variables as in part (a) and make the same assumptions as in part (b) to determine the time-dependence. Is there a simple DA-related connection between the time-dependence of $R_v(t)$ and $R_\omega(t)$? The long-time tails in angular momentum correlations (Lowe et al. 1995) you've found here have been demonstrated using computer simulations, which also confirmed earlier theoretical predictions by Huage and Martin-Löf (1973) and Chow (1973).

(d) The results above are valid for long-times, while the initial values of $R_v(0)$ and $R_\omega(0)$ are determined by the thermal environment ($k_B T$) and by the particle properties, m and I, respectively. Can you determine how, by just using dimensions, and perhaps the equipartition theorem?

P8.24 Liquid–gas foams I: (a)† Assuming a relation of the form $\Pi_1 = K\Pi_2^n$, use the data plotted in Fig. 8.4 and confirm that $K \approx 2900$ and $n \approx -1.8$, as derived in Pilon et al. (2001).
(b) Using the definitions of the dimensionless ratios $\mathcal{C}a$, $\mathcal{R}e$, and $\mathcal{F}r$, show that the scaling result $\mathcal{C}a(H_\infty/r_0) = K(\mathcal{R}e/\mathcal{F}r)^{-1.8}$ can be rewritten in the form

$$H_\infty = K\frac{\sigma}{r_0^{2.6}}\frac{(\mu\Delta j)^{0.8}}{(\rho_m g)^{1.8}} \tag{8.88}$$

and confirm that the dimensions on the right-hand side do actually give you a length.

P8.25 Liquid–gas foams II: In their paper "Modeling Study of Metallurgical Slag Foaming via Dimensional Analysis", Wang et al. (2021) obtain three different expressions for the foam height, H_∞, which they call Δh, in terms of the same parameters in Eqn. (8.88), namely,

$$\Delta h = K_1\frac{g^{3.53}\rho_m^{3.29}r_0^{7.82}(\Delta j)^{1.27}\mu^{1.75}}{\sigma^{5.04}} \tag{8.89}$$

$$\Delta h = K_2\frac{g^{1.35}\rho_m^{1.62}r_0^{3.97}(\Delta j)^{0.01}}{\mu^{0.53}\sigma^{1.09}} \tag{8.90}$$

$$\Delta h = K_3\frac{g^{0.91}\rho_m^{1.55}r_0^{3.46}\sigma^{0.18}}{\mu^{1.73}(\Delta j)^{0.45}}, \tag{8.91}$$

where $K_{1,2,3}$ are dimensionless constants. Show that despite the array of fitted exponents, that these three expressions all still respect dimensionality constraints, namely that the right-hand sides of each have dimensions of length.

P8.26 More dimensionless numbers in fluid dynamics: Two other Buckingham Pi ratios, dimensionless combinations of material and system parameters, which appear in hydrodynamics are

$$\text{Morton number} = \mathcal{M}o \;\propto\; g^{\alpha}\mu^{\beta}\rho_m^{\gamma}\sigma^{\delta} \tag{8.92}$$

$$\text{Weber number} = \mathcal{W}e \;\propto\; \rho_m^{\alpha}v^{\beta}l^{\gamma}\sigma^{\delta}, \tag{8.93}$$

where ρ_m, μ, σ are the fluid density, viscosity, and surface tension, g is the acceleration of gravity, and v, l are the characteristic speed and size of the object moving in the fluid. These quantities can be relevant for droplets, but also for the motion of gas bubbles in liquids.

(a) For each new named number, find the values of β, γ and δ, in terms of α, which make both sides dimensionless. The standard definitions of $\mathcal{M}o$ and $\mathcal{W}e$ both use $\alpha = 1$, which should then determine all of the exponents.

(b) Alongside the Reynolds ($\mathcal{R}e$), Froude ($\mathcal{F}r$), and Capillary ($\mathcal{C}a$) numbers discussed in **Example 8.7**, there are enough dimensionless combinations that not all of the named ratios are independent of each other. For example, write

$$\mathcal{M}o = (\mathcal{W}e)^a\,(\mathcal{R}e)^b\,(\mathcal{F}r)^c \tag{8.94}$$

and find values of a, b, c which make the expressions agree. Instead of doing this by matching M, L, T dimensions (since all four quantities are already dimensionless, that doesn't help), match powers of all physical quantities, namely $g, \mu, \rho_m, \sigma, v, l$, and see if there is a self-consistent solution.

Note: In "History and Significance of the Morton Number in Hydraulic Engineering" by Pfister and Hager (2014) describes the connection you've just found, and also reiterates the important role that DA can play in fluid engineering problems. It also alludes to (yet) another example of the "Zeroth Theorem of the History of Science"[23] as discussed in the Preface.

References for Chapter 8

Amador, C., et al. (2011). "Shear Elastic Modulus Estimation from Indentation and SDUV on Gelatin Phantom," *IEEE Transactions on Biomedical Engineering* 58, 1706–14.

Anderson, P. W. (1972). "More is Different," Science 177, 393–6.

Anderson, P. W., and Rowell, J. M. (1963). "Probable Observation of the Josephson Superconducting Tunneling Effect," *Physical Review Letters* 10, 230–2.

Ashcroft, N. W., and Mermin, N. D. (1976). *Solid State Physics* (New York: Holt, Rinehart, and Winston).

Banerjee, M., et al. (2017). "Observed Quantization of Anyonic Heat Flow," *Nature* 545, 75–79.

Banerjee, M., et al. (2018). "Observation of Half-Integer Thermal Hall Conductance," *Nature* 559, 205–10.

Bartalucci, S., et al. (VIP Collaboration). (2006). "New Experimental Limit on the Pauli Exclusion Principle Violation by Electrons," *Physics Letters B* 641, 18–22.

Blencowe, M. P., and Vitelli, V. (2000). "Universal Quantum Limits on Single-Channel Information, Entropy, and Heat Flow," *Physical Review A* 62, 052104.

Bohm, D., and Pines, D. (1951). "A Collective Description of Electron Interactions. I. Magnetic Interactions," *Physical Review* 82, 625–34.

[23] Part of the title by an article by J. D. Jackson (2008), Am. J. Phys. 76 704–19)

Bohm, D., and Pines, D. (1952). "A Collective Description of Electron Interactions. II. Collective vs Individual Particle Aspects of the Interactions," *Physical Review* 85, 338–53.

Chow, T. S. (1973). "Simultaneous Translational and Rotational Brownian Movement of Particles of Arbitrary Shape," *Physics of Fluids* 16, 31–34.

Cohen, M. L., and Louie, S. G. (2016). *Fundamentals of Condensed Matter Physics* (Cambridge: Cambridge University Press).

Costantino, D. J., Bartell, J., Scheidler, K., and Schiffer, P. (2011). "Low-Velocity Granular Drag in Reduced Gravity," *Physical Review E* 83, 011305.

Daoud, M., and Van Damme, H. (1999). "Fractals". In *Soft Matter Physics*, edited by M. Daoud and C. E. Williams, 47–86. (Berlin: Springer).

Davis, J. C., et al. (1991). "Observation of Quantized Circulation in Superfluid 3He-B," *Physical Review Letters* 66, 329–32.

Deaver, B. S., Jr., and Fairbank, W. M. (1961). "Experimental Evidence for Quantized Flux in Superconducting Cylinders," *Physical Review Letters* 7, 43–46.

Delahaye, G., et al. (2000). "Comparison of Quantum Hall Effect Resistance Standards of the NIST and the BIPM," *Metrologia* 37, 173–6.

Des Cloiseaux, J., Conte, R., and Jannink, G. (1985). "Swelling of an Isolated Polymer Chain in a Solvent," *Journal de Physique Lettres* 46, L595–L600.

Doi, M. (2013). *Soft Matter Physics* (Oxford: Oxford University Press).

Doi, M., and Edwards, S. F. (1986). *The Theory of Polymer Dynamics* (Oxford: Clarendon Press).

Doll, R., and Näbauer, M. (1961). "Experimental Proof of Magnetic Flux Quantization in a Superconducting Ring," *Physical Review Letters* 7, 51–52.

Dugdale, S. B. (2016). "Life on the Edge: A Beginner's Guide to the Fermi Surface," *Physica Scripta* 91, 053009.

Dyson, F. J. (1967). "Ground-State Energy of a Finite System of Charged Particles", *Journal of Mathematical Physics* 8, 1538–45.

Dyson, F. J., and Lenard, A. (1967). "Stability of Matter. I," *Journal of Mathematical Physics* 8, 423–34.

Dyson, F. J., and Lenard, A. (1968). "Stability of Matter. II," *Journal of Mathematical Physics* 9, 698–711.

Edwards, S. F. (1965). "The Statistical Mechanics of Polymers with Excluded Volume," *Proceedings of the Physical Society* 85, 613–24.

Ehrenfest, P. (1959). *Collected Scientific Papers*, edited by M. J. Klein, 617–22. (Amsterdam: North-Holland). This citation is in an Address delivered in 1931 on the awarding of the Lorentz medal to Prof. W. Pauli and is in German. It appears in English translation in both Dyson (1967) and Lieb (1976).

Ezawa, Z. F. (2013). *Quantum Hall Effects: Recent Theoretical and Experimental Developments*, 3rd ed. (Hackensack, NJ: World Scientific).

Feynman, R. P., Leighton, R. B., and Sands, M. (1964). *The Feynman Lectures on Physics Vol II: Mainly Electromagnetism and Matter* (Reading: Addison-Wesley). See also the (2011) The New Millennium Edition, which is freely available online at https://www.feynmanlectures.caltech.edu/.

Florry, P. J. (1949). "The Configuration of Real Polymer Chains," *The Journal of Chemical Physics* 17, 303–10.

Frenkel, D. (2002). "Soft Condensed Matter," *Physica A* 313, 1–31.

Gennes, P-G. de (1979). *Scaling Concepts in Polymer Physics* (Ithaca, NY: Cornell University Press).

Gennes, P-G. de (1992). "Soft Matter," *Reviews of Modern Physics* 64, 645–8.

Gennes, P-G. de (2005). "Soft Matter: More than Words," *Soft Matter* 1, 16.

Girvin, S. M., and Yang, K. (2019). *Modern Condensed Matter Physics* (Cambridge: Cambridge University Press).

Gomes, M. A. F. (1987). "Fractal Geometry in Crumpled Paper Balls," *American Journal of Physics* 55, 649–50.

Hall, E. H. (1879). "On a New Action of the Magnet on Electric Currents," *American Journal of Mathematics* 2, 287–92.

Hartnoll, S. A., Kovtun, P. K., Müeller, M., and Sachdev, S. (2007). "Theory of the Nernst Effect near Quantum Phase Transitions in Condensed Matter and in Dyonic Black Holes," *Physical Review B* 76, 144502.

Hauge, E. H., and Martin-Löf, A. (1973). "Fluctuating Hydrodynamics and Brownian Motion," *Journal of Statistical Physics* 7, 259–81.

Jezouin, S., et al. (2013). "Quantum Limit of Heat Flow across a Single Electronic Channel," *Science* 342 601–4.

Jones, R. A. L. (2002). *Soft Condensed Matter* (Oxford: Oxford University Press).

Josephson, B. D. (1962). "Possible New Effects in Superconducting Tunnelling," *Physics Letters* 1, 252–3.

Josephson, B. D. (1965). "Supercurrents through Barriers," *Advances in Physics* 14, 419–51.

Kantor, Y., Kardar, M., and Nelson, D. R. (1986). "Statistical Mechanics of Tethered Surfaces," *Physical Review Letters* 57, 791–4.

Karn, P. W., Starks, D. R., and Zimmerman, W., Jr. (1980). "Observation of Quantization of Circulation in Rotating Superfluid ^{4}He," *Physical Review B* 21 1797–1805.

Kekez, D., Ljubičièc, A, and Logan, B. A. (1990). "An Upper Limit to Violations of the Pauli Exclusion Principle," *Nature* 348, 224.

Kim, Y., et al. (2023). "Evidence for the Utility of Quantum Computing before Fault Tolerance," *Nature* 618, 500–5.

Kittell, C. (1971). *Introduction to Solid State Physics* (New York: Wiley and Sons).

Kleman, M., and Lavrentovich, O. D. (2003). *Soft Matter Physics: An Introduction* (Berlin: Springer).

Lamb, H. (1945). *Hydrodynamics*, reproduction of 6th ed. (New York: Dover).

Landau, L. (1930). "Diamagnetismus der Metalle (Diamagnetism of Metals)," *Zeitschrift für Physik* 64, 629–37.

Langenberg, D. N., et al. (1966). "Microwave-Induced D.C. Voltages across Josephson Junctions," *Physics Letters* 20, 563–5.

Laughlin, R. B. (1981). "Quantized Hall Conductivity in Two Dimensions," *Physical Review B* 23, 5632–3.

Le Guillou, J. C., and Zinn-Justin, J. (1977). "Critical Exponents for the *n*-Vector Model in Three Dimensions from Field Theory," *Physical Review Letters* 39, 95–98.

Li, B., Madras, N., and Sokal, A. D. (1995). "Critical Exponents, Hyperscaling, and Universal Amplitude Ratios for Two- and Three-Dimensional Self-Avoiding Walks," *Journal of Statistical Physics* 80, 661–754.

Lieb, E. H. (1976). "The Stability of Matter," *Reviews of Modern Physics* 48, 553–69.

Lowe, C. P., Frenkel, D., and Masters, A. J. (1995). "Long-Time Tails in Angular Momentum Correlations," *The Journal of Chemical Physics* 103, 1582–7.

Makhlin, Y., Schön, G., and Shnirman, A. (2001). "Quantum-State Engineering with Josephson-Junction Devices," *Reviews of Modern Physics* 73, 357–99.

Marder, M. P. (2000). *Condensed Matter Physics* (New York: Wiley and Sons).

Matthews, M. R., et al. (1999). "Vortices in a Bose–Einstein Condensate," *Physical Review Letters* 83, 2498–2501.

McLeish, T. (2020). *Soft Matter: A Very Short Introduction* (Oxford: Oxford University Press).

Meschke, M., Guichard, W., and Pekola, J. P. (2006). "Single-Mode Heat Conduction by Photons," *Nature Letters* 444, 187–90.

Nagel, S. R. (2017). "Experimental Soft-Matter Science," *Reviews of Modern Physics* 89, 025002.

Newburgh, R., Peidle, J., and Ruckner, W. (2006). "Einstein, Perrin, and the Reality of Atoms: 1905 Revisited," *American Journal of Physics* 74, 478–81.

Nienhuis, B. (1982). "Exact Critical Point and Critical Exponents of *O(n)* Models in Two Dimensions," *Physical Review Letters* 49, 1062–1065.

Nittrouer, C. A., et al. (2021). "Amazon Sediment Transport and Accumulation along the Continuum of Mixed Fluvial and Marine Processes," *Annual Review of Marine Science* 13, 501–36.

Osman, A., et al. (2021). "Simplified Josephson-Junction Fabrication Process for Reproducibly High-Performance Superconducting Qubits," *Applied Physics Letters* 118, 064002.

Paalanen, M. A., Tsui, D. C., and Gossard, A. C. (1982). "Quantized Hall Effect at Low Temperatures," *Physical Review B* 25, 5566–9.

Peierls, R. (1955). *Quantum Theory of Solids* (New York: Oxford University Press).

Pendry, J. B. (1983). "Quantum Limits to the Flow of Information and Entropy," *Journal of Physics A: Mathematical and General* 16, 2161–71.

Perrin, M. J. (1910). *Brownian Movement and Molecular Reality* (London: Taylor and Francis). Translated from the *Annales de Chimie et de Physique*, 8th Series, September 1909 by F. Soddy.

Pfister, M., and Hager, W. H. (2014). "History and Significance of the Morton Number in Hydraulic Engineering," *Journal of Hydraulic Engineering* 140, 02514001.

Pilon, L., Fedorov, A. G., and Viskanta, R. (2001). "Steady-State Thickness of Liquid–Gas Forms," *Journal of Colloid and Interface Science* 242, 425–36.

Pines, D. (1956). "Collective Energy Losses in Solids," *Reviews of Modern Physics* 28, 184–98.

Poon, W. C. K. (2015). "Colloidal Suspensions." In *The Oxford Handbook of Soft Condensed Matter*, edited by E. M. Terentjev and D. A. Weitz, 1–50. (Oxford: Oxford University Press).

Poon, W. C. K., and Andelman, D. (2006). *Soft Condensed Matter Physics in Molecular and Cell Biology* (New York: Taylor & Francis).

Powell, C. J., and Swan, J. B. (1959a). "Origin of the Characteristic Electron Energy Losses in Aluminum," *Physical Review* 115, 869–75.

Powell, C. J., and Swan, J. B. (1959b). "Origin of the Characteristic Electron Energy Losses in Magnesium," *Physical Review* 116, 81–83.

Preskill, J. (2018). "Quantum Computing in the NISQ Era and Beyond," *Quantum* 2, 79.

Ramberg, E., and Snow, G. A. (1990). "Experimental Limit on a Small Violation of the Pauli Principle," *Physics Letters B* 238, 438–41.

Rayfield, G. W., and Reif, F. (1964). "Quantized Vortex Rings in Superfluid Helium," *Physical Review* 136(5A), A1194–A1208. See also Rayfield, G. W., and Reif, F. (1965). "Erratum," *Physical Review* 137, AB4.

Rego, L. G. C., and Kirczenow, G. (1998). "Quantized Thermal Conductance of Dielectric Quantum Wires," *Physical Review Letters* 81, 232–5.

Ryan, M., Gurian, J., Shandera, S., and Jeong, D. (2022). "Molecular Chemistry for Dark Matter," *The Astrophysical Journal* 934, 120.

Schwab, K., et al. (2000). "Measurement of the Quantum of Thermal Conductance," *Nature* 404 974–7.

Slade, G. (2019). "Self-Avoiding Walk, Spin Systems and Renormalization," *Proceedings of the Royal Society of London. Series A, Mathematical and Physical Sciences* 475, 20180549.

Snoke, D. W. (2009). *Solid State Physics: Essential Concepts* (San Francisco: Addison-Wesley).

Taylor, R. N., Parker, W. H., and Langenberg, D. N. (1969). "Determination of e/h, Using Macroscopic Quantum Phase Coherence in Superconductors: Implications for Quantum Electrodynamics and the Fundamental Constants," *Reviews of Modern Physics* 41, 375–496. See also Taylor, R. N., Parker, W. H., and Langenberg, D. N. (1973). "Erratum", *Reviews of Modern Physics* 45 109.

Terentjev, E. M., and Weitz, D. A. (2015). *The Oxford Handbook of Soft Condensed Matter* (Oxford: Oxford University Press).

Trachenko, K., and Brazhkin, V. V. (2020). "Minimal Quantum Viscosity from Fundamental Physical Constants," *Science Advances* 6, eaba3747.

Trachenko, K., and Brazhkin, V. V. (2021). "The Quantum Mechanics of Viscosity," *Physics Today* 74(12), 66–67.

Vinen, W. F. (1961). "The Detection of Single Quanta of Circulation in Liquid Helium II," *Proceedings of the Royal Society of London Series A: Mathematical and Physical Sciences* 260, 218–36.

von Klitzing. K. (1986). "The Quantized Hall Effect," Reviews of Modern Physics 58, 519–31. This reference contains his lecture from December 9, 1985, on the occasion of his accepting the Nobel Prize.

von Klitzing, K. (2017). "Metrology in 2019," *Nature Physics* 13, 198.

von Klitzing, K. (2019). "Essay: Quantum Hall Effect and the New International System of Units," *Physical Review Letters* 122, 200001.

von Klitzing, K., Dorda, G., and Pepper, M. (1980). "New Method for High-Accuracy Determination of the Fine-Structure Constant Based on Quantized Hall Resistance," Physical Review Letters 45, 494–7.

Wang, R., et al. (2021). "Modeling Study of Metallurgical Slag Foaming via Dimensional Analysis," *Metallurgical and Materials Transactions B* 52B, 1805–17.

Weisskopf, V. F. (1965). "In Defense of High-Energy Physics." In *Nature of Matter: Purposes of High-Energy Physics*, edited by L. C. L. Yuan, 24–27. (Upton, NY: Brookhaven National Laboratory Publications).

Whitmore, S. C., and Zimmerman, W., Jr. (1968). "Observation of Quantized Circulation in Superfluid Helium," *Physical Review* 166, 181–96.

Wiederseiner, S., et al. (2011). "Refractive-Index and Density Matching in Concentrated Particle Suspensions: A Review," Experiments in Fluids 50, 1183–1206.

Yoshioka, D. (2002). *The Quantum Hall Effect* (Berlin: Springer-Verlag).

Zhang, B., et al. (2011). "The Largest Synthetic Structure with Molecular Precision: Towards a Molecular Object," *Angewandte Chemie International Edition* 50, 737–40.

Zhu, J. X., et al. (1992). "Scaling of Transient Hydrodynamic Interactions in Concentrated Suspensions," *Physical Review Letters* 68, 2559–62.

Zieve, R. J. (2023). "Sixty Years of Quantized Circulation," *Journal of Low Temperature Physics* 212, 155–67.

Chapter 9

Astrophysics and gravitation

In the last two chapters, we focused on the nature of the fundamental constituents of nature and their interactions in quite different ways. In Chapter 7, using the lens of quantum mechanics, we examined electrons and protons at the atomic and molecular level (i.e., the very small) focusing on electromagnetic (EM) interactions, including treatments of the electricity and magnetism (E&M) interactions themselves using quantum theory, up to relativistic energies (i.e., the very fast.) We also (briefly) explored the impact of the strong and weak interactions at even smaller length scales, to acknowledge three of the four fundamental forces of nature. In Chapter 8, we concentrated on the relevance of quantum mechanics to the development of emergent phenomena in condensed matter systems, especially at the extremes of low temperatures.

We now turn our attention to topics in astrophysics, where we focus more on the gravitational interaction (both via Newton's law and Einstein's extensions to general relativity) in settings on larger scales (solar system to cosmological distances) and applied to more extreme parts of the phase diagrams of Nature (especially at high temperatures and huge density), requiring both classical physics and quantum mechanics. While physics and astronomy/astrophysics are sometimes organized in different academic units (here at my institution, for example, the departments are formally separate programs), we both use the same basic physical principles, just studying different aspects of our universe. Perhaps recalling two quotes from the Feynman lectures is a good reminder of the essential interplay of our two fields:

> Astronomy is older than physics. In fact, it got physics started by showing the beautiful simplicity of the motion of the stars and planets, the understanding of which was the beginning of physics. But the most remarkable discovery in all of astronomy is that *the stars are made of atoms of the same kind as those on the earth.*
>
> Even though we cannot reproduce the conditions on the earth, using the basic physical laws we often can tell precisely, or very closely, what will happen. So it is that physics aids astronomy (Feynman et al. 1963, section 3-4).

We begin, however, with an extended discussion of the physics of hot ionized gases (plasmas), which is of relevance to both disciplines, in applied and more fundamental ways.

9.1 Plasma physics and magneto-hydrodynamics

While plasmas now appear in many everyday situations (you may be reading this book with the help of a fluorescent light, with a plasma TV on in the background, and later going to a bar with a neon sign), and have important applications in industrial settings (see, e.g., Bonizzoni and Vassalo (2002) and Domonkos et al. (2021)), plasma physics is perhaps most closely associated with physical phenomena in geophysical or astrophysical contexts. Examples from our own planet include lightning, the aurorae (either *borealis* or *australis*), and the ionosphere, while stellar systems run

Dimensional Analysis Across the Landscape of Physics. Richard W. Robinett, Oxford University Press. © Richard W. Robinett (2024).
DOI: 10.1093/oso/9780192867551.003.0009

on an engine of hot ionized gas, with plasmas also emitted in the solar wind, and existing in the interplanetary and even interstellar media.

A related topic is that of **M**agneto-**H**ydro**D**ynamics (**MHD**), which covers the study of electrically conducting fluids (including plasmas) and their interactions with magnetic fields. Technologically important applications in engineering and metallurgy include the manipulation (magnetic stirring) of liquid metals and, of course, plasma confinement in support of nuclear fusion devices, while naturally occurring systems include the Earth's rotating molten core giving rise to the geomagnetic dynamo, planetary magnetospheres, and sunspots. For those reasons, we find it useful to include a focused discussion on plasma physics and related MHD phenomena in this chapter. Excellent discussions of both topics at a very accessible level include Clemmow and Dougherty (1969), Davidson (2001), Kulsrud (2005), Fitzpatrick (2015), and Chen (2018). Three very useful references, all self-described as *formularies*, have been produced by Diver (2001), Huba (2013), and Richardson (2019), each of which has a wealth of formulae and technical information, but not in a standard textbook form. We also note that both plasma and MHD systems often require the use of physical quantities "spelled" with all five basic M, L, T, Q, Θ dimensions, further illustrating the power of dimensional analysis (DA) methods.

In one of our first applications of DA, in Section 2.3 we analyzed plasma oscillations, deriving the plasma frequency as

$$\omega_P = C_P \left(\frac{n_e e^2}{m_e \epsilon_0} \right)^{1/2}, \tag{9.1}$$

which, as we show below, is the complete answer, since $C_P = 1$.

To confirm this, let us reproduce the standard derivation of the oscillation process, which results when a plasma is displaced by a small amount δx from equilibrium, as shown in Fig. 9.1. The $\pm\sigma_e$ (area) charge density excess on either side is related to the average electron number density[1] as $\sigma_e = \pm e n_e \delta x$, giving rise to an effective parallel-plate capacitor electric field (explored using DA in Section 4.2.3) as $|\boldsymbol{E}| = \sigma_e/\epsilon_0$. The resulting restoring force on any individual electron is given by

$$m_e \frac{d^2\delta x(t)}{dt^2} = m_e a = F = qE = (-e)\left(\frac{\sigma_e}{\epsilon_0}\right) = (-e)\left(\frac{e n_e \delta x(t)}{\epsilon_0}\right)$$

$$\frac{d^2\delta x(t)}{dt^2} = -\left(\frac{e^2 n_e}{m_e \epsilon_0}\right)\delta x(t) = -(\omega_p)^2 \delta x(t) \qquad \text{where} \qquad \omega_p = \sqrt{\frac{e^2 n_e}{m_e \epsilon_0}}, \tag{9.2}$$

which is the equation for simple harmonic motion, as expected. We find it useful to review this explicit example, where again for oscillatory motion the angular frequency ω (as opposed to $f = \omega/2\pi$) is the natural quantity (to which ordinary differential equations (ODE) or DA can be applied equally well), and that for electro-static applications where Gauss's law is invoked (as done here), there are often no factors of 4π accompanying the ϵ_0. The judicious use of such intuition can help inform DA predictions on whether or not to include factors of 2π in one case and 4π in the second.

This derivation shows that there are plasma oscillations, but since there is no k (wave-number) dependence to ω_P, the wave velocity given by $d\omega_P/dk$ vanishes, so no plasma waves. There is another contribution to the plasma oscillation frequency due to thermal effects (hence ω_{TH}) that does have a wavelength dependence, which we can explore by writing

[1] As it's the much lighter electrons that can move in response to any external perturbation, compared to the heavier ions.

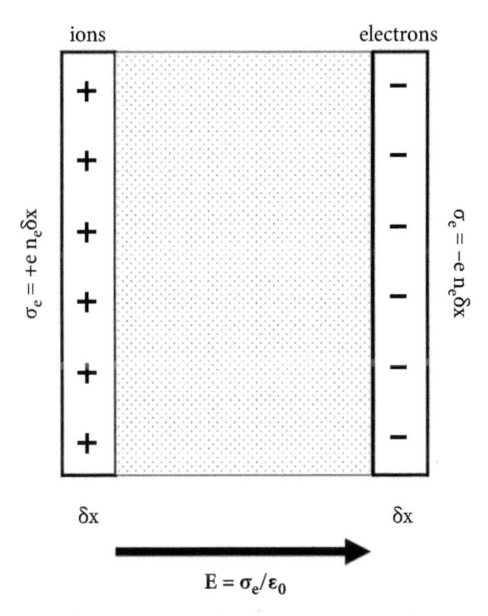

Fig. 9.1 A small displacement of electron density in a plasma leading to a restoring force, giving rise to plasma oscillations.

$$\omega_{TH} \propto k_B^\alpha T^\beta m_e^\gamma k^\delta$$
$$\frac{1}{\mathsf{T}} = \left(\frac{\mathsf{ML}^2}{\Theta\mathsf{T}^2}\right)^\alpha \Theta^\beta \mathsf{M}^\gamma \left(\frac{1}{\mathsf{L}}\right)^\delta, \tag{9.3}$$

which gives $\alpha = \beta = 1/2$, $\gamma = -1/2$, and $\delta = 1$, or $\omega_{TH} = (3k_BT/m_e)^{1/2}k = v_{TH}k$, where $v_{TH}^2 = 3k_BT/m_e$ is the usual thermal speed in three dimensions, hence the dimensionless 3. We thus have two distinct possible contributions to $\omega(k)$, and it turns out that the correct combination is to add them in **quadrature**, namely,

$$\omega^2 = \omega_P^2 + \omega_{TH}^2 = \omega_P^2 + 3\left(\frac{k_BT}{m}\right)^2 k^2, \tag{9.4}$$

which is called the **Bohm–Gross** (1949a, 1949b) **dispersion relation.** (A similar case of two distinct physical contributions to a dispersion relation being added in quadrature is that of gravity and capillary waves, as in Eqn. (3.78).)

In addition to this important time/frequency scale, we have already introduced in **Example 5.2** the possible thermal length scales (l_P) in plasmas, where we found in Eqn. (5.56) a general expression for the dependence of l_P on all of the likely relevant physical quantities, namely

$$l_P \propto k_B^\alpha T^\beta e^\gamma \epsilon_0^\delta n_e^\rho m_e^\sigma = \left(\frac{4\pi\epsilon_0 k_BT}{e^2}\right)^\alpha n_e^{-(1+\alpha)/3} m_e^0 = \left(\frac{4\pi\epsilon_0 k_BT}{e^2\, n_e^{1/3}}\right)^\alpha n_e^{-1/3} \equiv (\mathcal{R}_P)^\alpha n_e^{-1/3}, \tag{9.5}$$

where we **have** included the 4π factor this time. We now want to systematically explore the physical interpretations of several relevant combinations, and the meaning of the new dimensionless combination $\mathcal{R}_P$ that naturally occurs.

The simplest case is $\alpha = 0$ corresponding to

$$l_P = n_e^{-1/3} \equiv l_n, \tag{9.6}$$

which is simply the average inter-particle distance and so would be relevant for any gas. Another physically obvious combination corresponds to $\alpha = -1$ where

$$l_P = \frac{e^2}{4\pi\epsilon_0 k_B T} = l_{ca} = l_L, \tag{9.7}$$

which we associate with the **distance of closest approach** (l_{ca}) or the **Landau** distance (l_L) for two particles determined by when their kinetic energy $k_B T$ equals their mutual Coulomb electro-static repulsion.

The dimensionless ratio $\mathcal{R}_P$ is then seen to encode information on the ratio of the average (l_n) to closest ($l_{ca} = l_L$) distances between electrons, since

$$\mathcal{R}_P = \left(\frac{4\pi\epsilon_0 k_B T}{e^2\, n_e^{1/3}}\right) = \frac{l_n}{l_L}. \tag{9.8}$$

As one numerical example, for the Earth's ionosphere, one has $n_e \sim 10^{12}\ m^{-3}$ and $T \sim 10^3\ K$, which gives $\mathcal{R}_P \sim 6 \times 10^3$, and many plasmas are similarly characterized by $\mathcal{R}_P >> 1$, as you can confirm in **P9.1.**

An important combination that seemingly has the most "symmetric" use of all of the variables corresponds to $\alpha = 1/2$, giving

$$l_P = \left(\frac{4\pi\epsilon_0 k_B T}{e^2 n_e}\right)^{1/2} \propto \left(\frac{\epsilon_0 k_B T}{e^2 n_e}\right)^{1/2} \equiv l_D, \tag{9.9}$$

which is the **Debye length.** It's instructive to review the derivation leading to this length scale by solving (as usual, somewhat schematically) the basic electro-static equations in a region with a non-zero charge density, ρ_e. We also assume that the thermal kinetic energy is much larger than electric effects, namely that $e\phi(\mathbf{r}) << k_B T$.

Using the Laplace formulation for the electric potential, and the same Maxwell–Boltzmann thermal distribution approach as for **Example 5.3,** we write

$$\nabla^2\phi(\mathbf{r}) = \frac{\rho_e(\mathbf{r})}{\epsilon_0} = \frac{1}{\epsilon_0}\left[en_e\left\{e^{e\phi(\mathbf{r})/k_B T} - 1\right\}\right] \approx \left(\frac{e^2 n_e}{\epsilon_0 k_B T}\right)\phi(\mathbf{r}) = \left(\frac{1}{l_D}\right)^2 \phi(\mathbf{r}), \tag{9.10}$$

where

$$l_D \equiv \sqrt{\frac{\epsilon_0 k_B T}{e^2 n_e}} \quad \text{so that for a point charge Q inside the plasma} \quad \phi(\mathbf{r}) = \frac{Q}{4\pi\epsilon_0 r} e^{-r/l_D}. \tag{9.11}$$

The Debye length therefore governs the **largest** distance scale over which charged particles can effectively interact, since for $r >> l_D$, the respective charges are shielded (exponentially so) from one another as the plasma is polarized by the external charge and rearranges itself in response.

At the other extreme, the **shortest** relevant length scale is the **distance of closest approach** or $l_{ca} = l_L$, where the Coulomb and thermal energies balance, so the dynamic range of lengths scales (max to min ratio) relevant for many plasma interactions is

$$\frac{l_{max}}{l_{min}} = \frac{l_D}{l_{ca}} = \frac{l_D}{l_L} \sim \mathcal{R}_P^{3/2} >> 1. \tag{9.12}$$

Another of the oft-cited variables is the **plasma parameter,** which is simply the number of particles confined within a spherical region with the Debye radius, namely

$$N_{Debye\,sphere} \equiv \Lambda = n_e\left(\frac{4\pi}{3}l_D^3\right) \sim n_e\left(\frac{\epsilon_0 k_B T}{e^2 n_e}\right)^{3/2} = \left(\frac{\epsilon_0 k_B T}{e^2 n_e^{1/3}}\right)^{3/2} = \mathcal{R}_P^{3/2}, \tag{9.13}$$

giving another hint as to the importance of the $\mathcal{R}_P$ parameter.

A final case of Eqn. (9.5) is that of $\alpha = 2$, where

$$l_P = \left(\frac{4\pi\epsilon_0 k_B T}{e^2 n_e^{1/3}}\right)^2 n_e^{-1/3} = \frac{1}{n_e}\left(\frac{4\pi\epsilon_0 k_B T}{e^2}\right)^2 \sim \frac{1}{n_e(l_L)^2} \sim \frac{1}{n_e\sigma_{ca}} \sim l_{mfp}, \tag{9.14}$$

where we associate $(l_{ca})^2$ with a scattering cross-section (σ_{ca}), which in turn gives a length scale equivalent to a **mean free path,** l_{mfp}, as in Eqn. (5.90).

Collecting all of these results, we find a hierarchy of relevant length scales for plasma physics characterized by powers of $\mathcal{R}_P$ as

$$\begin{array}{ccccccc} l_L & : & l_n & : & l_D & : & l_{mfp} \\ (\mathcal{R}_P)^{-1} & : & 1 & : & (\mathcal{R}_P)^{1/2} & : & (\mathcal{R}_P)^2, \end{array}$$

and we see that the Buckingham Π theorem ratio $\mathcal{R}_P$ that "pops out" is useful in spanning multiple decades of distances, over a range $(\mathcal{R}_P)^3$. We also note the similarity to the EM length scales involving powers of $\hbar, m_e, e, \epsilon_0$ discussed in Section 7.1, all related to each other by powers of the fine-structure constant α_{FS}.

Early in Chapter 4, we considered the impact that time-dependent and inhomogeneous electric fields can have in generating forces on charged particles. We found that the **ponderomotive force** $\boldsymbol{F}_p$ of an oscillating electric field of the form $\boldsymbol{E}(\boldsymbol{r}, t) = \boldsymbol{E}_0(\boldsymbol{r})\cos(\omega t)$ was

$$\boldsymbol{F}_p = -\frac{e^2}{2m_e\omega^2}\nabla\left(\boldsymbol{E}_0(\boldsymbol{r})\cdot\boldsymbol{E}_0(\boldsymbol{r})\right) = -\nabla U_p(\boldsymbol{r}), \qquad \text{where} \qquad U_p \equiv \frac{e^2|\boldsymbol{E}_0(\boldsymbol{r})|^2}{2m_e\omega^2} \tag{9.15}$$

is sometimes called the **ponderomotive potential.** We note that a similar type of effect giving a net force due to an inhomogeneous oscillating <u>acoustic</u> field is also possible, with clear similarities; see **P9.4.**

A related application is **<u>l</u>aser-driven plasma-based <u>w</u>ake<u>f</u>ield <u>a</u>cceleration**[2] (or LWFA) whereby electrons that find themselves in a plasma wave can experience forces larger than traditional radiofrequency (RF) electric field driven methods. The original proposal by Tajima and Dawson (1979) has led to many experimental realizations (see, e.g., Buck et al. 2011) and forefront research continues (see, e.g., Götzfried et al. 2020, or Miao et al. 2022), often focusing on

[2] For reviews, see Esarey et al. (2009) or Tajima et al. (2020).

the development of such techniques for future ultra-high-energy particle accelerators, and some authors[3] cite plasmas as the "... accelerator medium of the future ...". All of this work relies on a few novel concepts involving the interaction of the laser and plasma parameters, many of which are amenable to a DA analysis.

Example 9.1 Laser driven electron acceleration—Electric field values

An important baseline electric field (E_0) value related to the plasma parameters in an LWFA system is given by the one-parameter family constrained by DA as

$$E_0 \propto m_e^{\alpha} c^{\beta} n_e^{\gamma} e^{\delta} \epsilon_0^{\sigma}$$
$$\frac{\mathsf{ML}}{\mathsf{QT}^2} = \mathsf{M}^{\alpha} \left(\frac{\mathsf{L}}{\mathsf{T}}\right)^{\beta} \left(\frac{1}{\mathsf{L}^3}\right)^{\gamma} \mathsf{Q}^{\delta} \left(\frac{\mathsf{Q}^2\mathsf{T}^2}{\mathsf{ML}^3}\right)^{\sigma}$$
$$E_0 \propto m_e^{2-3\gamma} c^{4-6\gamma} n_e^{\gamma} e^{-3+6\gamma} \epsilon_0^{1-3\gamma}, \tag{9.16}$$

and we note that the electron mass and speed of light suggestively appear in the combination $m_e c^2$. To better constrain the value of γ, we can make some assumptions about how E_0 depends on various parameters. We might expect that the larger the value of $m_e c^2$ the higher the allowed field, which suggests that $2 - 3\gamma > 0$ or $2/3 > \gamma$. If the strength of the fundamental electro-static force is larger, meaning that $1/\epsilon_0$ increases, we could expect E_0 to increase, and that implies that $1 - 3\gamma < 0$ or $\gamma > 1/3$. Combining these we find the limits $2/3 > \gamma > 1/3$, which does bracket the correct solution corresponding to $\gamma = 1/2$, or

$$E_0 = \sqrt{\frac{n_e(m_e c^2)}{\epsilon_0}}. \tag{9.17}$$

A perhaps useful mnemonic device that can be associated with this relationship is to rewrite it in the form $\epsilon_0 E_0^2 = n_e(m_e c^2)$, where the quantities on either side are energy densities, that stored in the electric field (on the left) and in the rest (mass) energy of the electrons (on the right). Another version of Eqn. (9.17), this one in terms of the plasma frequency, is

$$E_0 = \sqrt{\frac{n_e(m_e c^2)}{\epsilon_0}} = \frac{m_e c}{e}\sqrt{\frac{n_e e^2}{m_e \epsilon_0}} = \frac{(m_e c)\,\omega_P}{e}, \tag{9.18}$$

which can be expressed as $eE_0 = (m_e c)\omega_P$, where both sides now have the dimensions of force (electric on the left, and a "dp/dt" form on the right).

Turning now to an important result in magneto-hydrodynamics, we can explore the possible wave motion of charged particles in external magnetic fields.

Example 9.2 Alfvén waves

It can sometimes be useful to bring the tools of DA to bear as a first step when trying to understand background material one finds about a new physical problem, perhaps a subject that one is interested

[3] Miao et al. (2023).

in exploring in more depth or rigor. Such prefatory comments might be found in the introduction of a research paper, or even in a very general "public science" presentation on the internet. As just such an example, we note that Wikipedia has the following description[4] of **Alfvén waves**:

> An Alfvén wave in a plasma is a low-frequency (compared to the ion cyclotron frequency) traveling oscillation of the ions and the magnetic field.
>
> - The ion mass density provides the inertia and the magnetic field line tension provides the restoring force.
> - The wave propagates in the direction of the magnetic field, although waves exist at oblique incidence and smoothly change into the magnetosonic wave when the propagation is perpendicular to the magnetic field.
> - The motion of the ions and the perturbation of the magnetic field are in the same direction and transverse to the direction of propagation.
> - The wave is dispersionless.

Let us explore to what extent we can flesh out some of these statements, using just DA methods.

To first explore the speed of the wave (v_A), let us consider using the quantities B_0 (the ambient magnetic field), the basic constant of magnetism (μ_0), the ion mass density (ρ_m), and the fundamental electric charge (e), and then write

$$v_A \propto B_0^\alpha \mu_0^\beta \rho_m^\gamma e^\delta$$

$$\frac{\mathsf{L}}{\mathsf{T}} = \left(\frac{\mathsf{M}}{\mathsf{QT}}\right)^\alpha \left(\frac{\mathsf{ML}}{\mathsf{Q}^2}\right)^\beta \left(\frac{\mathsf{M}}{\mathsf{L}^3}\right)^\gamma \mathsf{Q}^\delta, \tag{9.19}$$

and we easily find that $\alpha = 1$, $\beta = \gamma = -1/2$, and $\delta = 0$. We can then write

$$v_A = \frac{B_0}{\sqrt{\mu_0 \rho_m}} = \sqrt{\frac{B_0^2/\mu_0}{\rho_m}} = \sqrt{\frac{2U_B}{\rho_m}}, \tag{9.20}$$

where $U_B = B_0^2/(2\mu_0)$ is the energy density of the magnetic field. This relation can also be expressed as $\rho_m v_A^2/2 = U_B$, with U_B balancing the kinetic energy density. This type of association helps confirm the first bullet, especially if we compare to the expression for sound velocity of a gas, given by $v_s = \sqrt{\gamma P/\rho_m}$ where P is the ordinary gas pressure, so that U_B is indeed providing the restoring force, in the form of **magnetic pressure.** (It's perhaps interesting to rewrite Eqn. (9.17) yet again, this time in the form $c = \sqrt{(\epsilon_0 E_0^2)/(n_e m_e)} = \sqrt{2U_E/\rho_m}$ for comparison to Eqn. (9.20), where in this case the electric field energy density/pressure, U_E, appears.)

To explore whether the waves exhibit dispersion (i.e., if the wave speed depends on k or λ), you can repeat the steps in Eqn. (9.19), but substitute $e \to k$, and find that δ still vanishes, so v_A does **not** depend on wavelength. One can also see this by exploring the dimensions of the wave frequency (ω_A) and instead of assuming that ω_A depends on e (which it seems not to), assume instead that it might depend on the wave-number, $k = 2\pi/\lambda$, which then gives you the dimensions of k if you need them. Writing

$$\omega_A \propto B_0^\alpha \mu_0^\beta \rho_m^\gamma k^\delta$$

$$\frac{1}{\mathsf{T}} = \left(\frac{\mathsf{M}}{\mathsf{QT}}\right)^\alpha \left(\frac{\mathsf{ML}}{\mathsf{Q}^2}\right)^\beta \left(\frac{\mathsf{M}}{\mathsf{L}^3}\right)^\gamma \left(\frac{1}{\mathsf{L}}\right)^\delta, \tag{9.21}$$

[4] See https://en.wikipedia.org/wiki/Alfvén_wave. Accessed 7/31/2023.

we actually find the same α, β, γ as in Eqn. (9.20), but now with $\delta = 1$, namely

$$\omega_A = \left(\frac{B_0}{\sqrt{\mu_0 \rho_m}} \right) k = v_A k, \tag{9.22}$$

which is the **linear dispersion relation** for a wave with a k-independent velocity $v_A = d\omega(k)/dk = \sqrt{B_0^2/\mu_0 \rho_m}$, also confirming that it's dispersionless.

A complete derivation of Alfvén waves at an accessible undergraduate level is given by Reitz, Milford, and Christy (2008), but the original paper by Alfvén (1942) is very readable, but in CGS units. Some hint of the wide relevance of magneto-hydrodynamics to astrophysical phenomena can be glimpsed in *Cosmical Electrodynamics* (Alfvén 1950), which discusses (but again in CGS units) applications in such areas as solar physics, magnetic storms and the aurora, and cosmic ray physics.

Experimental observations of Alfvén waves in terrestrial lab settings, in a variety of physical systems and materials, have been described by authors such as Jephcott (1959), Motz (1966), Alboussiére et al. (2011), and Schroeder et al. (2021). Evidence for Alfvén waves *"strong enough to power the solar wind"* were presented in three back-to-back-to-back papers in Science by de Pontieu et al. (2007), Okamoto et al. (2007), and Cirtain et al. (2007).

9.2 **Stellar structure**

Starting as early as in introductory level courses in earth sciences, the use of classical mechanics, gravitation, and thermal physics to describe natural phenomena such as hydrodynamics, geosciences, and environmental physics, in a quantitative manner, for the "physics under our feet," is an important extension of those topics outside the laboratory. Looking upwards, the same concepts can be applied to atmospheric science and meteorology, although to gases, instead of solids or (very viscous) liquids. Beyond the Earth, the rocky or gaseous planetary neighbors in our own solar system provide a natural extension of the study of our own planet. But perhaps the most obvious candidate for the complex physical modeling using such ideas is the study of our own Sun, and by extension, other stellar objects.

Many texts on stellar structure start with four fundamental equations that encode most of the important physical constraints relevant for determining the observable properties of stars. One can imagine measuring (by remote observation, of course) quantities such as the radius and total mass of a star, and its energy production (total luminosity or power output, fueled by complex nuclear fusion reactions), and infer through modeling the internal structure leading to the external observables.

The simplest such models assume spherical symmetry and that the system is in hydrostatic equilibrium (i.e., no time-dependence), so the obvious properties, such as (i) $M(r)$ or the mass within the radius r, (ii) the pressure $P(r)$, (iii) the luminosity $L(r)$, and (iv) the temperature profile $T(r)$, are all functions of radius alone. Imagining that we're starting with only that level of background, we note that the four basic equations of stellar structure reduce to the following:

$$\textbf{1: Mass balance} \quad \frac{dM(r)}{dr} = 4\pi r^2 \rho_m(r) \tag{9.23}$$

$$\textbf{2: Pressure balance} \quad \frac{dP(r)}{dr} = -\frac{GM(r)\rho_m(r)}{r^2} \tag{9.24}$$

$$\textbf{3: Energy production} \quad \frac{dL(r)}{dr} = 4\pi r^2 \rho_m(r)\,\epsilon \tag{9.25}$$

$$\textbf{4: Thermal transport} \quad \frac{dT(r)}{dr} = -\frac{3\chi\rho_m(r)}{64\pi\sigma T(r)^3 r^2} L(r). \tag{9.26}$$

The first relation in Eqn. (9.23) describes the small amount of mass, $dM(r)$, in a thin spherical shell of thickness dr in terms of the local mass density, $\rho_m(r)$, via

$$\begin{aligned} dM(r) &= dV\rho_m(r) = (4\pi r^2\, dr)\rho_m(r) \\ \mathsf{M} &\stackrel{!}{=} \mathsf{L}^3(\mathsf{M}/\mathsf{L}^3), \end{aligned} \tag{9.27}$$

and so is simply related to the definition of density. The second relation in Eqn. (9.24) represents **hydrostatic equilibrium**, with the net outward pressure, whether due to hot gas or photon radiation (or even quantum zero-point energy, as in the next section) on that thin shell, balanced by the gravitational attraction of the mass inside the shell, namely

$$\begin{aligned} (dP)(4\pi r^2) = F_P^{(out)} = \quad & F_G^{(in)} = \frac{GM(r)dM(r)}{r^2} = \frac{GM(r)[(4\pi r^2 dr)\rho_m(r)]}{r^2} \\ (\mathsf{M}/\mathsf{LT}^2)(\mathsf{L}^2) = \mathsf{ML}/\mathsf{T}^2 \stackrel{!}{=} \quad & \mathsf{ML}/\mathsf{T}^2 = (\mathsf{L}^3/\mathsf{MT}^2)\mathsf{M}[\mathsf{L}^3(\mathsf{M}/\mathsf{L}^3)]/\mathsf{L}^2. \end{aligned} \tag{9.28}$$

So far, we have used familiar symbols/notations, concepts, and of course dimensions, to review these two relations.

The energy production equation in Eqn. (9.25) can be written as

$$dL(r) = [(4\pi r^2 dr)\rho_m(r)]\epsilon = dM(r)\epsilon \qquad \text{or} \qquad \frac{dL(r)}{dM(r)} = \epsilon, \tag{9.29}$$

which corresponds to ϵ being the **power per unit mass** generated by whatever nuclear fusion reactions are taking place in the local environment. In terms of dimensions, this requires that $[\epsilon] = [dL/dM] = \{(\mathsf{ML}^2)/\mathsf{T}^3\}/\mathsf{M} = \mathsf{L}^2/\mathsf{T}^3$. This term is sometimes written as $\epsilon = \epsilon_{nuc} - \epsilon_\nu$, where e_ν is the power production that goes into neutrinos, and those particles escape the star without much/any interaction and so don't contribute to the energy transport in the final equation. Those neutrinos <u>do</u> reach earth and can be detected (see **P9.23**) and can be used to probe the solar energy production mechanisms in detail, as well as finding new physics about the particles[5] themselves!

The final relation in Eqn. (9.26) contains some familiar quantities such as the Stefan–Boltzmann constant σ and explicit temperature ($T(r)$) dependence, but also a new quantity called the **opacity**, here labeled κ (but sometimes cited as χ in the literature), which we review first. For many scattering processes, such as particles traversing matter from which they can scatter or be absorbed (with a cross-section for either process labeled as σ, not to be confused with the Stefan–Boltzmann constant), the loss of intensity as such a beam travels through a sample can be described in two ways, either in terms of the number density (n) or mass density (ρ_m) of the target material, namely

[5] Including the phenomena of neutrino oscillations and providing bounds on neutrino masses; see **P9.24**.

$$dI = -(\sigma n\, dx)I \qquad\qquad dI = -(\kappa \rho_m\, dx)I$$
$$I(x) = I_0 e^{-\sigma n x} \qquad\qquad I(x) = I_0 e^{-\kappa \rho_m x}. \tag{9.30}$$

The opacity then has dimensions given by $[\kappa] = 1/[\rho_m x] = (\mathsf{L}^3/\mathsf{M})(1/\mathsf{L}) = \mathsf{L}^2/\mathsf{M}$. In both cases, the typical length after which the beam is attenuated can be associated with the mean free path given by $l_{mfp} = 1/(\sigma n) = 1/(\kappa \rho_m)$.

To connect Eqn. (9.26) to earlier discussions of thermal transport, we recall the Fourier expression for the heat flow due to a temperature gradient and connect it to the spherical geometry considered here, via

$$\frac{dQ}{dt} = KA\frac{dT}{dx} \qquad \longrightarrow \qquad L(r) = K(4\pi r^2)\frac{dT(r)}{dr}, \tag{9.31}$$

where K is the coefficient of thermal conductivity. We can then use results from Chapter 5 and kinetic theory (Eqn. (5.97)) to associate K with

$$K = \frac{1}{3}\overline{v} C \lambda_{mfp}, \tag{9.32}$$

where for photons we use $\overline{v} = c$ as the average speed and $\lambda_{mfp} = 1/(\kappa \rho_m)$. For the specific heat C, we assume a gas of photons for which the energy density (from Section 6.2) is (here σ **is** the Stefan–Boltzmann constant)

$$U_T = aT^4 = \frac{4\sigma}{c}T^4 \qquad \text{implying} \qquad C = \frac{\partial U_T}{\partial T} = 4aT^3 = \frac{16\sigma}{c}T^3 \tag{9.33}$$

with $a = 4\sigma/c$, and both forms are used in the astrophysical literature. Putting these together, we have

$$K = \frac{16}{3}\frac{\sigma}{\kappa \rho_m}T^3 \qquad \longrightarrow \qquad \frac{L(r)}{4\pi r^2} = \left(\frac{16}{3}\frac{\sigma}{\kappa \rho_m}T^3\right)\frac{dT}{dr} \tag{9.34}$$

or

$$\frac{dT(r)}{dr} = -\frac{3}{64\pi}\left(\frac{\kappa \rho_m}{\sigma T^3}\right)\frac{L(r)}{r^2}$$
$$\frac{\Theta}{\mathsf{L}} \stackrel{?}{=} \left(\frac{\mathsf{L}^2}{\mathsf{M}}\right)\left(\frac{\mathsf{M}}{\mathsf{L}^3}\right)\left(\frac{\mathsf{M}}{\mathsf{T}^3\Theta^4}\right)^{-1}(\Theta)^{-3}\left(\frac{\mathsf{ML}^2}{\mathsf{T}^3}\right)(\mathsf{L})^{-2}$$
$$\Theta\,\mathsf{L}^{-1} \stackrel{!}{=} \mathsf{M}^{-1+1-1+1}\,\mathsf{L}^{2-3+2-2}\,\mathsf{T}^{3-3}\,\Theta^{4-3} = \mathsf{M}^0\,\mathsf{L}^{-1}\,\mathsf{T}^0\,\Theta^1. \tag{9.35}$$

There are then four important coupled differential equations determining the structure of stars, with boundary conditions at both $r = 0$ and $r = R_*$, which govern the mass, pressure, luminosity, and temperature profiles as functions of the radial distance, all which have to be solved numerically. In addition, many of the material properties required for a physical description of the system, namely P, ϵ, κ, are not just r-dependent, but also require information on the local density, temperature, and even chemical composition, as

$$P = P(\rho_m, T, X_i), \qquad \epsilon = \epsilon(\rho_m, T, X_i), \qquad \text{and} \qquad \kappa = \kappa(\rho_m, T, X_i), \tag{9.36}$$

where X_i describes the amount of each element, with X_1 for hydrogen, X_2 for helium, etc. The first relation is equivalent to specifying an equation of state for the gas, while the second encodes information on the nuclear fusion processes, with their interaction cross-sections highly dependent on temperature via the Coulomb barrier terms, as in Section 6.4.1.

Given all of this information, it would seem that a solution is determined, and there is a "folk theorem" due to Russell and Vogt that suggests that[6]

> ... the structure of a star, in hydrostatic and thermal equilibrium with all energy derived from nuclear reactions, is uniquely determined by its mass and the distribution of chemical elements throughout its interior.

While this is not a rigorous result, as there are a number of counter-examples related to both uniqueness or existence of a solution, computer solutions of these equations (supplemented by even more physics) do form the basis of most stellar structure calculations. Given that we're certainly not going to pursue detailed numerical modeling, let us instead use these four equations to see if we can get some rough sense of how stellar properties might scale.

Example 9.3 Scaling laws

Instead of solving the important differential equations directly, let us use them in the spirit of matching dimensions for the final stellar quantities, such as M_*, R_*, P_*, T_*. For example, from Eqn. (9.24) we can write

$$\frac{P}{R} \sim \frac{GM\rho_m}{R^2} \sim \frac{GM^2}{R^5} \qquad \text{or} \qquad P_* \sim \frac{GM_*^2}{R_*^4}, \tag{9.37}$$

which can be used to estimate a lower bound on the central pressure of the star (recall **P3.15**).

If we assume an equation of state, such as the ideal gas law, we can equate the pressure with temperature via

$$PV = n\mathcal{R}T \quad \text{where} \quad n = \frac{m}{\mu} \qquad \longrightarrow \qquad P = \frac{\rho_m \mathcal{R}T}{\mu}, \tag{9.38}$$

where n and m are the number of moles and mass of gas, related by the molecular weight μ, and $\mathcal{R} \equiv N_A k_B$ is the gas constant, with a non-standard notation to distinguish it from radius. This gives approximate scaling relations via

$$\left(\frac{M_*}{R_*^3}\right)\frac{\mathcal{R}T}{\mu} = \underbrace{\frac{\rho_m \mathcal{R}T}{\mu}} = P = \frac{GM_*^2}{R_*^4} \qquad \text{or} \qquad T_* \sim \left(\frac{\mu G}{\mathcal{R}}\right)\frac{M_*}{R_*}, \tag{9.39}$$

which can also be used to estimate central stellar temperatures. The thermal transport relation in Eqn. (9.26) can be scaled as

$$\frac{T}{R} \sim \left(\frac{\kappa\rho_m}{\sigma}\right)\frac{1}{T^3}\frac{L}{R^2} \qquad \text{or} \qquad L \sim \frac{T^4 R}{\rho_m}\left(\frac{\sigma}{\chi}\right) \sim \frac{(TR)^4}{M}\left(\frac{\sigma}{\chi}\right). \tag{9.40}$$

---

[6] https://en.wikipedia.org/wiki/Vogt-Russell_theorem. Accessed 8/18/2023.

Combining this with Eqn. (9.39) we have

$$\begin{aligned} L &= \frac{(TR)^4}{M}\left(\frac{\sigma}{\kappa}\right) \\ &= \left(\frac{R^4}{M}\right)\left\{\left(\frac{\mu G}{\mathcal{R}}\right)^4 \frac{M^4}{R^4}\right\}\left(\frac{\sigma}{\kappa}\right) \\ L_* &\sim \left[\frac{\sigma}{\kappa}\left(\frac{\mu G}{\mathcal{R}}\right)^4\right] M_*^3 \\ L_* &\propto M_*^3, \end{aligned} \tag{9.41}$$

which is a **mass–luminosity** relation. No single power-law spectrum correctly describes the $L_* \propto (M_*)^p$ behavior, but values in the range $p \sim 3.5$–4 cover much of the range of possible masses for stars on the main-sequence. It's then an interesting question about how the dimensionalities of the various factors like σ, κ, μ, G, and $\mathcal{R}$ enter into such a scaling relation when the exponent is not an integer.

9.3 White dwarf and neutron stars

Once the nuclear fusion processes which power a star during its active lifetime are exhausted, it's natural to ask about the fate of what some texts (such as Longair 2011) call **dead stars.** We'll see that in such systems there is an interplay of different mass scales (M_*, m_p, m_e), with classical gravitation attraction, now balanced by quantum mechanical effects, in both non-relativistic and relativistic limits. Besides involving what we might call $\hbar$ physics (de Broglie wavelength and uncertainty principle-type arguments), such systems provide dramatic examples of the central role played by the Pauli principle for both electrons and neutrons, which even touch on the number of space-time dimensions. We do not describe the detailed calculations of the type needed for real stellar structure results for such collapsed systems (just as we avoided them for "active" stars), but focus instead on results from DA to the extent we can.

Instead of focusing on the "big four" equations of stellar structure in Eqns. (9.23)–(9.26), we use energy methods, starting with the total gravitational energy and how it scales with G, M_* and R_*, namely

$$E_G \sim -\frac{GM_*^2}{R_*}. \tag{9.42}$$

As counter-intuitive as it may sound, we note that electrons will exhibit a zero-point energy when confined in a region even of a size as large as $L = R_*$, with individual quantum states (labeled by the quantum number n) scaling as

$$E_Q(n) \sim \frac{p_n^2}{2m_e} \sim \frac{\hbar^2 n^2}{m_e R_*^2} \qquad \longrightarrow \qquad E_Q(N_e) \sim +\frac{\hbar^2}{m_e R_*^2} F(N_e), \tag{9.43}$$

giving a quantum contribution to the overall energy balance for which the dimensions are dictated by $\hbar$-physics, but with a dimensionless function of the total number of electrons (N_e), $F(N_e)$, which (as we'll see) plays an absolutely crucial role. The positive contribution of this term gives a radially outward force called the **electron degeneracy pressure**, which can balance the (attractive) gravitational force/energy.

To estimate the total number of electrons contributing to this effect, we assume for simplicity that the star is initially only hydrogen (so equal numbers of electrons and protons), and that the mass of the star is due to the protons, giving $N_e = N_p = M_*/m_p$. To set scales, we'll often use a solar mass $M_* = M_\odot \equiv M_S = 2 \times 10^{30}\ kg$, in which case $N_e \sim 10^{57}$.

Before pursuing the form of $F(N_e)$, let us continue to focus on dimensional quantities and write the total energy, noting that the different R_* dependences ensure that there will be a (stable) minimum, namely

$$E(R_*) \sim \frac{\hbar^2}{m_e R_*^2} F(N_e) - \frac{GM_*^2}{R_*}$$

$$0 = \frac{\partial E(R_*)}{\partial R_*} \sim -\frac{2\hbar^2}{m_e R_*^3} F(N_e) + \frac{GM_*^2}{R_*^2}$$

$$\downarrow$$

$$\frac{GM_*^2}{R_*^2} \sim \frac{\hbar^2}{m_e R_*^3} F(N_e)$$

$$\downarrow$$

$$R_* \sim \left(\frac{\hbar^2}{GM_*^2 m_e}\right) F(N_e) = R_0 F(N_e), \tag{9.44}$$

where $R_0 \sim 5 \times 10^{-89}\ m$ (an astonishingly small number!) comes from the physical quantities with explicit dimensions.

The simplest assumption on the behavior of $F(N_e)$ would be to have $F(N_e) = N_e$, implying that all of the electrons are allowed to be in their ground state, which gives a white dwarf radius of $R_{WD} = R_0 N_e \approx 5 \times 10^{-32}\ m$, a value much closer to the Planck length than any physically relevant astrophysical distance scale. This assumption would be valid for bosons, where any number can exist in the lowest/ground state, as for Bose–Einstein condensates (BECs), but not relevant for fermions (either the electrons we're considering here first, or the neutrons we consider next). DA, using just $\hbar$ physics, can only take us so far in this problem and we need to implement the spin-statistics theorem or Pauli principle, even if only in an approximate manner, respecting all of the constraints of DA along the way, to evaluate $F(N_e)$.

To correctly take into account the Pauli principle that only two spin-1/2 particles are allowed in any quantum state (one spin-up, one spin-down), we have to carefully enumerate the number of allowed "spaces," so, more counting. We start by doing a one-dimensional example, using only "particle-in-a-box" notions. To see what the maximum quantum number (n_F for Fermi) allowed is for a given total number of spin-1/2 fermions (here N_e for electrons), we first write

$$N_e = \sum_{n=1}^{n_F} 2 \quad \rightarrow \quad N_e = 2\int_0^{n_F} dn = 2n_F \qquad \text{or} \qquad n_F = \frac{N_e}{2}, \tag{9.45}$$

where we use the fact that $N_e, n_F >>> 1$ to approximate the discrete sums as continuous integrals. This is just the statement that two electrons (neutrons) can be in a given state (spin-up and spin-down).

The total quantum energy, with two states for each allowed energy eigenvalue, is then

$$E_Q = \sum_{n=1}^{n_F} E_n = \sum_{n=1}^{n_F} 2\left(\frac{\hbar^2\pi^2 n^2}{2mL^2}\right) \to 2\left(\frac{\hbar^2\pi^2}{2mL^2}\right)\int_0^{n_F} (n^2)\, dn$$
$$\sim \frac{\hbar^2\pi^2}{mL^2}\left(\frac{n_F^2}{3}\right)$$
$$E_Q(N_e) \sim \left(\frac{\hbar^2\pi^2}{24mL^2}\right) N_e^3, \tag{9.46}$$

so that for a one-dimensional infinite well system, we have $F(N_e) \propto N_e^3$.

To extend this to a more realistic three-dimensional system, we also model the spherical geometry by a three-dimensional cubical "particle-in-a-box" and will eventually associate $L \sim R_*$. We first write

$$N_e = \sum_{n_x}\sum_{n_y}\sum_{n_z} 2 \quad \rightarrow \quad N_e = 2\int d\Omega_3 \int_0^{n_F} n^2\, dn = \frac{8\pi}{3} n_F^3 \quad \text{or} \quad n_F = \left(\frac{3N_e}{8\pi}\right)^{1/3}, \tag{9.47}$$

where we translate freely from the three-dimensional (abstract) number-space description of $\boldsymbol{n} = (n_x, n_y, n_z)$ to (n, θ_n, ϕ_n) in order to do the generalized radial integrals in $\boldsymbol{n}$-space. This dependence increases far less rapidly with N_e than does Eqn. (9.45), since the number-states can extend into three dimensions. The total energy as one fills up three-dimensional energy levels is then

$$E(N_e) = \sum_{n_x}\sum_{n_y}\sum_{n_z} 2\left(\frac{\hbar^2\pi^2}{2mL^2}\right)(n_x^2 + n_y^2 + n_z^2)$$
$$\downarrow$$
$$E(N_e) \sim 2\left(\frac{\hbar^2\pi^2}{2mL^2}\right)\int d\Omega_3 \int_0^{n_F} (n^2)\, n^2\, dn$$
$$\sim \left(\frac{\hbar^2\pi^2}{mL^2}\right) 4\pi \left(\frac{n_F^5}{5}\right)$$
$$E(N_e) \sim \frac{4\pi}{5}\left(\frac{3}{8\pi}\right)^{5/3}\left(\frac{\hbar^2\pi^2}{mL^2}\right) N_e^{5/3}. \tag{9.48}$$

Thus, in this more realistic three-dimensional enumeration of states, we find that $F(N_e) \sim N_e^{5/3}$, which increases far less rapidly with N_e than for the one-dimensional case ($F(N_e) \propto N_e^3$), but much more so than the naive $F(N_e) \propto N_e$. (You're asked in **P9.19** to repeat these arguments for two dimensions, and the general d-dimensional case, to see how the fact that we live in three dimensions makes a huge difference in this dimensionless numerical factor.)

With this result in hand, we find for a **quantum star** where gravitational pressure is balanced by electron degeneracy pressure, namely a **white dwarf star**, the radius a value of

$$R_{WD} = R_0 (N_e)^{5/3} = (5 \times 10^{-89}\, m)(10^{57})^{5/3} \approx 5 \times 10^6\, m \approx 5{,}000\, km, \tag{9.49}$$

comparable to the radius of the Earth, which is indeed of the right order of magnitude, especially given that we've neglected many dimensionless factors of $\mathcal{O}(0.1-10)$ along the way.

We note that while DA provided us with the approximate value of R_0 in Eqn. (9.44), which does set the length scale for the radii of collapsed objects, the difference between assuming bosons ($F(N_e) \propto N_e$) and fermions ($F(N_e) \propto N_e^{5/3}$) makes a numerical difference of a factor of $N_e^{2/3} \approx 10^{38}$! (The difference between being in one dimension versus three dimensions, or N_e^3 versus $N_e^{5/3}$ gives an even bigger error of $N_e^{4/3} \approx 10^{76}$, in the other direction.) It's hard to imagine ignoring any other physical effect/constraint that would have such a dramatic impact on getting close to the "right answer." We (hopefully) already know the crucial impact that the Pauli principle has on the structure of nuclei and atoms, and therefore molecules, solids, and chemistry—recall the quote by Ehrenrest in Section 8.1—but here the effect is huge.

To evaluate the zero-point energy contributions of the electrons, we have made one important assumption, namely that their energies are such that the non-relativistic expression $E_n = p_n^2/2m_e$ is valid. We can see if our results are self-consistent by evaluating the average electron energy and compare to the m_ec^2 rest energy. Using expressions for the total electron energy and the R_{WD} radius, we can write

$$E_Q = \frac{\hbar^2}{m_e(R_{WD})^2}F(N_e) \qquad \text{and} \qquad R_{WD} = \frac{\hbar^2}{G(M_{WD})^2 m_e}F(N_e) \tag{9.50}$$

giving

$$E_Q(N_e) = \frac{G^2 M_{WD}^4 m_e}{\hbar^2}\frac{1}{N_e^{5/3}}, \tag{9.51}$$

so that the average electron energy is

$$\langle E_e(n)\rangle \sim \frac{E_Q}{N_e} = \frac{G^2 M_{WD}^4 m_e}{\hbar^2}\frac{1}{N_e^{8/3}} \approx 4\times 10^{-14}\,J \approx 0.2\,MeV. \tag{9.52}$$

More detailed calculations give a somewhat smaller value (like 0.05 MeV) that we can compare to the electron rest energy (0.5 MeV) to see that to first approximation the non-relativistic ($E_e << m_ec^2$) expression is a reasonable first choice. Using $N_e = M_{WD}/m_p$, we find that this average energy scales with the white dwarf mass M_{WD} as

$$\langle E\rangle \sim \frac{G^2 m_e m_p^{8/3}}{\hbar^2}(M_{WD})^{4/3}, \tag{9.53}$$

so that for much larger initial stellar masses, one has to assume a relativistic energy-momentum relation for the electrons.

Making that transition generates two important changes in the dependence of E_Q on $\hbar, R_*, n$ and ultimately N_e, namely

$$E_Q^{(non-rel)} \sim \frac{p_n^2}{2m_e^2} \rightarrow \frac{\hbar^2 n^2}{2mR_*^2} \qquad \text{and} \qquad F(N_e) \propto N_e^{5/3} \tag{9.54}$$

$$\downarrow \qquad\qquad\qquad\qquad \downarrow$$

$$E_{QM}^{(rel)} \sim p_n c \rightarrow \frac{\hbar n c}{R_*} \qquad \text{and} \qquad F(N_e) \propto N_e^{4/3}. \tag{9.55}$$

You're asked in **P9.22** to use the same strategies as in Eqns. (9.47) and (9.48) to confirm the $N_e^{4/3}$ dependence in Eqn. (9.55) in the relativistic case.

The energy balance between the classical gravitational energy and quantum zero-point energy in this case is

$$E_{tot} = E_{QM} + E_G \sim \frac{\hbar c N_e^{4/3}}{R_*} - \frac{GM_*^2}{R_*}, \tag{9.56}$$

where now both terms have the same $(R_*)^{-1}$ dependence, so there is no local minimum as a function of R_*. Rather, there is now a critical point where the two balance, and for which larger values of M_* dominate leading to continuing collapse. This boundary is given by

$$GM_*^2 \sim \hbar c N_e^{4/3} \quad \text{or} \quad M_*^2 = \left(\frac{\hbar c}{G}\right)\left(\frac{M_*}{m_p}\right)^{4/3} \tag{9.57}$$

$$M_*^6 = \left(\frac{\hbar c}{G}\right)^3 \frac{M_*^4}{m_p^4} \quad \text{or} \quad M_{CH} \equiv M_* = \frac{(\hbar c/G)^{3/2}}{m_p^2} = \frac{M_{PL}^3}{m_p^2} \approx 1.7\, M_\odot, \tag{9.58}$$

where $M_{PL} = \sqrt{\hbar c/G}$ is the Planck mass and $M_S = M_\odot$ is the **solar mass.** This type of dependence is indeed seen in what is called the **Chandrasekhar limit** (hence M_{CH}), and close to the accepted value of $1.4\, M_\odot$. It's interesting to see that such an important cut-off mass value (between stability as a white dwarf or neutron star) can be expressed so simply in terms of four dimensional quantities, $G, \hbar, c$, and m_p.

If the white dwarf stage is not stable, the continuing gravitational pressure can squeeze the electrons and protons close enough that they can undergo inverse beta decay. Free neutrons are unstable due to the process $n \to p + e^- + \overline{\nu}_e$ since the energy budget for this decay is $Q \approx (m_n - m_p - m_e)c^2 \approx 0.78\, MeV > 0$, as neutrinos are effectively massless on this scale. If the additional relativistic quantum bound state energy becomes comparable or, greater than, this amount, then the inverse process $p + e^- \to n + \nu_e$ becomes favored, and the weakly interacting neutrinos (see Chapter 7) escape (often in an explosive way via a supernova; see **P9.24**) leaving a neutron-rich stellar remnant.

Given that neutrons are also spin-1/2 particles, just with a larger mass, all of the results obtained from the white dwarf analysis can be taken over (so long as the neutrons remain non-relativistic) by using the simple substitution $m_e \to m_n$, with now **neutron degeneracy pressure** balancing the system against further gravitational collapse. Assuming that every proton/electron pair has been compressed into a neutron, we have $N_n = N_p = N_e = M_*/m_p$, and the resulting **neutron star** radius can be estimated as

$$R_{NS} \sim \frac{\hbar^2}{GM_*^2 m_n} F(N_n) \propto R_{WD}\left(\frac{m_e}{m_n}\right) \approx \frac{R_{WD}}{2000} \sim 3\, km. \tag{9.59}$$

For even heavier initial stellar masses still, the neutrons themselves may become relativistic leading to further collapse, and the final state of the system can be a **black hole** and we encourage readers to explore the research literature or popular science accounts for more details. The gravitational fields of such compact objects can become strong enough to require physics beyond the Newtonian limit, and we explore some aspects of general relativity in the next section.

9.4 Gravitational physics

Most students of physics have a well-deserved (and perhaps hard-earned) appreciation of how important the basic field equations describing electromagnetism are, and learn how to interpret and manipulate the Maxwell equations shown in Eqns. (1.49)–(1.52) or (1.54)–(1.57), at least to some extent. The related mathematical constructs governing gravitational physics in the form of general relativity are perhaps less often studied in detail at the undergraduate or even graduate level, but they can be expressed in one of the most "densely packed" relationships in all of physics, namely the Einstein field equation, given by

$$G_{\mu\nu} = \kappa_G T_{\mu\nu}, \tag{9.60}$$

which has, perhaps, the most physical content/implications per unit symbol of any relation in physics.

9.4.1 Einstein field equations

The mathematics which can be brought to bear in understanding Eqn. (9.60) can range in level of formalism from differential geometry to numerical relativity and the purpose of this section is not to pursue either extreme of abstract rigor or computational power, but to understand at least some of the relevant dimensional connections. The first such relation involves the definition of κ_G, namely

$$\kappa_G \equiv \frac{8\pi G}{c^4} \qquad \text{where} \qquad [\kappa_G] = \frac{[G]}{[c^4]} = \left(\frac{\mathsf{L}^3}{\mathsf{MT}^2}\right)\left(\frac{\mathsf{T}^4}{\mathsf{L}^4}\right) = \frac{\mathsf{T}^2}{\mathsf{ML}}. \tag{9.61}$$

We note that $[\kappa_G] = [F]^{-1}$ has the dimensions of an inverse force, which you can confirm in **P9.30.**

The numerical value of $\kappa_G = 2 \times 10^{-43}\, s^2/(m \cdot kg)$ is another reminder of how extremely weak an interaction gravity is. The impact of the gravitational interaction is perhaps then surprising, given the fact that the electro-static force, which has the same functional inverse-square form as Newton's law, namely

$$F_E = \frac{q_1 q_2}{4\pi\epsilon_0 r^2} \qquad \text{versus} \qquad F_G = \frac{G m_1 m_2}{r^2}, \tag{9.62}$$

has a much larger magnitude, at least for elementary particles, as shown in **P9.26.** Ultimately however, F_E is restricted in its long-distance impact, not by the $1/r^2$ nature of Coulomb's law, but rather by the fact that matter is, to a remarkable degree, electrically neutral in bulk, with $|Q_p + Q_e|/|e| \leq 10^{-21}$, while there is no such restriction on the huge masses of astrophysical objects.

The two-index quantity on the right-hand side (RHS) of Eqn. (9.60), $T_{\mu\nu}$, is called the **stress-energy tensor** and encodes information on the energy and momentum densities and pressures/stresses in a system. This quantity is useful for both classical hydrodynamics and electromagnetism (see **P9.27** for an example in E&M) and for our purposes is simply a 4×4 matrix, given by

$$\left(\begin{array}{c|ccc} T^{00} & T^{01} & T^{02} & T^{03} \\ \hline T^{10} & T^{11} & T^{12} & T^{13} \\ T^{20} & T^{21} & T^{22} & T^{23} \\ T^{30} & T^{31} & T^{32} & T^{33} \end{array}\right).$$

We have intentionally added lines to "guide the eye" to reinforce the notion that the 0 and $i = 1, 2, 3$ components are connected in important ways to other relativistic constructs, such as the space-time and energy-momentum four-vectors, $\boldsymbol{\mathcal{X}}$ and $\boldsymbol{\mathcal{P}}$, in Eqns. (1.93) and (1.94), or those for the four-current and four-potentials, $\boldsymbol{\mathcal{J}}$ and $\boldsymbol{\mathcal{A}}$, in Eqns. (1.97) and (1.98), where factors of c or $1/c$ appear to ensure that all components/entries in $T^{\mu\nu}$ are dimensionally consistent.

As an example to help specify its physical dimensions, we cite the non-relativistic form of $T_{\mu\nu}$ for a fluid, in a notation where the indices correspond to 0 (for the time or ct dimension) and $i = 1, 2, 3$ (for the space dimensions). In that case we have

$$\begin{aligned} T^{00} &= \rho_m c^2 \\ T^{0i} = T^{i0} &= \rho_m c v_i \\ T^{ij} &= \rho_m v^i v^j + P\delta_{ij}, \end{aligned} \tag{9.63}$$

where ρ_m and P are the mass density and pressure, and v_i being the velocity components. The T^{00} term is the rest-energy density (since we've included the c^2), while the T^{0i} are the components of the momentum (actually $\boldsymbol{pc}$) density. The dimensions of the $\rho_m v_i v_j$ and P combinations should be familiar from fluid dynamics and the Bernoulli equation.

In our notation, where four-vectors and tensors in special/general relativity all share the same dimensions, with explicit factors of c as required, we have the dimensions for the stress-energy tensor being

$$\begin{aligned} [T^{00}] &= [\rho_m c^2] = \left(\frac{\mathsf{M}}{\mathsf{L}^3}\right)\left(\frac{\mathsf{L}}{\mathsf{T}}\right)^2 = \frac{\mathsf{M}}{\mathsf{LT}^2} \\ \left[T^{0i}\right] &= \left[\rho_m c v^j\right] = \left(\frac{\mathsf{M}}{\mathsf{L}^3}\right)\left(\frac{\mathsf{L}}{\mathsf{T}}\right)\left(\frac{\mathsf{L}}{\mathsf{T}}\right) = \frac{\mathsf{M}}{\mathsf{LT}^2} \\ \left[T^{ij}\right] &= [P] = \frac{[F]}{[A]} = \left(\frac{\mathsf{ML}}{\mathsf{T}^2}\right)\left(\frac{1}{\mathsf{L}^2}\right) = \frac{\mathsf{M}}{\mathsf{LT}^2}, \end{aligned} \tag{9.64}$$

which all agree, giving $[T_{\mu\nu}] = \mathsf{M}/\mathsf{LT}^2$ overall.

The quantity on the left-hand side (LHS) of Eqn. (9.60) is the **Einstein tensor**, which, in turn, is related to the **Ricci tensor** $R_{\mu\nu}$ (and its trace R) and the **metric tensor** $g_{\mu\nu}$ via

$$G_{\mu\nu} = R_{\mu\nu} - \frac{R}{2} g_{\mu\nu}. \tag{9.65}$$

The metric tensor $g_{\mu\nu}$ describes the structure of space-time by determining generalized distances in $\boldsymbol{\mathcal{X}} = (ct, \boldsymbol{x})$ space-time via[7]

$$ds^2 = g_{\mu\nu} dx^\mu dx^\nu, \tag{9.66}$$

where $[dx^\mu] = [dx^\nu] = [ds] = \mathsf{L}$, but the main result we require is that $g_{\mu\nu}$ is itself dimension**<u>less</u>**.

---

[7] Repeated upper and lower $\mu, \nu = 0, 1, 2, 3$ indices are to be interpreted as being summed over.

The elements of the Ricci tensor are obtained from the **Christoffel** symbols defined as

$$\Gamma^k_{ij} = \frac{1}{2} g^{kl} \left(\partial_i g_{jl} + \partial_j g_{il} - \partial_l g_{ij} \right), \tag{9.67}$$

where each partial space-time derivative (∂_i) gives one factor of L^{-1}, so that $[\Gamma^k_{ij}] = 1/\mathsf{L}$. Finally, the Ricci tensor elements are given by

$$R_{jk} = \partial_i \Gamma^i_{jk} - \partial_j \Gamma^i_{kl} + \Gamma^i_{ip} \Gamma^p_{jk} - \Gamma^i_{jp} \Gamma^p_{kl}. \tag{9.68}$$

In terms of dimensions, it's then clear that $[R_{jk}] = \mathsf{L}^{-2}$, which is all we need as we can now confirm that Einstein's field equation is dimensionally correct, namely

$$\begin{aligned} [G_{\mu\nu}] &\stackrel{?}{=} [\kappa_G] [T_{\mu\nu}] \\ \frac{1}{\mathsf{L}^2} &\stackrel{?}{=} \left(\frac{\mathsf{T}^2}{\mathsf{ML}} \right) \left(\frac{\mathsf{M}}{\mathsf{LT}^2} \right) \\ \frac{1}{\mathsf{L}^2} &\stackrel{!}{=} \frac{1}{\mathsf{L}^2}. \end{aligned} \tag{9.69}$$

Putting together the dimensions of $G_{\mu\nu}$, κ_G, and $T_{\mu\nu}$ we can write schematically

$$G_{\mu\nu} = \kappa_G T_{\mu\nu} \qquad \Longleftrightarrow \qquad \frac{1}{area} = \frac{1}{force} \times pressure. \tag{9.70}$$

One can then proceed to make use of Eqn. (9.60), in a variety of approximations, to examine many implications of the theory of general relativity, including the three classic tests, namely (i) the precession of the perihelion of Mercury, (ii) deflection of light by the Sun (or gravitational lensing in general), and (iii) gravitational red shift, the first two of which we explore in **P9.28** and **P9.29**.

We cite one example in passing to make a reassuring connection to classical Newtonian gravity. To recover Newton's law, in a static limit about flat-space, one writes $g_{00} = 1 + 2\Phi_G(\mathbf{r})/c^2$, where $\Phi_G(\mathbf{r})$ is the gravitational, potential, which gives $\mathbf{g} = -\nabla\Phi_G$ as the gravitational field, in much the same way that the electro-static potential is related to the electric field by $\mathbf{E} = -\nabla\phi$. The relevant entry for $T_{\mu\nu}$ is then $T_{00} = \rho_m c^2$ and the manipulations of the various derivatives leading to $G_{\mu\nu} = G_{00}$ give

$$\nabla^2 \left(\frac{2\Phi_G}{c^2} \right) = \left(\frac{8\pi G}{c^4} \right) (\rho_m c^2) \qquad \text{or} \qquad \nabla^2 \Phi_G = 4\pi G \rho_m, \tag{9.71}$$

which is the Poisson form of the Newton's law as discussed in Section 3.3 and Eqn. (3.47). As two final dimensional checks, we know that

$$\frac{\mathsf{L}}{\mathsf{T}^2} = [\mathbf{g}] = [\nabla][\Phi_G] = \frac{1}{\mathsf{L}}[\Phi_G] \quad \rightarrow \quad [\Phi_G] = \left(\frac{\mathsf{L}^2}{\mathsf{T}^2} \right) \quad \text{giving} \quad \left[\frac{\Phi_G}{c^2} \right] = 1 \tag{9.72}$$

is dimensionless, so that $g_{00} = 1 + 2\Phi(\mathbf{r})/c^2$ is dimensionally consistent, as is

$$[\nabla^2 \Phi_G] \stackrel{?}{=} [4\pi G \rho_m] \quad \rightarrow \quad \frac{1}{\mathsf{L}^2} \frac{\mathsf{L}^2}{\mathsf{T}^2} = \frac{1}{\mathsf{T}^2} \stackrel{!}{=} \frac{1}{\mathsf{T}^2} = \left(\frac{\mathsf{L}^3}{\mathsf{MT}^2} \right) \left(\frac{\mathsf{M}}{\mathsf{L}^3} \right). \tag{9.73}$$

Each generation of physicists seem to find excellent discussions of gravitational physics, including three fifty-year-old classics, by Weinberg (1972), Misner, Thorne, and Wheeler (1973), and Landau and Lifshitz (1975), and more recent, undergraduate-accessible, treatments by Hartle (2003) and Carroll (2004), to which the reader is referred for more thorough treatments of general relativity.

We explore below, in much more detail, one additional, very timely topic, which makes connections to ideas in E&M. Just as Maxwell's equations support the existence of EM waves traveling at the speed of light, so too do the Einstein equations allow for gravitational radiation, and that is the topic of the next section.

9.4.2 **Gravitational radiation**

Given that EM radiation is one of the most important outcomes of the unification codified in Maxwell's equations, and the seemingly obvious similarities between electromagnetism and gravitation (advertised in Section 3.3), even at the level of Eqn. (9.62), it's instructive to apply the same methods of DA as we did in **Example 4.9** and **P4.31** and **P4.32** to the study of gravitational waves.

Given that the production of EM waves can be described in terms of the time-dependence of the **electric dipole moment** of a system of charges, let us begin with the electro-static/gravitational analogy

$$\boldsymbol{p}_E = \sum_{k=1}^{N} q_k \boldsymbol{r}^{(k)} = \int \rho_e(\boldsymbol{r})\, \boldsymbol{r}\, dV \qquad \text{with} \qquad [\boldsymbol{p}_E] = \left(\frac{\mathsf{Q}}{\mathsf{L}^3}\right)(\mathsf{L})(\mathsf{L}^3) = \mathsf{QL} \tag{9.74}$$

$$\boldsymbol{p}_G = \sum_{k=1}^{N} m_k \boldsymbol{r}^{(k)} = \int \rho_m(\boldsymbol{r})\, \boldsymbol{r}\, dV \qquad \text{with} \qquad [\boldsymbol{p}_G] = \left(\frac{\mathsf{M}}{\mathsf{L}^3}\right)(\mathsf{L})(\mathsf{L}^3) = \mathsf{ML}, \tag{9.75}$$

where we sum over the $k = 1, \ldots N$ point masses or charges, with locations given by $\boldsymbol{r}^{(k)}$, or integrate over a continuous body with charge density $\rho_e(\boldsymbol{r})$ or mass density $\rho_m(\boldsymbol{r})$. We can then call the quantity $\boldsymbol{p}_G$ the **gravitational** or **mass dipole moment**, by analogy with the **electric dipole moment** $\boldsymbol{p}_E$.

Let us assume that we have yet to explore how the power radiated by a time-dependent electric dipole depends on the relevant physical quantities, not even knowing which time-derivative is relevant for the process, and so define

$$\boldsymbol{P}_E^{(n)} \equiv \frac{d^n}{dt^n}\boldsymbol{p}_E \qquad \text{with dimensions given by} \qquad [\boldsymbol{P}_E^{(n)}] = \frac{[\boldsymbol{p}_E]}{[dt^n]} = \frac{\mathsf{QL}}{\mathsf{T}^n}. \tag{9.76}$$

Knowing that ϵ_0, μ_0, c are not independent, we can then propose that the power ($\mathcal{P}_E$) radiated in E&M waves depends on the combination

$$\mathcal{P}_E \propto \left(\boldsymbol{P}_E^{(n)}\right)^\alpha \epsilon_0^\beta c^\gamma$$

$$\frac{\mathsf{ML}^2}{\mathsf{T}^3} = \left(\frac{\mathsf{QL}}{\mathsf{T}^n}\right)^\alpha \left(\frac{\mathsf{Q}^2\mathsf{T}^2}{\mathsf{ML}^3}\right)^\beta \left(\frac{\mathsf{L}}{\mathsf{T}}\right)^\gamma, \tag{9.77}$$

which give the dimensional constraint equations

$$\begin{aligned} \mathsf{M}: &\quad 1 = \quad -\beta \\ \mathsf{L}: &\quad 2 = \quad \alpha - 3\beta + \gamma \\ \mathsf{T}: &\quad -3 = -n\alpha + 2\beta - \gamma \\ \mathsf{Q}: &\quad 0 = \quad \alpha + 2\beta. \end{aligned} \tag{9.78}$$

We've so far considered only truly linear equations for the various unknown exponents $(\alpha, \beta, \ldots)$, but in this case, it's useful to see an example where one can also obtain constraints on the number of derivatives of a given quantity, incorporating the $[d^n/dt^n]$ or T^{-n} dimensions as a separate unknown. In this case, we do have a unique solution, namely $n = 2$ (thus requiring $\ddot{\boldsymbol{p}}_E$), $\alpha = 2$, $\beta = -1$, and $\gamma = -3$, all of which reproduce the Larmor formula $\mathcal{P}_E \propto (\ddot{\boldsymbol{p}}_E)^2/\epsilon_0 c^3$, as expected.

We can then extend this analysis to explore to what extent we can determine (or at least constrain) the dependence on the power radiated due to gravitational radiation ($\mathcal{P}_G$) in a similar way, by writing

$$\mathcal{P}_G \propto \left(\boldsymbol{P}_G^{(n)}\right)^\alpha G^\beta c^\gamma \qquad \text{where} \qquad \boldsymbol{P}_G^{(n)} \equiv \frac{d^n}{dt^n}\boldsymbol{p}_G \qquad \text{giving}$$

$$\frac{\mathsf{ML}^2}{\mathsf{T}^3} = \left(\frac{\mathsf{ML}}{\mathsf{T}^n}\right)^\alpha \left(\frac{\mathsf{L}^3}{\mathsf{MT}^2}\right)^\beta \left(\frac{\mathsf{L}}{\mathsf{T}}\right)^\gamma \tag{9.79}$$

or

$$\begin{aligned} \mathsf{M}: &\quad 1 = \quad \alpha - \beta \\ \mathsf{L}: &\quad 2 = \quad \alpha + 3\beta + \gamma \\ \mathsf{T}: &\quad -3 = -n\alpha - 2\beta - \gamma. \end{aligned} \tag{9.80}$$

Because there are now only three $\mathsf{M}, \mathsf{L}, \mathsf{T}$ constraints, we don't find a unique solution, but rather two very distinct options, with quite different physical interpretations.

Simply summing the three equations, we find the constraint $0 = \alpha(2 - n)$, which can be satisfied in two ways. If $\alpha = 0$ (with n then undetermined, but also not relevant) we have the solution $\beta = -1$ and $\gamma = 5$, or

$$\mathcal{P}_G \propto \frac{c^5}{G} = \mathcal{P}_{Pl}, \tag{9.81}$$

which doesn't involve the time-dependent motion of the massive object(s) at all, but rather is the **Planck luminosity**,[8] which is a relation imposed by special and general relativity, and one can find this combination in several ways. For example, given only the variables that would be relevant for a massive object in general relativity, namely M, G, c, if we write $\mathcal{P}_G = M^\alpha c^\beta G^\gamma$, matching dimensions gives $\alpha = 0$, $\beta = 5$, and $\gamma = -1$. This is also seen if we taken the Planck name literally and construct a power from $\hbar^\alpha c^\beta G^\gamma$ (as in Section 6.5.2 and **P6.32**) and find the same results for $\mathcal{P}_{Pl}$.

[8] So-called, despite not having any dependence on $\hbar$, but still involving c and G, as with other Planckian quantities as in Section 6.5.2.

This is another of Nature's "limiting values," with no object allowed to have an instantaneous power output larger than this value. For a discussion of luminosities of some astrophysical objects, and their comparison to this maximum value, see, for example, Hogan (1999) or Cardoso et al. (2018).

The other solution corresponds to $n = 2$ (so again, two derivatives), but now with a general form given by

$$\mathcal{P}_G \propto \left(\ddot{p}_G\right)^{\alpha} G^{-1+\alpha} c^{5-4\alpha}. \tag{9.82}$$

Given the vector nature of $\boldsymbol{p}_G$, and several obvious analogies with the EM result, we might well choose $\alpha = 2$ to find

$$\mathcal{P}_G \propto \frac{(\ddot{p}_G)^2 G}{c^3}, \tag{9.83}$$

which is indeed dimensionally correct, but turns out to be physically irrelevant! This is because we can write the mass dipole moment as

$$\boldsymbol{p}_G = \sum_{k=1}^{N} m_k \boldsymbol{r}^{(k)} = \left(\sum_{k=1}^{N} m_k\right)\left(\frac{\sum_{k=1}^{N} m_k \boldsymbol{r}^{(k)}}{\sum_{k=1}^{N} m_k}\right) \equiv M_{tot}\,\boldsymbol{R}_{cm}, \tag{9.84}$$

where $\boldsymbol{R}_{cm}$ is the **center-of-mass** coordinate of the extended system (the multi-particle generalization of the two body version in Eqn. (7.36)) and M_{tot} being the total mass. Given this connection, the second derivative appearing in Eqn. (9.83) satisfies

$$\ddot{\boldsymbol{p}}_G = M_{tot}\ddot{\boldsymbol{R}}_{cm} = \boldsymbol{F}_{ext} = 0, \tag{9.85}$$

since for an isolated system (such as a binary star gravitational wave source) the total external force vanishes.

Continuing to use analogies with electro-static fields, we know that when the total electric charge <u>and</u> the dipole moment both vanish, then the next relevant charge distribution metric is the **quadrupole moment**, which is a tensor, and so has two indices $(\overleftrightarrow{Q})_{ij}$ instead of one for a dipole moment $(\boldsymbol{p}_{E,G})_i$. In our discussions in Chapter 4, we only noted the dimensions of the higher-multipole moments for electro-static problems, so we cite here that the parallel results for the respective quadrupole moments for E&M and gravity are

$$\left(\overleftrightarrow{Q}_E\right)_{ij} \equiv \sum_k q_k\left(3r_i^{(k)}r_j^{(k)} - |\boldsymbol{r}^{(k)}|^2\delta_{ij}\right) \quad \text{or} \quad \left(\overleftrightarrow{Q}_G\right)_{ij} \equiv \sum_k m_k\left(3r_i^{(k)}r_j^{(k)} - |\boldsymbol{r}^{(k)}|^2\delta_{ij}\right) \tag{9.86}$$

with $[Q_E] = \mathsf{QL}^2$ and $[Q_G] = \mathsf{ML}^2$.

We can then posit that the power radiated in gravitational waves depends on some time-derivative of $\overleftrightarrow{Q}$ via

$$\mathcal{P}_G \propto \left(\frac{d^n}{dt^n}\overleftrightarrow{Q}\right)^{\alpha} G^{\beta} c^{\gamma}$$

$$\frac{\mathsf{ML}^2}{\mathsf{T}^3} = \left(\frac{\mathsf{ML}^2}{\mathsf{T}^n}\right)^{\alpha}\left(\frac{\mathsf{L}^3}{\mathsf{MT}^2}\right)^{\beta}\left(\frac{\mathsf{L}}{\mathsf{T}}\right)^{\gamma}. \tag{9.87}$$

Matching dimensions, we once again find two classes of solutions, since $(3 - n)\alpha = 0$, and when $\alpha = 0$ we recover the Planck luminosity. For $n = 3$, we have

$$\mathcal{P}_G \propto \left(\frac{d^3}{dt^3}\overleftrightarrow{Q}\right)^{\alpha} G^{\alpha-1}c^{5-5\alpha}, \tag{9.88}$$

and since we again require the power to be a scalar, $\alpha = 2$ is still a natural choice, giving

$$\mathcal{P}_G = C_{\mathcal{P}}\frac{(\dddot{Q})^2 G}{c^5}, \tag{9.89}$$

where a detailed calculation gives $C_{\mathcal{P}} = 1/5$.

This type of calculation is perhaps most relevant for the power emitted in the form of gravitational waves by orbiting binary systems, either neutron stars and/or black holes. For two such bodies in a circular orbit of radius a, with reduced mass $\mu = m_1 m_2/(m_1 + m_2)$, rotating with an angular frequency ω, the relevant quadrupole moment and its derivative scale as $Q \propto \mu a^2$ and $\dddot{Q} \propto (\mu a^2)\omega^3$, and including appropriate numerical factors, the emitted power is

$$\mathcal{P}_G = \frac{32}{5}\frac{G\mu^2 a^4 \omega^6}{c^5}, \tag{9.90}$$

with a characteristic ω^6 dependence due to the $(\dddot{Q})^2$ quadrupole dependence, analogous to the Rayleigh scattering ω^4 ("blue sky") factor from the $(\ddot{p})^2$ factor for EM dipole radiation. We can now use this relation to explore in more detail the information that can be extracted from the signals observed in gravitational wave detectors.

Example 9.4 Chirp mass in GR wave detection

P1.21 briefly introduced a very specific combination of the two masses (m_1, m_2) of an inspiraling binary pair that could be directly measured by data of the time-dependent frequency, $f(t)$, of the gravitational radiation observed in terrestrial gravitational wave detectors such as LIGO or VIRGO, namely

$$\frac{(m_1 m_2)^{3/5}}{(m_1 + m_2)^{1/5}} \equiv \mathcal{M}_c = \frac{c^3}{G}\left\{\left(\frac{5}{96}\right)^3 \pi^{-8} f^{-11} \dot{f}^3\right\}^{1/5}, \tag{9.91}$$

where $\mathcal{M}_c$ is the **chirp mass.** This relation can be simply rewritten as

$$\dot{f}(t) = \frac{df(t)}{dt} = \frac{96}{5}\pi^{8/3}\left(\frac{G\mathcal{M}_c}{c^3}\right)^{5/3} f(t)^{11/3} \tag{9.92}$$

and we wish to explore reproducing this result now that we have Eqn. (9.90), given that the various 1/3 and 1/5 power exponents are far from obvious.

To see how far a DA approach can get us (not all that far, as it turns out) we can write $\dot{f} \propto G^{\alpha}\mathcal{M}_c^{\beta}c^{\gamma}f^{\delta}$ where $\mathcal{M}$ is <u>some</u> mass and we find only that $\dot{f} \propto G^{\alpha}\mathcal{M}_c^{\alpha}c^{-3\alpha}f^{2+\alpha}$, which does reproduce the dimensional dependence in Eqn. (9.92) if we use the non-obvious exponent $\alpha = 5/3$. To reproduce that result in more detail, we can combine familiar results from gravitational physics (Kepler orbits) and the newer (to us) gravitational power radiation relation in Eqn. (9.90).

We first recall the definitions

$$M_T = m_1 + m_2 \qquad \text{and} \qquad \mu = \frac{m_1 m_2}{(m_1 + m_2)} \qquad \text{so that} \qquad m_1 m_2 = \mu M_T \tag{9.93}$$

for the total and reduced masses of the two-body system. To get a sense of how the chirp mass combination might arise, we note that

$$(\mathcal{M}_c)^5 = \frac{(m_1 m_2)^3}{(m_1 + m_2)} = \left(\frac{(m_1 m_2)^3}{(m_1 + m_2)^3}\right)(m_1 + m_2)^2 = \mu^3 M_T^2. \tag{9.94}$$

The total energy of such a binary system as a function of the radius a (we'll always assume circular orbits) can then be written as

$$E(a) = -\frac{G m_1 m_2}{2a} = -\frac{G\mu M_T}{2a}, \tag{9.95}$$

so that as the system in-spirals, the rate of energy loss, given by Eqn. (9.90), is related to the change in radius via

$$\underbrace{\frac{G\mu M_T}{2a^2}\dot{a} = \frac{dE(a)}{dt}}_{\text{derivative of E(a)}} = \underbrace{P_G = \frac{32}{5}\frac{G\mu^2 a^4 \omega^6}{c^5}}_{\text{gravitational waves}}. \tag{9.96}$$

(Note that we have canceled two minus signs, one from differentiating $E(a) \propto a^{-1}$ with respect to t, and one from the fact that the change in energy due to GR wave emission is negative.) In order to relate $\dot{a} = da(t)/dt$ to the change in frequency, we seek we use Kepler's law in the form

$$a^3 = \frac{GM_T}{\omega^2} \qquad \text{or} \qquad a = (GM_T)^{1/3}\omega^{-2/3} \qquad \text{so that} \qquad \dot{a} = -\frac{2}{3}(GM_T)^{1/3}\omega^{-5/3}\dot{\omega}. \tag{9.97}$$

Using the expressions for a and $\dot{a}$ from Eqn. (9.97) in Eqn. (9.96) we find that

$$\dot{\omega} = \frac{96}{5}\left\{\frac{\mu(GM_T)^{5/3}}{M_T c^5}\right\}\omega^{11/3}. \tag{9.98}$$

The expression in curly brackets is seen to be

$$\left(\frac{\mu^3 G^5 M_T^5}{M_T^3 c^{15}}\right)^{1/3} = \left(\frac{G^5 \mu^3 M_T^2}{c^{15}}\right)^{1/3} = \left(\frac{G\mathcal{M}_c}{c^3}\right)^{5/3} \tag{9.99}$$

as expected, showing how the chirp mass $\mathcal{M}_c$ and the other factors arise.

The final step is to replace ω with f, which would seem trivial, simply substituting $\omega = 2\pi f$. However, because the gravitational radiation is governed by a quadrupolar process, the symmetry dictates that the system will repeat itself every π rotations, and not just every 2π, so we are told to use $\omega = \pi f$. (To visualize this, you can imagine the electric dipole and quadrupole charge configurations in Fig. 4.17 spinning around an axis.) Doing so, we find that Eqn. (9.98) reduces to

$$\dot{f} = \frac{96}{5}\pi^{8/3}\left(\frac{G\mathcal{M}}{c^3}\right)^{5/3} f^{11/3} \tag{9.100}$$

as advertised. While DA has certainly been helpful "along the way," to guide judgment about the dependence of the power radiated in gravitational radiation in Eqns. (9.89) and (9.90), a detailed analysis is required to obtain the actual functional form and the particular combination of masses giving $\mathcal{M}_c$. One can integrate Eqn. (9.100) to obtain the explicit t-dependence for $f(t)$ (see **P9.35**) and similar methods can be used to derive the time-variation of $a(t)$ (as in **P9.36**) and other processes (**P9.37** and **9.38**) involving GR radiation.

We have only touched on some of the physics, astrophysics, and cosmology with gravitational waves[9] that are possible with the rapidly increasing amounts of data from GR wave observatories across the globe, including novel tests of the **second law of black hole mechanics** or **Hawkings area theorem**; see Isi et al. (2021). We explore some aspects of black hole physics in the problems.

9.5 Problems for Chapter 9

Q9.1 Reality check: (a) What was the most important, interesting, engaging, or novel thing you learned about how to use DA, or about anything else, in this chapter?
(b) Have you previously encountered (or learned) many (or any) of the topics covered in this chapter?

P9.1 $\mathcal{R}_P$ values and Debye lengths for plasmas of interest: Using the definitions in Eqns. (9.8) and (9.11), complete the table below for a large array of lab-based and naturally occurring plasmas. The last entry has the relevant values of n_e and T for the free-electrons in a room-temperature metal for comparison.

System	$n_e\ (m^{-3})$	$T\ (K)$	$\mathcal{R}_P$	$l_D\ (m)$	$\ln(\mathcal{R}_P^{3/2})$
Gas discharge	10^{16}	10^4			
Tokamak	10^{20}	10^8			
Ionosphere	10^{12}	10^3	6×10^3		
Magnetosphere	10^7	10^7			
Solar core	10^{32}	10^7			
Solar wind	10^6	10^5			
Interstellar medium	10^5	10^4			
Intergalactic medium	1	10^6			
Metal	$10^{28.5}$	$10^{2.5}$			

P9.2 Wakefield acceleration I—Unpacking a research result: The authors of a paper on laser-plasma wakefield acceleration (Götzfried et al. 2020) make the following statement:

> The force responsible for setting up the wakefield is thus the ponderomotive force
>
> $$F_{pond} = -\frac{m_e c^2}{2\langle\gamma\rangle}\nabla(a)^2$$

[9] The title of an excellent review article on the subject by Sathyaprakash and Schutz (2009).

with m_e as the electron mass, c the speed of light in vacuum, $\langle\gamma\rangle$ is the cycle-averaged Lorentz factor, and

$$\boldsymbol{a} = \frac{eA}{m_e c} = \frac{eE_0}{m_e c^2} \times \left(\frac{\lambda_0}{2\pi}\right) \approx 0.31 \times E_0 \left[TVm^{-1}\right] \lambda_0 \left[\mu m\right]$$

is the laser's local normalized vector potential, where λ_0 denotes the wavelength of the laser pulse and E_0 stands for the electric field strength of the laser.

(a) Recalling that the Lorentz factor is given by $\gamma = 1/\sqrt{1-(v/c)^2}$ (with $[\gamma] = 1$), use the first equation to show that $\boldsymbol{a}$ is dimensionless, hence the "normalized" description.

(b) In the second equation, show that the combinations involving $\boldsymbol{A}$ and $\boldsymbol{E}$ proportional to $\boldsymbol{a}$ are also dimensionless. Recall that $\boldsymbol{A}$ is the EM vector potential.

(c) In the second equation, show that the last equality, generating a numerical value for $\boldsymbol{a}$ is correct, once values or m_e, c, e are inserted and the stated values of $\boldsymbol{E}_0$ and λ_0 are used. Note that *TV* stands for *teravolt.*

P9.3 Wakefield acceleration II—Numerical examples: (a) Esarey et al. (2009) translate the result in Eqn. (9.17) into a general "working version" involving actual numbers, namely

$$E_0\left(\frac{V}{m}\right) \approx 96\sqrt{n_e\,(cm^{-3})}.$$

Insert values of m_e, c and ϵ_0 to confirm this relation.
(b) Using wakefield acceleration techniques, Miao et al. (2022) have produced an "... acceleration gradient as high as 25 *GeV/m* ...." What does this imply about the number density of the plasma in their device?
(c)† Buck et al. (2011) measured the period of plasma oscillations "... as a function of electron density while all other experimental parameters were kept constant ...", which "... corroborates that the oscillations originate from the plasma wave ..." and their results are shown in Fig. 9.2. In their notation the period is

$$T_{plasma} = 2\pi\sqrt{\frac{\epsilon_0 m_e}{e^2 n_e}}$$

and the solid curve represents that prediction. Use values of ϵ_0, m_e, and e to compare at least two values of n_e on that plot with Eqn. (9.5), and to also confirm th $T_{plasma} \propto n_e^{-1/2}$ dependence.

Note: It's a regular occurrence in science that a major discovery (e.g., plasma oscillations by Tonks and Langmuir (1929)), can become a routine experimental methodology or diagnostic tool as research advances in that field.

9.4 Acoustic analog of ponderomotive force: A time-dependent acoustic field, say a sound wave of frequency ω and pressure p_0, can generate a **ponderomotive potential** (U_p) akin to that in Eqn. (9.15), giving a net force via $\boldsymbol{F}_p = -\boldsymbol{\nabla} U_p$. Assume that the oscillating pressure field is incident on a spherical object of radius a and mass m, so that the potential might depend on $U_p \propto a^\alpha p_0^\beta m^\gamma \omega^\delta$.

(a) Match M, L, T dimensions and determine the exponents to the extent you can.

(b) Compare the result from part (a) to Eqn. (9.15) and use analogy to see if you can constrain the exponents completely.

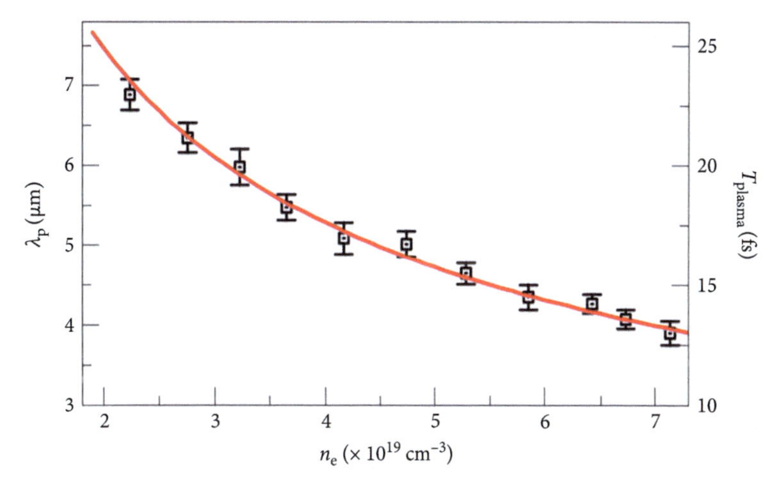

Fig 9.2 Wavelength (left) and period (right) versus electron density for laser-driven electron acceleration. Reproduced from Buck et al. (2011) with permission from Springer Nature. Note the units used for all quantities, especially *fs* = *femtosecond* for the period.

Note: A very nice theoretical discussion and experimental realization of *acoustic levitation and the acoustic radiation force* using these methods has been described by Jackson and Chang (2021).

P9.5 Spitzer resistivity: The electric **resisitivity** (here to be called η_e to avoid confusion with ρ_e or charge density) of an ionized plasma was first described by Spitzer and Härm (1953).

(a) Assuming that the relevant quantities are e, ϵ_0, m_e, k_B, and T, write $\eta_e = C_\eta e^\alpha \epsilon_0^\beta m_e^\gamma k_B^\delta T^\sigma$ and solve for all of the exponents. You may have to look up the fact that $[\eta_e] = [1/\sigma_e] = (\mathrm{ML}^3)/(\mathrm{Q}^2\mathrm{T})$. where σ_e is the electrical conductivity.

(b) A detailed calculation finds that the dimensionless constant is given by $C_\eta = (4\sqrt{2\pi}/3)\ln(\Lambda)$, where the last term is often described as the **Coulomb logarithm**. It arises from an integral of the form

$$\ln(\Lambda) = \ln\left(\frac{b_{max}}{b_{min}}\right) = \int_{b_{min}}^{b_{max}} \frac{db}{b}, \tag{9.101}$$

where b is an impact parameter in a given Coulomb collision. Considering the discussion following our definitions of the Debye length and distance of closest approach, leading to Eqn. (9.12), can you relate Λ to $\mathcal{R}_P$? And then use the last column of the table from **P9.1** to confirm the conventional wisdom that $4 < \ln(\Lambda) < 40$ for many plasmas?

P9.6 Dreicer field: In a thermal plasma, there is a critical electric field (E_D, after Dreicer 1959, 1960) above which value electron collisions aren't sufficient to damp the acceleration due to the applied field, leading to runaway behavior. Assume that this field depends on the plasma properties via $E_D \propto e^\alpha \epsilon_0^\beta n_e^\gamma k_B^\delta T_e^\sigma$, match dimensions, and determine the exponents to the extent you can. A detailed calculation (see, e.g., Longair 2011) finds that E_D depends linearly on the electron density. Does that fix all of the other exponents?

Note: You might not be surprised to find that the final form also depends on the Coulomb logarithm $\ln(\Lambda)$ discussed in **P9.5** as well.

P9.7 MHD parameters I—Magnetic Reynolds number: There is an analog of the familiar Reynolds number (from classical fluid dynamics) in magneto-hydrodynamics that describes how "frozen" magnetic field lines are in a conducting fluid. Just like Re, the parameter Re_m is a dimensionless number and in this case can depend on the permeability of free space (μ_0), the conductivity of the material (σ_e), the speed and size (v and l) of the system, and the magnetic field B_0.

(a) Write $R_m \propto \mu_0^\alpha \sigma_e^\beta v^\gamma l^\delta B_0^\epsilon$, match dimensions, and solve for as many exponents as you can.

(b) Recall (or look up) how the regular Reynolds number (Re) depends on v and l and see if that analogy can help you determine the exponents more completely.

(c) Recall the skin-depth problem in **P4.37** where the differential equation for the $\boldsymbol{E}$ (or $\boldsymbol{B}$) fields in a particular limit was

$$\nabla^2 \boldsymbol{E}(\boldsymbol{r}, t) = \mu_0 \sigma_e \frac{\partial \boldsymbol{E}(\boldsymbol{r}, t)}{\partial t}. \tag{9.102}$$

If we use l and $t = l/v$ as the appropriate scales for the ∇ and $\partial/\partial t$ operators, show that the dimensionless magnetic Reynolds number you found from part (a) or (b) naturally appears.

(d) Rewrite Eqn. (9.102) so that it has the form of the diffusion equation, and identify what plays the role of the **magnetic diffusion constant.**

(e) The **Lundquist number** is yet another dimensionless ratio, also relevant to magneto-hydrodynamics, given by

$$Lu = \sqrt{\frac{\mu_0}{\rho_m}} \sigma_e B\, l \tag{9.103}$$

where ρ_m is the mass density. Can you rewrite this expression in terms of the Alfvén velocity and compare to the results from parts (a) and (b)?

Note: For examples of the use of the magnetic Reynolds number in astrophysical settings, see Davidson (2001) or Kulsrud (2005).

P9.8 MHD parameters II—Magnetic damping time: One important time scale for the dissipation of energy via Joule heating in magneto-hydrodynamics processes might depend on the conductivity (σ_e), the ambient magnetic field (B_0), the mass density of the fluid (ρ_m), and the permeability of free space (μ_0). Write $\tau \propto \sigma_e^\alpha B_0^\beta \rho_m^\gamma \mu_0^\delta$ and solve for the exponents to the extent you can.

P9.9 Magnetospheres and the Chapman–Ferraro distance: We often hear that the Earth's magnetic field helps protect us from the possibly harmful effects of cosmic rays from the solar wind, and schematic images of the magnetosphere with its distorted dipole field configuration are familiar. The distance at which a planet's magnetic field starts to provide such protection can be denoted R_M, and is often given in terms of the planet's (surface) equatorial radius R_S in a relation such as

$$\frac{R_M}{R_S} = F(\Pi), \qquad \text{with } \Pi \text{ depending on} \qquad \Pi \propto B_S^\alpha \mu_0^\beta \rho_m^\gamma v^\delta, \tag{9.104}$$

where B_S is the magnetic field at the planet's surface (assumed to be a dipole field) and ρ_m, v being the mass density and speed of the incident particles.

(a) Evaluate β, γ, δ in terms of α so that Π is dimensionless. Do the combinations you find seem familiar? The exponent turns out to be $\alpha = 1/3$ and we can try to understand why in part (b).

(b) The energy/force balance involved in the magnetosphere is between the dynamical hydrodynamical pressure of the solar wind and that provided by the magnetic pressure of the planet's field, evaluated at the distance R_M. Recalling that the units of energy density and pressure are the same, use $\rho_m v^2/2$ and $B_M^2/2\mu_0$, and what you know about how the magnetic field varies with distance (recall, it's a dipole field), to evaluate B_M in terms of B_S to find an expression for R_M/R_S as in Eqn. (9.104) and thereby evaluating the exponent α.

Notes: The relationship you're finding was proposed by Chapman and Ferraro (1930, 1931). See also '"Solar System Magnetospheres" by Blanc, Kallenbach, and Erkaev (2005).

P9.10 Alfvén radius: A special case of the magnetospheric phenomena discussed in **P9.10** is when the influx of particles giving the $\rho_m v^2$ **ram pressure** is due to accretion with a mass rate of $\dot{m} = dm/dt$ on a star of mass and radius M_*, R_*. This most often happens in a binary system where an "active" star accretes matter onto a collapsed object, like a white dwarf or neutron star. Longair (2011) has an expression for the distance at which the ram pressure and magnetic field density balance given by

$$R_A = \left(\frac{2\pi^2}{G\mu_0^2}\right)^{1/7} \left(\frac{B_*^4 R_*^{12}}{M_* \dot{m}^2}\right)^{1/7} \tag{9.105}$$

where B_* is the magnetic field at the surface which has an interesting array of 1/7 powers!

(a) Show that Eqn. (9.105) is dimensionally correct.

(b) Try to reproduce that expression to the extent you can by identifying

$$(4\pi R_A^2)\rho_m v = \dot{m} = \frac{dm}{dt} \quad \text{and} \quad v \sim \sqrt{\frac{GM_*}{R_A}} \tag{9.106}$$

to calculate the incoming particle flux and ram pressure and use $B(R_A) = B_*(R_*/R_A)^3$ in the expression for the magnetic pressure

P9.11 Spin-up rate due to accretion: (a) If the angular momentum (L) of a spinning object is changed, its period (P) will also change, and a simple relation connecting $\dot{L}$ and $\dot{P}$ is

$$\dot{P} = -\frac{\dot{L}P^2}{2\pi I}, \tag{9.107}$$

where I is the moment of inertia. Show that Eqn. (9.107) is dimensionally correct.

(b) Derive the expression in Eqn. (9.107) by differentiating the relation $L = I\omega$, where $\omega = 1/(2\pi P)$.

(c) Longair (2011) discusses how matter accreting onto a compact object can exert a torque on the magnetosphere of the star, leading to an decrease in its rotational period. Using the simple relation above, they cite an expression for the fractional rate of change of the period P as

$$\frac{\dot{P}}{P} = -\frac{1}{2\pi I}\left(\frac{2\pi^2}{\mu_0^2 G^6}\right)^{1/14}\left(B_*^{2/7} R_*^{12/7} M_*^{-3/7}\right)\left(L^{6/7} P\right), \tag{9.108}$$

where $I \propto M_* R_*^2$ is the moment of inertia of the compact object and L is its luminosity (power output). Show that this relation is dimensionally correct! (Let us not try to reproduce this complicated expression!) See Bonanno and Urpin (2015) for a similar (but equivalent) relation, using different variables, which you can also confirm is dimensionally correct.

P9.12 Diffusion in plasmas: One of the most important physical processes that determines the ability to control and maintain thermonuclear reactions via magnetic confinement is the diffusion from regions of high to low density. The parameter that chiefly describes this is another **diffusion constant**, D, a concept we've encountered in many other systems, always with the same dimensionality, $[D] = \mathsf{L}^2/\mathsf{T}$. In a plasma, the most likely properties that would determine this would be k_B and T, e and B_0 from the E&M side, and the electrical resistivity (like from **P9.5**) and number density of the plasma, η_e and n_e.

(a) Explore this by writing $D \propto k_B^\alpha T^\beta e^\gamma B^\delta \eta_e^\sigma n_e^\rho$, match dimensions, and see to what extent you can constrain the exponents.

(b) Are any of the exponents completely determined? The theory of so-called **classical diffusion** predicts that $\delta = -2$, so what is your result in that case? The theory of **Bohm diffusion**[10] finds that $D \propto B^{-1}$, and what is your final result then? For a discussion of both limits, see Chen (2018).
Note: Bohm diffusion turns out to play an important role in theories that try to explain the acceleration mechanisms of cosmic rays; see, for example, Stage et al. (2006).

(c) One standard reference on plasma physics (Chen 2018) cites the Bohm diffusion constant as $D_B = (k_B T)/(16eB)$, while Bohm himself (1949) states that

> This diffusion coefficient D_e is
>
> $$D_e = \frac{10^4\, T_e}{H} \tag{8}$$
>
> where T_e is the electron temperature in volts, and H is the magnetic field strength in thousands of Gauss.

Compare the two expressions, by using $k_B T_e = 1\ eV$ and $B = 10^3\ Gauss$ in the formula for D_B, while noting that Bohm uses CGS units in evaluating his version, so that the units of his D_e are cm^2/s.

P9.13 Dissipation in Alfvén waves: (a) Magneto-hydrodynamic waves in a conducting material (one with a finite conductivity σ_e) will also exhibit attenuation, with the wave amplitude decreasing exponentially as $e^{-\kappa z} = e^{-z/l_\kappa}$. To see how the **attenuation length** l_κ depends on the physical parameters, write $l_\kappa \propto \omega^\alpha \sigma_e^\beta \mu_0^\gamma v_A^\delta$, match dimensions, and see to what extent you can constrain the exponents. What if you knew that $l_\kappa \propto \sigma_e$?

(b) Compare your result for part (a) to the general expression for skin depth for E&M waves propagating in a conducting medium from Eqn. (4.121) and comment on any similarities and differences, especially the dependences on σ_e and ω. Do things look even more similar if one replaces ϵ_0 with c^2/μ_0?

---

[10] See Bohm (1949c).

P9.14 Faraday rotation: The plane of polarization of an EM wave can be rotated as it passes through a medium with free electrons (charge e, mass m_e) and spatially varying magnetic field, $B_0(z)$. The angle of rotation of the wave $\Delta\phi$ (so dimensionless) is given by

$$\Delta\phi = \frac{e^3\lambda^2}{8\pi^2\epsilon_0 m_e^2 c^3}\int_{path} n_e(z)B_0(z)\,dz, \tag{9.109}$$

where $n_e(z)$ is the (path-dependent) number density and λ is the wavelength.

(a) Show that Eqn. (9.109) is dimensionally consistent.

(b) What if you tried to apply DA "from scratch" by assuming that $\Delta\phi$ depended on powers of $e, \epsilon_0, \lambda, m_e, c, n_e, B_0$ and path length dz. What constraints would you get among the possible exponents?

(c) Say you read that $\Delta\phi$ was proportional to $1/\omega^2$, and that it could be used to measure the "*... line integral of n_eB_0 along the straight line path ...*" connecting the observer to an astronomical object. With that extra information, can you uniquely determine all of the other parameters? What if you were told that it was directly proportional to the *line integral* quantity mentioned?

P9.15 Central stellar pressure and temperature minima from scaling: Use the two scaling relations in Eqn. (9.37) and (9.39) to estimate lower bounds on the pressure and temperature at the center of the Sun, using solar mass values $M_* = M_\odot$ and $R_* = R_\odot$.

P9.16 Mass–luminosity relation: Some treatments of the approximate scaling $L_* \propto M_*^3$ relation derived in Eqn. (9.41) drop the dependence on the important physical constants $\sigma, \kappa, \mu, G, \mathcal{R}$ and only keep track of the factors of M_*, R_*, T_*. Let us see to what extent we can determine them by DA.

(a) If we are told that $T_* \propto M_*/R_*$, add enough powers of $\mu, G, \mathcal{R}$ to make this dimensionally correct, and compare to Eqn. (9.39).

(b) If we're given that $L_* \propto (TR_*)^4/M_*$, add enough factors of σ and κ to make it dimensionally correct, and compare to Eqn. (9.40).

(c) Finally, if we're only given that $L_* \propto M_*^3$, write $L_* = (M_*)^3\sigma^\alpha\kappa^\beta\mu^\gamma G^\delta\mathcal{R}^\epsilon$, match dimensions to find all exponents in terms of α. Is there a simplest choice? And does that one reproduce the final result in Eqn. (9.41)?

P9.17 Radiation versus gas pressure: For stars of relatively low mass, ordinary gas pressure is much larger than that due to radiation pressure from the photons. For high enough temperatures, P_{rad} can begin to dominate.

(a) Estimate the critical temperature where the two pressures are equal by writing $T_c \propto \mathcal{R}^\alpha\rho_m^\beta a^\gamma\mu^\delta$, where $\mathcal{R} = Nk_B$ is the gas constant, μ is the average molecular mass of the gas, and we're including the **radiation constant**, $a = 4\sigma/c$, as it's often used in the astrophysical literature. Match dimensions, and solve for exponents to the extent you can.

(b) If we look up that the pressure due to radiation is given by $P_{rad} = aT^4/3$, we can balance this with an ideal gas law result in Eqn. (9.38) to find the temperature at which the two are equal, giving T_c. Does that value agree with your result from part (a)?

(c) What if you had tried this problem by using σ (the Stefan–Boltzmann constant) instead of a?

(d) Find the ratio of the thermal gas pressure to that of radiation using your estimates of P_c and T_c from **P9.15** and the expression for P_{rad} from part (b).

P9.18 Stellar time scales: There are many important time scales for stellar processes that depend on the star's parameters and fundamental constants in quite different ways.

(a) The **dynamical time scale** corresponds to gravitationally driven oscillations. Write $\tau_{dyn} \propto R_*^{\alpha} M_*^{\beta} G^{\gamma}$, match dimensions, and solve for the exponents. Evaluate τ_{dyn} for solar parameters.

(b) The **thermal** or **Kelvin–Helmholtz time** is obtained from assuming that the gravitational energy is used up in powering the star at a constant luminosity L_*. Write $\tau_{th} = \tau_{KH} \propto G^{\alpha} M_*^{\beta} R_*^{\gamma} L_*^{\delta}$, match exponents, and use your intuition about the physics involved to finalize your answer. Evaluate τ_{th} for solar parameters and decide if this is a likely mechanism to have powered our own Sun over its lifetime.

(c) The **nuclear** time scale assumes that some (dimensionless!) fraction f of the Sun's rest mass can be converted to energy and to power its luminosity. Write $\tau_{nuc} = f M_*^{\alpha} c^{\beta} L_*^{\gamma}$, match exponents, and evaluate τ_{nuc} for solar parameters, using $f \approx 10^{-3}$.

P9.19 Two-dimensional electron energies and generalization to d-dimensions: We've seen from Eqns. (9.45)–(9.46) and Eqns. (9.47)–(9.48) how to do the counting of fermions of mass m as they fill up energy levels in one-dimensional and three-dimensional infinite well potentials, of size L and L^3, respectively.

(a) Generalize those results to the case of two dimensions for a box of area L^2 to see how n_F and E_Q depend on N_e. This is relevant to two-dimensional electron gas (2DEG) systems, as in the quantum Hall effect and in many two-dimensional devices.

(b) Generalize this to d-dimensions, but you don't have to explicitly evaluate the d-dimensional angular integrals (the analogs of $\int d\theta$ in two dimensions and $\int d\Omega$ in thee dimensions), just focusing on the generalized "radial" integration over powers of n. Does your result depend on d in a way that reproduces the one-dimensional, two-dimensional, and three-dimensional results?

Note: If you're really curious, the solid angle in d dimensions is $\Omega_d = 2\pi^{d/2}/\Gamma(d/2)$, where $\Gamma(z)$ is the generalized factorial function. You can then check that this gives $\Omega_3 = 4\pi$, $\Omega_2 = 2\pi$, and $\Omega_1 = 2$. Using this expression, the volume of the d-sphere of radius R is then $V_d = \Omega_d R^d/d$, but the R dependence is as far as DA takes us.

P9.20 White dwarf and neutron star scaling: Consider the following quantities, and how they scale with R_* for fixed stellar mass, to obtain approximate values relevant for a white dwarf (WD) and neutron star (NS) star, starting a one solar mass ($M_* = M_{\odot} = M_S$) object. Assume our rough estimates of $R_{WD} \approx 5 \times 10^3\ km$ and $R_{NS} \approx 3\ km$.

(a) How does the density depend on R_*? What are ρ_{WD} and ρ_{NS} in kg/m^3?

(b) If you know that the **magnetic flux** of a star is a constant, how does the magnetic field depend on R_*? What are B_{WD} and B_{NS} if we assume that the progenitor star had a magnetic field of, say 100 G.

(c) Assuming that angular momentum is conserved in the collapse process, how does the rotational period depend on R_*? If the original star had a period of roughly a month (like our own Sun), what are τ_{WD} and τ_{NS} predicted by this argument.

Note: Neutron stars don't typically have their periods predicted by this simple model, as they can lose mass and angular momentum in the often cataclysmic collapse process.

P9.21 Quantum degeneracy pressure stars and their mass dependence: Let us assume that the radius of any collapsed object supported by degeneracy pressure can depend on only five quantities, $\hbar$ and G, and one or more of the fundamental masses in the problem M_*, m_e, m_p, where for simplicity we invoke the fact that the neutron and proton are almost degenerate in mass.

(a) Write $R_* \propto \hbar^\alpha G^\beta M_*^\gamma m_e^\delta m_p^\sigma$, match dimensions, and solve to the extent you can.

(b) Using $F(N_E) \sim N_e^{5/3}$, write the expressions for both the WD and NS masses in terms of the five quantities above and confirm that the general result from part (a) includes them as special cases.

9.22 Relativistic electron total energy calculations: Following the workflow for enumerating the counting of states and total energy for the three-dimensional non-relativistic infinite well in Eqns. (9.47) and (9.48), confirm that for relativistic particles, where $E(n) \propto p_n c = n\hbar c/L$, that the total electron energy does scale like $N_e{}^{4/3}$, instead of $N_e{}^{5/3}$ as in the non-relativistic case.

P9.23 Astrophysical ν's I—Solar neutrinos: In **P9.24** we discuss the detection of neutrinos from a cataclysmic event like a supernova explosion, but detectors on Earth have already been measuring the flux of neutrinos from the Sun as part of the solar energy generation cycle for decades. The main fusion mechanism in the Sun is the *pp* reaction $4p \rightarrow 4He + 2e^+ + 2\nu_e$, which releases about 27 *MeV*, with about 2–3% of the total energy is released as neutrinos. The average neutrino energy from these reactions is roughly 0.5 *MeV*. If the total solar energy production (luminosity) is $L_{sun} \approx 4 \times 10^{26}\ W$, how many neutrinos are emitted per second? What is their flux (number per cm^2 per *sec*) on Earth?

Note: Neutrinos from the Sun have been detected on Earth in dedicated underground experiments, but with event rates about 1/3 of that expected from the standard solar model. The discrepancy turns out to be due to the fact that three neutrinos flavors can "oscillate" from one form to another on their way to Earth, and only one type is visible in the detector. The neutrino mass problem is still an ongoing subject of research and one data-point on the limits on neutrino mass (a necessary ingredient for oscillations to occur) also comes from astrophysics, as in the next problem.

P9.24 Astrophysical ν's II—Neutrino mass limits from SN1987A: Massless particles all move at the speed of light and stringent bounds have been placed on the mass of the photon from a variety of methods; see Section 4.8.2. A bound on a possible mass for the neutrino was obtained from data on the arrival times of a pulse of neutrinos from SN1987A (the closest supernova since the time of Kepler, SN1604) measured[11] in large underground detectors designed to search for proton decay. Some of the observable data from that historic observation include the average neutrino energy and the spread in values ($\langle E_\nu \rangle \sim 30 \pm 10\ MeV$), the distance to the source ($D \sim 50\ kpc$ where *kpc* is *kiloparsec*), and the pulse arrived within a time window of roughly $\Delta t_{obs} \sim 2\ s$. Previous limits

[11] See Hirata et al. (Kamiokande Collaboration) (1987) and Bionta et al. (1987).

on a possible neutrino mass or rest energy, $m_\nu c^2$ were already much smaller than $\langle E_\nu \rangle$ and so the neutrinos were known to be ultrarelativistic and one can use Eqn. (1.90) to write

$$1 - \frac{v}{c} \approx \frac{1}{2}\frac{(m_\nu c^2)^2}{E_\nu^2} \quad \longrightarrow \quad \frac{\Delta v}{c} \sim \frac{(m_\nu c^2)^2}{E_\nu^3}\Delta E_\nu \tag{9.110}$$

to estimate the velocity spread Δv of the neutrino burst. If all of the neutrinos were emitted at the same time, then we can use

$$t_{arr} = \frac{D}{v} \quad \longrightarrow \quad \Delta t_{arr} = \frac{D}{v^2}\Delta v \leq \Delta t_{obs} \tag{9.111}$$

to compare to the observed spread in times.

(a) Use this logic to find an approximate upper bound on the neutrino mass, in *eV*.
Note: An avalanche of papers appeared almost immediately after these events were announced, with a similar analysis to what you're asked to reproduce here, such as Arnett and Rosner (1987), Ellis and Olive (1987), and Schaeffer et al. (1987). An exploration of the probability of determining the upper limit of the neutrino mass from the time of flight from a nearby supernova was actually done by Zatsepin as early as 1968.

(b) If you were going to try to explore this effect using only DA, you could write $m_\nu c^2 \propto E_\nu^\alpha (\Delta E_\nu)^\beta c^\gamma D^\gamma (\Delta t_{obs})^\epsilon$. Try matching dimensions and see how far you get. Can you write your result from part (a) in such a way that the bound on $m_\nu c^2$ is in this form and see if it's at least consistent?

P9.25 Cosmological constant: Einstein originally postulated an additional term on the LHS of Eqn. (9.60), namely one of the form $\Lambda g_{\mu\nu}$ where Λ is the **cosmological constant.**

(a) What are the dimensions of $[\Lambda]$?

(b) If Λ is related to the Hubble parameter (H_0), which describes the expansion of the universe, the speed of light (c), and G, what can you tell about such a dependence from writing $\Lambda \propto H_0^\alpha c^\beta G^\gamma$? Recall that the Hubble parameter is the slope of the roughly linear relation between the speed of galaxies versus their distance, so a plot of v versus d, with $v \sim H_0 d$.

(c) If the Hubble constant is roughly $H_0 \approx (60\text{–}70)(km/s)/Mpc$, where Mpc is *megaparsec*, find the value of Λ.

(d) The **critical mass density** to close the universe is often cited as $\rho_c = 3 \times H_0^2/(8\pi G)$. Evaluate this in kg/m^3 and show that it's roughly five hydrogen atoms per cubic meter. Show that the corresponding **critical energy density** is roughly 3 keV/cm^3.

P9.26 Electro-static to gravitational strength for the hydrogen atom: (a) Evaluate the ratio of the gravitational to electro-static force for the electron and proton in a hydrogen atom, noting that it's independent of distance, and given by

$$\frac{F_G}{F_E} = \frac{G m_e m_p}{e^2/4\pi\epsilon_0} = \frac{G m_e m_p}{K_e}. \tag{9.112}$$

(b) One (now long outdated) explanation[12] for the expansion of the universe is to assume that there is a small difference, Δe, between the electron and proton charges (but that the neutron is still neutral), so that each atom has a net charge of Δe. The gravitational force between say the Earth and Moon would be $F_G = GM_E M_M/r^2$, while the electro-static force would now be $F_E = Q_E Q_M/4\pi\epsilon_0 r^2$, where

$$Q_{E,M} \approx \Delta e N_{E,M} = \Delta e\left(\frac{M_{E,M}}{m_p}\right) \tag{9.113}$$

if we assume that the mass of a macroscopic object is due to the protons. If these were to balance, what would the value of $\Delta e/e$ be? Does your limit depend on which two macroscopic bodies are being considered? If we know that $F_E/F_G < R$, where $R < 1$ is some (presumably small) number, how does the bound you obtain depend on R?

Note: Recall the discussion in Section 1.2.2 (between Eqns. (1.47) and (1.48)), which included experimental bounds on $\Delta e/e$.

P9.27 Stress-energy tensor for the EM field: Besides being useful for gravitational physics, many energy–momentum relationships in electromagnetism can be nicely described via the stress-energy tensor formalism in Sec. 9.4.1. The most obvious component for the E&M stress tensor is

$$T^{00} = \frac{\epsilon_0}{2}|\mathbf{E}|^2 + \frac{1}{2\mu_0}|\mathbf{B}|^2, \tag{9.114}$$

which are just the energy densities U_E, U_B of the EM fields and is the clear analog of Eqn. (9.63). The T^{ij} components given by

$$T^{ij} = \epsilon_0 E_i E_j + \frac{1}{\mu_0} B_i B_j - \frac{1}{2}\left(\frac{\epsilon_0}{2}|\mathbf{E}|^2 + \frac{1}{2\mu_0}|\mathbf{B}|^2\right)\delta_{ij}, \tag{9.115}$$

which are similar to those for the fluid case in Eqn. (9.63). Given that long introduction, the question is, what do you think are the relevant entries for the T^{0i} components, which for the fluid case, represented the momentum densities? What combination(s) of $\mathbf{E}, \mathbf{B}$ values and c would make sense?

P9.28 Classic tests of GR I—Deflection of light: Light from distant objects can be deflected by the gravitational field of intervening objects, with an apparent angular shift in the position on the sky given by $\theta = C_\theta G^\alpha M_*^\beta R_*^\gamma c^\delta$, where M_* is the mass of the object and R_* the grazing distance, often being close to the edge of the star.

(a) Match dimensions and determine the exponents to the extent you can. The simplest assumption is that this effect is proportional to G, which turns out to give the correct scaling with the variables.

(b) An analysis of this problem based purely on Newtonian mechanics gives $C_\theta = 2$, while a complete analysis using general relativity finds $C_\theta = 4$. Use your results from part (a) and $C_\theta = 4$ to evaluate θ relevant for starlight passing by our own Sun. Compare your value to that observed by the famous 1919 expedition which was $(1.6 \pm .3)''$, where the units used are $''$ = *arcseconds*.

[12] See Lyttleton and Bondi (1959).

Notes: The problem of the deflection of light goes back to Newton and two nice history of science papers (Sauer 2021, and Lotze and Simionato 2022) explore the development of predictions for this effect and include many original references. Will (2015) provides a historical description of the 1919 solar eclipse expedition by Dyson, Eddington, and Davidson (1920), which confirmed Einstein's prediction.

P9.29 Classic tests of GR II—Precession of perihelia of planets: Point-like, two-body systems interacting by only an inverse square law of gravity are described by elliptical orbits, where the "long" axis remains fixed in space, and in many cases we consider only one large-mass star and a smaller-mass planet. Other effects, such as a quadrupole moment of the star, and especially the interactions with other planets, can cause the semi-major axis (and point of perihelion) to precess, and values of the precession rate for many of the inner planets are well-measured.

The orbit of Mercury is known to exhibit such a precession, and a recent review article[13] cites it as roughly $d\Phi/dt \approx 575''\ /century$ where again the" units are again in *arcsecond*. The effects of all of the other planetary orbits (and of the sun's quadrupole moment) are estimated to account for a total of $532''\ /century$, leaving about $43''\ /century$ unaccounted for when all classical gravitational effects are included.

(a) General relativity predicts just such a precession, and all textbook presentations calculate the precession angle due to GR per revolution, so predict $d\Phi_G$ for one cycle. Given that this should depend on G, M_*, c and now the relevant length being the semi-major axis of the orbit a, one could simply repeat **P9.28**, making the same assumptions about the dependence on G. Instead, we note that one famous textbook cites a value of

$$d\Phi_G = \frac{6\pi M_*}{a(1-e^2)}, \tag{9.116}$$

where e is the (dimensionless) eccentricity of the orbit. Add enough factors of G and c to make this dimensionally correct.

(b) The table below presents values of a, e and the orbital period τ for the five inner most planets and ask you to use Eqn. (9.116) to estimate the general relativistic correction to $d\Phi/dt$ for each body. For Mercury, compare to the experimentally well-measured "deficit" noted above.

Planet	a	e	τ
Mercury	$5.8 \times 10^{10}\ m$	0.205	88 *days*
Venus	$1.1 \times 10^{11}\ m$	0.0068	224 *days*
Earth	$1.49 \times 10^{11}\ m$	0.0167	365 *days*
Mars	$2.28 \times 10^{11}\ m$	0.0934	687 *days*
Jupiter	$7.78 \times 10^{11}\ m$	0.049	11.86 *years*

P9.30 Planck force and the rigidity of space-time: (a) Using only the dimensional parameters M, G, c for any massive object, find the combination that produces a **force** (with subscript ST for space-time) by writing $F_{ST} \propto M^\alpha G^\beta c^\gamma$ and solving for the exponents. Repeat, but use only $\hbar, G, c$ as the relevant physical parameters.

[13] See Park et al. (2017).

(b) Do you see any connections between F_{ST} and κ_G as defined in Eqn. (9.61) and the associations suggested in Eqn. (9.70)?
(c) You may have seen demonstrations designed to visualize how massive objects curve space-time, with a rubber sheet stretched tightly over a two-dimensional form, and a heavy object causing a "dimple." Small objects released along the surface roll in what often look like elliptical orbits, determined not by their gravitational interaction, but by the curvature of the two-dimensional surface.

Say we want to construct the analog of a stress, similar to say **Young's modulus** Y for an ordinary material that would express, in some sense, the **rigidity of space-time** itself. In the context of gravitational waves, the only other dimensional quantity besides G and c would be the frequency f of the wave. Write $Y_{ST} \propto G^\alpha c^\beta f^\delta$, match dimensions (recall that $[Y] = [P]$ has the same units as pressure), and solve for the exponents.
(d) For the GR waves observed at LIGO or VIRGO, the frequencies at which events are detected are in the $10^2-10^3\ Hz$ range. Evaluate Y_{ST} for these values and compare to Y for diamond, which is $10^3\ GPa$.
(e) For a gravitational wave, the LHS of the Einstein equation is roughly $\nabla^2 h_{\mu\nu}$ or $\partial^2 h_{\mu\nu}/\partial(ct)^2$, where $g_{\mu\nu} = g_{\mu\nu}^{(flat)} + h_{\mu\nu}$ and h represents the wave-like amplitude on top of the constant (flat) background space-time. For a wave-like solution of the form $h_{\mu\nu} \propto \sin(kx - \omega t)$, the dimensions of the LHS are then determined by k^2 or $(\omega/c)^2$, which if we divided by κ_G would give the dimensions of stress from $T_{\mu\nu}$. Does this analysis agree with your result from part (a)?

P9.31 Planck luminosity: (a) Using only the dimensional parameters M, G, c, find combinations that given an energy and time (separately), then divide the two to find the possible power, and note the mass dependence.
(b) Repeat this exercise, but use $\hbar, G, c$ instead and compare results.
(c) Evaluate the Planck luminosity in units of *erg/s* (typical units used in astrophysics) and compare to two well-known astrophysical events, namely

- SN1987A (the first supernova to produce neutrinos measured on Earth) with $P_\nu \approx (3 \times 10^{53}\ erg)/(3\ sec) \approx 10^{53}\ erg/sec$, and
- GW170729 (one of the largest black hole merger events detected by LIGO/VIRGO), which had an estimated luminosity in gravitational waves of $P_{GW} = 4 \times 10^{56}\ erg/sec$.

P9.32 Gravitational radius: For a compact enough massive (M) object, the escape velocity exceeds the speed of light, giving an event horizon, and the size of a collapsed object for which this occurs is the **Schwarzschild radius**, r_S. This is a special and general relativity effect, but doesn't rely on quantum mechanics. Explore the dependence of r_S on the physical parameters by writing $r_S = C_r G^\alpha M^\beta c^\gamma$ and solve for the exponents. Detailed calculations show that $C_r = 2$.

Note: Some books on general relativity[14] actually use units where $G = c = 1$, so that the Schwarzschild radius and the associated mass are used interchangeably.
(b) Use DA to find the acceleration of gravity g_S of an object of mass M at a distance r_S by writing $g_S = C_g G^\alpha M^\beta c^\gamma$ and matching exponents. Calculations show that $C_g = 1/4$ and you can confirm this by writing $g_S = GM/r_S^2$.

[14] See, e.g., Hartle (2003), their Appendix A.

P9.33 Hawking radiation—Temperature and luminosity: Black holes aren't actually "black" in that they can radiate due to quantum mechanical effects, and have an effective temperature T_H and therefore a luminosity/power output L_H.

(a) Since this is now a quantum **and** a thermal process, we explore the dependence of the Hawking temperature by writing $T_H = C_T \hbar^\alpha c^\beta G^\gamma M^\delta k_B^\sigma$, including both the Planck and Boltzmann constants. Match dimensions and evaluate the exponents to the extent you can. Since it's natural to assume that G and M appear together as GM to some power, does this help you evaluate all of the exponents?

(b) Can you rewrite your final expression for T_H in terms of the Schwarzschild radius r_s? Can you express your result in terms of the acceleration of gravity, g_s, of an object at the Schwarzschild radius that you found in **P9.32**(b)?

(c) Repeat part (a), but find an expression for the luminosity or power L_H, using only $\hbar, c, G, M$ as dimensional variables. Does the same observation about GM appearing together help here?

(d) Repeat parts (a) and (c), but find an expression for the power per unit area, or intensity I_H, using only $\hbar, c, G, M$ as dimensional variables. Does the same observation about GM appearing together help here?

(e) Is your result from part (d) expressible in the form $I_H = L_H/(4\pi r_S^2)$, where r_S is the Schwarzschild radius?

(f) Given your results from parts (a) and (d), is there a (dimensional, at least) connection between I_H and T_H via the Stefan–Boltzmann relation, namely, are they related by $I_H \propto \sigma T_H^4$ where σ is the Stefan–Boltzmann constant, defined in Eqn. (6.10)?

P9.34 Davies–Unruh effect: A theoretical prediction about the thermal impact of an object moving in space-time was made by Davies (1975) and Unruh (1976). The former summarized his main finding in the abstract by saying

> The result is that an observer who undergoes uniform acceleration κ apparently sees a fixed surface radiate with a temperature of $\kappa/2\pi$ (Davies 1975).

In this case, the answer already includes the correct dimensionless factor $(1/2\pi)$, but not any of the possible fundamental constants that might contribute to the effect, all of which have seemingly been set to unity or "assumed."

(a) To see which of those constants, and which types of physics, determine this effect, write $T_{UD} = (a/2\pi)\hbar^\alpha c^\beta G^\gamma k_B^\delta$ (where we use the more traditional notation of a for acceleration instead of κ), match dimensions, and determine all four exponents.

(b) If you've done **P9.33**(b), compare your answer to (a) and comment on any similarities. Could your answer to this question be framed in the context of the **equivalence principle**? (Since we haven't talked about that, you may have to look it up.)

P9.35 Time-dependence of gravitational wave frequency: In a very readable review, Abbot et al. (2017) state that the time-dependence of the frequency of the waveforms they observe due to gravitational radiation is given by

$$f_{GW}^{-8/3}(t) = \frac{(8\pi)^{8/3}}{5}\left(\frac{G\mathcal{M}_c}{c^3}\right)^{5/3}(t_c - t), \tag{9.117}$$

where t_c is the coalescence or collapse time at which the merge is complete which corresponds to $f(t \to t_c) \to \infty$.

(a) Show that this relation is dimensionally correct.

(b)* Integrate the result for $\dot{f}(t)$ from Eqn. (9.100) by writing in the form

$$\int_{f(t)}^{\infty} f^{-11/3}\, df \propto \int_{t}^{t_c} dt \tag{9.118}$$

to reproduce Eqn. (9.117).

(c) Does the relation in Eqn. (9.117), when written in the form $f_{GW}(t) \propto (t - t_c)^{-3/8}$, agree qualitatively with the data in Fig. 1.2?

P9.36 Time-dependence of inspiraling binaries I—Radial separation: In one of the first papers to discuss extracting the chirp mass directly from gravitational wave frequency dependence, Cutler and Flanagan (1994) use units where $G = c = 1$ and you're asked here to "translate" their results into ones with those explicit factors in place.

(a) For Kepler's third law, they write $\Omega = M^{1/2} r^{-3/2}$, where Ω is the angular velocity (we used ω), M is the total mass of the binary, and r is the separation. Add enough powers of G and/or c to make this consistent with our notation.

(b) Using the same gravitational wave power expressions we discussed in Eqn.(9.96), they give an expression for the time-dependence of the radius (here $r(t)$) as

$$\frac{dr}{dt} = -\frac{64}{5}\frac{\mu M^2}{r^3}. \tag{9.119}$$

Add enough powers of G and/or c to the RHS to make this dimensionally correct. Reproduce this expression yourself (so it'll have the G, c already in place) by writing Eqn. (9.96) in terms of just r and $\dot{r}$ by eliminating ω.

(c)* Finally, they present a result for $r(t)$, namely

$$r(t) = \left(\frac{256}{5}\mu M^2\right)^{1/4} (t_c - t)^{1/4}, \tag{9.120}$$

where t_c is again the collapse or coalescence time defined in **P9.35**. Can you reproduce this result by integrating Eqn. (9.119)?

P9.37 Time-dependence of inspiraling binaries II—Period: One of the first (albeit indirect) confirmations of the existence of gravitational waves was the interpretation (by Taylor and Weisberg 1982) of the decrease in the orbital period (P) of the binary pulsar PSR B1913+16 (discovered by Hulse and Taylor (1975)) as being due to energy lost due to gravitational radiation. The formula they used to compare data to the predictions of general relativity was

$$\dot{P} = -\frac{192\pi}{5}\left(\frac{P}{2\pi}\right)^{-5/3}\left(\frac{m_1 m_1}{(m_1 + m_2)^{1/3}}\right) F(e), \tag{9.121}$$

where $F(e)$ is a function of the dimensionless eccentricity, e, a quantity we discuss in more detail in the next problem.

(a) We have left out important factors of G and c in Eqn. (9.121), so add powers of $G^{\alpha}c^{\beta}$ to the RHS, match dimensions, and evaluate the exponents.

(b)* Once you have added factors of G, c to Eqn. (9.121), use that result for $\dot{P}$ to find the rate of increase of the frequency of the emitted radiation, $\dot{f}$, by using the relations

$$\frac{2}{P} = f \quad \text{or} \quad P = \frac{2}{f} \quad \text{and} \quad (\mathcal{M}_c)^5 \equiv \frac{(m_1 m_2)^3}{(m_1 + m_2)}. \tag{9.122}$$

The first two formulae are once again different from the typical $f = 1/P$ frequency–period relation, due to the quadrupolar nature of the gravitational waves. Finally, compare your result to that in Eqn. (9.100), where $f(e) = 1$ is the relevant dimensionless function of the eccentricity.

Notes: Using more than three decades of pulsar timing measurements for PSR B1913+16, Weisberg, Nice, and Taylor (2010) compared theory to experimental data and found that

> The system's orbital period has been decreasing at a rate 0.997 ± 0.002 times that predicted as a result of gravitational radiation damping in general relativity.

We see that the formulae for $\dot{P}$ and $\dot{f}$ are equivalent, with the former being more useful if one is monitoring the period over long intervals of time, and the latter being invaluable in analyzing the direct detection of GR waves near the end of their in-spiral history.

P9.38 Time-dependence for inspiraling binaries III—Angular momentum and eccentricity: In addition to expressions for the rate of energy loss (dE/dt) and the radial separation (dr/dt) for binary systems, Peters (1964) derived expressions for dL/dt and de/dt, where L is the angular momentum and e is the eccentricity (for a non-circular orbit). Recall the definition of e (calculable in terms of the energy E and angular momentum L of an elliptical orbit) as

$$e^2 = \frac{r_a - r_p}{r_a + r_p}, \tag{9.123}$$

where r_a, r_p are the largest (shortest) distances in the elliptical orbit from the focus at the center of mass, meaning that e is dimensionless.

Note: The fact that gravitational waves radiate away angular momentum implies that the inspiraling orbits become increasingly circular.

(a) Peters (1964) expression for dL/dt, without factors of either G or c, is

$$\frac{dL}{dt} = -\frac{32}{5}\frac{\mu^2 M_T^{5/2}}{a^{7/2}} f(e), \tag{9.124}$$

where $f(e) = (1 + 7e^2/8)/(1 - e^2)^2$ is a dimensionless function of the eccentricity. Add enough powers of G and/or c to make this dimensionally correct.

(b) The corresponding expression for the rate of change of the eccentricity itself is

$$\frac{de}{dt} = -\frac{304}{15}\frac{\mu M_T^2}{a^4}(e\, g(e)), \tag{9.125}$$

with $g(e) = (1 + 121e^2/304)/(1 - e^2)^{5/2}$ and we've kept the explicit linear dependence on e separate. Add enough powers of G and/or c to make this dimensionally correct.

(c) Peters (1964) also makes the connection

$$\frac{da}{de} = \frac{12}{19}\left(\frac{a}{e}\right)h(e), \tag{9.126}$$

where $h(e) \approx 1$ for small e. Using the results of **P9.36**(b) and part (b) just above, confirm this result and show that it implies the relation $a(e) \propto e^{12/19}$.

Notes: I wanted to include this final result, as it's one of the least obvious of all power-law dependences that I've ever encountered in exploring this topic! The result of part (b) implies that the eccentricity of an inspiraling binary system decreases in time, becoming more nearly circular, and driving $e \to 0$ in other formulae.

P9.39 Dimensionless spin vector for binary systems: Another measurable property of a binary system emitting gravitational radiation is its spin angular momentum, **S**. This is most frequently given in terms of a dimensionless combination called χ, which depends on **S**, but also includes powers of G, c, and the mass m. Write $\chi \propto S^{\alpha} c^{\beta} G^{\gamma} m^{\delta}$, match dimensions and solve for the exponents to the extent you can. Given that both χ and **S** are vectors, does this help you completely constrain all of the exponents?

P9.40 Density of gravitational wave sources from frequency data: (a) Recalling a result from classical physics already discussed in **P2.15** and **P3.14**, explore the likely frequency dependence of oscillations due to gravitational attraction on Newton's constant G, the density ρ_m, and the size of the system d, by writing $f \propto G^{\alpha} \rho_m^{\beta} d^{\gamma}$ and solving for the exponents.
(b) Inserting a value for G and knowing that gravitational waves are observed in the *kHz* range, what does this imply about the densities of the material in the system producing them? Does this seem to be consistent with some of the neutron star values from **P9.20**?

References for Chapter 9

Abbot, B. P., et al. (LIGO and VIRGO Collaborations). (2017). "The Basic Physics of Binary Black Hole Merger GW150914," *Annalen der Physik* 529, 1600209.

Alboussiére, T., et al. (2011). "Experimental Evidence of Alfvén Wave Propagation in a Gallium Alloy," *Physics of Fluids* 23, 096601.

Alfvén, H. (1942). "Existence of Electromagnetic-Hydrodynamic Waves," *Nature* 150, 405–6.

Alfvén, H. (1950). *Cosmical Electrodynamics* (Oxford: Oxford University Press).

Arnett, W. D., and Rosner, J. L. (1987). "Neutrino Mass Limits from SN1987A," *Physical Review Letters* 58, 1906–9.

Bionta, R. M., et al. (IMB Collaboration). (1987). "Observation of a Neutrino Burst in Coincidence with Supernova 1987A in the Large Magellanic Cloud," *Physical Review Letters* 58, 1494–6.

Blanc, M., Kallenbach, R., and Erkaev, N. V. (2005). "Solar System Magnetospheres," *Space Science Reviews* 116, 227–98.

Bohm, D. (1949). "Qualitative Description of the Arc Plasma in a Magnetic Field." In *The Characteristics of Electrical Discharges in Magnetic Fields*, edited by A. Guthrie and R. K. Wakerling, 1–12. (New York: McGraw-Hill).

Bohm, D., and Gross, E. P. (1949a). "Theory of Plasma Oscillations. A. Origin of Medium-Like Behavior," *Physical Review* 75, 1851–64.

Bohm, D., and Gross, E. P. (1949b). "Theory of Plasma Oscillations. B. Excitation and Damping of Oscillations," *Physical Review* 75, 1865–76.

Bonanno, A., and Urpin, V. (2015). "The Accretion Rate and Minimum Spin Period of Accreting Pulsars," *Monthly Notices of the Royal Astronomical Society* 451, 2117–22.

Bonizzoni, G., and Vassalo, E. (2002). "Plasma Physics and Technology; Industrial Applications," *Vacuum* 64, 327–36.

Buck, A., et al. (2011). "Real-Time Observation of Laser-Driven Electron Acceleration," *Nature Physics* 7, 543–8.

Cardoso, V., et al. (2018). "Remarks on the Maximum Luminosity," *Physical Review D* 97, 084013.

Carroll, S. M. (2004). *Spacetime and Geometry: An Introduction to General Relativity* (San Francisco: Addison-Wesley).

Chapman, S., and Ferraro, V. C. A. (1930). "A New Theory of Magnetic Storms," *Nature* 126, 129–30.

Chapman, S., and Ferraro, V. C. A. (1931). "A New Theory of Magnetic Storms," *Terrestrial Magnetism and Atmospheric Electricity* 36, 77–97 and 171–86.

Chen, F. F. (2018). *Introduction to Plasma Physics and Controlled Fusion*, 3rd ed. (Cham: Springer).

Cirtain, J. W., et al. (2007). "Evidence for Alfvén Waves in Solar X-Ray Jets," *Science* 318, 1580–2.

Clemmow, P. C., and Dougherty, J. P. (1969). *Electrodynamics of Particles and Plasmas* (Reading: Addison-Wesley).

Cutler, C., and Flanagan, É. E. (1994). "Gravitational Waves from Merging Compact Binaries: How Accurately Can One Extract the Binary's Parameters from the Inspiral Waveform?," *Physical Review D* 49, 2658–97.

Davidson, P. A. (2001). *An Introduction to Magnetohydrodynamics* (Cambridge: Cambridge University Press).

Davies, P. C. W. (1975). "Scalar Particle Production in Schwarzschild and Rindler Metrics," *Journal of Physics A: Mathematical and Theoretical* 8, 609–16.

de Pontieu, B., et al. (2007). "Chromospheric Alfvénic Waves Strong Enough to Power the Solar Wind," *Science* 318(5856), 1574–7.

Diver, D. A. (2001). *A Plasma Formulary for Physics, Technology, and Astrophysics* (New York: Wiley-VCH).

Domonkos, M., et al. "Applications of Cold Atmospheric Pressure Plasma Technology in Medicine, Agriculture, and Food Industry," *Applied Sciences* 11, 4809.

Dreicer, H. (1959). "Electron and Ion Runaway in a Fully Ionized Gas. I," *Physical Review* 115, 238–49.

Dreicer, H. (1960). "Electron and Ion Runaway in a Fully Ionized Gas. II," *Physical Review* 117, 329–42.

Dyson, F. W., Eddington, A. S., and Davidson, C. (1920). "A Determination of the Deflection of Light by the Sun's Gravitational Field from Observations Made at the Total Eclipse of May 29, 1919," *Philosophical Transactions of the Royal Society of London. Series A, Containing Papers of a Mathematical or Physical Character* 220, 291–333.

Ellis, J., and Olive, K. A. (1987). "Constraints on Light Particles from Supernova 1987A," *Physics Letters B* 193, 525–30.

Esarey, E., Schroeder, C. B., and Leemans, W. P. (2009). "Physics of Laser-Driven Plasma-Based Electron Accelerators," *Reviews of Modern Physics* 81, 1229–85.

Feynman, R. P., Leighton, R. B., and Sands, M. (1964). *The Feynman Lectures on Physics Vol II: Mainly Electromagnetism and Matter* (Reading: Addison-Wesley). See also the (2011) The New Millennium Edition, which is freely available online at https://www.feynmanlectures.caltech.edu/.

Fitzpatrick, R. (2015). *Plasma Physics: An Introduction* (Boca Raton, FL: CRC Press).

Götzfried, J., et al. (2020). "Physics of High-Charge Electron Beams in Laser-Plasma Wakefields," *Physical Review X* 10, 041015.

Hartle, J. B. (2003). *Gravity: An Introduction to Einstein's General Relativity* (San Francisco: Addison-Wesley).

Hirata, K., et al. (Kamiokande II Collaboration). (1987). "Observation of a Neutrino Burst from the Supernova SN1987A," *Physical Review Letters* 58, 1490–3.

Hogan, C. J. (1999). "Energy Flow in the Universe." In *Structure Formation in the Universe*, edited by R. G. Crittenden and N. G. Turok, 283–93. (Dordrecht: Kluwer Academic Publishers).

Huba, J. D. (2013). *NRL Plasma Formulary* (Washington, DC: Naval Research Laboratory).
Hulse, R. A., and Taylor, J. H. (1975). "Discovery of a Pulsar in a Binary System," *The Astrophysical Journal* 195, L51–53.
Isi, M., et al. (2021), "Testing the Black-Hole Area Law with GW150914," *Physical Review Letters* 127, 011103.
Jackson, D. P., and Chang, M-H. (2021). "Acoustic Levitation and the Acoustic Radiation Force," *American Journal of Physics* 89, 383–92.
Jephcott, D. F. (1959). "Alfvén Waves in a Gas Discharge," *Nature* 183, 1652–4.
Kulsrud, R. M. (2005). *Plasma Physics for Astrophysics* (Princeton: Princeton University Press).
Landau, L. D., and Lifshitz, E. M. (1975). *The Classical Theory of Fields*, 4th rev. English ed. (New York: Pergamon Press).
Longair, M. S. (2011). *High Energy Astrophysics*, 3rd ed. (Cambridge: Cambridge University Press).
Lotze, K-H., and Simionato, S. (2022). "Henry Cavendish on Gravitational Deflection of Light," *Annalen der Physik* 534, 2200102.
Lyttleton, R. A., and Bondi, H. (1959). "Physical Consequences of a General Excess of Charge," *Proceedings of the Royal Society of London. Series A, Mathematical and Physical Sciences* 252, 313–33.
Miao, B., et al. (2022). "Multi-GeV Electron Bunches from an All-Optical Laser Wakefield Accelerator," *Physical Review X* 12, 031038.
Miao, B., Shrock, J., and Milchberg, H. (2023). "Electrons See the Guiding Light," *Physics Today* 76(8), 54–55.
Misner, C. W., Thorne, K. S., and Wheeler, J. A. (1973). *Gravitation* (San Francisco: W. H. Freeman and Co.).
Motz, R. O. (1966). "Alfvén Wave Generation in a Spherical System," *Physics of Fluids* 9, 411–12.
Okamoto, T. J., et al. (2007). "Coronal Transverse Magnetohydrodynamic Waves in a Solar Prominence," *Science* 318, 1577–80.
Park, R. S., et al. (2017). "Precession of Mercury's Perihelion from Ranging to the *MESSENGER* Spacecraft," *The Astronomical Journal* 153, 121.
Peters, P. C. (1964). "Gravitational Radiation and the Motion of Two Points Masses," *Physical Review* 136, B1224–32.
Reitz, J. R., Milford, F. J., and Christy, R. W. (2008). *Foundations of Electromagnetic Theory*, 4th ed. (San Francisco: Addison-Wesley).
Richardson, A. S. (2019). *2019 NRL Plasma Formulary* (Washington, DC: U.S. Naval Research Laboratory).
Sathyaprakash, B. S., and Schutz, B. F. (2009). "Physics, Astrophysics, and Cosmology with Gravitational Waves," *Living Reviews in Relativity* 12, 2.
Sauer, T. (2021). "Soldner, Einstein, Gravitational Light Deflection, and Factors of Two," *Annalen der Physik* 533, 2100203.
Schaeffer, R., Declais, Y., and Jullian, S. (1987). "The Neutrino Emission from SN1987A," *Nature* 300, 142–4.
Schroeder, J. W. R., et al. (2021). "Laboratory Measurements of the Physics of Auroral Electron Acceleration by Alfvén Waves," *Nature Communications* 12, 3103.
Spitzer, L., Jr., and Härm, R. (1953). "Transport Phenomena in a Completely Ionized Gas," *Physical Review* 89, 977–81.
Stage, M. D., et al. (2006). "Cosmic-Ray Diffusion near the Bohm Limit in the Cassiopeia A Supernova Remnant," *Nature Physics* 21, 614–19.
Tajima, T., and Dawson, J. M. (1979). "Laser Electron Accelerator," *Physical Review Letters* 43, 267–70.
Tajima, T., Yan, X. Q., and Ebisuzaki, T. (2020). "Wakefield Acceleration," *Reviews of Modern Plasma Physics* 4, 7.
Taylor, J. H., and Weisberg, J. M. (1982). "A New Test of General Relativity: Gravitational Radiation and the Binary Pulsar PSR 1913+16," *The Astrophysical Journal* 253, 908–20.
Tonks, L., and Langmuir, J. (1929). "Oscillations in Ionized Gases," *Physical Review D* 33, 195–210.
Unruh, W. G. (1976). "Notes on Black-Hole Evaporation," *Physical Review D* 14 870–892.

Weinberg, S. (1972). *Gravitation and Cosmology: Principles and Applications of the General Theory of Relativity* (New York: Wiley).

Weisberg, J. M., Nice, D. J., and Taylor, J. H. (2010). "Timing Measurements of the Relativistic Binary Pulsar PSR B1913+16," *The Astrophysical Journal* 722, 1030–1034.

Will, C. M. (2015). "The 1919 Measurement of the Deflection of Light," *Classical and Quantum Gravity* 32, 124001.

Zatsepin, G. I. (1968). "On the Possibility of Determining the Upper Limit of the Neutrino Mass by Means of the Flight Time," *Zhurnal Eksperimental'noi i Teroreticheskoi Fiziki* 8, 333–4 (translation in *Journal of Experimental and Theoretical Physics Letters* 8, 305–6.)

Part IV

Resources for exploration

Chapter 10

Ancillary material

Outside the world of publishing, the word *appendix* may well bring to mind images of a human internal organ, of uncertain function, and which most people can safely ignore, and possibly even remove with little or no ill effect. On the other hand, the word *ancillary* can be defined using phrases such as ". . . providing necessary support to the main work or activities . . ." and that is the spirit in which we collect the topics included in this final chapter as a series of additional sections.

It might well seem redundant to include a collection of "facts" in the form of traditional appendices, given that Google and Wikipedia are just a few keystrokes away, using the keyboard of any computer, tablet, or smart phone, or that voice recognition software can access much of the same information.[1] But to allow this book to be as self-contained (and hopefully as useful) as possible, we include here a collection of resources that might facilitate checking most of the statements in the text, verifying some of the worked out **Examples**, or help in solving the homework problems, without having to refer to outside sources.

Besides including reminders about the dimensional spelling of many quantities, the values of some important physical constants, and statements of some basic calculus results, we also stress again the utility of symbolic manipulation (more Mathematica© code) and fitting data to confirm many of the scaling laws and parameter dependences that can be derived using dimensional analysis (DA). We end with a brief discussion of some of the history of DA and the foundational Buckingham Pi theorem, as well as some personal comments.

10.1 DA "dictionary"

In this section, we include the name, typical mathematical symbol(s), M, L, T, Q, Θ dimensions, and *MKSA* units for some quantities relevant to mechanics, electricity and magnetism (E&M), and thermal physics. We recall that no new base dimensionful variables are needed for quantum mechanics/relativity.

[1] On July 8, 2023, I had the following "conversation" with Alexa. **Me:** "Alexa, what is the value of Planck's constant?" **Alexa:** "The Planck constant is about 6.6 times 10 to the power of -34 joule-seconds."

Dimensional Analysis Across the Landscape of Physics. Richard W. Robinett, Oxford University Press. © Richard W. Robinett (2024).
DOI: 10.1093/oso/9780192867551.003.0010

Table 10.1 Dimensional "spelling" of, and MKS units for, quantities in mechanics in terms of length (M), mass (K), and time (T) base dimensions, and m = *meter*, kg = *kilogram*, and s = *second* units. We refer here to MKS, as we have yet to introduce a base dimensionful unit corresponding to E&M. Any dimensionless quantity (such as angle or solid angle) corresponds to dimension 1. We recall that the dimensions of any vector quantity are the same as those of its magnitude, hence $[\mathbf{v}] = [v]$ are both included.

Quantity	Symbol(s)	Dimensions	MKS units
Mass		M	kg
Length		L	m
Time		T	s
Area	A, a	L^2	m^2
Volume	V	L^3	m^3
Speed/velocity	$v, \mathbf{v}$	L/T	m/s
Acceleration	$a, \mathbf{a}$	$\mathrm{L/T}^2$	m/s^2
Force	$F, \mathbf{F}, \mathcal{F}$	$\mathrm{ML/T}^2$	$(kg \cdot m)/s^2 = Newton = N$
Spring constant	k	$\mathrm{M/T}^2$	$kg/s^2 = N/m$
Energy/work	E, W	$\mathrm{ML}^2/\mathrm{T}^2$	$(kg \cdot m^2)/s^2 = N \cdot m = Joule = J$
Energy density	$U = E/V$	$\mathrm{M/LT}^2$	$kg/(m \cdot s^2) = J/m^3$
Power	$P = dE/dt$	$\mathrm{ML}^2/\mathrm{T}^3$	$(kg \cdot m^2)/s^3 = J/s = Watt = W$
Intensity	$I = P/A$	$\mathrm{M/T}^3$	$W/m^2 = kg/s^3$
Momentum	$p, \mathbf{p}$	ML/T	$(kg \cdot m)/s$
Torque	$\tau, \boldsymbol{\tau}, \mathcal{T}$	$\mathrm{ML}^2/\mathrm{T}^2$	$(kg \cdot m^2)/s^2 = N \cdot m$ (but not *Joule*!)
Moment of inertia	I	ML^2	$kg \cdot m^2$
Angular momentum	$L, \mathbf{L}, \mathcal{L}$	ML^2/T	$(kg \cdot m^2)/s$
Wavelength	λ	L	m
Wave-number	$k = 2\pi/\lambda$	1/L	$1/m$
Frequency	$f = \omega/2\pi$	1/T	$1/s$
Angular frequency	$\omega = 2\pi f$	1/T	$1/s$
Angle	θ	1	*dimensionless*
Angular velocity	$\omega, \boldsymbol{\omega}$	1/T	$1/s$
Angular acceleration	$\alpha, \boldsymbol{\alpha}$	$1/\mathrm{T}^2$	$1/s^2$
Mass density	ρ_m	$\mathrm{M/L}^3$	kg/m^3
Pressure	P	$\mathrm{M/LT}^2$	$kg/(m \cdot s^2) = N/m^2 = Pascal = Pa$
Bulk modulus	B	$\mathrm{M/LT}^2$	Pa
Young's modulus	Y, B	$\mathrm{M/LT}^2$	Pa
Shear modulus	S, B	$\mathrm{M/LT}^2$	Pa
Viscosity	η, μ	M/LT	$kg/(m \cdot s) = (N \cdot s)/m^2 = Pa \cdot s$
Kinematic viscosity	$\nu = \eta/\rho_m$	L^2/T	m^2/s
Surface tension	γ, S	$\mathrm{M/T}^2$	$kg/s^2 = N/m$
Diffusion coefficient	D	L^2/T	m^2/s
Solid angle	Ω	1	*dimensionless*

Table 10.2 Dimensional "spelling" of, and MKSA units for, quantities in E&M. The (A) in the MKS(A) column is in parenthesis to emphasize that we express quantities using $C = Coulomb$ for charge Q instead of the MKSA system of units, which uses the *Ampere* as the base quantity for E&M. We discuss the reasoning behind using SI units in Section 10.6. We will not make use of the various auxiliary E&M fields often encountered when dealing with materials, including the **displacement vector** ($\boldsymbol{D}$) or **polarization field** ($\boldsymbol{P}$), or their magnetic analogs, $\boldsymbol{H}$ and $\boldsymbol{M}$, for **magnetic induction** and **magnetization vector**, nor the (dimensionless) **electric** and **magnetic susceptibilities** (χ_e, χ_m). We do include them here for completeness, in case the interested reader encounters them in their exploration of applications of E&M. **Note:** We have sometimes switched symbols for magnetic moment from m, m_0 to μ, especially when discussing fundamental particles, to avoid confusion with their mass (also m).

Quantity	Symbol	Dimensions	MKS(A) units
Charge	q	Q	$Coulomb = C$
Charge density	ρ_e	Q/L^3	C/m^3
Current	$I = dq/dt$	Q/T	$C/s = Ampere = A$
Current density	J_e	Q/TL^2	$C/(s \cdot m^2) = A/m^2$
Electric potential/voltage	V	ML^2/QT^2	$(kg \cdot m^2)/(C \cdot s^2) = J/C = Volt = V$
Electromotive force	$\mathcal{E}$	ML^2/QT^2	$(kg \cdot m^2)(C \cdot s^2) = J/C = Volt = V$
Electric field	$\boldsymbol{E}$	ML/QT^2	$(kg \cdot m)/(C \cdot s^2)) = N/C = V/m$
Electric dipole moment	$p, \boldsymbol{p}$	QL	$C \cdot m$
Electric quadrupole moment	$\mathcal{Q}$	QL^2	$C \cdot m^2$
Polarization density	P	Q/L^2	C/m^2
Capacitance	C	Q^2T^2/ML^2	$(C^2 \cdot s^2)/(kg \cdot m^2) = C/V = Farad = F$
Electric permittivity	ϵ_0, ϵ	Q^2T^2/ML^3	$(C^2 \cdot s^2)/(kg \cdot m^3) = F/m$
Resistance	R	ML^2/Q^2T	$(kg \cdot m^2)/(C^2 \cdot s) = Ohm = \Omega$
Conductance	$G = 1/R$	Q^2T/ML^2	$(C^2 \cdot s)/(kg \cdot m^2) = 1/\Omega = siemens = S$
Electrical resistivity	ρ, η	ML^3/Q^2T	$(kg \cdot m^3)/(C^2 \cdot s) = \Omega \cdot m$
Electrical conductivity	$\sigma_e = 1/\rho$	Q^2T/ML^3	$(C^2 \cdot s)/(kg \cdot m^3) = siemens/m = S/m$
Magnetic field (induction)	$\boldsymbol{B}$	M/QT	$kg/(C \cdot s) = Tesla = T$
Magnetic flux	Φ_B	ML^2/QT	$(kg \cdot m^2)/(C \cdot s) = T \cdot m^2$
Inductance	$L, \mathcal{L}$	ML^2/Q^2	$(kg \cdot m^2)/C^2 = Henry = H$
Magnetic permeability	μ_0	ML/Q^2	$(kg \cdot m)/C^2 = H/m$
Magnetic dipole moment	m, m_0 (or μ)	QL^2/T	$(C \cdot m^2)/s$
Scalar potential	ϕ	ML^2/QT^2	$(kg \cdot m^2)/(C \cdot s^2)$
Vector potential	$\boldsymbol{A}$	ML/QT	$(kg \cdot m)/(C \cdot s)$
Poynting vector	$\boldsymbol{S} = \boldsymbol{E} \times \boldsymbol{B}/\mu_0$	M/T^3	$kg/s^3 = W/m^2$
Auxiliary electric field	$\boldsymbol{D} = \epsilon \boldsymbol{E}$	Q/L^2	C/m^2
Polarization field	$\boldsymbol{P} = \chi_e \epsilon \boldsymbol{E}$	Q/L^2	C/m^2
Auxiliary magnetic field	$\boldsymbol{H} = \boldsymbol{B}/\mu$	Q/TL	$C/(s \cdot m)$
Magnetization	$\boldsymbol{M} = \chi_m \boldsymbol{H}$	Q/TL	$C/(s \cdot m)$
Electric susceptibility	χ_e	1	*dimensionless*
Magnetic susceptibility	χ_m	1	*dimensionless*

Table 10.3 Dimensional "spelling" of, and MKSA units for, thermal quantities.

Quantity	Symbol	Dimensions	MKS(A) units
Temperature (degree)	T	Θ	$Kelvin = K$
Heat (energy, work)	$Q\ (E, W)$	ML^2/T^2	$(kg \cdot m^2)/s^2 = J$
Entropy	S	$ML^2/T^2\Theta$	$(kg \cdot m^2)/(s^2 \cdot K) = J/K$
Specific heat (per unit mass)	C	$L^2/T^2\Theta$	$m^2/(s^2 \cdot K)$
Latent heat (fusion or vaporization)	$\mathcal{L}_F, \mathcal{L}_V$	L^2/T^2	m^2/s^2
Thermal conductivity	κ	$ML/T^3\Theta$	$(kg \cdot m)/(s^3 \cdot K)$

Table 10.4 Dimensional "spelling" of, and MKSA units for, quantities related to medical physics and radiation as cited in current SI units.

Quantity	Symbol	Dimensions	MKSA units
Absorbed quantity (gray)	Gy	L^2/T^2	J/kg or m^2/s^2
Dose equivalent (sievert)	Sv	L^2/T^2	J/kg or m^2/s^2
Activity of radionuclide (becquerel)	Bq	$1/T$	$1/s$

Table 10.5 Dimensional "spelling" of some fundamental (universal) physical constants.

Physical quantity	Symbol	Dimensions
Newton's gravitational constant	G	L^3/MT^2
Fundamental charge	e	Q
Fundamental electric constant (permittivity)	ϵ_0	Q^2T^2/ML^3
Electrostatic interaction "strength"	$K_e \equiv e^2/(4\pi\epsilon_0)$	ML^3/T^2
Fundamental magnetic constant (permeability)	μ_0	ML/Q^2
Boltzmann's constant	k_B	$ML^2/\Theta T^2$
Avogadro's number	N_A	*dimensionless*
Gas constant	$R = N_A k_B$	$ML^2/\Theta T^2$
Planck's constant	$\hbar$ (or h)	ML^2/T
Speed of light	$c = 1/\sqrt{\epsilon_0\mu_0}$	L/T
Stefan–Boltzmann constant	$\sigma \equiv (2\pi^5 k_B^4)/(15c^2h^3)$	$M/T^3\Theta^4$
Radiation constant	$a = 4\sigma/c$	$M/LT^2\Theta^4$
Bohr magneton	$\mu_B = e\hbar/2m_e$	QL^2/T
Quantum of flux	$\Phi_0 = h/2e$	ML^2/QT
Quantum of circulation	$C_0 = h/m$	L^2/T
Quantum of resistance (von Klitzing constant)	$R_K = h/e^2$	ML^2/Q^2T
Quantum of conductance	$G_K = 2/R_K = 2e^2/h$	Q^2T/ML^2

10.2 **SI prefixes**

Table 10.6 SI prefixes for "powers of ten".

Power	Name	Prefix	Power	Name	Prefix
10^{24}	yotta	Y	10^{-1}	deci	d
10^{21}	zetta	Z	10^{-2}	centi	c
10^{18}	exa	E	10^{-3}	milli	m
10^{15}	peta	P	10^{-6}	micro	μ
10^{12}	tera	T	10^{-9}	nano	n
10^{9}	giga	G	10^{-15}	femto	f
10^{6}	mega	M	10^{-18}	atto	a
10^{3}	kilo	k	10^{-21}	zepto	z
10^{1}	deka	da	10^{-24}	yocto	y

10.3 **Mini-handbook of physical constants**

Table 10.7 Table of important physical constants and their SI numerical values and units. See also Table 7.2 for values of rest masses of some elementary particles in MeV/c^2 or GeV/c^2 units.

Symbol	Physical quantity	MKSA numerical value
c	Speed of light	$3.0 \times 10^8\ m/s$
G	Newton's constant	$6.67 \times 10^{-11}\ N \cdot m^2/kg^2$ or $m^3/(kg \cdot s^2)$
g_E	Acceleration of gravity (Earth)	$9.8\ m/s^2$
M_E	Mass of Earth	$5.97 \times 10^{24}\ kg$
R_E	Radius of Earth	$6.36 \times 10^6\ m$
D_{ES} (or AU)	Earth–Sun distance	$1.50 \times 10^{11}\ m$
M_S (or $M_\odot$)	Mass of Sun	$1.99 \times 10^{30}\ kg$
R_S (or $R_\odot$)	Radius of Sun	$6.96 \times 10^8\ m$
L_S (or $L_\odot$)	Luminosity of the Sun	$3.8 \times 10^{26}\ W$
$\hbar$	Planck's constant $(h/2\pi)$	$1.05 \times 10^{-34}\ J \cdot s$ or $6.58 \times 10^{-16}\ eV \cdot s$
h	Planck's constant $(2\pi\hbar)$	$6.63 \times 10^{-34}\ J \cdot s$ or $4.14 \times 10^{-15}\ eV \cdot s$
$\hbar c$		$3.16 \times 10^{-26}\ J \cdot m$ or $1.97 \times 10^{-7}\ eV \cdot m$
m_p	proton mass	$1.67 \times 10^{-27}\ kg$ or $938.3\ MeV/c^2$
m_n	neutron mass	$1.67 \times 10^{-27}\ kg$ or $939.6\ MeV/c^2$
m_e	electron mass	$9.11 \times 10^{-31}\ kg$ or $0.511\ MeV/c^2$
e	electron/proton charge	$1.6 \times 10^{-19}\ C$
ϵ_0	permittivity constant	$8.85 \times 10^{-12}\ Farad/m$ or $(C^2 s^2)/(kg \cdot m^3)$
$K_e \equiv e^2/(4\pi\epsilon_0)$	electrostatic force "strength"	$2.3 \times 10^{-28}\ kg \cdot m^3/s^2$ or $(N \cdot m^2)$
μ_0	permeability constant	$1.26 \times 10^{-6}\ H/m$ or $(kg \cdot m)/C^2$

Continued

Table 10.7 *Continued*

Symbol	Physical quantity	MKSA numerical value
k_B	Boltzmann's constant	$1.38 \times 10^{-23}\ J/K$ or $8.62 \times 10^{-5}\ eV/K$
T_r	room temperature	$\sim 295\ K$
$k_B T_r$	at room temp	$4 \times 10^{-21}\ J$ or $1/40\ eV$
N_A	Avogadro's constant	6.02×10^{23} (*dimensionless*)
R	Gas constant ($N_A \cdot k_B$)	$8.31\ J/(K \cdot mole)$
σ	Stefan–Boltzmann constant	$5.67 \times 10^{-8}\ W/(m^2 \cdot K^4)$ or $(kg \cdot m^2)/(s^3 \cdot K^4)$
a_0	Bohr radius	0.53 Å or $5.3 \times 10^{-11}\ m$
R_∞	Rydberg constant	$1.10 \times 10^7\ m^{-1}$
μ_B	Bohr magneton	$9.27 \times 10^{-24}\ J/T$ or $(C \cdot m^2)/s$

10.4 Conversion factors

Table 10.8 Conversion factors, from SI to Gaussian, and others.

$1\ km$	$10^3\ m$
$1\ cm$	$10^{-2}\ m$
$1\ mm$	$10^{-3}\ m$
$1\ \mu m$	$10^{-6}\ m$
$1\ nm$	$10^{-9}\ m$
$1\ parsec = 1\ pc$	$3.09 \times 10^{16}\ m$
$1\ light\text{-}year = 1\ ly$	$9.46 \times 10^{15}\ m$
1 Å	$10^{-10}\ m$
$1\ Fermi = F$	$10^{-15}\ m$
$1\ barn$	$10^{-28}\ m^2$
$1\ eV$	$1.6 \times 10^{-19}\ J$
$1\ gr$	$10^{-3}\ kg$
$1\ dyne = 1\ (gr \cdot cm)/s^2$	$10^{-5}\ N$
$1\ erg = 1\ (gr \cdot cm^2)/s^2$	$10^{-7}\ J$
$1\ Gauss$	$10^{-4}\ Tesla$
$Farad$ = F	$(C^2 \cdot s^2)/(kg \cdot m^2)$
$Henry = H$	$(kg \cdot m^2)/C^2$
$1\ inch$	$2.54\ cm$
$1\ foot$	$12\ inch$
$1\ mile$	$1.61\ km$
$1\ lb$	$4.45\ N$
$1\ cal$	$4.18\ J$

10.5 The Greek alphabet

Table 10.9 Upper and lower case Greek alphabet

Alpha	A	α	Nu	N	ν
Beta	B	β	Xi	Ξ	ξ
Gamma	Γ	γ	Omicron	O	o
Delta	Δ	δ	Pi	Π	π
Epsilon	E	ϵ	Rho	R	ρ
Zeta	Z	ζ	Sigma	Σ	σ
Eta	H	η	Tau	T	τ
Theta	Θ	θ	Upsilon	Υ	υ
Iota	I	ι	Phi	Φ	ϕ
Kappa	K	κ	Chi	X	χ
Lambda	Λ	λ	Psi	Ψ	ψ
Mu	M	μ	Omega	Ω	ω

10.6 Units for electromagnetism: SI/MKSA versus CGS/Gaussian

In order to maximize the utility of this presentation on DA, we have made the very conscious (and we feel obvious) choice to use SI E&M units to describe electromagnetism. The author of one of the most popular and long-standing graduate level textbooks on the subject, J. D. Jackson, after having consistently used Gaussian units in the first two editions, noted in the preface to the third edition (1999) of his famous *Classical Electrodynamics* text that he had chosen to use a hybrid approach, citing

> The most visible change is the use of SI units in the first 10 chapters. Gaussian units are retained in the later chapters, since such units seem more suited to relativity and relativistic electrodynamics than SI.

He then added that

> My tardy adoption of the universally accepted SI system is a recognition that almost all undergraduate physics texts, as well as engineering books at all levels, employ SI units throughout.

It is indeed true that the vast majority of books adopted for algebra-trig or calculus-based first- or second-year intro-level courses, and those used in the undergraduate physics curriculum, are SI-based, as are almost all texts used in every engineering discipline, including, but not restricted to, electrical engineering. In fact, the only popular undergraduate text of which I'm aware that used Gaussian units was one by Purcell, but in the most recent (third) edition of Purcell and Morin (2013), the same switch to SI has been implemented.[2] A popular textbook on plasma physics by Chen (2018) also underwent a similar change in notation in its most recent edition. A few

[2] Jackson (1999) cites a long-term pact between himself and Purcell to "... support each other in the use of Gaussian units ..." and then having "... betrayed him!" in making the switch to SI. Purcell and Morin (2013) in discussing the decision to change to SI units note that "Of course, there are differing opinions as to which system of units is 'better' for an introductory course. But this issue is moot, given the reality of these courses."

Table 10.10 Conversion factors from SI to Gaussian (G) units for many E&M source and field quantities.

(electric charge)	$q_{SI} = q\sqrt{4\pi\epsilon_0}$	$\boldsymbol{E}_{SI} = \boldsymbol{E}/\sqrt{4\pi\epsilon_0}$	(electric field)
(charge density)	$\rho_{SI} = \rho\sqrt{4\pi\epsilon_0}$	$V_{SI} = V/\sqrt{4\pi\epsilon_0}$	(voltage)
(current)	$I_{SI} = I\sqrt{4\pi\epsilon_0}$	$\phi_{SI} = \phi/\sqrt{4\pi\epsilon_0}$	(scalar potential)
(current density)	$J_{SI} = J\sqrt{4\pi\epsilon_0}$	$\boldsymbol{B}_{SI} = \boldsymbol{B}\sqrt{\mu_0/4\pi}$	(magnetic field)
(electric dipole moment)	$\boldsymbol{p}_{SI} = \boldsymbol{p}\sqrt{4\pi\epsilon_0}$	$\boldsymbol{A}_{SI} = \boldsymbol{A}\sqrt{\mu_0/4\pi}$	(vector potential)
(magnetic dipole moment)	$\boldsymbol{m}_{SI} = \boldsymbol{m}\sqrt{4\pi/\mu_0}$	$\Phi_{SI} = \Phi\sqrt{\mu_0/4\pi}$	(magnetic flux)

authors have chosen to present results in both units, such as Kittel (1971) in his *Introduction to Solid State Physics*, where he notes that "Nearly every important equation is repeated in SI and in CGS-Gaussian units, wherever they differ."

We do recognize that there can be an "impedance mismatch" for students who explore physics topics at the graduate and/or research level, where CGS-Gaussian units are more popular, and that the resulting E&M equations can have different forms (no ϵ_0 or μ_0 and factors of 4π and c in different places), so we start with some of the conversions that allow one to translate from one system to another, in case your journeys across the "landscape of physics" take you across borders into regions where this language is used.

All quantities in this text are assumed to be in SI (unless stated otherwise), which can then be substituted using the "translations" included in Table 10.10 into any of the equations we've cited to obtain the corresponding formulations in Gaussian (but with no additional subscript). We encourage readers who are interested in the logical structure leading to different choices of E&M units to consult Kowalski (1986), Zimmerman (1998), or the appendices in Jackson (1999) and in Purcell and Morin (2013). A series of interesting discussions by Birge (1934, 1935a, 1935b) are still very readable and each provides some valuable background and historical references. Maxwell (1892/1953) in his *Treatise on Electricity and Magnetism* dedicates a chapter (pp. 620-9) to "Dimensions of Electric Units" exploring different choices such as having a "unit of magnetic pole" as the base E&M unit.

As an example, consider the Lorentz force for a charge q in electric and magnetic fields, namely

$$\begin{aligned}
&\boldsymbol{F} \stackrel{SI}{=} q(\boldsymbol{E} + \boldsymbol{v} \times \boldsymbol{B}) \\
&\downarrow \\
&\boldsymbol{F} = q[\sqrt{4\pi\epsilon_0}\,]\left(\left[\frac{1}{\sqrt{4\pi\epsilon_0}}\right]\boldsymbol{E} + \boldsymbol{v} \times \left[\sqrt{\frac{\mu_0}{4\pi}}\,\right]\boldsymbol{B}\right) \\
&\downarrow \\
&\boldsymbol{F} \stackrel{G}{=} q\left(\boldsymbol{E} + \frac{\boldsymbol{v}}{c} \times \boldsymbol{B}\right),
\end{aligned} \tag{10.1}$$

since $\sqrt{\epsilon_0\mu_0} = 1/c$. One clear benefit of using the Gaussian system of units is that the dimensions of the electric and magnetic fields are the same!

Coulomb's law for the electric field is translated as

$$\boldsymbol{E} \overset{SI}{=} \frac{q\hat{\boldsymbol{r}}}{4\pi\epsilon_0 r^2}$$

$$\downarrow$$

$$\left[\frac{1}{\sqrt{4\pi\epsilon_0}}\right]\boldsymbol{E} = \left[\sqrt{4\pi\epsilon_0}\,\right]\frac{q\hat{\boldsymbol{r}}}{4\pi\epsilon_0 r^2}$$

$$\downarrow$$

$$\boldsymbol{E} \overset{G}{=} \frac{q\hat{\boldsymbol{r}}}{r^2}. \tag{10.2}$$

Perhaps more fundamentally, the related expression for the Coulomb force between two like charges, q, is

$$|\boldsymbol{F}_C| \overset{G}{=} \frac{q^2}{r^2}, \tag{10.3}$$

which is then used to define the *statCoulomb* in the CGS-Gaussian system as the charge which gives a 1-*dyne* force at a 1-*cm* separation. The *dimensions* (and value) of the statCoulomb are then determined by Eqn. (10.3) as

$$[q^2] = [F_C][r^2] = \left(\frac{\mathsf{ML}}{\mathsf{T}^2}\right)\left(\mathsf{L}^2\right) = \frac{\mathsf{ML}^3}{\mathsf{T}^2} \quad\longrightarrow\quad [q] = \left(\frac{\mathsf{ML}^3}{\mathsf{T}^2}\right)^{1/2} = \left(\frac{gr\cdot cm^3}{s^2}\right)^{1/2}. \tag{10.4}$$

The translation between Coulomb and statCoulomb (or perhaps one should say "association", as the equations are different in the two systems) is $1\ C \sim 3\times 10^9\ statC$ and the fundamental electric charge has a value of $e = 1.6\times 10^{-19}\ C = 4.8\times 10^{-10}\ statC$, which you're asked to confirm in **P10.1**.

The Biot–Savart law for the magnetic field due to a small current element is

$$d\boldsymbol{B} \overset{SI}{=} \frac{\mu_0}{4\pi}\frac{I d\boldsymbol{l}\times\hat{\boldsymbol{r}}}{r^2}$$

$$\downarrow$$

$$\left[\sqrt{\frac{\mu_0}{4\pi}}\right]d\boldsymbol{B} = \frac{\mu_0}{4\pi}\frac{\left[\sqrt{4\pi\epsilon_0}\,\right]I d\boldsymbol{l}\times\hat{\boldsymbol{r}}}{r^2}$$

$$\downarrow$$

$$d\boldsymbol{B} \overset{G}{=} \frac{1}{c}\frac{I d\boldsymbol{l}\times\hat{\boldsymbol{r}}}{r^2}, \tag{10.5}$$

where as always we use the fact that $\sqrt{\epsilon_0\mu_0} = 1/c$. The fact that $\boldsymbol{E}, \boldsymbol{B}$ have the same Gaussian dimensions is again clear, given that $[I d\boldsymbol{l}\times\hat{\boldsymbol{r}}/c] = (\mathsf{Q}/\mathsf{T})(\mathsf{L})(1)/(\mathsf{L}/\mathsf{T}) = \mathsf{Q} = [q]$. The common dimensionality of, and symmetry between, electric and magnetic fields is seen perhaps most obviously in their respective energy densities,

$$U_E \overset{SI}{=} \frac{\epsilon_0}{2}|\mathbf{E}|^2 \; U_B \overset{SI}{=} \frac{1}{2\mu_0}|\mathbf{B}|^2$$

$$\downarrow$$

$$U_E = \frac{\epsilon_0}{2}\left[\frac{1}{\sqrt{4\pi\epsilon_0}}\right]^2 |\mathbf{E}|^2 \; U_B = \frac{1}{2\mu_0}\left[\sqrt{\frac{\mu_0}{4\pi}}\right]^2 |\mathbf{B}|^2$$

$$\downarrow$$

$$U_E \overset{G}{=} \frac{1}{8\pi}|\mathbf{E}|^2 \; U_B \overset{G}{=} \frac{1}{8\pi}|\mathbf{B}|^2. \tag{10.6}$$

The forms for Gauss's law and Faraday's law are obtained as

$$\nabla \cdot \mathbf{E} \overset{SI}{=} \frac{1}{\epsilon_0}\rho_e$$

$$\downarrow$$

$$\left[\frac{1}{\sqrt{4\pi\epsilon_0}}\right](\nabla \cdot \mathbf{E}) = \frac{1}{\epsilon_0}\left[\sqrt{4\pi\epsilon_0}\,\right]\rho_e \tag{10.7}$$

$$\downarrow$$

$$\nabla \cdot \mathbf{E} \overset{G}{=} 4\pi\rho_e$$

and

$$\nabla \times \mathbf{B} \overset{SI}{=} \mu_0 \mathbf{J}_e + \mu_0\epsilon_0 \frac{\partial}{\partial t}\mathbf{E}$$

$$\downarrow$$

$$\left[\sqrt{\frac{\mu_0}{4\pi}}\,\right](\nabla \times \mathbf{B}) = \mu_0\left[\sqrt{4\pi\epsilon_0}\,\right]\mathbf{J}_e + \mu_0\epsilon_0\left[\frac{1}{\sqrt{4\pi\epsilon_0}}\right]\frac{\partial}{\partial t}\mathbf{E}$$

$$\downarrow$$

$$\nabla \times \mathbf{B} \overset{G}{=} \frac{4\pi}{c}\mathbf{J}_e + \frac{1}{c}\frac{\partial}{\partial t}\mathbf{E}, \tag{10.8}$$

and other examples are left to the problems.

From **P4.8**, which explored the so-called Hillas bound in cosmic ray physics, you may have found that the maximum energy of a particle of charge q in a region of magnetic field B and size R is

$$E_{max} \overset{SI}{=} qBRc$$
$$\downarrow$$
$$E_{max} = \left(q\sqrt{4\pi\epsilon_0}\right)\left(\sqrt{\frac{\mu_0}{4\pi}}B\right)Rc$$
$$\downarrow$$
$$E_{max} \overset{G}{=} qBR, \tag{10.9}$$

where once again we use $\sqrt{\epsilon_0\mu_0} = 1/c$. As an example of a numerical calculation of a physical quantity in the two systems of units, let us use these results to evaluate the energy for a charge e in a magnetic field $B = 10^2\ G = 10^{-2}\ T$ in a region of space of size $R = 10^{15}\ cm = 10^{13}\ m$. We find

$$E_{max} \overset{SI}{=} (1.6 \times 10^{-19}\ C)(10^{-2}\ T)(10^{13}\ m)(3 \times 10^8 m/s) = 4.8\ J$$
$$E_{max} \overset{G}{=} (4.8 \times 10^{-10}\ statC)(10^2\ G)(10^{15}\ cm) = 4.8 \times 10^7\ erg, \tag{10.10}$$

which do agree.

Finally, some important dimensionful quantities in fundamental physics, such as the fine-structure constant (Section 7.1), the Bohr magneton (Section 6.3.4), and the quantum of magnetic flux (Section 8.2) look slightly different in the two schemes, namely

$$\alpha_{FS} \overset{SI}{=} \frac{e^2}{4\pi\epsilon_0\hbar c} \quad \rightarrow \quad \alpha_{FS} = \left[\sqrt{4\pi\epsilon_0}\right]^2 \frac{e^2}{4\pi\epsilon_0\hbar c} \overset{G}{=} \frac{e^2}{\hbar c} \tag{10.11}$$

$$\mu_B \overset{SI}{=} \frac{e\hbar}{2m} \quad \longrightarrow \quad \left[\sqrt{\frac{4\pi}{\mu_0}}\right]\mu_B = \left[\sqrt{4\pi\epsilon_0}\right]\frac{e\hbar}{2m} \quad \longrightarrow \quad \mu_B = \sqrt{\epsilon_0\mu_0}\,\frac{e\hbar}{2m} \overset{G}{=} \frac{e\hbar}{2mc} \tag{10.12}$$

$$\Phi_0 \overset{SI}{=} \frac{\hbar}{2e} \quad \longrightarrow \quad \left[\sqrt{\frac{\mu_0}{4\pi}}\right]\Phi_0 = \frac{\hbar}{2e}\left[\frac{1}{\sqrt{4\pi\epsilon_0}}\right] \quad \longrightarrow \quad \Phi_0 = \frac{\hbar}{2e}\frac{1}{\sqrt{\epsilon_0\mu_0}} \overset{G}{=} \frac{\hbar c}{2e}. \tag{10.13}$$

10.7 **Used mathematics**

In the same way that values of physical constants and dimensions are easily obtainable from virtual resources, most of the (we hope relatively simple) mathematics used (or cited) in this work can be found in textbooks, or by the use of symbolic math programs. We do think it can be useful to recall some of the simplest results that appear in this book.

For example, the binomial formula can be used in a variety of applications, including a series expansion.

$$(a+b)^n = a^n + na^{n-1}b + \frac{n(n-1)}{2}a^{n-2}b^2 + \cdots + \frac{n(n-1)}{2}a^2b^{n-2} + nab^{n-1} + b^n. \tag{10.14}$$

Properties of logarithms (here referring specifically to natural logs) include:

$$\ln(a\,b) = \ln(a) + \ln(b) \tag{10.15}$$

$$\ln(a/b) = \ln(a) - \ln(b) \tag{10.16}$$

$$\ln(a^c) = c\ln(a) \tag{10.17}$$

$$\ln(ax^b) = \ln(a) + b\ln(x) \tag{10.18}$$

$$\ln(ae^{bx}) = \ln(a) + bx. \tag{10.19}$$

Converting from natural to base-10 logs involves assuming

$$10^x = e^y \quad \rightarrow \quad x = x\log_{10}(10) = y\log_{10}(e) = 0.4343\,y \qquad \text{or} \qquad y = x\ln(10) = 2.303\,x. \tag{10.20}$$

Complex exponentials can be written in terms of trig functions via

$$e^{\pm i\theta} = \cos(\theta) \pm i\sin(\theta) \qquad \text{or} \qquad \cos(\theta) = \frac{1}{2}\left(e^{+i\theta} + e^{-i\theta}\right),\ \sin(\theta) = \frac{1}{2i}\left(e^{+i\theta} - e^{-i\theta}\right). \tag{10.21}$$

Series expansions for $|x| << 1$ include

$$e^x = 1 + x + \frac{x^2}{2!} + \cdots \tag{10.22}$$

$$\ln(1+x) = x - \frac{x^2}{2!} + \frac{x^4}{3!} + \cdots \tag{10.23}$$

$$\sin(x) = x - \frac{x^3}{3!} + \frac{x^5}{5!} + \cdots \tag{10.24}$$

$$\cos(x) = 1 - \frac{x^2}{2} + \frac{x^4}{4!} + \cdots \tag{10.25}$$

$$(1+x)^n = 1 + nx + \frac{n(n-1)}{2}x^2 + \cdots, \tag{10.26}$$

where the last entry can be used for fractional values of n, such as for $\sqrt{1+x}$. We note that these formulae assume that any/all dimensional dependence of the variable x has been taken into account so that $[x] = 1$ is dimensionless.

A few basic integrals are included below:

$$\int \cos(x)\, dx = \sin(x) \tag{10.27}$$

$$\int \sin(x)\, dx = -\cos(x) \tag{10.28}$$

$$\int e^x\, dx = e^x \tag{10.29}$$

$$\int_{-\infty}^{+\infty} e^{-x^2}\, dx = \sqrt{\pi} \tag{10.30}$$

$$\int_{-\infty}^{+\infty} e^{-ax^2-bx}\, dx = \sqrt{\frac{\pi}{a}}\, e^{b^2/4a} \quad \text{(if a>0)} \tag{10.31}$$

$$\int_{-\infty}^{+\infty} x^2 e^{-x^2}\, dx = \frac{\sqrt{\pi}}{2}. \tag{10.32}$$

A useful approximation for the factorial function, $n! \equiv n \cdot (n-1) \cdot (n-2) \cdots 2 \cdot 1$, is called the **Stirling approximation** and can be written as

$$n! \sim \sqrt{2\pi n}\left(\frac{n}{e}\right)^n \qquad \text{so that for large n} \qquad \ln(n!) \sim n\ln(n) - n + \mathcal{O}(\ln(n)). \tag{10.33}$$

This is useful in calculations of entropy, as in Section 6.1.2.

Simple second-order differential equations, and their solutions, appear throughout the physics curriculum, and familiar examples are

$$\mathcal{M}\ddot{x}(t) = -\mathcal{K}x(t) \quad \longrightarrow \quad x(t) = A\cos\left(\sqrt{\frac{\mathcal{K}}{\mathcal{M}}}t\right) + B\sin\left(\sqrt{\frac{\mathcal{K}}{\mathcal{M}}}t\right) \tag{10.34}$$

$$\mathcal{M}\ddot{x}(t) = +\mathcal{K}x(t) \quad \longrightarrow \quad x(t) = \tilde{A}e^{+\sqrt{\mathcal{K}/\mathcal{M}}\,t} + \tilde{B}e^{-\sqrt{\mathcal{K}/\mathcal{M}}\,t}, \tag{10.35}$$

corresponding to oscillatory or exponential time-dependence. The coefficients A, B and $\tilde{A}, \tilde{B}$ are determined by imposing the initial conditions, $x(0)$ and $\dot{x}(0)$.

A more general approach to solving ordinary differential equations with constant coefficients represented by the problem

$$\mathcal{M}\ddot{x}(t) + \mathcal{B}\dot{x}(t) + \mathcal{K}x(t) = 0 \tag{10.36}$$

is to assume an exponential time dependence of the form $x(t) = A\exp(rt)$, which gives (after canceling the remaining common $A\exp(rt)$ factors) a quadratic constraint on the exponent r, namely

$$\mathcal{M}r^2 + \mathcal{B}r + \mathcal{K} = 0 \quad \longrightarrow \quad r_{(\pm)} = \frac{-\mathcal{B} \pm \sqrt{\mathcal{B}^2 - 4\mathcal{M}\mathcal{K}}}{2\mathcal{M}}, \tag{10.37}$$

so that $x(t) = Ae^{r_{(+)}t} + Be^{r_{(-)}t}$, which can correspond to pure oscillatory, pure exponential, or combined solutions using Eqn. (10.21). Examples of such behavior are the damped harmonic oscillator in **Example 3.1** or the RLC circuit discussed in Section 4.5. The notations of $\mathcal{M}$ and $\mathcal{K}$ are designed to be reminiscent of physical parameters representing inertia and spring constant (for the harmonic oscillator), but have analogs in the LC circuit system. Whatever the physical realization leading to an equation of the form in Eqn. (10.36), the coefficients must be dimensionally consistent, as in Eqn. (4.85).

At the other extreme from the formula-driven formal mathematics used in all physics textbooks, the ability to perform "back-of-the-envelope" calculations is another "used math" skill also prized by physicists. Just as DA works to derive the power-law dependences of physical quantities, hoping that any dimensionless prefactor(s) will be within a power of 10 (or so) on either side of unity, so too do "order-of-magnitude" problems try to use physical intuition and general background knowledge/experience to quickly estimate the answers to questions requiring numerical values. These problems can arise in STEM disciplines, but also in almost any other aspect of everyday life, and a number of (rightly) popular treatments on the subject have appeared for quite general audiences. Examples include:

- Harte, J. (1998). *Consider a Spherical Cow: A Course in Environmental Problem Solving* (Melville, NY: University Science Books).
- Harte, J. (2001). *Consider a Cylindrical Cow: More Adventures in Environmental Problem Solving* (Melville, NY: University Science Books).
- Swartz, C. (2003). *Back-of-the-Envelope Physics* (Baltimore: Johns Hopkins University Press).
- Santos, A. (2009). *How Many Licks?: Or, How to Estimate Damn Near Anything* (Philadelphia: Running Press).
- Mahajan, S. (2010). *Street-Fighting Mathematics: The Art of Educated Guessing and Opportunistic Problem Solving* (Cambridge, MA: MIT Press).
- Weinstein, L., and Adam, J. A. (2008). *Guesstimation: Solving the World's Problems on the Back of a Cocktail Napkin* (Princeton: Princeton University Press).
- Weinstein, L. (2012). *Guesstimation 2.0: Solving Today's Problems on the Back of a Napkin* (Princeton: Princeton University Press, Princeton).
- Mahajan, S. (2014). *The Art of Insight in Science and Engineering* (Cambridge, MA: MIT Press) has a section on dimensions.
  - This treatment grew out of Mahajan's (1998) Caltech PhD thesis *Order of Magnitude Physics: A Textbook with Applications to the Retinal Rod and the Density of Prime Numbers*; it's available at https://resolver.caltech.edu/CaltechTHESIS:10302009-110303489.

In the context of physics culture, such "puzzlers" are often referred to as **Fermi problems** after Enrico Fermi, who posed many such challenges, to colleagues and students, some of which were included in the *University of Chicago Graduate Problems in Physics* (Cronin, Greenberg, and Telegdi 1979). One of the more famous of these might be considered the back-of-the-envelope analog of the Taylor–von Neumann–Sedov blast wave calculation considered in Section 2.2. Numerous authors (including von Baeyer 1993, or Burgess 2005) have discussed the quick-and-dirty calculation attributed to Fermi, which Katz (2021) describes as follows:

> The official history of Los Alamos and many popular accounts describe how Enrico Fermi, during Trinity, the first nuclear test conducted July 16, 1945, estimated its explosive yield with a simple experiment. He dropped some small scraps of paper before the blast wave passed by and observed how far they were displaced. From this, he was able to estimate its explosive yield as 10 *kt*. This is about 40% of the modern estimate . . .

Katz (2021) attempts to "reconstruct his reasoning" using still fairly sophisticated arguments, while Weinstein (2012), just cited above, provides an even "quicker and dirtier" estimate.

We won't pursue any further such back-of-the-envelope calculations, save only to quote what is sometimes jokingly referred to the **fundamental theorem of engineering**, and the content of a number of memes, namely

$$\begin{array}{ccccccccc} \pi & = & e & = & \sqrt{10} & = & \sqrt{g} & = & 3 \\ 3.14... & & 2.72... & & 3.16... & & 3.13 & \approx & 3 \end{array}$$

which is not dimensionally consistent, but also not exactly wrong in the spirit for such calculations!

10.8 Sample Mathematica© code for DA

The vast majority of the mathematics actually used in the examples and homework in this text involves solving systems of linear equations. As early as Section 1.4, we have encouraged readers to consider the use of symbolic math programs to automate these processes, and included several examples written in Mathematica©. We continue to lobby for such tools here by presenting three more examples of such code to illustrate several different possible outcomes, namely

- a 5×5 example where $\mathsf{M}, \mathsf{L}, \mathsf{T}, \mathsf{Q}, \Theta$ are all used,
- a 4×4 example where there is actually no solution to the linear algebra problem, and
- a 3×4 example where there are four parameters, but only three constraints from $\mathsf{M}, \mathsf{L}, \mathsf{T}$.

We use `M,L,T,Q,THETA` as names of the dimensional variables in the code, being lucky that Mathematica© does not already have pre-defined variables/constants with any of those symbols! The first example is the Mathematica© version of the Wiedemann-Franz problem in Section 2.4.

```
(* Wiedemann_Franz.nb *)
(* This does the Wiedemann-Franz law problem *)
(* LHS has the thermal and electrical conductivities and the temperature *)
(* RHS has powers of e, n_e, m_e, tau, and k_B *)
```

```
(* This example uses all five base dimensions, namely M, L, T, Q, and THETA *)
(* This is what we need on the LHS *)
sigma = Q^2*T/(M*L^3);                  (* electrical conductivity *)
kappa = (M*L)/(T^3*THETA);              (* thermal conductivity *)
temperature = THETA;                    (* temperature *)
LHS = kappa/sigma/temperature;          (* This is the Lorenz number *)
(* Now do the RHS *)
ee = Q;                                 (* electron charge *)
me = M;                                 (* electron mass *)
edensity = 1/L^3;                       (* carrier density *)
tau  = T;                               (* collision time *)
kB = (M*L^2)/(THETA*T^2);               (* Boltzmann's constant *)
RHS = kB^(alpha) * ee^(beta) * edensity^(gamma) * me^(delta) * tau^(epsilon);
ratio = RHS/LHS;
(* Ratio of RHS to LHS should be dimensionless, as dimensions should match *)
aa = Exponent[ratio, M];
bb = Exponent[ratio, L];
cc = Exponent[ratio, T];
dd = Exponent[ratio, Q];
ee = Exponent[ratio, THETA];
Solve[{0 == aa, 0 == bb, 0 == cc, 0 == dd, 0 == ee}, {alpha, beta, gamma, delta, epsilon}]
(* Press SHIFT+ENTER to compile *)

{{alpha -> 2, beta -> -2, gamma -> 0, delta -> 0, epsilon -> 0}}
```

We see that the output agrees with the "pencil-and-paper" approach from Eqns. (2.17)–(2.20).

The next example is a version of the Larmor problem considered in **Examples 1.2** and **4.9** (and **P4.31**), but asking about the power radiated by a moving charge, one with speed v, and not one that is accelerating.

```
(* Larmor_Speed.nb *)
(* Derives the dependence of the power radiated by charge moving at speed v,
on q, v, epsilon_0, and mu_0 *)
(* This is a variation of the Larmor problem for an accelerating charge *)
(* This one assumes instead that the particle is moving with constant SPEED, v *)
(* All quantities are given in terms of their M, L, T, and Q base dimensions *)
(* This is an example where there is no solution to the system of linear equations
and we want to see what that looks like -- the output is simple, just two empty
'curly brackets' like {} *)
(* This is what we need for the RHS *)
q = Q;                                  (* particle charge dimensions *)
v = L/T;                                (* particle speed dimensions *)
epsilon0 = (Q^2*T^2)/(M*L^3);           (* permittivity dimensions *)
mu0 = (M*L)/Q;                          (* permeability dimensions *)
RHS = q^(alpha) * v^(beta) * epsilon0^(gamma) * mu0^(delta);
(* This is what we need on the LHS *)
power = (M*L^2)/T^3;                    (* power dimensions *)
LHS = power;
```

```
ratio = RHS/LHS;
(* Ratio of RHS to LHS should be dimensionless, as dimensions should match *)
aa = Exponent[ratio, M];
bb = Exponent[ratio, L];
cc = Exponent[ratio, T];
dd = Exponent[ratio, Q];
Solve[{0 == aa, 0 == bb, 0 == cc, 0 == dd}, {alpha, beta, gamma, delta}]
(* Press SHIFT+ENTER to compile *)

{}
```

This is an example where the output is (rightly) the empty set (represented as {} in Mathematica©), as there is no valid solution to this set of constraints, as noted by Larmor himself!

As our final example, we consider the Reynolds number problem of **Example 3.4**, which discusses the drag force F_D on a sphere of diameter D moving through a fluid medium with density ρ_m and viscosity η. In that case, there are four dimensionful variables, and only three M, L, T constraints, leading to the identification of a useful dimension**less** combination, which turns out to be the Reynolds number (or its inverse). We show here how solving for the exponents in two different orders gives the same dimensionless ratio, but with different prefactors.

```
(* Reynolds_Number.nb *)
(* Dimensional analysis for the force on a sphere due to a moving fluids *)
(* Sphere has diameter Dsphere, moving at speed vsphere, through a fluid medium
with density and viscosity *)
density = M/L^3;                    (* density of fluid *)
viscosity = M/(L*T);                (* viscosity of fluid *)
Dsphere = L;                        (* diameter of sphere *)
vsphere = L/T;                      (* velocity of sphere *)
RHS = density^(alpha) * viscosity^(beta) * Dsphere^(gamma) * vsphere^(delta);
(* Force is what we need on the LHS *)
Fdrag = (M*L)/T^2;                  (* drag force *)
LHS = Fdrag;
ratio = RHS/LHS;
aa = Exponent[ratio, M];
bb = Exponent[ratio, L];
cc = Exponent[ratio, T];
(* Solve for the exponents in terms of alpha *)
Solve[{0 == aa, 0 == bb, 0 == cc}, {alpha, beta, gamma, delta}]
(* Press SHIFT+ENTER to compile *)

{{beta -> 1 - alpha, gamma -> 1 + alpha, delta -> 1 + alpha}}

(* Or solve for things in terms of beta *)
Solve[{0 == aa, 0 == bb, 0 == cc}, {beta, alpha, gamma, delta}]

{{alpha -> 1 - beta, gamma -> 2 - beta, delta -> 2 - beta}}
```

These two approaches correspond to equally valid solutions, namely

$$F_D \propto \eta v D \left(\frac{\rho_m v D}{\eta}\right)^{\alpha} \qquad \text{or} \qquad F_D \propto \rho_m v^2 D^2 \left(\frac{\eta}{\rho_m v D}\right)^{\beta}, \tag{10.38}$$

and both results single out the relevant Buckingham Pi theorem dimensionless ratio!

We emphasize that automating DA problems can be done using almost any modern software. For example, Karam and Saad (2021) have produced *BuckinghamPy: A Python Software for Dimensional Analysis*, which is designed to generate the possible dimensionless Π ratios in symbolic terms; see also Dumka et al. (2022). Bakarji et al. (2022) propose numerical techniques (including machine learning tools) that use the Buckingham Pi theorem as a constraint to identify appropriate dimensionless groups which collapse data to lower-dimensional sets in an optimal way.

10.9 **Overlap of data/DA**

In the great feedback loop that is the scientific enterprise, the collection, organization, and analysis of data may be followed by attempts to encode that information in compact ways, perhaps described by functional relationships or mathematical models informed by theoretical expectations, which can then, in turn, be used to make predictions about new experiments, accompanied by estimates of the likely errors arising from all sources of uncertainty along the way. Physicists, with a wide variety of skills and from many disciplines, can play a role in any aspect of this cycle; helping perhaps to extend the boundaries of the landscape of physics, even if only in small ways, or in helping others understand the processes involved and results so obtained. At each stage of the process, dimensions can play an important role in ensuring that any purported relationships are logically consistent, and in this section we briefly review two important aspects of this cycle: the fitting of data (especially to power-law and exponential functional forms) and what role dimensional constraints play in the propagation of uncertainty in error analysis calculations.

10.9.1 **Data fitting**

One of the desired outcomes of DA is to predict the scaling behavior of physical quantities on parameters in the problem, often via a power-law dependence. In addition to trying to do so proactively, by imposing dimensional constraints, we often wish to compare predictions to experimental data, or have such data suggest the functional form of the behavior itself. In either case, having the ability to "match data" to various functional forms is an important tool.

One simple version of such techniques results in the standard definition of the **mean value** for a set of data points. For example, if we ask what value a is the "best fit" to a set of data $\{y_1, y_2, \ldots, y_N\}$ points in an objective way, one might consider the function

$$S(a) \equiv \sum_{i=1}^{N} \delta_i = \sum_{i=1}^{N} (a - y_i), \tag{10.39}$$

where $\delta_i \equiv a - y_i$ measures the spread of the y_i values about the average, with some δ_i being positive, some negative. If we impose the condition that $S(a) = 0$, namely that "on average" the deviations are equally positive or negative, we have the condition

$$S(a_0) = \sum_{i=1}^{N}(a_0 - y_i) = 0 \qquad \text{implying} \qquad a_0 \sum_{i=1}^{N}(1) = \sum_{i=1}^{N} y_i \qquad \text{or} \qquad a_0 = \frac{1}{N}\left(\sum_{i=1}^{N} y_i\right), \tag{10.40}$$

which is the standard definition of the **mean value** of a data set. Beyond reproducing this well-known result, however, this approach provides no information on the actual spread of the data points about the mean value, since $S(a_0) = 0$ could arise from either very small or very large values of the individual $|\delta_i|$.

An only very slightly different approach, one that still reproduces the mean value, is to consider instead

$$S(a) \equiv \sum_{i=1}^{N}(\delta_i)^2 = \sum_{i=1}^{N}(a - y_i)^2 \tag{10.41}$$

and find the value of a that minimizes this quantity, hence the phrase **least squares fit.** Doing so implies that

$$0 = \frac{\partial S(a)}{\partial a} = 2\sum_{i=1}^{N}(a_0 - y_i) = 0 \quad \text{or} \quad a_0 \sum_{i=1}^{N}(1) = \sum_{i=1}^{N} y_i \qquad \text{giving} \qquad a_0 = \frac{1}{N}\left(\sum_{i=1}^{N} y_i\right), \tag{10.42}$$

which still gives the familiar mean or average value, but now the value of $S(a_0)$, or perhaps more appropriately $\sqrt{S(a_0)}$, does give a well-defined measure of the spread about the "best fit."

Following this least squares approach, the next most straightforward example would be to try to find the most appropriate straight line fit to data of the form (x_i, y_i) for N data points by assuming a relationship of the form $y(x) = bx + a$ where the a, b are constants (let us not worry about dimensions here, you've hopefully taken care of that) to be determined. One measure of the global "closeness" of such a line to the data is obtained by generalizing Eqn. (10.41) to

$$S(a, b) \equiv \sum_{i=1}^{N}(d_i)^2 = \sum_{i=1}^{N}(bx_i + a - y_i)^2, \tag{10.43}$$

which uses the (vertical) "distance" d_i between the actual y_i values and a putative linear fit value, $(bx_i + a)$. We can then find the values of a, b, which minimize this two-parameter function, by writing

$$\begin{aligned} 0 &= \frac{\partial S(a,b)}{\partial a} = 2\sum_{i=1}^{N}(bx_i + a - y_i) = 0 \\ 0 &= \frac{\partial S(a,b)}{\partial b} = 2\sum_{i=1}^{N}(bx_i + a - y_i)x_i = 0. \end{aligned} \tag{10.44}$$

These two constraints correspond to linear equations in the variables a, b

$$\begin{aligned} b\left(\sum_{i=1}^{N} x_i\right) + aN &= \left(\sum_{i=1}^{N} y_i\right) \\ b\left(\sum_{i=1}^{N} x_i^2\right) + a\left(\sum_{i=1}^{N} x_i\right) &= \left(\sum_{i=1}^{N} x_i y_i\right), \end{aligned} \tag{10.45}$$

which can be solved[3] for the "best fit" values of a and b. This method has the benefit that it can easily be extended to higher-order polynomial fits, and to include error bars (call them σ_i) on the y_i data, by writing

$$S(a, b, c, \ldots) \equiv \sum_{i=1}^{N} \frac{(a + bx_i + cx_i^2 + \cdots - y_i)^2}{(\sigma_i)^2}. \tag{10.46}$$

Such methods can be extended even further to evaluate the likely error bars $\sigma_a, \sigma_b, \sigma_c, \ldots$ on the fitted parameters $a, b, c, \ldots$ themselves, to allow for comparison of experimental data to theory.

To finally make connections to power-law dependences of the form $y(x) = ax^b$, or exponential ones of the form $y(x) = ae^{bx}$, (again, ignoring dimensions), we can turn these formulae into linear relations (in other variables) by taking logarithms of both sides, giving

$$\ln(y) = b\ln(x) + \ln(a) \qquad \text{or} \qquad \ln(y) = bx + \ln(a). \tag{10.47}$$

By taking logs of either one or both values in an (x_i, y_i) data set, we can then apply the method behind Eqn. (10.43) to find the best fit for these forms.

Instead of doing data fits "from scratch" by using such methods, it's possible to use many existing symbolic math programs which already contain powerful data-fitting functions, and we include below one such example using Mathematica©. In the first example of code below, we generate "fake data" from a simple power-law relation ($y = 7x^3$), evaluate the logarithms of both x_i, y_i, use the `Fit[ ]` command, and check the results against the original "data." (**Note:** A line of code ending in a semi-colon generates no output, just to save space.)

```
(* Fit.nb shows a simple use of the Fit[ ] function in Mathematica *)
(* Generates power-law data which are then fit to check parameters *)
F[x_] := 7*x^3;
data = Table[{x, F[x]}, {x, 1, 10, 1}];
(* The Log[ ] command in Mathematica is the natural log *)
logdata = Log[data];
Fit[logdata, {1, z, z^2}, z]
1.9459101490553095' + 3.000000000000004' x -  8.481124372248901`*^-16 x^2
(* Coefficient of the z = Log[x] term is very close to 3 *)
(* Coefficient of the z^2 term found to be very close to zero *)
(* Check the prefactor by taking inverse log or exponential *)
Exp[1.94591]
6.999998956612885'
(* This is very close to 7 *)
```

Such methods can, of course, be implemented in many programs (including Excel© and others) and we encourage readers to make use of such tools when comparing data to predictions (from DA or otherwise obtained), such as in **P2.4**, where data from the Trinity bomb explosion is provided to compare to the DA prediction in Eqn. (2.9). But we also trust that all due diligence has been used in terms of making sure dimensions are appropriately taken into account before doing purely numerical manipulations!

[3] For the final form of the a, b and many more details on methods of fitting/modeling techniques, see any standard book on data analysis methods, such as Hansen et al. (2013) or Strutz (2010).

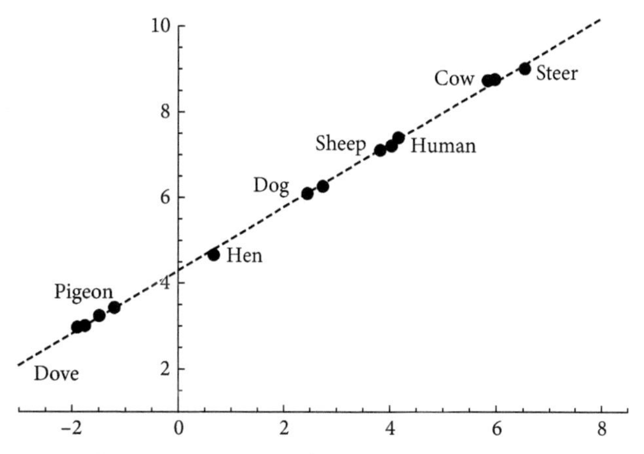

Fig. 10.1 Data analysis of animal metabolism (M in calories per day) versus mass (W, in kg) from Klieber (1932), taken from his Table 1. The data are plotted on a log–log scale and fit to a power-law of the form $M = AW^{\alpha}$ using the Mathematica© code in Section 10.9.1. The data values (dots) and "best fit" curve (dashed line) are shown on the plot. The fitted value of the scaling exponent is $\alpha \sim 0.74$, consistent with the original Kleiber analysis of $\alpha = 3/4$, and many other "quarter power" scaling results.

As an example of fitting real data, we include below the animal metabolic data from Kleiber (1932) (his table 1) discussed in Section 5.1.2 and displayed in Fig. 5.7. The author cites the data is "best-fit" by a 3/4 power law exponent and the log–log "slope" we find is indeed close to that, namely ~ 0.74.

```
(* Kleiber_Fit.nb *)
(* Original Klieber (1932) data on animal metabolism *)
(* Fits to a power-law dependence *)
(* Exponent close to 3/4 instead of 2/3 ! *)
KlieberData = {{679, 8274}, {342, 6255}, {388, 6421}, {64.1, 1632},{56.5, 1349},
{45.6, 1219}, {15.5, 525}, {11.6, 443}, {1.96, 106}, {0.3, 30.8}, {0.226, 25.5},
{0.173, 20.2}, {0.15, 19.5}};
logdata = Log[KlieberData];
Fit[logdata, {1, x}, x]
4.298821555742978` + 0.737540314622882` x
```

Showing the data and this least-squares fit line together in Fig. 10.1 helps visually confirm the relatively good fit obtained.

Even if we choose to not perform any actual fitting ourselves, we can still make use of published data plots to extract power-law exponents or exponential scaling parameters. For example for data fit to the form $y = ax^b$, plotted on log–log scales, we have from Eqn. (10.47), $\ln(y) = b\ln(x) + \ln(a)$ so that the value of b can be obtained by using any two points taken from the fitted line, giving the generalized slope as

$$b = \frac{\ln(y_2) - \ln(y_1)}{\ln(x_2) - \ln(x_1)} = \frac{\ln(y_2/y_1)}{\ln(x_2/x_1)}. \tag{10.48}$$

We note that b is dimensionless, as it must, being a power-law exponent. We can then obtain the value of $a = y_i x_i^{-b}$ using any data point (x_i, y_i), where a may well have dimensions, since the (x_i, y_i) might.

As an example, consider Fig. 5.7, which shows the allometric scaling described as having a best-fit exponent of roughly 3/4. Picking two points from the graph that seem to fall on the fitted straight line, say the dove and sheep, we have $(W(kg), M(Cal))$ values of $(0.15, 19.5)$ and $(45.6, 1220)$ and Eqn. (10.48) gives $b = \ln(1220/19.5)/\ln(45.6/0.15) \approx 4.136/5.717 \sim 0.72$.

Finally, for data fit to an exponential form $y = ae^{bx}$, we plot $\ln(y) = bx + \ln(a)$ giving

$$b = \frac{\ln(y_2) - \ln(y_1)}{x_2 - x_1} = \frac{\ln(y_2/y_1)}{x_2 - x_1} \tag{10.49}$$

with $[b] = 1/[x]$ as required. Then $a = y_i e^{-bx_i}$ for any (x_i, y_i) pair. Let us use this method to confirm an analysis from a research paper.

Example 10.1 Residence time for adsorbed atoms

Frenkel (1924), in paper titled (in translation) "The Theory of Adsorption and Related Phenomena," noted that when gas atoms impinge on a surface, they can condense and remain on the surface for a time determined by a Arrhenius-type relation of the form

$$\tau = \tau_0 e^{Q/k_B T}, \tag{10.50}$$

where τ_0 is a typical molecular vibrational period and Q is the threshold energy for release. As an example of this relation, Smith, Wolleswinkel, and Los (1970) obtained data on $\tau\,(sec)$ versus $1/T\,(K^{-1})$ which we reproduce in Fig. 10.2, with τ plotted on a log scale. They fit the upper (A) and lower (B) curves to the form in Eqn. (10.50) and obtained values

Fitted values	*A* data	*B* data
$Q\,(eV)$	2.36 ± 0.06	2.35 ± 0.06
$\tau_0\,(sec)$	$(2.4 \pm 0.9) \times 10^{-13}$	0.6×10^{-13}

To verify these values, let us try to read off data from the *A* line, namely

$$\begin{aligned}(\tau_2, 1/T_2) &= (10^{-3}\,s, 8.05 \times 10^{-4}\,K^{-1})\\ (\tau_1, 1/T_1) &= (10^{-4}\,s, 7.2 \times 10^{-4}\,K^{-1})\end{aligned} \tag{10.51}$$

and then use

$$\ln(\tau) = \ln(\tau_0) + \left(\frac{Q}{k_B}\right)\frac{1}{T} \quad \longrightarrow \quad \frac{Q}{k_B} = \frac{\ln(\tau_2/\tau_1)}{1/T_2 - 1/T_1} = \frac{\ln(10)}{0.85 \times 10^{-4}\,K^{-1}} = 2.7 \times 10^4\,K. \tag{10.52}$$

This gives

$$Q = k_B(2.7 \times 10^4\,K) \sim 3.74 \times 10^{-19}\,J \sim 2.3\,eV, \tag{10.53}$$

consistent with the fitted value. The dimensionful prefactor, τ_0, is then given by using either of the two $(\tau, 1/T)$ values, for example,

$$\tau_0 = \tau_i e^{-Q/k_B T_i} \sim (10^{-3}\,s)\, e^{-21.8} \sim 3 \times 10^{-13}\,s. \tag{10.54}$$

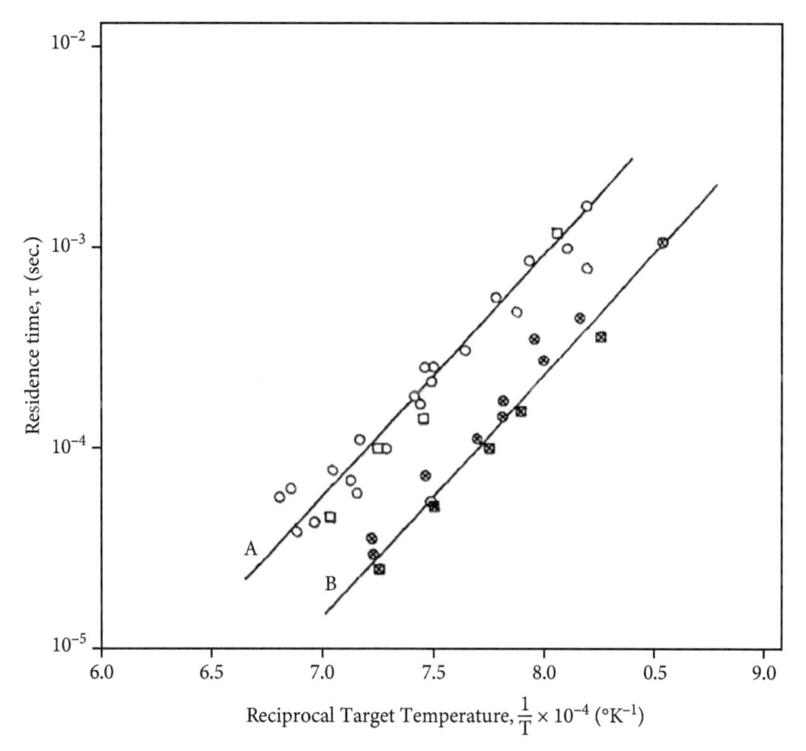

Fig. 10.2 Residence time (*sec*) versus inverse temperature $1/T \times 10^{-4}\,(K^{-1})$ for adsorbed atoms. Reprinted from Smith, Wolleswinkel, and Los (1970) with permission of North-Holland Publishing. Note the log scale for the vertical axis, and the units used for the horizontal axis!

We see that the "slopes" of the two log–linear plots appear to be the same within errors, while the "intercept" of the *B* data is smaller, which you can confirm by using values from the lower curve.

10.9.2 **Dimensions in error analysis**

As we've just emphasized, many of the quantities we have discussed in the context of DA have yielded products, powers, and/or functions of dimension**ful** variables. When each component variable has an associated experimental error, it's important to respect the dimensionality of all quantities and relations involved when implementing any **propagation of uncertainty** analyses.

One of the easiest examples of such connections is when the quantity in question is the simple sum (or difference) of two variables, namely $x = a \pm b$, where each quantity on the RHS has their own estimated error, call them σ_a and σ_b. We are always required by dimensional consistency to have $[\sigma_a] = [a]$ and $[\sigma_b] = [b]$, whatever the dimensions of a, b might be. In this case, given the sum/difference connection assumed, we must also have $[\sigma_x] = [x] = [a] = [b]$ as well. The well-known result on how the errors in a, b combine to give an error in x is that they add in **quadrature**, namely

$$\sigma_x^2 = \sigma_a^2 + \sigma_b^2 \qquad \text{or} \qquad \sigma_x = \sqrt{\sigma_a^2 + \sigma_b^2}, \tag{10.55}$$

which is clearly dimensionally consistent.[4] A small extension of this is to consider linear combinations of the variables with (let us assume dimensionless) constants, α, β, whose values are precisely known (so no error included), namely

$$x = \alpha a \pm \beta b \qquad \text{with associated error} \qquad \sigma_x^2 = \alpha^2 \sigma_a^2 + \beta^2 \sigma_b^2. \tag{10.56}$$

This can be extended further to account for the case of a mean (or average) value of a measurement, in which case the linear combination has $\alpha_1 = \alpha_2 = \cdots = 1/N$ for all variables, giving

$$x = \sum_{i=1}^{N} \alpha_i a_i = \frac{1}{N} \sum_{i=1}^{N} a_i \qquad \text{with error} \qquad \sigma_x^2 = \underbrace{N}_{\text{sum}} \underbrace{\left(\frac{1}{N}\right)^2}_{\text{the } \alpha_i^2} \sigma_a^2 \qquad \text{or} \qquad \sigma_x = \frac{\sigma_a}{\sqrt{N}}. \tag{10.57}$$

This implies that repeating an experiment multiple times decreases the uncertainty in an average value, but only at a "rate" of $1/\sqrt{N}$.

For a product relation of the form $x = a \cdot b$, each of the three quantities involved can clearly have different dimensions, but must together satisfy the constraint $[x] = [a \cdot b] = [a]\,[b]$. The appropriate combination of errors in this case is known to be

$$\left(\frac{\sigma_x}{x}\right)^2 = \left(\frac{\sigma_a}{a}\right)^2 + \left(\frac{\sigma_b}{b}\right)^2 \qquad \text{or} \qquad \frac{\sigma_x}{x} = \left(\left(\frac{\sigma_a}{a}\right)^2 + \left(\frac{\sigma_b}{b}\right)^2\right)^{1/2}, \tag{10.58}$$

which is again dimensionally appropriate, given that all three terms are dimension**less**! In this case it is the **fractional errors** that add in quadrature. This relation also works for the ratio of two variables, namely $x = a/b$.

A general result that is useful when the relation between the quantities is more complex, say $x = f(a, b)$, is that

$$\sigma_x^2 = \left(\frac{\partial f(a,b)}{\partial a}\right)^2 \sigma_a^2 + \left(\frac{\partial f(a,b)}{\partial b}\right)^2 \sigma_b^2, \tag{10.59}$$

which is also seen to be dimensionally consistent for any functional relation, and can be used to confirm the specific results in Eqns. (10.55), (10.56), and (10.58). For example, for the product rule where $x = a \cdot b$ we have

$$\sigma_x^2 = \left(\frac{\partial f(a,b)}{\partial a}\right)^2 \sigma_a^2 + \left(\frac{\partial f(a,b)}{\partial b}\right)^2 \sigma_b^2 = b^2 \sigma_a^2 + a^2 \sigma_b^2 \tag{10.60}$$

or

$$\left(\frac{\sigma_x}{x}\right)^2 = \frac{\sigma_x^2}{(ab)^2} = \frac{b^2 \sigma_a^2 + a^2 \sigma_b^2}{a^2 b^2} = \left(\frac{\sigma_a}{a}\right)^2 + \left(\frac{\sigma_b}{b}\right)^2. \tag{10.61}$$

The general result in Eqn. (10.59) can be used for powers of variables (integral or fractional) such as $x = f(a) = a^n$, where $\partial f(a)/\partial a = n a^{n-1}$, to give

$$\sigma_x^2 = \left(n a^{n-1}\right)^2 \sigma_a^2 \quad \rightarrow \quad \left(\frac{\sigma_x}{x}\right)^2 = \left(\frac{\sigma_x}{a^n}\right)^2 = \left(\frac{n a^{n-1}}{a^n}\right)^2 \sigma_a^2 \qquad \text{or} \qquad \frac{\sigma_x}{x} = |n| \frac{\sigma_a}{a}. \tag{10.62}$$

[4] In all of the cases we consider here, we assume that the variables a, b are uncorrelated to each other, otherwise there are important "cross-terms" in σ_x, which take any such connections into account.

We have been careful to write $\sqrt{n^2} = |n|$ to accommodate negative exponents, say $n = -2$, and this derivation also includes fractional powers (positive or negative). This relation is consistent with the fact that power-law relations with $|n| < 1$ imply less sensitivity of the final result to changes in a, while an $|n| > 1$ relation magnifies the dependence, thereby increasing the fractional error. You're asked to explore specific cases in **P10.7**.

10.10 Connections to metrology

Human cultures all over the world have found (often very diverse) ways of expressing themselves in terms of food, dress, housing, social interactions, and organization, and perhaps most especially in the ways they communicate with each other. Such information exchange can involve the spoken or written word, in everyday conversation or more formal discourse, but it can also include the knowledge encoded about either mathematics or science in their own specialized notations. Notions of "how far," "how heavy," or "how long did it take" can play an important role in commerce, economics, and of course science, and not surprisingly various civilizations have settled upon different sets of basic units to measure such quantities.

Many human-sized units of length, for example, were directly tied to the size of body parts, for example, finger close to an inch, outstretched palm a span, a foot like a foot, and even the supposed definition of the **yard** as a unit of length as the distance from the King's nose to the tip of his outstretched thumb. With the more scientific approach arising from the time of the French Revolution, definitions based more on reproducible physical phenomena were developed, for example, with the definition of the **meter** as one ten-millionth of the distance from the pole to the equator.[5]

The history of metrology (the science of measurement) has many treatments describing its development and impact on society, including Gupta (2020),[6] as well as very readable popular accounts by Cooper and Grozier (2017) and especially Crease (2011). The landscape of the **Système International** (SI) system of units changed dramatically in 2019 when there were redefinitions of the seven standardly cited quantities in metrology. Five of these fundamental units are, of course, the familiar T, L, M, Q, and Θ ones we've cited as "sufficing" for most practical problems in physical science and engineering and forming the five-dimensional space of dimensions used throughout this text. We chose to focus attention on the first five dimensionful quantities encoded by these units, since the last two (mole and candela) are encountered far less frequently in the physics curriculum, at either the undergraduate or graduate level.[7]

In "A More Fundamental International System of Units," Newell (2014) presented this information at a pedagogical level, as did Schlamminger (2018) more recently. It's interesting to note that many of the new definitions included below rely on high-precision atomic physics experiments (using the hyperfine transition of Cesium as a "standard clock") and involve fundamental physics constants (as in the repeated appearance of c, $\hbar$, and e). Two short articles on the importance of the quantum Hall effect (Section 8.2) on the new SI units were written by the discoverer of this phenomena, von Klitzing (2017, 2019), who notes the extreme reproducibility of measurements of $R_K = h/e^2$ at the roughly 10^{-9} level. Other precision measurements from Josephson junctions (Section 8.2) (complementary information on $\hbar$ and e) and the Kibble–Watt balance (see **P4.43**,

---

5 Try it!

6 A very exhaustive study, including background on non-Western units of measurement!

7 In somewhat the same way that the Q and X tiles/choices in Scrabble© or Wordle© are rare.

Robinson and Schlamminger (2016), or Schlamminger (2018)) for the mass standard, can all be explored using DA, as we've seen.

Two quotations, from authors who work at the British and French equivalents of the National Institute of Standards and Technology (NIST), respectively, describe some important aspects of the recent progress in metrology:

> The revision of the International System of Units (SI) in 2019 was a triumph of scientific endeavour, perseverance, and ingenuity. It was the result of decades of progress in metrology and is one of the greatest stories of international scientific collaboration in recent times . . . The current system of seven SI base units reflects the technical, political, and cultural compromises made over many years to achieve a global consensus on a practical measurement system (Brown 2024).
>
> . . . the new definitions use the rules of nature to create the rules of measurement and tie measurements at the atomic and quantum scales to those at the macroscopic level (Stock et al. 2019).

The official definitions of the 2019 revised units can be found on almost all government (United States or otherwise) metrological organizations websites, and one example from NIST[8] is reproduced below. An "infographic" version of this is also provided in Fig. 10.3. Note that despite listing the **Ampère** as the basic unit of charge, its value is obtained in terms of the fundamental electric charge e and the newly revised definition of the **second**, so our emphasis on the Q dimension is perhaps not unwarranted.

Second: The second is defined by taking the fixed numerical value of the cesium frequency $\Delta\nu_{Cs}$, the unperturbed ground-state hyperfine transition frequency of the cesium-133 atom, to be 9,192,631,770 when expressed in the unit Hz, which is equal to s^{-1}.

Meter: The meter is defined by taking the fixed numerical value of the speed of light in vacuum c to be 299,792,458 when expressed in the unit $m\,s^{-1}$, where the second is defined in terms of $\Delta\nu_{Cs}$.

Kilogram: The kilogram is defined by taking the fixed numerical value of the Planck constant h to be $6.62607015 \times 10^{-34}$ when expressed in the unit $J\,s$, which is equal to $kg\,m^2\,s^{-1}$, where the meter, and the second are defined in terms of c and $\Delta\nu_{Cs}$.

Ampere: The ampere is defined by taking the fixed numerical value of the elementary charge e to be $1.602176634 \times 10^{-19}$ when expressed in the unit C, which is equal to $A\,s$, where the second is defined in terms of $\Delta\nu_{Cs}$.

Kelvin: The kelvin is defined by taking the fixed numerical value of the Boltzmann constant k to be 1.380649×10^{-23} when expressed in the unit $J\,K^{-1}$, which is equal to $kg\,m^2\,s^{-1}\,K^{-1}$, where the kilogram, meter, and second are defined in terms of h, c, and $\Delta\nu_{Cs}$.

Mole: One mole contains exactly $6.02214076 \times 10^{23}$ elementary entities. This number is the fixed numerical value of the Avogadro constant, N_A, when expressed in the unit $mole^{-1}$ and is called the Avogadro number. The amount of substance, symbol n, of a system is a measure of the number of specified elementary entities. An elementary entity may be an atom, a molecule, an ion, an electron, or any other particle or specified group of particles.

Candela: The candela is defined by taking the fixed numerical value of the luminous efficacy of monochromatic radiation of frequency 540×10^{12} $Hz\,Kcd$, to be 683 when expressed in the unit $lm\,W^{-1}$, which is equal to $cd\,sr\,W^{-1}$, or $cd\,sr\,kg^{-1}\,m^{-2}\,s^3$, where the kilogram, meter, and second are defined in terms of h, c, and $\Delta\nu_{Cs}$.

[8] https://www.nist.gov/si-redefinition/definitions-si-base-units. Accessed October 17, 2023.

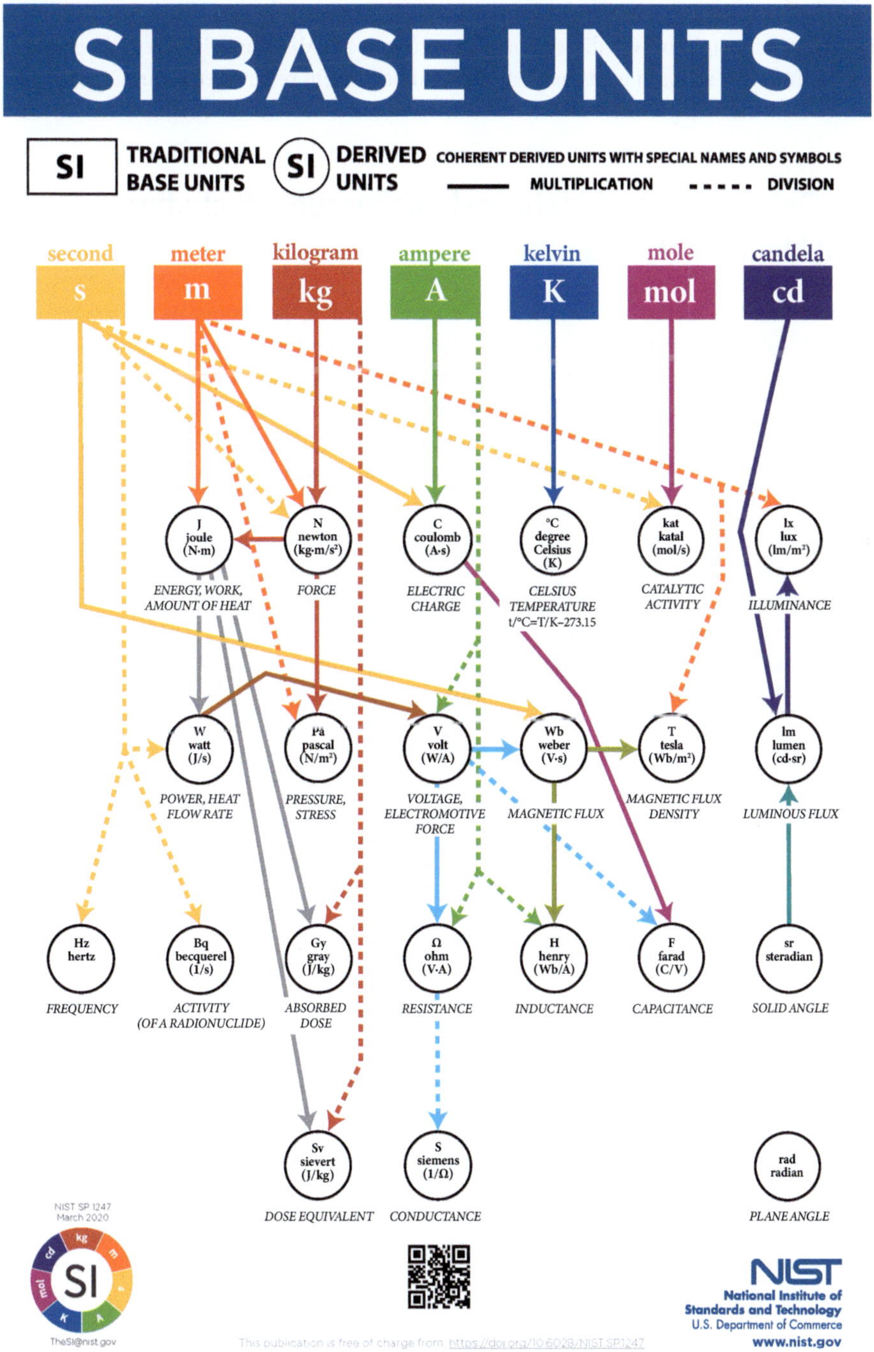

Fig. 10.3 NIST (National Institute of Science and Technology) educational poster on SI units. Downloaded from https://www.nist.gov/pml/owm/metric-si/si-units on 19 August, 2023. Reprinted courtesy of NIST. All rights reserved, US Secretary of Commerce.

While not part of the official metrological "canon" of defined base quantities, one still sees discussions of **plane angle** in two dimensions (in *radians*) or **solid angle** in three dimensions (in *steradians*) and their status in measurement theory. For a recent review, see Kalinin (2019), which has some historical background.

10.11 References and historical background, including the Buckingham Pi theorem

We reiterate that this text is not intended to be an exercise in the history of the subject of DA, even though we are happy to quote classic results in the subject as examples of its long-standing (namely, along the "time" axis in the four-dimensional space-time "landscape of physics") use in physics. We have quoted several prominent scientists who've cited DA as an important tool, including Fourier (1878) (in our Chapter 5) and Rayleigh (1915) (in our Chapter 2) to which we can add Bertrand (1878) and note that other standard texts have provided some historical references, including Birkhoff (1950). In his serious *Historico-Critical Review of Dimensional Analysis*, Macagno (1971) provides a nice discussion about "... the first notions of dimensions in old civilizations to the powerful methods of recent times." That treatment also cites overlooked early contributions by Vaschy (1892) and Riabouchinsky (1911), along with a substantial set of original references, so interested readers can consult that resource.[9]

In terms of textbook level treatments of the subject, we include below (in chronological order of publication) a number of the most relevant/accessible ones (many still in print), namely

- Bridgeman, P. W. (1922). *Dimensional Analysis* (New Haven, CT: Yale University Press).
  - Available at https://archive.org/details/dimensionalanaly00bridrich. Accessed 19 January, 2024).
- Langhaar, H. L. (1951). *Dimensional Analysis and Theory of Models* (New York: Wiley).
- Huntley, H. E. (1955). *Dimensional Analysis* (New York: Rinehart and Company, New York). Reprinted 1967 by Dover, New York.
- LeCorbeiller, P. (1966). *Dimensional Analysis* (New York: Appleton-Century Crofts). This is a textbook designed to accompany a module of *programmed instruction* (written in the Basic language) coded by A. V. Lukas.
- Goldenfeld, N. (1992). *Lectures on Phase Transitions and the Renormalization Group* (Reading: Addison-Wesley).
  - Goldenfeld's excellent treatment of the subject outlined in its title has a short introductory review of DA, but then includes the **very** important reminder that

> ... in many scaling laws, the power law, or the **exponent** is a rational function, often deduced from simple dimensional considerations. This partly accounts for the fact that the phenomena described are so well understood, and are taught in elementary physics courses. However, there is a broader class of phenomena where power-law behavior occurs, but the exponent is not a simple fraction (as far as is known). This class of phenomena includes, but is by no means restricted to, phase transitions where there is a **critical point.**

[9] But see also the comments on that paper made by Martins (1981), who begins his work with the statement "The complete history of dimensional analysis has not yet been written".

- Such power-law behaviors are not restricted to approximate ones, like the scaling of cosmic ray fluxes with energy, or stellar mass–luminosity relations, but include ones where the critical exponent can be predicted with field theory techniques (see, e.g., Zinn-Justin 2001) and measured with extreme accuracy. One such example is the measurement of the heat capacity of superfluid helium, where Lipa et al. (1996, 2003) performed experiments in Earth orbit to minimize systematic errors and found a value of $\alpha = -0.0127 \pm 0.0003$ for one critical exponent.
- Discussion of this important topic is beyond the scope of this work, and we encourage interested readers to review the vast literature on this subject, including reviews by Fisher (1998) and Pelisseto and Vicari (2002).

- Sedov, L. I. (1993). *Similarity and Dimensional Methods in Mechanics*, 10th ed. (Boca Raton, FL: CRC Press).
- Szirtes, T., and Rozsa, R. (1997). *Applied Dimensional Analysis and Modeling*, 2nd ed. (Amsterdam: Elsevier).
- Hornung, H. G. (2006). *Dimensional Analysis: Examples of the Use of Symmetry* (New York: Dover).
- Palmer, A. C. (2008). *Dimensional Analysis and Intelligent Experimentation* (Hackensack, NJ: World Scientific).
- Lemon, D. S. (2017). *A Students Guide to Dimensional Analysis* (Cambridge: Cambridge University Press).
- Santiago, J. G. (2019). *A First Course in Dimensional Analysis* (Cambridge, MA: MIT Press).
- Williams, J. H. (2021). *Dimensional Analysis: The Great Principle of Similitude* (Bristol: IOP Publishing Ltd).

A very incomplete set of references from journals (ones not already included in earlier chapters on individual content topics), some at a more pedagogical level, is included below.

- Tolman, R. C. (1914). "The Principle of Similitude," Physical Review 3, 244–55. This was the first in a series of papers that proposed a much deeper meaning to DA, based on a "novel" physical principle, namely that

  > The fundamental entities of which the physical universe is constructed are of such a nature that from them a miniature universe could be constructed exactly similar in every respect to the present universe.

  In contrast to the famous paper by Rayleigh (1915), which shares the same title, this was an ill-fated dead end in terms of the proposed physical principles, but the approach does provide more practice in DA, in an unfamiliar context. Other examples by the same author[10] (and comments by others) include:
  - Tolman, R. C. (1914). "The Specific Heat of Solids and the Principle of Similitude," *Physical Review* 4, 145–53.
  - Tolman, R. C. (1915). "The Principle of Similitude and the Principle of Dimensional Homogeneity," *Physical Review* 6, 219–33.

[10] Who did otherwise make major contributions to special relativity, statistical mechanics, and cosmology.

- Tolman, R. C. (1916). "Note on the Homogeneity of Physical Equations," *Physical Review* 8, 8–11.
- Tolman, R. C. (1921). "The Principle of Similitude and the Entropy of Polyatomic Gases," *Journal of the American Chemical Society* 43, 866–875.
- Bridgman, P. W. (1916). "Tolman's Principle of Similitude," *Physical Review* 8, 423–31
- Ehrenrest-Afanassjewa, T. (1916). "On Mr. R. C. Tolman's 'Principle of Similitude,'" *Physical Review* 8, 1–7.

- Murphy, N. F. (1949). "Dimensional Analysis," *Bulletin of the Virginia Polytechnical Institute* XLII(6), 1–41.
- Corrsin, S. (1951). "A Simple Geometrical Proof of Buckingham's π-Theorem," *American Journal of Physics* 19, 180–1.
- Stahl, W. R. (1961). "Dimensional Analysis in Mathematical Biology I. General Discussion," *Bulletin of Mathematical Biophysics* 23, 355–76.
- Stahl, W. R. (1962). "Dimensional Analysis in Mathematical Biology II," *Bulletin of Mathematical Biophysics* 24, 81–108.
- Evans, J. H. (1972). "Dimensional Analysis and the Buckingham Pi Theorem," *American Journal of Physics* 40, 1815–22.
- Price, J. F. (2003). "Dimensional Analysis of Models and Data Sets," *American Journal of Physics* 71, 437–47.
- Bohren, C. F. (2004). "Dimensional Analysis, Falling Bodies, and the Fine Art of *Not* Solving Differential Equations," *American Journal of Physics* 72, 534–37.
- Bolster, D, Hershberger, R. E., and Donnelly, R. J. (2011). "Dynamic Similarity, the Dimensionless Science," *Physics Today* 42(9), 42–47. This review provides a brief history of dimensional analysis, citing contributions from Newton, Euler, Fourier, and Rayleigh.
- Robinett, R. W. (2015). "Dimensional Analysis as the *Other* Language of Physics," *American Journal of Physics* 83, 353–61.
  - Based on modules included in a Physics first year seminar and junior "skills" courses for Physics majors taught by the author over the years, and what got me excited about this project!

Example 10.1. Rayleigh example

Before our final discussion of the Buckingham (1914) Pi theorem, let us reproduce one of the more extended examples from the original Rayleigh (1915) paper "The Principle of Similitude", one which includes the identification of an important dimensionless ratio. Rayleigh says:

> As a last example, let us consider, somewhat in detail, Boussinesq's problem of the steady passage of heat from a good conductor immerse in a stream of fluid moving (at a distance from the solid) with velocity v. The fluid is treated as incompressible and for the present as inviscid, while the solid has always the same *shape* and presentation to the stream. In these circumstances the total heat (h) passing in unit time is a function of the linear dimensions of the solid (a), the temperature-difference (θ), the stream-velocity (v), the capacity for heat of the fluid per unit volume (c), and the conductivity κ. The density of the fluid clearly does not enter into the question. We now have to consider the "dimensions" of the various symbols.

Those of a are	(Length)^1
.............v....	(Length)^1 (Time)$^{-1}$
.............θ....	(Temperature)^1
.............c....	(Heat)^1 (Length)$^{-3}$ (Temp.)$^{-1}$
.............κ....	Time)$^{-1}$
.............h....	(Heat)^1 (Time)$^{-1}$.

Hence if we assume

$$h = a^x \theta^y v^z c^u \kappa^v$$

we have

by heat	$1 = u + v$
by temperature	$0 = y - u - v$
by length	$0 = x + 2 - 3u - v$
by time	$-1 = -z - v$

so that

$$h = \kappa a \theta \left(\frac{avc}{\kappa}\right)^z.$$

Since z is undetermined, any number of terms of this form may be combined, and all that we can conclude is that

$$h = \kappa v \theta \, F(avc/\kappa)$$

where F is an arbitrary function of the one variable avc/κ.

You should be able to reproduce the dimensional matching to find that indeed $x = 1 + z$, $y = 1$, $u = z$ and $v = 1 - z$ as advertised. This example can remind us that (i) it's possible to use independent base dimensional quantities other than the standard M, L, T, Q, Θ set, and (ii) that it's often the case that dimensionless ratios of the physical parameters appear in such an analysis, which leads us to our final topic.

Instead of starting this book with a formal discussion of the foundational aspects of DA, especially as codified in the Buckingham Pi theorem, we've chosen to describe its main results closer to the end, after we've had a chance to see many of the possible outcomes of a DA approach to problems. In what follows, we adopt the notation of Buckingham (1914) to the extent possible.

We start by assuming that we've selected a number of physical variables (with well-defined and identified dimensions) given by $Q_1, Q_2, \ldots, Q_n$, as well as any possible dimensionless ratios that are apparent in the problem, $r_1, r_2, \ldots, r_m$, examples of which could be ratios of lengths (L_x/L_y) or masses (m_e/M) or energies ($E/k_B T$). It may seem like a tautology to say that all of these variables can be connected in the form of an equation of the form

$$f(Q_1, Q_2, \ldots, Q_n; r_1, r_2, \ldots, r_m) = 0, \tag{10.63}$$

since any relation/equation can be written in the form $LHS = RHS$, and we can simply write $f(Q_i; r_i) = LHS - RHS = 0$. Buckingham then assumes "... that the ratios do not vary[11] during the

[11] For example, m_e/M stays fixed.

phenomenon described by the equation . . ." so that the constraint in Eqn. (10.63) reduces to one only involving the Q_i,

$$F(Q_1, Q_2, ..., Q_n) = 0 \tag{10.64}$$

and notes

> If none of the quantities involved in the relation has been overlooked, the equation will give a complete description of the relation subsisting among the quantities represented in it, and will be a complete equation (Buckingham 1914.)

One then assumes that there is a set of the Q_i that are independent of each other in terms of dimensional "spelling" that will be used as bases or fundamental units or building blocks (different from the M, L, T, Q, Θ basic dimensions) in terms of which the remaining variables can be expressed. One picks k of them, a value that almost always represents the applicable number of constraint equations allowed by the basic dimensions, and we use their dimensional spellings as a starting point, namely $[Q_1], [Q_2], ..., [Q_k]$.

With k constraints and $n > k$ variables, the Pi theorem states that there will be $n - k$ dimensionless ratios, the Π_i, among the n possible Q_i, an outcome that should be familiar from many of our DA examples. If we call the $N = n - k$ remaining (non-base) variables P_i, with $i = 1, N$, those ratios will be written in the forms

$$\begin{aligned}
\Pi_1 &= (Q_1^{\alpha_1} Q_2^{\beta_1} \cdots Q_k^{\epsilon_1}) P_1 \\
\Pi_2 &= (Q_1^{\alpha_2} Q_2^{\beta_2} \cdots Q_k^{\epsilon_2}) P_2 \\
&\vdots \\
\Pi_N &= (Q_1^{\alpha_N} Q_2^{\beta_N} \cdots Q_k^{\epsilon_N}) P_N
\end{aligned} \tag{10.65}$$

and given that they're dimensionless, we must also have

$$\begin{aligned}
[\Pi_1] &= \left[Q_1^{\alpha_1} Q_2^{\beta_1} \cdots Q_k^{\epsilon_1} P_1\right] = 1 \\
[\Pi_2] &= \left[Q_1^{\alpha_2} Q_2^{\beta_2} \cdots Q_k^{\epsilon_2} P_2\right] = 1 \\
&\vdots \\
[\Pi_N] &= \left[Q_1^{\alpha_N} Q_2^{\beta_N} \cdots Q_k^{\epsilon_N} P_N\right] = 1.
\end{aligned} \tag{10.66}$$

At this point, we apply our usual M, L, T, Q, Θ matching equations for each Π_i combination to extract the dimensionless ratios. This is very similar to our automated Mathematica© approach in Sections 1.5 and 10.8, where we set the dimensions of *LHS*/*RHS* equal to unity. We note that the assumption that all terms in any given equation must have the same set of basic dimensions is often referred to as **homogeneity**.

So far, this seems a bit less straightforward than our standard approach, but to see how this works, let us consider the most obvious example, where we pick say 3 Q_i to try to determine the dependence of a new variable $P_1 = X$ on those physical quantities. Buckingham notes that for such problems,

> If we wish to treat some one quantity X as the unknown . . . This means that in selecting the variables which are to act as Q's and P's. . .X must be one of the P's.

Let us see how this approach works, comparing to examples we've already encountered.

Example 10.2 Buckingham Pi theorem approach applied to earlier problems

Recalling our very first historical case-study, from Section 2.1, we there wanted to find the dependence of the oscillation frequency $X = f$ of a liquid drop on the surface tension (S), density (ρ_m), and diameter (D). We assume that $F(S, \rho_m, d; f) = 0$ with f the quantity to be constrained. We then have

$$[S^{\alpha}\rho_m^{\beta}d^{\gamma}f] = 1$$

$$\left(\frac{M}{T^2}\right)^{\alpha}\left(\frac{M}{L^3}\right)^{\beta}L^{\gamma}\left(\frac{1}{T}\right) = 1, \tag{10.67}$$

which gives $\alpha = -1/2$, $\beta = 1/2$, and $\gamma = 3/2$ yielding a dimensionless ratio

$$\Pi = S^{-1/2}\rho_m^{1/2}d^{3/2}f \qquad \text{implying} \qquad f \propto S^{1/2}\rho_m^{-1/2}d^{-3/2} = \sqrt{\frac{S}{\rho_m d^3}}. \tag{10.68}$$

In this type of application, the dimensionless Π factor is associated with the undetermined constant we've assumed all along.

One of the first examples we found of an important dimension**less** ratio was in the study of the harmonic oscillator when we explored its dependence of the total energy in terms of k, m, x_0, v_0. Given only the three M, L, T or mechanics constraints, if we write $F(k, m, x_0; v_0) = 0$ this requires

$$[k^{\alpha}m^{\beta}x_0^{\gamma}v_0] = 1 \qquad \text{or} \qquad \left(\frac{M}{T^2}\right)^{\alpha}M^{\beta}L^{\gamma}\left(\frac{L}{T}\right) = 1, \tag{10.69}$$

which gives $\alpha = -1/2$, $\beta = 1/2$, and $\gamma = -1$ or the ratio

$$\Pi = k^{-1/2}m^{1/2}x_0^{-1}v_0 = \sqrt{\frac{m}{k}}\frac{v_0}{x_0}, \tag{10.70}$$

which we found in Eqn. (3.17). This relation could also be used to connect $v_0 \sim x_0\sqrt{k/m}$ in such relations as

$$x(t) = x_0\cos\left(\sqrt{\frac{k}{m}}t\right) \quad \longrightarrow \quad v(t) = -x_0\sqrt{\frac{k}{m}}\sin\left(\sqrt{\frac{k}{m}}t\right). \tag{10.71}$$

As a final example, we can consider again the fluid flow problem in **Example 3.4** and ask how the frictional force $F = P_1$ depends on the variables ρ_m, v, D (density, speed, and diameter), but also include viscosity η as part of a second P_2 ratio. We then write $F(\rho_m, v, D; F, \eta) = 0$ and impose

$$[\rho_m^{\alpha_1}v^{\beta_1}D^{\gamma_1}F] = 1 \qquad \text{and} \qquad [\rho_m^{\alpha_2}v^{\beta_2}D^{\gamma_2}\eta] = 1 \tag{10.72}$$

to find the the two dimensionless ratios Π_1, Π_2. Doing so, we find that the solution must have the scaling form

$$0 = \psi(\Pi_1, \Pi_2) = \psi\left(\frac{F}{\rho_m v^2 D^2}, \frac{\eta}{\rho_m v D}\right) \quad \text{so that for fixed } \Pi_2 \quad F = \rho_m v^2 D^2 G\left(\frac{\eta}{\rho_m v D}\right), \tag{10.73}$$

and the Reynolds number $\mathcal{R}e \equiv \rho_m v D/\eta$ (or its inverse) is indeed naturally obtained. In addition to many examples dealing with fluid mechanics, Buckingham (1914) also considered the energy stored in the electric and magnetic fields (pp. 358–359) and the Larmor radiation equation (pp. 362–363), especially in the context of finding dimensionless Π ratios.

10.12 **Populating the landscape of physics: Census data**

Science, like most human endeavors, is ultimately organized and accomplished by real people, with, of course, the help of powerful tools, whether mathematical, computational, experimental/technological, financial, or social. We want to acknowledge here, however briefly, what types of research or applications professional physicists associate themselves with the most, at least in the United States. The American Physical Society (APS) maintains a number of **divisions** to which members can belong, along with various **topical groups** and **forums**. We reproduce here the 2023 division membership[12] data for the first two categories below, as they are the groups most focused on research and application related topics.

Division	Count	Division	Count
Atomic, Molecular, & Optical (DAMOP)	3,322	Materials Physics (DMP)	2,802
Astrophysics (DAP)	2,725	Nuclear Physics (DNP)	2,559
Biological Physics (DBIO)	2,160	Physics of Beams (DPB)	1,001
Condensed Matter Physics (DCMP)	6,568	Particles & Fields (DPF)	3,100
Computational Physics (DCOMP)	2,765	Polymer Physics (DPOLY)	1,371
Chemical Physics (DCP)	1,222	Plasma Physics (DPP)	2,422
Fluid Dynamics (DFD)	3,299	Quantum Information (DQI)	3,539
Gravitation (DGRAV)	1,526	Soft Matter (DSOFT)	2,147
Laser Science (DLS)	1,080		

Topical group	Count	Topical group	Count
Data Science (GDS)	1,399	Plasma Astrophysics (GPAP)	424
Energy Research & Applications (GERA)	721	Physics of Climate (GPC)	596
Few-Body Systems (GFB)	328	Physics Education Research (GPER)	747
Hadronic Physics (GHP)	576	Fundamental Constants (GPMFC)	547
Instrument & Measurement Science (GIMS)	592	Shock Compression (GSCCM)	543
Magnetism (GMAG)	1,218	Statistical & Non-Linear (GSNP)	1,302
Medical Physics (GMED)	486		

It's been one intention of this book to explore the use of DA across as many different physics (and physics-adjacent) disciplinary areas as possible. I hope that readers will have "seen themselves" in at least some of the topics covered, worked-out **Examples**, problems suggested for practice, or citations of the original literature (pedagogical, research, or otherwise) for additional study or future reference.

---

[12] Taken from https://www.aps.org/membership/units/statistics.cfm. Accessed 20 August, 2023.

While it's sometimes said that "The University exists to find and communicate the truth,"[13] science (even in the academy) is not necessarily done for its own sake. Increasingly, funding agencies (rightly) want to document the "broader impact" of the work they support, and so it's fitting to conclude this book with a nod to the APS forum groups, which help us all express those important social/societal outcomes.

Forum	Count	Forum	Count
Diversity & Inclusion (FDI)	3,115	Industrial & Applied (FIAP)	5,099
Early Career Scientists (FECS)	5,587	International Physics (FIP)	3,917
Education (FED)	3,865	Outreach & Engaging the Public (FOEP)	2,890
Graduate Student Affairs (FGSA)	3,177	Physics & Society (FPS)	5,228
History & Philosophy of Physics (FHPP)	3,930		

While I'm happy to have contributed as an author and collaborator on a number of projects in particle physics (i.e., the DPF), mathematical physics, and related areas, I was more than honored to have been selected as an APS Fellow (in 2003) from the Forum on Education (FED), and I hope to continue to be able to contribute to the educational mission of our exciting field!

10.13 **Problems for Chapter 10**

Q10.1 Reality check: (a) What was the most important, interesting, engaging, or novel thing you learned about how to use DA, or about anything else, in this chapter?
(b) Have you previously encountered (or learned) many (or any) of the topics covered in this chapter?

P10.1 SI versus Gaussian units—Numerical examples: (a) Find the value of the SI charge that would give a one dyne force between two identical q objects at a separation of one centimeter to confirm the association that $1\ C \sim 3 \times 10^9\ statC$.
(b) Two electrons are separated by a distance of $1\ Å = 10^{-8}\ m = 10^{-10}\ cm$ where $e_{SI} = 1.6 \times 10^{-19}\ C$ and $e_G = 4.8 \times 10^{-10}\ statC$. Evaluate the resulting force in N using the SI values/formulae and in *dyne* for the Gaussian ones and show that they agree.

P10.2 SI to Gaussian equation conversion I: (a) Extend the conversion of Gauss's and Faraday's laws from SI to Gaussian from Eqns. (10.7) and (10.8) to the "no monopole" and Ampère's law equations.
(b) Rewrite all of the Maxwell equations in integral form (from Eqns. (1.54)–(1.57) or Eqns. (4.96)–(4.99)) in Gaussian form.
(c) Rewrite the connections between the electromagnetic potentials $\boldsymbol{A}, \phi$ and the E&M fields $\boldsymbol{E}, \boldsymbol{B}$ from Eqn. (1.65) in Gaussian form.
(d) Rewrite the **power lost per unit volume** formula $\mathcal{P} = \boldsymbol{J}_e \cdot \boldsymbol{E}$ from Eqn. (4.72) in Gaussian form.
(e) Using the CGS dimensions of electric charge in Eqn. (10.4), evaluate the corresponding dimensions (in *gr*, *cm*, and *s*) of electric and magnetic fields, using the Lorentz force relation in Eqn. (10.1). With those results for the dimensions of $\boldsymbol{E}, \boldsymbol{B}$ in place, show that the expressions in Eqn. (10.6) do give energy densities.

[13] Attributed to Robert Maynard Hutchins.

P10.3 SI to Gaussian equation conversion II: (a) Rewrite the SI expressions for torque and energy for magnetic/electric dipoles $(\boldsymbol{m}, \boldsymbol{p})$ in $\boldsymbol{E}, \boldsymbol{B}$ fields, namely

$$\tau_m = \boldsymbol{m} \times \boldsymbol{B} \qquad \tau_e = \boldsymbol{p} \times \boldsymbol{E}$$
$$\mathcal{E}_m = -\boldsymbol{m} \cdot \boldsymbol{B} \qquad \mathcal{E}_e = -\boldsymbol{p} \cdot \boldsymbol{E}. \tag{10.74}$$

(b) Rewrite the SI expressions for Poynting vector, conservation of E&M energy, radiation pressure, and momentum density from Eqns. (4.76)–(4.79) in Gaussian notation.
(c) Using Ohm's law, $\boldsymbol{J}_e = \sigma_e \boldsymbol{E}$, find how the conductance σ_e changes from SI to Gaussian units.
(d) Use the definitions of inductance, resistance, and capacitance from Section 4.5 (Eqn. (4.81)) to show how L, R, C change from SI to Gaussian units. Are your results consistent with how much energy is stored in an L or C circuit element? Or how much power is dissipated in a resistor?

P10.4 SI to Gaussian equation conversion III: As a more open-ended problem, how do the Maxwell equations change when magnetic charge is included (as discussed in Section 4.8.1), assuming the conversion $g_{SI} = g\sqrt{4\pi/\mu_0}$ and related ones for magnetic charge and magnetic current densities in one dimension, two dimensions, and three dimensions. Can you think of arguments about why this is the appropriate *SI→CGS* conversion factor?

P10.5 SI to Gaussian equation conversion IV: Rewrite the following expressions, given in SI units, into Gaussian:

Eqn. (2.13)	Plasma oscillation frequency	$\omega_P = (n_e e^2/m_e \epsilon_0)^{1/2}$
Eqn. (4.13)	Cyclotron frequency	$\omega_C = qB_0/m$
Eqn. (4.13)	Larmor radius	$R_L = mv_\perp/qB_0$
Eqn. (4.16)	$\boldsymbol{E}, \boldsymbol{B}$ drift velocity	$\boldsymbol{v}_d = \boldsymbol{E} \times \boldsymbol{B}_0/\lvert\boldsymbol{B}_0\rvert^2$
Eqn. (4.113)	Larmor power radiated	$P = q^2a^2/6\pi\epsilon_0 c^3$
Eqn. (9.20)	Alfvén velocity	$v_A = \sqrt{B_0^2/(\mu_0 \rho_m)}$

P10.6 Translating equations from particle astrophysics: Ptitsyna and Troitsky's (2010) "Physical Conditions in Potential Accelerators of Ultra-High-Energy Cosmic Rays," quotes a result for a bound on the possible energy due to diffusive processes as

$$E_d \sim \frac{3}{2} \frac{m^4}{Rq^4B^2}, \tag{10.75}$$

where m, q are the mass and charge of a cosmic ray, and B, R are the magnetic field strength and size of the field region respectively. The result as it stands isn't dimensionally correct in our language for two reasons: the E&M quantities are expressed in Gaussian units, and there are missing factors of c ("$c = 1$"-type issues) as well.

(a) Use the conversions from Table 10.8 to switch to SI units and then add enough powers of c to make it dimensionally correct. Is it obvious just how many extra powers of c there should be because of the m dependence?

(b) The authors translate Eqn. (10.75) into a numerical expression as

$$E_d \sim \left[2.91 \times 10^{16}\, eV\right] \frac{A^4}{Z^4} \left(\frac{1\, kpc}{R}\right) \left(\frac{1\, Gauss}{B}\right)^2, \tag{10.76}$$

where the A, Z (atomic number and nuclear charge) dependences make sense since $m = Am_p$ and $q = Ze$ with m_p and e being the proton mass and charge. The distances are scaled in units of *kiloparsec* and the magnetic fields in *Gauss*. Use your result from part (a), values of the fundamental constants m_p, ϵ_0, c, e, and stated values of R, B to see if you can reproduce the prefactor in Eqn. (10.76) (in eV).

(c) The same authors find several other limiting energy values depending on different physical mechanisms, all of which can be encoded in a single parameter family (when written in Gaussian and $c = 1$ units), namely

$$E_{max} \propto m^{\alpha} q^{1-5\alpha/4} B^{1-3\alpha/4} R^{1-\alpha/2}. \tag{10.77}$$

Show that the case of $\alpha = 0$ corresponds to the Hillas result from Eqn. (10.9), while $\alpha = 4$ corresponds to the diffusive result in Eqn. (10.75). Generalize the results of part (a) by again switching from Gaussian to SI E&M units, and adding factors of c, translate this general expression to one we'd be more familiar with.

Note: This type of exercise can occur when reading research level results in many fields where $c = 1$ **and** Gaussian units are used, so it's good to have seen at least one such "practice problem." If it makes you feel better, this was one of the papers I found that I really struggled with getting into SI units, so stay strong!

P10.7 Error propagation: (a) The power dissipated in a resistor is given by $P = I^2 R$, and if the errors in measured values of the current and resistance are σ_I and σ_R, what is the likely error in P? (b) For the Rayleigh capillary oscillation example in Section 2.1, the frequency was given by $f = C_f \sqrt{S/\rho_m d^3}$ where C_f is a numerical constant. If there are errors σ_S, σ_{ρ_m}, and σ_d in all three values, what is the estimated fractional error in f, namely σ_f/f? Does your answer depend on the value of C_f?

P10.8 Buckingham Pi theorem—Von Kármán flow rate: "On the Turbulent Friction of Various Fluids," von Kármán (1911) examined existing experimental data on the pressure drop (ΔP) in a pipe flow, as a fixed volume (Q) of fluid was drained as a function of time (τ). Earlier authors had found that $\Delta P \sim A(1/\tau)^p$ where $p \sim 1.5-1.8$ for fluids of varying viscosity (μ) and density (ρ_m), but with differing prefactors A. Von Kármán proposed to take that dependence into account by presenting a non-dimensional representation of the data, by finding two Buckingham Pi dimensionless quantities.

(a) Verify his result by writing

$$\text{dimensionless} = (\Delta P)^{\alpha} \rho_m^{\beta} \tau^{\gamma} \mu^{\delta} Q^{\epsilon} = (\Pi_1)^{\alpha} (\Pi_2)^{\beta} \tag{10.78}$$

by matching dimensions, solving for γ, δ, ϵ in terms of α, β, thereby defining $\Pi_{1,2}$. In a later, very readable short paperback on aerodynamics, von Karman (1954) includes a figure showing how the data for the various fluids all "collapse" onto a single universal curve when one plots Π_1 versus Π_2, similarly to our Fig. 8.4.

(b) Show that your result for Π_2 is essentially the same as the Reynolds number, $\mathcal{R}e = \rho_m vl/\mu$ where v, l are typical velocity and length scales.

References for Chapter 10

Bakarji, J., et al. (2022). "Dimensionally Consistent Learning with Buckingham Pi," *Nature Computational Science* 2, 834–44.

Bertrand, J. (1878). "Sur l'homogénéité dans les formules de physique (On Homogeneity in Physics Formulas)," *Comptes Rendus* 86, 916–20.

Birge, R. T. (1934). "On Electric and Magnetic Units and Dimensions," *American Journal of Physics* 2, 41–48. Note that the title of this journal was *The American Physics Teacher* until 1940.

Birge, R. T. (1935a). "On the Establishment of Fundamental and Derived Units, with Special Reference to Electric Units. Part I," *American Journal of Physics* 3, 102–109. Note that the title of this journal was *The American Physics Teacher* until 1940.

Birge, R. T. (1935b). "On the Establishment of Fundamental and Derived Units, with Special Reference to Electric Units. Part II," *American Journal of Physics* 3, 171–9. Note that the title of this journal was *The American Physics Teacher* until 1940.

Birkhoff, G. (1950). *A Study in Logic, Fact, and Similitude* (Princeton: Princeton University Press).

Brown, R. J. C. (2024). "On the Distinction between SI Base Units and SI Derived Units," *Metrologia* 61, 013001.

Buckingham, E. (1914). "On Physically Similar Systems: Illustrations of the Use of Dimensional Equations," *Physical Review* 4, 345–76.

Burgess, M. (2005). "A Backward Glance: Eyewitnesses to History," *Nuclear Weapons Journal* 2, 47.

Chen, F. F. (2018). *Introduction to Plasma Physics and Controlled Fusion*, 3rd ed. (Heidelberg: Springer).

Cooper, M., and Grozier, J. (2017). *Precise Dimensions: A History of Units from 1791-2018* (Bristol: IOP Publishing).

Crease, P. (2011). *World in the Balance: The Historic Quest for an Absolute System of Measurement* (New York: Norton).

Cronin, J. A., Greenberg, D. F., and Telegdi, V. L. (1979). *University of Chicago Graduate Problems in Physics with Solutions*, rev. ed. (Chicago: University of Chicago Press).

Dumka P et al. (2022) "Implementation of Buckingham's Pi Theorem Using Python," *Advances in Engineering Software* 173, 103232.

Fisher M (1998) "Renormalization Group Theory: Its Basis and Formulation in Statistical Physics," *Reviews of Modern Physics* 70, 653–81.

Fourier, J-B. J. (1878). *The Analytical Theory of Heat*, translated by Alexander Freeman (London: Cambridge University Press).

Frenkel, J. (1924). "Theorie der adsorption und verwandter Erscheinungen (Theory of Adsorption and Related Phenomena)," *Zeitschrift für Physik* 26, 117–38.

Gupta, S. V. (2020). *Units of Measurement: History, Fundamentals, and Redefining the SI Base Units* 2nd ed. (Cham: Springer).

Hansen, P. C., Pereyra, V., and Scherer, G. (2013). *Least Squares Data Fitting with Applications* (Baltimore: Johns Hopkins University Press).

Jackson, J. D. (1999). *Classical Electrodynamics*, 3rd ed. (New York: Wiley and Sons).

Kalinin, M. I. (2019). "On the Status of Plane and Solid Angles in the International System of Units (SI)," *Metrologica* 56, 065009.

Karam, M., and Saad, T. (2021). "BuckinghamPy: A Python Software for Dimensional Analysis," *Software X* 16, 100851.

Katz, J. I. (2021). "Fermi at Trinity," *Nuclear Technology* 207(Suppl 1), S326–34.

Kittel, C. (1971). *Introduction to Solid State Physics*, 4th ed. (New York: Wiley and Sons).

Kleiber, M. (1932). "Body Size and Metabolism," *Hilgardia* 6, 315–53.

Kowalski, L. (1986). "A Short History of the SI Units in Electricity," *Physics Teacher* 24, 97–99.

Lipa, J. A., et al. (1996). "Heat Capacity and Thermal Relaxation of Bulk Helium Very near the Lambda Point," *Physical Review Letters* 76, 944–7.

Lipa, J. A., et al. (2003). "Specific Heat of Liquid Helium in Zero Gravity Very near the Lambda Point," *Physical Review B: Condensed Matter and Materials Physics* 68, 174518.

Macagno, E. O. (1971). "Historico-Critical Review of Dimensional Analysis," *Journal of the Franklin Institute* 292, 391–402.

Martins, R. D. A. (1981). "The Origin of Dimensional Analysis," *Journal of the Franklin Institute* 311, 331–7.

Maxwell, J. C. (1892/1953). *A Treatise on Electricity and Magnetism*, 3rd ed. (Stanford, CA: Academic Reprints).

Newell, D. B. (2014). "A More Fundamental International System of Units," *Physics Today* 67, 35–41.

Pelisseto, A., and Vicari, E. (2002). "Critical Phenomena and Renormalization-Group Theory," *Physics Reports* 368, 549–727.

Ptitsyna, K. V., and Troitsky, S. V. (2010). "Physical Conditions in Potential Accelerators of Ultra-High-Energy Cosmic Rays: Updated Hillas Plot and Radiation-Loss Constraints," *Advances in Physical Sciences* 53, 691–701.

Purcell, E. M., and Morin, D. J. (2013). *Electricity and Magnetism*, 3rd ed. (Cambridge: Cambridge University Press).

Rayleigh, L. (J. W. Strutt) (1915). "The Principle of Similitude," *Nature* 95, 66–68.

Riabouchinsky, D. (1911). "Méthode des variables de dimensions zéro et son application en aérodynamique (Method of Zero-Dimensional Variables and its Application in Aerodynamics)," *L'Aerophile* 19, 407–8.

Robinson, I. A., and Schlamminger, S. (2016). "The Watt or Kibble Balance: A Technique for Implementing the New SI Definition of the Mass," *Metrologia* 53, A46–74.

Schlamminger, S. (2018). *Refining the Kilogram and Other SI Units* (Bristol: IOP Publishing).

Smith, J. N., Jr., Wolleswinkel, J., and Los, J. (1970). "Residence Time Measurement for the Surface Ionization of K on W: The Effects of Surface Contaminants," *Surface Science* 22, 411–25.

Stock, M., et al. (2019). "The Revision of the SI—The Result of Three Decades of Progress in Metrology," *Metrologia* 56, 022001

Strutz, T. (2010). *Data Fitting and Uncertainty: A Practical Introduction to Weighted Least Squares and Beyond* (Berlin: Springer).

Vaschy, A. (1892). "Sur les lois de similitude en physique (On the Laws of Similarity in Physics)," *Annales Télégraphiques* 19, 25–28.

von Baeyer, H. C. (1993). *The Fermi Solution: Essays on Science* (New York: Random House).

von Kármán, Th. (1911). "Über die turbulenzreibung verschiedener Flüssigkeiten (On the Turbulent Friction of Various Fluids)," *Physikalische Zeitschrift* 12, 283–4.

von Kármán, Th. (1954). *Aerodynamics* (New York: Cornell University Press). See also the first paperback edition published in 1963 by McGraw-Hill.

von Klitzing, K. (2017). "Metrology in 2019," *Nature Physics* 13, 198.

von Klitzing, K. (2019). "Essay: Quantum Hall Effect and the New International System of Units," *Physical Review Letters* 122, 200001.

Zimmerman, N. M. (1998). "A Primer on Electrical Units in the Système International," *American Journal of Physics* 66, 324–31.

Zinn-Justin, J. (2001). "Precise Determination of Critical Exponents and Equation of State by Field Theory Methods," *Physics Reports* 344, 159–78.

Index